D0983297

Applied Biochemistry of Clinical Disorders

Edited by

Allan G. Gornall, PH.D., D.SC., F.R.S.C.

Professor Emeritus, Department of
Clinical Biochemistry
University of Toronto, Toronto, Ontario, Canada

with 29 contributors
and 140 illustrations

J. B. LIPPINCOTT COMPANY

Philadelphia
London Mexico City New York
St Louis São Paulo Sydney

Applied Biochemistry of Clinical Disorders

SECOND EDITION

Acquisitions Editor: Lisa A. Biello
Sponsoring Editor: Delois Patterson
Manuscript Editor: Marjory Fraser
Indexer: Angela Holt
Design Director: Tracy Baldwin
Design Coordinator: Earl Gerhart
Designer: Katharine Nichols
Production Manager: Kathleen P. Dunn
Production Coordinator: Susan Hess
Compositor: Tapsco, Inc.
Printer/Binder: R. R. Donnelley & Sons Co.

Second Edition

6 5 4 3 2 1

Library of Congress Cataloging-in-Publication Data

Applied biochemistry of clinical disorders.

Includes bibliographies and index.
1. Clinical biochemistry. I. Gornall, Allan G., 1914- . [DNLM: 1. Chemistry, Clinical. QY 90 A652]
BB112.5.A67 1986 616.07 86-2781
ISBN 0-397-50768-2

The authors and publisher have exerted every effort to ensure that drug selection and dosage set forth in this text are in accord with current recommendations and practice at the time of publication. However, in view of ongoing research, changes in government regulations, and the constant flow of information relating to drug therapy and drug reactions, the reader is urged to check the package insert for each drug for any change in indications and dosage and for added warnings and precautions. This is particularly important when the recommended agent is a new or infrequently employed drug.

This book is dedicated
to the vision of a
chair in clinical biochemistry
in the
faculty of medicine
of all major universities

Contributors

Lynn C. Allen, Ph.D., D.Cl.Chem. **Chapter 23**
Assistant Biochemist, Department of Clinical Biochemistry, Toronto General Hospital; Assistant Professor, Department of Clinical Biochemistry, University of Toronto, Toronto, Ontario

Kenning M. Anderson, M.D., Ph.D., F.R.C.P.(C) **Chapter 24**
Associate Professor, Departments of Biochemistry and Medicine, Rush Medical College, Chicago, Illinois

Andrew DeWitt Baines, M.D., Ph.D. **Chapter 8**
Professor, Departments of Clinical Biochemistry and Medicine, University of Toronto; Biochemist-in-Chief, Toronto General Hospital, Toronto, Ontario

Bhagu R. Bhavnani, Ph.D. **Chapter 14**
Professor, Department of Obstetrics and Gynecology, McMaster University Medical Center, Hamilton, Ontario

Philip D. Bonomi, M.D., M.S. **Chapter 24**
Assistant Professor, Department of Internal Medicine, Rush Medical College, Chicago, Illinois

W. Carl Breckenridge, Ph.D. **Chapter 20**
Professor, Department of Biochemistry, Dalhousie University, Halifax, Nova Scotia

Peter J. Brueckner, M.D., C.M., F.R.C.P.(C) **Chapter 6**
Staff Biochemist, Department of Biochemistry, Sunnybrook Medical Center, Associate Professor, Department of Clinical Biochemistry, University of Toronto, Toronto, Ontario

Stanislaw Dubiski, M.D., Ph.D. **Chapter 4**
Professor, Departments of Immunology, Medical Genetics and Clinical Biochemistry, University of Toronto, Toronto, Ontario

David M. Goldberg, M.D., Ph.D., F.R.I.C., F.R.C. Path. **Chapters 3 and 11**

Biochemist-in-Chief, Department of Biochemistry, The Hospital for Sick Children; Professor and Chairman, Department of Clinical Biochemistry, University of Toronto, Toronto, Ontario

Allan G. Gornall, Ph.D., D.Sc., F.R.S.C. **Chapters 1, 10, 11, 14, and 26**

Professor Emeritus, Department of Clinical Biochemistry, University of Toronto, Toronto, Ontario

Marc D. Grynpas, Ph.D. **Chapter 17**

Assistant Professor, Departments of Pathology and Medicine, University of Toronto and Connective Tissue Research Group, Mt. Sinai Hospital, Toronto, Ontario

J. Gilbert Hill, M.D., Ph.D. **Chapter 21**

Head, Service Division, Department of Biochemistry, The Hospital for Sick Children; Professor, Department of Clinical Biochemistry, University of Toronto, Toronto, Ontario

J. Thomas Hindmarsh, M.D., F.R.C.P.(C) **Chapter 10**

Chief, Division of Biochemistry, Department of Laboratory Medicine, Ottawa Civic Hospital, Ottawa, Canada

Joseph B. Houpt, M.D., F.R.C.P.(C) **Chapter 13**

Director, Trihospital Rheumatic Disease Unit, Mount Sinai Hospital; Associate Professor, Department of Medicine, University of Toronto, Toronto, Ontario

John A. Kellen, M.D., Ph.D., F.R.C.P.(C) **Chapter 16**

Professor, Department of Clinical Biochemistry, University of Toronto; Staff Physician, Sunnybrook Medical Center, Toronto, Ontario

Stephen J. Kish, Ph.D. **Chapter 15**

Assistant Professor, Departments of Psychiatry and Pharmacology, University of Toronto and Clarke Institute of Psychiatry, Toronto, Ontario

Choong-Chin Liew, Ph.D. **Chapters 27 and 28**

Professor, Departments of Clinical Biochemistry and Medicine, University of Toronto, Toronto, Ontario

Allan W. Luxton, M.D., F.R.C.P.(C) **Chapter 14**

Department of Laboratories, Clinical Chemistry Section, Henderson General Hospital, Hamilton, Ontario; Assistant Professor, Department of Clinical Biochemistry, University of Toronto, Toronto, Ontario

Michael D. D. McNeely, M.D., F.R.C.P.(C) **Chapters 18 and 22**

Medical Biochemist, Island Medical Laboratories, Victoria, British Columbia

Robert L. Patten, M.D., C.M., F.R.C.P.(C) **Chapters 9 and 20**

Clinical Biochemist-in-Chief, Department of Clinical Biochemistry, Saint Michael's Hospital; Associate Professor, Department of Clinical Biochemistry, and Professor, Department of Medicine, University of Toronto, Toronto, Ontario

Alan Pollard, M.B., M.A., M.R.C.P., F.R.C. Path. **Chapters 17 and 19**

Head, Division of Clinical Biochemistry, Department of Laboratories, Mount Sinai Hospital; Associate Professor, Departments of Clinical Biochemistry and Medicine, University of Toronto, Toronto, Ontario

Kenneth P. H. Pritzker, M.D., F.R.C.P.(C) **Chapter 17**

Professor, Department of Pathology, University of Toronto; Pathologist and Head, Connective Tissue Research Group, Mt. Sinai Hospital, Toronto, Ontario

Frank H. Sims, M.B., Ch.B., Ph.D., F.A.A.C.B. **Chapter 2**

Pathology Department, Medical School, University of Auckland, Auckland, New Zealand

Steven J. Soldin, Ph.D., D.Cl.Chem, F.A.C.B. **Chapter 25**

Associate Biochemist and Director of Therapeutic Drug Monitoring, Hospital for Sick Children; Professor, Department of Clinical Biochemistry, University of Toronto, Toronto, Ontario

Zulfikarali H. Verjee, Ph.D., D.Cl.Chem. **Chapter 18**

Clinical Chemist, Sunnybrook Medical Centre; Assistant Professor, Department of Clinical Biochemistry, University of Toronto, Toronto, Ontario

W. H. Chris Walker, M.B., F.R.C. Path., F.R.C.P.(C) **Chapter 12**

Head, Clinical Chemistry Service, and Professor, Department of Pathology, McMaster University Medical Center, Hamilton, Ontario

John R. Wherrett, M.D., C.M., Ph.D., F.R.C.P.(C) **Chapter 15**

Head, Division of Neurology, Department of Medicine, Toronto General Hospital; Professor of Neurology, Department of Medicine, University of Toronto, Toronto, Ontario

Colin R. Woolf, M.D., F.R.C.P.(C), F.R.C.P.(Lon.) **Chapter 7**

Professor, Department of Medicine, University of Toronto; Senior Staff Physician, Department of Medicine, Toronto General Hospital, Toronto, Ontario

Leebert A. Wright, Ph.D. **Chapter 5**

Director, Department of Clinical Biochemistry, The Wellesley Hospital; Associate Professor, Department of Clinical Biochemistry, University of Toronto, Toronto, Ontario

Preface

The success of *Applied Biochemistry of Clinical Disorders* and the rapid growth of new knowledge over the wide spectrum of human diseases have provided the challenge to prepare a second edition. The task of bringing the book up to date has placed heavy demands on most of the authors, which they have met commendably. Two chapters have been rewritten (4, 17). Chapter 15 has been expanded to include new knowledge of psychiatric disorders. Chapter 23 has been extensively revised and includes a new section on prenatal diagnosis of genetic defects. Three new chapters have been added. Chapter 25 emphasizes the importance of therapeutic drug monitoring. Chapter 26 summarizes the role of receptors in disease. Our new understanding of the molecular basis of many diseases and the rapid development of techniques that provide a definitive diagnosis has required a last-minute addition of Chapter 28, Molecular Diagnosis of Genetic Defects. A few changes in authorship were inevitable, but the objectives and style are essentially the same as those of the first edition.

Written in textbook format, this book encompasses the knowledge of clinical biochemistry that should be acquired between the third year of medicine and second year of residency. It is intended as a useful reference for professionals responsible for the biochemical aspects of laboratory medicine and should be on the desk of every clinical chemist.

Medical students need to understand basic statistical principles and must be impressed with the pervading uncertainties of the decision-making process in medicine. Postgraduate education should ensure that young physicians have sufficient knowledge of the laboratory to exercise good judgment in assessing the relevance and importance of 'abnormal' laboratory data.

Clinical biochemists must assume responsibility for reducing to a minimum the waste of valid laboratory information. They must be capable of 'flagging' results that are outside the reference range for the patient under study and of assessing the likelihood that it is significant. Responsibility then reverts to the clinician. A more effective use of laboratory services will come when clinicians and clinical biochemists are motivated to obtain and record reliable data in a computerized system, which will permit a printout of ranked-order probabilities of diseases that could be associated with the laboratory results.

The book could be subtitled "The ABCD of Effective Laboratory Utilization." It undertakes to present the central core of information that is the common ground of both the clinician and the clinical biochemist. Each has a much wider range of knowledge and experience in quite different areas, but in *utilizing* and *providing* laboratory services this book offers a basis for understanding, communication, and cooperation.

The growing sophistication of laboratory instrumentation and analytic technology, the advent of robotics, and the power of the computer, will mean exciting new developments in the next few years. The effective use of laboratory services will continue to depend on well educated and well trained clinicians and clinical biochemists, dedicated to a better health care delivery system.

Allan G. Gornall, PH.D., D.SC., F.R.S.C.

Preface to the First Edition

The Clinical Chemistry Laboratory Service is the practical expression of all that we know about the biochemistry of human diseases. The limited usefulness of this service has been due largely to the fact that neither the principles underlying the selection of laboratory tests, nor the interpretation of laboratory data, have been properly taught or understood. Complex and difficult problems have arisen as a byproduct of the effort to make effective use of the resources of analytical chemistry in support of the practice of medicine. These problems involve hospitals, universities, public and private laboratories, health professionals, and government, but the cost-effectiveness of all health services becomes ultimately a concern of the whole community.

This book addresses several aspects of these problems, and the incentive to edit such a volume came from different sources. There have been recurring requests for copies of the lecture synopses provided to students in our course in Clinical Biochemistry. Curriculum revisions have left medical students in many centers with a less-than-coherent account of the biochemistry of human diseases and the role of the laboratory in clinical decision making. Clinicians have shown an interest in updating their knowledge in this area. Clinical biochemists have come to recognize that at the core of their discipline is an understanding of the extent to which each test is sensitive (positive in disease) and specific (negative in health). They must be qualified to comment on the predictive value of the data they provide and to contribute to the continuing education of the practicing physician.

Based on our experience in the field, we present here a practical, systematic account of the biochemistry of human diseases. An effort has been made to steer a course between the Scylla of being superficial and the Charybdis of too much controversial detail. Each chapter has a reading list to facilitate further study. Literature references to support individual statements have been omitted, but can be obtained from the authors. An effort has been made to survey all the common, or particularly interesting, disorders in which biochemical changes contribute to an explanation of the clinical findings and pathologic changes. Special attention has been given to the potential value of laboratory data in understanding the disorder, confirming a diagnosis, monitoring treatment and making a prognosis. North America is in a transition period in the methods of reporting laboratory results. Looking to the future, most reference ranges have been expressed in système international (S.I.) units, with current terminology in parentheses.

The aim has been to create a better understanding of factors that govern the information gained from laboratory tests. The clinician and clinical chemist together must determine the extent to which each test is *sensitive* and *specific.* The clinician should reach a provisional diagnosis, which increases the *prevalence* of the suspected disorder to an acceptable statistical level. In situations of *low* prevalence specificity is more important, since the objective is to rule out the presence of disease. In situations of *higher* prevalence sensitivity assumes importance, since the objective is to establish that the patient has the suspected disease. The close relationship between prevalence and the *predictive value* of laboratory tests must become a natural component of our reasoning. A number of *algorithms* are included as a guide to reaching an appropriate decision in a minimum number of steps with a high degree of precision.

Finally, there has been a modest attempt to direct our attention to an obligation that should challenge all health professionals. We must continually question current practice, insist on adequate justification for the use of our technical skills, and establish a means of measuring with reasonable accuracy the benefits of what we accomplish in terms of the demands made on the economy.

Allan G. Gornall, PH.D., D.SC., F.R.S.C.

Acknowledgments

As Editor I wish to acknowledge here the contribution of my wife Sheila, whose intelligence, insight, and strength of character have been a source of support and inspiration over more than forty years.

The Editor would pay tribute to the Publishers for their cooperation and assistance throughout the preparation of this second edition. He is grateful also to Claire Brenner for her care and skill in typing many revisions of the manuscript.

Original illustrations, except in the few instances indicated, were prepared by Frederick Lammerich, Department of Art as Applied to Medicine, University of Toronto. New and revised figures were prepared by Jennifer A. Walker on the basis of material supplied by each author.

A number of colleagues have assisted the authors by providing critical appraisals of various chapters. Appreciation of their efforts is expressed by listing their names below.

J. Richard Hamilton, M.D., F.R.C.P.(C)
Chief, Division of Gastroenterology, The Hospital for Sick Children; Professor of Pediatrics, University of Toronto, Toronto, Ontario

A. Ralph Henderson, M.B., Ch.B., Ph.D., M.R.C. Path.
Chief of Clinical Biochemistry, University Hospital; Professor of Biochemistry, University of Western Ontario, London, Ontario

Connie Hoff, B.Sc., M.Sc.
Research Assistant, Island Medical Laboratories, Victoria, British Columbia

Robert G. Josse, M.D., F.R.C.P.(C), F.A.C.P., A.B.I.M.
Staff Physician and Endocrinologist, St. Michael's Hospital; Associate Professor of Medicine, University of Toronto, Toronto, Ontario

Stephen I. Kandel, Ph.D.
Professor, Faculty of Pharmacy, University of Toronto, Toronto, Ontario

Pang Nin Shek, Ph.D.
Research Scientist, Defence and Civil Institute of Environmental Medicine; Assistant Professor of Clinical Biochemistry, University of Toronto, Toronto, Ontario

Authors who have provided helpful appraisals of chapters other than their own are Dr. Joseph B. Houpt, Dr. Alan Pollard, and Dr. W. H. Chris Walker.

Contents

Part Four • Special Topics

In co-authored chapters, the sequence of names reflects, as a general rule, the portions of the chapter contributed by each author.

Applied Biochemistry of Clinical Disorders

Part One

General Topics

1. **Basic Concepts in Laboratory Investigation**
2. **Pathologic Processes**
3. **Diagnostic Enzymology**
4. **Diagnostic Immunology**
5. **Diagnostic Clinical Toxicology**

1

Basic Concepts in Laboratory Investigation

Allan G. Gornall

"Medicine is a science of uncertainty and an art of probability. One of the chief reasons for this uncertainty is the increasing variability of the manifestations of any one disease."
—Osler

HEALTH AND DISEASE

People usually seek the help of a physician because of some discomfort or disease. On average, about 80% of the population experiences some form of illness during any one year. What is it that distinguishes health from disease? An abnormality noted or suspected by a patient is called a *symptom* (e.g., dizziness, pain, cough, diarrhea). An abnormality observed by the physician is called a *sign.* Signs include information obtained by inspection (e.g., edema of the ankles, jaundiced sclera), by palpation (e.g., enlarged liver, a lump in the breast), by percussion (e.g., fluid in the lung or abdomen), and by auscultation (e.g., heart sounds, wheezy breathing). Patients may suspect that they have an illness and complain of real or imagined symptoms. A medical investigation may reveal no detectable disease, and the patient may be cured by effective reassurance. Alternatively, people may believe that they are free of any symptoms of ill-health and yet be found at a routine examination to have early signs of serious disease.

A ***disease*** is a composite of signs and symptoms associated with a specific pathologic process. Some disease states are relatively easy to identify because there are observable *lesions* (e.g., measles). In other cases only the consequences of the disorder are observed, the lesion perhaps being a genetic defect involving a single amino acid in a protein (e.g., sickle-cell anemia). Most diseases have a known cause or *etiology,* but some illnesses cannot be proved to be due to any known cause. When an abnormality is observed but the cause is unknown it is described as *idiopathic* (or *occult,* or *cryptic*); when function is disturbed but no organic cause can be found (e.g., when an abnormality is believed to be psychosomatic) it is often called *functional.* The *pathogenesis* of a disease is a description of the factors involved in its development. The definitive diagnosis of a disease usually requires objective evidence of the pathologic process, such as isolation of a bacillus or virus, demonstration of a specific (pathognomonic) biochemical abnormality, visualization of a kidney stone by radiography or a tumor by ultrasound, the finding of an inflamed appendix at surgery, or location of a lesion in the tissues by biopsy or at autopsy. Sometimes a disease may be difficult to define, or may have more than one cause; the combination of symptoms, signs, and lesions may then be referred to as a *syndrome.*

The problem of distinguishing between health and disease depends on a concept of what is ***normal.*** There are many definitions of normal, but in relation to health the choice lies between trying to define an 'ideal' state, which is difficult, or determining an 'average' state in people considered healthy. In terms of a particular biochemical parameter the latter method results in a more or less symmetric (Gaussian) distribution about a mean, producing the familiar bell-shaped curve. *Statistical theory* then defines, at two standard deviations (±2 SD), the limits within which 95% of the examined population will lie. This should encompass variations due to analytic, biologic, demographic, and environmental factors. A population having a specific disease will usually have an asymmetric (nonparametric) distribution, which in most cases overlaps the 'normal' bell curve (Fig. 1–1). Because of this asymmetry the frequency distributions of laboratory results in both normal and diseased populations can be listed in ranked order and decision lines drawn as the 2.5 and 97.5 percentiles.

DIAGNOSIS AND THE VALUE OF LABORATORY TESTS

A physician will have in mind three *primary* questions: What is the matter? What can I do? What will the outcome be? Three *secondary* questions are: Why did all this happen? Could it have been prevented? What is the pathogenesis?

The prime reasons for ordering laboratory tests are:

a. To confirm or exclude a diagnosis
b. To monitor progress and therapy
c. To establish or complete a data base
d. To assess severity of disease (e.g., rerisk of surgery)
e. To screen for unsuspected disease
f. To avoid malpractice claims

In coming to a diagnostic decision most clinicians appear to follow a recognized system (Fig. 1–2). Initial hypotheses are formulated from memory (experience) as soon as the patient has been questioned and examined. These hypotheses can be tested in various ways, including the acquisition and perusal of laboratory data, which often account for 25% of the new information. The data may lead to new hypotheses, which are 'recycled' several times, if necessary, until reduced to a diagnostic decision on which action can be taken. The patient's response to therapy may lead to clinical improvement or to new developments that may require the consideration of new hypotheses.

When a specific diagnostic problem can be identified (e.g., hypercalcemia, hyponatremia), the physician may switch to a tree-branching approach (see, for example, Fig. 6–3). Such diagnostic ***algorithms*** can assist in reaching an appropriate decision in a minimum number of steps with a high degree of confidence. They may help the clinician in some situations, but in general they have been used most effectively by nurse practitioners and paramedics. The diagnostic process is thus one of garnering information until, ideally, the probability

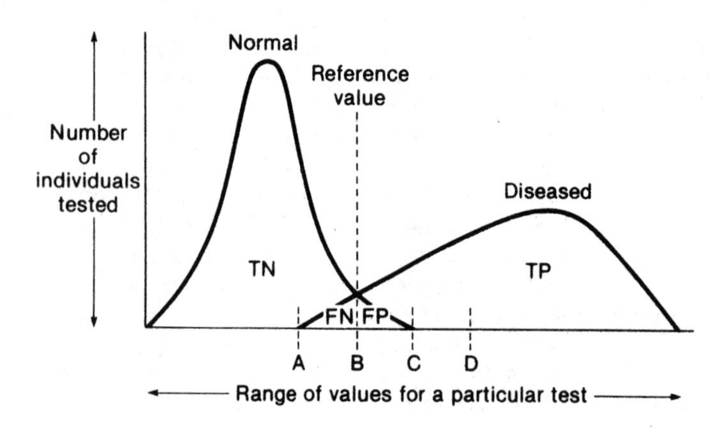

FIGURE 1–1. Hypothetical distribution of laboratory values for a particular biochemical test in 'normal' and 'diseased' populations. TN = True negative; TP = true positive; FN = false negative; FP = false positive. A, B, C, and D are decision points (or reference values) referred to in the text.

of the disease approaches 100%, but in practice this is often much lower before a decision has to be made.

How can one assess the *efficiency* of a biochemical test in discriminating between the presence and absence of disease? The value of a laboratory procedure is measured in terms of the information gained, which depends on the following factors:

PRECISION AND ACCURACY

Analytic test procedures are assessed for *precision* (reproducibility) and *accuracy* (proximity to the 'correct' or 'true' value). They may be referred to as sensitive (capable of measuring small quantities) or specific (free of interference by other substances). These latter terms are used in a quite different sense when describing the *clinical usefulness* of a test (discussed later).

REFERENCE VALUES

The distinction between a test result being a *true positive* (TP) or *true negative* (TN) is usually determined with reference to a selected 'normal range,' based on 95% confidence limits. Since some normal people will fall outside this range (false positive, FP), and some with disease will fall within it (false negative, FN), the term ***reference range*** (RR), or reference interval, is now preferred. For any particular disease, attention is usually focused at either the upper or the lower limit of this range, so we are really considering a ***reference value*** (RV). A major problem is to define, for each laboratory test, the RV that provides the greatest usefulness in the process of coming to a decision, or in discriminating between health and disease. It is sometimes called the *decision threshold* or *decision point* (see Fig. 1–1). In some situations the reference mean (RM) may prove more useful.

Also important is the concept of the ***reference state*** (RS), the conditions under which the RV can be assumed to apply. *Ideally,* for example, the RS might be defined as male (or female), white (or black), aged 20 to 40, not obese, off drugs, on a normal diet (with a limited intake of caffeine, alcohol, and nicotine), at rest and fasted 10 hours. In *practice,* reference ranges are often applied to much wider populations, and only the more significant variants are considered. These are usually

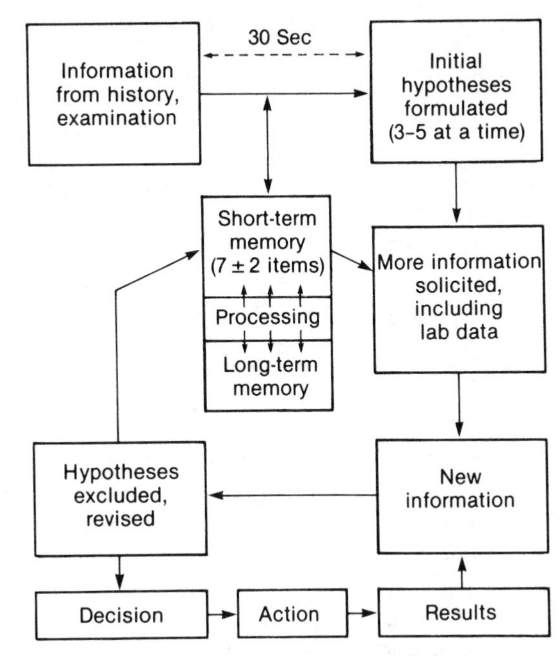

FIGURE 1–2. The hypotheticodeductive process of reaching a diagnosis or a decision. (Modified from Fabb WE et al [eds]: Focus on Learning in Family Practice. Melbourne, Australia, Royal Australian College of General Practitioners, 1975)

based on experience. In neonate, pediatric, obstetric, and geriatric practices the RR will differ for such parameters as calcium, phosphate, alkaline phosphatase, creatinine, cholesterol, urate, and some hormones.

To *establish* a reference range properly at least 100 subjects are required, even when the distribution of results is symmetrical. For nonparametric tests larger numbers are needed. The RV for each new test must be validated and its efficiency assessed under three broad sets of conditions:

1. In 'healthy' persons under as nearly comparable conditions as possible
2. In patients *with* disease *x,* who have no symptoms, as well as those who encompass the full range of signs and symptoms that may accompany the disease
3. In patients who do *not* have disease *x,* but whose signs and symptoms can mimic the disease, or in whom there is pathologic involvement of the same organ system.

SENSITIVITY AND SPECIFICITY

Laboratory test statistics are best shown as a decision matrix (Fig. 1–3).

The ***sensitivity*** (SENS) of a test is its 'positivity,' the percentage of patients with the disease that falls beyond the RV. This would be 100% at point *A* in Figure 1–1.

$$\text{SENS} = \frac{\text{TP}}{\text{TP} + \text{FN}} \times 100$$

= % of disease cases that give a positive result with the RV selected

The ***specificity*** (SPEC) of a test is its 'negativity,' the percentage of people who do not have the disease that falls within the reference range. It would be 100% at point *C* in Figure 1–1.

$$\text{SPEC} = \frac{\text{TN}}{\text{TN} + \text{FP}} \times 100$$

= % of nondisease cases that give a negative result with the RV selected

It can be noted that moving the RV toward the 'normal' range increases *sensitivity* for ruling *out* disease (when negative). Moving the RV toward the 'disease' range increases *specificity* for ruling *in* disease (when positive). The particular clinical application of a test will govern the best choice of RV.

LABORATORY TESTS AND DISEASE

The accuracy with which a disease can be defined varies considerably, and its relationship to laboratory data has to be expressed in probabilities. Laboratory tests may be used to discover, confirm, exclude, or monitor disease.

Discovery tests are often applied to asymptomatic patients, and should have good *sensitivity.* Many clinicians will order an initial profile of tests to establish a data base on each patient. Selective screening with such tests as urinalysis, serum calcium, thyroxin, iron (or ferritin), creatinine, gammaglutamyl transferase, alkaline phosphatase, and lipids can be justified in a family practice when clinical suspicion is aroused. Neonatal screening for hypothyroidism, or certain genetic diseases, is

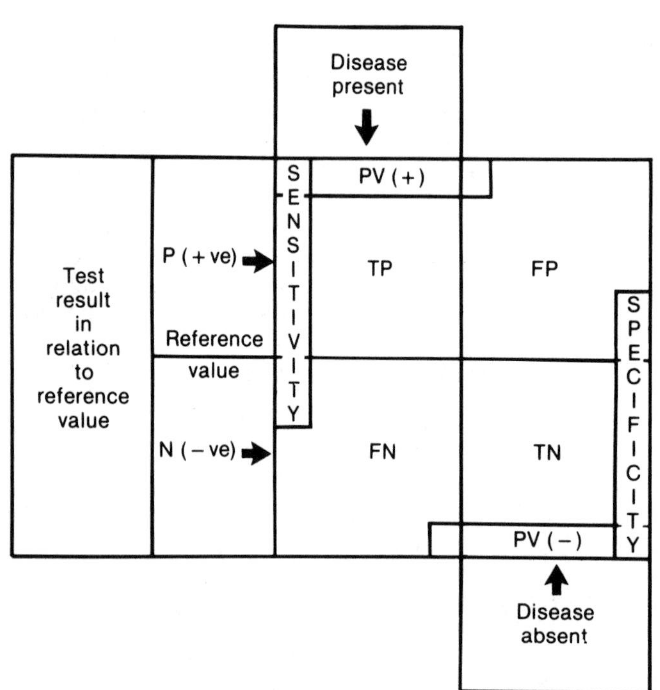

FIGURE 1–3. A decision matrix illustrating the meaning of sensitivity, specificity, and the predictive value of laboratory tests.

$$\text{SENS} = \frac{\text{TP}}{\text{TP} + \text{FN}} \times 100$$

$$\text{SPEC} = \frac{\text{TN}}{\text{TN} + \text{FP}} \times 100;$$

predictive value of a positive test,

$$\text{PV}(+) = \frac{\text{TP}}{\text{TP} + \text{FP}} \times 100$$

of a negative test,

$$\text{PV}(-) = \frac{\text{TN}}{\text{TN} + \text{FN}} \times 100$$

TP = true positive; FP = false positive; TN = true negative; FN = false negative; P = positive (+ve); N = negative (−ve) clearly of value. Blanket screening of the general population can rarely be condoned.

Confirmatory tests need to have high *specificity,* in order to establish that a disease is present. A combination of two tests, the first specific and the second sensitive, has the best chance of ruling a disease *in* or *out.*

Tests to *exclude* disease must have *very high sensitivity,* with virtually no false negatives. An example is the glucose tolerance test.

Tests used for *monitoring* are of two types, those that reflect the course of a disease (e.g., serum bilirubin), and those that are required to follow blood levels of a drug used in treatment (e.g., digoxin). In an office practice about 20% of tests may be for monitoring, in hospitals the figure can be as high as 80%.

Very few biochemical tests have such high sensitivity *and* specificity that they are unique or *pathognomonic* for a particular disease; the new techniques of 'molecular diagnosis' based on DNA polymorphism hold great promise in this regard. In many cases a pathognomonic test may involve a biopsy (e.g., of liver, or prostate). To avoid morbidity and expense, reliance is often placed on secondary manifestations of the disease (e.g., increased serum aminotransferases, or acid phosphatase) and these are called *surrogate* tests. Here there is no absolute relationship, only a greater or lesser likelihood that the test abnormality results from a particular disease. A knowledge of SENS and SPEC is essential, and the underlying concept is one of probability.

PREVALENCE, PREDICTIVE VALUE, AND EFFICIENCY

The ***prevalence*** of a disease is the number of cases in the population being tested at a designated point in time. It is usually expressed as 1 per *x* number of people, or number per 100,000, or as a percentage. The prevalence is not the same as the ***incidence*** of the disease, which refers to the number of cases occurring in a series of patients over a period of time, generally one year, or the *epidemiology,* which is concerned with both. Diabetes is a high-prevalence, low-incidence disease; the common cold has a low prevalence but a high incidence.

The essential characteristic of any test is its ***predictive value*** (PV) which depends not only on specificity and sensitivity but also on the prevalence of the condition in the population being tested (Bayes' theorem). In Figure 1–3 it can be seen that the PV is affected by the reference value and depends on the percentage of tests that are true positives or true negatives, relative to the false positives or false negatives. A laboratory test should answer a specific 'clinical question' of some consequence. The PV indicates the information that can be gained by performing the test under appropriate conditions. The predictive value of a positive test

$$PV(+) = \frac{TP}{TP + FP} \times 100$$

The predictive value of a negative test

$$PV(-) = \frac{TN}{TN + FN} \times 100$$

It is important to understand that laboratory data are often more useful in ruling *out* than ruling *in* a diagnosis. A test with a high PV+ works best in a high-prevalence population such as a specialty ward in a hospital. It will not be as helpful in an ambulatory outpatient service. In the low-prevalence situation of an office practice, tests with a high PV− are more useful. Therefore, different test strategies and laboratory profiles are needed for ambulatory (low-prevalence) and hospitalized (high-prevalence) situations. As the prevalence increases, the PV of a positive test increases and the PV of a negative test decreases. As the diagnosis becomes less and less likely, the PV of a negative test increases while that of a positive test decreases. It is useful to note that 100 − PV(−) is the probability that a patient has disease *x* in spite of a negative test result.

The ***efficiency*** of a laboratory test is the percentage of *all* results that are *true* results, whether TP or TN. It is expressed as $\frac{TP + TN}{\text{Total tested}} \times 100$, which is the same as $\frac{SENS + SPEC}{2}$. The efficiency of flipping a coin is 50%. If a perfect test existed the efficiency would be 100%. As a general rule a test is probably not worth doing if its efficiency is less than 80%.

RECEIVER OPERATING CHARACTERISTICS AND LIKELIHOOD RATIOS

A so-called *receiver operating characteristic* (ROC) curve is based on the relation of the true positive to the false positive rate. It is obtained by plotting the TP ratio [TP/(TP + FN) = SENS] against the FP ratio [FP/(FP + TN) = 100 – SPEC] for a series of reference values. This can be useful not only in assessing the clinical value of different tests, but also in selecting the best RV for the test chosen. It is also possible to calculate the ***likelihood ratio,*** which is the ratio of these two conditional probabilities; that is, the ratio of TP to FP for detecting disease, and the ratio of FN to TN for excluding disease.

For a *positive* test:

$$\frac{\%\ \text{TP}}{\%\ \text{FP}} = \frac{\text{SENS}}{100 - \text{SPEC}}$$

For a *negative* test:

$$\frac{\%\ \text{FN}}{\%\ \text{TN}} = \frac{100 - \text{SENS}}{\text{SPEC}}$$

These ratios have been incorporated in a useful nomogram (Fig. 1–4).

FIGURE 1–4. Information gained from a laboratory test: A nomogram showing an example of the effect of clinical input (which increases prevalence from 1% to 30%) on diagnostic probability, with a single test that has a likelihood ratio of 19 (95% sensitivity, 95% specificity). TP = True positives; FP = false positives; FN = false negatives; TN = true negatives; SENS = sensitivity; SPEC = specificity. (Modified from Fagan TJ: Letter to the Editor: Nomogram for Bayes's theorem. N Engl J Med 293:257, 1975)

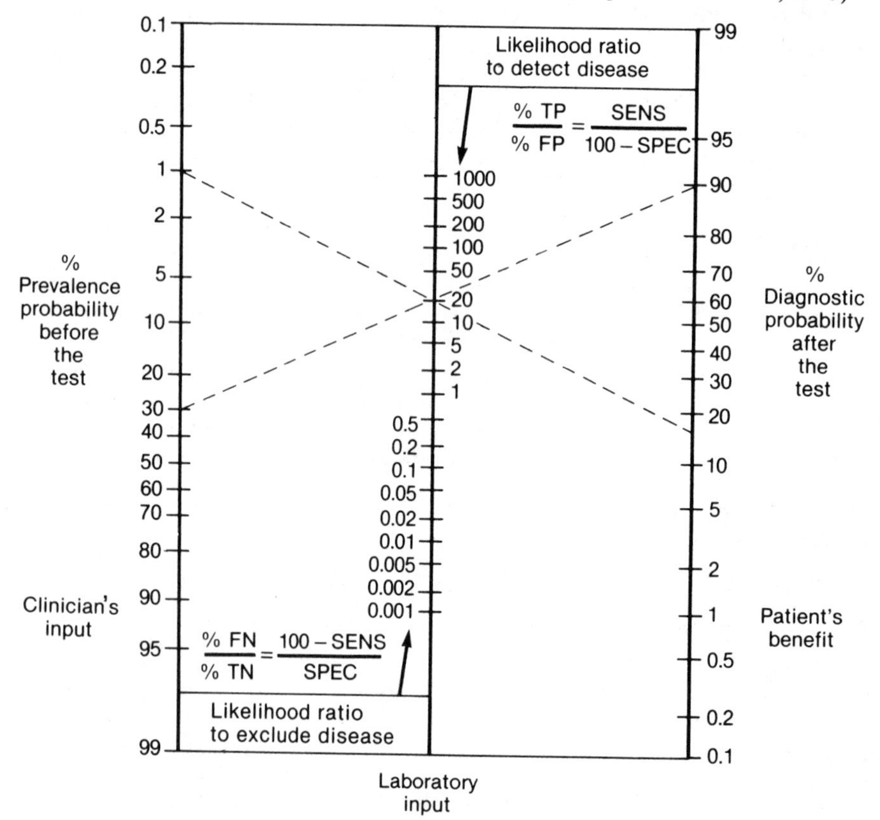

CHOICE OF A REFERENCE VALUE

The *choice of a reference value* is influenced also by *clinical* factors.

If a disease is serious but is treatable without risk, so that an incorrect diagnosis due to an FP test will not harm the patient, then the reference value may approach *A* (in Fig. 1–1) to increase ***sensitivity*** (e.g., phenylketonuria).

If a disease is serious and incurable, so that diagnosis based on an FP test could be psychologically damaging, and a TN test can be reassuring, then the reference value may approach *C* (in Fig. 1–1) to increase ***specificity*** (e.g., many cancers).

If the treatment of the disease may have serious consequences and accuracy of diagnosis is important, then conditions that yield the highest ***predictive value*** should be sought. This often involves using a combination of tests having good sensitivity and specificity (e.g., illnesses involving major surgery).

When a disease is serious but treatable, and the risks of FP and FN tests are about equal, then reference values offering maximum ***efficiency*** are required (e.g., in myocardial infarction).

THE IMPORTANCE OF PREVALENCE

Diagnosis is a process of reducing hypotheses and increasing the probability that the patient has a particular disease; medical education is directed toward improving this deductive capability. The laboratory is one means of extending the physician's powers of observation. The importance of establishing a fairly *high prior probability* of the disorder is illustrated by the following example, plotted on the nomogram in Figure 1–4 (dashed lines).

Suppose a laboratory test (e.g., a urine 24-hour vanillylmandelic acid) is a fairly good one and in relation to the selected reference value (7.5 mg) has a SENS of 95% and a SPEC of 95%. This gives a likelihood ratio of 95/5, or 19. In 100 patients with *pheochromocytoma* there will be 95 true positives (TPs) and 5 false negatives (FNs). In a population recognized by the clinician as having elevated blood pressure, the prevalence of this tumor is only 1% (or less). Table 1–1 shows the results for 10,000 hypertensive patients. The predictive value (PV) is only 17.5%.

TABLE 1–1. Results of Test (Prevalence = 1%)

	Positive	Negative	Totals
Disease present	TP 95	FN 5	100
Disease absent	FP 445	TN 9455	9900
Totals	540	9460	10,000

The predictive value (PV) of a positive test = $\frac{95}{540} \times 100$ = 17.5%

But if the physician has noted episodic bouts of hypertension, accompanied by symptoms often associated with pheochromocytoma, and has found no evidence of cardiac, renal, or gastrointestinal disease, the probability (prevalence) would have increased to about 30%. Table 1–2 shows the results for 1000 such patients. The PV of a positive test is now 89%.

TABLE 1–2. Results of Test (Prevalence = 30%)

	Positive	Negative	Totals
Disease present	TP 285	FN 15	300
Disease absent	FP 35	TN 665	700
Totals	320	680	1000

The PV of a positive test = $\frac{285}{320} \times 100 = 89\%$

EFFECTIVE LABORATORY UTILIZATION—A SHARED RESPONSIBILITY

Figure 1–4 serves also to clarify the roles and responsibilities of the clinician and the clinical biochemist in effective laboratory utilization. The clinician initiates and completes the decision-making process. In 80% of patients the history and physical examination should result in a diagnosis, although intuitive pattern recognition has become less frequent as tests have multiplied. In most cases additional information is sought, either to confirm the diagnosis or to screen for other possibilities as a precautionary measure. In about 20% of cases clinical chemistry will provide a major data base, and for half of these the decision will rely on this data. It can be stated that unless the sensitivity and specificity of a laboratory test average 80% or better and the probability (prevalence) of the disease is at least 5%, the information gained from a *positive* laboratory test will be very limited.

It is the ***clinician's responsibility*** to take a careful history, examine the patient, and raise the level of probability to the maximum that is possible. A sequence of well-chosen *tests* can play a significant role in this process. However, without a working understanding of the characteristics of the tests employed, the physician will have no basis for confidence in the probability of the diagnosis. Where there is no clinical input, as in mass screening or admission profiles, the predictive value of a *positive* test is usually very low, no matter how good the likelihood ratio. Only *negative* results are clearly of value in this situation. It is the joint responsibility of the clinician and the clinical biochemist to establish test characteristics based on a reference value selected to suit the problem at hand.

The ***clinical biochemist*** is concerned with the explanation of pathologic phenomena at the molecular level, and with the search for manifestations that can be measured and that are helpful in differentiating the presence or absence of disease. A fundamental consideration is that a suitable specimen must be obtained, at the proper time, and under appropriate conditions. Quality control of laboratory services is a very complex process:

Analytic factors include the precision, accuracy, and calibration of methods and instruments; interfering factors; and the stability of specimens during their transport and storage.

Biologic factors include the patient's age, sex, body weight (proportion of muscle and fat), organ size, metabolic status, genetic background, physical condition, posture, biologic variation and circadian rhythms, and life-style.

Pathologic factors include toxic effects of tobacco, alcohol, and drugs, the effects of stress, and the effects of one disease on another.

Interpretation factors include test characteristics (SENS, SPEC, PV, Efficiency), the prevalence of the supposed diagnosis, and the methods of recording, processing, and reporting laboratory data.

Reliable data on test sensitivity, specificity, predictive values, and disease prevalence are not easy to find. Value judgments, which depend on intellect and experience, underlie virtually all clinical assessments and decisions. The need for research in this area is being recognized gradually and better data are beginning to become available.

Sophisticated data processing techniques have been tried with the help of the computer. *Cluster analysis* has some merit, but its practical utility is in doubt. *Discriminant function analysis* has considerable potential in spite of rather stringent test requirements. It does permit some improvement in the interpretation of laboratory data and can be used to identify the best three of perhaps twelve tests for a profile. However, no amount of statistical or computer manipulation of laboratory results can correct errors of analysis or of inadequate or inaccurate clinical observation.

A simple yes (+) or no (−) answer may obscure valuable information in a test result. The PV(+) increases in proportion to the extent that a test result deviates from the RV (e.g., exceeds ±3 or more standard deviations). *Repeating* a test will eliminate most false positives, but only if analytical errors are independent from test to test. *Multiple* tests are best run in sequence, because the PV of one test becomes the prevalence for the next. For practical reasons they are usually run in parallel and must be interpreted as a group. A second or third test has value only when the results are conditionally independent of the first (e.g., creatinine and albumin, but not creatinine and urea).

Appropriate sequential testing is more effective than 'blanket' testing, because specificity decreases as the number of tests is increased. It can be shown that to be effective most *profiles* of tests should be kept to no more than three, selected for their efficiency in discriminating between the presence and absence of the suspected disease.

COMPUTERS AND THE LABORATORY

The relationship between laboratory data and a putative diagnosis is statistical and must be expressed as a probability. The extent and complexity of disease variations and laboratory data have outstripped human mental capacity but can be resolved effectively by use of the computer. Not only does it have the ability to deal with thousands of manifestations of hundreds of diseases, but it can be programmed to follow the hypotheticodeductive approach. This power will soon be exploited to convert laboratory data into a more informative message, in some cases with a ranked order of diagnostic probabilities. It remains the physician's responsibility to combine this information with first-hand knowledge of the patient, to make the diagnosis that has the greatest likelihood of being correct, and to initiate treatment.

TREATMENT, TREND ANALYSIS, AND PROGNOSIS

Laboratory data usually form part of the primary clinical investigation of a patient and often contribute to a decision between hospitalization or management at home. It does not follow that establishing a high probability of a correct diagnosis will automatically call for ***treatment*** of the disorder. For certain diseases an effective remedy can be prescribed with perfect safety. For some diseases there is no known cure and only supportive management can be offered to reduce morbidity. In many instances there is a reasonably effective method of treatment that involves a recognized risk of an adverse reaction, for example to a prescribed drug or to surgery. In such situations the clinician has to establish a new reference value, which we may call RV_T (point *D* in Fig. 1–1): the degree of abnormality where the risk of therapy (T) is outweighed by the risk of allowing a mild form of the disease to go untreated. For example, the RV for 'diagnosis' of diabetes mellitus may be a fasting blood glucose above 6 mmol/liter (110 mg/dL), whereas the RV_T (for treatment) may be around 10 mmol/liter (180 mg/dL).

Whether or not the diagnosis calls for some form of treatment, the laboratory may have a role to play in *monitoring* the course of the patient's illness. The serum creatinine in renal failure, bilirubin in hepatitis, and glucose (and now HbA_{IC}) in diabetes are examples. A growing application of laboratory services is to monitor blood levels of prescribed drugs, especially where the range between effectiveness and toxicity is rather narrow (see Chap. 25).

Trend analysis is useful in situations where the direction in which test results are changing may be important. When time factors follow a predictable pattern in the course of a disease, the day-to-day changes in laboratory results help to confirm the diagnosis. Examples are serum enzymes in myocardial infarction and acute pancreatitis, and bile pigments in hepatitis.

In some situations, small changes are important. For example, following a renal transplant relatively small changes in serum creatinine may signal the onset of rejection. How certain can we be that today's value is significantly different from yesterday's? One must have a knowledge of ***variance*** (the standard deviation squared), both *individual* (SD_{IV}^2) and *analytic* (SD_{AV}^2). The former varies with the conditions under which the patient is being observed; the latter depends on laboratory precision. When both can be calculated then the ***expected variability*** (EV) can be expressed in SD units:

$$SD_{EV} = \sqrt{SD_{IV}^2 + SD_{AV}^2}$$

The probability of there being a real difference (*d*) between the two results is then:

$$50\% \text{ at } d = 0.95 \times SD_{EV}$$

$$75\% \text{ at } d = 1.65 \times SD_{EV}$$

$$85\% \text{ at } d = 2.05 \times SD_{EV}$$

$$95\% \text{ at } d = 2.80 \times SD_{EV}$$

Tables for SD_{IV}^2 and SD_{AV}^2 have been published for the more common laboratory tests, but their utility in specific situations may be limited.

Laboratory evaluation of the biochemical status of a patient will sometimes suggest the probable outcome of the illness. Decisions ranging from the need for intensive care or major surgery to discontinuing drugs or discharge from hospital may be made on the basis of laboratory data. In certain illnesses the long-range outcome, or ***prognosis,*** of the patient's disease can sometimes be forecast.

THE COST-EFFECTIVENESS OF LABORATORY SERVICES

If costs were unimportant we could apply our technologic capabilities to the development of profiles of laboratory data for every person in the country. Theoretically, the value of so huge a data bank would be that a patient's laboratory results during an illness could be compared with his or her 'normal' values. However, although *intrapatient* data (i.e., one patient's test results at different times) are remarkably constant, methods change and results from different laboratories at different times are rarely comparable. The predictive value of positive results obtained by *multiphasic screening* in low-prevalence situations is so poor that such data are generally useless or misleading. It is only for populations sorted into higher-prevalence categories by clinical examination that special laboratory work is justified. The clinician's cursory investigation of nonrelevant areas adds very little to the cost of a physical examination. By contrast, a request for a cursory screen of nonrelevant laboratory tests can add considerably to the cost of making a diagnosis, especially if a spurious result prolongs the investigation. It is probably fortunate that the clinician, in most cases, tends to ignore the false positives generated by multichannel profiles, because they do not fit any of the hypotheses that are being reduced to a diagnosis.

Cost effectiveness depends on efficient data utilization, not on data production. Appropriate test ordering and perceptive data interpretation are skills that should be perfected during postgraduate training. The young clinician must understand why tests ordered for different reasons require different characteristics based on the selected reference values.

Cost containment of laboratory services has been attempted in several ways. The greatest improvement thus far has come from the use of problem-oriented order forms, developed in collaboration between clinician and clinical chemist. At a more advanced level the clinician indicates what the *problem* is; the laboratory then sets in motion a predetermined protocol to provide an *answer.*

Finally, one must consider the cost in effort and dollars in relation to the extent that laboratory services improve the physician's ability to reduce morbidity and extend useful life; will it influence the course of the disease in a way that imparts a significant benefit relative to the risks taken.

PRACTICAL CONSIDERATIONS IN REQUESTS FOR LABORATORY ASSISTANCE

The analytic function of the laboratory represents less than half the effort required to provide the clinician with useful information, and this function is usually under good quality control. When the relevant laboratory tests have been ordered the following points must be considered:

1. The *identification* of patients and their specimens must be clear and as free of clerical error as possible.
2. The *collection* and *preservation* of specimens requires special knowledge and efficient organization.
3. In some cases, control of *physiologic* or *emotional* factors is mandatory (e.g., diet, exercise, posture, worry).
4. In some situations, *temporal* and *environmental* factors are important (e.g., fasting, biologic rhythms, stress).
5. Certain *drugs,* including patent medicines, may distort biochemical values or interfere with the analytic procedure.
6. There must be good *communication* between the clinician and the laboratory staff if the patient is to derive maximum benefit from the information provided.
7. The *interpretation* of laboratory data must include an understanding of test characteristics, that is, the sensitivity, specificity, and predictive value of a test in the situation in which it has been used.

Clinical chemists have led the way in identifying analytic variance and its causes. Clinicians are aware of the effects that biologic factors have on the ranges of biochemical values. Both are now placing more emphasis on test characteristics and prevalence, and on the impact these have on the information content of laboratory reports and the effective use of laboratory services.

SUGGESTED READING

BENSON ES, RUBIN M: Logic and Economics of Clinical Laboratory Use. New York, Elsevier, 1978

CONNELLY DP, BENSON ES, BURKE MD, FENDERSON D: Clinical Decisions and Laboratory Use. Minneapolis, University of Minnesota Press, 1982

FEINSTEIN AR: Clinical Epidemiology: The Architecture of Clinical Research. Philadelphia, WB Saunders, 1985

GALEN RS: The normal range. Arch Pathol Lab Med 101:561, 1977

GALEN RS, GAMBINO SR: Beyond Normality—The Predictive Value and Efficiency of Medical Diagnosis. New York, John Wiley & Sons, 1975

MCNEELY MDD: Computerized interpretation of laboratory tests: An overview of systems, basic principles and logic techniques. Clin Biochem 16:141, 1983

PATRICK EA: Decision Analysis in Medicine. Florida, CRC Press, 1979

ROBERTSON EA, ZWEIG MH: Use of receiver operating characteristic curves to evaluate the clinical performance of analytical systems. Clin Chem 27:1569, 1981

SACKETT DL, HAYNES RB, TUGWELL P: Clinical Epidemiology. A Basic Science for Clinical Medicine. Boston, Little Brown, 1985

WEINSTEIN MC, FINEBERG HV: Clinical Decision Analysis. Philadelphia, WB Saunders, 1980

WULFF HR: Rational Diagnosis and Treatment, 2nd ed. Oxford, Blackwell, 1981

2

Pathologic Processes

Frank H. Sims

Pathology is the branch of medicine that examines and describes the effects of disease on the structure and function of the tissues of the body and their reactions to such injury. It is the foundation on which 20th century scientific medicine has been built. The more recent chemical examination of cells and body fluids has introduced new ways of studying the effects of disease on the body at metabolic and cellular levels.

As a result of electron microscopy, microchemical analysis, and the use of immunologic methods, the astounding complexity of ***living cells*** has become increasingly evident. The separate functions of the nucleus, cell organelles, and cytoplasm have been more clearly defined, and complex details of the activities within the cell have been described. A unicellular organism undergoes cell movement by *diapedesis,* a flowing of the cytoplasm from one area to another; it ingests food particles by enveloping them with cytoplasm, the process known as *phagocytosis,* or *pinocytosis* in the case of liquid materials. It reproduces by cell division, the process of *mitosis,* creating two daughter cells of identical genetic constitution.

In the transition from single cells to more complex living systems, many of these basic properties of cells have been lost. Most of the cells of the body have become highly specialized to perform only restricted tasks, a process known as *differentiation.*

THE CELLS OF THE HUMAN BODY

Unicellular organisms are composed mostly of water (70%) containing protein (25%) and dissolved organic compounds and inorganic salts (5%). The human body has approximately the same composition, because it is made up of myriads of cells supported by connective tissue elements. The unicellular organism cannot exist if the surrounding aqueous medium does not contain appropriate food materials and mineral salts in concentrations lying within narrow limits. Human beings carry with them the aqueous medium that constantly bathes the cells of the body. This carefully regulated environment is preserved from loss or excessive evaporation by the enveloping skin, a structure especially suited for this purpose.

Cells of the same type, which occur in a relatively fixed structural relationship to one another and together perform a special function, are said to form a *tissue.* Other groups of cells move freely throughout the body, such as the red cells of the circulating blood and the leukocytes which may migrate through the walls of the blood capillaries into the interstitial tissue. Special tissues have developed for digesting and absorbing food, for transporting nutrients and gases, for removing waste products, and as support structures for the body. Examples of special tissues are

1. The ***skin*** and 'external' ***mucosa*** (Fig. 2–1), with their many layers of cell regenerating from an underlying basal layer, the skin becoming increasingly keratinized and protective.
2. ***Glandular cells*** (Fig. 2–2), which have the special function of producing a secretion, either *exocrine,* to a free surface, or *endocrine,* into the circulating blood.
3. ***Bone,*** which is composed of cells surrounded

by a calcified matrix of great mechanical strength.

4. ***Heart muscle,*** for pumping blood, ***skeletal muscle,*** producing voluntary movement, and ***smooth muscle*** (Fig. 2–3), providing involuntary contractile functions such as peristalsis and vascular resistance.
5. ***Connective tissue*** (see Fig. 2–1), composed of a matrix of proteoglycans, containing collagen and elastic fibers having pliability and strength. It imparts toughness to the skin, provides connection between muscle and bone, supports organ structures, and serves many other purposes.
6. ***Red blood cells,*** carrying oxygen to and some carbon dioxide from the tissues.
7. ***Platelets,*** forming an essential part of the blood clotting mechanism.
8. ***Leukocytes,*** which play a major role in body defenses by their capacity to recognize invading organisms and foreign proteins as 'not self.' In response to suitable stimuli they migrate through the blood vessel walls into the area where a defense role is required (Figs. 2–4 through 2–6). The monocytes of the blood become, in the tissues, wandering histiocytes with phagocytic activity (Fig. 2–7).

NORMAL CHANGES IN BODY CELLS

For an understanding of the changes that occur in the cells and tissues of the body in disease, some knowledge is required of the changes that occur under physiologic circumstances.

Epithelium. The protective stratified squamous epithelium is constantly regenerated by multiplication of the cells of the basal layer and by progressive flattening and keratinization of the cells, with ultimate shedding of the superficial squames. Any change in the nature of the epithelial cells is thus reflected in the shape and appearance of the cells shed from the surface. This is the basis of the *Papanicolaou smear* to detect possible cancer of the uterine cervix. Similarly, the columnar epithelium of the respiratory passages and of the intestine is constantly being shed and renewed. Cytologic examination of the cells present in washings from the lungs or stomach may be of considerable help in detecting sinister alterations in these cells.

Another property of epithelium is its *secretory* capability. The columnar cells of the bronchi of the lung and of the intestinal lining secrete mucus, which protects the surface and also tends to act as a lubricant (see Figs. 2–2 and 2–5). These secretory cells may show increased or reduced activity when conditions are abnormal. For example, the irritation caused by bacterial and viral infections, and the resultant inflammatory response, may cause an increased production of mucus by the cells of the respiratory tract, as in acute bronchitis or in the running nose of the common cold.

Glands. Glands such as the sebaceous glands that are associated with the external skin may undergo hyperplasia, or increased growth and activity, in response to androgenic hormonal influences, as in the facial acne of puberty. Endocrine glands (e.g., the adrenal gland) may also undergo increased activity in response to physiologic stimuli such as stress. The cells of the adrenal cortex produce steroid hormones, particularly cortisol, aldosterone, and androgens. Cortisol is essential for normal functioning of virtually all the cells of the body.

Connective Tissue. Connective tissue has the property of showing increased growth in response to stimuli. For example, the specialized connective tissue of the female breast, as well as the duct system, undergoes increased growth under hormonal influence at the time of puberty. The capacity of connective tissue to proliferate is important in the body's reaction to injury, as will be described later.

Reproduction

Most of the cells of the body have the capacity for reproduction. The life of each is limited, and replacement of the dead cells is accomplished by cell division of neighboring cells as long as the structure or framework of the tissue remains intact. The cells of the mucosa of the intestine are an example. The red blood cells, however, because they have lost their nuclei in the process of development, cannot regenerate by cell division and are replenished by new cells formed in the bone marrow.

In some highly specialized tissues, such as brain, heart, and certain glandular organs, the cells have

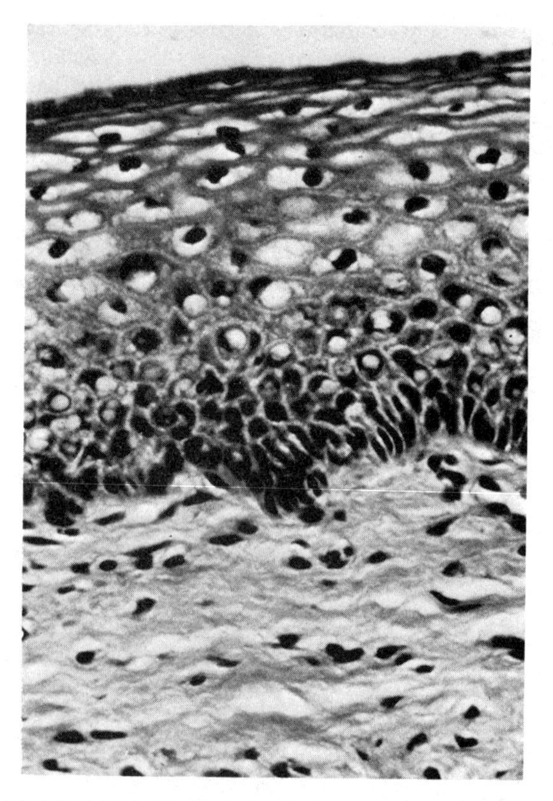

FIGURE 2–1. Normal stratified squamous epithelium from the cervix. Note the regular arrangement of the epithelial cells, the progressive flattening of cells toward the free surface, the clearly defined basal layer, and the adjacent supporting connective tissue. The skin would show, in addition, a horny layer of keratinized epithelial cells of variable thickness on the free surface.

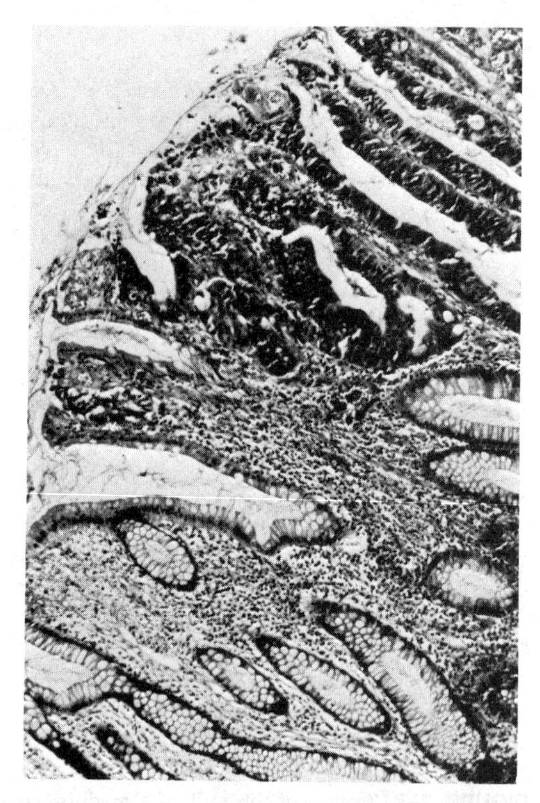

FIGURE 2–2. A portion of mucosa from the colon showing malignant changes in the glandular epithelium. Note the change from regularly arranged mucus-secreting columnar cells to heaped-up epithelium, irregularly arranged and with diminished or absent secretory activity. In other areas (not shown) the cells were freely invading the deeper portions of the mucosa and the muscular wall.

lost their ability to divide and thus cannot regenerate.

PATHOLOGIC CHANGES AFFECTING INDIVIDUAL CELLS

CELL DAMAGE AND CELL DEATH (NECROSIS) AND THEIR CAUSES

In health, the cells of the body are protected mechanically by the surrounding tissues and the aqueous fluid of the interstitial space, and are supplied with nutrient materials by the bloodstream. They are maintained at a constant temperature in a medium of nearly constant composition, and are damaged by any serious departure from this protective and controlled environment. The ways in which cells may be damaged and their response to injury are outlined as follows.

External Violence; Chemical and Physical Agents

Direct *trauma* applied to any body tissue is likely to damage or destroy cells. This obviously can be caused by external forces. It may, however, also result from abnormalities within the body, such as blocking of a duct by a calculus; for example, salivary duct obstruction can cause atrophy of the secretory cells of the gland. Another example is the excessive turbulence of blood flow through

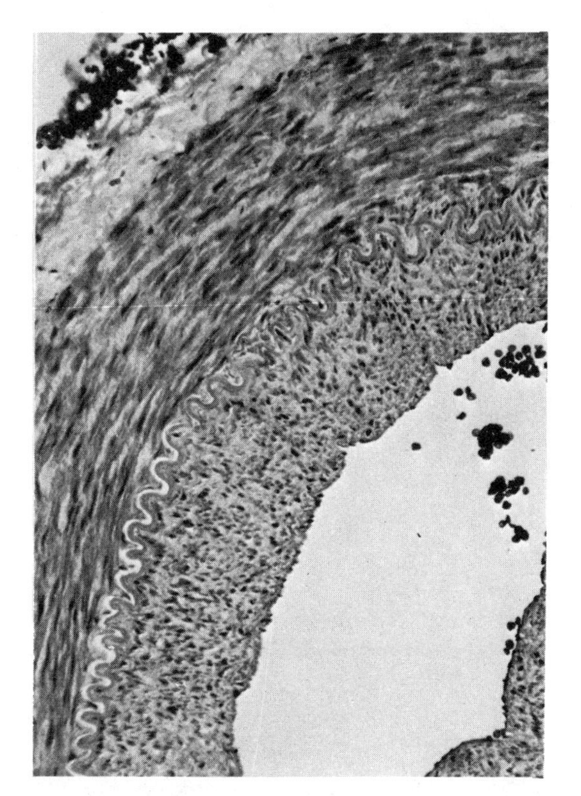

FIGURE 2–3. Section of the wall of a medium-sized muscular artery showing the changes occurring in arteriosclerosis. The smooth-muscle wall and connective tissue of the adventitia are essentially normal, and the internal elastic membrane is clearly seen. The intima is grossly thickened by proliferating cells, most of which are smooth muscle cells arranged longitudinally in the direction of blood flow. The inner surface is covered by normal endothelial cells. In this section there is no evidence of necrosis or abnormal lipid deposition in the thickened intima.

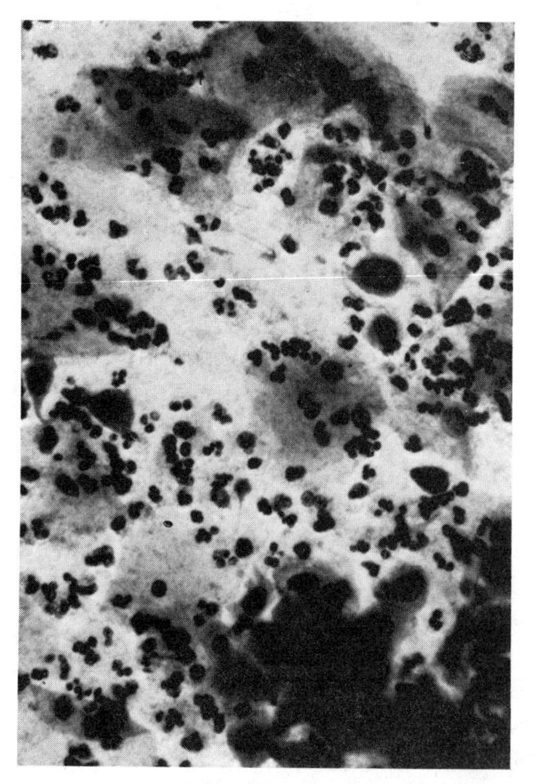

FIGURE 2–4. A Papanicolaou smear showing evidence strongly suggestive of malignancy. Some normal squames are present, but there are numerous small rounded cells (with large densely staining nuclei) which tend to occur in clumps. The area is obviously infected since the smear contains very large numbers of leukocytes (polymorphonuclear neutrophils).

some types of artificial heart valves, resulting in destruction of a proportion of the red cells.

Chemical agents can damage or destroy cells, either by altering the chemical composition of the surrounding medium (e.g., acids or alkalis producing major *p*H changes), or by a direct toxic effect on the cells. For example, in severe uncontrolled diabetes, there is a marked lowering of the blood *p*H due to excessive production of keto-acids from the metabolism of fatty acids. This stimulates the respiratory center, producing 'air hunger' or hyperpnea, and causes cerebral confusion, and ultimately loss of consciousness or coma. Another example is the damage to the liver cells from excess alcohol, which, if recurrent, will lead to their replacement by fibrous tissue and the development of 'cirrhosis' of the liver. As a result there is obstruction to the portal blood flow, effusion of fluid into the peritoneal cavity (ascites), and engorgement of the esophageal veins (esophageal varices). Inhaled chlorinated hydrocarbons may also seriously damage the liver.

Physical agents can damage cells, as for example excessive radiation with x-rays or γ-rays. Radiology staff have to wear lead aprons to protect them from scattered radiation, especially the cells of the go-

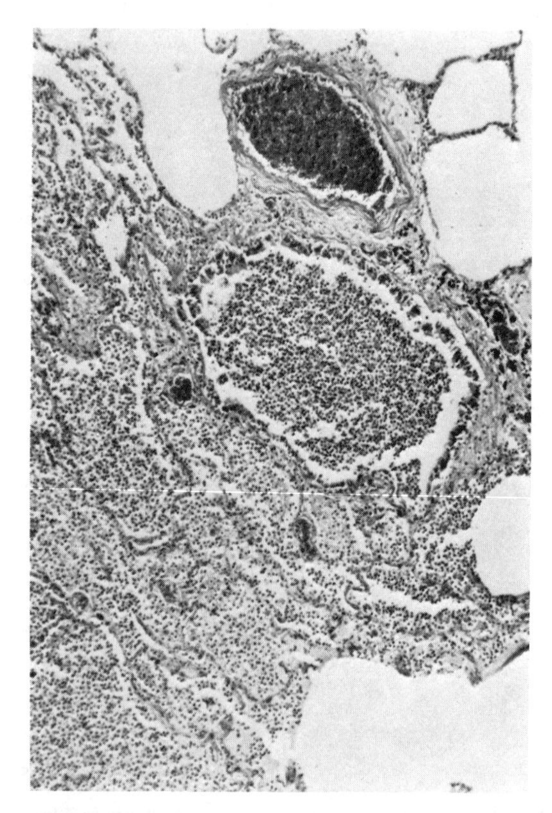

FIGURE 2–5. Lung showing bronchopneumonic changes. Note the bronchus, filled with inflammatory cells, and the neighboring lung alveoli also filled with edema fluid containing large numbers of inflammatory cells. The alveolar walls are still preserved, and resolution of the infection would restore the lung to normal.

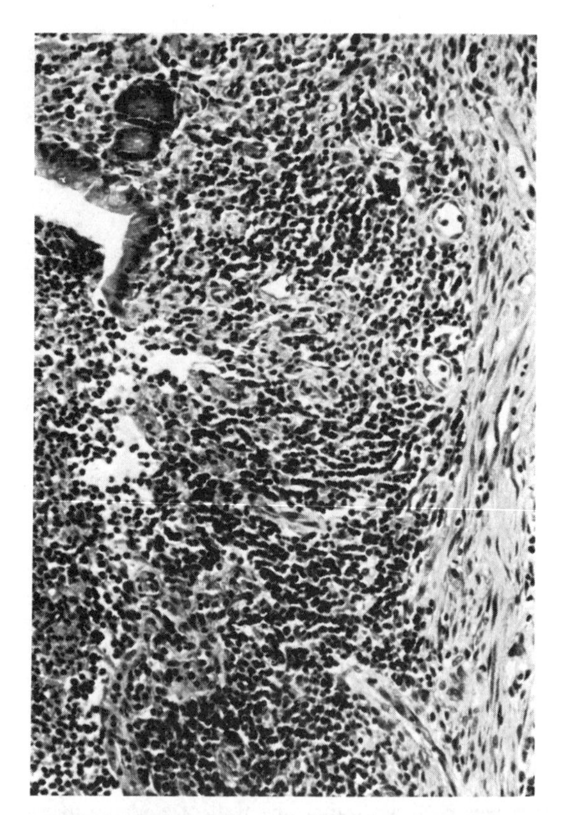

FIGURE 2–6. Portion of the base of an ulcer of the mucosal surface of the colon. Notice the destruction of the epithelium and the heavy infiltration of the base of the ulcer by inflammatory cells. At a deeper level there is increased fibrosis. The surface of the ulcer is covered by an inflammatory exudate containing large numbers of inflammatory cells.

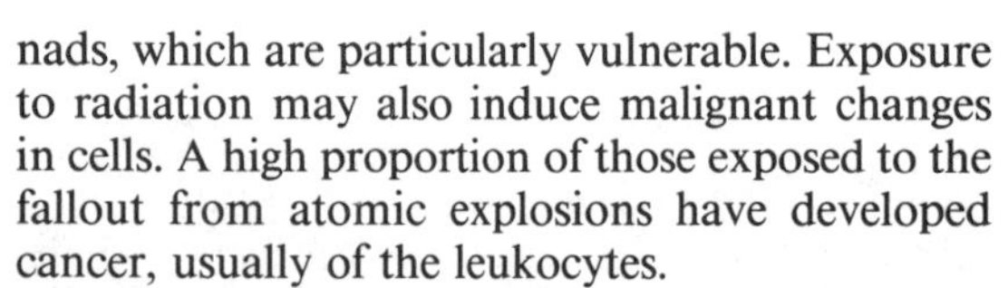

nads, which are particularly vulnerable. Exposure to radiation may also induce malignant changes in cells. A high proportion of those exposed to the fallout from atomic explosions have developed cancer, usually of the leukocytes.

The action of the ultraviolet rays of sunlight in damaging the cells of the skin (sunburn) is well known. There is an increased incidence of skin cancer in individuals exposed to high levels of sunlight for long periods, particularly those without much protective skin pigment.

Biologic Agents

Of major importance in human pathology is the cell injury produced by biologic agents. The cells may be injured by *toxins* produced by foreign organisms that have in some way penetrated the first line of defense of the body (the epithelial covering of all the free surfaces) and have begun to live and multiply within the body. A typical example is the common boil, in which the organism *Staphylococcus aureus* enters the tissues through a hair follicle or some minute skin abrasion and sets up an acute inflammatory mass. Viruses not only penetrate the tissue spaces between the cells but also enter the cells themselves and damage or destroy them (Fig. 2–8). A prime example is viral *hepatitis* B, in which the virus may cause the death of a high proportion of the cells of the liver. Another example is *poliomyelitis,* in which the virus enters

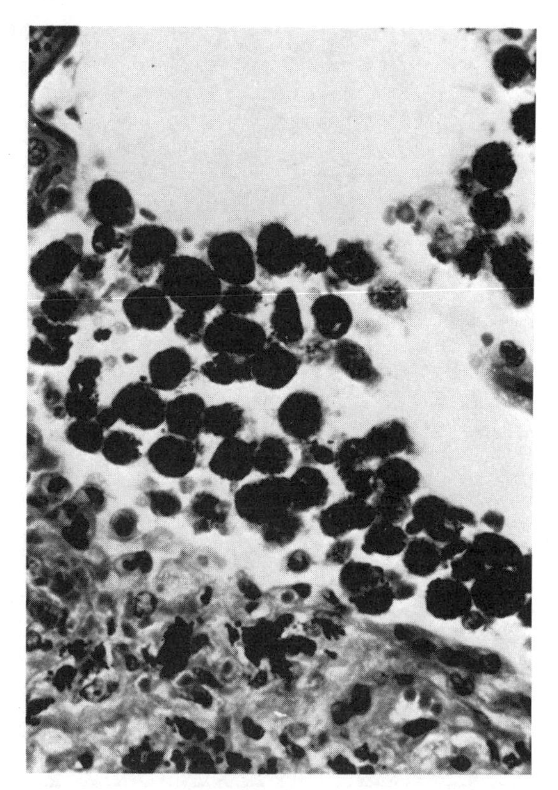

FIGURE 2–7. A portion of a lung alveolus containing numbers of phagocytic cells filled with blood pigment. The patient had congestive heart failure. More pigment-filled phagocytes are present in the interstitial tissue of the adjacent lung.

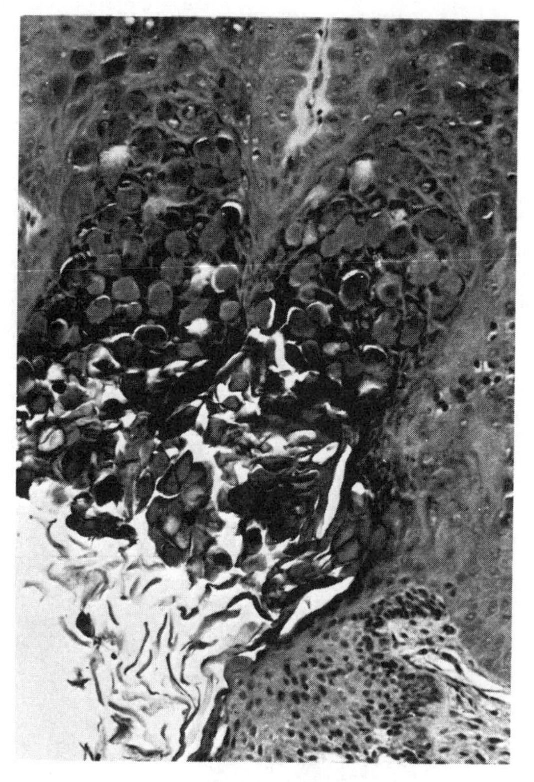

FIGURE 2–8. Molluscum contagiosum. This benign skin lesion shows the effect of a virus on epithelial cells. These are swollen, and under higher power can be seen to be necrotic. A small area of unaffected stratified squamous epithelium can also be seen.

and destroys the motor nerve cells of the spinal cord, resulting in permanent paralysis of the muscle fibers controlled by these nerve cells. If a virus (e.g., Herpes simplex) enters the nerve cells of the brain to produce an *encephalitis,* with associated congestion and swelling of the brain tissue, the patient usually becomes comatose, and even if the patient survives there may be severe permanent damage to the nerve cells.

Foreign protein molecules can themselves have a toxic effect on the cells of the body and may be absorbed in small amounts through an intact epithelial surface. Such absorption of intact protein is responsible for the manifestations of *allergy,* such as the reaction that may follow the ingestion of shellfish in an individual sensitive to this food. An urticarial skin rash and possibly severe respiratory difficulty may develop. Foreign proteins may be introduced in other ways, such as by bee or wasp stings.

Normally we do not produce ***antibodies*** to our tissues, because the immune mechanism recognizes as 'self' the wide range of proteins produced by our own body cells. At times, however, the defense forces of the body can cause damage to one's own cells. If a foreign protein becomes coupled to one of the body's tissue components, or has antigenic determinants in common with one of the body's own proteins, the antibodies produced by the defense cells against the foreign protein may react also with the body's tissue. If this happens the *autoantibody* mechanism may produce damage and destruction of the cells that were the source of the antigen and cause their replacement by fi-

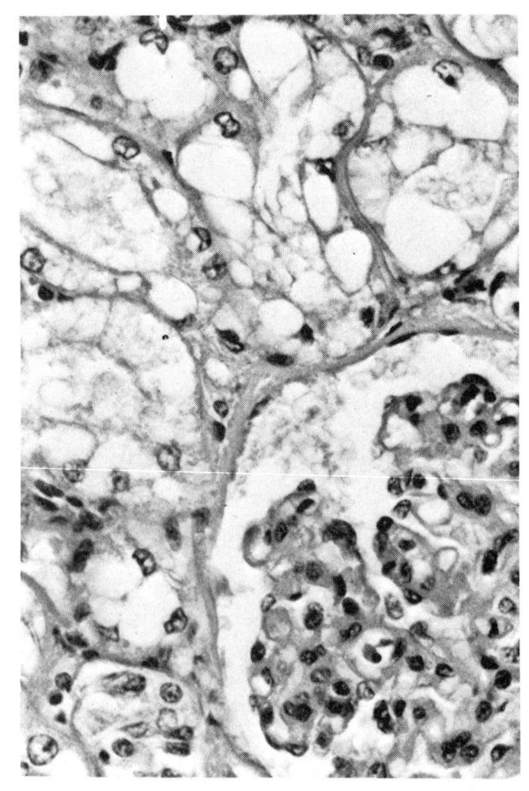

FIGURE 2–9. Kidney tissue from a patient with severe potassium depletion. The portion of glomerulus shown is essentially normal, but the adjacent proximal tubular cells are swollen and vacuolated; cells in the distal tubule would be affected as well.

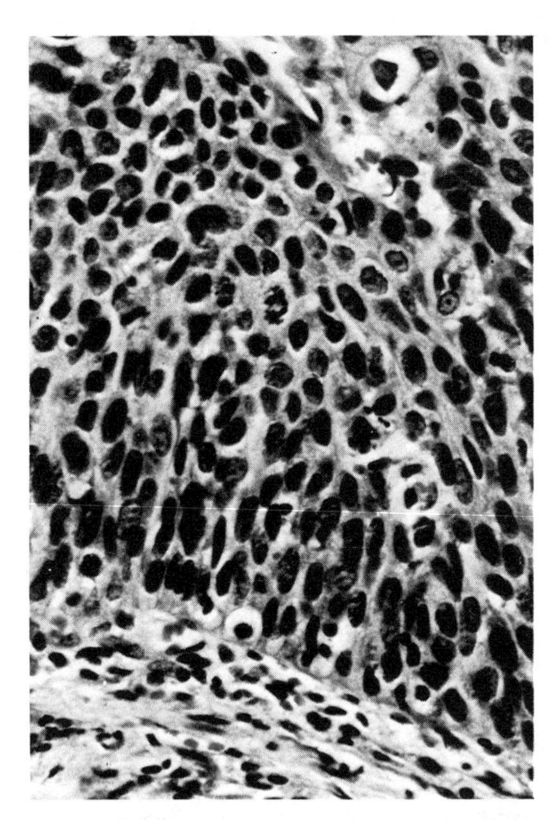

FIGURE 2–10. Stratified squamous epithelium from the cervix, showing severe dysplasia. There is irregularity in the arrangement of the epithelial cells, an absence of layered flattened cells, evidence of active growth (mitotic figures), and early disorganization of the basal layer.

brous tissue; an example is the Hashimoto type of *thyroiditis.* Another example is the damage caused by *rheumatic fever.* In this situation, sensitization to *β-hemolytic streptococci* of Lancefield group A, infecting the pharynx of the patient is thought to produce a reaction directed against the connective tissue of the body, particularly to the heart valves, which may suffer severe permanent scarring and distortion.

Nutritional Deficiency

A reduction in blood flow to an area of the body will have no effect on the health of the cells so long as the flow is adequate for the nutritional and gaseous exchanges required. A progressive reduction, however, will reach a point at which the requirements of the cells can no longer be met, and at this point cellular function will become impaired. More severe circulatory impairment will cause serious cell damage, and will ultimately produce cell death. This ***ischemia,*** or circulatory 'starving' of the tissues, is one of the important changes that occurs in the aging process. As we will see later, there is no escape from it; the end result is failure of critical organs or tissues, and finally death of the individual. This type of cell damage is an important underlying factor in many human diseases.

Many other types of nutritional deficiency occur, only a few of which can be mentioned. For example, in some patients, owing to prolonged abnormal loss of *potassium* in feces or urine, a state of potassium deficiency may develop. In this situation skeletal muscle may become so weak that the patient cannot stand up, and there may be impairment of respiratory or cardiac function. Vacuolar damage to the cells of the renal tubules can

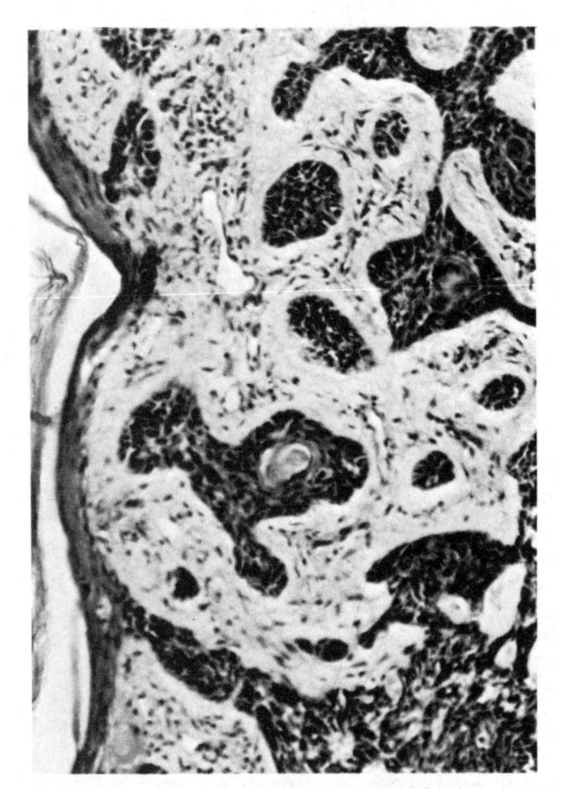

FIGURE 2–11. The typical structure of a basal cell carcinoma. The photomicrograph shows the tumor cells growing inward from the epithelial surface (at the left) into the adjacent connective tissue. In this case no ulceration of the overlying epithelium has occurred. This type of tumor is locally invasive, but only with extreme rarity forms metastases. It is therefore easily cured.

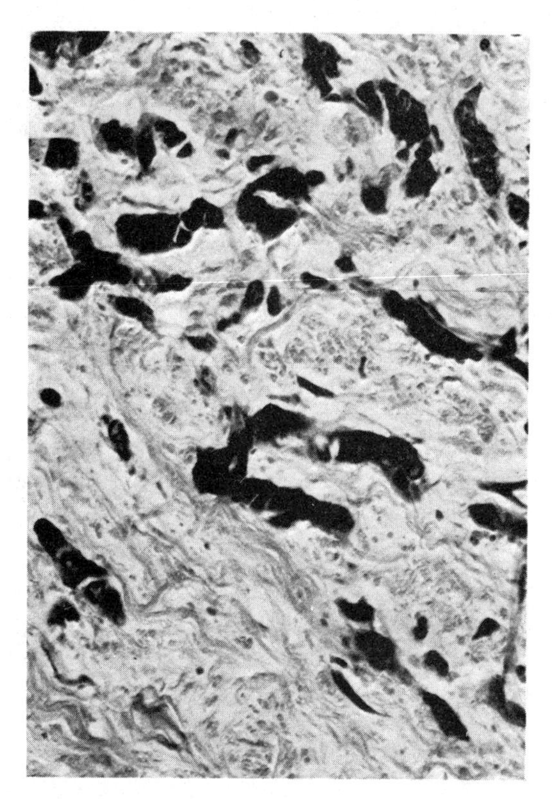

FIGURE 2–12. Section from an infiltrating breast carcinoma. The darkly staining tumor cells are present singly and in small groups growing along the connective tissue planes, and may set up metastatic tumor growths in other sites in the body. Notice in this case an apparent absence of defense cells or inflammatory reaction.

also occur (Fig. 2–9). These changes are reversible upon correction of the potassium deficiency.

When the diet lacks vitamin D, particularly in the absence of sufficient sunlight to permit synthesis in the skin, there is impairment of calcium absorption from the food and of calcium deposition in the bones. The result is soft bones, inadequate in mechanical strength, which bend under stress and give rise to the clinical findings of *rickets*—bow legs, pigeon breast. In vitamin C deficiency (the basic cause of scurvy) the body is unable to hydroxylate proline, and the production of collagen is imperfect. This results in friable tissues that allow the rupture of small blood vessels. Bleeding gums occur, with petechial hemorrhages elsewhere and defective bone formation, and the patient will die in the absence of treatment, as did many of the early explorers and their ship's crews. Vitamin B deficiency causes alterations of nerve conduction, and alterations of function result from trace metal deficiencies. Multiple deficiencies can occur in malnutrition, as shown by the severe liver damage in children suffering from kwashiorkor, and the advanced fatty change of the liver cells seen in ulcerative colitis.

THE EFFECTS OF CELL DAMAGE

Recovery with No Residual Impairment of Function

The effects of cell damage may be transient, with recovery of the cell upon restoration of a normal environment. If the damage is too severe, however, the end result is cell death. When ***cell death*** occurs there is a reaction on the part of the surrounding tissue, directed toward regeneration or repair. if the basic structure of an organ or tissue is not damaged, and if the source of the trauma is removed, the remaining cells may divide and multiply to replace the dead cells and reconstruct the tissue. This happens, for instance, when the renal tubular cells die as a result of toxic damage or a transient hypoxia from a severe fall in blood pressure. If the patient can be kept alive, by peritoneal dialysis or with an artificial kidney, for a period of some 10 to 14 days, there is a good chance that regeneration of the tubular epithelium will occur, with restoration of the kidney to essentially normal function. Similarly, if a lobe of the lung is the site of an inflammatory process such as lobar pneumonia, or is damaged by a blockage of its blood supply, as in pulmonary embolus, the patient's respiratory capacity will be impaired. If, however, the patient can be kept alive in spite of the reduced ventilatory capacity, and if any coexistent infection can be overcome, the lung tissue will be restored to normal by phagocytic removal of the debris within the lung alveoli. No permanent disability will then result.

Permanent Loss and Fibrous Replacement of the Cells

If, on the other hand, damage to cells is more severe and involves the basement membrane of the tissue, regeneration is no longer possible, and the repair process is accomplished by filling in the damaged area with fibrous connective tissue. Cells with a specialized function will have been permanently destroyed, and the connective tissue replacing them has no corresponding activity. Thus, the tissue or organ is permanently damaged. The area of fibrous tissue replacement is referred to as ***scar tissue.*** Such scarring is not of serious concern when it affects structures such as the skin or skeletal muscle, but it may cause serious impairment that can limit life when it occurs in vital structures such as heart muscle, kidney tissue, liver, or brain.

METAPLASIA

Metaplasia is the term applied to the transformation of one type of specialized cell to a form that is abnormal in that particular situation. For example, chronic irritation of the columnar cells of the bronchi of the lungs leads to a transformation of these cells to stratified squamous epithelium. Similar changes can occur in the columnar epithelial cells of the glands of the uterine cervix. The initiating factor may be mechanical or chemical trauma, and in some cases the transformation may constitute a premalignant change.

DYSPLASIA

In dysplasia, highly specialized cells show evidence of incomplete or abnormal development. In the case of the stratified squamous epithelium of the cervix, for example, the epithelial cells may cease to grow in the regular fashion, (larger cells in the basal layer and successively flatter cells toward the surface); instead, they may be of irregular size and shape, arranged in a haphazard fashion. These abnormal cells will be seen in the Papanicolaou smear. This is prepared by scraping the surface of the female cervix with a blunt edge (e.g., a wooden applicator stick), spreading the material on the surface of a glass slide, and staining. The smear contains a selection of superficial epithelial cells that are readily dislodged by this simple procedure. Abnormal cells and cells of sinister appearance may be identified readily by a skilled observer (Fig. 2–10; see Figs. 2–1 and 2–4). Similarly, in the alimentary tract the columnar cells may lose their capacity for mucus secretion and grow in a heaped-up fashion that is quite abnormal in appearance. This change in the character of the cells is an indication of unregulated multiplication and is regarded as a premalignant transformation.

ANAPLASIA

Anaplasia is the sequel to dysplasia. It represents the situation in which cells that are normally highly differentiated have become much less differentiated, and thus have an appearance of comparatively unregulated growth. If the growth of the cell is still under control, the changes are of the same type as those constituting dysplasia but are more severe. When applied to tumor cells the term refers to cells that are much more primitive in appearance than the cells from which they arose.

NEOPLASIA

Neoplasia refers to the situation in which a certain type of cell grows and multiplies by cell division in an uncontrolled fashion. Such cells form by their growth a mass that compresses and pushes aside the surrounding tissue and may destroy it. Alternatively, the cells may infiltrate widely into the surrounding tissue and set up proliferating tumor masses elsewhere in the body. The cell mass is called a ***benign tumor*** if the cells remain a local mass and do not move through the body tissues to set up metastatic deposits elsewhere (Fig. 2–11). The tumor is said to be ***malignant*** if a characteristic feature of the cells is their ability to infiltrate the surrounding tissues and to be carried by the lymphatic channels or vascular system to distant sites, there to set up satellite growths or metastases (Fig. 2–12).

It is to be noted that a benign tumor may still be locally destructive as a result of the pressure of its expanding mass, and if it occurs in a critical area of the body it may damage or destroy some vital structure. For example, a meningioma, a tumor growing from the fibrous connective tissue membrane covering the brain, may compress and destroy vital brain tissue. A malignant tumor, if not successfully treated, will eventually kill the patient. The capacity of tumor cells to survive and grow in other locations throughout the body means that inevitably, in the course of time, vital body structures will be irreparably damaged by the metastatic tumor deposits.

PATHOLOGIC CHANGES AFFECTING TISSUES

THE NATURE OF THE REACTION OF TISSUES TO INJURY

The Inflammatory Process

The reaction of the tissues to injury is a complex process that is still not fully understood. The principal steps in inflammation are essentially the same whether the injury is due to mechanical trauma, chemical or physical agents, invading organisms, or some antigen–antibody reaction. The cardinal signs were formulated in the succinct but still apt words of Celsius in the first century A.D.; they are heat, redness, pain, and swelling, leading to loss of function.

The first phase, ***heat*** and ***redness,*** results from a reflex dilatation of arterioles and capillaries due to stimulation of the nerve endings, perhaps associated with histamine release at the site of the tissue injury. The increased blood flow carries defense cells to the area. The ***swelling*** is due partly to the increased vascularity and partly to the fact that the injury leads to increased permeability of the capillary walls, with an effusion of protein-containing fluid into the surrounding tissue spaces. It is in this way that antibodies arrive at the site of the injury. The ***pain*** may result from direct damage to nerve endings and from the increased tissue tension caused by the swelling. Blood and tissue contain precursors of proteolytic enzymes such as kallikrein, believed to be formed at the site of injury. These enzymes act on plasma and tissue proteins to produce vasoactive polypeptides, such as bradykinin, which increase capillary permeability, are chemotactic to white cells, and cause pain. They may be the prime mediators of acute inflammation.

Blood flow is slower in the dilated capillaries and venules and is accompanied by changes in the cells lining these vessels. The changes cause adhesion of leukocytes and platelets to the endothelial surface. This facilitates the ***migration*** of white blood cells, which pass between the endothelial cells of the capillary wall into the surrounding tissue. In the *initial phase,* the cells moving into the inflammatory area are nearly all polymorphonuclear leukocytes (polymorphs), which are the acute defense cells of the body. Later, monocytes, lymphocytes, and plasma cells move out of the vessel into the area. The monocytes become tissue macrophages; the lymphocytes and plasma cells preside over cell-mediated immunity and antibody production (see Chap. 4).

If bacteria have gained access to the area, the defense cells, aided by antibodies present in the ***effusion fluid,*** will take up and destroy the organisms. Some of the tissue components and many of the polymorphs may be destroyed, but this inflammatory debris will be removed by remaining phagocytic cells. The edema fluid contains fibrinogen, which is converted by the products of tissue damage to fibrin, strands of which serve to wall off the affected zone from the surrounding tissues and aid in the movement of fibroblasts into the area during the stage of repair. During this phase there is replacement of the dead tissue by fibrous tissue. If highly specialized cells have been lost and cannot regenerate to restore normal function, there will be permanent and sometimes serious consequences of the tissue damage.

When the cause has not been removed, or healing is delayed, a later or *chronic phase* of an inflammatory process occurs. The stimulus for this phase is a complex mixture of abnormal tissue constituents, including vasoactive amines, short-chain polypeptides, and probably prostaglandins. Although the tissues tend to become refractory to histamine and to bradykinin, the release of lysosomal enzymes by leukocytes leads to continued production of these substances, to cleavage products of complement, and possibly to formation of serotonin (5-hydroxytryptamine). Prostaglandins may be formed by the platelets; they are potent vasodilators and cause smooth-muscle contraction.

The Effects of Inflammation

The effects of the inflammatory process are both good and bad. The pain and swelling promote rest and protection of the damaged area. The inflammatory process is an important *defense* mechanism of the tissues against infection and plays a vital part in the repair of any area damaged by trauma. There are, however, a number of undesirable effects, which in some circumstances may be serious. It has already been pointed out that the damaged tissue is often replaced by fibrous tissue. Thus damage to the heart or kidney, as a result of infection or a serious reduction in blood supply, will result in an area of ***fibrosis*** having no ability to function like the original tissue. When brain damage occurs through trauma, cerebral hemorrhage, or thrombosis, the damaged brain tissue is permanently replaced by fibrous tissue, sometimes with serious and permanent loss of neuronal function.

In some circumstances the ***swelling*** of the damaged tissue can have serious consequences. For example, swelling of damaged brain tissue within the confined space of the cranial cavity can lead to such a rise in pressure that the blood supply to vital brain centers is impaired, at times fatally. In osteomyelitis, in which there is an acute bacterial infection of the bone, the rise in pressure within the marrow cavity produces acute pain and sometimes further destruction of marrow tissue and bone, owing to deprivation of the blood supply as a result of the pressure. When a fractured limb is placed in a rigid plaster cast the circulation has to be watched with great care, since the subsequent swelling may interfere with the circulation of blood and could, in extreme circumstances, result in gangrene and death of the whole limb.

When ***fibrous tissue*** is first formed it is red owing to its high vascularity, but in time it tends to shrink and eventually becomes white, having a very scanty blood supply. This means that major vessels, nerves, ducts, and other essential structures that pass through the damaged tissue will, in the course of time, become constricted, with the possibility of serious loss of function. For example, a kidney stone impacted in a ureter may cause sufficient fibrosis to obstruct the ureter, even though the stone has been passed or removed surgically. Traumatic damage to the skin can produce extensive scars that contract in the course of time and may produce limitation of movement or deformities. For example, after extensive burns, plastic surgery may be necessary to rectify the disfigurement produced by this contraction of scar tissue; after radical mastectomy scar tissue may cause some limitation of movement of the arm on the affected side.

INFECTION

The defense system of the body is highly effective in keeping the tissues free of invading organisms, in spite of an environment that is crowded with potentially hostile life, ranging in size from viruses through bacteria and fungi to parasitic worms and flukes.

When air containing dust and microorganisms is breathed in, the convoluted nasal passages, covered by mucus-secreting ciliated *epithelium,* mechanically entrap a good portion of the foreign particles and organisms as the air passes over these surfaces before entering the trachea and bronchi. More of the foreign material is deposited in the mucus covering the surface of these larger respiratory passages, and very few organisms reach the lung alveoli. Any that do, however, are engulfed by phagocytic cells (see Fig. 2–7).

Organisms entering the alimentary tract are less

likely to find their way into the body tissues, because it would be necessary for them to survive gastric acidity and to traverse the intact epithelium of the intestinal wall. Any organisms that do gain access to the body through this route meet with the cellular defenses of the deeper layers of the mucosa of the alimentary canal, and they are destroyed in the normal individual.

If entering organisms are not immediately destroyed, but multiply in a particular area, they may cause tissue necrosis. There will be a surrounding zone of inflammation crowded with defense cells. The central area will be filled with cellular debris from the necrotic tissue and the defense cells that have died. Such an area is called an ***abscess,*** and the turbid fluid from the central cavity containing cell debris is termed ***pus.***

Occasionally the infection may spread throughout an area of the body *without* localization. In the lung this is a common phenomenon and is the general situation in diffuse bronchopneumonia. There is an outpouring of fluid and inflammatory cells into the lung alveoli surrounding the bronchi (see Fig. 2–5). Therefore these alveoli become temporarily useless for the interchange of oxygen and carbon dioxide between the inspired air and the circulating blood. If the infection is overcome, and provided there is no destruction of lung tissue, the lung alveoli are cleared of inflammatory debris by phagocytic action, and the lung tissue is restored to normal.

ULCERATION

When the epithelium of body surfaces is lost as the result of severe damage (whether mechanical, chemical, or due to some other factor), the underlying tissue is directly exposed to the external environment. Infection of the denuded area is virtually certain, and an acute inflammatory reaction will develop in the surrounding tissue. Usually the trauma and infection will cause necrosis of the superficial zone of the exposed tissue, which will be covered by an exudate containing (like the necrotic center of an abscess) numerous dead defense cells and cell debris. This is the basic structure of an ***ulcer*** (see Fig. 2–6). The healing of such a lesion is dependent upon a successful defense of the underlying tissue from bacterial invasion, an adequate blood supply to the area, and favorable conditions for the growing in of new epithelium from living cells at the periphery. The *depth* of the ulcer crater depends on the balance between trauma from the damaging agent and resistance of the underlying tissue. One of the normal defense mechanisms of the body is a proliferation of fibrous tissue, and this occurs in the base of the ulcer, walling off the area of damage from the surrounding normal tissue.

Malignant disease frequently results in severe intractable ulceration when it involves a free surface. The reason for this is that the tumor cells destroy the surface epithelium, and their presence in the underlying tissue seriously hinders the normal reaction against infection. Even if the infection is contained, the cancer cells prevent the ingrowth of normal epithelium to cover the defect, and a *permanent malignant ulcer* results.

ISCHEMIA

We have indicated how function is impaired and cell death may occur if the cell is deprived of nutrients. Because similar cells are grouped together to form tissues and organs that have special functions, such structures depend for their supply of oxygen and other nutrient materials on the circulating blood of the area. This also carries away water and small metabolic waste products of the cells. Substances of large relative mass that escape from the blood capillaries are drained from the area by the lymphatic system.

The ***blood supply*** of any tissue or organ may be ***impaired*** in several ways. With the passage of years the arteries of the body undergo degenerative changes, which ultimately affect their capacity to carry blood to the areas they supply. These changes constitute a major cause of the impairment of function with advancing age and share with malignant disease a leading place as a cause of death.

Arterial Disease

The arterial wall of the larger vessels (down to arteriolar size) is composed of smooth-muscle cells with variable amounts of elastic connective tissue, lined by a thin inner layer called the *intima,* and strengthened on the outer surface by collagenous connective tissue (see Fig. 9–2). The intima layer in the newborn infant is composed of a layer of flattened endothelial cells, separated from the in-

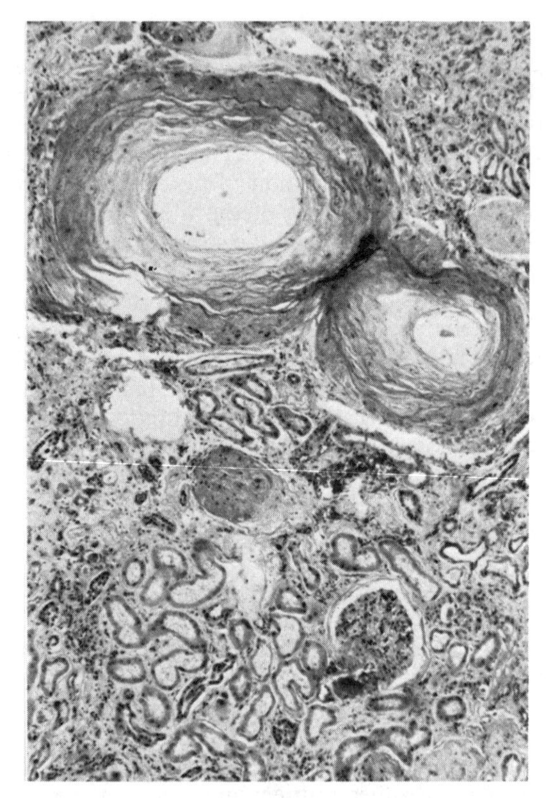

FIGURE 2–13. The end result of severe arteriosclerosis, seen in the small renal arteries. The thickened intima here contains much fibrous tissue and shows areas of degeneration. The severe narrowing of the lumen can be seen clearly. The diminished blood supply causes pronounced degenerative changes in the surrounding kidney tissue, indicated by fibrosis in the glomeruli and increased fibrosis in the interstitial tissue between the renal tubules.

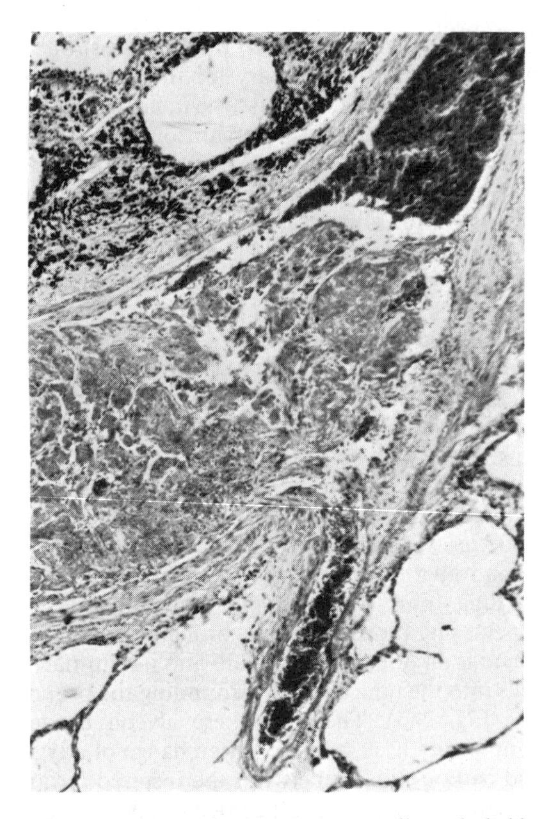

FIGURE 2–14. A pulmonary artery totally occluded by an embolus. The removal of the embolus by phagocytic activity is well advanced. The embolus is pale in color owing to the removal of the red cells by phagocytes; other phagocytic cells are résolving the fibrin and precipitated protein. A freshly formed thrombus is present close to the embolus. Note the heavy pigmentation of the adjacent lung tissue, due to large numbers of pigment-filled phagocytes.

ternal elastic lamina (which forms the inner surface of the musculoelastic wall) by a few scanty strands of connective tissue that are scarcely recognizable in microscopic sections. Normal circulatory trauma and the permeability of the endothelial lining of the ***arteries*** allow diffusion of protein-containing fluid and cells through the endothelial surface. This sets up a reactionary repair process, resulting in the proliferation of smooth-muscle cells and fibrous connective tissue in the intima between the endothelium and the internal elastic lamina (see Fig. 2–3).

At first the ***intimal thickening*** has no significant effect on the internal diameter of the vessel, but in the course of time the process seriously reduces the size of the lumen, particularly in medium-sized arteries, resulting in a pronounced reduction in the blood supply to the tissues they serve. Serious degenerative changes then begin to take place in these tissues (Fig. 2–13).

The rate of intimal thickening varies from one individual to another, and from one artery to another in the same individual, and is often complicated by the deposition of lipid in the thickened

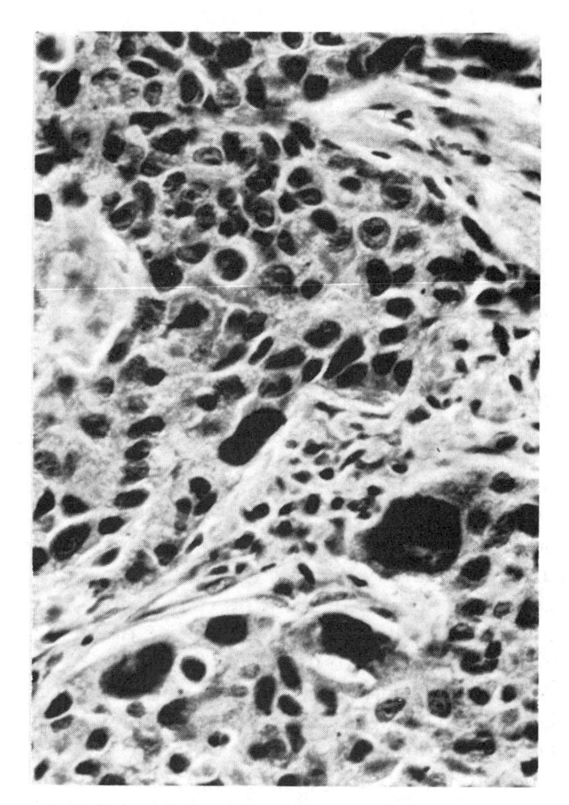

FIGURE 2–15. A section of breast carcinoma. It shows the irregular arrangement of the tumor cells, the presence of mitotic figures, and a number of large, irregularly shaped darkly staining masses, which are calcium deposits in small areas of necrosis.

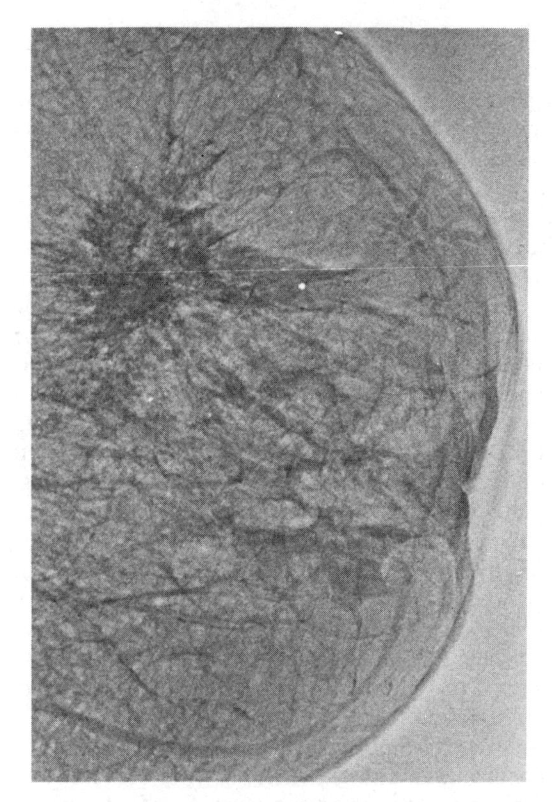

FIGURE 2–16. Mammogram of a breast carcinoma showing the irregular extension of the tumor into the surrounding breast tissue and its fibrous reaction. The contraction of the fibrous tissue connecting the tumor and the nipple has caused retraction of the nipple. A higher magnification would show fine dense stippling of the tumor due to calcium deposition.

intima. Initially this takes the form of large cells with foamy cytoplasm containing lipid. Many of these cells break down and lipids accumulate in the connective tissue to form the degenerative deposit called ***atheroma.*** Cholesterol, freed by hydrolysis of its esters, separates out in characteristic crystals that are extremely resistant to re-solution and removal. Areas of necrosis and hemorrhage may occur, probably aggravated by the hypoxia that exists in the deeper layers of the thickened intima. There is almost always calcium deposition as well. The rigidity of the vessel, due to the thickened and calcified intima, leads to atrophy of the smooth muscle of the media of the arterial wall. The end result is severe ***atherosclerosis*** with its thickened wall, narrowed lumen, and lipid deposition. The reduction in the cross-sectional area of the vessel may vary from slight to a constriction so severe that a lumen scarcely exists at all. The mechanism of this intimal thickening and lipid deposition is still not clearly understood (see Chap. 9 for further discussion).

Thrombosis. One of the dangerous complications of the process of intimal thickening and arterial narrowing is *complete obstruction* by the development of a blood clot in a narrow portion of the vessel. This event is made more likely by damage to the endothelial surface of the vessel, perhaps by increased turbulence of flow or by the breakdown of an atheromatous plaque. Platelets, fibrin, and entrapped red cells may collect on the damaged surface, and the mural clot so formed can enlarge to occlude the lumen of the vessel, particularly if this is small. In some anatomic sites, such as heart muscle or kidney, the area supplied by each of the major branches of the arterial tree may be poorly supplied by collaterals from neighboring vessels. In this case, the sudden obstruction of such an artery may result in the death of tissue in the area supplied, with eventual replacement of the specialized tissue by fibrous connective tissue. This type of lesion is termed an ***infarct.*** It can be appreciated readily that such a process of infarction, if it involves a substantial volume of tissue, or if it affects a critical area, may produce serious disability or even death. The heart muscle is particularly vulnerable, because its blood supply is carried by only three main arterial branches, it is in constant activity, and it is the vital circulatory pump of the body.

When sudden obstruction of a *cerebral* artery occurs, the patient has suffered a form of *stroke,* and the neurons in the affected area will die and can never be replaced. The patient will die at once if this area of damage includes vital centers such as are present in the brain stem. Survival is probable if the damaged area involves the cortical region of the brain, but the patient may suffer muscular paralysis caused by the destruction of nerve cells with motor function.

Although the process of intravascular thrombosis may occur at any time within the lumen of arteries, the walls of which have been damaged by degenerative processes, the rapidity of flow of the arterial blood tends to ensure that it is a relatively uncommon event. In the case of the ***veins,*** however, occlusion by thrombosis is much more common, especially if the wall of the vein is damaged or inflamed, as in *phlebitis.* Venous walls are not subjected to the same stresses as arterial walls and hence do not suffer the same type of intimal damage with all of its sequelae. On the other hand, the flow of blood through the veins is relatively sluggish, and this tends to favor the clotting of blood within the vessels. Such an event is more likely to occur in lower portions of the body (e.g., the legs), where the flow in the veins is particularly slow unless facilitated by muscular exercise. It thus occurs most frequently when the patient is lying in bed, especially if movement is restrained, or when there is sufficient pressure on the legs to obstruct some of the veins. Venous thrombosis in the leg will prevent the return of blood from the limb and will thus increase capillary pressure, causing an effusion of fluid from the capillaries into the surrounding tissue. The result is a swollen and edematous limb that at times causes little discomfort, but which may be painful and tender.

Embolism. If venous thrombosis should occur, the patient is exposed to another serious danger, that of embolism. This is the dislodgement of a portion or the whole of the blood clot, its transport to the right side of the heart, and from there into the *pulmonary* circulation (Fig. 2–14). If the dislodged blood clot is a large one, such an accident can almost totally occlude the pulmonary circulation, and sudden death will occur. If the blood clot is smaller, it will occlude a portion of the pulmonary bed and produce an infarct of the lung tissue. The supply of blood from any branch of the pulmonary artery to its particular segment of the lung is so richly joined, or anastomosed, with that from neighboring arteries that the lung tissue usually does not undergo necrosis. However, the involved area does become seriously deprived of oxygen and nutrient substances, which results in damage to the capillary walls and an effusion of blood into the damaged tissue. In the course of time the embolus is removed by phagocytic cells; the blood in the lung alveoli is also removed by phagocytes, and in about 2 months the infarcted area of the lung tissue is restored to normal. In contrast, infarction of the heart muscle or other organs supplied by the systemic circulation generally does not end with complete restoration in this manner, but with varying degrees of replacement fibrosis or scarring.

Although in clinical practice venous emboli resulting in pulmonary embolism are seen most commonly, *other types* of embolism occur. In bacterial endocarditis, vegetations composed of platelets, fibrin, and entrapped bacteria may be dislodged from the heart valves and may be carried in the arterial blood to produce infarcts in the kidneys, brain, or bowel. When the atria of the heart are in fibrillation and fail to contract normally, blood clots may form in the atrial appendages and produce emboli that will be liberated into the arterial circulation.

The term *embolus* refers to any particle or foreign material carried by the bloodstream that may block a blood vessel, and includes gas emboli and fat emboli. Air emboli can be formed by the accidental introduction of air into the veins (e.g., from a deep neck wound or an intravenous needle). Nitrogen emboli may form when a person has been breathing air at high pressure and returns too rapidly to an environment at normal pressure. Because the solubility of a gas in the circulation is proportional to its pressure, reduction of the external pressure will cause the nitrogen to separate out of solution in the form of minute bubbles, which may take a considerable amount of time to absorb. Carried in the circulation these bubbles will block small vessels, causing severe pain (the 'bends') and sometimes fatal damage to the brain. Divers, caisson workers, and occasionally high altitude pilots are at risk.

A fat embolus occurs when traumatic injury to fatty tissue (usually the fatty tissue of bone) dislodges fat from the damaged tissue, which then enters the circulating blood. Cholesterol and lipid emboli may arise from atheromatous plaques on the inner surface of diseased arteries.

HEMORRHAGE

In health the circulating blood, with a volume of approximately 5 liters, forms a special compartment within a total of roughly 50 liters of body water. It carries the red blood cells, which are the source of oxygen for the body tissues; the leukocytes of the body's defense system; and a very wide range of macromolecules. These include a complex mixture of plasma proteins, the immunoglobulins, and various binding proteins that transport substances (e.g., cortisol and thyroxin) that otherwise could not remain in aqueous solution.

Cells and *macromolecules* remain in the vascular compartment because they cannot readily traverse the endothelial lining of the vessels. If this lining is damaged or destroyed, however, these constituents may pass into the tissues. If the damage is confined to very small vessels, the effusion of blood into the tissues is small, as for example the bruise resulting from local trauma. In the course of days phagocytes move into the area, transforming the hemoglobin into bilirubin (the bruise becomes yellow), and the repair process is completed by replacement of the damaged tissue with newly formed fibrous connective tissue. If the damage is more severe and larger vessels are involved, there may be a considerable effusion of blood into the tissues, forming a substantial blood clot. Such a clot, unless aspirated, may take months or even years to resolve, and may critically damage important tissue. This happens with a ***stroke,*** when the hemorrhage occurs in the soft substance of the brain. The term 'stroke' is used to describe a cerebrovascular accident which may consist in either hemorrhage into the brain tissue or a thrombosis of one of the cerebral vessels. In both cases there is permanent destruction of the surrounding brain tissue with similar clinical manifestations. The patient may be comatose and may have muscular paralysis caused by the destruction of the corresponding motor nerve cells. Hemorrhage may also occur into body cavities, for example, the pleural or peritoneal cavities, sometimes resulting in interference with organ function.

A large loss of blood in a short time will produce a state of ***shock*** owing to a dangerous fall in blood pressure, with impairment of the circulation to such important structures as the heart, brain, and kidneys. Unless treated promptly this situation becomes irreversible. When an injury is the result of severe trauma, the hemorrhage may be a visible outward loss of blood from the body through an open wound or laceration. The volume of blood loss that an individual can survive depends on many factors, such as the rate of bleeding, the use of supportive fluids, and the temperature of the injured patient.

The body has an effective system for minimizing the loss of blood from ruptured vessels. Tissue damage will set in train an elaborate ***clotting mechanism*** resulting in the conversion of the fibrinogen in plasma to fibrin. The meshwork of fibrin entraps red cells to form a blood clot, thus blocking the lumen of the damaged vessels. In addition, the muscular walls of the small arterioles contract, narrowing their lumens and reducing the rate of blood loss. With the rupture of major vessels, the rate of flow may be too great for any such mechanism to be effective in the short period of survival. Penetrating wounds of the lung are likely to sever large vessels in this very vascular organ. Another example is the disastrous tearing of highly vascular muscle tissue, such as occurs in shark bite of the limbs; most victims die from severe hemorrhage in a very short time.

DEHYDRATION, EDEMA, AND CONGESTION

Alterations in body fluid compartments are considered in Chapter 6. The interstitial fluids amount to about 11 liters in a 70-kg human, and much of this water is in the layers of loose connective tissue and fat that underlie the skin. In the normal state of hydration, the skin has an elastic feel, and folds disappear at once when released. With serious reduction of the extracellular fluid volume, as in states of ***dehydration,*** this elasticity is lost and skin folds do not quickly flatten out. Conversely, when salt and water retention leads to swelling and *edema* of subcutaneous tissues, firm finger pressure causes pitting that returns slowly to normal. In tissue sections these changes are seen microscopically as reduced or increased spaces between the collagen bundles of the connective tissue.

Tissue ***edema*** occurs frequently in congestive heart failure, in which increased venous pressure causes accumulation of fluid in the subcutaneous tissues, particularly in the dependent parts of the body, where hydrostatic pressure is greatest. Interstitial edema is also a constant finding in an area of inflammation; here there is engorgement of the blood vessels, with increased permeability of the capillary walls as a secondary phenomenon.

Congestion in any part of the body, that is, dilatation of the blood vessels of the area, may be a consequence of an inflammatory process, as described earlier, or it may be caused by increased venous pressure, as in congestion of the lungs from left heart failure.

DEPOSITION OF ABNORMAL SUBSTANCES IN TISSUES

The detection of abnormal substances in tissues, by their appearance in biopsy sections or with special stains, can in many instances help in making a diagnosis or assist in treatment. Some examples are as follows:

Uric Acid. A common example is the recognition of sodium urate crystals in joint fluid or tissue sections submitted for examination. These fine, needle-shaped crystals are highly characteristic and offer firm evidence for the diagnosis of *gout.* Deposition of urate in the tissues causes the severe pain of clinical gout and leads to tissue damage that may occur progressively in the kidneys of the gouty subject.

Iron. An abundance of iron in the tissues indicates an intake of iron in excess of the capacity of the body to excrete it. The iron may come from transfusions of blood containing large quantities of red cells (a calculated risk when repeated blood transfusions are necessary). Excessive absorption of ingested iron can take place in two situations. (1) It can occur when there is an abnormality of the control mechanism for the intestinal absorption of iron, and far more is absorbed than the body requires. This is the situation in the rare condition of *hemochromatosis,* in which iron accumulates among other sites in the skin, liver, and pancreas, resulting in damage in both of the latter organs. Replacement of the damaged liver cells by fibrous tissue leads to cirrhosis of the liver, and damage to the islet cells of the pancreas may cause diabetes. The patient develops an unusual slate-colored skin, giving rise to the old term "bronzed diabetes." (2) Excessive absorption can also occur when an otherwise normal individual has an iron intake so great that it overwhelms the regulatory system of absorption. These individuals may also develop damage to liver and pancreas because of the same widespread deposition of iron, but the condition is referred to as *hemosiderosis.* This condition occurs with excessive intake of wine in some persons and in those who drink large quantities of beer brewed in iron vessels. The iron is present in the tissues as hemosiderin, which is a polymer of ferritin with excess iron, and the iron is readily detected by the Prussian blue reaction with potassium ferricyanide.

Cholesterol. Although accumulation of cholesterol in the intima of arterial walls is a common phenomenon in the aging process, certain persons have a greater-than-normal tendency toward cholesterol deposition. This may be a primary disorder relating specifically to cholesterol metabolism, or it may be secondary to some other disorder such as diabetes or chronic biliary tract obstruction. The serum cholesterol may or may not be elevated, but when it is chronically and markedly elevated the cholesterol will be deposited in yellowish xanthomatous plaques in the skin, tendon sheaths, and other sites throughout the body.

Amyloid. *Amyloid disease* is the deposition in the tissues of a fibrillary protein called amyloid; it is associated with prolonged intense stimulation of immunoglobulin synthesis. Thus it is formed most commonly in chronic infective processes, such as leprosy, in which an antibody stimulus persists over a number of years. The abnormal protein appears to be a polymer of the immunoglobulin light chains (see Fig. 4–2) and is seen also in association with abnormal immunoglobulin synthesis such as occurs in myeloma. The abnormal protein is deposited in the tissues and can cause impairment of function and ultimate destruction of important structures such as renal glomeruli or arteriolar walls.

Carbon. A variety of foreign materials may be inhaled with varying degrees of damage to the lung tissues. One of the commonest is ordinary city dust, composed largely of fragments of carbon from incompletely burned fuel. The carbon particles that escape entrapment in the upper respiratory passages find their way into the lung alveoli. There they are ingested by the macrophages of the lung and carried into the interstitial tissue of the lung (see Fig. 2–14) or carried to the mediastinal lymph nodes, where they remain for the life of the individual. In mild cases they cause no serious loss of lung function. When exposure to dust is very severe, as in the case of coal miners or foundry workers, the excessive deposition of carbon causes increased fibrosis of the lung tissue with some impairment of expansion and of respiratory exchange with the blood flowing through the lung capillaries (***anthracosis***).

Silica. If the inhaled dust contains particles of silica, as is possible when quartz is being mined, the effects are more serious. The fragments of silica set up a severe inflammatory reaction in the surrounding tissue, and the resulting progressive fibrosis of the lung has a most deleterious effect on the efficiency of the respiratory exchange (***silicosis***).

Talcum and Starch. Talcum powder is a form of magnesium silicate. Its access to the tissues will provoke a vigorous fibrous reaction that may be harmful in some situations. For example, in former years talcum powder used on surgeons' gloves was sometimes released into the peritoneal cavity during operations. It set up a chronic inflammatory reaction resulting in fibrosis, in some cases causing undesirable adhesions between loops of bowel. Talcum powder has now been replaced with starch powder, but even starch granules will cause a fibrous reaction in some cases. Starch is a common diluent in pill manufacture and in recent years, during the intravenous use of drugs by the ignorant and unskilled addict, starch granules have been introduced into the body. These lodge in the lungs and may produce serious impairment of lung function.

Asbestos. Another type of damaging inhaled particle is asbestos. These fragments not only cause a fibrous reaction in the surrounding lung tissue, but after the passage of years may also be associated with *mesothelioma,* a malignant change of the layer of cells (mesothelium) covering the surface of the lung.

Calcium. One of the normal body constituents that may become deposited in abnormal sites in the body is calcium. In disturbances of serum calcium and phosphate levels such as occur in hyperparathyroidism, there is an increase in the concentration of calcium in the plasma and of calcium and phosphate in the urine. When the solubility product is exceeded, calcium may be deposited over a period of time in the tissues or, as renal calculi, in the urinary tract.

Some types of tumor are associated with calcium deposition in areas of central necrosis that, in small aggregates, have a characteristic appearance on an x-ray film and can usually be recognized microscopically within the tumor mass (Fig. 2–15). Such a finding on an x-ray film of the breast (mammogram; Fig. 2–16) is obviously a useful diagnostic aid when a breast lump is felt. In some cases, also, a suspicious calcified area is of considerable diagnostic help when no discrete lump is palpable. On x-ray of the abdomen, calcium deposition in the pancreas or adrenal glands is an indication of probable past inflammatory damage to these structures and is a warning of possible deficiency of function. Calcium deposition in arterial walls as seen in x-ray photographs is also an indication of probable severe vascular damage.

SUGGESTED READING

BOYD W, SHELDON H: An Introduction to the Study of Disease, 8th ed. Philadelphia, Lea & Febiger, 1980

ROBBINS SL, ANGELL M, KUMAR V: Basic Pathology, 3rd ed. Philadelphia, WB Saunders, 1981

ROBBINS SL, COTRAN RS, KUMAR V: Pathologic Basis of Disease, 3rd ed. Philadelphia, WB Saunders, 1984

WALTER JB: An Introduction to the Principles of Disease, 2nd ed. Philadelphia, WB Saunders, 1982

3

Diagnostic Enzymology

David M. Goldberg

Enzymes are proteins composed of 200 or more amino acids covalently linked in a sequence specified by the cell's genetic code. This sequence dictates the extent of interactions, such as hydrogen bonding, between different amino acid residues and thus determines the three-dimensional *conformation* that the protein assumes. The unique nature of this conformation is responsible for both the catalytic activity and the substrate specificity of an enzyme. Many enzymes exist in multiple molecular forms, known as ***isoenzymes.*** These may consist of entirely different peptides determined by unique structural genes. Commonly, they are caused by molecular aggregates of one or two different polypeptides. In other instances isoenzymes may result from the formation of complexes of the catalytically active peptide with other proteins, carbohydrates, or lipids. Genetic variants may result from the mutational substitution of one or more amino acids in the peptide sequence.

The *active site* of an enzyme is that part directly involved in binding and catalysis. The activity of some enzymes depends only on conformation; others need additional nonprotein cofactors. Such cofactors may be a metal, or organic substances called *coenzymes.* These are usually carrier intermediates involving protons, specific atoms, or functional groups. Altering the pH or temperature can affect an enzyme's conformation and hence its activity; beyond certain limits the protein will be denatured.

The commonest ***enzyme units*** ('international' units [IU]) indicate micromoles (μmol) of substrate transformed per minute under standard conditions. A new unit of enzyme activity, the *katal* (kat), is the amount of activity that converts one mole of substrate per second (1 katal = 6×10^7 IU).

Tools of the Enzymologist

A variety of analytic techniques have been used to measure product formed or substrate consumed during a fixed time and temperature of incubation. Automated instrumentation has enabled laboratories to keep pace with increased demands and to implement 'kinetic' enzyme assays. Demand has also led to the development of packaged reagent kits, the commonest of which are for enzyme assays. At times the temptation to sacrifice standards for convenience has been responsible for analytic errors and some abuse of the laboratory service. The standardization of reference methods has proved to be difficult, and few enzyme assays are sufficiently sound that one set of conditions can be advocated as superior to all others. The Expert Panel on Enzymes of the International Federation of Clinical Chemistry (IFCC) has published Reference Methods for a number of enzymes measured frequently in routine diagnostic laboratories. The methods, as such, are not always applicable for routine use, but the recommended formulations are now being provided by reagent manufacturers in kit form, so that optimal and standardized substrate concentrations can be used in all laboratories. They will improve accuracy in assigning catalytic activities to enzyme preparations. Wide adoption of these methods will reduce the variance between different laboratories, although complete uniformity is some way off. It is likely that the recommended temperature of 30°C will be adopted because the gallium cell provides exquisite control in this region. The quality control of enzyme assays is plagued by the problems of trace contaminants, the scarcity of reference enzyme preparations, the instability of enzymes, and species variations in enzymes whose characteristics differ from those for the corresponding human enzymes. The IFCC Panel on Reference Materials is making strenuous efforts to develop appropriate quality control preparations for enzyme assays.

Mechanisms of Elevated Enzyme Activities

The activity of an enzyme in blood serum represents a balance between its rate of liberation into the extracellular space and its rate of clearance or uptake from the extracellular space. Factors such as age, sex, and body weight seem to affect this balance in individual subjects. Serum enzymes in ***health*** almost certainly are derived from the metabolic breakdown and turnover of cells and tissues. Because the gradient between intracellular and serum levels exceeds 10^3 for most enzymes, it is probable that the *cell membrane* performs a crucial function in retaining enzymes within the cell.

When isolated cells are incubated *in vitro,* the liberation of enzymes, either spontaneously or under provocation by phospholipase A or by a high potassium ion concentration in the medium, can be prevented by adding ATP or ATP-generating compounds. In the absence of added ATP, the rate of escape of intracellular enzymes is strictly related to the rate of depletion of intracellular ATP. It follows that ***anoxia,*** occurring in pathologic states such as severe cardiac or respiratory disease, may allow a significant escape of enzymes from cells even though no structural damage is evident. Some of the highest serum enzyme activities occur in patients with *status asthmaticus,* a largely functional though potentially fatal condition in which anatomic damage to the lung and other body tissues is rarely present. Even relative hypoxia, such as that occurring during severe exercise, may allow a transient escape of enzymes from skeletal muscle with a consequent elevation of serum activities.

Drugs and ***poisons*** are frequently responsible for raised serum enzyme activities and may act as inhibitors of energy-yielding processes within cells. Some drugs, notably chlorpromazine and promethazine, act directly upon the cell membrane to accelerate enzyme release. Bacterial toxins probably act within the cell as metabolic poisons and externally upon the cell membrane; generalized septicemia can cause dramatic increases in serum enzyme activities.

Another point of view, based on a predominance of *catabolic* over *anabolic* processes, such as would result from cell damage and stress, emphasizes the role of electrolytes and osmotically active molecules. When the cell swells, the membrane becomes stretched, cellular pores become enlarged, and molecules such as enzymes are able to escape at a greatly increased rate.

These mechanisms explain why serum enzyme elevations are relatively nonspecific with regard to the tissue involved or the nature of the injury that occurred. The fact that serum enzyme elevations may reflect transient *functional* impairment of the cell from which they originate—even though this impairment is consequent upon a serious generalized pathologic process—has stimulated a search for indices more directly related to actual ***cell necrosis.*** Enzymes released from cells during anoxia, or as a result of changes in membrane permeability, are mainly those present in the cytosol. Mitochondrial enzymes are liberated during cell necrosis. Their presence in the serum is thus evidence for a more severe form of cell injury. Glutamate dehydrogenase is one such enzyme, and aspartate aminotransferase exists in a mitochondrial form which is electrophoretically, chromatographically and immunochemically distinct from the cytosolic form. Detection of such enzymes can be a help in assessing the extent of tissue injury.

Other mechanisms may lie behind the elevation of microsomal or membrane-associated enzymes found in the serum. *Lipid peroxidation* in biomembranes is one such mechanism; hepatic ***enzyme induction*** is another. The latter can cause an increase in the activity of microsomal enzymes in the serum of patients receiving enzyme-inducing drugs. Alkaline phosphatase, γ-glutamyl transferase, and 5′-nucleotidase are frequently elevated in epileptic patients treated with anticonvulsants, even though no impairment of the liver or biliary tree can be demonstrated in these patients. So far as the liver and biliary tree are concerned, obstruction of the latter causes an increased synthesis of alkaline phosphatase in the liver which, in turn, is reflected in increased serum alkaline phosphatase activity. Evidence also suggests that bile acids play a role in promoting enzyme liberation from damaged hepatic cells and organelles when the bile ducts are obstructed.

Another variable in determining the level of activity of an enzyme in the serum is its rate of ***clearance*** from the serum. A measure of clearance can be obtained by determining the half-life of the enzyme in clinical subjects, ideally under conditions in which the termination of enzyme release is clearly demarcated, although few such situations exist clinically. Figure 3–1 presents an analysis of the clearance rates of serum enzymes in a patient with *paroxysmal myoglobinuria,* a condition associated with an acute but sharply delimited liberation of enzymes from damaged muscle. Variations in the half-life of the different enzymes measured are apparent; lactate dehydrogenase (LDH) demonstrates an initially fast and subsequently slower half-life, consistent with the fact that the two primary polypeptide units of this enzyme are cleared at different rates. However, results obtained after the infusion of enzymes in humans suggest that the clearance mechanisms are far more complex than can be described by a single half-life measurement.

The rate of ***enzyme elimination*** from the serum appears to have an upper limit and different mechanisms apply to the clearance of different enzymes. Thus, amylase and lipase are cleared mainly in the urine, while alkaline phosphatase is excreted partly in the bile, but neither of these routes contributes significantly to the clearance of lactate or isocitrate dehydrogenases. The reticuloendothelial system is actively concerned in the clearance of some enzymes. Other removal mechanisms are thought to involve either nonbiologic decay of the type occurring when serum specimens are allowed to stand without refrigeration, or gradual degradation by proteolytic enzymes, or combination with specific inhibitors or antibodies. Indeed, the persistence in the serum of enzymes conjugated with antibodies has given rise to unusual forms that are detectable on electrophoresis and that generally turn out to have no clinical significance; these have been documented particularly with creatine kinase and lactate dehydrogenase. These immunoreactive forms of enzymes persist in the blood beyond the point at which their biologic potency is lost. Half-lives determined by immunochemical procedures will therefore be much longer than those based upon catalytic assays. Enzymes frequently demonstrate K_m (Michaelis constant) values in serum that differ from those in their tissues of origin. The changes may result in increased or decreased substrate affinity, suggesting conformational changes following release from cells rather than degradation of the enzymes in the serum.

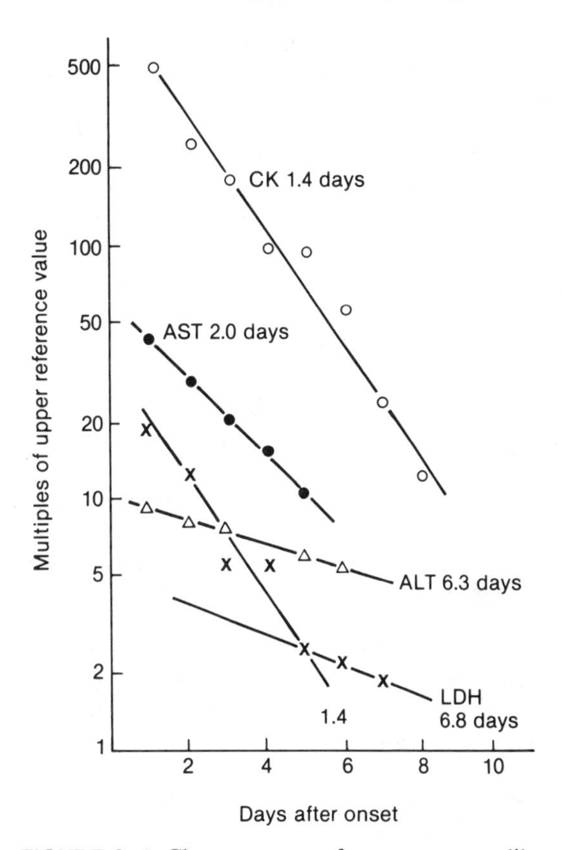

FIGURE 3–1. Clearance rates of serum enzymes liberated from damaged cells in a patient with paroxysmal myoglobinuria. CK = creatine kinase; AST = aspartate aminotransferase; ALT = alanine aminotransferase; LDH = lactate dehydrogenase. (Wilkinson JH: Principles and Practice of Diagnostic Enzymology. London, Arnold, 1976. Redrawn)

With the use of isotopically labeled enzymes, it is apparent that the *degradation* of enzymes *within tissues* is quantitatively an important phenomenon. Moreover, tissues vary enormously in the rate at which they degrade enzymes. If enzymes are taken up as well as released by cells, the balance of these processes, together with autodegradation and clearance, determines the serum activities encountered. Current knowledge is insufficient to explain all of the serum enzyme alterations seen in disease.

SOME IMPORTANT ENZYMES

ACID PHOSPHATASE (ORTHOPHOSPHORIC MONOESTER PHOSPHOHYDROLASE, ACP, EC 3.1.3.2*)

The term *acid phosphatase* is applied to a family of enzymes catalyzing the hydrolysis of a wide range of phosphate esters at acid pH. Assay methods measure the hydrolysis of a suitable phosphate ester, and a number of substrates permit easy determination of ACP activity because their nonphosphate (alcohol) moiety yields a colored product when the reaction pH is adjusted from the optimum (around pH 5.0) to an alkaline environment. The activity of ACP decreases quite rapidly at room temperature unless the pH is reduced below 6.0.

Isoenzymes

A series of acid phosphatases exists in virtually all cells of the body, and although soluble forms are found in the cytosol, the lysosome is the predominant intracellular location of the enzyme. Clinically, the important tissue sources are the prostate, red blood cells (RBC), and platelets. The prostate enzyme is sensitive to L-tartrate and the RBC enzyme to formaldehyde. The latter will therefore inhibit ACP released in hemolyzed samples. Unfortunately, other tissue forms of ACP are sensitive to L-tartrate, because serum levels of L-tartrate–sensitive enzyme are detectable in females, and in males after prostatectomy.

Clinical Applications

The main use of ACP lies in the diagnosis of ***prostatic cancer.*** When the disease has spread to involve adjacent bones, the serum activity of this enzyme is greatly elevated. Spread to the soft tissues without bone involvement may also be associated with high activity, but in a lower proportion of patients. When the tumor has remained as a discrete nodule, or has not extended beyond the capsule of the prostate, it can be anticipated that half the patients will have abnormal values for serum ACP activity. The test is of limited diagnostic value and has been disappointing as a screening test for prostatic cancer.

Although a specific *radioimmunoassay* (RIA) for prostatic ACP has been developed, the sensitivity of this test in the diagnosis of prostatic cancer has been no better than those published previously for catalytic assays (Table 3–1). It remains to be seen whether the test has a role in monitoring patients with established prostatic cancer. Continued use of these tests is due more to habit than to utility.

Prostatic massage during rectal examination was reported to cause a transitory elevation in serum ACP in some patients. Although this is rarely so, it is nevertheless advisable to draw blood for the enzyme test before performing the examination.

In view of the predominance of ACP in *lysosomes* it is not surprising that certain lysosomal storage diseases (e.g., Gaucher's disease and Niemann–Pick disease) are associated with raised serum ACP activity. Its association with formed *blood elements* is responsible for elevated ACP activities in thromboembolic and myeloproliferative disorders.

ALKALINE PHOSPHATASE (ORTHOPHOSPHORIC MONOESTER PHOSPHOHYDROLASE, ALP, EC 3.1.3.1)

The term *alkaline phosphatase* is applied to a family of enzymes with low substrate specificity that catalyze the hydrolysis of a wide range of phosphate esters at alkaline pH. All ALP enzymes so far purified have proved to be zinc metalloproteins with a serine residue at the active center. The enzymes are located in the *brush borders* of both the proximal convoluted tubule of the kidney and the small intestinal mucosa. Many other *membranes,* including the sinusoidal and canalicular surfaces of the hepatocyte, are rich in ALP. In these locations the enzyme is believed to play a role in absorption and transport processes. ALP of *bone,* on the other hand, has long been suspected of hav-

* International Enzyme Commission classification number.

ing a role in bone mineralization, since its activity correlates with changes in osteoblastic function.

Serum ALP activity shows a striking dependency upon ***age,*** values being high in infancy and declining quite dramatically after puberty, but with occasional fleeting elevations well into the pathologic range occurring during prepubertal growth spurts in normal children. The growth rate of bone is frequently altered as a result of disease in infancy, and these changes in bone growth will be reflected in an altered serum ALP activity. During adult life the values in men are significantly higher than in women, but after the menopause this difference is eliminated owing to a sharp increase in ALP activity among women and a slower rise among men of comparable age.

Isoenzymes

Distinct molecular forms of ALP are present in the human placenta and the small intestinal mucosa; these are under separate genetic control. The isoenzymes in bone, liver, and kidney may be products of the same gene locus but subject to posttranscriptional modification within each tissue, the differences between them being relatively small. Various electrophoretic media have been used to separate these isoenzymes in the serum of patients with elevated ALP activity (Fig. 3–2). The liver isoenzyme has the fastest rate of migration toward the anode, with bone only slightly slower, followed by placental and intestinal isoenzymes. A slow-moving isoenzyme with a high relative mass is present in the serum of patients with bile duct obstruction and hepatic metastases. Other methods of distinguishing ALP isoenzymes include the following:

1. ***Heat stability.*** When heated at 65°C for 5 min, placental ALP is virtually unaffected, whereas all other ALP isoenzymes suffer a total loss of activity. When heating is performed at 56°C for 10 min, intestinal ALP loses 20% of its activity, liver ALP 60%, and bone ALP 80%.
2. ***Urea denaturation.*** When exposed to 2 M urea, ALP isoenzymes show a stability pattern similar to that seen with heat, placental ALP being the most and bone ALP the least resistant of the four.
3. ***Inhibition by amino acids.*** L-Phenylalanine inhibits intestinal and placental ALP but has no effect on the bone and liver isoenzymes. L-Homoarginine has little effect on intestinal or placental ALP but inhibits the bone and liver isoenzymes.

The higher activity of serum ALP in infants is due to predominance of the *bone* isoenzyme. This disappears after puberty so that the *liver* isoenzyme accounts for most of the activity in healthy adults. *Intestinal* ALP appears in the serum only if persons are of blood groups O or A, are secretors of ABH red cell antigens, and are positive for Lewis antigen.

TABLE 3–1. Sensitivities of Prostatic Acid Phosphatase Radioimmunoassays from Eight Published Studies

Study	Specificity (%)	Sensitivity (%) Stage of Disease			
		A	*B*	*C*	*D*
1	94	33	79	71	92
2	97.5	8		35	
3	100	0	30	55	80
4	97.5	13	26	30	62
5	94	12	32	47	86
6	94	22	29	52	87
7	94	75	75	77	86
8	95	0	14	16	63

* In the context of this table, these stages are defined as follows: A = not palpable; B = palpable, but within the prostatic capsule; C = spread beyond the prostatic capsule but not to bone or beyond pelvic cavity; D = spread to bone or beyond pelvis.

FIGURE 3–2. Isoenzymes of alkaline phosphatase separated by gel electrophoresis. + = anode; − = cathode. (Wilkinson JH: Principles and Practice of Diagnostic Enzymology. London, Arnold, 1976. Redrawn)

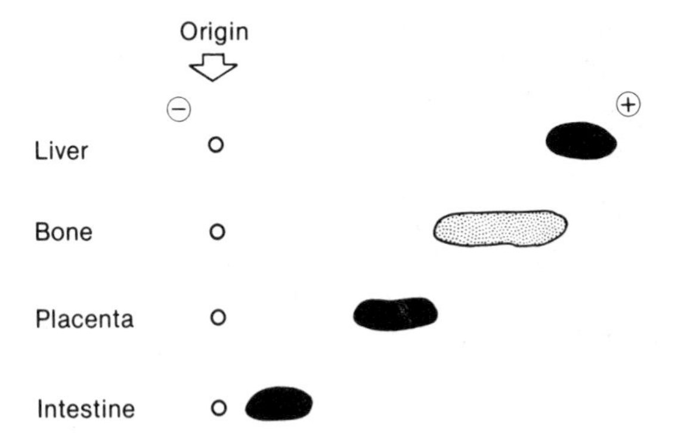

Placental ALP is present in the serum of pregnant women, but only during the second and third trimesters of pregnancy, and accounts for the marked increase in total serum ALP in these patients. A sharp reduction in placental ALP often heralds placental insufficiency and imminent death of the fetus.

Clinical Applications

Hepatobiliary Disease. Patients with biliary obstruction usually have high levels of ALP, or even very high levels, whereas those with primary parenchymal liver disease have, as a rule, only moderately increased activity. The assay is thus of considerable importance in the differential diagnosis of *hepatic* and *posthepatic (obstructive) jaundice.* In extrahepatic obstruction, serum bilirubin and ALP usually rise and fall in parallel. Patients with *acute hepatic necrosis* often show little change in ALP even when the serum bilirubin concentration is rising dramatically; if icterus fails to clear and the patient enters a cholestatic phase, serum ALP may rise substantially, but this is usually a late manifestation. *Portal cirrhosis* is generally associated with normal or only slightly elevated values for serum bilirubin and ALP activity, but primary *biliary cirrhosis* is characterized by striking increases in ALP activity, even when the bilirubin is normal. Elevated serum ALP may also occur in liver disease without icterus. It occurs frequently in patients with alcoholic *fatty liver,* as well as in patients with metastatic *carcinoma* or *lymphomatous* infiltration of the liver.

The increased serum ALP activity in the above conditions is due to increase of the *liver* isoenzyme. An unusual isoenzyme (Regan isoenzyme), which has all the properties of placental ALP, has been detected in the serum of patients with various carcinomas metastasizing to the liver. Its incidence in most studies is 1% to 3% of such patients, and its appearance seems to be caused by derepression of the genome. This phenomenon allows the reexpression of a gene product whose synthesis is suppressed in normal individuals. At least three other tumor-associated ALP isoenzymes have been described which differ among each other in their sensitivities to heat and amino acids, and in electrophoretic mobility. Their importance at present is in providing insight into genetic changes occurring during neoplasia, rather than as aids to diagnosis.

Bone Disease. High serum ALP activity occurs in many skeletal disorders. Primary *bone tumors* (e.g., osteogenic sarcomas) and secondary osteoblastic bone tumors (e.g., those originating from the prostate) can cause very high levels. However, osteolytic 'secondaries' are often associated with only slight to moderate increases. *Paget's disease* of bone is invariably a cause of high activity. So, too, are the *vitamin D deficiency* states of rickets and osteomalacia. Primary *hyperparathyroidism* is often associated with increased serum ALP activity when bone involvement is present. Secondary hyperparathyroidism, as in renal osteodystrophy, causes dramatic changes in bone metabolism and structure, and high serum ALP activities are usual. Healing *fractures* also cause elevated levels. A rare hereditary disease of bone is associated with hyperphosphatasia. The converse abnormality—hypophosphatasia—also exists as a congenital disease, associated with retarded skeletal development; such patients excrete excessive amounts of ethanolamine phosphate and inorganic pyrophosphate in the urine and exhibit a very low serum ALP.

Intestinal Disease. High ALP activities are quite often found in patients with intestinal disease and after gastrectomy. The predominant isoenzyme in the serum in these situations is not the intestinal but the *bone* isoenzyme, and its presence is due to metabolic bone disease, often subclinical, which occurs in these conditions as a consequence of impaired intestinal absorption of calcium, phosphate, and vitamin D.

Cancer. A number of ALP isoenzymes have been identified in patients with cancer (Table 3–2). In the main, they are variants of the placental isoenzyme. They are uncommon and occur predominantly in hepatocellular cancer or when the liver is the site of metastatic spread. Their sensitivity is thus low, but they are virtually cancer specific. Their existence has provided a strong stimulus to the concept of gene derepression as a cancer-associated phenomenon, and they have provoked a continued search for more useful oncofetal gene products. A high-relative-mass form of ALP occurs

TABLE 3-2. Properties of Cancer-Specific Alkaline Phosphatase Isoenzymes

Isoenzyme	Heat Stable	L-PheA (5 mM)	L-Leu (10 mM)	EDTA (4 mM)	Type
Regan	+++	≡	0	0	Placental
Nagao	+++	≡	≡	≡	D-Variant
Regan variant	+	≡	=	−	Hybrid
Kasahara*	+	−	−	−	Fogh-Lund
Fetal intestinal*	+	−	−	−	As stated
Timperley	−	0	0	0	Non-Regan

* Distinguished by electrophoresis and reaction with Concanavalin A.

+, Resistant; +++, strongly resistant; −, inhibited; ≡, strongly inhibited; 0, unaffected.

in a much higher proportion of patients with hepatic cancer (primary or secondary) but is seen also in patients with nonmalignant biliary tract disease.

CREATINE KINASE (ATP:CREATINE PHOSPHOTRANSFERASE, CK, EC 2.7.3.2)

This enzyme has assumed major importance in the diagnosis of myocardial infarction, although it first attained prominence because of its high serum activities in Duchenne muscular dystrophy.

Creatine kinase catalyzes the following equilibrium reaction:

$$\text{Creatine} + \text{ATP} \rightleftharpoons \text{Creatine Phosphate} + \text{ADP}$$

CK is a magnesium-dependent enzyme. It loses activity on storage, but can be reactivated by thiol reagents. Hemolysis of the sample may interfere with certain methods of CK assay. Dilution may yield higher activity than expected. Controversy rages as to whether this phenomenon is a methodologic quirk or is caused by the presence of dissociable *inhibitors* of CK in human serum. There is now good evidence that uric acid is a powerful inhibitor of CK in the serum, and a peptide inhibitor has been isolated but not yet characterized. Another source of difficulty is the elevated serum value that can follow an intramuscular injection. This must always be ruled out before one attempts to interpret elevated serum CK activities.

The levels of CK in *healthy* subjects are prone to various demographic influences. Males have higher values than females, and activities increase with age and with body weight. High levels are found in the neonate but fall rapidly during the first weeks of life, and by 6 months are in the range of values found in adolescents and young adults.

Isoenzymes

There are three principal molecular forms of CK; these can be recognized by electrophoresis, by immunotitration, and by column chromatography. Brain and most other nonmuscle tissues in which CK can be detected contain a single band which moves fairly rapidly toward the anode. This isoenzyme is a homodimer (BB) of a peptide known as the B chain. Skeletal muscle contains predominantly one slow-moving band, a homodimer (MM) of the M peptide chain. This form comprises the bulk of CK in cardiac muscle, but in addition this tissue contains 20% to 40% of the enzyme in the form of a heterodimer (MB) not found in any other organ, except skeletal muscle, in which small amounts are present in red fibers. Three variant isoenzymes designated X, Y, and Z have been described. Their nature and clinical significance are not yet elucidated although CK-Z appears to be present in patients only following a myocardial infarction. The status of these variants is confounded by the fact that nonidentity of the M-peptide and B-peptide chains in their respective homodimers has been demonstrated. Macro-CK complexes of uncertain or benign clinical significance have also been demonstrated and occur in 2% to 8% of hospital populations when a serious search is made; the commonest form is the result of a complex between CK-BB and IgG. In some tissues a mitochondrial form of CK is present; this moves more cathodally than the MM dimer on electrophoresis. Its presence in the serum of cancer patients is being increasingly reported.

Clinical Applications

Myocardial Infarction (See also Chap. 9). Measuring the serum activity of CK is the most useful biochemical test in establishing or refuting a diagnosis of myocardial infarction. After such an infarct the serum CK level rises more promptly, and more specifically than that of any other serum enzyme utilized thus far, and the elevation above the reference limit is greater. Values become abnormal about 6 hours after the onset of symptoms and rise, crescendo fashion, to a peak around 36 hours, after which they decline, reaching reference values after the third day. The onset of cardiac failure does not usually alter the response pattern of CK; but reinfarction, thromboembolic phenomena, or arrhythmias may cause a delay in the return toward reference values or secondary spikes of increased activity detectable only through frequent monitoring. Figure 3–3 shows the time course of a number of serum enzyme activities after myocardial infarction in a typical case.

FIGURE 3–3. Changes in serum enzyme levels with time, following myocardial infarction. CK = creatine kinase; AST = aspartate aminotransferase; LDH = lactate dehydrogenase; LDH_1 = most anodal isoenzyme of LDH; URV = upper reference value. (Wilkinson JH: Principles and Practice of Diagnostic Enzymology. London, Arnold, 1976. Redrawn)

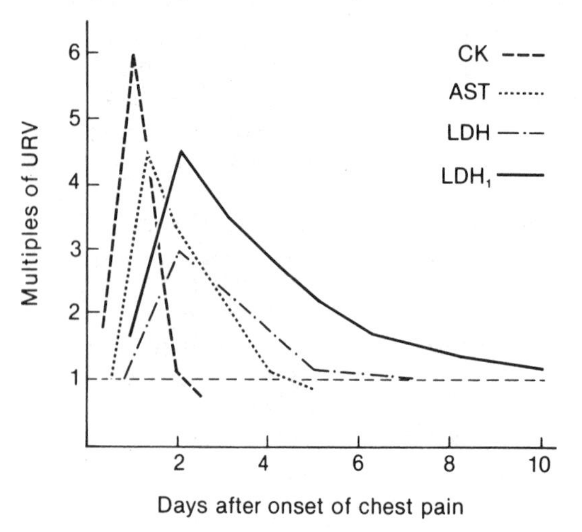

The following cardiac conditions, which have to be considered in the *differential diagnosis* of acute transmural myocardial infarction, also can give rise to increased serum CK activity:

Prolonged chest pain with transient ST-segment and T-wave changes (≃40% of cases)
Prolonged chest pain without electrocardiographic abnormalities (≃20% of cases)
Classic angina pectoris (occasionally)

However, the extent of the elevation in these conditions is usually slight. *Procedures* carried out by the cardiologist, such as cardiac catheterization, can cause elevated serum CK activity. Cardiac surgery and electric cardioversion almost invariably cause elevated serum CK, although the latter is mainly associated with damage to skeletal muscle of the chest and MM is the form of CK liberated. The feature which characterizes the increase in CK following a myocardial infarct is the proportion (10% to 30%) of this activity represented by the ***MB isoenzyme.*** The co-occurrence of chest pain and a rise in serum CK-MB is virtually pathognomonic of infarction. With sufficiently sensitive methods, one finds that up to 2% of CK present in normal serum is of the MB type, and it is therefore necessary to think of a diagnostic *increase* rather than a diagnostic *presence* of this isoenzyme. The time course of CK-MB after an infarct is similar to that of total serum CK, except that the elevation may be detectable within a few hours; the increase is even more dramatic, but of shorter duration, as it is rapidly cleared from the vascular space. Although total serum CK activity is raised in a wide spectrum of muscle diseases, few are accompanied by an increase in the MB component.

Attempts are being made to relate the extent of CK or CK-MB release to the mass of infarcted heart muscle. This would enable the prognosis to be predicted and would also provide a means of assessing the efficacy of therapeutic measures aimed at protecting threatened myocardial tissue. Such estimates require frequent serum assays for CK activity and use a range of constants for CK release and decay which are the subject of dispute among investigators. It will be some time before these techniques can be used clinically.

Muscle Disease. Very high activities of CK are found in the serum of patients with X-linked

Duchenne muscular dystrophy, a genetic disease transmitted by female carriers to their affected sons. The primary biochemical lesion in Duchenne muscular dystrophy is not known, although the membrane of the muscle cell and the actomyosin complex are prime candidates. Leakage of intracellular enzymes, predominantly CK-MM, to the vascular compartment is currently held to be the major factor contributing to elevated serum CK activity in affected patients. Although other enzymes found in skeletal muscle, such as aldolase, lactate dehydrogenase, and the aminotransferases, all may show high serum activities, none are raised to an extent comparable with CK, and it is unnecessary to pursue their determination. Highest CK activities are encountered at the earliest stages of the disease (pseudohypertrophy), when the muscle mass is expanded because of fat infiltration. As the disease progresses inexorably, with muscle wastage, the serum CK activity falls in parallel, and in the terminal stages of the disease the lowest CK values are obtained (Fig. 3–4).

The *inheritance pattern* of Duchenne muscular dystrophy is such that half the female children of a female carrier will themselves be carriers. Abnormal CK levels have been reported in as many as 75% of these obligate carriers, and when a healthy female relative of a known patient shows consistently raised serum levels she will generally also show electromyographic or histochemical evidence of latent muscle abnormality. The problem of carrier detection is made difficult because of the factors affecting reference values, such as age and body weight. *Physical exercise* may also increase serum CK activity in healthy subjects; therefore, samples of blood for detecting carriers should always be taken in the early morning, or after full recovery from any physical exertion.

Another form of x-linked muscular dystrophy (*Becker type*) is more benign: much lower serum CK activities occur in affected male subjects, and 50% of female carriers demonstrate abnormal values. Lesser elevations of serum CK also occur in other forms of muscular dystrophy, such as the limb-girdle, facioscapulohumeral, and myotonic varieties.

Certain other primary diseases of muscle, such as *polymyositis,* are associated frequently with raised serum CK. In most secondary diseases of muscle, as in neurogenic atrophy, values remain normal; however, increased activities have been reported in motor neuron disease. *Malignant hyperpyrexia* develops in response to general anesthesia as widespread muscular rigidity, severe metabolic acidosis, and rapidly rising body temperature. The trait, which affects skeletal muscle, is inherited in a mendelian dominant pattern. Very high serum CK activities are encountered during the episode, and some authors have detected the presence of MB and BB isoenzymes, although MM predominates. Serum CK determination is not a reliable means of detecting carriers in this condition.

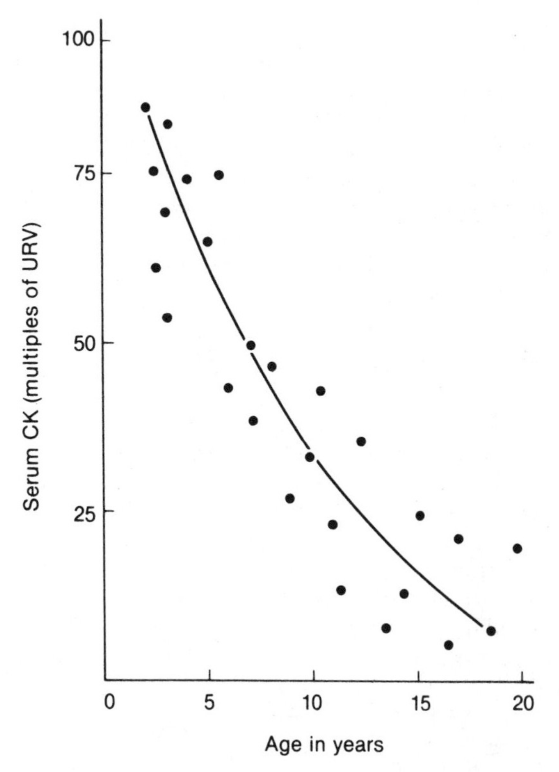

FIGURE 3–4. Fall in serum creatine kinase (CK) during the muscle wasting of Duchenne muscular dystrophy. URV = upper reference value.

Cerebral Disease. High serum levels of CK occur in patients after *epileptic seizures* and are caused by muscular activity rather than an abnormal brain focus. A similar elevation has been reported in patients with *tetanus.* Raised levels have also been recorded in *organic neurologic* disease such as cerebral infarction, meningitis, and encephalitis, occurring some days after the onset of symptoms and showing little relationship to prognosis. Cerebrospinal fluid levels of CK show no consistent change in these conditions, and the isoenzyme in the serum is predominantly of the MM type. It thus appears that the source of the elevation is skeletal muscle and not cerebral tissue. After neurosurgery, the total CK activity of the serum may be increased, and in many such instances a significant proportion is CK-BB.

Other Conditions. Most other diseases that are associated with a raised CK activity of serum can cause secondary muscle disease, and it is now widely believed that this is the causal mechanism in most circumstances, the elevated activity being due exclusively to an increase in the MM isoenzyme. During acute *alcohol intoxication* in chronic alcoholics, high levels can occur as the result of a toxic myopathy. *Hypothyroidism* is another disease in which high serum CK activity is encountered; it has been advocated as a useful diagnostic test for the younger patient presenting with a suspected thyroid disorder. High serum CK activity is encountered during *labor* and the *puerperium;* a significant proportion of this activity is due to the BB isoenzyme, which is the predominant form found in uterine muscle. Reports are now appearing of increased CK-BB in the serum of patients with *cancer;* the incidence is low and the exact diagnostic significance of this phenomenon needs to be defined.

LACTATE DEHYDROGENASE (L-LACTATE: NAD OXIDOREDUCTASE, LDH, EC 1.1.1.27)

This zinc-containing enzyme is distributed widely throughout all body tissues, where it is present predominantly in the cytosol. By regulating the interconversion of lactate and pyruvate, it exercises a control function over the balance between respiration and glycolysis.

The enzyme catalyzes the following reversible reaction:

$$\text{Pyruvate} + \text{NADH} + H^+ \rightleftharpoons \text{lactate} + \text{NAD}^+$$

High LDH activity is present in erythrocytes; therefore, when samples have hemolyzed or when separation of the serum has been delayed, they are unsuitable for analysis.

Isoenzymes

Each of the five isoenzymes of LDH comprises a tetramer of four subunit polypeptide chains. The subunits consist of two types of polypeptides—the H chain characteristic of heart muscle LDH, and the M chain typical of hepatic and skeletal muscle LDH. Each polypeptide is the product of a single gene locus, and polymorphism may occur at one or other locus to yield up to 15 distinct bands on electrophoresis. The isoenzymes are numbered LD_1 to LD_5 in order of their rate of migration toward the anode. LD_1 has the composition H_4 (four H chains) and LD_5 the composition M_4. The intermediate isoenzymes have the following composition: LD_2, H_3M; LD_3, H_2M_2, LD_4, HM_3. Occasionally, abnormal bands that do not correspond to the position of any of these five isoenzymes have been described; one of the commoner causes of their existence is the combination of an immunoglobulin with one of the peptide chains. A variant, termed LD_6, which is cathodal to LD_5 on electrophoresis and may be a posttranslational modification of this isoenzyme, has been reported in terminally ill patients and is a grave prognostic index.

Clinical Applications

Myocardial Infarction (see also Chap. 9). Elevations of serum LDH activity commence 12 to 18 hours after the onset of symptoms and a peak is reached around the third day, after which, in uncomplicated cases, the activity declines, returning to normal by the 10th day (see Fig. 3–3). Abnormal values may persist beyond this point, so that this assay is more valuable in a case presenting several days after the suspected infarction. Heart muscle is especially rich in LD_1 and LD_2 (in that order), whereas LD_2 predominates over LD_1 in normal serum. The typical LDH isoenzyme pattern of serum from a case of myocardial infarction reflects this predominance of LD_1 (Fig. 3–5) and is referred to as the 'flipped LD pattern.' A return to reference values is delayed if complications arise.

velop icterus although their aminotransferases may be grossly abnormal. Aminotransferase determinations have been useful in studying the epidemiology of the disease, and in its prophylaxis through the detection and isolation of infected contacts of known cases. Early in the genesis of the disease, AST activity exceeds that of ALT since AST occurs in higher concentration in liver tissue, but owing to the slower clearance of ALT, this enzyme accumulates and ultimately exceeds the activity of AST. This ALT > AST ratio is maintained as the aminotransferases, following a crescendo-like rise, fall dramatically toward reference values, ALT persisting at abnormal levels for a longer time than AST.

If, during the recovery period, a recrudescence of the disease occurs, the downward trend in the aminotransferases is reversed and a further transient rise occurs. Continuing elevation of the aminotransferases is indicative of the presence of chronic *active* hepatitis. If values are moderate and the AST:ALT ratio is less than unity (i.e., ALT > AST), chronic *persistent* hepatitis is more likely. If the values are higher than moderate and the AST:ALT ratio is greater than unity (i.e., AST > ALT) chronic *aggressive* hepatitis is likely, the prognosis is less favorable, and treatment with steroids and immunosuppressants may be required. The aminotransferases are not helpful in distinguishing type A from type B viral hepatitis, nor are they useful in detecting asymptomatic carriers of hepatitis B antigen (see Viral Hepatitis, Chap. 11). However, they will continue to be powerful tools in the diagnosis of non-A, non-B hepatitis until immunologic markers for this disease become available.

Elevated serum aminotransferase levels occur in most types of hepatobiliary disease, because the tests correlate with biliary regurgitation as well as with hepatic cell necrosis. Except in prehepatic hyperbilirubinemia, the jaundiced patient will almost invariably have raised aminotransferase activities in the serum. If attention is devoted to the *degree* of elevation and the activity *ratio* of the two aminotransferases, a series of criteria can serve as a general guide to the type of liver disease and its probable mechanism. In *portal cirrhosis,* both micro- and macronodular, aminotransferase levels may be normal if jaundice is absent, and only slightly raised if it is present. Elevations are rarely more than three times normal, and the activity of AST usually exceeds that of ALT. Primary and secondary *biliary cirrhosis* cause moderately pronounced and equal increases in the activity of both aminotransferases to approximately four times normal.

Raised aminotransferases occur in virtually all patients with hepatobiliary *obstruction,* and the levels usually range from two to eight times the upper reference value, depending upon the degree of icterus. It is uncommon for the activity of AST to exceed that of ALT when the obstruction is caused by nonmalignant disease. In a proportion of cases in which the obstruction is caused by a tumor, AST activity is the higher of the two, and it is the dominant aminotransferase in about half those cases manifesting hepatic metastases.

The major disadvantage of the aminotransferases in the diagnosis of hepatobiliary disease is their lack of *specificity.* Values outside the reference range are found in many patients with diseases of other organs, and this has stimulated many workers to experiment with alternative tests of possibly greater specificity for hepatobiliary disease. None has been found with sufficient advantages to warrant displacing the aminotransferases from their position of diagnostic primacy in this area, although some are useful adjuncts.

Muscle Disease. High levels of both aminotransferases are seen in Duchenne muscular dystrophy. Abnormal levels are also quite frequent but less pronounced in dermatomyositis and myoglobinuria.

Pulmonary Disease. Various lung diseases may be associated with a raised serum aminotransferase activity, although the abnormalities are rarely of value in diagnosis. Bronchial carcinoma is one such condition and pulmonary embolus another. The increased AST activity in the latter may derive from extrapulmonary sources such as the liver and erythrocytes.

5′-NUCLEOTIDASE (5′-RIBONUCLEOTIDE PHOSPHOHYDROLASE, 5NT, EC 3.1.3.5)

Distribution and Action

The main clinical interest in this enzyme is in the differential diagnosis of hepatobiliary disease. In a sense this is surprising, for 5NT enjoys a wide tissue distribution, high activities being found in endocrine organs and the vascular endothelium, while the liver is not especially notable for its 5NT content. It was widely believed that 5NT was restricted to the plasma membrane, and its activity has been employed to determine the purity of plasma membrane fractions. This view is now untenable. Four distinct enzymes with 5NT activity have already been purified from rat liver, differing in catalytic properties and in the cell locus to which they are confined.

The fundamental reaction in most methods for determining serum 5NT activity is the following:

$$\text{Adenosine 5'-monophosphate (AMP)} \rightarrow \text{adenosine} + \text{Pi}$$

Early methods measured the inorganic phosphate (Pi) liberated and were not sufficiently specific to contend with interference by high ALP activity. Alternative methods are based upon the measurement of adenosine, using the enzyme adenosine deaminase. This permits the suppression of AMP hydrolysis due to ALP by adding another substrate for ALP.

Clinical Applications

Hepatobiliary Disease. Many reports testify to the fact that serum 5NT activity is raised predominantly in diseases of the *biliary tree;* activities in patients with primary hepatic parenchymal diseases generally show only modest elevation. Whereas a high activity is strongly suggestive of biliary tract disease in a jaundiced patient, there is no reference value that can reliably segregate primary biliary from hepatic diseases. *Bile ductules* are the major source of the elevated serum 5NT. The assay is a sensitive indicator of diseases which induce proliferation or destruction of ductular cells, or inflammatory cell infiltration within the portal tracts.

The age- and growth-related changes in ALP activity (see Alkaline Phosphatase) are disadvantages that do not apply to 5NT. Being actually lower in serum during infancy than during adulthood, 5NT is superior to ALP in the frequency and magnitude of the elevations occurring in hepatobiliary disease during childhood, and in the specificity of such elevations, since they are rarely encountered in other childhood illnesses.

Liver vs. Bone Disease. ALP activity is raised in diseases of the bone and the liver; serum 5NT is raised only in diseases of the liver. Therefore if both are raised ALP is probably of hepatic origin, and if only ALP is raised the abnormality usually signifies bone disease. 5NT may occasionally be elevated in other than hepatobiliary disorders; the existence of ALP isoenzymes other than those of bone and liver, and the possibility of raised serum activity in diseases affecting other tissues, must be considered in interpreting the relationship between the two enzymes. Leucine aminopeptidase and γ-glutamyl transferase have also been assayed in attempts to define the cause of raised serum ALP activity, and there is no reason to think of 5NT as superior in this regard.

Cancer. Raised serum 5NT activity occurs in a high proportion of patients with metastatic spread of a tumor to the liver. This often occurs before the onset of jaundice, probably owing to blockage of small bile ducts. Although most such patients also have raised serum ALP activity, this is not invariable, and 5NT activity seems to be the better test. These two enzymes have also been used to monitor the treatment of cancer affecting the liver, 5NT being more reliable in indicating response to therapy and subsequent recurrence.

γ-GLUTAMYL TRANSFERASE (γ-GLUTAMYLTRANSFERASE, GGT, EC 2.3.2.1)

Distribution and Assay

This enzyme is distributed in many tissues of the body, highest activity being located in the kidney, followed by the pancreas, liver, spleen, and placenta. It catalyzes a reaction involving glutathione and an amino acid:

$$\gamma\text{-Glutamyl-cysteinylglycine (glutathione)} + \text{amino acid} \rightarrow \gamma\text{-glutamyl-amino acid} + \text{cysteinylglycine}$$

GGT may play a role in the reabsorption of amino acids from the glomerular filtrate within the renal tubule and is present in the endoplasmic reticulum of the hepatocyte where its activity is increased in situations leading to microsomal enzyme induction. In this locus it may function in detoxification mechanisms through conversion of drugs to their mercapturic acid derivatives, which are then highly water soluble. The walls of the bile canaliculi are also richly endowed with GGT activity. Many reports testify to the microheterogeneity of GGT in tissues and serum. Differences in relative mass and in carbohydrate composition provide the molecular basis for these isoenzymes of GGT. Although characteristic isoenzyme patterns have been described by some authors in patients with cholestasis, there is insufficient agreement about their incidence and the most appropriate techniques for their visualization to recommend these diagnostic procedures at the present time.

The *assay* of GGT activity is dependent upon the availability of synthetic substrates. The commonest substrate used at the present time is γ-L-glutamyl-*p*-nitroanilide (or its carboxy derivative, which is much more soluble than the parent compound and is the substrate recommended by several national regulatory bodies). Transfer of the γ-glutamyl residue to glycylglycine releases *p*-nitroaniline which, at the pH for optimal activity, absorbs strongly at 405 to 420 nm. There is confusion over the appropriate *reference range* for GGT. This stems from its skewed distribution and the influence of environmental factors such as sedatives and alcohol which, even when consumed in moderation, may elevate serum GGT activity without being associated with overt disease. It is widely agreed that GGT activity is higher in males than in females, and different reference values should be used for each sex.

Clinical Applications

Hepatobiliary Disease. The GGT activity of serum is raised in a variety of diseases affecting the liver, biliary tract, and pancreas. The enzyme appears to display the highest activity in those diseases in which *biliary stasis* is a dominant feature, and a high activity is also observed in association with space-occupying lesions such as hepatic metastases. GGT levels tend to parallel those shown by other enzymes such as ALP and 5NT, which are also 'markers' for biliary obstruction. However, the relative elevation of GGT is much greater, conferring superior sensitivity in the diagnosis of biliary tract disorders, although diseases of the pancreas are also associated with high activity.

Values in *hepatitis* are generally lower than those in biliary stasis, but there is too much overlap to allow an accurate discrimination between these conditions to be made on the basis of serum GGT assays alone. The activity is quite variable in *cirrhosis*. Occasionally it may be very high, and may be the only serum enzyme activity to be elevated. At other times it may be normal. The serum GGT level tends to decline as the condition reaches a terminal phase and hepatic synthesis of GGT becomes impaired.

The utility of GGT in detecting hepatic *metastases* appears to be well documented. However, it is if anything too sensitive in this regard, and reliance on this test alone would lead to errors in classifying many patients subsequently found at laparotomy to be free of liver cancer. For this reason, and also because increases in GGT activity may reflect a biologic response to therapy rather than spread of the tumor, it is of limited value in monitoring treatment in cancer patients. GGT is a more powerful tool in *excluding* hepatobiliary disease than in confirming it, although it can provide the answer to some clearly defined problems so long as its lack of specificity is recognized (see section on Other Conditions).

Chronic Alcoholism and Enzyme Induction. Elevated levels of serum GGT activity have been described in alcoholics and are frequently the only enzyme abnormality encountered. But raised values are seen almost as frequently in those without tangible evidence of liver disease as in those with clear stigmata of hepatic involvement. A high proportion of abnormal results occur among persons who are defined as heavy drinkers but not alcoholics, a lesser proportion among moderately heavy drinkers. This is partly explainable in terms of hepatic microsomal *enzyme induction,* which occurs after exposure to alcohol or to certain drugs, especially barbiturates, antidepressants, and anticonvulsants. It has now been established that raised serum GGT activities occur in patients to whom these drugs are administered therapeutically, for example, those with epilepsy or psychiatric illness. Raised serum GGT activity in a suspected alcoholic may therefore arise as a consequence of alcohol-induced hepatic damage, or drug- or alcohol-induced microsomal enzyme synthesis in an otherwise healthy liver. A note of caution is necessary: it has now been shown that the plasma membrane, and not smooth endoplasmic reticulum, is the richest source of the enzyme and that parenchymal cells have a low content of GGT relative to other hepatic cell types (Kupffer cells and biliary epithelium).

Myocardial Infarction. Serum GGT activity increases after myocardial infarction; the increase reaches a peak 7 to 11 days after the onset of symptoms and persists for about a month. Because human myocardium is low in GGT activity, the enzyme may originate from the lysosomes of invading leukocytes, or from the capillary endothelium of regenerating blood vessels. An increased serum LD_5 content in infarct patients often accompanies raised GGT activity, and the hypothesis that the postinfarct increase in GGT originates from the liver (as a consequence of anoxic damage) appears as plausible as the suggestion that it arises as a result of repair processes within the myocardium.

Other Conditions. Serum GGT elevations occur during rejection of kidney transplants and in certain neurologic diseases. A normal GGT activity generally prevails in bone diseases, in which ALP levels are usually elevated; the test has therefore been used to determine whether bone or liver is the probable source of raised ALP activity. Raised levels are not encountered in diseases of muscle, but threefold elevations have been reported in 65% of cases suffering from angina, and increases up to sixfold in some diabetics, especially those with vascular complications.

α-AMYLASE (AMYLASE: α-1,4-GLUCAN-4-GLUCANOHYDROLASE, EC 3.2.1.1)

Origins and Assay

The most significant α-amylase functionally is that of the exocrine pancreas. The enzyme is secreted into the duodenum where it participates in the hydrolysis of macromolecular carbohydrates such as starch and glycogen. A related enzyme is secreted by the salivary glands, but although it initiates the digestion of carbohydrates, its action is terminated on contact with gastric acid and pepsin. Other α-amylases are found in the small intestine, kidney, testes, and fallopian tubes.

The action of α-amylase is restricted to hydrolysis of α-1,4-glucosidic bonds. The residual complex therefore preserves its α-1,6-glycosidic links intact and the other products are essentially maltose units. Earlier methods for *measuring* α-amylase activity used a determination of the rate of appearance of reducing sugar (saccharogenic method) or the rate of disappearance of starch (amyloclastic method). Substrates have now been synthesized in which a dye is covalently linked to an insoluble or easily precipitable carbohydrate polymer. Hydrolysis of the substrate leads to the liberation of water-soluble fragments containing the dye. More recently, direct optical rate methods have become available. These methods are based on the hydrolysis by amylase of small-relative-mass glucose polymers of defined chain length containing *p*-nitroanilide residues that are cleaved by action of the enzyme. Coupled reactions have also been developed in which reduced pyridine nucleotides are generated and produce the final optical signal through their high absorbance at 340 nm.

Isoenzymes

The heterogeneity of amylase isoenzymes in human tissues and fluids has been documented in various publications. At least two zones of amylase activity exist in human serum, corresponding to the mobility of human pancreatic and salivary amylases. Clinical studies have established the pancreas as the usual source of pancreatic amylase in human serum; however, other human tissues contain amylases with properties similar to those of the pancreas and salivary glands. The enzyme is especially rich in genital tissues.

Clinical Applications

Serum Amylase. The principal application of amylase determinations has been in the diagnosis of *acute pancreatitis* (see Chap. 12). This may exist in edematous or hemorrhagic forms; the latter carries a more serious prognosis and is usually, though not invariably, associated with more pronounced increases in serum amylase activity (*hyperamylasemia*). Certain conditions which may mimic acute pancreatitis are also associated with raised amylase activity, notably perforated peptic ulcer, intestinal obstruction, cholecystitis, and ruptured ectopic pregnancy. The administration of analgesics such as morphine may cause transient amylase elevation, ascribed to spasm of the sphincter of Oddi. Mumps parotitis and orchitis also generate raised amylase activities, although here clinical confusion with acute pancreatitis is unlikely.

The elevation of serum amylase in acute pancreatitis occurs promptly after the onset of symptoms, reaches a peak within 24 hours, and in uncomplicated cases returns to normal by the fourth day (see Fig. 3–6). Although no level is absolutely diagnostic of acute pancreatitis, it is generally held that the probability of the disease increases with the *magnitude* of the serum amylase elevation. The incidence of hyperamylasemia following operations on the biliary tree varies from 6% to 30% in different series, but it is only rarely accompanied by clinical manifestations of pancreatitis. Another condition associated with hyperamylasemia is *diabetic ketoacidosis,* in which up to 60% of cases may show this feature during the initial 24 hours. The *predictive value* of serum amylase elevations for acute pancreatitis is disappointingly low. It is frequently useful in monitoring resolution of the disease; in the presence of acute pancreatitis prolonged and persistent elevations of serum amylase herald the onset of complications such as pseudocyst formation.

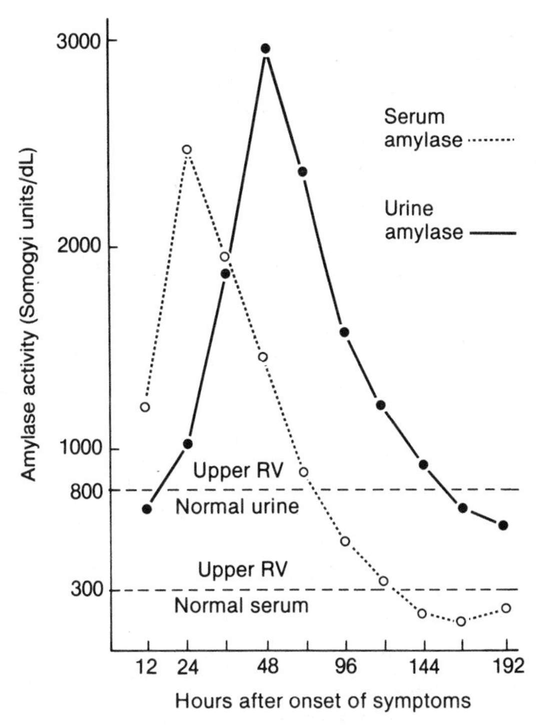

FIGURE 3–7. Serum and urine amylase values during an attack of acute pancreatitis. RV = reference value. Pain and vomiting were present for the first 36 hours.

Urine Amylase. The renal clearance of amylase is directly proportional to the creatinine clearance and is about 3% of the glomerular filtration rate in man. Urine amylase will therefore be increased whenever the serum amylase is raised. The clinical role of urinary amylase determinations is not clearly established, partly because there is no unanimity as to whether measurement should be as a concentration, as an output, or as an excretory rate. However, simple concentration measurements can often be useful (Fig. 3–7) as long as the time course of altered urine activity is appreciated. The elevation of urinary amylase *follows* the increase in activity of the serum enzyme, but the return to reference values is delayed; thus the test is misleading if performed too early in the course of the disease, but valuable when presentation is 48 hours or more after the onset of symptoms.

A number of reports have advocated measurement of the ratio of *amylase clearance* to *creatinine clearance,* values for which are increased in acute pancreatitis. This clearance ratio is also increased in a number of conditions besides acute pancreatitis, such as diabetic ketoacidosis, severe burns, and renal failure. The mechanism seems to involve the effect of kinins, which are released by the acute inflammatory process within the pancreatic gland or other tissues and which block the reabsorption of filtered amylase by the renal tubular cells. In routine use the sensitivity and specificity of the test do not match the claims of the original reports.

Macroamylasemia. In this condition the elevated serum amylase activity is caused by the presence of *macroamylase,* a circulating high-relative-mass complex of the enzyme with carbohydrate or with another protein. There is no evidence that specific clinical features and the macroamylasemia are necessarily related, and tests of pancreatic function have consistently yielded normal results in such patients. The clearance of amylase relative to that of creatinine is greatly reduced in patients with macroamylasemia, in contrast to renal disease, in which both clearances are reduced in parallel, and acute pancreatitis, in which the amylase clearance is greatly increased. The condition will usually be suspected in persons manifesting raised serum amylase and low urinary amylase in the presence of normal renal function; this occurs because the large molecule cannot pass through the glomerular membrane.

Isoamylases. The clinical utility of ***isoamylase determinations*** has not so far proved impressive. A relative absence of pancreatic isoenzymes occurs in the serum of patients with cystic fibrosis and reflects decreased pancreatic function in this disease. Patients with mumps have elevations of serum salivary isoamylases; patients with acute pancreatitis have elevated levels of pancreatic-type isoamylases. However, isoenzyme differentiation cannot unambiguously identify the tissue source of hyperamylasemia, because some patients with neoplasia have hyperamylasemia shown to be caused by salivary-type isoamylases. Indeed, *ectopic tumor* amylases are being increasingly reported in cancers, such as those of the bronchus, arising in tissues which are normally devoid of or low in amylase activity.

LIPASE (GLYCEROL-ESTER HYDROLASE, EC 3.1.1.3)

Properties

This enzyme hydrolyzes triglycerides. The products are mainly monoglycerides and free fatty acids, with some diglycerides and a little glycerol. Pancreatic lipase acts primarily at the interface of an aqueous triglyceride emulsion. A protein cofactor, called *colipase,* is required for full activity. Bile acids activate the enzyme in the presence of colipase, and inhibit the enzyme in its absence.

Clinical Application

Opinions concerning the value of serum lipase estimations in acute pancreatitis have until recently been prejudiced by technical difficulties with the assay. Many reports were based upon fallacious methods using nonspecific water-soluble substrates readily hydrolyzed by esterases other than lipase. The *diagnostic superiority* of lipase over amylase in acute pancreatitis seems increasingly to be established, especially as regards the late stages of the disease, and lipase is more specific for pancreatic disease. It has been claimed that serum lipase elevations also precede those of amylase in acute pancreatitis, but this is controversial. Despite reports that serum lipase may be raised in pancreatic carcinoma, the changes do not seem sufficiently consistent to warrant diagnostic use. Increased serum lipase has been reported to follow severe bone fractures, the highest values being found when fat embolism occurs as a complication.

TRYPSINOGEN AND TRYPSIN

A number of pancreatic proteases have been measured in serum in an attempt to assess the integrity or functional state of the pancreas. Greatest interest has centered on trypsin, although normal serum contains only its precursor trypsinogen. In certain disease states, and particularly in acute pancreatitis or necrotic episodes of chronic pancreatitis, the active enzyme is also liberated into the plasma but is rapidly bound by various protease inhibitors such as α_1-antitrypsin and α_2-macroglobulin. Catalytic assays for trypsin therefore show very low activities.

Renewed interest in ***trypsin*** as a diagnostic tool has been due to the availability of radioimmunoassays for the enzyme. There are actually two, quite distinct proteins: one is anionic and the other is cationic. The latter has been shown to have diagnostic value in the following clinical states:

Acute Pancreatitis: High concentrations of trypsin occur in this condition. The test appears to be sensitive but takes an unduly long time to perform and is not suited to dealing with a clinical emergency. Elevations have been reported in cirrhosis of the liver, diabetic ketoacidosis, and renal failure, as well as in patients with cholecystitis, a condition which it is important to distinguish from acute pancreatitis.

Chronic Pancreatitis: Increased concentrations of trypsin reflect subacute episodes of pancreatic necrosis. With the decrease in viable pancreatic tissue, the concentrations fall progressively until they are below the *lower* reference limit. This usually occurs when >90% of the exocrine pancreatic tissue has been destroyed and steatorrhea is present. Immunoreactive trypsin assays are therefore useful in distinguishing between intestinal and pancreatic causes of *steatorrhea.* Most patients with >10% residual pancreatic function have normal or low-normal serum trypsin concentrations, and so the test is unreliable in detecting even moderate chronic pancreatic damage.

Cystic Fibrosis: Serum immunoreactive trypsin assays may have a special role in screening for this disease. Paradoxically high values are encountered in newborn infants with cystic fibrosis; the test has a sensitivity of >90% and a specificity >99%. It is cost effective, but the impact of earlier detection of the disease on the duration and quality of life has yet to be determined. When the disease progresses, pancreatic function undergoes steady impairment to the point where steatorrhea develops. Deteriorating function is accompanied by a reduction in the serum trypsin concentrations, initially to normal and subsequently to subnormal levels. We now have a better understanding of the pathologic mechanism underlying pancreatic destruction in this disease. At birth, pancreatic ductal obstruction due to inspissated mucous plugs, in the presence of normal acinar tissue, leads to regurgitation of trypsin into the bloodstream and hence to high concentrations. In the face of persisting obstruction, the acini eventually undergo necrosis and fibrosis, and consequently little or no trypsin passes to the blood. It is now apparent that ductal obstruction leading to high serum trypsin concentrations is almost invariable in cystic fibrosis, but in approximately 15% of cases the obstruction does not lead to pancreatic necrosis.

SUGGESTED READING

ADOLPH L, LORENZ R: Enzyme Diagnosis in Diseases of the Heart, Liver, and Pancreas. Basel, Karger, 1982

GOLDBERG DM: The diagnostic efficiency of enzymes and isoenzymes in myocardial infarction. Mod Med Can 34:454–461, 1979

GOLDBERG DM: Structural, functional and clinical aspects of γ-glutamyltransferase. Crit Rev Clin Lab Sci 12:1–58, 1980

GOLDBERG DM: A perspective of diagnostic cancer biochemistry with special reference to enzymes. In Kaiser E, Gabl F, Muller MM, Gayer M (eds): Topics in Clinical Chemistry, pp 383–394. Berlin, de Gruyter, 1982

GOLDBERG DM: Enzymes and Isoenzymes in the Evaluation of Diseases of the Pancreas. In Homburger HA (ed): Clinical and Analytical Concepts in Enzymology, pp 31–55. Skokie, Illinois, College of American Pathologists, 1983

HENDERSON AR: Advances in Clinical Enzymology. In Goldberg DM (ed): Annual Review of Clinical Biochemistry, Vol 1, pp 233–270. New York, Wiley, 1980

HENDERSON AR: Clinical Enzymology. In Goldberg DM (ed): Clinical Biochemistry Reviews, Vol 3, pp 187–234. New York, Wiley, 1982

LANG H: Creatine Kinase Isoenzymes: Pathophysiology and Clinical Application. Berlin, Springer–Verlag, 1981

LANG H, WURZBURG H: Creatine kinase, an enzyme of many forms. Clin Chem 28:1439–1447, 1982

MCCOMB RB, BOWERS GN, POSEN S: Alkaline Phosphatase. New York, Plenum, 1979

MOSS DW: Acid Phosphatase: An Archetype of the Role of Isoenzyme Analysis in Clinical Enzymology. In Goldberg DM, Werner M (eds): Progress in Clinical Enzymology, Vol 2, pp 113–123. New York, Masson, 1983

WILKINSON JH: The Principles and Practice of Diagnostic Enzymology. London, Arnold, 1976

4

Diagnostic Immunology

Stanislaw Dubiski

Diagnostic immunology is concerned primarily with detecting and characterizing diseases of the immune system, which can be classified into four major categories:

1. *Genetic* defects of the immune apparatus (e.g., *immunodeficiency diseases*).
2. *Malignancies* of the immune system (e.g., *monoclonal gammopathies*).
3. Malfunctioning of the mechanisms that *regulate* the immune system; failure to distinguish 'self' from 'nonself' may result in development of the ***autoimmune diseases.*** Malfunctioning of the mechanism that determines the class and amount of antibody formed to a given antigen may lead to the formation of excessive amounts of IgE antibodies and may result in ***atopic*** (allergic) diseases. Excessive suppressive activity within the immune system is believed to lead to the development of acquired immune deficiency syndrome (AIDS).
4. Conditions caused by a properly functioning immune response directed against a mismatched *graft* (e.g., transplant rejection, hemolytic transfusion reaction, and hemolytic disease of the newborn. The ***graft-versus-host*** reaction, a process during which a transplanted tissue 'rejects' the immunologically incompetent host, also belongs in this category.

A brief account of the normal functions of the immune system is necessary before immunologic disorders can be discussed.

DEFINITIONS

Immunity is the ability of the organism to recognize and destroy 'nonself' macromolecules or cells. These functions are performed by a few interacting systems, some of which are of low specificity (*constitutive*) and serve as a first line of defense; others are the highly specific *adaptive* systems that provide protection against repeated invasions of the same macromolecules or cells. The main ***constitutive*** systems are (1) the complement system, a series of enzymes present in the circulation in their inactive forms and activated as a result of the invasion, (2) phagocytic cells, which ingest foreign material and may be helped in their activity by other constitutive and adaptive systems, (3) interferon, which is induced by viruses and triggers the formation of a nonspecific antiviral substance, and (4) 'natural' killer cells, large granular lymphocytes, derived from the bone marrow, that destroy foreign and some tumor cells.

The ***adaptive*** systems improve their performance on repeated encounter with the invading macromolecules or cells. They are highly specific, that is, they recognize the invading agents by structures, called determinants. The specificity of the adaptive systems allows it not only to recognize self and nonself, but also to distinguish among different nonself macromolecules.

The ability of the immune system to adapt implies a capacity to 'remember' previously encountered macromolecules. *Immunologic memory* was the first manifestation of the immune apparatus discovered by man; individuals who survived epidemics were known to be immune to a second outbreak of the same disease. Today an understanding of immunologic memory allows us to prepare safe and effective vaccines, to immunize millions of people, and to make significant attempts at complete eradication of certain infectious diseases.

Nonself substances that can stimulate the adaptive part of the immune system are called ***antigens*** or ***immunogens.*** The ability of an antigen to stimulate an immune response is called *immunogenicity,* and an *antigen* can be defined as an immunogenic macromolecule capable of reacting specifically with the effectors of the immune system. *Antigenic determinants* (sometimes called *epitopes*) are the portions of the antigen molecule that are recognized by, and react with, the immune system. *Paratopes* are the structures capable of recognizing *epitopes.* Antigenic determinants define the *specificity* of the antigens. If two distinct antigens share some determinants, these antigens will *cross-react,* that is, the immune system will not be able to distinguish fully between them. A *hapten* is usually a small molecule that contains one or more antigenic determinants but is not itself immunogenic. Haptens usually become immunogenic if attached to a strongly immunogenic macromolecule called a 'carrier.'

THE IMMUNE SYSTEM

The emergence of multicellular organisms created a need to distinguish between self and nonself. This ability is a pivotal regulatory mechanism of the immune system of today's vertebrates. Traditionally, *humoral* and *cellular* arms of the immune system have been recognized. This distinction is fully justified only at the effector level; the recognition and regulatory mechanisms are shared by the two arms.

Anatomically, the immune functions are carried by the ***lymphoid system*** which consists of bone marrow, thymus, spleen, appendix, Peyer's patches of the small intestine, the adenoids, the tonsils, and a large number of peripheral lymph nodes scattered throughout the body. These organs are interconnected by a network of lymphatic and blood vessels through which the lymphoid cells constantly circulate. Lymphoid cells originate in the bone marrow from pluripotential *stem cells.* Cells that are involved in regulatory functions, and those that later become effectors of the cellular arm, migrate to the thymus, undergo complex differentiation processes, and become ***T cells*** (thymus-derived cells). Cells that are going to secrete humoral effector molecules (antibodies) differentiate in the bone marrow, and become ***B cells*** (bone-marrow–derived cells). B and T cells populate other lymphoid organs, where they coexist with each other. A third category, *accessory* cells (macrophages, monocytes, and others) intercept, process, and present the antigens to the lymphocytes.

EVENTS THAT LEAD TO THE IMMUNE RESPONSE

Prior to antigenic stimulation, T and B cells undergo *antigen-independent maturation* and acquire antigen-specific cell-surface receptors, thus becoming *committed* for responding to a given antigen. The receptors of B cells are antibody molecules (IgD and monomers of IgM). The receptors of T cells are also antigen specific, but they are not identical with those of the B cells.

Antigens that enter the body are intercepted and processed by the accessory cells, which present such processed antigen to the immunocompetent cells. After combining with the antigen, committed immunocompetent cells undergo *antigen-driven differentiation.* Because one antigen-sensitive cell can carry receptors of only one specificity and because the number of possible specificities is almost unlimited, only a very small proportion of the receptor-carrying cells will carry the receptors of the right configuration, and only those cells will become stimulated.

The immune response (Ir) is triggered only if a *third* cell, a ***T-helper cell*** (***T_H***), is involved. If a hapten, rendered immunogenic by conjugation to a carrier, is used for immunization, B cells will combine with the haptenic determinant, and T_H cells with the carrier portion of the molecule. It is not clear yet what the sequence of these events is and whether the process requires direct contact between the cells involved. At least some of the information is transferred by soluble substance(s) released by accessory and/or T_H cells. Such soluble molecules appear to contain not only the processed antigen, but also the product of the *Ir* locus of the major histocompatibility complex (MHC). Both these determinants are recognized by the T_H cell. As a result, the message is *antigen specific* and *genetically restricted,* that is, can be recognized only by T cells that carry both the antigen-specific receptor and the appropriate *Ir* gene product. Alternatively, the antigen and the *Ir* gene product are presented directly to the T_H cell by the macrophage. In general, in order to be recognized by the T_H cell and to trigger the immune response, the antigenic 'message' must contain both self and nonself determinants.

A number of antigen-nonspecific and genetically unrestricted humoral substances are also involved in communication between the cells. *Interleukin 1* is produced by macrophages and stimulates maturation and activation of T_H and ***T-cytotoxic*** (***T_C***) cells. *Interleukin 2* is produced by activated T cells as a result of the action of interleukin 1. It appears that the principal function of interleukin 2 is the enhancement of antigen-specific T and B cellular responses.

In addition to the T_H cells, there are also ***T-suppressor cells*** (***T_S***), which inhibit the immune response. Like the T_H cells, the T_S cells are also antigen specific.

The final outcome of these cellular interactions is the stimulation of the B-cell, which after a number of divisions matures into an effector, antibody-secreting *plasma cell.* Initially, particularly during the primary response, the antibody produced is of the IgM class, but later the cell may undergo a 'switch' and may start making antibody of another class, but of the same specificity. Eventually, *one* cell secretes antibody of only *one* class. Some stimulated B cells, instead of differentiating into plasma cells, transform into *memory* cells that are responsible for the phenomenon of *immunologic memory.* Memory cells are already committed to a given class of antibodies and do not have to undergo the switch from IgM.

Immunocompetent T cells mature through pathways analogous to those of B-cell maturation: they are 'helped' by T_H cells, are suppressed by T_S cells, and differentiate into T_C and T_D effector cells. The former are responsible for *cytotoxic* reactions, the latter for *delayed hypersensitivity* reactions.

CLONAL SELECTION THEORY OF IMMUNITY

The *preformed* cell surface ***receptors*** play a central role in the initiation of the immune response. The formation of these receptors is *not* the result of an encounter with the antigen. This fundamental precept was foreseen by Paul Ehrlich who, in 1900, formulated the first *selective* theory of immunity. Because the role of antigen in the immune processes is so apparent, and because one can induce formation of antibodies against almost any chemical configuration, the concepts of preformed receptors and the consequent selective mechanism of immunity seemed unacceptable. A number of *instructive* hypotheses were proposed that postulated an active role of antigen in the determination of quaternary or primary structure of antibody molecules.

The dawn of modern immunology has seen the resurrection, by Niels Jerne in 1955, of the concept of a selective mechanism of immunity. Two years later, Macfarlane Burnet formulated the *modern theory of immunity* by introducing a fundamental modification of Ehrlich's concept. Burnet postulated that *one* cell can express receptors of only *one* specificity. Further maturation of such a cell is possible only after the combination with an antigen molecule. The maturing cell divides and generates an entire ***clone*** of identical daughter cells, all of them capable of synthesizing antibody molecules of specificity identical to the specificity of the mother cell's receptors. Clonal precursor receptor-carrying cells are generated *without* intervention of antigen by a random process equivalent to somatic mutations. If, during this process, an *immature* cell displaying autoreactive receptors is generated, it is eliminated as a ***forbidden clone.*** On the other hand, exposure of a *mature* antigen-sensitive cell to an antigenic determinant triggers clonal expansion and an immune response.

Logical extensions of the clonal selection hypothesis were the concepts of ***immune surveillance*** and ***immunologic tolerance.*** Self cells that have acquired new antigenic determinants as a result of a somatic mutation or virus-induced changes in its genome, would be eliminated as nonself. Such surveillance mechanism may be capable of eliminating potentially malignant cells.

IMMUNOLOGIC TOLERANCE

In his concept of tolerance, Burnet postulated the elimination, as a forbidden clone, of any receptor-carrying cell that has prematurely been exposed to an antigen, be it self or nonself. The intervening research has shown that the mechanism of this phenomenon is much more complex than originally proposed. The inability to respond to self antigens while responding normally to all other immunogens is called *naturally induced tolerance.* Under some conditions, nonself immunogens can mimic self substances and can induce *acquired tolerance.* This phenomenon was first observed in dizygotic twins; in a proportion of cattle twins placental anastomoses are established allowing exchange of blood between the twins *in utero.* Such twins become *chimeras,* that is, their blood cells are a mixture of their own and of their twin sibling. In spite of the persistent chimerism, the calves never form antibodies against the grafted cells. Natural chimeras have been described in humans; experimental chimerism can be induced by injecting spleen cells into newborn mice, or in birds by joining two incubating eggs.

Allophenic (tetraparental) mice are a special example of tolerance induced by fusion of two 8-cell murine embryos. The resulting 16-cell embryos are implanted into the uterus of a foster mother. All organs of such allophenic mice are chimeric to a different degree. The chimerism of skin tissue produces spectacular color patterns.

Tolerance to soluble, nonviable antigens can be induced by injecting tolerogen (an antigen that induces tolerance) to newborn animals, by injecting excessive doses of antigen (so-called immunologic paralysis), by injecting antigen in its nonimmunogenic form (e.g., the form in which it is free of any aggregated material), or by injecting antigen into immunosuppressed animals. Persistence of tolerogen in the circulation is essential for the maintenance of tolerance.

Tolerance affects both B and T cells, but they become tolerant at different rates and under different conditions. Because B and T cells cooperate to produce an immune response, tolerance of either of them will make the animal tolerant, and consequently, different mechanisms may operate in different forms of tolerance.

REGULATION OF THE IMMUNE RESPONSE THROUGH IDIOTYPE—ANTI-IDIOTYPE NETWORKS

In 1974, Jerne formulated his *idiotype network hypothesis.* ***Idiotype*** is a term used for an antigenic determinant associated with the antigen-combining site (or the hypervariable part) of the antibody molecule. One can, under certain experimental conditions, raise an antibody directed to an idiotypic determinant. This 'antiantibody' (or more commonly, anti-idiotype antibody) may compete for binding with the antigen. If it does so, it is a proof that it actually combines with the antigen binding site of its target antibody. Thus, the antigen-combining site of this anti-idiotype antibody (1) must be a sort of *negative image* of the combining site of the target antibody and (2) must resemble the configuration of the antigenic determinant, with which the target antibody reacts. To avoid confusion we must designate the antigen ***Ag,*** the first antibody ***Ab1,*** and the second (anti-idiotype) antibody ***Ab2.*** Higher ranks of anti-idiotypes, for example Ab3 and Ab4, were raised by appropriate immunization and it was shown that Ab1, Ab3, and so forth can be compared to a photographic *negative* of the Ag, whereas Ab2, Ab4, and so forth resemble a photographic *positive* of the Ag.

In such experiments each Ab has been raised in a different animal. Jerne postulated that an analogous process is taking place in the individual's immune system; there is a continuous formation of auto-Ab2 against the cell surface idiotypes. Subsequently, Ab2 may be recognized as an immunogen and an auto-Ab3 is formed, and so forth. The idiotypes on the cell surface and the anti-idiotypes in the circulation recognize each other and, through this recognition, *regulate* the immune response. Small amounts of anti-idiotype may stimulate cellular proliferation; high concentrations may arrest the proliferation. Errors in this recognition system may result in an autoimmune disorder.

The ***idiotype–anti-idiotype network*** may also provide a stimulus to keep the immune system

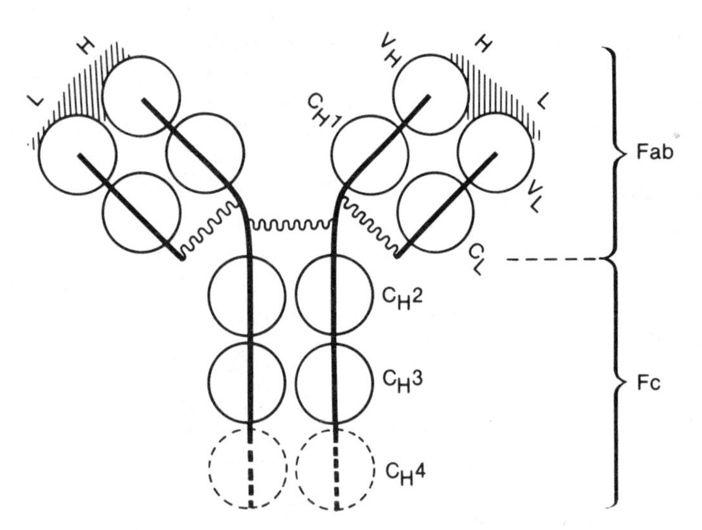

FIGURE 4–1. Schematic and simplified drawing of a basic immunoglobulin unit composed of two light (L) and two heavy (H) chains. Domains are represented as circles; C = constant; V = variable. Wavy lines represent interchain disulfide bonds; the number of H-H interchain bonds varies in different classes and subclasses; intrachain disulfide bonds are not shown. Shaded areas indicate the location of the antigen combining sites. C_H4 domains, drawn in broken lines, are present only on μ and ϵ chains. V_H = variable heavy; V_L = variable light; C_H1, etc. = constant domains of the heavy chain; C_L = constant light; Fab and Fc = antigen-binding and constant fragments formed by cleaving the molecule with papain.

fully alert. The immune system is always ready to manufacture antibodies of almost *any* specificity, be it against a very common pathogen or against never-to-be encountered synthetic antigen. One would expect that the ability to synthesize never-used antibodies should be lost, either during the individual's lifetime, or even more likely, during evolution. Because this is not the case, perhaps the immune system keeps itself in shape by making auto-Ab1, Ab2, Ab3, and so forth, which resemble the antigens that the immune system expects to encounter in the future. The immune networks may one day be used to manipulate the biochemistry and immunity of the body; Ab2 may possess some of the properties of the Ag, for example, Ab2s in the insulin–anti-insulin system may resemble insulin and combine with insulin receptors. Viral Ab2s may be used as vaccines to immunize against the respective antigens without exposing the body to potentially dangerous viruses.

HUMORAL IMMUNITY

STRUCTURE OF ANTIBODY MOLECULES

The *antibodies,* secreted by plasma cells, are the effector molecules of the humoral arm of the immune system. They belong to a heterogeneous group of proteins called ***immunoglobulins.*** All immunoglobulins are either monomers or polymers of a basic unit consisting of two pairs of polypeptide chains, *heavy* (***H***) and *light* (***L***), the relative masses (M_r*) of these chains being 52,000 (or 70,000) and 22,500 respectively (Fig. 4–1, Table 4–1). These chains are joined together by noncovalent bonds and by disulfide bonds. The location and number of disulfide bonds depends on the class (or subclass) of immunoglobulins, of which five classes (IgG, IgA, IgM, IgD, and IgE) and several subclasses (e.g., $\gamma 1$, $\gamma 2$) have been described (Table 4–1). Immunoglobulin classes are defined by a particular type of heavy chain (γ, α, μ, δ, and ϵ).† There are two types of light chains (κ and λ),‡ and either may be associated with any of the heavy chain types. One molecule, however, can have only one type of heavy chain and one type of light chain.

Amino terminal halves of the L chains and amino terminal one-third or one-fourth of the H chains vary considerably from one molecule to another and are thus designated *variable* (***V***) regions. Amino acid sequences of the remaining portions of the chains are characteristic for a given chain class or type, but do not vary significantly within each chain type; these portions are thus designated *constant* (***C***) regions.

Closer analysis of the amino acid sequences of the H and L chains reveals the existence of some

* M_r = relative mass ≃ molecular weight.

† γ = gamma, α = alpha, μ = mu, δ = delta, ϵ = epsilon.

‡ κ = kappa, λ = lambda.

TABLE 4–1. Properties and Functions of Antibody Classes

Class	Heavy Chains*	Number of Basic Units	Valence	Relative Mass ($\times 10^{-3}$)	Serum (mg/mL) (g/L)	Percentage of Serum Immunoglobulins	Main Functions
IgG	$\gamma 1$	1	2	150	10	50	Serum antibody, produced in increased amounts after secondary stimulation (see Immunologic Memory) activates complement (except IgG2), is transported through human placenta and provides antibacterial, antiviral, and antitoxic defenses. Facilitates phagocytosis by combining with Fc receptors on phagocytic cells. Activates antibody-dependent cell-mediated cytotoxicity.
	$\gamma 2$					15	
	$\gamma 3$					6	
	$\gamma 4$					4	
IgM	μ	1	2	190 (?)	—		Cell surface determinant of B cells
		5†	10	900	1	15	'Natural' antibody, produced early in humoral response; amounts produced after secondary stimulation do not differ significantly from amounts produced in primary stimulation; excellent activator of complement. Because of its multivalency, even low-affinity antibodies combine firmly with antigen ('zipper effect').
IgA	$\alpha 1$, $\alpha 2$	1	2	160	2	10	Serum antibody
		2†	4	330			
		2‡	4	400	—		Antibody in external secretions (saliva, tears, nasal mucosa, bronchial secretions, mucous secretion of the small intestine, milk and colostrum, prostatic fluid, and vaginal secretions). Protects the body against the invasion of foreign microorganisms; protects the nursing infant.
IgD	δ	1	2	180	.03	0.2	IgD (as well as monomeric IgM) is the cell surface determinant of B cells.
IgE	ϵ	1	2	190	.0003	0.004	Binds with very high affinity to mast cells. Upon combination with antigen stimulates the mast cell to release pharmacologically active substances (histamine, leukotrienes, and bradykinin) thus mediating allergic reactions (see type I immune injury). May play a role in immunity to parasites.

* Each heavy-chain class can occur with κ or λ light-chain type.

† Contains one J-chain per polymeric molecule.

‡ May occur also in larger polymeric molecules, consisting of three to five basic units.

TABLE 4–2. Main Functions of Immunoglobulin Domains

Function	Domains Involved	Requires Activation
Antigen binding	$V_H + V_L$	—
Subunit interaction		
H – L	$V_H + V_L$, $C_H1 + C_L$	No
H – H	C_H3	No
Complement (C1q) binding	$C_\gamma2$*, $C_\mu2$	Yes
Interaction with cellular receptors of		
macrophages	$C_\gamma3$	Yes
neutrophils	$C_\gamma2 + C_\gamma3$	Yes
K cells	$C_\gamma2 + C_\gamma3$	Yes
trophoblast	$C_\gamma2 + C_\gamma3$	No
mast cells	$C_\epsilon4$	No
Control of catabolic rate	C_H2	No

* C_H2 of IgG2 subclass cannot bind complement because of its inaccessibility (steric hindrance).

degree of periodicity of amino acid sequences. This periodicity defines *domains* consisting of approximately 100 amino acid residues each. Domains are folded tightly and are held in place by intrachain disulfide bonds. They are linked by short stretches of chains that are less compactly folded. Sequences of the domains show a considerable degree of homology, suggesting that they might have originated by gene duplication. Each light chain possesses 2 domains; μ and ϵ chains have 5; all other H chains have 4 (Table 4–2, Fig. 4–1).

The variability of both V_H and V_L regions is confined to certain 'hot spots' called *hypervariable* regions. The remainder of the domain, called the *framework,* shows much less variability. Because of the folding of polypeptide chains in the assembled immunoglobulin molecule, the hypervariable regions of the neighboring L and H chains find themselves near each other and together form the *combining site* of the antibody.*

The basic antibody unit resembles the letter 'Y'; its upper arms (the ***Fab*** portions) carry the antigen combining sites and are comprised of an L chain and two aminoterminal domains of the H chain (V_H and C_H1) each. The stem of the letter (the ***Fc*** portion) contains the remaining constant domains of both H chains. Fc is the biologically active portion of the antibody molecule. A longitudinal axis of symmetry divides each antibody unit into two identical halves, each composed of one H and one L chain. The two halves are joined together by one or more disulfide bonds at the *hinge region,* located approximately in the middle of the molecule. Hinge region is the point of considerable segmental flexibility between the Fab and Fc regions, essential for the function of the antibody molecule (Fig. 4–1).

Immunoglobulins G, D, and E are *monomers* of the basic four-chain unit, whereas IgM is a *pentamer* (Fig. 4–2). IgA may occur as a monomer (in blood) or as a dimer or polymer (the so-called *secretory* IgA in external secretions such as milk, saliva, tears, intestinal secretions). Polymeric im-

* Each L-chain has 3, and each H-chain 4 hypervariable regions, of which 2 and 3, respectively, always participate in the formation of the combining site.

FIGURE 4–2. Pentameric molecule of IgM; note the presence of J-chain (center) and of the disulfide bonds (wwww) connecting the units; the presence of a second set of disulfide bonds between C_H3 domains (not shown) is controversial. For other explanations, see Fig. 4–1 (Zaleski MB, Dubiski S, Niles ES, Cunningham R. Immunogenetics. Belmont, Pitman, 1983. Reprinted with permission.)

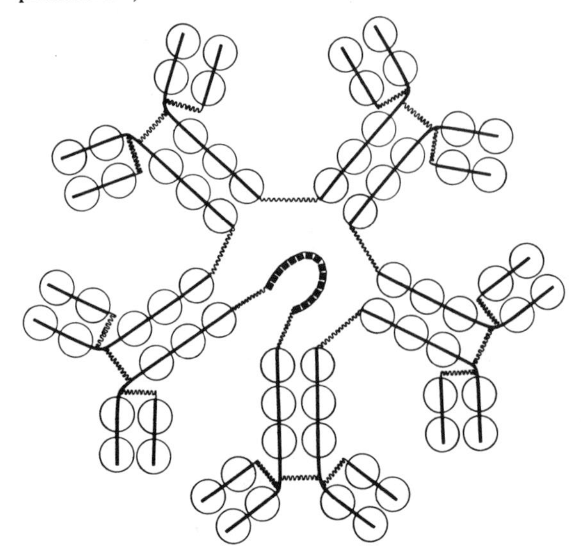

munoglobulins contain an additional component, ***J*** chain ($M_r \simeq 15,000$), which joins the units together by means of disulfide bonds. Secretory IgA is also associated with another polypeptide chain *secretory component* ($M_r \simeq 70,000$), synthesized by epithelial cells and added to the IgA molecule during its passage through the mucous membrane.

FUNCTION OF ANTIBODY MOLECULES

Each basic antibody unit is *divalent,* that is, it has *two combining sites.* The combining site, a crevice whose configuration is complementary to that of the antigenic determinant, measures approximately 3.4 × 1.2 × 0.7 nm.* Both H and L chains cooperate in the formation of combining sites through their hypervariable regions. The *specificity* of a combining site is determined by the primary structure of the respective hypervariable regions. One of the goals of contemporary immunology is to relate amino acid sequences of these regions to the specificities of the resulting antibodies, thus compiling a 'dictionary' of primary structures and their respective specificities. Recent advances in protein and DNA sequencing and in monoclonal antibody techniques place this goal within reach.

Biologic functions of the antibody molecule are performed by the *constant* domains. Properties, functions, and biologic significance of various antibody classes have been summarized in Table 4–1. Binding of antigen by the combining site activates the constant portion to perform its biologic functions. Some functions, however, do not require such activation (Table 4–2).

IMMUNOGLOBULIN GENES

The information necessary for the synthesis of immunoglobulins is stored in the genome in a very compact manner. Storing a complete copy of each immunoglobulin gene would exceed the capacity of the genome. Instead, the genome contains gene segments, which are to a certain degree 'interchangeable' and may be used for assembling many different genes. The 'interchangeability' of the immunoglobulin building blocks is manifested on three levels: (1) the existence of *separate* ***V*** and ***C*** genes, (2) the *cooperation* between H and L chains in the formation of the combining site, and (3) the existence of *gene segments* from which the ***V*** genes are being assembled.

There are three independently segregating *linkage groups* of immunoglobulin genes: the κ and λ light-chain groups, and the heavy-chain groups. Within each group there are genes controlling various classes and subclasses, genes controlling subtypes (if any), and genes for the variable portions. The genes can rearrange themselves during the 'switch' from IgM to IgG production: A given ***V*** gene is initially expressed together with the $\boldsymbol{C\mu}$ gene to form μ chains and IgM antibodies. Later, the same ***V*** gene may express itself with another ***C*** gene (e.g. $C\gamma$ or $C\epsilon$) to form IgG or IgE molecules. A similar mechanism controls the synthesis of κ and λ light chains. As a result, any ***V*** gene may associate itself with a large number of ***C*** genes within its own linkage group, and any ***C*** gene with an even larger number of ***V*** genes. This switching and splicing mechanism ensures that any antibody class can be associated with any antibody specificity available in the genome, without the need to store DNA sequences of *all* complete immunoglobulin chains.

A given L or H chain may, in cooperation with different H or L chains, form different combining sites, so that a set of x***VL*** and ***VH*** genes may generate more than x combining sites.

Finally, each ***V*** gene is assembled from ***V, J,*** and ***D*** segments, several of which are positioned in tandem on a given chromosome (light-chain genes do not seem to possess the ***D*** segments). During the maturation of the B-cell, the required ***V, J,*** and ***D*** segments are brought together and are spliced, and the intervening (not needed) nucleotides are excised. The ***V*** gene thus assembled is then spliced with the ***C*** gene, which is assembled from four or five segments, each ***C*** segment coding for a separate domain.

Besides the aforementioned mechanisms, *somatic mutations,* and possibly *somatic recombinations,* also play a role in the generation of the diversity of antibodies.

* Dimensions of combining sites of different specificities vary; those given above were estimated by Kabat for antibodies to dextran.

CELL MEDIATED IMMUNITY

The *T-cytotoxic cells* (T_C) are involved in killing of some neoplastic cells, self cells infected by viruses, and incompatible transplanted cells. The cells are killed by direct contact, probably by a potent nonspecific humoral substance that is quickly inactivated and capable of acting only across a short distance. In order to be destroyed, cells infected by viruses must first be recognized as self. Antigens of the K and D regions of the major histocompatibility complex are essential for this recognition.

T_D cells are primarily responsible for *delayed hypersensitivity (DH).* DH is a complex reaction in response to stimulation with certain antigens. Usually soluble substances (e.g., toxins) activate a mainly *humoral* immune response, whereas the exposure to particular and cellular antigens results predominantly in the development of a *cellular* immune response. DH is especially pronounced in fungal, some bacterial (*Mycobacterium, Brucella*) and protozoal (*Leishmania, Toxoplasma*) infections.

Experimentally, DH can be produced by intradermal injection of an antigen with which the subject has been previously immunized. The reaction begins 4 to 6 hours after the injection, reaches a peak at 24 to 72 hours and then slowly disappears. The characteristic features are erythema (due to vasodilation) and induration (due to increased vascular permeability and cellular infiltration). In severe reactions, necrosis may occur. The cells infiltrating the site are originally polymorphonuclear and later mononuclear cells, predominantly lymphocytes and macrophages.

The events leading to DH can be summarized as follows: T_D antigen-sensitive lymphocytes or T-memory cells, upon exposure to antigen, become activated and release humoral substances known under the collective term of ***lymphokines.*** Lymphokines *attract* monocytes from the circulation and *inhibit* their further movements. At the site, the monocytes are activated and converted into tissue macrophages. Activated macrophages kill bacteria by releasing lymphotoxin and by increasing phagocytosis. Increased vascular permeability facilitates the infiltration of antibodies, which neutralize the soluble toxins, activate the ADCC (antibody-dependent cell-mediated cytotoxicity), and enhance phagocytosis by opsonization. T_C lymphocytes are also present at the site and may be actively killing some cells.

Soluble Factors That Mediate Cellular Immunity

The number of antigen-reactive T cells present at the site of the DH reaction is actually very small. These cells recruit large numbers of immunologically nonspecific cells by secreting *lymphokines.* Most lymphokines are known only by their function; very few have been characterized in molecular terms. Thus, it is not known how many distinct substances carry the described functions. The better known lymphokines are migration inhibition factor (MIF), macrophage activation factor, macrophage chemotactic factor, leukocyte inhibitory factor, chemotactic factors affecting cells other than macrophages, histamine-releasing factor, and the mysterious transfer factor (TF). TF can specifically transfer DH to a nonimmune or even immunologically deficient host. It is a dialyzable substance ($M_r < 4000$) produced by activated peripheral blood lymphocytes. TF is resistant to the action of RNase and DNase. It is difficult to understand how the information about both the antigen and the species specificity can be transferred by such a small molecule. TF has been used successfully to treat some conditions (e.g., *Candidiasis*) resulting from deficiency of the cellular arm of the immune system.

THE MAJOR HISTOCOMPATIBILITY COMPLEX (MHC)

Although clinically the most noticeable role of the MHC is in eliciting the rejection of incompatible transplants, this function is in fact only incidental to the main physiologic function of the system, that is, the recognition of, and the distinction between, self and nonself. In humans, the MHC is encoded in a segment of chromosome 6; the complex itself is called ***HLA*** (for *human lymphocyte antigens*). Several polymorphic genes have been described that control four different types (classes) of products (Table 4–3).

The *Class I* products are membrane-bound glycoproteins composed of two polypeptide chains. The heavy chain ($M_r \simeq 39{,}000$) is embedded by its carboxyterminal in the cell membrane, with 25

TABLE 4–3. Genetic Control of Human Major Histocompatibility Complex; HLA

Chromosomal Region (Class of Products)	Loci	Characteristic Features (Number of Alleles)
I	*A*	4 + 13 nfc* alleles
	B	7 + 25 nfc alleles
	C	8 nfc alleles
II†	*D*	12 nfc alleles
	DR	6 + 4 nfc alleles *D* = *DR* (?)
	DC	nfc
	SB	nfc
III	*C2* *C4A* *C4B* *BF*	C2, C4 and B components of the complement system and antigens of Rodgers and Chido 'blood groups'

* nfc = not fully characterized.

† This chromosomal region controls also the **Ir** (immune response) genes.

to 30 amino acid residues penetrating into the cell's interior. Its primary amino acid sequence is controlled by the Class I (HLA-A, HLA-B or HLA-C) genes and is extremely polymorphic. It carries a carbohydrate side chain and is noncovalently associated with the light chain (M_r = 11,800), which contains 100 amino acids and is known as *β_2-microglobulin.* It is worth noting that β_2-microglobulin shows about 35% sequence homology with immunoglobulin domains. Some degree of homology exists also between immunoglobulin domains and parts of the Class I heavy chains. This may indicate that the immunoglobulin and the MHC genes have a common origin. Class I products are present on all nucleated cells.

Class II products are not so ubiquitous as the Class I products. They also consist of two noncovalently linked polypeptide chains, α (M_r = 32,000) and β (M_r = 28,000). Both units are embedded in the cytoplasmic membrane of most B cells, some T cells, macrophages, and some other cells.

Class III products are components of the complement system.

Biologic Function of the MHC Determinants. As outlined earlier (see Cellular Interactions), in order to trigger the immune response, the antigen must be presented to the T-cell together with one of the self MHC molecules. Class I molecules are recognized by the T_C cells, Class II molecules by the T_D and by the regulatory T cells. The only antigenic determinants that do not require the *self* component to be included in the immunogenic message are the MHC molecules themselves. The significance and the exact mechanism of this linked recognition remain the subject of extensive research and speculation.

The Class II genes function as *immune response* (Ir) genes, controlling the responses against a variety of antigenic determinants. It has been shown that animals carrying certain alleles are 'low responders' when immunized with a given hapten, whereas some other animals that carry other Class II alleles are 'high responders.' The same allele may confer low respondency to one determinant and high respondency to another. The effect of the ***Ir*** genes on the response to complex multideterminant antigens is not so noticeable, because the animal may be a low responder to one determinant of a given antigen and a high responder to another determinant of the same antigen. The difference between high and low responders may arise from the ability or inability to recognize a given combination of antigenic determinant and Class II molecule.

Whatever the *primary physiologic role* of the MHC genes and their products, it is obvious that some malfunctions of the system may lead to overt pathologic manifestations. It has been shown that persons expressing an HLA–B27 gene are at much greater risk of developing *ankylosing spondylitis* than those expressing other HLA–B alleles. The B27 determinant is present in only about 7% of healthy people, whereas approximately 90% of patients suffering from ankylosing spondylitis are B27 positive. Similar but less pronounced associations between various HLA molecules and diseases have also been described: Reiter's syndrome, uveitis, juvenile rheumatoid arthritis (all associated with HLA–B27); psoriasis (HLA–B13); celiac disease (HLA–DR3); multiple sclerosis (HLA–DR2); juvenile diabetes mellitus (HLA–DR2,3,4); Graves' disease (HLA–DR3), and others. Most of these diseases have an immune component; some are autoimmune, and some are thought to result from an infection with an unknown viral agent. Thus, the associations may be the result of possessing a *low response* ***Ir*** allele for the immune response against a given pathogen. Some other diseases may be caused by genetic errors of metabolism; in these conditions the association between the disease and the MHC determinants may be caused by a linkage disequilibrium between genes controlling the metabolic pathway in question, and one of the MHC alleles. The two hypotheses are not mutually exclusive and may explain different associations.

THE COMPLEMENT SYSTEM*

The *humoral* immune system can inactivate foreign antigens in cellular form only in cooperation with mechanisms of *constitutive* immunity such as the complement system or phagocytic cells.

The *complement system* is composed of several structurally and functionally distinct proteins that are present in fresh serum in their precursor, inactive forms. The *activation* of the complement system leads to (1) release of pharmacologically active peptides that cause local inflammation (i.e., vasodilation, enhanced vascular permeability, and chemotaxis of phagocytes), (2) opsonization, that is, enhancement of phagocytosis, and (3) cell death resulting from perforation of the cell wall.

Most of the complement components become activated as the result of an association with one or more earlier components, or by a limited proteolytic cleavage effected by an earlier component in its activated, enzymatic form. The *larger* cleavage products generally serve in the reaction sequence as activators of the next component, and the *small* fragments are often mediators in the inflammatory response. There are two separate but converging pathways of complement activation, the ***'classic' pathway,*** which is activated by antigen–antibody complexes and the ***'alternate' pathway,*** activated by microbial cell wall polysaccharides (endotoxin, inulin, zymosan). The alternate pathway may be older in an evolutionary sense, since it does not require antibody for its activation.

* This section was written jointly with J. O. Minta. Dr. Minta is Associate Professor, Department of Pathology, University of Toronto.

During each step of the complement sequence, considerable amplification occurs; one molecule of an earlier component may activate many molecules of the next component (Fig. 4–3). Overamplification is prevented by a number of specific *inactivators* and by fast decay of some activated components. Usually a component loses its activity, unless it forms a solid phase by binding to a cell or to antigen–antibody aggregates.

The ***first component*** of the classic pathway, C1† is a complex of three distinct proteins, C1q, C1r, and C1s, held together by calcium ions. C1 becomes activated by a single molecule of IgM (C_H4 domain) or by two molecules of IgG, combined with an antigen in proximity ($\simeq$700 nm apart, C_H2 domains). The activated C1 initiates the sequential activation of eight more components (C1 → C4 → C2 → C3 → C5 → C6 → C7 → C8 → C9) by cleaving C4 and C2, whose major fragments fuse in the presence of magnesium ions, forming the *classic* pathway C3 convertase enzyme (C4b2a). The C3 convertase, in turn, cleaves C3 into C3a and C3b.

The conversion of C3 into C3b can be accomplished through both classic and alternate pathways and is a *central event* in complement activation. The activation of C3 via the alternate pathway utilizes the slow spontaneous conversion of C3 into C3b. Normally, nascent C3b binds to cell surfaces covalently. Cell-bound C3b can form a complex with Factor H and, in this form, be cleaved and inactivated (C3bi) by a serum endopeptidase, Factor I. The binding of Factor H to cell-bound C3b is favored by the presence of membrane sialic acid residues. Some microbial cell surface constituents, containing zymosan, inulin, or endotoxin can also bind the nascent C3b, but the C3b bound to these surfaces has a diminished affinity for Factor H. Cell-bound C3b has affinity also for Factor B, with

† Notation of complement components: the components of the *classic* pathway are denoted by a number prefixed with a capital C; the factors of the *alternate* pathway are denoted by a capital letter, for example, B or D. Enzymatically *active* components are designated by a bar over the numeral(s) of the component involved, for example, $C\overline{1}$ is an activated C1. Fragments of a cleaved component are designated by lowercase letters, for example, a, b, and c. Usually (except for C2) the letter 'a' is assigned for the smaller fragment and the letter 'b,' for the larger one. The suffix 'i' may be used for *inactivated* components. Inactivators are designated by the symbol of the target component followed by the abbreviation 'INA.'

which it forms complexes. Serum enzyme D then cleaves the bound factor B to generate the *alternate* pathway convertase (C3bBb), which, like C4b2a, cleaves more C3. C3bBb, on interaction with a multiplicity of C3b molecules, is transformed into C5 convertase. It has to be noted that the C3bs formed through either pathway are 'interchangeable,' so that C3b formed via the *classic* pathway can participate in the last stages of the *alternate* pathway, if it can bind to the 'activating' surfaces. C3bBb has a short half-life due to inactivation by Factors H and I; however, Factor P (previously known as *properdin*) stabilizes these complexes.

Antibodies to 'new' antigens of the classic and alternate pathways' C3 convertase enzymes have been demonstrated in sera of patients suffering from systemic lupus erythematosus, partial lipodystrophy and membrane proliferative glomerulonephritis. Interaction of these antibodies with the convertases stabilizes the enzymatic activities and causes *hypo*complementemia.

The *membrane attack complex* (***MAC***) begins to assemble with the formation of C5b. This labile fragment must bind a molecule of C6 or decay. C5b and C5b6 may remain in the fluid phase or bind noncovalently to membranes. C5b6 binds C7 and, at this point, must attach to membrane, because the complex in the fluid phase decays rapidly by the binding of the serum glycoprotein known as protein S and subsequently loses the ability to bind to cell membranes. C5b67 can attach to membranes that did not combine with antibodies and thus is able to attack 'innocent bystander' cells. Hydrophobic peptides become exposed as these terminal components bind to one another, thus permitting them to insert themselves into biomembranes. The cell-bound C5–7 complex on interaction with C8 initiates partial membrane permeability. Attachment of three to six C9 molecules to this complex results in the formation of a 23S MAC of an apparent relative mass of 10^6. MAC forms a doughnut-shaped ring (external diameter $\simeq$ 20 nm) on the cell surface, with a central pore having a diameter of 10 nm and penetrating the entire thickness of the cell membrane. It is through this transmembrane pore that the osmotic lysis of the cell takes place. Nucleated cells, due to intrinsic metabolic and repair processes, are not readily lysed by the MAC. Bacterial cells sustain damage to the outer cell wall, but are not killed by the complement action alone. After the damage of the outer cell wall, the inner cell wall becomes susceptible to the attack of lysozyme, which leads to cell death.

Biologic Significance of the Complement System

The cell lysis by the C5–9 MAC, with or without the participation of lysozyme, is an important mechanism of immunity to viruses, protozoa, and some bacteria. When antibody concentrations are low, the fixation of C3b is needed for improved neutralization. Cells coated with C4b, C3b, and C5b bind to specific receptor sites on macrophages, lymphocytes, neutrophils, and erythrocytes. This is the *immune adherence phenomenon,* which may enhance phagocytosis. C3b has also been shown to facilitate antibody and lymphokine production.

Activation of the complement sequence leads to the release of a number of pharmacologically active substances, which usually are low-relative-mass products of the cleavage of the complement precursors. These substances, in general, can be classified as kinins, anaphylatoxins and chemotactic agents. Kinins resemble histamine in their activity, anaphylatoxins are histamine-liberating substances, and chemotactic agents cause the migration of cells along the concentration gradient of the agent. The generation of C2- or C4-derived kinin-like material has been described; C3a and C5a have anaphylatoxin activity; C5a and C567 are chemotactic for macrophages, leukocytes, and monocytes (Fig. 4–3).

PRINCIPLES OF IMMUNOLOGIC METHODS

METHODS BASED ON ANTIGEN–ANTIBODY REACTION

The binding of antigen (Ag) by antibody (Ab) is the basis of most immunologic reactions that utilize or explore the humoral arm of the immune system. In these reactions either the antigen or the antibody can be detected qualitatively or quantitatively, provided that the nature or the quantity of the other reagent is known. The technical approach will differ in accordance with the form of the antigen (soluble, cellular, biologically, or biochemically active), the expected concentration of

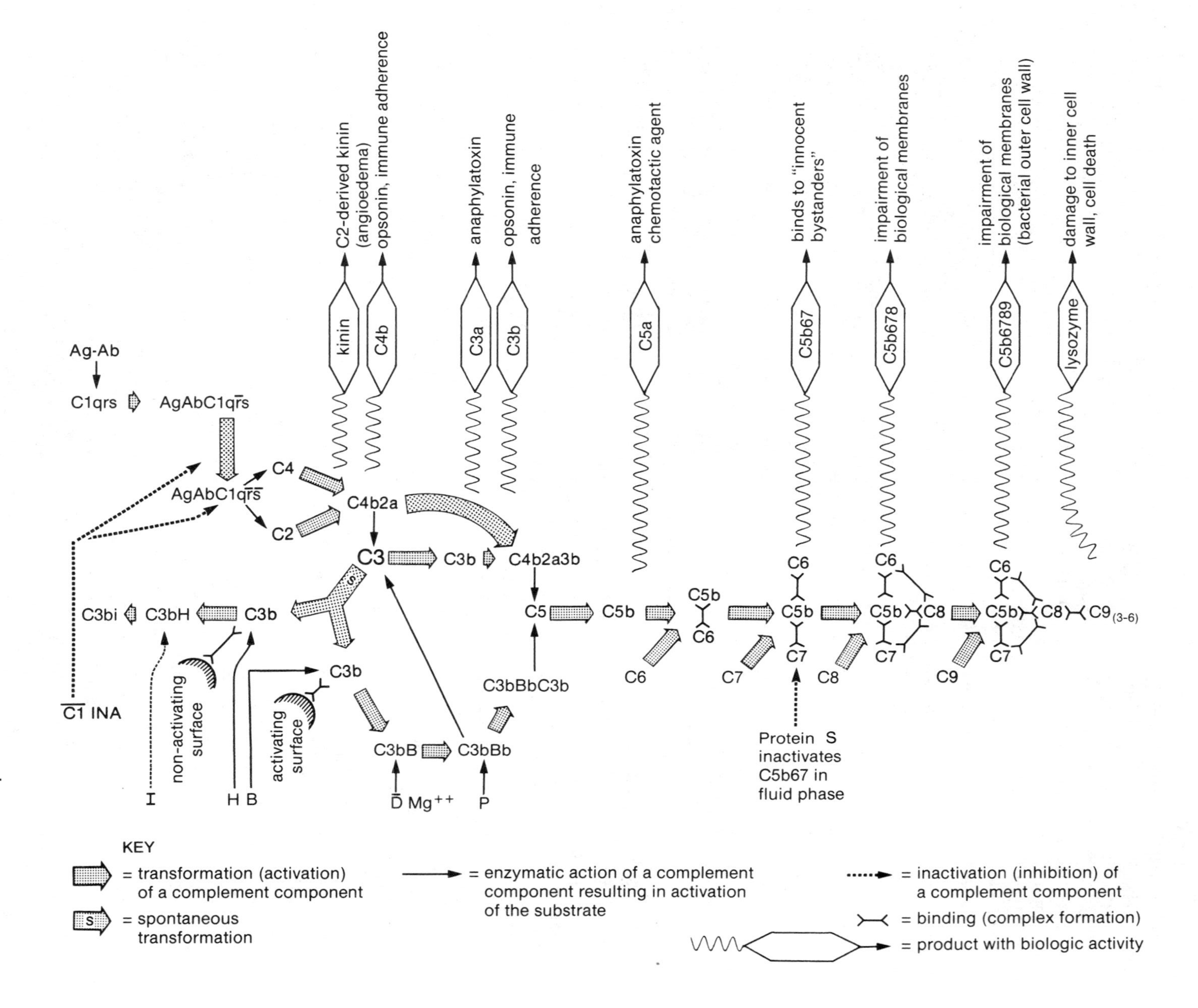
Ag-Ab
C1qrs
AgAbC1qrs
AgAbC1qrs
C4
C2
C4b2a
C3
s
C3b
C4b2a3b
C5
C5b
C5b
C6
C6
C6
C5b
C7
C7
C8
C6
C5b
C8
C7
C9
C6
C5b
C8
C9 (3-6)
C7
C1 INA
C3bi
C3bH
C3b
I
non-activating surface
H B
activating surface
C3b
C3bB
C3bBb
D Mg++
P
C3bBbC3b
Protein S inactivates C5b67 in fluid phase
kinin
C2-derived kinin (angioedema)
C4b
opsonin, immune adherence
C3a
anaphylatoxin
C3b
opsonin, immune adherence
C5a
anaphylatoxin chemotactic agent
C5b67
binds to "innocent bystanders"
C5b678
impairment of biological membranes
C5b6789
impairment of biological membranes (bacterial outer cell wall)
lysozyme
damage to inner cell wall, cell death
KEY
= transformation (activation) of a complement component
= spontaneous transformation
= enzymatic action of a complement component resulting in activation of the substrate
= inactivation (inhibition) of a complement component
= binding (complex formation)
= product with biologic activity

FIGURE 4–3. Interaction of complement components and their biologic activities. A bar over a component indicates the activated state. Ag-Ab = Antigen-antibody complex; C1, etc. = complement components; INA = inactivator; B = factor B; D = factor D; H = factor H; I = factor I; P = properdin.

the Ag or Ab, and the nature of antibodies (class, polyclonal, monoclonal). Ag or Ab can also be manipulated so as to become insoluble. This usually brings about a significant increase in sensitivity. Commonly employed techniques are outlined briefly in the following section; for details the reader is referred to specialized textbooks.

Tests with Antigen in Soluble Form

Precipitation may occur if the Ag–Ab complexes formed during the interaction are so large that they can no longer stay in solution. In order to form such complexes with divalent Ab molecules, the valency of the Ag must be >2. The largest complexes are formed at equivalence, that is, when the proportions between the Ag and Ab are optimal. In Ag excess, the complexes are trimolecular (Ag–Ab–Ag) and retain their solubility. Complexes formed at Ab excess are usually insoluble.*

Formation of Ag–Ab precipitates can be detected by a simple ***ring precipitation*** test or be quantified by measuring the amount of nitrogen in the centrifuged and washed precipitates. Automated assays measure light scattering by the precipitate in suspension.

Immunologic techniques based on precipitation have been revolutionized by the introduction of gelified media. Concentration gradients of Ag and Ab are formed by diffusion through gel; precipitates form in the area in which optimal proportions of the two reactants have been established. In the *Ouchterlony* technique, Ag and Ab diffuse toward each other from reservoirs punched in the gel. If more than one Ag–Ab system is reacting, separate equivalence zones are likely to be formed in slightly different areas of the gel, forming *multiple precipitation lines* (Fig. 4–4). The technique allows also a comparison of two antigens so as to show antigenic similarity (cross-reaction), nonidentity, or apparent identity (Fig. 4–5).

* Antithyroglobulin antibodies found in some patients with Hashimoto's thyroiditis may form soluble complexes with thyroglobulin at Ab excess.

In the ***single diffusion*** method only one concentration gradient is formed. Antigen in excess is allowed to diffuse into the gel, which contains a constant concentration of antibody. As a result, a well-defined precipitation zone is formed at antibody excess and at equivalence. The precipitate may gradually dissolve, as the excess antigen diffuses further from the interface and a larger precipitation zone is formed. Two modifications of this technique (***Oudin's*** and ***Mancini's***) are in use. In the former, the antigen is placed on top of the agar-antibody layer in tubes; the precipitation band descends through the tube from the interface. In Mancini's (radial diffusion) technique, the antigen diffuses radially from a central reservoir in an antibody-containing agar plate; the precipitation zone forms a disc around the reservoir. The length (or diameter) of the precipitation zone is measured and is compared with zones formed at standard concentrations of the antigen, thus allowing quantification of the antigen.

Immunoelectrophoresis. The resolving power of the *double diffusion* method can be greatly increased by combining it with an electrophoretic separation of antigens (Fig. 4–6). In immunoelectrophoresis the antigens are applied to the gel and are separated according to their mobility in an

FIGURE 4–4. Multiple precipitation lines in a double diffusion Ouchterlony test.

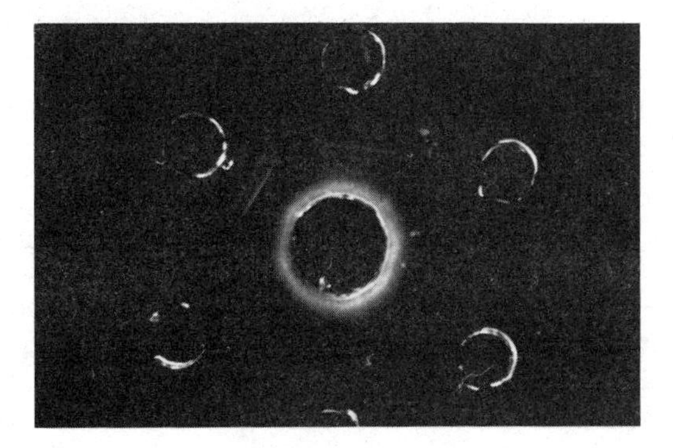

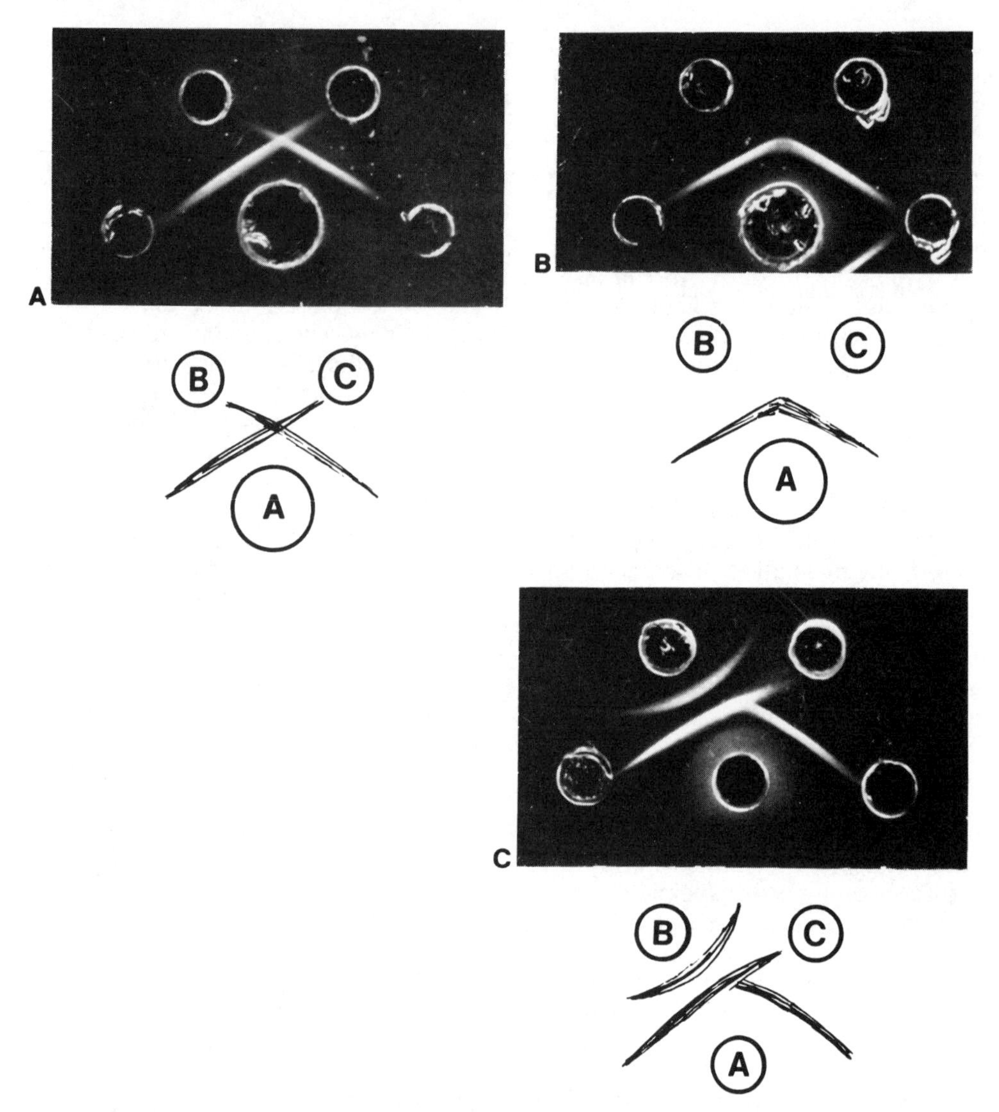

FIGURE 4–5. Comparisons of two antigen–antibody systems by double diffusion (Ouchterlony) method. A = Antiserum: B, C = antigens. ***A.*** Precipitation lines cross: the two antigen–antibody systems are different: antiserum-A must contain two different antibodies, anti-B and anti-C. ***B.*** Precipitation lines fuse: both antigens (B and C) appear to be identical when examined with antibody-A. ***C.*** Spur formation as an indication of a cross reaction. Antiserum-A was raised by immunization with antigen-B. Antigen-B has at least two types of antigenic determinants: some shared with C and some present on B molecules only. Antiserum-A contains antibody molecules against both types of determinants. Since both types of determinants are on the same molecule, one precipitin line is formed between A and B and this line is then split in two: one part fuses with the line between A and C and the second part, specific for B, spurs over. Preparation B contains another antigen against which the antiserum A contains antibodies. This antigen–antibody system forms a second precipitation line. The line is curved around the antigen well, which indicates that the antigen is of high molecular weight.

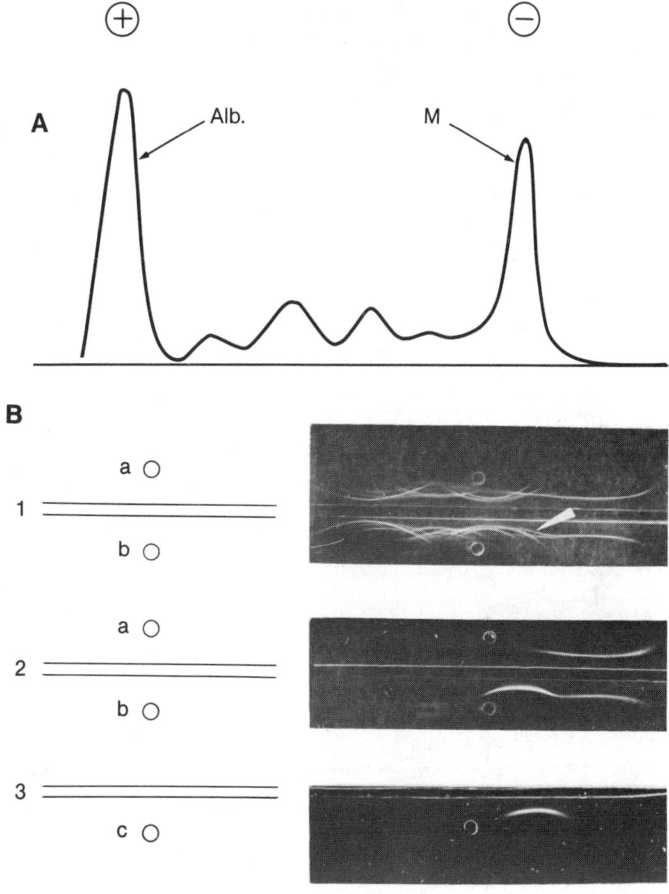

FIGURE 4–6. Monoclonal IgG in serum and Bence Jones protein in urine. ***A.*** Densitometer tracing of serum electrophoresis on a cellulose acetate strip. Alb. = Albumin; M = monoclonal 'spike.' ***B.*** Immunoelectrophoresis of serum and urine. 1. Antiserum directed to whole human serum. 2. Anti-IgG immune serum. 3. Anti-kappa immune serum. a = Normal human serum; b = patient's serum; c = patient's urine (concentrated 100×). Monoclonal arc on the first slide indicated by an *arrow.* The patient's serum contains monoclonal IgG of kappa type. The anode (+) is on the left.

electric field. Subsequently, the antibody is applied, causing arc-shaped precipitation lines to form. Antigens so precipitated can often be identified by the location and shape of the arcs.

Counterimmunoelectrophoresis is a modification of double diffusion in which differences in the electrophoretic mobilities of antigen and antibody are used to accelerate the process of diffusion and to increase the sensitivity of the test by minimizing unwanted diffusion. The test can be performed only with negatively charged antigens which, when placed in agar gel in an electric field, will move in the direction opposite to that of the positively charged immunoglobulins. Two wells are punched in agar along the line parallel to the direction of the current flow; the antibody is placed closer to the positive electrode and the antigen closer to the cathode, so that the two components are accelerated towards each other.

Flocculation. Lipid antigens, which form fine suspensions in water, may aggregate upon addition of antibody and form visible floccules. This technique is used for the detection of antibodies in syphilis.

Three Component Systems. Biochemical activity of antigens (enzymes, toxins) may be neutralized by antibodies. In order to *inhibit* the enzyme activity, the antibody has to be directed to a determinant located on the active site of the enzyme. Antibodies directed to determinants adjacent to the active site may *protect* the enzyme from neutralization by antibody. Combination with antibodies directed against determinants remote from the active site does not affect the enzyme activity. Some antibodies may distort the enzyme molecule in such a way as to facilitate its interaction with the substrate, thus *enhancing* the enzyme activity.

Complement Fixation Texts. Complement is activated and consumed during Ag–Ab reactions. Consumption (fixation) of complement can be used in assays for antigens or antibodies. This method is particularly sensitive in the detection of IgM antibodies; it can detect as little as 0.1 μg of antibody nitrogen. The indicator for the presence of active complement is sheep red cells combined with antierythrocyte antibody. The cells lyse in the presence of active complement; the reaction can be quantified by measuring the hemoglobin released from the lysed indicator cells. This technique was widely used in the diagnosis of syphilis, but has been replaced by more convenient methods.

Tests with Antigen in Cellular Form

Agglutination of red cells or bacteria was one of the first *in vitro* immunologic reactions described. Today, these techniques are used almost exclusively in blood grouping and bacteriology laboratories.

Complement-mediated hemolysis is widely used for the detection of single antibody-forming cells. A mixture of lymphoid cells and erythrocytes is incubated to allow the lymphoid cells to release the antibody. The subsequent addition of complement results in the hemolysis of the red cells that have combined with the antibody during incubation. Each antibody-forming cell can be identified by a clear zone (plaque) of hemolysis visible around it. Since the hemolytic capacity of IgG antibodies is much lower than that of the IgM ones, plaques due to IgG antibodies (indirect plaques) can be visualized only after the addition of an anti-immunoglobulin antibody. In the assay originally described by Jerne, cells are embedded in agarose gel; in the modification by Cunningham the reaction takes place in a narrow space between two microscope slides.

Complement-Mediated Cytotoxicity. Unlike erythrocytes, nucleated cells (e.g., lymphocytes killed by complement) do not disintegrate. Dead lymphocytes can be identified by changes in morphology and light transmission and by the uptake of vital stains such as trypan blue, eosin, or propidium iodide (fluorescent, especially useful in flow cytometry). Complement-mediated cytotoxicity of nucleated cells is the principle for MHC determination.

Labeled Antibodies

Antibody molecules can be labeled, without destruction of their combining activity, by conjugating them with other molecules. The attached label may be detectable by its (1) fluorescence, (2) enzyme activity, (3) radioactivity or (4) high electron density. Fluorescent antibodies are used for the detection of antigens in tissue sections or on single cells by fluorescent microscopy. Flow cytometry and cell sorting are powerful new techniques that utilize fluorescent antibodies. Most solid phase assays require enzyme- or radionuclide-labeled antibodies. Ferritin, an iron-containing protein, provides an electron-dense label for electron microscopy; hybrid antibodies are prepared by recombination of Fab fragments of antiferritin and anti-immunoglobulin antibodies. Such hybrid antibodies combine with unlabeled antibodies attached to tissues and, at the same time, to ferritin, making the site of attachment electron dense. An indirect method of depositing high-density label uses peroxidase-conjugated antibodies; the products of the reaction between the enzyme and its substrate (hydrogen peroxide and diaminobenzidine) have a high affinity for osmium tetroxide. Among the radioactive labels, ^{131}I is one of the most widely used.

Solid-Phase Tests

Techniques using Ag or Ab on an *insoluble support* are approximately 1000 times more sensitive than tests based on precipitation. Solid-phase tests are also capable of detecting nonprecipitating systems, for example, monoclonal antibodies or hapten-antihapten systems. Most solid-phase tests involve the use of labeled antibodies.

Passive agglutination and ***passive hemolysis*** do not require the use of labeled antibody. Ag is conjugated to erythrocytes, which are then agglutinated upon adding the Ab. In passive agglutination, latex particles can substitute for the erythrocytes. Passive agglutination is used for the detection of rheumatoid factor, of antibodies directed against thyroglobulin, and of antibodies against thyroid microsomal antigen. ***Inhibition of passive agglutination*** detects antigen (e.g., chorionic gonadotropin in 'pregnancy tests') in the presence of a limiting concentration of antibody.

Red cells are *passively hemolyzed* if exposed, in the presence of complement, to an antibody directed against the antigen conjugated with the erythrocytes. Passive hemolysis is widely used in single cell plaque assays for the detection of Ab-forming cells (see p 68).

Enzyme-Linked Immunosorbent Assay (ELISA). The solid phase in this test is provided by plastic microtiter plates. Proteins incubated in the trays bind to the plastic. In order to *measure antibodies,* plates are first incubated with the antigen, then with the unknown sample, and finally with an anti-immunoglobulin antibody labeled with an enzyme. The trays are then washed and the enzyme activity is assayed colorimetrically by adding an appropriate substrate. The amount of the enzyme-labeled anti-immunoglobulin antibody bound to the plate is a measure of the amount of the antibody present in the tested sample. When *antigen* is *measured* by the ELISA technique, the antibody is first combined with the plates. Next, the tested antigen is added and allowed to bind to the antibody. Finally, the enzyme-labeled antibody is added, followed by the substrate. Enzymes most frequently used for labeling the antibodies are alkaline phosphatase and peroxidase.

In ***radioimmunoassays,*** as in other solid-phase techniques, one of the reactants has to be attached to an insoluble support or rendered insoluble by precipitation. There are many variants of these two basic principles, as well as innumerable applications in research and diagnostic laboratories. A detailed description of these techniques is beyond the scope of this chapter. Two assays, the *radioallergosorbent test (RAST)* and the *Farr assay,* should be mentioned as examples of the use of insoluble support and precipitation approaches, respectively.

The ***RAST*** test is used for the detection of antibody activity associated with IgE, the immunoglobulin class responsible for allergic reactions. The antigen, for instance an extract from ragweed pollen, is covalently linked to cellulose fibers of a small filter-paper disc. The tested serum sample is incubated with the disc. Incubation with labeled anti-IgE antibody follows. Radioactivity retained on the disc reflects the presence of antibodies of IgE class directed to the antigen used in the assay.

The ***Farr*** assay for antibodies to DNA is used in the diagnosis of *lupus erythematosus.* The test serum is incubated with a radioactively labeled DNA preparation. Ammonium sulfate is used to precipitate immunoglobulins, since it does not affect DNA. The presence of labeled DNA in the precipitate is therefore an indication that this DNA has combined with antibody.

In Vivo Assays

In general, *in vivo* assays using experimental animals are rarely performed for diagnostic purposes. ***Passive cutaneous anaphylaxis (PCA)*** is one of the more important research techniques. Immune serum is injected intradermally into a normal guinea pig; this results in attachment of antibodies to mast cells in the skin. Three to six hours later, a mixture of antigen and Evans blue is injected intravenously. Localized edema and increased vascular permeability at the injection site is made visible by blueing, especially well seen on the inner side of the skin. Human IgE antibodies do not combine with guinea pig mast cells. They can be detected in the ***Prausnitz–Küstner (P–K)*** reaction, in which both the antibody and the antigen are injected intradermally into humans or primates. A positive result manifests itself as local edema and erythema.

Skin allergy tests are rapid, sensitive and inexpensive *in vivo* tests that detect IgE antibodies to a variety of antigens. Antigenic extract is allowed to contact the skin through a small puncture in the epidermis. A positive reaction manifests itself as a local edema and erythema. By testing many, in some cases dozens of extracts, it is possible to identify the environmental allergens that the patient must avoid.

DETECTION OF IMMUNE COMPLEXES

Immune complexes form *in vivo* whenever the Ag–Ab reaction takes place. Small amounts of complexes formed under physiologic conditions are quickly eliminated without any detrimental effect. However, chronic exposure to antigen may lead to the formation of large amounts of complexes and to considerable damage (see p 77). The source of this antigen may be an infectious agent in chronic infections, drugs or serum introduced for therapeutic reasons, or autologous antigens in autoimmune disorders. Immune complexes tend to be eliminated from the circulation, so their detection there is difficult. The reader is referred to specialized literature for the description of the many available methods.

Detection of Immune Complexes in Tissues. Ag–Ab complexes can become trapped in glomeruli or deposited in other tissues. Immune complexes may also form in tissues in the presence of autoantibodies directed against these tissues. Complexes, whether trapped or formed *in situ* can readily be detected in biopsy material by staining with fluorescent antibodies; either anti-IgG or anti-C3 may be used for this purpose.

ASSESSMENT OF THE IMMUNE SYSTEM

Functional tests of all the component parts of the immune system are essential for the accurate diagnosis of immunodeficiencies. The armamentarium of available methods is growing constantly.

Skin testing is an inexpensive yet extremely valuable method for the assessment of immune functions. Intracutaneous injection of diphtheria toxin (***Schick test***) provides information on the capability to produce antibodies. If the subject has been immunized against diphtheria toxin, the circulating antibodies will neutralize the injected toxin and prevent local erythema. A positive skin reaction after injection of the toxin suggests a deficiency of humoral immunity.

Intradermal tests with various environmental antigens (PPD,* *Candida,* streptokinase-streptodornase, *Trichophyton*) help to assess the components of the immune system responsible for the delayed hypersensitivity (DH). Lack of reaction (anergy) can be attributed to the deficiency of either lymphocytes or macrophages. Conversion from a negative to a positive DH reaction is an important sign of clinical improvement following transfer factor treatment or bone marrow transplantation.

* PPD = purified protein derivative, a protein extracted from *Mycobacterium tuberculosis.*

Enumeration of T and B Lymphocytes. T and B lymphocytes carry a number of surface markers. These markers are characteristic for different cell types and subsets, as well as for different differentiation stages, and can be detected by polyclonal and monoclonal antibodies. Cells bearing the markers can be found in peripheral blood, lymph nodes, and bone marrow biopsies. Immunoglobulin molecules are present either on the cell surface or in the cytoplasm of the B cells and can serve as markers of these cells. These cell markers are detected by fluorescent techniques, with the help of either a fluorescence microscope or the cell flow analyser.

Functional Assays of T and B Lymphocytes. B-cell functions can be assessed by an *in vitro* synthesis of immunoglobulins (see p 68) and by mitogen-induced proliferation. B-cell mitogens are usually substances derived from bacterial cell walls. Antibodies to the cell-surface immunoglobulin receptor may also function as B-cell mitogens. T cells proliferate, when cultured in the presence of T-cell mitogens (phytohemagglutinin, concanavalin A) or nonself cells that carry incompatible MHC (major histocompatibility complex) antigens. This is called the mixed lymphocyte reaction. Functions of T cells can also be assessed by measuring the ability to generate cytotoxic cells in response to nonself target cells (CML = cell-mediated lympholysis) and the ability to release lymphokines (MIF = migration inhibition factor; interferon; interleukin 2).

Evaluation of Phagocytic Cells. Appropriate tests can be used to evaluate the following functions of phagocytic cells: motility, recognition, ingestion, degranulation, and intracellular killing. Impairment of one or more of these functions may lead to serious clinical consequences.

HYBRIDOMA TECHNOLOGY

A great deal of research in immunology has been devoted to the development of high-resolution tools, in the form of antibodies capable of distinguishing between molecules showing only minute differences in their structure. The ultimate goal was to raise antibody directed to a single antigenic determinant. Such ‘fine tuning’ of antibody specificity was unattainable, because of the inherent heterogeneity of the antibody response and of the multideterminant nature of all antigens. Antibodies produced by different clones differ in their specificity and affinity, so that even immunization with a simple hapten still results in a very heterogeneous antibody response.

An extremely important breakthrough in antibody production has been achieved by adopting the cell fusion technique. Antibodies produced by each clone are absolutely homogeneous and are directed against only one antigenic determinant. ***Monoclonal antibodies*** are rapidly becoming indispensable high-resolution tools of modern immunology.

In this technique, pioneered by Milstein and Köhler, myeloma cells are fused with antibody-producing B cells (obtained from spleens of mice immunized with the appropriate antigen) to form antibody-producing 'hybridoma' cell lines. The success of this technique is due to the very efficient and simple selection method, which discriminates against the unfused cells and favors the growth and propagation of fused cells. After fusion, the cells are cloned and the many clones obtained in one fusion experiment are screened for the production of antibody of the desired specificity. Established hybridoma cell clones may be grown either *in vitro* or *in vivo* (passaged in the peritoneal cavity of BALB/c mice). Both the culture supernatant and the ascites fluid are good sources of monoclonal antibodies. Once established, the hybridoma cell line may be propagated or stored in liquid nitrogen indefinitely.

Hybridoma technology can be used to obtain antibodies directed to virtually any antigenic determinant. Monoclonal antibodies to some self determinants have also been obtained. *Analytical applications* of monoclonal antibodies are increasing rapidly (see Chap. 28). Once production of human hybridomas is accomplished, *therapeutic use* of monoclonal antibodies will become feasible; monoclonal antibodies directed against the determinant of a malignant cell may be conjugated with a toxin (diphtheria toxin, ricin) and used as 'guided missiles' to destroy the malignant tissue. It has been shown that the toxicity of such 'immunotoxin' for the target cells is several thousand times higher than the toxicity for 'innocent bystander' cells. Monoclonal antibodies might also be used to *regulate* the immune system either by interference with the idiotype regulatory network or with the helper or suppressor cells.

DISORDERS OF THE IMMUNE SYSTEM

IMMUNODEFICIENCIES

Immunodeficiencies are classified according to the part of the immune system affected. A hereditary defect or an acquired condition may affect the T or B arm of the immune apparatus, phagocytosis or the complement system. Some deficiencies affect more than one of these systems; the etiology and classification of others is still not clear. Some immunodeficiencies are correlated with a higher incidence of autoimmune diseases. The mechanism of these associations is unknown.

T-CELL DEFICIENCIES

The classic T-cell deficiency is the ***nude*** mutation in the laboratory mouse. Besides the lack of thymus, the mice also lack hair follicles, hence the name of the autosomal recessive gene responsible for this condition. *Clinically,* isolated T-cell deficiencies are rare. The patients are susceptible to viral, fungal, and protozoal infections, and to bacterial intracellular infections. Vaccination with live attenuated viral vaccines leads to a fatal generalized infection.

DiGeorge *syndrome,* the best characterized human T-cell deficiency results from an intrauterine accident, perhaps an infection, affecting the development of the third and fourth pharyngeal pouches around the 12th week of gestation. DiGeorge syndrome is associated with hypoparathyroidism, congenital heart defects, and abnormalities in facial features, all these being ontogenically associated with the pharyngeal pouches affected. Because of these abnormalities, it can be diagnosed relatively early. Replacement therapy with cells from fetal thymus (less than 14 weeks' old to avoid a graft-*versus*-host reaction) has been reported effective.

Other, less well known T-cell deficiencies include *episodic lymphopenia with lymphocytotoxins,* in which T cells are lysed by autoantibodies in the presence of complement, and *T-cell deficiency* in the course of *Hodgkin's disease,* in which otherwise normal T cells are blocked with serum factors.

B-CELL DEFICIENCIES

Complete failure to develop the B-cell arm of the immune system occurs in an *X*-linked recessive condition known as *infantile sex-linked* ***agammaglobulinemia*** or ***Bruton's disease.*** Its prevalence is approximately 1:100,000. The disease is characterized by a complete absence or deficiency of all classes of immunoglobulins and absence of circulating B cells. The symptoms consist of recurrent bacterial infections, bronchitis, pneumonia, otitis media, meningitis, and so forth, *Streptococcus pneumoniae* and *Hemophilus influenzae* usually being responsible. The infections begin 5 to 6 months after birth, when the protective effect of passively acquired maternal antibody wears out, and usually do not respond to treatment with antibiotics. Regular treatment with human immunoglobulins is essential for the survival of these patients. There is an increased incidence of collagen diseases, autoimmune disorders, and leukemias among people affected with Bruton's disease.

Selective Immunoglobulin Deficiencies. In contrast to Bruton's disease, B-cell deficiencies may affect only one immunoglobulin class or subclass. The most common of these disorders (prevalence ≃ 1:700) is IgA deficiency. Since the main function of the IgA is the protection of mucous membranes, a deficiency of IgA is associated mainly with recurrent pulmonary and sinus infections. Patients with IgA deficiency very often develop anti-IgA antibodies. This makes replacement therapy impossible. Even blood or plasma transfusions may result in severe anaphylactic shock.

COMBINED T- AND B-CELL DEFICIENCIES

The total absence of both T- and B-cell immunity is due to a defect in *stem cell maturation* and occurs in three forms: *X*-linked recessive, autosomal recessive with adenosine deaminase deficiency, and autosomal recessive. The symptoms are a combination of symptoms characteristic for B- and T-cell deficiencies and begin at about 6 months of age. The prevalence of these deficiencies is hard to calculate because many patients die during the first year of life, before the diagnosis is made. The affected infants are not only extremely susceptible to any infection, but are also vulnerable to graft-*versus*-host attack by maternal cells transferred during gestation or delivery, or to other viable lymphoid cells given to them in the form of blood transfusion. If a histocompatible donor can be found, the patient's immune system can be repopulated by the transfer of bone marrow cells. Some success has been described after the transfer of fetal liver cells (<8 weeks' gestation) or fetal thymus cells (<14 weeks' gestation), but these techniques have not resulted in complete reconstitution of immune functions.

Ataxia Telangiectasia with Immunodeficiency. This is a multisystem disorder of unknown etiology, inherited in an autosomal manner. It is primarily a neurologic disorder with endocrine, vascular, and immune abnormalities, which become worse with age. Immunologically, most of the patients lack both serum IgA and secretory IgA; some have antibodies to IgA, and in some cases there is a deficiency of IgE. All patients show some degree of deficiency of T-cell immunity.

Immunodeficiency with Eczema and Thrombocytopenia (Wiskott–Aldrich syndrome). The prevailing symptoms of this rare *X*-linked trait are related to thrombocytopenia. T-cell immunity is initially unaffected, but may deteriorate later. Numbers of B cells and IgG levels are normal, concentrations of IgA and IgE are elevated, and IgM is decreased, with marked inability to make antibodies against polysaccharide antigens. The etiology is unknown.

Acquired Immune Deficiency Syndrome (AIDS). This 'new' disease, first described in the United States, is now seen also in Canada and Europe, and the number of cases is on the rise. Since 1979, approximately 12,000 cases of AIDS have been reported to the Center for Disease Control (CDC) in Atlanta. At present, 100 new cases are reported each week to the CDC. The overall case fatality rate is 65%, most of these patients dying within 2 years of the diagnosis. The disease is characterized by a general loss of immunity and subsequent development of opportunistic infection (most frequently *Pneumocystic carinii* pneumonia) or a rare tumor such as Kaposi's sarcoma. A syndrome termed the *AIDS-related complex* is also seen in at-risk individuals. Persistent generalized unexplained lymphadenopathy, fevers, night sweats, extreme weight loss and intense fatigue occur in these patients in association with immune deficiency. Some (15%) of these patients will go on to develop AIDS.

Persons at risk are homosexual or bisexual men, intravenous drug abusers, and hemophiliacs. Epidemiologically, AIDS is similar to hepatitis B, which suggests that the etiologic agent may be a virus transferred by sexual contact, blood or blood products, or maternal–fetal contact (infants of mothers with AIDS may also acquire the condition). Recent studies have demonstrated an association between AIDS and the human T-cell leukemia virus-III (HTLV-III), or the lymphocyte-associated virus (LAV) isolated by a different research group. This virus preferentially infects the T-helper (T_H) cell population and is cytolytic, leading to T_H cell depletion. The factors responsible for the development of AIDS in individuals exposed to HTLV–III or LAV remain to be determined.

DISORDERS OF PHAGOCYTIC CELLS

Any step in phagocytosis and any enzyme instrumental in the process of intracellular killing of bacteria may underlie these disorders. Their common symptoms are chronic and acute infections with normally nonpathogenic or unusual bacteria (*Staphylococcus epidermis, Serratia marcescens, Aspergillus, Candida, Pseudomonas,* etc.) The two conditions, thought to be caused by a defect in lysosomal enzymes, are ***Chronic granulomatous disease,*** an *X*-linked recessive disorder, and ***Job's syndrome,*** an autosomal recessive disorder. In the ***Chediak-Higashi syndrome,*** a multisystem autosomal recessive disorder, the lysosomal enzymes seem to be normal, but the lysosomal structure is defective.

GENETIC DEFECTS OF THE COMPLEMENT SYSTEM

Inherited deficiencies of all the classic complement components and of properdin, Factors I and H, and C1INA have been described. Because the complement system can be activated by two pathways, the deficiency in one pathway may be bypassed by way of the other. Thus, deficiencies of single complement components are not necessarily detrimental to the individual affected, unless the missing component plays a central role in both pathways (e.g., C3). Similarly, because some biologic effects of complement activation are already expressed at earlier stages of the sequence, the deficiency of a late-acting component may have few or no clinical consequences.

Inhibitors of various stages of the complement sequence are essential in preventing damage to self tissues by activated complement components. A deficiency of *C1 inactivator* (C1INA) is associated with the dramatic symptoms of ***hereditary angioedema.*** This single-gene autosomal trait is characterized by local subepithelial edema of the skin and mucous membranes. The condition may be life threatening if it occurs in the respiratory tract. A kinin, probably derived from C2, is responsible for the pathogenesis. An attack of edema may be triggered by local trauma or may occur spontaneously. After the attack the serum concentration of C4 is decreased, this being a result of uncontrolled activation of the classic pathway. In angioedema either the *synthesis* of C1INA is affected or (in 15% of cases) an *inactive form* of the factor is synthesized. The former can be diagnosed by a quantitative immunoassay, the latter by a functional assay.

Deficiencies of Factor I or H lead to hypercatabolism and low serum concentrations of C3. Serum C3 is present mostly in the form of C3b. Constant activation of the alternate pathway results in depletion of the terminal components (those common to both pathways). The consequences of such generalized ***hypocomplementemia*** are impaired serum phagocytic, chemotactic, and bactericidal activities, resulting in recurrent infections, especially meningitis, septicemia, pneumonia, or sinusitis. The addition of fresh serum replaces the missing Factor I and restores the regulation of the alternate pathway. Deficiency of C6, C7, or C8 is associated with recurrent *Neisseria* infections. Persons deficient in components C1, C4, or C2 often suffer from systemic lupus erythematosus or lupus-like syndromes, rheumatoid diseases, and vasculitis syndromes. The mechanism of this association is unknown.

MALIGNANCIES OF THE LYMPHOID CELLS

All categories of lymphoid cells at all stages of differentiation are susceptible to malignant transformation; most of these malignancies, with the exception of B-cell tumors, are investigated by the hematologists.

Disorders associated with immunoglobulin-secreting *B-cell tumors* are known as ***monoclonal gammopathies, plasma-cell dyscrasias,*** or ***plasmacytomas.*** These disorders must be considered in terms of the homogeneity as well as the concentration of the immunoglobulin produced. A challenge with an antigen normally results in the stimulation of many clones of antibody-producing cells and, consequently, in a *polyclonal,* heterogeneous antibody response. Tumor cells usually originate from a single cell and form a single clone of malignant cells. The immunoglobulins synthesized by these cells are thus of *monoclonal* origin. Immunoglobulins of monoclonal origin are sometimes called paraproteins.

In investigating *paraproteinemias* we must determine whether or not the serum contains a monoclonal population of immunoglobulin molecules. The question is quantitative; one must look for a homogeneous population of immunoglobulin molecules in a concentration that significantly exceeds the concentration of any other population. This means that any malignant clone will escape detection unless its product constitutes a sizeable proportion of the total immunoglobulins. How does one distinguish between monoclonal and polyclonal immunoglobulins? By *definition,* monoclonal immunoglobulins must have an identical primary structure. In practice, one uses two criteria: electrophoretic mobility and immunoglobulin class. An electrophoretically homogeneous population of molecules can be detected by acetate strip electrophoresis (from the presence of a monoclonal 'spike'; Fig. 4–6*A*), and by immunoelectrophoresis. If this population also belongs to one immunoglobulin class, and carries only one type of light chain, one can be reasonably confident of its monoclonal origin. Usually antisera for light-chain types κ or λ are used. Within the IgG class the analysis can be extended further by the use of antisera to IgG subclasses.

The *diagnosis* of an ***IgG*** of monoclonal origin is relatively straightforward. The IgG class has a wide range of electrophoretic mobilities, so the presence of a homogeneous population can be detected without major difficulties (Fig. 4–6*B*). Because of their high relative mass and the consequent slow diffusion through the gel, as well as relatively low normal concentrations, the investigation of ***IgM*** is not always straightforward. Sometimes the abnormal IgM is in monomeric rather than in pentameric form; consequently, its diffusion rate through gel is faster, and this may cause higher readings in tests based on single diffusion. The determination whether the IgM is monoclonal can usually be made by comparison of electrophoresis patterns developed by anti-κ and anti-λ antisera. The detection and characterization of monoclonal immunoglobulins of other classes, IgA, IgD, and IgE may often require procedures that can be carried out only in a specialized protein chemistry laboratory.

The detection of immunoglobulins (or their fragments) in the ***urine*** is an important diagnostic procedure. Immunoelectrophoretic analysis of urine (concentrated 100 to 200 times by ultrafiltration) shows the presence of immunoglobulin light chains. Usually immunoglobulin heavy chains are absent, and the light chains form dimers, known as Bence–Jones proteins. Monoclonal Bence–Jones proteins belong, of course, to only one light-chain type (κ or λ; Fig. 4–6*B*).

Polyclonal hypergammaglobulinemias are due to prolonged antigenic stimulation, as in chronic bacterial or parasitic infections, or autoimmune diseases. In certain chronic diseases with hypergammaglobulinemia (rheumatoid arthritis, Sjögren's syndrome), antigenic stimulation is suspected, but the identity of the immunogen is uncertain. The mechanism of hypergammaglobulinemia in parenchymal liver disease is unknown.

Multiple Myeloma. This is the most common malignant gammopathy (prevalence $\simeq$ 4:100,000). Multiple myeloma rarely occurs before the age of

30; some 80% of myeloma cases occur after the age of 50. The prevalence of monoclonal proteins, not necessarily associated with multiple myeloma, has been estimated at 0.9% in the population over 25 years of age, and at 3% in the over-70 age group.

The uncontrolled proliferation of malignant plasma cells causes not only the appearance of monoclonal immunoglobulin in the serum, but also multiple osteolytic bone lesions (hence the term 'multiple'), and a decrease in the concentrations of other immunoglobulins owing to crowding out and destruction of lymphoid tissue and to the activity of suppressor T cells. This results in increased susceptibility to infections. The immunoglobulin light chains are sometimes synthesized in asynchrony with the production of the heavy chains, which lags behind. As a result, free light chains (Bence–Jones protein) are found in the serum and, more frequently, in the urine. Less frequently, an overproduction of defective heavy chains is observed (***heavy-chain disease***). Symptoms include those caused by bone lesions and by kidney damage. Symptomatic cryoglobulinemia, hyperviscosity syndrome, coagulation abnormalities, and amyloidosis are commonly observed.

Solitary plasmacytoma is regarded as an initial, localized stage of plasma-cell myeloma. In this state a monoclonal immunoglobulin may not yet be detectable. The tumor may eventually develop into a generalized multiple myeloma; it may also recur as a solitary plasmacytoma after irradiation and surgical treatment.

Waldenström's Macroglobulinemia. This gammopathy involves IgM-producing cells. Monoclonal IgM is found in a pentameric form, in larger aggregates, or sometimes in a monomeric form. Clinical symptoms are produced mainly by the increased viscosity of the blood, by the formation of complexes with the coagulation factors or other serum proteins, and by cold-induced insolubility of the macroglobulins and complexes (***cryoglobulinemia***). Bence–Jones proteins are found in only 10% of cases.

Benign monoclonal gammopathies can be defined as conditions in which monoclonal protein appears in the serum, and sometimes also in urine, without any indication that a malignant tumor, either multiple or solitary, is present. Laboratory investigations cannot distinguish between a monoclonal immunoglobulin which is a product of malignant cells and one synthesized by a clone of benign plasma cells. Rabbits immunized with streptococcal antigens sometimes synthesize antibody of *restricted heterogeneity,* that is, practically monoclonal. It is conceivable that a similar antibody response resulting from an unknown antigenic stimulation may cause the synthesis of a monoclonal immunoglobulin in man. Another origin of these benign gammopathies may be a plasma-cell tumor, similar to myeloma, but nonprogressive. Benign gammopathies occur more frequently in higher age groups; the concentration of monoclonal immunoglobulin is usually lower than in the malignant diseases, and the urine is less likely to contain Bence–Jones protein. Other symptoms associated with myeloma (bone lesions, clotting defects) are absent, and the cellular composition of the bone marrow is normal. The *tentative diagnosis* of a benign gammopathy can be made only after careful exclusion of all possibilities of malignancy and should be reviewed periodically.

AUTOIMMUNITY AND DISEASE

Autoimmune damage may be inflicted by autoantibodies, (e.g., in *autoimmune hemolytic anemias*), by mechanisms of cellular immunity (e.g., in *demyelinating diseases*), or by both arms of the immune system. In autoimmune thyroiditis (***Hashimoto's disease***) circulating autoantibodies are formed, but most of the tissue damage seems to be done by the cellular mechanism.

Because the basic function of the immune system is the distinction between *self* and *nonself,* autoimmune diseases may be regarded as examples of a fundamental failure of the immune apparatus. This is only partly true, because in autoimmunity this failure is limited to one, or at most a few, antigens. Nevertheless, a complete elucidation of the mechanism or mechanisms causing autoimmunity would be a major breakthrough in immunology. The factors instrumental in development of autoimmune diseases may be divided into *extrinsic* and *intrinsic* factors.

Extrinsic Factors. Some tissue antigens never come in contact with immunocompetent cells and thus never induce natural tolerance. Such *secluded* autologous antigens can cause immunization if the organ (e.g., brain, testes, eye lens, thyroid) becomes damaged as the result of either a pathologic process or a surgical procedure, causing the antigens to enter the circulation. These antigens can stimulate autoantibody formation in the course of a number of diseases, for example, *orchitis* and *sympathetic ophthalmia.*

Experimentally acquired tolerance can be *circumvented* by stimulation with an antigen that cross-reacts with the *tolerogen.* An animal tolerant to *bovine* serum albumin, immunized with *human* serum albumin will subsequently produce antibodies to *bovine* serum albumin. Clinically, certain forms of ***nephritis*** may be caused by a similar mechanism of tolerance circumvention. Certain types of hemolytic streptococci are known to possess an antigen that is structurally similar to human heart antigen. Invasion of such organisms results in the production of antibodies that cross-react with the autologous tissue antigen and may cause a breakdown of tolerance to the heart antigen in ***rheumatic fever.***

In another category of pathologic conditions, antibodies are formed to hapten molecules (e.g., simple drugs) attached to autologous macromolecules or cells. The combining of the hapten with the antibodies will cause the destruction of cells to which the hapten molecules are attached. A consequence may be the development of *thrombocytopenia, agranulocytosis,* or *hemolytic anemia,* depending on the type of cells destroyed. Drugs that can be responsible for these reactions include aminopyrine, quinidine, apronalide (allylisopropylacetylurea, Sedormid), and stibophen. These reactions are relatively rare, and when administration of the drug is stopped, the symptoms disappear without any further treatment.

In some autoimmune diseases of *unknown etiology,* processes similar to those responsible for the drug-associated cytopenias, or to a circumvention of tolerance, may take place. An infectious agent (e.g., a virus) may become attached to autologous cells. Antiviral antibodies or T_C cells would cause the destruction of the autologous cells. Another mechanism that is possible and not mutually exclusive might act through the creation of 'neoantigens,' by combination of an infectious agent with self-macromolecules. Antibodies against this 'neoantigen' may also cross-react with the unchanged self-macromolecules, causing the *circumvention,* and eventually a *breakdown,* of tolerance, resulting in a full-fledged autoimmunization. Autoimmunization by circumvention of tolerance may also occur if an autologous macromolecule becomes antigenically altered as a result of somatic mutation.

Intrinsic Factors. A mutation of an immunocompetent cell into an autoreactive one may give rise to an autoimmune clone of T or B cells. In terms of the *immune surveillance* hypothesis, such autoreactive cells do arise, but are eliminated. Thus, failure of the surveillance mechanism is one of the possible intrinsic mechanism of autoimmunity. Unfortunately, this hypothesis is of little practical value, because the surveillance mechanism itself remains hypothetical. Nevertheless, it must be stressed that the incidence of both autoantibodies and tumors increases with age, which may be taken as an indication that the efficiency of the surveillance mechanism deteriorates with age.

Some *genetic* factors appear to play a role in autoimmune processes, since certain strains of animals regularly develop autoimmune diseases (a generalized autoimmune process in New Zealand black mice; autoimmune thyroiditis in the obese strain of chicken). The association of the major histocompatibility genes with some human diseases, including those of autoimmune etiology, has already been mentioned.

An important intrinsic condition leading to autoimmunity seems to be the *imbalance of the regulatory system,* that is, loss of T-suppressor functions or increase in the T-helper functions. There are examples in which an infectious agent acts directly on the lymphoid system, causing a loss of immunoregulation. Infections with the ***Epstein-Barr*** virus, the etiologic agent of *infectious mononucleosis,* result in proliferation of B cells and in a subsequent reaction of T cells, which presumably are attempting to limit the B-cell proliferation. An infectious agent causing proliferation of T-helper or damaging the T-suppressor cells may thus be responsible for some autoimmune diseases, for example, lupus erythematosus. An infectious agent that has an opposite effect on the immune system, that damages the T_H cells, has already been mentioned in connection with the *acquired immune deficiency syndrome (AIDS)*.

In considering the etiology and pathogenesis of human autoimmune diseases, one should remember two things. First, several of the aforementioned mechanisms may cooperate in the pathogenesis of a given autoimmune disease. For instance, the emergence of an autoreactive cell by somatic mutation may have to coincide with the presence of a virus affecting the balance between the helper and suppressor cells. Furthermore, an invasion by the virus may be possible only if the host cells possess certain MHC determinants. Second, autoantibodies are not necessarily the *cause* of the diseases classified as autoimmune. Production of antibodies may be secondary to the tissue damage inflicted by an infectious agent, which may expose a secluded antigen or may change a self-antigen into a cross-reactive one. Whatever the etiology, detection of autoantibodies in these conditions is of considerable diagnostic importance.

INJURY MEDIATED BY IMMUNOLOGIC MECHANISMS

The immune apparatus, in carrying out its normal or abnormal functions, can inflict damage to own body tissues. Traditionally, the immune injury is classified into four categories.

Type I: Anaphylactic, Immediate, IgE-Dependent Reaction. As discussed earlier, antibody molecules of the IgE class (reagin) attach themselves to the surface of mast cells (through the C_H4 domain of the heavy chain). Reaction of the mast-cell–bound IgE antibody with the antigen (bridging of two adjacent IgE molecules by a molecule of the antigen is required) results in *degranulation* of the mast cell. During this process a number of biologically active substances (histamine, serotonin, leukotrienes), stored in the granules, are released. These substances cause contraction of the smooth muscles, increased vascular permeability, and vasodilation, and are responsible for the pathologic symptoms of hypersensitivity in such diseases as *hay fever, asthma,* and *allergic rhinitis.* The most common allergens causing hypersensitivity are pollens of higher plants (ragweed, grasses, trees). Allergies to a variety of substances, including food, cosmetics, bee and wasp venom, and penicillin, are not uncommon. The tendency to make antibodies of the IgE class is genetically controlled, but the physiologic functions of these antibodies are not well understood. IgE may play an important role in protection against helminthic and perhaps other parasitic diseases. By triggering smooth-muscle contraction, IgE may facilitate expulsion of worms from the digestive system.

Type II: Cytotoxic Reaction. In this category, the damage is caused by cytotoxic antibodies. Complement is not necessarily involved. The pathogenesis involves self antibodies directed against foreign cells (blood transfusion), transferred antibodies directed against self erythrocytes (hemolytic disease of the newborn), self antibodies directed against simple chemical substances (aminopyrine, apronalide) that have become attached to self cells, and finally, autoantibodies inflicting damage to self cells. Diseases in which self tissues are damaged by cytotoxic autoantibodies include *autoimmune hemolytic anemia* (antibodies to one's own red cells), *thrombocytopenic purpura* (antibodies to platelets), and *Goodpasture's syndrome.* In Goodpasture's syndrome, autoantibodies to lung and kidney basement membranes inflict severe damage to these organs. The disease can be life threatening. A similar mechanism is responsible for the damage to the basement membrane of the skin in *pemphigus.*

Type III: Damage by Antigen–Antibody Complexes. When an antigen–antibody reaction takes place *in vivo,* soluble and insoluble Ag–Ab complexes lodge in the capillaries of various organs and especially in the glomeruli of the kidney. The damage may be inflicted to a certain degree by physical obstruction of capillaries and small blood vessels, but results mainly from activation of the complement sequence, which triggers various effector mechanisms and causes the release of biologically active substances. This results in damage to the tissue lodging the complexes. Examples of this type of immunologic injury follow.

The *Arthus reaction* is an experimentally induced local inflammatory reaction to the intradermal injection of an antigen into an animal that

TABLE 4–4. Offending Antigens and Forms of Extrinsic Allergic Alveolitis

Forms of Disease	Source of Offending Antigen	Suspected Offending Antigen
Farmer's lung	Moldy hay	*Microspora faeni*
Mushroom worker's lung	Mushroom compost	*Thermoactinomyces vulgaris*
Bagassosis	Moldy sugar cane	*Thermoactinomyces vulgaris*
Maple bark disease	Dry moldy bark	*Cryptostroma corticale*
Woodworker's disease	Moldy sawdust	*Pullularia pullulans*
Mill worker's lung	Mill dust	*Sitophilus granarius*
Malt worker's lung	Germinating barley	*Aspergillus clavatus*
Air conditioning lung	Contamination of air conditioning, heating, or humidification systems	Thermophylic actinomycetes
Bird breeder's lung	Feathers and droppings	Antigens specific for a given species (chicken, pigeon, and parakeet)*
Bronchopulmonary aspergillosis	Colonization of bronchial tree by fungus	*Aspergillus fumigatus*
	Inhaled organic antigens	Enzyme detergents, cotton antigens, etc.

* Blood serum of a suspected bird species may be used as an antigen in tests for antibodies.

already has circulating antibodies directed against this antigen.

Serum sickness is a generalized reaction to Ag–Ab complexes. It develops 8 to 15 days after the injection of an antigen, in quantities sufficient to form large amounts of complexes. Serum sickness was a common complication of passive immunotherapy (injection of heterologous antitetanus or antidiphtheria antibodies). Today, most of the antibodies for passive immunotherapy are prepared from human sera, so that the risk of serum sickness has been practically eliminated. The symptoms of serum sickness include swelling at the site of injection, enlargement of regional lymph nodes, rash, joint pain, fever and, in severe cases, glomerulonephritis, arthritis, and involvement of the nervous, gastrointestinal, and circulatory systems.

Ag–Ab complexes may form in a number of pathologic conditions, with the antigen being of either extrinsic or autologous origin:

Extrinsic allergic alveolitis (immune complex pneumonitis) is a syndrome caused by an Ag–Ab reaction in the lung tissue. Patients who suffer from this disease have detectable precipitating antibodies against antigens present in organic dust to which they have been chronically exposed (Table 4–4). Symptoms appear within 6 to 8 hours after exposure to the offending antigen and consist of dry cough, shortness of breath, fever, and general malaise. Antibodies responsible for this condition can be detected and identified in a double-diffusion test of the patient's serum with extracts of various suspected microorganisms (Table 4–4).

Cryoglobulinemia is another condition in which severe damage is inflicted by Ag–Ab complexes. Mixed (polyclonal) cryoglobulins are complexes of IgG and IgM, or IgA and IgM. These complexes are reversibly precipitable in the cold. Their formation was found to be associated with some autoimmune processes and infections. The IgM component is usually a rheumatoid factor (see below) directed against IgG or IgA. The complexes are responsible for renal damage (similar to that seen in chronic glomerulonephritis or chronic serum sickness), Raynaud's phenomenon, vasculitis, and often, lung damage. Monoclonal cryoglobulins are frequently found in patients who suffer from Waldenström's macroglobulinemia or multiple myeloma. Cryoglobulins may be overlooked because they may precipitate during the normal processing of blood samples. Blood samples tested for cryoglobulins have to be processed at 37°C, and the serum placed at 4°C to allow the cryoglobulins to precipitate.

Antigen–antibody complexes are also responsible for at least some of the symptoms in so-called ***collagen diseases,*** among which rheumatoid arthritis and systemic lupus erythematosus are the most representative. In *lupus erythematosus,* the predominant immunologic feature is the presence of antibodies to double-stranded (native) DNA. Antibodies to single-stranded DNA are not specific to lupus erythematosus and are frequently found in other collagen diseases.

In *rheumatoid arthritis,* the most prominent immunologic feature is the presence of *rheumatoid factor,* which is an antibody of IgM, IgG, or IgA class directed against immunoglobulins. The specificity of rheumatoid factor in various patients differs widely. It can react with self heavy-chain determinants and sometimes with allotypic* determinants; in most cases it reacts best with immunoglobulins that have undergone slight denaturation (aggregation) following heating or reaction with an antigen. It can also react with slightly aggregated immunoglobulins of other species (rabbit antibodies combined with sheep red cells). Rheumatoid factor can be demonstrated by agglutination of latex particles coated with human immunoglobulins or by agglutination of sheep red cells coated with rabbit antisheep antibodies. Complexes of rheumatoid factor and immunoglobulins often circulate in the blood and can be detected as aggregates having a sedimentation coefficient of 22S. These complexes will dissociate at low pH into 19S IgM and 7S IgG. High titers of rheumatoid factor are characteristic of rheumatoid arthritis, but antibodies indistinguishable from rheumatoid factor can be detected also in other collagen diseases, such as *lupus erythematosus* and *Sjögren's syndrome,* and, as mentioned previously, in cryoglobulinemia.

Type III immune injury is responsible for the damage to the glomeruli that occurs in *chronic progressive glomerulonephritis.* It is generally believed that a chronic infection of an unknown nature evokes an antibody response to the antigen involved; formation of antigen–antibody complexes and kidney damage follow. In this disease immunoglobulins are readily detectable by fluorescent antibody techniques in sections of the kidney, on the epithelial side of the glomerular membrane. C3 is also deposited along with the immunoglobulins. The activation of the complement sequence by the antigen–antibody complexes and an increased utilization of complement components are reflected in a low concentration of C3 in the serum.

Type IV: Cell-Mediated Injury. Cell-mediated immunity, through its T_D and T_C effector cells, is an efficient mechanism in fighting viral, bacterial, and fungal infections. This efficiency can easily be appreciated if one compares immune competency of normal and of T-cell–deficient persons.

Delayed hypersensitivity reactions are of considerable value as *diagnostic skin tests.* The classic example of such tests is the tuberculin test in which an extract of *Mycobacterium tuberculosis* is injected intradermally or placed on the surface of the skin. A typical delayed reaction (see p 60) develops within 24 to 48 hours in the presence of cellular immunity against the organism. Extracts of other microorganisms or fungi can be used for diagnosis of cellular immunity in *brucellosis, tularemia, leprosy,* and *histoplasmosis.* The skin test with an extract of *Candida* is important in investigating T-cell competency.

Delayed hypersensitivity is the main cause of immune damage in some autoimmune diseases, such as *allergic encephalomyelitis, Hashimoto's thyroiditis, aspermatogenesis,* and *sympathetic ophthalmia.* It is also the chief mechanism in the rejection of *transplants.* ***Encephalomyelitis*** is a rare (about 1:1,000) complication of vaccination with attenuated viruses and was also a complication of immunization with crude rabies vaccine (which, in addition to inactivated rabies virus, contained a variety of antigens of the nervous tissue). The main pathogenetic process in encephalomyelitis is demyelinization. Encephalomyelitis can be produced experimentally by immunization with brain extracts, or with a purified myelin basic protein. Autoimmune pathogenesis is also suspected in *multiple sclerosis,* because demyelinization is a characteristic feature of this disease.

Contact Dermatitis. Many substances contained in jewelry, cosmetics, clothing, leather, and rubber can cause allergy of the delayed type; formaldehyde, nickel, chromium, copper, or other metals, potassium dichromate, glycol, phenylenediamine, and many others may be responsible. Presumably, these simple haptens attach themselves to tissue proteins and become antigenic. Repeated topical application results in delayed hypersensitivity lesions of the skin. *Catechols* present in poison ivy, poison oak, and primrose are notorious for causing contact dermatitis. The chemistry of these catechols and of their attachment to proteins is fairly well known.

* Allotypes are genetically determined polymorphic markers of immunoglobulin molecules.

SUGGESTED READING

BASIC AND CLINICAL IMMUNOLOGY

ATASSI MZ, VAN OSS CJ, ABSOLOM DR: Molecular Immunology: A Textbook. New York, Marcel Dekker, 1984

KLEIN J: Immunology: The Science of Self-Nonself Discrimination. New York, John Wiley & Sons, 1982

MILGROM F, ABEYOUNIS CJ, KANO K: Principles of Immunological Diagnosis in Medicine. Philadelphia, Lea & Febiger, 1981

PAUL WE: Fundamental Immunology. New York, Raven Press, 1984

ROITT IM, BROSTOFF J, MALE DK: Immunology. St. Louis, CV Mosby, 1985

STITES DP, STOBO JB, FUDENBERG HH, WELL JV: Basic and Clinical Immunology, 5th ed. Los Altos, Lange Medical Publications, 1984

ZALESKI MB, DUBISKI S, NILES EG, CUNNINGHAM R: Immunogenetics. Boston, Pitman, 1983

IMMUNOLOGIC TECHNIQUES

ALOISI RM, HYUN J (eds.): Immunodiagnostics. Laboratory and Research Methods in Biology and Medicine, vol 8. New York, Alan R. Liss, 1983

HUDSON L, HAY FC: Practical Immunology. Oxford, Blackwell Scientific, 1980

MISHELL BB, SHIIGI SM: Selected Methods in Cellular Immunology. San Francisco, WH Freeman & Co, 1980

ROSE NR, FRIEDMAN H: Manual of Clinical Immunology. Washington, American Association for Microbiology, 1980

WEIR DM (ed): Handbook of Experimental Immunology, 3rd ed. Oxford, Blackwell Scientific, 1978

5

Diagnostic Clinical Toxicology

Leebert A. Wright

Drug and chemical poisonings, whether accidental or deliberately induced, as well as adverse reactions to drugs administered for therapeutic reasons, have become a major health problem in the Western world, often resulting in hospitalization and death. These drug effects can mimic a variety of disease syndromes, leading in some instances to misdiagnosis, mismanagement, and the associated social and economic burdens imposed on the individual and the health care system. The problem is exaggerated by the fact that a significant number of persons, not receiving medication, present with symptomatic complaints that could be attributed to adverse drug reactions. In particular, easy access to nonprescription (over-the-counter) drugs increases the possibility of multiple drug ingestion, and therefore of drug interaction and toxicity that may be disproportionate to the quantity of individual drugs taken. All too often, notably in drug abuse cases, the substance is impure, containing chemical or biologic materials that may individually or collectively produce a variety of toxic clinical conditions.

Implicit in these considerations is a greater need for up-to-date information on the epidemiologic and pharmacologic basis of drug and chemical intoxication, as well as an awareness of the diagnosis and appropriate management of such cases. Although there is much interest in the areas of industrial, environmental, and forensic toxicology, this chapter will be limited to a discussion of clinical toxicology, with particular emphasis on drug reactions that lead to or prolong hospitalization, drug interference with clinical chemistry tests, and some aspects of drug analysis.

EPIDEMIOLOGY OF DRUG AND CHEMICAL TOXICITY

Approximately 4% of all admissions to general hospitals are attributable to drugs or chemicals, and 30% of these patients develop a second reaction as a result of drugs taken during their stay in hospital. Further, it is estimated that about 15% of *all* hospital patients experience adverse drug reactions resulting in prolonged hospitalization. Where *drug-related* admissions have been analyzed, a breakdown shows that 75% of them were associated with therapeutic dosages, 15% with drugs of abuse, and 10% with accidental poisoning. Approximately 60% of the accidental poisoning cases occur in children less than 5 years of age. In adults the drugs accounting for 45% of all adverse reactions from *normal dosages* are digitalis, acetylsalicylic acid (aspirin), prednisone, warfarin, insulin, diuretics, and antibiotics. More than 100 drugs account for the other 55% of adverse reaction cases.

The drugs and chemicals that are primarily responsible for emergency care and hospitalization can be listed in order of importance for different age groups. Such substances for infants and children under 5 years of age are

Cleaning fluid	Sedatives
Furniture polish	Analgesics

Solvents
Aspirin
Vitamins

Those important in 5- to 24-year olds are

Alcohol
Aspirin
Sedatives
Analgesics
Hallucinogens
Amphetamines
Inhalants
Narcotics

Finally, for those over 25 years of age the responsible drugs are

Alcohol
Barbiturates
Aspirin
Prednisone
Warfarin
Insulin
Methaqualone
Analgesics
Phenothiazines
Digitalis
Antihypertensives
Diuretics
Antibiotics
Tranquilizers
Antidepressants
Narcotics
Diethylpropion

Similarly, drugs accounting for 95% of drug abuse cases in the adolescent and adult population can be listed as follows:

Acetaminophen
Amitriptyline
Amobarbital
Amphetamine
Butabarbital
Cannabis
Chloral hydrate
Chlordiazepoxide
Chlorpromazine
Cocaine
Codeine
Desipramine
Diazepam
Diphenhydramine
Diphenylhydantoin
Doxepin
Ephedrine
Ethanol
Ethchlorvynol
Flurazepam
Imipramine
Maprotiline
Meperidine
Meprobamate
Methaqualone
Methyprylon
Morphine/heroin
Nortriptyline
Pentobarbital
Phencyclidine
Phenobarbital
Phenylpropanolamine
Promazine
Propoxyphene
Salicylate
Secobarbital
Thioridazine

Although the order of preference for drugs of abuse will vary according to place, age, and sex, ethanol accounts for 32% of drug intoxication cases in adults. Cannabis and ethanol, depending on the locality, are the drugs preferred by the adolescent. Of all acute drug intoxication cases reported, 28% have involved a combination of 2 to 17 drugs; thus, drug interactions can be a major factor in acute intoxication.

TOXIC REACTIONS TO DRUGS

Absorption of Drugs

Drug toxicity depends not only on the chemical composition, concentration, and formulation of the drug, but also on its mode of administration, its rate of absorption; its distribution, metabolism, and excretion; its interaction with other drugs; intestinal factors—pH, motility, and the type of intestinal microbial enzymes—and the clinical status of the individual. The toxicity of a drug may be a function of its relative insolubility, the type of complexes it forms with other drugs (e.g., tetracycline and iron), or the availability of binding proteins in the plasma.

In the *gastrointestinal tract* the absorption of *acid* drugs (salicylates, barbiturates) will be enhanced by an alkaline environment, whereas the absorption of *basic* drugs (amphetamines, quinine) will be enhanced by an acid environment. Drugs such as salicylates, phenylbutazone, and amphetamines are retarded by an unfavorable environmental pH. The anticholinergic drugs, which delay gastric emptying, may contribute to mucosal damage or promote the development of carcinogenic nitrosamines through the interaction of secondary amines with nitrite from foods.

Metabolic Clearance of Drugs

Drugs (and other chemical substances) may be biotransformed into metabolites that are pharmacodynamically active, some of which are beneficial, depending on their concentrations, while others may be toxic. Some drug metabolites that are known to be pharmacologically active and potentially toxic are listed in Table 5–1. The elimination of drugs and their metabolites from the body is a prime responsibility of the liver and kidneys. Any decrease in the functional status of these organs may decrease the metabolic clearance of drugs and other chemical substances. Higher circulating levels will then increase the possibility of drug toxicity reactions.

In the ***liver*** many larger molecules ($M_r > 300$) are eliminated in the bile. Any obstruction of bile

TABLE 5-1. Some Pharmacologically Active or Toxic Metabolites

Compound	Reaction	Metabolite
Amitriptyline	*N*-Demethylation	Nortriptyline
Morphine	O-Methylation	Codeine
Codeine	O-Demethylation	Morphine
Methanol	Oxidation	Formaldehyde, formic acid
Imipramine	*N*-Demethylation	Desipramine
Phenacetin	Deethylation	Acetaminophen
Primidone	Alicyclic hydroxylation	Phenobarbital
Chlorpromazine	7-Hydroxylation	7-OH-Chlorpromazine
Acetylsalicylic acid (aspirin)	Deacetylation	Salicylic acid
	Hydroxylation	Gentisic acid
Chloral hydrate	Reduction	Trichloroethanol
Diazepam	*N*-Demethylation	*N*-Desmethyldiazepam
	+3-Hydroxylation	Oxazepam
Carbamazepine	Arene oxidation	Carbamazepine-10,11-epoxide
Procainamide	*N*-Acetylation	Acetylprocainamide
Methsuximide	*N*-Demethylation	Desmethylmethsuximide
Propoxyphene	*N*-Demethylation	Norpropoxyphene

flow or interference with bile production will raise drug and metabolite levels and increase the risk of toxicity. Almost all drugs are reversibly bound to proteins; thus, defective synthesis of albumin, glycoproteins, or other carrier proteins will result in an increase in the circulating level of free drugs.

If the risk of toxicity is to be minimized, drugs that are metabolized extensively by the hepatic microsomal enzymes (amobarbital, diazepam, chlordiazepoxide, glutethimide, diphenylhydantoin, isoniazid, meperidine, morphine, phenylbutazone, and thiopental) should be administered in low dosages during *liver disease.* The degree of hypoalbuminemia and other evidence of liver dysfunction correlate directly with the liver's handling capacity for many drugs.

The induction or inhibition of ***drug-metabolizing enzymes,*** particularly the microsomal enzymes of the liver, can significantly influence drug toxicity. Certain drugs, when administered together, compete for the initially low levels of drug-metabolizing enzymes, resulting in a transient inhibition of the metabolism of both drugs. Some drugs are active enzyme *inducers* (e.g., barbiturates) and, should they be given with another drug (e.g., diphenylhydantoin), the dosage of the latter may have to be increased in order to achieve a therapeutic level. Should the barbiturate be stopped, the possibility of diphenylhydantoin toxicity will be increased. The same situation may arise if phenobarbital (an inducer) is administered with phenacetin, warfarin, or prednisone. Another consequence of the induction of certain metabolizing enzymes is a raised level of metabolites, such as those from halothane and chlorpromazine, which are known to be hepatotoxic. Drug *inhibition* of enzymes is another cause of toxicity. Monoamine oxidase inhibitors (e.g., tranylcypromine) not only may inhibit the metabolism of dietary amines (a possible cause of hypertensive crisis) but also may inhibit those enzymes required for the metabolism of barbiturates and narcotics.

Microbial enzymes in the gastrointestinal tract appear to play a significant role in the detoxification of some drugs (imipramine) by reductive and hydrolytic processes; thus, the administration of antibiotics with these drugs may decrease their microbial degradation and increase the risk of drug toxicity. In addition, antibiotics may inhibit the tissue metabolism of certain drugs, as happens when erythromycin is given with theophylline.

In the ***kidney*** certain drugs and their metabolites are eliminated through a process of active secretion. Drugs that share the same secretory mechanism, such as salicylates, sulfonamides, and phenylbutazone, will inhibit each other competitively when administered together. Drug toxicity is a major hazard for patients with *poor renal function,* and has been reported in up to 45% of such patients.

Drug toxicity may also be affected by the *pH of the urine.* Thus, the excretion of weakly acidic drugs such as phenylbutazone, salicylates, barbiturates, and sulfonamides is decreased in *acidic* urine, whereas excretion of weak bases such as the amphetamines, quinine, quinidine, methaqualone, amitriptyline, imipramine, and meperidine is decreased in *alkaline* urines. Changes in urine pH that favor the formation of ionized compounds will decrease tubular reabsorption and increase excretion. The lipophilic properties of a drug will also affect its tubular reabsorptivity. The greater the lipid solubility of a drug the greater the proportion reabsorbed by the renal tubular cells. Biotransformation of drugs results usually in metabolites that are more polar, less lipid soluble, and therefore more easily excreted than the parent drug.

Classification of Toxic Reactions to Drugs

Reactions to drugs account for many biochemical abnormalities, which are documented in "Compendium of Pharmaceuticals and Specialties" (20th ed., Canadian Pharmaceutical Association, 1985). The major clinical manifestations of drug toxic reactions are summarized in Tables 5–2 and 5–3.

Neurotoxicity can occur with a wide variety of drugs, which are listed in Table 5–3.

Other Organs and Tissues. Toxic reactions to drugs may also be pulmonary (propoxyphene, oral contraceptives), dermatogenous (chlorpromazine, disulfiram, furosemide, meprobamate, sulfonamides, carbamazepine, tricyclic antidepressants), ocular (methanol), mutagenic and carcinogenic (chlorpromazine, LSD, diethylstilbestrol) or dysmorphogenic (thalidomide, cyclophosphamide). Some of these toxicities have been mentioned in other sections.

Fetal, Maternal, and Neonatal Toxicity. Certain drugs will pass through the placenta and are likely to produce abnormalities in the *fetus.* Some of these drugs are aspirin, alcohol, androgens, barbiturates, cortisone, bishydroxycoumarin, cyclophosphamide, estrogens, heroin, lithium, LSD, methotrexate, morphine, phenacetin, and sulfonamides.

TABLE 5-2. Drug Toxicity Reactions

Drug	Toxic Effect
Hypersensitivity Reactions	
Penicillin	Skin-sensitizing IgE antibodies, particularly to metabolite; rhinitis, urticaria
Sulfonamides	Skin reaction—urticaria, purpura
Methsuximide	Skin rash
Meprobamate	Skin rash
Digitoxin	Hemolytic anemia
Phenindione	Renal tubular necrosis
Halothane	Hepatic necrosis
Hydralazine	Lupus erythematosus syndrome
Glutethimide	Acute hypersensitivity reaction
Chlorpromazine	Agranulocytosis, hepatitis, skin rash
Phenacetin	Disruption of erythrocyte membrane, hemolytic anemia
Hematopoietic Toxicity	
Primaquine and its hydroxylated metabolite Sulfonamides Phenacetin	Disruption of cellular membranes, hemolytic anemia
Chloramphenicol	Inhibition of protein synthesis, aplastic anemia
Diphenylhydantoin (Phenytoin) Primidone Phenobarbital	Inhibition of folate metabolism, megaloblastic anemia
Chlorpromazine	Inhibition of DNA synthesis, agranulocytosis

TABLE 5-2. *Continued*

Drug	Toxic Effect
Hepatotoxicity	
Acetaminophen	Acute hepatocellular necrosis
Barbiturates	Metabolites cause liver necrosis, hypertrophy, and cirrhosis
Phenothiazines	Allergic reaction, cholestasis, necrosis
Oral contraceptives (17-alkyl steroids or their derivatives)	Cholestasis, jaundice, occasionally cirrhosis
Imipramine	Cholestatic jaundice (rare)
Halothane	Allergic reaction to metabolites; damage to mitochondrial membranes, cell necrosis
Phenylbutazone	Allergic reaction, cholestasis, cytotoxicity
Carbon tetrachloride	Allergic reaction, hepatic necrosis
Other organic solvents	Hepatic necrosis
Gastrointestinal Toxicity	
Glucocorticoids	Decrease in synthesis of glycoproteins; gastritis, peptic ulceration
Phenothiazines	Gastroenteritis
Phenylbutazone Aspirin	Peptic ulcer, metabolic acidosis (in infants), respiratory alkalosis (in adults)
Digitalis	Vomiting, nausea
Nephrotoxicity	
Acetaminophen Caffeine Phenacetin Salicylates	Possible toxicity to aminophenol metabolites; analgesic nephropathy—papillary necrosis, pyelonephritis, renal failure
Amphotericin B Ampicillin Tetracycline Bacitracin	Renal tubular damage—necrosis, fibrosis
Phenindione	Allergic reaction, glomerular damage, tubular necrosis
Sulfonamides	Insoluble crystals, tubular necrosis
Primaquine	Hemoglobinuria, which may cause acute renal failure
Phenylbutazone	Acute renal failure, nephrotic syndrome
Organic solvents	Acute renal failure, necrosis
Cardiovascular and Cerebrovascular Toxicity	
Acetaminophen	Myocardial necrosis
Imipramine	Arrhythmias, coma
Amitriptyline	Seizures, hypotension, myocardial damage (rare)
Chlordiazepoxide	Hypotension
Disulfiram + alcohol	Myocardial infarction, acute circulatory failure
Estrogen Antiovulatory drugs	Myocardial infarction, increase in serum glycerides
Lithium	Interstitial myocarditis
Phenothiazines	Hypotension
Digitalis	Hypokalemia can precipitate toxicity. Atrioventricular block.
Tetrahydrocannabinol	Tachycardia
Lidocaine Amphetamines Halothane Propoxyphene	Cardiovascular collapse

TABLE 5–3. Drugs Causing Neurotoxicity

Drug	Effects
Narcotic Analgesics	
Heroin, morphine, codeine, demerol, methadone, propoxyphene	CNS depression, pinpoint pupils, respiratory paralysis, brain catecholamine depletion (morphine), hypotension
Cocaine	CNS stimulation, dilated pupils, convulsions, respiratory arrest
Hallucinogens	
Lysergic acid diethylamide (Lysergide, LSD)	Visual or tactile hallucinations, hyperthermia, nausea, vomiting, hyperglycemia, tachycardia, 5-hydroxytryptamine inhibition, convulsions
Tetrahydrocannabinol (marihuana, hashish)	CNS depression or stimulation, hallucinations, illusions, dual personality. Aggressive action and toxic psychosis are dose dependent
Antidepressants	
Amphetamines	CNS stimulation, loss of appetite, fatigue, delusions, hallucinations, psychosis, exhaustion, pneumonia, coma, cardiac failure, death. Effects depend on dose
Tricyclic Compounds	
Imipramine, amitriptyline, desipramine, protriptyline	Ataxia, convulsions, confusion, parkinsonian syndrome, hallucinations. Arrhythmias, hypotension, hyperthermia, coma; cardiac failure may occur at high dosage levels
Sedatives, Hypnotics	
Barbiturates, methyprylon, glutethimide, methaqualone	CNS depression, incoherence; constricted pupils, reactive to light (except glutethimide); respiratory paralysis and death can occur. Synergistic with alcohol
Tranquilizers	
Diazepam, oxazepam, chlordiazepoxide, meprobamate	Similar to the sedatives but less intense
Organic Solvents	
Toluene, acetone, carbon tetrachloride, gasoline, turpentine	CNS depression, dilated pupils, dizziness, euphoria, drunken appearance, sometimes hallucinations; brain damage and death from suffocation
Alcohol	
Ethanol, methanol,* isopropyl alcohol	Effects depend on dose and rate of consumption. Incoordination, aggressiveness, vomiting, malnutrition. Brain, liver, and kidney damage can occur
Metabolites of methanol (formic acid, formaldehyde)	May produce acidosis, ocular damage, vertigo, headache, epigastric pain, coma, death

* The CNS depressant effect of methanol is less than that of ethanol. Its oxidation rate is about one-seventh that of ethanol.

Toxic reactions resulting from drug intake by mothers during *pregnancy* have elicited much concern in recent years. Normally, the placenta has a very limited capacity to metabolize drugs, and because most drugs can diffuse passively through its membrane to some extent, the fetus is quite vulnerable to drugs during pregnancy. It is well documented that pregnant women are likely to have a decreased ability to metabolize and conjugate drugs, presumably in part owing to high levels of progesterone and pregnanediol in the blood and to a deficiency of cytochrome P450; thus, the level of free drugs in these patients may be relatively high. This will not only impose a higher level of drug transfer on the fetus but also increase the possibility of maternal toxic reactions. Drugs administered during the first trimester can cause mutations, malformations, and impaired growth, whereas drugs given during the second and third trimesters are unlikely to be teratogenic but may impair growth and body functions.

Both the *fetus* and the *neonate* have limited metabolic and excretory functions. The hepatic metabolizing enzymes and renal glomerular and tubular functions are immature. There are relative deficiencies of glucuronyl transferases and oxidases required to conjugate drugs and their metabolites, and higher levels in the circulation must be transported by plasma proteins. This increases the possibility of bilirubin displacement, jaundice, kernicterus, and elevated free drug levels. Also, the blood–brain barrier of newborn infants is not fully operative and hence there is a low tolerance for many drugs, as for example, morphine. In spite of the immaturity of the oxidation system of the fetus, its level of hepatic and adrenal oxidation during the second trimester of gestation is about 20% to 50% of that of the adult. This is of major concern because oxidation is the primary mechanism for the conversion of drugs and other chemicals to highly reactive toxic metabolites.

Drug toxicity may develop in infants as a result of drug intake via *breast-feeding* as well as by direct administration. Some of the drugs known to be secreted into breast milk are aspirin, alcohol, barbiturates, and morphine.

Geriatric Toxicity. Drug toxicity represents a greater hazard in geriatric patients who find it increasingly difficult to metabolize and eliminate drugs. Furthermore, some take many different drugs in a given day and are thus highly susceptible to drug interactions and toxicity. For example, long-acting insulin or oral hypoglycemic agents may cause hypoglycemic confusion, particularly during the night and early morning. Even a suboptimal level of diuretics can cause dehydration, hypokalemia, and confusion, whereas the abrupt withdrawal of corticosteroids may induce acute adrenal insufficiency.

DRUG INTERFERENCE WITH CLINICAL CHEMISTRY TESTS

Several reports have indicated that clinical chemistry assay values are significantly altered by drugs. However, pooled serum samples, each containing one of 39 commonly prescribed drugs, were analyzed for 22 standard serum components, and only glucose, carbon dioxide (CO_2), uric acid, aspartate aminotransferase (AST), and bilirubin levels were significantly altered (Table 5–4). The effect of the drugs was investigated at the upper limit of the therapeutic range (A in Table 5–4) and at five times this level (B in Table 5–4). The study did not exclude possible interference by metabolites of these drugs.

The laboratory tests investigated were serum sodium, potassium, chloride, creatinine, and glucose (Trinder's reagent) on Technicon SMA 6/60; total protein, albumin, bilirubin, uric acid, urea nitrogen, phosphorus, calcium, alkaline phosphatase (ALP), lactate dehydrogenase (LDH), and AST on Technicon SMA 12/60; AST and alanine aminotransferase (ALT) kinetically; and γ-glutamyl transpeptidase (GGT), amylase (Phadebas test), cholesterol (Enzymatic, CHOD-PAP), ALP (SMA 12/60, *p*-nitrophenylphosphate).

On the basis of these and other findings it would appear unnecessary to monitor patients routinely for a possible drug interference with the results of clinical chemistry tests, although there should be appropriate communication between the laboratory and the physician concerning the administration of particular medications.

DRUG ANALYSIS

The role of the toxicology laboratory should encompass the evaluation of acute or chronic poisonings, testing for drugs of abuse, and the therapeutic monitoring of specific drugs. Significant contributions to diagnosis, prognosis, and optimal management of drug cases depend on the provision of prompt and reliable qualitative and quantitative data to the physician. Often forgotten is that the metabolites of some drugs and chemicals are active and that the determination of the parent compound alone may produce misleading or incomplete information.

Acute Poisonings

The extent to which analytic toxicology may be required in different centers or by various physicians can be a vexing problem for the clinical chemist. Specialists in clinical toxicology fre-

TABLE 5-4. Changes in Clinical Chemistry Test Results Due to Drug Interference

	Percent of Correct Values									
	Glucose		*CO_2*		*Uric Acid*		*AST*		*Total Bilirubin*	
Drug	*A*	*B*	*A*	*B*	*A*	*B*	*A*	*B*	*A*	*B*
Ascorbic acid	↓ 5.5	↓ 50			↑ 6	↑ 20				
Aspirin			↓ 5	↓ 27						
Methyldopa	↓ 2.5	↓ 9.5			↑ 1	↑ 19				↑ 40
Nitrofurantoin							↓ 9	↓ 35		
Tetracycline	↓ 3.5	↓ 10								

Drugs Investigated

Acetaminophen	Codeine	Penicillin G
Allopurinol	Diazepam	Perphenazine
Amitriptyline	Digoxin	Phenobarbital
Amobarbital	Diphenylhydantoin	Prochlorperazine
Ampicillin	Flurazepam	Promethazine
Ascorbic acid	Furosemide	Propranolol
Aspirin	Gentamicin	Secobarbital
Cefazolin	Hydralazine	Tetracycline
Cephalothin	Meprobamate	Theophylline
Chlordiazepoxide	Methicillin	Tolbutamide
Chlorpromazine	Methotrimeprazine	Trichloroethanol
Chlorpropamide	Methyldopa	Trifluoperazine
Cloxacillin	Nitrofurantoin	Warfarin

AST = aspartate aminotransferase; A = tested at upper limit of therapeutic range; B = tested at 5 × upper limit; ↓ = decrease; ↑ = increase in test value. For methods used, see text. (Wright LA, Foster G: Effect of some commonly prescribed drugs on certain chemistry tests. Clin Biochem 13:249, 1980)

quently request blood levels of drugs for each patient known to have taken some toxicant, regardless of the symptoms. Some physicians want an actual identification of the drugs taken; others are interested primarily in whether the drug taken was acetaminophen, amphetamines, barbiturates, benzodiazepines, digoxin, ethylene glycol, lithium, salicylate, theophylline, the tricyclics, or the alcohols.

The four *principles* of good *management* of the acutely poisoned patient are

1. Identification of the toxicant as quickly as possible
2. Evacuation of the poison from the stomach when necessary and possible
3. Administration of an antidote, if available
4. Symptomatic and supportive therapy

It should be understood, however, that in most instances *treatment* cannot await laboratory results, even though the correct diagnosis, specific antidote administration, or other therapy, may depend on the laboratory findings. In some instances the drug will have disappeared even though its effects are present (e.g., chloral hydrate, busulfan, anticholinesterases). Ordinarily, symptoms of drug poisoning are manifested within 30 minutes, but in some cases they may be delayed for up to 6 hours. In others, the drug may be localized in a tissue rather than in the blood, and therefore its level may be low, as for example thiopental (brain, adipose tissues), amitriptyline (liver, brain, lungs), LSD (liver, kidney), and imipramine (liver, brain, lungs). The presence of coma in some patients with an uncharacteristically low level of a particular drug may be due to the presence of another drug or to medical complications. In drug abuse cases some patients are recalcitrant, paranoid, or misleading as to the name and the amount of the drug taken. Thus, in order to save time, a good deal of cooperation must exist between the laboratory, the physician, and those who were in contact with the patient. Any suspected drug(s) or containers found should be presented to the laboratory. In any case, it is imperative that the analytic procedures chosen be rapid, specific, and suitably accurate.

Drug Monitoring (See Also Chap. 25)

Drug dosage should be adjusted according to *individual requirements* as indicated by steady-state blood concentrations of the drug. This is one example of the benefits to be obtained from drug monitoring. Other examples follow.

In order for some drugs to achieve therapeutic effectiveness they must be given in amounts that approach toxic levels, and so they will come within the toxic range for some patients. The *therapeutic index* of certain drugs such as diphenylhydantoin, primidone, carbamazepine, digoxin, procainamide, lidocaine, ethosuximide, phensuximide, disopyramide, and theophylline is low, so that their effective minimal therapeutic levels are close to their toxic levels, and therefore a knowledge of their blood levels could be important. Accordingly, various analyses may be required of the laboratory as the physician attempts to reverse or prevent serious untoward effects. Often, the toxic reactions to a drug appear slowly, but appropriate function tests of vital organs such as the liver, kidney, and hematopoietic system may serve as warning indicators of the impending symptoms. The incidence of adverse drug reactions becomes disproportionately greater as the number of drugs taken increases.

When a drug is known to cause such untoward effects as leukopenia, anemia, cholestasis, or hepatocellular or renal damage in some patients, it is often prudent to establish the patient's *baseline* test levels for later comparisons. When symptoms such as anorexia, discomfort, or malaise, and signs such as dark urine or jaundice have been observed, the performance of appropriate laboratory tests is clearly indicated. In the case of *nephrotoxicity,* apart from a few cases of acute toxic reactions, damage usually occurs subtly, well in advance of the symptoms, and routine urine analysis can be very useful.

Although information on the therapeutic and toxic blood levels of drugs and chemicals is incomplete, the data presented in Table 5–5 will enable the physician and the clinical chemist to make valuable judgments. It must be understood that this table is only a guide and that individuals will vary in their ability to absorb, distribute, metabolize, and eliminate drugs. In addition, age, sex, and weight; chemical, dietary, and genetic factors; and the clinical status of the patient will influence both the drug level and the pharmacologic response.

Analytic Techniques

During the last decade many new techniques have emerged for the analysis of drugs in biologic fluids. These procedures have shortened analysis times and in some cases have increased the sensitivity of the assay compared with the traditional methods of spot tests, spectrophotometry, colorimetry, thin-layer chromatography, and gas–liquid chromatography.

Thin-Layer Chromatography (TLC). For routine purposes TLC is a comprehensive system that is suitable for screening. It provides for the testing of many specific drug standards and samples simultaneously, as well as the application of several detection systems. A commercial TLC kit packaged with standardized materials and simplified steps for extraction, concentration, inoculation, development, and detection of drugs is available and deserves consideration. However, the sensitivity and specificity of TLC are low; therefore, drugs at low levels may be undetected, and positive results may require confirmation by another method.

Immunoassays. Immunoassays are fast becoming the major tool for the quantitative and qualitative measurements of drugs in therapeutic drug monitoring and clinical toxicology. These assays include isotopic (radioimmunoassay) and nonisotopic immunoassay. The nonisotopic assays, particularly the homogeneous immunoassay techniques, are rapidly gaining in popularity. This group includes enzyme mediated immunoassay

(*Text continues on p. 93*)

TABLE 5–5. Significant Blood Concentrations of Some Common Drugs and Chemicals

Tests		Serum Concentrations	
		Conventional Units	*SI Units*
Acetaminophen	Therapeutic	2–13 mg/L	13–86 μmol/L
	Toxic <4 h	>100 mg/L	>660 μmol/L
	Toxic <12 h	>30 mg/L	>200 μmol/L
Acetone	Toxic	>400 mg/L	>6.9 mmol/L
Acetylsalicylic acid	Analgesic/Antipyretic Therapeutic	<200 mg/L	<1100 μmol/L
	Toxic	>400 mg/L	>2200 μmol/L
	Anti-inflammatory Therapeutic	<400 mg/L	≤2200 μmol/L
	Toxic	>400 mg/L	>2200 μmol/L
Alcohols			
Ethanol	Therapeutic	—	—
	Legal	<800 mg/L	<17.4 mmol/L
	Toxic	>1800 mg/L	>39 mmol/L
Isopropanol	Therapeutic	—	—
	Toxic	>400 mg/L	>6.7 mmol/L
Methanol	Therapeutic	—	—
	Toxic	>200 mg/L	>6.2 mmol/L
Amikacin	Therapeutic peak	15–30 mg/L	15–30 mg/L
	Therapeutic trough	1–8 mg/L	1–8 mg/L
	Toxic peak	>35 mg/L	>35 mg/L
	Toxic trough	>10 mg/L	>10 mg/L
Barbiturates			
Amobarbital	Therapeutic	0.3–5.0 mg/L	1–22 μmol/L
	Toxic	>10 mg/L	>40 μmol/L
Barbital	Therapeutic	5–10 mg/L	27–54 μmol/L
	Toxic	>60 mg/L	>320 μmol/L
Butabarbital	Therapeutic	1.0–5.0 mg/L	5–24 μmol/L
	Toxic	>10 mg/L	>45 μmol/L
Glutethimide	Therapeutic	0.2–0.8 mg/L	0.9–3.7 μmol/L
	Toxic	>10 mg/L	>45 μmol/L
Meprobamate	Therapeutic	5–20 mg/L	23–92 μmol/L
	Toxic	>100 mg/L	>450 μmol/L
Methyprylon	Therapeutic	5–10 mg/L	27–55 μmol/L
	Toxic	>30 mg/L	>160 μmol/L
Pentobarbital	Therapeutic	0.3–5.0 mg/L	1–22 μmol/L
	Toxic	>10 mg/L	>40 μmol/L
Phenobarbital	Therapeutic	15–30 mg/L	65–130 μmol/L
	Toxic	>40 mg/L	>170 μmol/L
Secobarbital	Therapeutic	0.1–1.0 mg/L	0.4–4.0 μmol/L
	Toxic	>7 mg/L	>30 μmol/L
Bromide	Therapeutic	75–100 mg/L	0.9–1 mmol/L
	Toxic	>500 mg/L	>6 mmol/L
Carbamazepine	Therapeutic	4–12 mg/L	17–50 μmol/L
	Toxic	>12 mg/L	>50 μmol/L
Chloral hydrate	Therapeutic	7–12 mg/L	42–73 μmol/L
	Toxic	>100 mg/L	>600 μmol/L
Chloramphenicol	Therapeutic	10–20 mg/L	10–20 mg/L
	Toxic	>25 mg/L	>25 mg/L
Chlordiazepoxide	Therapeutic	0.5–1.0 mg/L	1.7–3.3 μmol/L
	Toxic	>2 mg/L	>6.7 μmol/L

TABLE 5-5. *Continued*

Tests		Serum Concentrations	
		Conventional Units	*SI Units*
Diazepam	Therapeutic	0.1–1.5 mg/L	0.4–5.3 μmol/L
	Toxic	⩾2 mg/L	⩾7.0 μmol/L
Digoxin	Therapeutic	0.8–2.0 μg/L	1.0–2.5 nmol/L
	Toxic	>2.0 μg/L	>2.5 nmol/L
Diphenhydramine	Therapeutic	0.01–0.11 mg/L	0.04–0.43 μmol/L
	Toxic	>0.2 mg/L	>0.8 μmol/L
Dextropropoxyphene	Therapeutic	0.05–0.20 mg/L	0.15–0.60 μmol/L
	Toxic	>0.50 mg/L	>1.5 μmol/L
Diphenylhydantoin	Therapeutic	10–20 mg/L	40–80 μmol/L
	Toxic	>20 mg/L	>80 μmol/L
Disopyramide	Therapeutic	2–5 mg/L	6–15 μmol/L
	Toxic	>6 mg/L	>18 μmol/L
Ethchlorvynol	Therapeutic	2–8 mg/L	14–55 μmol/L
	Toxic	>20 mg/L	>135 μmol/L
Ethosuximide	Therapeutic	40–100 mg/L	280–710 μmol/L
	Toxic	>150 mg/L	>1050 μmol/L
Flurazepam	Therapeutic	0.005–0.020 mg/L	0.01–0.05 μmol/L
	Toxic	>0.2 mg/L	>0.50 μmol/L
Gentamycin	Therapeutic peak	5–10 mg/L	5–10 mg/L
	Therapeutic trough	0.5–2 mg/L	0.5–2 mg/L
	Toxic peak	>10 mg/L	>10 mg/L
	Toxic trough	>2 mg/L	>2 mg/L
Kanamycin	Therapeutic peak	15–30 mg/L	
	Therapeutic trough	1–8 mg/L	
	Toxic peak	>35 mg/L	
	Toxic trough	>10 mg/L	
Lidocaine	Therapeutic	1.5–5.0 mg/L	6–21 μmol/L
	Toxic	>7 mg/L	>30 μmol/L
Lithium	Therapeutic	0.5–1.5 mEq/L	0.5–1.5 mmol/L
	Toxic	>2.0 mEq/L	>2.0 mmol/L
Methaqualone	Therapeutic	1.0–5.0 mg/L	4–20 μmol/L
	Toxic	>5.0 mg/L	>20 μmol/L
Methotrexate	Therapeutic	Variable	Variable
Without citrovorum	Toxic > 42 h	$>8 \times 10^{-8}$ M	>0.08 μmol/L
With citrovorum	Toxic 24 h	$>10^{-5}$ M	⩾10.0 μmol/L
With citrovorum	Toxic 48 h	$>10^{-6}$ M	⩾1.0 μmol/L
Methsuximide	Therapeutic	2.5–7.5 mg/L	12–37 μmol/L
	Toxic	>18 mg/L	>88 μmol/L
Oxazepam	Therapeutic	0.2–1.4 mg/L	0.7–4.9 μmol/L
	Toxic	—	—
Phencyclidine	Therapeutic	—	—
	Toxic	>0.1 mg/L	>0.4 μmol/L
Phenothiazines			
Chlorpromazine	Therapeutic	0.1–0.5 mg/L	0.3–1.6 μmol/L
	Toxic	>1.0 mg/L	>3.1 μmol/L
Fluphenazine	Therapeutic	0.1–0.5 mg/L	0.2–1.1 μmol/L
	Toxic	>1.0 mg/L	>2.3 μmol/L
Methotrimeprazine	Therapeutic	0.1–0.5 mg/L	0.3–1.5 μmol/L
	Toxic	>1.0 mg/L	>3.0 μmol/L
Pericyazine	Therapeutic	—	—
	Toxic	—	—

TABLE 5-5. *Continued*

Tests		Serum Concentrations	
		Conventional Units	*SI Units*
Perphenazine	Therapeutic	—	—
	Toxic	>1.0 mg/L	>2.5 μmol/L
Procyclidine	Therapeutic	—	—
	Toxic	—	—
Prochlorperazine	Therapeutic	0.1–0.5 mg/L	0.3–1.3 μmol/L
	Toxic	>2.0 mg/L	>5 μmol/L
Promazine	Therapeutic	—	—
	Toxic	—	—
Thioridazine	Therapeutic	1.0–1.5 mg/L	2.7–4.0 μmol/L
	Toxic	>10 mg/L	>27 μmol/L
Trifluoperazine	Therapeutic	0.1–0.5 mg/L	0.2–1.2 μmol/L
	Toxic	>2.0 mg/L	>5 μmol/L
Trimeprazine	Therapeutic	0.1–0.5 mg/L	0.3–1.7 μmol/L
	Toxic	>2.0 mg/L	>6.5 μmol/L
Phensuximide	Therapeutic	10–19 mg/L	53–100 μmol/L
	Toxic	>80 mg/L	>420 μmol/L
Primidone	Therapeutic	5–12 mg/L	23–55 μmol/L
	Toxic	>15 mg/L	>68 μmol/L
Procainamide	Therapeutic	4–8 mg/L	17–34 μmol/L
	Toxic	>12 mg/L	>50 μmol/L
Procainamide + *N*-acetylprocainamide	Therapeutic	8–16 mg/L	34–68 μmol/L
	Toxic	—	—
Propranolol	Therapeutic	0.050–0.100 mg/L	0.19–0.39 μmol/L
	Toxic	—	—
Quinidine	Therapeutic	2–5 mg/L	6–15 μmol/L
	Toxic	>7 mg/L	>21 μmol/L
Theophylline	Therapeutic	10–20 mg/L	55–110 μmol/L
	Toxic	>20 mg/L	>110 μmol/L
Tobramycin	Therapeutic peak	5–10 mg/L	
	Therapeutic trough	0.5–2 mg/L	
	Toxic peak	>10 mg/L	
	Toxic trough	>2 mg/L	
Tricyclic Antidepressants			
Amitriptyline + nortriptyline	Therapeutic	0.125–0.250 mg/L	0.45–0.90 μmol/L
	Toxic	>0.500 mg/L	>1.8 μmol/L
Desipramine	Therapeutic	0.040–0.160 mg/L	0.15–0.60 μmol/L
	Toxic	>0.500 mg/L	>1.9 μmol/L
Doxepin + desmethyldoxepin	Therapeutic	0.150–0.300 mg/L	0.55–1.10 μmol/L
	Toxic	>0.500 mg/L	>1.8 μmol/L
Imipramine + desipramine	Therapeutic	0.150–0.250 mg/L	0.55–0.90 μmol/L
	Toxic	>0.500 mg/L	>1.8 μmol/L
Nortriptyline	Therapeutic	0.050–0.150 mg/L	0.20–0.60 μmol/L
	Toxic	>0.500 mg/L	>1.9 μmol/L
Protriptyline	Therapeutic	0.100–0.250 mg/L	0.38–0.95 μmol/L
	Toxic	>0.500 mg/L	>1.9 μmol/L
Maprotiline	Therapeutic	0.050–0.200 mg/L	0.18–0.72 μmol/L
	Toxic	>0.800 mg/L	>2.9 μmol/L
Valproic acid	Therapeutic	50–100 mg/L	350–700 μmol/L
	Toxic	>200 mg/L	>1400 μmol/L

(EMIT), nephelometric inhibition immunochemistry (ICS), substrate-labeled fluorescence immunoassay (SLFIA), foster energy transfer or fluorescence quenching (FETIA), and fluorescence polarization immunoassay (FPIA). The homogenous assay systems do not require separation, thus allowing for speed and ease of determination, although this is at the expense of background 'noise.' On the other hand, the heterogeneous assays require separation, a factor that poses a problem for automation.

Immunoassays permit rapid analysis of a large number of samples but sensitivity is highly variable and should be ascertained for each test. The specificity of the method depends on the antibody available. Antibodies for most drugs of similar chemical structure tend to show some degree of cross-reactivity, except for the antibody of diphenylhydantoin, which shows a high degree of specificity.

Gas–Liquid Chromatography (GLC). GLC is still widely used for both quantitative and qualitative (screening) drug analysis. Again, however, confirmation is required by an independent technique.

High-Performance Liquid Chromatography (HPLC). HPLC, a recent technique made possible by improvements in the manufacture of column packings, shares many of the attributes of GLC but can be applied to compounds of lower volatility without derivative formation.

Spectrophotometry and Colorimetry. Ultraviolet (UV), infrared (IR) spectrophotometry, and colorimetry have generally been used for the screening and measurement of drugs. UV spectrophotometric and colorimetric assays tend to lack specificity, and interference can occur from other substances. IR spectrometry is highly specific on pure samples, but the impurities present in extracts of biologic samples make it impractical for routine use.

Other Analytic Systems. In the future, mass spectrometry in conjunction with GLC or HPLC will probably be more widely used. In the immunologic field, there are many possible labels for drugs that have not yet been widely tested. These include bioluminescence, chemiluminescence, and bacteriophages. Currently, a novel approach is the use of specific properties of enzymes in association with methods based on enzyme inhibition by analyte, or enzyme degradation of the analyte, and the monitoring of the substrate or product. For example, an enzyme from a strain of *Pseudomonas fluorescens* can degrade paracetamol, producing aminophenol, which can then be detected colorimetrically.

RECOMMENDED ANALYTIC CAPABILITY

Toxic reactions to drugs can mimic a variety of disease syndromes and lead to inappropriate requests for laboratory tests. The clinical chemist needs to develop an epidemiologic, pharmacologic, and diagnostic consultative service. Good communication between the physician and the laboratory can affect the analytic approach and save valuable time. Although most toxic reactions are treated symptomatically without a knowledge of the drug taken, for correct diagnosis and specific modes of treatment (such as dialysis or antidotal therapy) the rapid identification and perhaps the quantitative measurement of the drug can be crucial.

Because of the large number of prescribed and "street" drugs in use, few laboratories can be expected to provide the full range of analytic capability. Therefore, taking into consideration the drugs most frequently abused (listed earlier), as well as drugs that require therapeutic monitoring (e.g., anticonvulsants, antiarrhythmics, antiasthmatics, antidepressants, and antineoplastics), the clinical chemist and local physicians should reach an agreement on the analyses that will be required routinely for emergency and special cases and for therapeutic monitoring.

The following suggestions for the establishment of an analytic service, based on hospital size, may be useful.

Hospitals with Fewer than 400 Beds

Laboratories in these hospitals should have the capacity to provide, or should have rapid access to, screening tests for the following compounds:

Acetaminophen (TLC; colorimetric, *o*-cresol, or sulfamic acid coupling; EMIT)

Acetone (Acetest)
Barbiturates (TLC; UV spectrophotometry; EMIT)
Ethanol (enzymatic, alcohol dehydrogenase)
Methanol (colorimetric, coupling of formaldehyde with chromotropic acid)
Phenothiazines (FPN reagent)
Salicylates (Trinder's reagent)

Each laboratory should be guided by the type of drugs commonly abused in that locality, as well as the life-threatening effects of certain drugs.

Laboratories required to provide monitoring of therapeutic drugs may wish to employ the enzyme multiplied immunoassay test (EMIT) system for the determination of such drugs as diphenylhydantoin, phenobarbital, theophylline, and lithium by flame emission photometry.

Hospitals with 400 to 600 Beds

Laboratories supporting a hospital of this size should be capable of providing the tests listed above, as well as carbon dioxide, iron and drug screens (EMIT) for opiates, cocaine, benzodiazepines, amphetamines, cannabinoids, and phencyclidine. In addition, these laboratories, with the aid of homogeneous immunoassay techniques, should provide a wider spectrum of drug monitoring services such as phenobarbital, diphenylhydantoin, primidone, carbamazepine, valproic acid, ethosuximide, digoxin, lidocaine, theophylline, gentamycin, tobramycin, and methotrexate.

Major Medical Centers

These centers should be capable of providing a full range of screening techniques and quantitative measurement of drugs suspected in acute poisoning cases, as well as procedures needed for therapeutic drug monitoring (TDM). These include the less frequently requested drugs in TDM as well as the volatiles, ethylene and propylene glycol, thiocyanate, lead, arsenic, organophosphates (insecticides), paraquat (a herbicide), cyanide, and carbon monoxide.

Frequently the laboratory will have difficulty in identifying or detecting a number of drugs. Therefore, all possible data on the patient and the drugs, including any empty containers found, should be ascertained. It is essential that biologic samples be sparingly and properly utilized and stored. It cannot be overemphasized that confirmation by alternative procedures is essential for a number of routine methods used for drug analyses. The various factors that will influence the analysis and interpretation of the analytic data have been discussed earlier (see Toxic Reactions to Drugs and Drug Analysis).

SUMMARY AND CONCLUSIONS

From an overall health and economic point of view there is a need for more extensive drug analyses for a variety of reasons:

1. Approximately 4% of all admissions to hospitals are due to drug syndromes, and therefore drug analyses on appropriate patients may serve to decrease hospitalization. Some of the drugs that are primarily involved are digitalis, salicylates, prednisone, warfarin, insulin, diuretics, antibiotics, alcohol, barbiturates, tricyclic antidepressants, and tranquilizers.
2. Hospitalization can be seriously prolonged as a result of the side-effects of medication provided in the hospital.
3. The incidence of adverse drug reactions becomes greater in proportion to the number of drugs the patient receives. Drug interactions may develop not only with prescribed and over-the-counter drugs but with 'environmental drugs' such as ethanol, caffeine, nicotine, and heavy metals.
4. The dosage of some drugs should be adjusted according to individual requirements as indicated by the steady-state blood level of the drug.
5. The therapeutic index of medications such as the antiepileptic and cardioactive drugs is low; therefore, the therapeutic level tends to approach the toxic level.
6. Abuse of drugs and accidental poisonings make up approximately 25% of drug-related admissions to hospital. Although alcohol, barbiturates and other sedative–narcotic drugs, tranquilizers, and analgesics are the primary drugs abused, many physicians are interested in knowing whether the drugs taken were acetaminophen, methanol, a salicylate, a barbiturate, or ethanol, since these drugs are often associated with severe clinical episodes that are amenable to specific treatment.

The clinical chemist should endeavor to provide a basic service and employ methods of analysis that are accurate, sensitive, specific, and rapid. In view of the limited number of tests requested from each laboratory, the wide range of abused drugs and drug toxicity, and the high degree of expertise required for reliable drug analyses, many centers will find it appropriate to set up a central toxicologic service.

SUGGESTED READING

BASELT RC: Analytical Procedures for Therapeutic Drug Monitoring and Emergency Toxicology. Davis, California, Biomedical Publications, 1980

BASELT RC: Disposition of Toxic Drugs and Chemicals in Man, 2nd ed. Biomedical Publications, 1982

BAER DM, DITO WR: Interpretations in Therapeutic Drug Monitoring. Chicago, American Society of Clinical Pathology, 1981

CLARKE EGC: Isolation and Identification of Drugs in Pharmaceuticals, Body Fluids, and Post-Mortem Materials, Vols 1, 2. London, Pharmaceutical Press, 1975

DE WOLFF FA, MATTIE H, BREIMER DD: Therapeutic Relevance of Drug Assays. Leiden University Press, 1979

DOULL J, KLAASSEN CD, AMDAR MO: Toxicology. The Basic Science of Poisons, 2nd ed. Toronto, MacMillan, 1980

DRAYER DE: Pharmacologically active metabolites of drugs and foreign compounds. Drugs 24:519–542, 1982

FLANAGAN RJ, CALDWELL R, LEWIS RR, CORLESS D: Toxicological investigations in the detection of drug-induced disease in elderly patients. Hum Toxicol 2:371–380, 1983

GOUGH TA, BAKER PB: Identification of major drugs of abuse using chromatography: An update. J Chromatog Sci 21:145–153, 1983

HASSAN FM, PESCE AJ, RITSCHEL WA: Pitfalls and errors in drug monitoring: Analytical aspects. Meth Find Exp Clin Pharmacol 5:567–573, 1983

HEPPLER BR, SUTHEIMER CA, SUNSHINE I: The role of the toxicology laboratory in emergency medicine. J Clin Toxicol 19:353–365, 1982

LANE RJM, ROUTLEDGE PA: Drug-induced neurological disorders. Drugs 26:124–147, 1983

PADMORE GRA, WEBB SF: The laboratory, the clinician, the poisoned patient. Practitioner 224:81, 1980

SADEE W, BEELEN GCM: Drug Level Monitoring. Analytical Techniques, Metabolism and Pharmacokinetics. New York, John Wiley & Sons, 1980

STEAD AH, MOFFAT AC: A collection of therapeutic, toxic and fatal blood drug concentrations in man. Human Toxicol 3:437–464, 1983

SUNSHINE I: Methodology for Analytical Toxicology. Cleveland, CRC Press, 1975

WOOD LJ, POWELL LW: Liver disease: When drugs may be the cause. Prac Therapeut August 1983, pp 53–56

Part Two

Organ System Diseases

6

Water, Electrolyte, and Hydrogen Ion Disorders

Peter J. Brueckner

CHEMICAL ANATOMY

Analogous to morphologic anatomy is a well-defined distribution of chemical constituents within the body—the *chemical anatomy.* Its integrity is largely dependent on physicochemical processes that maintain the composition and volume of the various compartments. The importance of these physiologic and chemical mechanisms, along with the fact that the concentration of a substance is not a measure of the total amount in a compartment, is often not fully appreciated. Fluid and electrolyte disorders and, to a lesser extent, hydrogen ion disorders are disturbances of chemical anatomy.

Compartmentalization

The sequestration of molecular or ionic species into specific regions may be described as compartmentalization (Table 6–1). Two important features of such compartments are (1) their dependence on a variety of *energy*-consuming processes that are governed by hormonal and other factors and are subject to alteration by changes in metabolism; and (2) the *equilibrium* of compartments with one another, so that an alteration in the state of one causes alterations in the states of others.

Space

Most solutes are contained within several compartments, at different concentrations. When solute is administered, it will enter these various compartments, at rates that depend on its permeability across the membrane boundaries, until equilibrium is reached. The *hypothetical volume* that would contain the total solute at the same concentration as in plasma is termed the ***distribution space*** for that substance. This concept can be used to quantify total body deficiencies of the substance in question and thereby provide an estimate of the amount required for therapeutic replacement.

Compartment Boundaries

The boundary of a chemical–anatomic compartment generally consists of a *membrane,* or layer of cells, whose behavior is governed by physiologic processes. The movement of molecules and ions across the boundary depends on at least two identifiable mechanisms:

1. ***Diffusion***—the passive movement of molecules or ions across the boundary. Rates of diffusion are dependent on the nature of the barrier ('pore' size, polarity, and charge) and the nature of the particle (size, shape, polarity, and charge). Diffusion rates (or coefficients) are measurable and may be altered by physiologic or hormonal stimuli. Although membranes are often referred to as permeable or impermeable, there is no sharp distinction. Rather, there is a spectrum of differing degrees of permeability.

TABLE 6–1. Physiologic Body Compartments That Form the Basis of Chemical Anatomy

Total body (42 L)*	
Intracellular (23 L)	*Extracellular (19 L)*
Specific organs	Interstitial (8 L)
Specific cell types	Plasma (3 L)
Cytoplasmic	Lymph (1 L)
Mitochondrial	Connective tissue (3 L)
Nuclear	Cerebrospinal fluid (0.2 L)
Other organelles	Bone (3 L)
	Body cavities—pleural, pericardial, peritoneal, etc. (1 L)

* Amounts = Volume of water per compartment in a lean 70-kg male.

2. ***Active transport***—the active transfer of molecules or ions across the boundary. Energy-consuming enzymatic mechanisms are localized in cell and organelle membranes. They are usually not absolutely specific for one molecular or ionic species. Other molecules or ions of similar configuration may be transported, but with lesser efficiency. Such processes are generally subject to physiologic controls. Passive diffusion always occurs in the presence of active transport, and both are affected by the concentration gradient across the boundary. Net transport is the algebraic sum of all movements.

In summary, the composition of a compartment is governed by passive and active processes that are characteristics of the compartment boundary, that respond to hormonal and metabolic factors, and that may be altered by pathologic processes.

Osmolality

Osmolality is a measure of the number of osmotically active particles (molecules and ions) in solution in the solvent (water) phase. It is related to the osmotic pressure that would be generated across a membrane totally impermeable to such particles but freely permeable to water.

Osmolality is preferred to *osmolarity* as a measure of osmotic activity because practical measurements, such as freezing point depression, of this colligative property of the solution give values in terms of solute per unit of solvent (not solution). Thus, the results are expressed in milliosmols (mosm) per kilogram of water (not per liter of solution).

Physiologic fluids contain many solutes, each with its own permeability coefficient, and each generating an osmotic pressure at the boundary of the compartment. The algebraic sum of all the osmotic pressures resulting on the two sides of the boundary is the *net* osmotic pressure. This is one of the factors that governs fluid movement between compartments. The net osmotic pressure is really only a small fraction of, and thus cannot be determined accurately from, the total osmotic pressure. Under special circumstances, however, it can be measured directly.

The net osmotic pressure between the circulatory compartment (plasma) and interstitial fluid is termed the ***oncotic pressure*** (see footnote in Chap. 8). A low oncotic pressure is an important cause of edema.

Balance

In a physiologic context, the term *balance* refers to the relationship between total input and total output of some system. *Positive balance* means an excess of input over output and results in a net gain by the system. *Negative balance* is the reverse. It is important to remember that a system may be in balance (input = output) and yet have a great excess or deficiency of one or more of its components. Thus, a measure of balance tells us nothing about total amount.

LABORATORY ASSESSMENT

In the evaluation of fluid and electrolyte status, four important factors must be kept in mind:

1. Although the laboratory provides valuable data, the *history* of the disease process and the *clinical status* of the patient are essential components of interpretation.
2. Standard laboratory measurements are usually carried out on serum and urine, which represent relatively small compartments of body fluid. The extrapolation of this information to processes in the entire body is often tenuous.
3. Most laboratory analyses are measures of *con-*

centration and do not quantify the total amount of a substance.

4. Except in acute illnesses, when sudden alterations in electrolyte balance may occur, the body is usually in a state of *dynamic equilibrium.* Thus, the quantity of electrolyte excreted in the urine generally reflects electrolyte intake in the diet. When the appropriateness of electrolyte excretion is being assessed, the dietary intake as well as the serum concentration of that electrolyte must be known.

Table 6–2 lists the commonly available measures of electrolyte concentration. The ***reference range*** is given for adult populations; specific groups will vary from this, and the range to be expected in disease or during therapy may be quite different. The degree of ***precision*** is dependent on method and instrumentation. Most laboratories will be able to provide this information on request.

General Comments

In most disease processes the *general direction* of fluid, electrolyte, or acid–base disturbance is predictable. However, the *magnitude* of the disturbance is often difficult to estimate, because there is individual response to the disease process and variations in the amount of fluid and electrolyte intake and loss during the evolution of the disorder.

The facts (1) that water is quite freely permeable across most physiologic barriers and (2) that compartment volume is determined by solute content, have important implications concerning the distribution of administered fluid and electrolyte. For example, if *water* is administered (as 5% glucose and water, the glucose being metabolized in minutes) it will enter intra- and extracellular compartments—the osmolalities of all compartments remaining equal. *Isotonic saline,* however, will remain in the extracellular space, since Na^+ does not readily enter cells—again, the osmolalities of all compartments remaining equal. Furthermore, if fluid administration is excessive, a water load can be eliminated more rapidly because the kidney can excrete excess water more quickly than excess sodium. The administration of *hypertonic saline* will cause a shift of water from the intra- to the extracellular space, because osmotic equilibrium is always preserved. The effect of such a shift on the cardiovascular system can be quite severe.

TABLE 6–2. Serum Electrolytes: Methods, Reference Ranges, and Precision of Measurement

Electrolyte	Method of Measuring	Reference Range, Adult (mmol/L)*	Expected Precision (± SD)
Sodium (Na^+)	Flame photometer	136–148	±2
	Ion-specific electrode	136–148	±2
Potassium (K^+)	Flame photometer	3.6–4.7	±0.1
	Ion-specific electrode	3.6–4.7	±0.05
Chloride (Cl^-)	Coulometric	98–108	±1.5
	Colorimetric	98–108	±2
Bicarbonate (HCO_3^-)	Colorimetric	22–26	±1.5
	Gasometric	23–29	±2

* As for all laboratory analyses, the reference ranges may vary according to the population surveyed. For example, the limits for serum Na^+ are likely to be slightly higher in ambulatory and slightly lower in recumbent patients.

DISTURBANCES OF WATER AND ELECTROLYTE BALANCE

ABNORMAL STATES OF HYDRATION

Abnormalities of hydration can be quantitative or qualitative; that is, there may be a deficit or excess, or an abnormal distribution of body water. In practice, four types of disturbances may be identified:

1. Dehydration, in which water loss occurs from all body compartments
2. Water intoxication, in which water accumulates in all body compartments
3. Edema, in which water accumulates in the extracellular space
4. An abnormal distribution of a normal total volume of water, with a deficit in some compartments and an excess in others

Assessment of State of Hydration

It is important to recognize that a patient's state of hydration must be assessed on a *clinical* and

TABLE 6–3. Principal Features of Disordered Hydration

	Dehydration*	Overhydration*
Pulse	Increased	Normal
Blood pressure	Decreased	Normal or increased
Skin turgor	Decreased	Increased
Eyeballs	Soft, sunken	Normal
Mucous membranes	Dry	Normal
Hematocrit	Increased†	Decreased†
Serum sodium	Increased	Decreased
BUN	Increased	Normal
Urine output	Decreased	Normal or decreased ‡

BUN = Blood urea nitrogen.

* The data in this table are applicable only in a general sense: there is considerable variation, depending on electrolyte status, excess fluid distribution, etc.

† Changes may not be apparent, if premorbid values were abnormal.

‡ There is usually some impairment of renal water excretion, which causes or contributes to the disorder.

not on a *laboratory* basis, although test results are often suggestive or confirmatory. The most informative tests are usually measures of the concentration of substances whose levels are normally maintained within narrow limits, so that small changes are readily apparent. The pertinent features of disordered hydration are listed in Table 6–3. Note that in dehydration, because of renal reabsorptive mechanisms, urea rises more than would be anticipated on the basis of hemoconcentration.

Dehydration

Once the presence of dehydration has been established, the differential diagnosis (Fig. 6–1) consists of determining the reason for either the impairment of water intake or the route of excessive water loss. In most circumstances this will be obvious from the clinical history and examination. The appropriateness of the renal response is an important differentiating feature. Normally, dehydration of even mild degree will result in a reduced volume of concentrated urine. On this basis, the following categorization is possible.

Appropriate Renal Response. This suggests an extrarenal etiology, with the following possible causes:

1. ***Insufficient intake of water,*** of which the commonest cause is nausea. Impaired consciousness also frequently results in dehydration; this is most often secondary to excessive drug ingestion in younger people and to stroke in the elderly. Several miscellaneous causes, such as the unavailability of water, or esophageal obstruction, may also come under consideration.
2. ***Excessive water loss*** through nonrenal routes. Gastrointestinal losses (vomiting and diarrhea) are common and may be particularly hazardous in infants. Losses through respiration or perspiration may be significant in unusual circumstances, while losses from fistulas or burns are usually immediately obvious.

Abnormal Renal Response. This is often manifested as the absence of response to dehydration and suggests a renal cause involving excessive loss of water through the kidney. However, it must be remembered that in severe dehydration, even with impaired renal function, the output of urine may fall and its concentration rise. Excessive loss of water through the kidney is an important entity, recognized clinically as *polyuria,* which will be discussed briefly here, although it does *not* always result in dehydration (see also Polyuria, Chap. 8).

Polyuria. By definition, polyuria is an excessively large volume of urine (>3 liters/24 hours) and does not refer to the frequency of micturition. Its various causes and features are summarized in Table 6–4. Polyuria may be the result of excessive water ingestion, abnormalities in the control of water excretion, or intrinsic renal disease. *Polydipsia* (the excessive drinking of water) is present in all three instances, since, in the last two, water loss results in thirst. The classification of disorders leading to polyuria is:

1. ***Primary polydipsia.*** Excessive water intake is usually the result of compulsive psychogenic problems, although rarely organic brain disorders may be the cause. The physiologic response to the large volume of ingested water is a diminution of vasopressin (antidiuretic hormone, ADH) secretion, mediated by the lowered plasma osmolality. There is no dehydration.

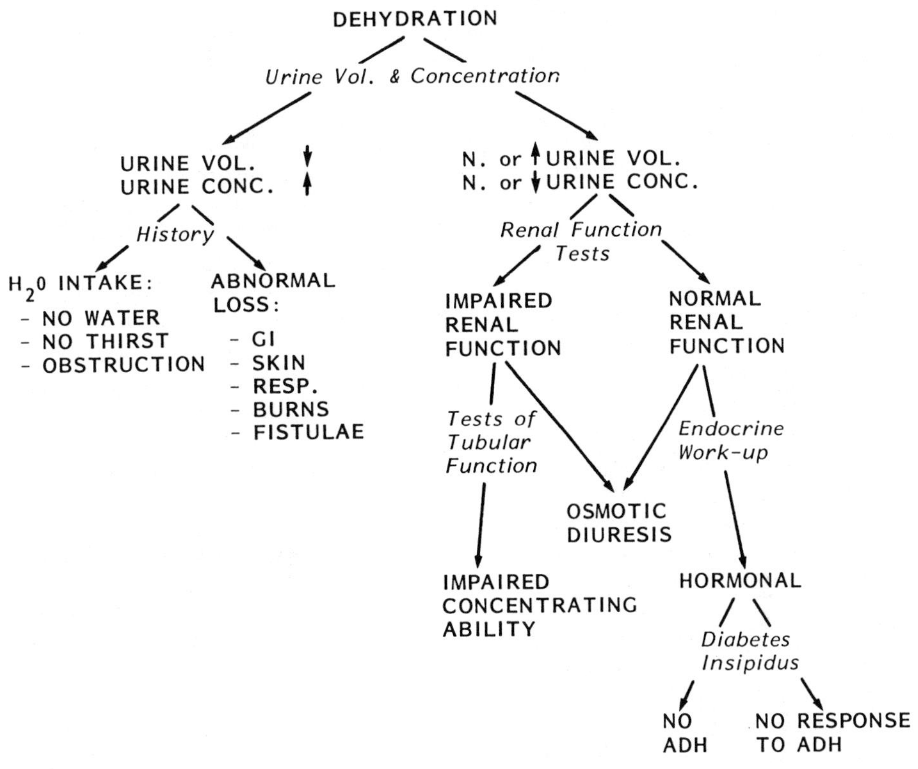

FIGURE 6–1. An approach to the differential diagnosis of dehydration. N = normal; GI = gastrointestinal; ADH = antidiuretic hormone (vasopressin); RESP = respiration; VOL = volume; CONC = concentration.

2. ***Diabetes insipidus,*** also an uncommon condition, results from the impaired production or release of pituitary ADH. In about 40% of cases the etiology is idiopathic, in 20% it is secondary to a neurosurgical procedure, and in 40% it is due to other intracranial lesions such as tumor, trauma, stroke, or infection. Patients with diabetes insipidus are usually

TABLE 6–4. Differential Diagnosis of Polyuria

				Response to H_2O Restriction			Response to ADH		
		Osmolality			*Osmolality*			*Osmolality*	
Disorder	**ADH Level**	*Serum*	*Urine*	*Urine Volume*	*Serum*	*Urine*	*Urine Volume*	*Serum*	*Urine*
Primary polydipsia	L	N, L	L	↓	↑	↑	↓	↓	↑
Diabetes insipidus	L	N, H	L	↔	↑	↔	↓	↓	↑
Nephrogenic diabetes insipidus	H	H, N	L	↔	↑	↔	↔	↔	↔
Osmotic diuresis	N, H	N, H	N	Sl ↓	↑	↑	—	—	—
Renal disease	N, H	N, H	N, L	Sl ↓	↑	↑	—	—	—

N = normal, L = low, H = high, ↑ increase, ↓ = decrease, ↔ = unchanged, — = not applicable, Sl = slight.

able to compensate for their water loss by maintaining a proportionately large intake, but are highly susceptible to dehydration if unable to drink. Because excessive water loss is the primary problem, and polydipsia is secondary, serum osmolality tends to be slightly increased.

3. ***Nephrogenic diabetes insipidus*** is a rare disorder, caused by unresponsiveness of the renal tubules to ADH. The symptoms are similar to those seen in pituitary diabetes insipidus.
4. ***Osmotic diuresis*** results from the presence of large amounts of nonreabsorbable solute in the renal tubular fluid. Such a solute may be nonphysiologic, such as mannitol, or a normal metabolite, such as glucose, which, if present in a quantity large enough to exceed the reabsorptive capacity of the tubule is, in effect, a nonreabsorbable solute. This solute decreases the osmotic pressure gradient beween the capillary and the tubule and thus impairs the reabsorption of water and electrolyte. Secondarily, the larger volume of tubular fluid that then remains diminishes the efficiency of the countercurrent concentrating mechanism.
5. ***Renal disease.*** This is a heterogeneous group of disorders. Polyuria is usually seen in the diuretic phase of acute tubular necrosis, and in hypokalemic or hypercalcemic nephropathy. All have in common damage to tubular cells, which impairs the transport of water and other substances. The specific characteristics of each condition should be sought.

Useful Tests in Polyuria. Although polyuria is one of the causes of dehydration, this occurs only if urinary water loss is not replaced. Biochemical procedures that may be useful in the differential diagnosis of polyuria (Table 6–4) are as follows:

1. ***Serum and urine osmolality.*** Normally, if serum osmolality is high, urine osmolality will also be high, and if one is low the other will be low. This represents an appropriate renal response, and is a fair test of the mechanism regulating ADH secretion.
2. ***Response to dehydration.*** If water intake is restricted, the normal response is a fall in urine volume, a rise in urine concentration, and an increase in plasma osmolality, all within 12 hours. It is important to remember that if renal concentrating ability is severely impaired, water loss continues and the resulting dehydration may be life threatening. Thus, the patient must be closely observed.
3. ***Response to the administration of ADH*** (or its synthetic analog). The normal response is a diminution in urine volume, an increase in urine concentration, and, if the patient continues to drink, a fall in plasma osmolality. If polyuria has been present for a long time, the response may be delayed by hours or a day and may be sluggish when it does occur.
4. The administration of ***drugs to stimulate ADH*** release has not been commonly used because of the availability of synthetic hormone; however, the response would be similar to that seen when ADH is administered, if the pituitary release mechanism is not defective.
5. ***Assay of ADH.*** With currently available methods, the limit of detectability of ADH is such that low normal values cannot be distinguished from the absence of hormone. Thus, this test is useful only in situations where excessive secretion is expected to be present.
6. ***Serum creatinine*** should be measured as a test of renal function, and the possibility of an osmotic diuresis must be kept in mind. Specific assays for many such osmotically active substances are available.
7. Most solutes that cause osmotic diuresis can be detected in urine or plasma by specific tests or assays.

Overhydration

The diagnosis of ***water overload*** is also based on clinical data. It is important to ascertain the *distribution* of excess fluid in order to distinguish between water intoxication and edema. Water intoxication is primarily a disorder in the handling of a water load, whereas edema is a problem of water retention secondary to sodium retention. Overhydration is nearly always associated with a dilutional hyponatremia, the differential diagnosis of which is given in the section entitled Disorders of Sodium Metabolism.

Water Intoxication. This is a state of overhydration in which fluid accumulates in all body compartments. It may cause some impairment of physiologic functions, especially those of the cardiovascular system (*circulatory overload*). Hyponatremia is nearly always present, resulting in additional symptoms. The excess water may, in milder form, simply result in hypoosmolality of body fluids, as will be discussed under hypoosmolar (hyponatremic) syndromes. The underlying mechanism is the inappropriate retention of water by the kidney (or, rarely, the administration of excessive volumes of water, usually parenterally). Such renal water retention may be the result of intrinsic renal disease or of the syndrome of inappropriate ADH secretion (SIADH). The administration of excessive quantities of sodium will also result in fluid retention, with symptoms that may resemble water intoxication, but osmolality does not decrease.

Edema. This is also a state of overhydration, but the excess water is localized in specific extracellular compartments instead of being global. In most instances abnormal oncotic or hydrostatic pressure plays a role in its development. The location of edema fluid varies with the disease process and thus provides diagnostic information.

Abnormal Distribution of Body Water

This occurs when transudates (or exudates) accumulate in body cavities, as in ascites. The electrolyte composition of transudates is essentially the same as that of plasma. Except for the large volume of fluid that may accumulate in the gut, this class of disorders generally does not have a significant effect on fluid and electrolyte balance. Most symptoms arise as the result of local mechanical effects due to the presence of the accumulated fluid. If fluid and electrolyte problems do occur, their nature is the same as if the transudated fluid had been lost outside the body.

Physiologic Consequences of Abnormal Hydration

Dehydration has two important effects:

1. *Volume depletion* occurs, with a consequent impairment of tissue perfusion. Initially this results in decreased urine output (*oliguria*), which may be followed by alterations in heart and brain functions. The central nervous system effect is an altered level of consciousness, which may progress to coma and death.
2. There is a *concentration* of all constituents in body fluids, which in general impairs metabolic processes and is manifested clinically as tiredness, lethargy, and altered reflexes.

Overhydration has two principal effects:

1. The abnormally large fluid load places a burden on *cardiac function* and may precipitate congestive failure or pulmonary edema, especially in patients with limited cardiac reserve.
2. The abnormally low electrolyte concentrations may result in *neuromuscular* problems, although these are rarely severe. Cellular overhydration is usually not a problem, although *convulsions* may be a late complication. These are probably the result of a combination of overhydration and low electrolyte concentrations within the brain.

DISORDERS OF OSMOLALITY

An abnormal serum osmolality means an abnormal (total) amount of solute per volume of serum water. It does not reveal whether the problem is the result of disordered hydration, a quantitative abnormality of one of the normal constituents of serum, or the presence of an abnormal constituent. The *measurement* of osmolality is a rather nonspecific test; under controlled circumstances it is useful in the assessment of mechanisms regulating water balance, allowing comparisons between the osmolalities of plasma and urine.

Under normal circumstances the principal contributors to plasma osmolality are sodium (and its associated anions), urea, and glucose. Investigation of an abnormal osmolality should include the determination of these constituents. A ***calculated*** osmolality may then be obtained by the following formula:

$$\text{Osmolality (mosm/kg)}$$
$$= 2 \times \text{Na (mmol/L)} + \text{urea (mmol/L)} + \text{glucose (mmol/L)}$$
$$= 2 \times \text{Na (mmol/L)} + \frac{\text{urea nitrogen (mg/dL)}}{2.8} + \frac{\text{glucose (mg/dL)}}{18}$$

A difference of 10 mosm/kg between the *calculated* and *measured* values strongly suggests the presence of some other substance. If necessary, the appropriate specific assays may then be carried out. Ethanol is probably the most common of these solutes. Other substances (e.g., mannitol) are less frequently encountered.

Hyperosmolality

There are three principal causes of an elevated serum osmolality (>300 mosm/kg). However, more than one mechanism may contribute to the disorder, because high concentrations of nonreabsorbable solute not only raise serum osmolality, but also cause osmotic diuresis, which then leads to dehydration. The causes are as follows:

1. ***Dehydration.*** Water loss usually results in an increase in the concentration of all body fluids. In serum, the most striking abnormality is hypernatremia. The differential diagnosis of dehydration was discussed earlier. Rarely, the dehydration may be isotonic, or even hypotonic, if the water loss is accompanied by a proportionate or greater loss of solute—primarily electrolyte. Serum osmolality will then not be elevated. Hypotonic dehydration may be seen in infants, but is quite uncommon in adults.
2. ***Abnormal concentration of a normal plasma constituent.*** The resulting hyperosmolality is most frequently due to hyperglycemia or uremia, and less often to ketosis. Such metabolites can easily be measured. It is important to remember that an abnormally high level of one of these metabolites, in the presence of a normal osmolality, implies the existence of hyponatremia.
3. ***The presence of an abnormal substance.*** An administered agent or abnormal metabolite may be suspected on clinical grounds or be suggested by the discrepancy between measured and calculated osmolalities. Specific assays can then be confirmatory.

Hypoosmolality

Sodium is the only component of serum present in sufficiently large concentration that a significant diminution in its level results in hypoosmolality. Thus, for practical purposes *hypoosmolality* is synonymous with *hyponatremia,* which is discussed later.

Physiologic Effects of Abnormal Osmolality

Three factors are of particular importance in considering the effects of osmolality on metabolism:

1. The magnitude of the abnormality
2. The rate of development of the abnormality
3. The permeability of compartment boundaries to the causative, osmotically active substance(s)

The deleterious effects of disordered osmolality appear to be due not so much to the osmolality itself as to the ***fluid shifts*** that may result, causing alterations in the concentration and volume of various body compartments. For example, a metabolite such as *urea,* when present in high concentrations, will result in marked hyperosmolality; but because it is quite freely permeable across most membranes, it does not cause a significant fluid shift. Such hyperosmolality may be well tolerated, provided the uremia has developed slowly enough to allow for equilibration. *Ethanol* falls into a similar category.

In contrast, however, substances such as *glucose* or *mannitol,* which by and large remain extracellular, will cause a shift of fluid out of the cell mass, resulting in intracellular dehydration, with its attendant decrease in cell volume and increased concentration of normal cell metabolites—primarily electrolytes. In *severe dehydration* this change in volume reduces the size of the brain, which may cause the rupture of small blood vessels in that organ and in the meninges, as these structures pull away from the rigid cranium. The result is an intracranial hemorrhage.

Osmolalities up to 400 mosm/kg can be fairly well tolerated when caused by permeative metabolites, provided that adequate time (approximately 24 hours) is allowed for equilibration. In contrast, osmolalities in the 330 mosm/kg range, when caused by nonpermeative metabolites, may be hazardous, especially if the onset is acute. The *clinical effects* are primarily those due to cerebral dehydration, namely, impaired consciousness, progressing to stupor and death. If the cause is the administration of a nondiffusible osmotic load (e.g., mannitol), rapid expansion of the circulatory volume should be anticipated, as fluid shifts out of the cells. This can be hazardous to patients with compromised cardiac function.

DISORDERS OF SODIUM METABOLISM

Sodium is primarily a cation of *extracellular* fluid, in which it constitutes the bulk of the osmotically active substances. The major features of its metabolism are illustrated in Figure 6–2. The ratio of intracellular to extracellular Na^+ concentration is important physiologically, but the intracellular concentration is not easily measured.

The actual ***function*** of sodium is somewhat obscure, but it is known to play a significant role in at least the following processes:

1. Volume maintenance, of both intracellular and total body fluid
2. Maintenance of osmolality and ionic strength in body fluids
3. Neuromuscular excitation processes
4. Hydrogen ion metabolism

Although the physiologically important measure of sodium is its concentration in body *water,* it is more practical to obtain its concentration in body *fluid.* In the case of serum, for example, the difference is about 8%—a figure that is approximately equal to the total solute content. (If serum Na^+ is 140 mmol/L and serum contains 92% water, the concentration of Na^+ in the *water* is 152 mmol/L). This fact assumes clinical importance in unusual circumstances, such as severe hyperlipidemia, in which plasma water content is reduced, giving rise to an artifactual hyponatremia.

Hyponatremia

By definition, hyponatremia is a serum Na^+ concentration < 136 mmol/L. This may arise either by a depletion of body sodium or by the excessive accumulation of water, resulting respectively in a ***depletional*** or a ***dilutional*** hyponatremia (Fig. 6–3).

These alternatives can usually be distinguished by an assessment of the patient's state of hydration, although other clinical data are important. Depletional hyponatremias are generally associated with normal fluid volume or dehydration, whereas dilutional hyponatremias are, by definition, accompanied by overhydration and are therefore disturbances of both sodium and water metabolism.

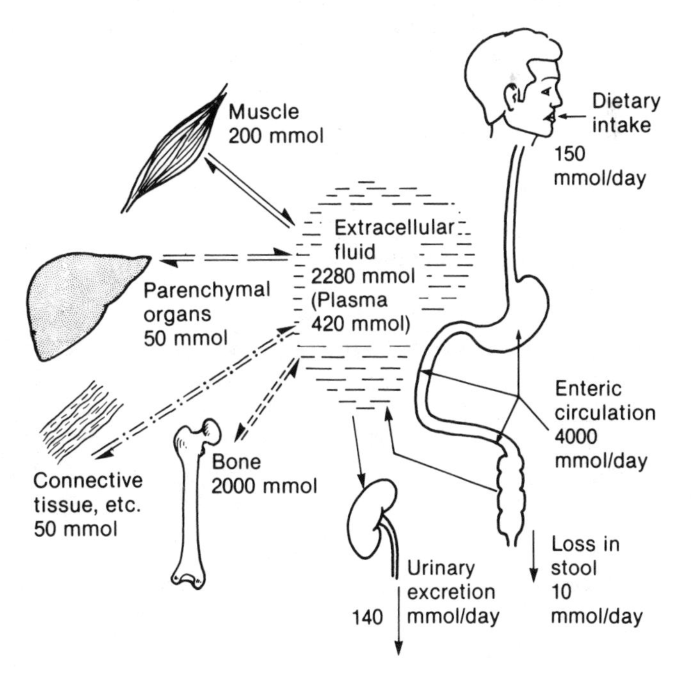

FIGURE 6–2. Principal features of sodium metabolism. In an adult, slightly over half of the 4600 mmol of sodium is exchangeable. Of this, nearly 90% is in the extracellular fluid and 16% in the plasma. Sodium in bone is not readily exchangeable. Note the magnitude of the enteric circulation.

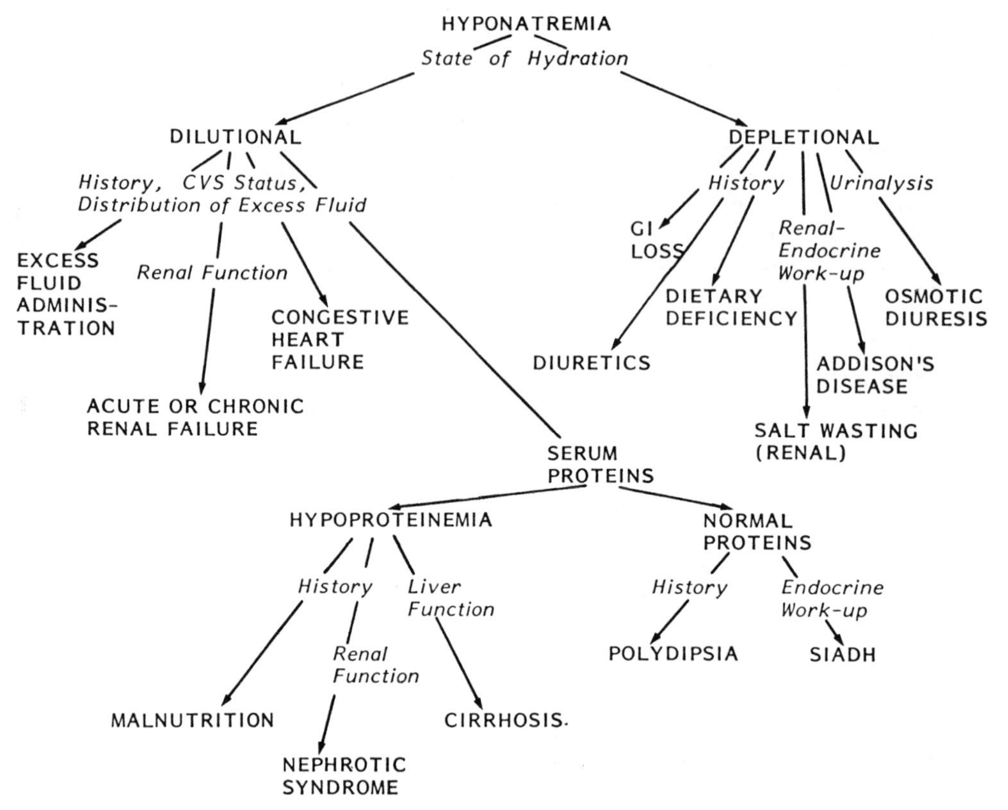

FIGURE 6–3. An approach to the differential diagnosis of hyponatremia. CVS = Cardiovascular system; GI = gastrointestinal; SIADH = syndrome of inappropriate antidiuretic hormone secretion.

Dilutional Hyponatremias

Five broad areas should be considered in the differential diagnosis. Although specific laboratory tests may be useful, the differentiation is primarily clinical; examination of the cardiovascular system is of special importance.

Congestive Heart Failure. Impaired cardiac function results in a diminished cardiac output and reduced renal perfusion. This is interpreted by renal and other homeostatic mechanisms as hypovolemia, activating stimuli for sodium and water retention. Water is usually retained in greater excess than sodium, giving rise to the dilutional hyponatremia. Although hyperaldosteronism and excessive secretion of ADH are common features of this condition, there is no specific biochemical test for congestive heart failure. It is essentially a clinical diagnosis.

Acute or Chronic Renal Failure. These disorders may be accompanied by an impaired ability to handle a water load. If water intake exceeds the capacity of the kidney to excrete it, a dilutional hyponatremia will ensue. Renal function tests and, if applicable, the determination of free water clearance (see Urine Dilution and Concentration, Chap. 8) will usually confirm the presence of this problem.

Fluid Administration. The administration of large quantities of fluid of low sodium concentration, if carried out too rapidly, will result in a dilutional hyponatremia. Usually this outcome is associated with intravenous administration, although duodenal tube feedings and enemas have been identified as causes. The appropriate historical data must be obtained.

Hypoalbuminemia. This possibility should be considered if the foregoing disorders can be excluded. Albumin is in part responsible for the maintenance of oncotic pressure, and when its concentration is diminished a shift of fluid from the intravascular space to the interstitial space occurs. This stimulates a hypovolemic response of increased aldosterone and ADH secretion. Other proteins do not have sufficient osmotic effect to cause, by their diminished concentration, such shifts of fluid. (Paradoxically, however, patients with congenital hypoalbuminemia usually do not have edema.) The diagnosis is based on the measurement of serum albumin.

The principal *causes* of hypoalbuminemia that should be considered are:

1. ***Liver disease***—usually cirrhosis, in which inadequate quantities of albumin are synthesized
2. ***Malnutrition*** (rarely seen in North America in the absence of severe disease), again accompanied by inadequate albumin synthesis
3. ***Nephrotic disease,*** in which excess quantities of albumin are lost in the urine. A urinary protein determination is an important diagnostic procedure if this disorder is being considered

The accumulation of fluid in the interstitial space is clinically manifested as *dependent edema* when it occurs in the limbs (e.g., in heart disease), as *ascites* when it collects in the peritoneal cavity (characteristic of cirrhosis), or as *facial edema* (characteristic of nephrotic disease). These conditions are accompanied by secondary increases in aldosterone and ADH secretion.

Excessive ADH Secretion. If the foregoing conditions have been excluded, and the serum albumin concentration is normal, primary excessive secretion of ADH (see also Chap. 14) must be considered as a possible cause of overhydration. This is a rare disorder in which the rate of ADH secretion is inappropriately high, resulting in water retention, hypoosmolality, and hyponatremia, although total body sodium may be increased. The etiology can be either a malfunction of the ADH-releasing mechanism (central nervous system disorder), pulmonary disorders, ectopic production of ADH by a tumor, or drugs that probably act by increasing ADH release. The associated clinical and laboratory findings are important in establishing the diagnosis, and it may be helpful to investigate the renal handling of water loads and, if possible, ADH levels in plasma.

One very important feature of dilutional hyponatremias is that they are accompanied by an *excess* of *total body sodium.* Thus, correction of the abnormality must not involve the administration of Na^+, for this would overload the cardiovascular system. Except for situations in which the patient's condition is rapidly deteriorating, dilutional hyponatremia usually represents a new steady state in sodium balance, and urinary sodium excretion tends to reflect dietary intake. Thus, the measurement of urine electrolytes is seldom a useful assay.

Depletional Hyponatremias

The depletional hyponatremias have several causes whose distinguishing features are again primarily clinical, although renal function studies may be helpful.

Dietary Deficiency. Normally functioning kidneys are capable of restricting Na^+ excretion to essentially nil; thus, a normal individual, even with a very low sodium intake, is quite unlikely to develop hyponatremia. However, when renal function is impaired, Na^+ loss may ensue. This may be insidious, especially in elderly patients, so that the hyponatremia presents as one related to inadequate intake.

Gastrointestinal Loss. This most frequently arises as a result of vomiting or diarrhea, but may also be due to drainage through fistulas. These causes of hyponatremia are usually obvious, and no specific biochemical tests are necessary.

Renal Loss. Impaired renal function, due either to a disorder in the physiologic mechanisms that regulate renal function or to intrinsic renal disease, may result in substantial electrolyte loss. Four disease processes should be considered:

1. ***Osmotic diuresis.*** The impaired reabsorption of tubular fluid results in the formation of urine that contains not only the nonreabsorbable solute (which was causative), but also considerable amounts of Na^+ and K^+. Measurement of urine volume and specific tests for suspected causative solutes may be helpful.
2. ***Hormonal disorders.*** Aldosterone is necessary for the normal conservation of Na^+, thus hypoaldosteronism (as in Addison's disease) leads to Na^+ loss. There is evidence that sodium depletion from this cause is more severe in extracellular than in intracellular fluid, because aldosterone plays a role in the regulation of ionic balance across cell membranes.
3. ***Renal tubular disease.*** This may be either acute or chronic (see Chap. 8 for a differential diagnosis). In the present context, an inability of the tubular cells to reabsorb an adequate portion of the filtered Na^+ can lead to hyponatremia. An assessment of renal function is necessary.
4. ***Chronic renal failure.*** The reduced nephron population places an increased functional demand on the remaining nephrons, resulting in an increased solute (principally urea) load for each. This, in effect, is an osmotic diuresis. In most circumstances, because the glomerular filtration rate and the filtered load of Na^+ are proportionately reduced, Na^+ excretion is not greater than normal. However, the capacity to reduce Na^+ excretion is impaired, and the patient becomes susceptible to hyponatremia from any reduction in dietary intake.

Diuretics. The desired action of diuretics is to promote Na^+ excretion and thereby reduce body water content. This may result either intentionally or inadvertently in hyponatremia.

Consequences of Hyponatremia

The ***symptoms*** resulting from hyponatremia depend on the Na^+ concentration and on the rapidity with which the hyponatremia has developed. Symptoms are seldom experienced at concentrations greater than 125 mmol/L and occasionally levels as low as 100 mmol/L may be well tolerated as long as there has been adequate time (days) for adaptation. Furthermore, the diminished Na^+ concentration appears to be less of a problem in dilutional than in depletional states. When effects do occur, they are principally neuromuscular—weakness, lethargy, and confusion, progressing to convulsions, coma, and death. Anorexia, nausea, and abdominal cramps have also been reported. The mechanisms of these symptomatic manifestations appear to be a combination of altered neuromuscular excitability and fluid shifts, the latter causing an increased cell volume.

Hypernatremia

This disorder, by definition a serum $Na^+ > 148$ mmol/L, is nearly always the result of *dehydration,* although the reverse is not true; that is, if water loss is accompanied by electrolyte loss, dehydration may not be associated with hypernatremia (see the previous discussion on dehydration).

Although hyperaldosteronism is accompanied by Na^+ (and water) retention, an 'escape mechanism' generally prevents serum Na^+ from rising above the reference range. Rarely, hypernatremia may be produced iatrogenically by the administration of hypertonic saline.

Effects of Hypernatremia

The predominant ***symptoms*** are neurologic and their severity depends on both the degree of hypernatremia and the rate at which the disorder developed. Lethargy, paradoxically associated with neuromuscular irritability and convulsions, can be anticipated, and may progress to impairment of consciousness, coma, and death. These manifestations are probably due both to the elevated Na^+ concentration itself and to the dehydration. Because of the high extracellular fluid osmolality, intracellular dehydration will always be present, and can be sufficiently severe to cause anatomic distortion of the brain. The *plasma volume* depletion that usually accompanies this condition may have severe effects on the cardiovascular and renal systems. Serum Na^+ concentrations above 155 mmol/L may produce symptoms, and those above 165 mmol/L are potentially hazardous, but the degree of volume depletion must also be taken into consideration.

DISORDERS OF POTASSIUM METABOLISM

Some important aspects of K^+ metabolism are summarized in Figure 6–4. *Potassium* is the major intracellular cation; approximately 95% of the total body K^+ is normally within cells. This disequilibrium is maintained by active transport mechanisms that depend on the integrity of cellular metabolic processes, so that a 20-fold concentration gradient can be maintained between intracellular and extracellular fluids. As in the case of Na^+, the exact ***metabolic role*** of K^+ has not been fully elucidated, but it is known to

1. Play an important role in neuromuscular activity, in which both the extracellular concentration and the concentration gradient are significant factors
2. Be largely responsible, in response to regulatory stimuli, for the maintenance of intracellular ionic strength and cell volume
3. Participate in K^+-dependent metabolic processes
4. Function as a complementary cation for intracellular proteins, and probably for other large molecules.

Only about 0.3% of total body potassium is contained in the plasma. One important consequence of this is that factors that cause even a small but sudden shift of intracellular K^+ will result in a relatively large change in the extracellular K^+ content and concentration. With the available technology it is practical to measure K^+ only in serum (itself a fraction of the extracellular compartment).

A significant amount of total body H^+ is buffered intracellularly. In order to maintain ***electroneutrality*** H^+ entering the cell will be balanced by K^+ leaving the cell, and *vice versa.* Because intracellular buffering capacity is large, shifts of K^+ during buffering may also be large, resulting in marked changes in the concentration of serum K^+. Figure 6–5 correlates serum K^+ and blood H^+ concentrations with total body potassium. Although this information is useful, it should be interpreted cautiously because the blood H^+ concentration is not a good indicator of intracellular H^+ status. Furthermore, there are *additional factors* that play a role in intracellular K^+ metabolism. For example:

1. Increases in the rate of *glucose metabolism* and glycogenesis cause a shift of K^+ into the cells,

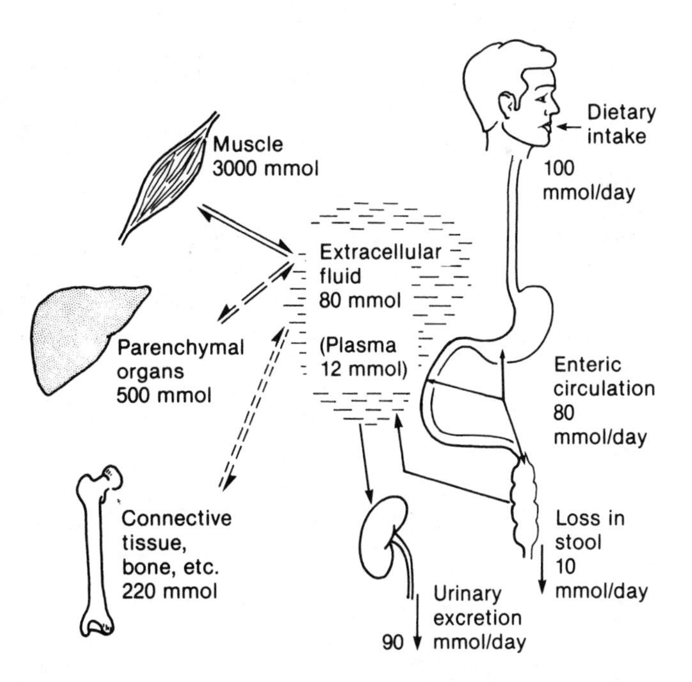

FIGURE 6–4. Principal features of potassium metabolism. Of the total body potassium, approximately 98% is intracellular, 2% is extracellular, and only 0.3% is in the plasma.

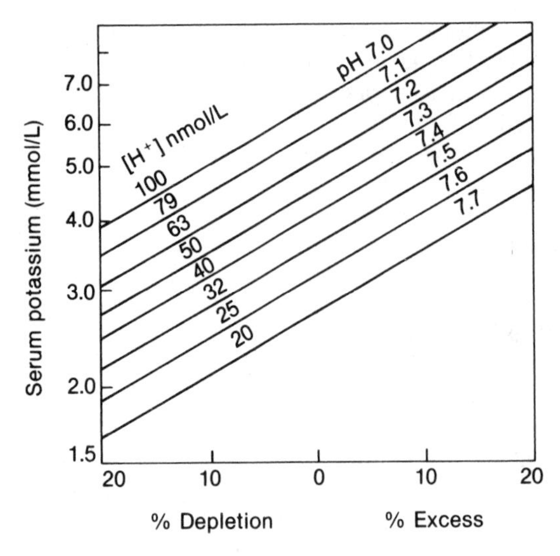

FIGURE 6–5. Relationship between serum K^+, blood H^+, and total body potassium.

resulting in hypokalemia, and *vice versa.* Insulin plays a regulatory role in this process.

2. *Aldosterone,* which is an important regulator of external K^+ balance, appears to facilitate an intracellular shift of K^+. Prolonged hyperaldosteronism, however, results in a depletion of total body potassium because it also causes a great increase in the renal excretion of this element.

The ***reference range*** of serum K^+ is 3.5 to 5.0 mmol/L. In a proportional sense, this is a wider range than that for sodium, so changes in hydration do not lead as readily to values outside the reference range for K^+ as they do for Na^+.

The regulation of 'external' potassium balance is incompletely understood, but the following points are worth noting:

1. Low plasma levels of K^+ suppress, and high plasma levels stimulate, aldosterone secretion.
2. Aldosterone is a powerful stimulant of renal K^+ excretion.
3. There is an obligatory renal excretion of K^+ of about 30 mmol/day, which must be replaced by dietary sources.
4. There is good evidence that the total body content of potassium (in addition to plasma concentration) plays a role in the maintenance of K^+ balance.

Hypokalemia

By definition, hypokalemia is a serum concentration of $K^+ < 3.5$ mmol/L, but other factors, particularly H^+ concentration, must be taken into account. Furthermore, some thought must be given to the expected level of intracellular potassium stores. Because these stores cannot be measured directly, the nature of the disease process provides valuable information. However, hypokalemia and potassium deficiency usually coexist and are discussed together in this section. The *differential diagnosis* (Fig. 6–6) is based principally on clinical data relating to the following conditions:

Dietary Deficiency. If the obligatory renal loss of 30 mmol/day is not replaced, K^+ deficiency and hypokalemia will develop. Because meats, vegetables, and fruit contain large amounts of K^+, a hypokalemic diet is one that consists primarily of cereals. This problem is most frequently encountered in elderly patients who subsist on 'tea and toast.'

Gastrointestinal Loss. Vomitus, intestinal fluids, and diarrhea fluid all contain relatively high concentrations of K^+. With their loss, a potassium deficit and hypokalemia may ensue.

Diuretics. Perhaps the commonest cause of hypokalemia in the adult population is the administration of diuretics, which impair the renal reabsorption of filtered Na^+ and K^+.

Hyperglycemia. The *osmotic diuresis* of hyperglycemia that results from uncontrolled diabetes causes a significant loss of Na^+ and K^+. Furthermore, the absence of insulin activity impairs intracellular glucose metabolism and facilitates glycogenolysis, resulting in a shift of K^+ extracellularly. The situation is compounded by the acidosis that usually develops, which also moves K^+ out of the cells. Potassium that has left the cell mass is rapidly lost in the urine. Thus, during the evolution of this disorder hyperkalemia may occur transiently in the presence of total body K^+ depletion.

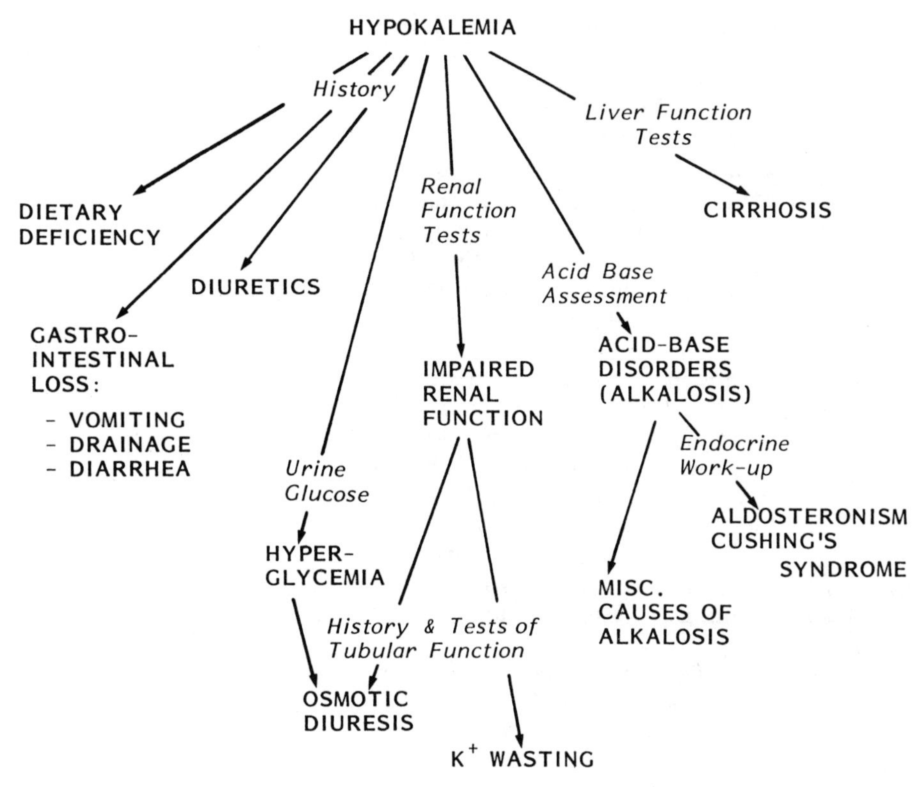

FIGURE 6–6. An approach to the differential diagnosis of hypokalemia. MISC = miscellaneous.

Cirrhosis. Here the mechanism for hypokalemia is not fully understood; however, the secondary hyperaldosteronism that often accompanies cirrhosis may play a role.

Alkalosis. In what may be described as 'reverse buffering,' a diminished H^+ concentration in plasma leads to an extracellular shift of this ion, which is compensated by movement of K^+ into the cells, thereby lowering plasma K^+ levels. Furthermore, the renal response to a low plasma H^+ concentration is an increase in the excretion of K^+.

Excess Steroid Hormone. Although aldosterone causes a shift of K^+ intracellularly, it also causes an increase in K^+ excretion, which ultimately results in a deficit of total body potassium. Because aldosterone acts by promoting Na^+ reabsorption in the kidney in exchange for K^+ or H^+ (excretion), a metabolic alkalosis (H^+ ion loss) can also develop in this condition. This, in turn, facilitates further K^+ loss. Although glucocorticoids have little *mineralocorticoid* activity, such activity may nevertheless be significant when high corticosteroid levels are present. Thus, hypercortisolism is frequently accompanied by hypokalemia; some synthetic analogs of cortisol have a similar effect.

A very unusual cause of hypokalemia is the excessive ingestion of licorice, which contains compounds that simulate the mineralocorticoid activity of aldosterone.

Renal Potassium Wasting. In this rare disorder there is an impairment of K^+ reabsorption in the proximal tubule. It is usually accompanied by the improper handling of other metabolites, such as glucose and amino acids, which also appear in the urine. The condition may be acquired genetically or caused by nephrotoxins. (Renal loss of

K^+ also occurs in osmotic diuresis, as described previously.)

An Approach to the Differential Diagnosis of Hypokalemia

The ***clinical assessment*** of potassium metabolism should include a dietary history, an enquiry into any recent illness that could result in fluid loss from the gastrointestinal tract, and a listing of all medications, particularly diuretics and corticosteroids. Such information will permit a decision on four common causes of hypokalemia (the first four above), for which there are no specific laboratory tests. Because serum K^+ concentration must be interpreted in the light of H^+ concentration, an *acid–base* assessment is often necessary. Frank diabetes mellitus causing K^+ depletion will usually be obvious, but an assessment of blood *glucose* may be necessary. Similarly, cirrhosis that is sufficiently advanced to result in hypokalemia will usually be evident, but *liver function* tests may be useful.

Exclusion of the foregoing entities leaves excess steroid *hormone* secretion or *renal K^+ wasting* as possible causes. The first requires an endocrinologic work-up (see Chap. 14), and the second should be suspected in the presence of continued renal loss of K^+ and other metabolites after the other causes listed above have been excluded.

In the presence of hypokalemia, a urinary K^+ concentration of >10 mmol/L suggests potassium loss via the kidney (provided the responsible pathologic mechanism remains active). A urinary K^+ concentration of <10 mmol/L suggests impaired intake or extrarenal loss (most commonly gastrointestinal).

An aspect of disorders of potassium metabolism that is worth reemphasizing is the *poor correlation* between *hypokalemia* and *potassium deficiency*. Because many factors may cause an extracellular shift of K^+, the diagnosis of potassium deficiency must occasionally be made on clinical grounds and in the presence of a normal, or, rarely, even an elevated, serum K^+ concentration. Similarly, an estimate of the magnitude of the potassium depletion must be based on clinical as well as laboratory information. Furthermore, when the mechanisms causing K^+ loss from the cells are reversed, catastrophically low levels of serum K^+ may be reached in a short time. For example, when a diabetic's disease has been out of control and has caused a K^+ deficit, the administration of insulin may rapidly (within one hour) result in hypokalemia, as K^+ moves back into the cells.

The Effects of Hypokalemia

The effects of hypokalemia and of total body K^+ depletion are similar. Both the *magnitude* of the deficit and the *rate* at which it has developed are important factors, lower levels being better tolerated if several days have been allowed for adaptation. A sudden fall to 2 mmol/L is probably hazardous, but patients may have few symptoms even at 1.5 mmol/L, if adaptation has occurred. However, it is prudent to treat hypokalemia, because if ventricular fibrillation occurs, this first symptom may be the last.

The manifestations are primarily *neuromuscular*—tiredness, weakness, and hyporeflexia. The most severe problem is a propensity to cardiac *arrhythmias,* particularly in patients who are digitalized. In this situation serum values of 2.5 mmol/L or less should be considered serious. Prolonged hypokalemia may also result in a renal tubular *nephropathy* characterized by impaired concentrating ability and polyuria. This effect can be expected if serum potassium levels of 2 mmol/L or less are present for a period of more than a few weeks.

Hyperkalemia

A serum concentration of $K^+ > 5$ mmol/L is, by definition, hyperkalemia. However, as in the case of hypokalemia, it must be interpreted in the light of blood H^+ concentration and the status of intracellular metabolic processes (primarily that of glucose). Because serum K^+ concentration is not a good index of total body potassium stores, hyperkalemia may coexist with intracellular K^+ depletion.

The various *causes* of hyperkalemia (Fig. 6–7) are discussed below.

Excess Intake. Persons with *normal* renal function are able to cope with substantial potassium loads; hence, only the administration of massive amounts of K^+ will lead to hyperkalemia. However, when renal function is *compromised* the ability to excrete K^+ is diminished and there is a proportionately greater hazard of hyperkalemia

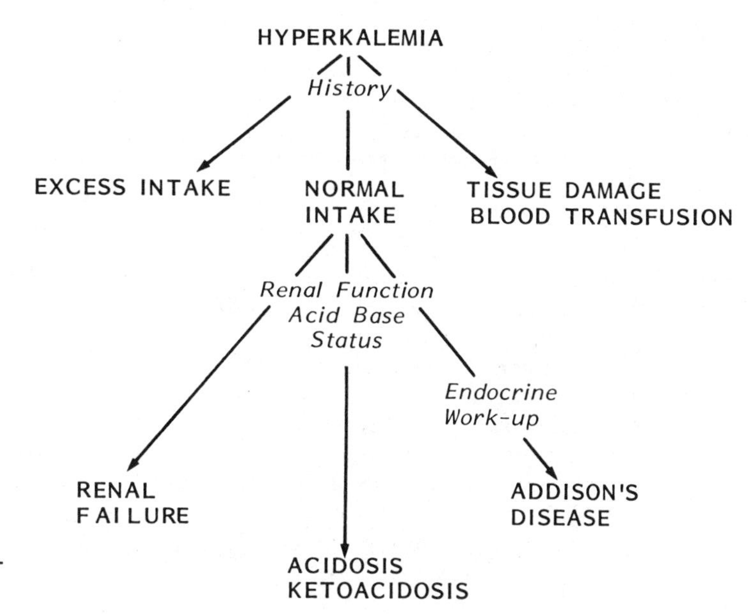

FIGURE 6–7. An approach to the differential diagnosis of hyperkalemia.

with dietary or therapeutic administration. A common clinical problem is the elderly patient on diuretic therapy who receives a potassium supplement in amounts that cannot be excreted adequately. Occasionally, the injudiciously rapid administration of intravenous K^+ may result in hyperkalemia if the ion cannot enter the cell mass sufficiently rapidly. Recently, there have been reports of hyperkalemia in infants who have received formula with an inappropriately high K^+ content, which could not be handled by the baby's immature kidneys.

Tissue Damage. A substantial portion of the large intracellular store of K^+ may be released from damaged tissue resulting from either physical (including thermal) or metabolic injury. A similar mechanism is operant when old banked blood is administered, because K^+ leaks from the cells during storage.

Renal Failure. Potassium excreted normally in the urine has been actively secreted by renal tubular cells. When destruction of functional renal tissue occurs, there is a loss of such secretory cells, and consequently the amount of K^+ that can be handled by the kidney is proportionately reduced. Renal reserve is substantial, so that hyperkalemia is usually not encountered until about 80% of renal function has been lost. Beyond this point, dietary intake of K^+ must be restricted, which is difficult because nearly all plant and animal cells have a high content of this element.

Acidosis. Intracellular buffering of H^+ causes an extracellular shift of K^+ from within the cell mass. This mechanism of hyperkalemia therefore results in intracellular K^+ depletion.

Addison's Disease. Hypoaldosteronism results in the renal retention of K^+ and an elevation of serum levels. The hyperkalemia is an important diagnostic clue for this relatively uncommon disorder. There is usually an accompanying hyponatremia, and there may be a mild metabolic acidosis.

An Approach to the Patient with Hyperkalemia

It is important to estimate the amount of potassium ingested, both in the diet and as medication. A determination of K^+ output is less important, but may be significant. Hydrogen ion status should also be considered. If excessive intake and tissue damage can be excluded (on clinical grounds), renal function should be investigated with appropriate tests. Addison's disease may be suspected on a clinical basis, or when electrolyte results suggest it, and must then be confirmed with the appropriate endocrinologic procedures.

Consequences of Hyperkalemia

Elevated plasma K^+ levels result in generalized neuromuscular irritability, characterized by hyperreflexia. The most serious effect is *myocardial irritability,* resulting in arrhythmias and, ultimately, cardiac arrest. A characteristic feature of hyperkalemia is peaked T waves on the electrocardiogram. In an emergency, this sign may allow one to suspect the presence of hyperkalemia before the test result is returned from the laboratory; it is, however, only a moderately reliable index. Serum K^+ levels above 7 mmol/L are hazardous, and levels above 8 mmol/L may be incompatible with life.

CHLORIDE AND BICARBONATE

Chloride is the major anion in plasma, although little is known about its specific role and regulation. It appears to be present to maintain electroneutrality, that is, as a complementary anion, together with bicarbonate. Thus, the level of Cl^- is related to both the total cation concentration and the HCO_3^- concentration. Plasma *bicarbonate* is an integral component of acid–base balance, and is discussed under disorders of H^+ metabolism in the next section.

FIGURE 6–8. Relationship between blood pH and nanomoles of H^+ ion per liter.

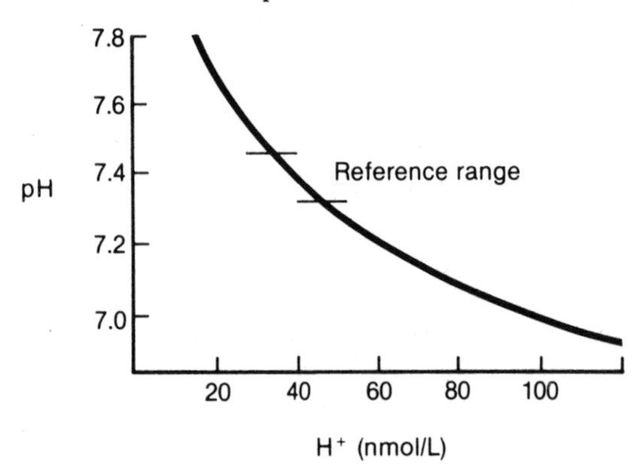

HYDROGEN ION REGULATION

Biochemical reactions are sensitive to H^+ concentration (pH). In the body, a complex mechanism maintains H^+ homeostasis, that is, ***acid–base balance.*** A H^+ disturbance represents a malfunction of this homeostatic system and is usually the result of overproduction or improper excretion of acids or bases. By convention, in the past, H^+ concentration has been expressed as ***pH.*** Now efforts are being made to introduce ***nanomoles/liter*** as the standard unit of concentration. It is thus necessary to understand both conventions. The relationship between them is illustrated in Figure 6–8. Note that pH is a *logarithmic* function, which decreases as H^+ concentration, a *linear* function, increases. A pH change from 7.5 to 6.5 represents a tenfold increase in H^+ concentration.

HYDROGEN ION METABOLISM

Hydrogen ion is a component of the chemical–anatomic structure, and its concentration varies from compartment to compartment. The major phases in its metabolism are the production of H^+, its buffering and its transport through cells and the circulatory system, and finally its excretion. The reference range in *plasma* is 45 to 35 nmol/L (pH 7.35–7.45), whereas *intracellular* levels are in the order of 125 nmol/L (pH 6.9). Intracellular H^+ concentration is difficult to measure, and consequently the practical assessment of acid–base status depends on the H^+ concentration in blood. This obviously imposes limitations on our knowledge about the intracellular aspects of disease processes.

Production of Hydrogen Ion

Hydrogen ions arise from two sources:

1. The metabolism of proteins, nucleic acids, and phospholipids results in the production of sulfuric and phosphoric acids, and the *incomplete* metabolism of fat and carbohydrates in the production of organic acids. The dissociation of these relatively strong acids forms about 40 mmol of H^+ daily. Because these substances are not metabolized further, they are termed *fixed acids.* Their only mechanism of disposal is excretion by the kidneys. During anoxia and other disease states, the incomplete metabo-

lism of fat or carbohydrate is increased, adding to the fixed acid load. This process is reversible, however, and the excess organic acids can be removed by further metabolism.

2. The *complete* metabolism of *fat* and *carbohydrate* results in the formation of CO_2. Carbon dioxide is not an acid. However, in solution some of it is hydrated to *carbonic acid,* which dissociates to H^+ (and bicarbonate). This is termed *volatile acid* because the mechanism by which it is produced is readily reversible and the H^+ from this source is utilized in the regeneration of CO_2, which in turn is excreted by the lungs.

Transport and Buffering

Normally, in a 24-hour period the entire *metabolic load* of 40 mmol of fixed acid and 20,000 mmol of CO_2 is transported by the circulatory system to the excretory organs. In order to maintain the concentration of *free* H^+ at 40 nmol/L (40×10^{-6} mmol/L), body fluids are extensively ***buffered.*** In blood there are extracellular (i.e., plasma) and intracellular (i.e., erythrocyte) buffers. The principal extracellular buffer is bicarbonate, with lesser contributions by phosphate, protein, sulfate, lactate, and so forth. The major intracellular buffer is hemoglobin, followed by other cellular proteins, phosphate, sulfate, and lactate. Within each compartment the various buffers are in equilibrium with each other; thus, it is necessary to examine the status of only one to assess the status of all. Assessment of the status of the buffer systems yields fundamental information about the acid–base status of the patient.

Excretion of Hydrogen Ion

Two mechanisms are involved in H^+ excretion:

1. ***Fixed acid*** is excreted almost entirely by the ***kidney.*** H^+ is actively secreted by the renal distal tubular cells. Although a nearly 300-fold concentration gradient can be achieved, extensive buffering is still needed to accommodate the daily complement of 40 mmol of H^+ in 2 liters of urine at a concentration of free H^+ not exceeding 0.04 mmol/L (pH 4.4). The required buffering capacity is provided in part by the sulfates, phosphates, and other anions normally excreted in the urine, and in part by ammonia produced by the renal tubular cells. Impairment of this latter process diminishes the ability of the kidney to excrete fixed acid.

   The reabsorption of filtered HCO_3^- conserves what was in the blood and does not contribute to net H^+ excretion. However, the renal loss of this HCO_3^- reduces its plasma level, diminishes buffering capacity, and results in an acid–base disturbance (acidosis).

2. ***Carbon dioxide*** is excreted only by the ***lungs.*** Since CO_2 is produced at a fairly constant rate, the level of blood and tissue CO_2 is regulated by the respiratory system (assuming normal circulatory function. See Chap. 7).

Regulation of Hydrogen Ion Metabolism

Factors governing H^+ metabolism are poorly understood. The plasma concentration of H^+ appears to be a function of its rates of production and excretion and of the body's ability to buffer. The rate of *production* is normally dependent on metabolic processes, which are controlled by hormonal and other mechanisms and are influenced by the dietary intake of foodstuffs. In disease, factors such as anoxia or the abnormal lipid metabolism that occurs in uncontrolled diabetes may greatly increase H^+ formation. Control of the *renal excretion* of fixed acid is not well understood beyond the fact that the kidney appears to excrete the fixed acid load presented to it in such fashion as to maintain normal plasma levels of H^+.

The excretion of CO_2, and hence the amount of volatile acid, is controlled by the *respiratory center,* in response to the plasma concentrations of CO_2 and H^+.

Extracellular buffering ability depends largely on an appropriate HCO_3^- concentration, maintained by renal regulation through a threshold mechanism. The amount of protein, particularly hemoglobin, is also important because of its transport and buffering functions, but its concentration does not appear to respond directly to acid–base homeostatic mechanisms.

CLASSIFICATION OF DISTURBANCES OF HYDROGEN ION METABOLISM

The clinical and laboratory classification of H^+ disorders is based on changes occurring in the extracellular fluid compartment, as observed by sampling blood. The change in H^+ concentration caused by disease processes is further modified by buffering and physiologic compensation. Thus the *assessment of acid–base status* is based on

1. The measurement of H^+ concentration (or pH)
2. The determination of blood buffer status by the measurement of HCO_3^-
3. The clinical history and examination

Although the extent of buffering could be estimated by the measurement of any of the buffers, HCO_3^- is chosen because it is the most abundant and because CO_2 is a component of the system.

Definitions

Acidosis is the accumulation of H^+ resulting from either fixed or volatile acid. Note that this term does not refer to free H^+ concentration, because all but a minute fraction has been taken up by buffers

Alkalosis is the opposite of acidosis, namely, a deficit of H^+. Again, buffering capacity alters the free H^+ concentration. The accumulation of HCO_3^- results in a deficit of H^+ because these ions combine and form CO_2 and H_2O.

Acidemia is an arterial blood concentration of free $H^+ > 45$ nmol/L (pH < 7.35).

Alkalemia is an arterial blood concentration of free $H^+ < 35$ nmol/L (pH > 7.45).

Metabolic (in acid–base terminology) refers to factors that govern the formation and excretion of fixed acid. The HCO_3^- concentration is referred to as the metabolic component.

Respiratory (in acid–base terminology) refers to the status of volatile acid formation (namely, that due to CO_2). This is usually measured as a partial pressure (see footnote at beginning of Chap. 7), and pCO_2 is termed the respiratory component.

Mechanisms of Hydrogen Ion Disturbances

The ***bicarbonate buffer system*** is described by the following equation:

$$H_2O + CO_2 \leftrightharpoons H_2CO_3 \leftrightharpoons HCO_3^- + H^+$$

There is doubt about the existence of carbonic acid, and the middle term is usually omitted. The equilibrium constant is then calculated on the basis of the overall reaction, and is generally taken to be approximately 7.9×10^{-7} (pK = 6.1). Using the appropriate logarithmic terms, the equation can then be expressed as

$$pH = pK + \log \frac{HCO_3^-}{\alpha\ pCO_2} \quad \textbf{(Henderson–Hasselbalch equation)}$$

where $\alpha = 0.03$, the factor required to convert pCO_2 in millimeters of mercury to the concentration of CO_2 in mmol/L. There are *two independent variables* in the equation, HCO_3^- and CO_2, each of which may be increased or decreased by disease. Thus, there are four possible primary disturbances, ***metabolic acidosis*** or ***alkalosis*** (where the *metabolic* component, HCO_3^-, is primarily disturbed), and ***respiratory acidosis*** or ***alkalosis*** (where the *respiratory* component, pCO_2, is primarily disturbed). Each such disturbance will be accompanied by a pH change, as follows:

Respiratory

Acidosis $$pH\downarrow = 6.1 + \log \frac{HCO_3^-}{0.03\ pCO_2\uparrow}$$

Alkalosis $$pH\uparrow = 6.1 + \log \frac{HCO_3^-}{0.03\ pCO_2\downarrow}$$

Metabolic

Acidosis $$pH\downarrow = 6.1 + \log \frac{HCO_3^-\downarrow}{0.03\ pCO_2}$$

Alkalosis $$pH\uparrow = 6.1 + \log \frac{HCO_3^-\uparrow}{0.03\ pCO_2}$$

This easily perceived concept is *complicated* by

1. A mass-action effect, which causes a shift in the equilibrium when any one factor is altered
2. Physiologic compensation, which is the organism's attempt to return a disordered H^+ concentration to normal
3. The simultaneous presence of more than one of the four possible primary acid–base disturbances, so that both of the independent variables in the Henderson–Hasselbalch equation are affected.

Thus, acid–base data obtained during a disease process will reflect the *combined effects* of the primary disturbance, the chemical shift in equilibrium (involving all physiologic buffers), and the body's compensatory response. ***Analysis*** of the ***data*** consists of assessing the role played by each of these mechanisms and arriving at

1. The nature of the primary disorder
2. Its magnitude
3. The appropriateness of compensation
4. In conjunction with (3), knowledge of the presence or absence of a *second* acid–base disorder

Two principal chemical and physiologic ***mechanisms*** are active in disorders of H^+ homeostasis:

1. Buffering. The degree to which acid–base parameters are disturbed, in response to a hydrogen ion disorder of a given severity, depends partly on the available buffering capacity of the body (e.g., hemoglobin concentration). Furthermore, it is difficult to assess intracellular buffering, which depends on cellular metabolism. These factors are important during therapy, in estimating quantities of buffer or other solutions to be administered. Unfortunately, since it is impractical to measure intracellular concentrations, such estimates are usually based on the medical history.

2. Physiologic Compensation. Under normal circumstances, a specific acid–base disturbance will evoke a compensatory response in the 'unaffected' component of the buffer pair, which is the physiologic opposite of the primary disturbance. For example, a respiratory acidosis ($\uparrow pCO_2$) will normally be followed by an elevation of HCO_3^- concentration, thereby diminishing the severity of the acidosis as the pH rises toward normal. Thus, the compensatory response to a respiratory acidosis is, in effect, a metabolic alkalosis; however, it is important to realize that this is a normal phenomenon, which is referred to as *compensation,* and is *not* a superimposed acid–base disorder. On the other hand, an absent or inadequate compensatory response, or an excessive alteration in the compensating component, represents a *second* acid–base disorder. Thus, with reference to the previous example, a respiratory acidosis without an elevated HCO_3^- (time having been allowed for compensation) suggests the presence of a coincident metabolic acidosis.

The *direction, rate of development,* and *magnitude* of the expected compensatory responses are detailed in Table 6–5. Estimation of the magnitude does not lend itself to computation by simple formulas,* and so the data are presented in tabular form. A rather wide range is observed, probably due to variation among individuals. It is evident from Table 6–5 that appropriate degrees of compensation, particularly for respiratory disorders, are achieved only after a period of time. This must be taken into consideration in the interpretation of results.

A Practical Approach

As a rule of thumb, the direction of the deviation of pH will indicate whether the problem is an acidosis or an alkalosis; the component (metabolic or respiratory) that is relatively farthest from normal, will represent the *primary* acid–base disorder. In the untreated patient this is a feature of the disease process; however, once treatment has been instituted, the primary disorder may be a feature of therapy. After the primary disorder has been identified, the data in Table 6–5 can be used to assess the *appropriateness of compensation.* Inappropriate compensation may represent a second acid–base disorder. Once the acid–base disorder has been classified in this fashion, a differential diagnosis can be constructed on the basis of this and other laboratory data, the clinical examination, and the history (as discussed later).

The Anion Gap

Electroneutrality is maintained in plasma, as in other solutions, by matching equivalents of positive and negative ions. Figure 6–9 shows a scalar representation of the ionic constituents of serum. The ***anion gap*** is obtained from the formula $Na^+ - (Cl^- + HCO_3^-)$. This algebraic sum represents the difference in the equivalence concentrations of the remaining positive ions (K^+, Ca^{2+}, Mg^{2+}) and the

* Expressions have been published that are said to describe primary acid–base changes and compensatory mechanisms with simple arithmetic relationships. In fact, the responses are not linear and vary within and among individuals, and the expected range of values increases with the magnitude of the abnormality. Such 'titration curves' are of questionable value.

TABLE 6–5. Compensation in Acid–Base Disturbances

	Metabolic Acid–Base Disorder (Appropriate Compensation Develops Over 15–30 Min)		Respiratory Acid–Base Disorder (Appropriate Compensation Develops Over 1–3 Days)		
	HCO_3^- (mmol/L)	*Expected pCO_2 with Compensation* (mm Hg)*	*pCO_2 (mm Hg)*	*Expected HCO_3^- without Compensation* (mmol/L)*	*Expected HCO_3^- with Compensation* (mmol/L)*
Acidosis	4	12–20	90	24–31	40–53
	8	20–27	80	23–30	37–48
	12	26–32	70	22–30	33–43
	16	31–37	60	23–30	29–37
	20	35–40	50	22–29	26–32
Reference Range	24	38–42	40	22–27	22–27
Alkalosis	28	39–47	30	21–26	20–26
	32	40–52	20	20–26	18–24
	36	41–55			
	40	42–59			
	44	43–63			
	48	44–67			

* Expected values refer to 95% confidence limits. Data are applicable only prior to the start of therapy.

remaining negative ions (protein, $H_2PO_4^-$, HSO_4^-, organic anions). The reference range for the anion gap is 12 to 18 mmol/L.

In practice, except for Na^+, changes in the *cation* content of plasma are so small that they can be ignored. An *increased* anion gap is nearly always due to an accumulation of organic acids, or sulfuric and phosphoric acids, either of which usually represents a metabolic acidosis. A small increase may also be caused by an elevation of serum protein concentration. A *decreased* anion gap is almost always the result of hypoproteinemia. It is important to remember that the anion gap is calculated from three measured concentrations; thus, laboratory imprecision may be compounded, and a significant error in any of the three components will result in an incorrect anion gap.

Methodology

As indicated in Table 6–6, ***blood gas measurements*** are usually made by using electrodes in contact with the blood sample. The Henderson–Hasselbalch equation is then used to calculate the HCO_3^- from the measured pCO_2 and pH values. In most laboratories it is this calculated bicarbonate concentration that is reported with blood gas measurements. An alternative method measures HCO_3^- directly, usually by a colorimetric or gasometric technique. This is the value that is generally reported in conjunction with other electrolyte values.

Even when determined on the same specimen, such *calculated* and *measured* HCO_3^- values differ by 2 to 3 mmol/L under normal conditions, and up to three or four times this amount when major acid–base disturbances are present. The discrepancy usually increases with the severity of the patient's disorder. These differences have not been satisfactorily explained. Two *contributing factors*

TABLE 6–6. Methods Used in Blood Gas Measurements

Test	Method	Reference Range (Adult)	Precision (± SD)
pCO_2	Electrode	38–42 mm Hg	±1.5
pH	Electrode	7.35–7.45	±0.01
H^+	Electrode	35–45 nmol/L	±1
HCO_3^-	Calculated	22–26 mmol/L	±1
HCO_3^-*	Colorimetric	22–26 mmol/L	±1.5
HCO_3^-*	Gasometric	23–29 mmol/L	±2

Values refer to arterial blood (or *venous serum).

are (1) that different methods of specimen collection and analysis (ionic activity *versus* color development) are used, and (2) in the calculation of the HCO_3^- value, the pK is set numerically at 6.1. There is evidence that pK varies from approximately 5.8 to 6.2 depending on pH, temperature, and other factors.

DIFFERENTIAL DIAGNOSIS OF HYDROGEN ION DISORDERS

Once the acid–base disorder has been classified, it is necessary to establish the *cause* of the disturbance. In most instances, the clinical history and examination will yield sufficient information for this to be done, but often in metabolic disorders further biochemical investigation is required.

METABOLIC ACIDOSIS

Etiology

Metabolic acidosis results from either an accumulation of fixed acid or a loss of extracellular buffers (when, in effect, the amount of fixed acid that is produced exceeds the buffering capacity). Four ***mechanisms*** are possible:

1. Impaired renal excretion of fixed acid
2. Overproduction (or administration) of fixed acid
3. Primary HCO_3^- loss, renal or gastrointestinal
4. Secondary HCO_3^- loss, due to elevated Cl^- levels

The first two of these mechanisms are characterized by anion (usually organic) accumulation and therefore the anion gap is *increased.* In the latter two, Cl^- is substituted for HCO_3^-, and so the anion gap remains *normal.*

Differential Diagnosis

Figure 6–10 schematically outlines a ***practical approach*** to differential diagnosis. The initial step is to ascertain, by measuring serum electrolytes, whether the anion gap is increased or normal.

Increased Anion Gap. Four classes of clinical disorders may be responsible for the accumulation of fixed acid:

1. ***Renal failure.*** The diminished number of functioning renal tubular cells results in a proportionate impairment of acid excretion, resulting in the accumulation of H^+ and its accompanying anions, which are mostly sulfates, phosphates, and other products of protein metabolism. Renal function tests will establish the diagnosis. Because the kidney has a substantial reserve, significant acidosis is usually not seen until renal function is less than 10% of normal. Supporting laboratory findings—hyperkalemia, anemia—can often be elicited. The development of this disorder may take months or years.
2. ***Diabetic ketoacidosis*** is a classic example of metabolic acidosis. Altered carbohydrate and lipid metabolism results in a great increase in ketoacid production, which overwhelms buffering and excretory mechanisms. Diagnosis

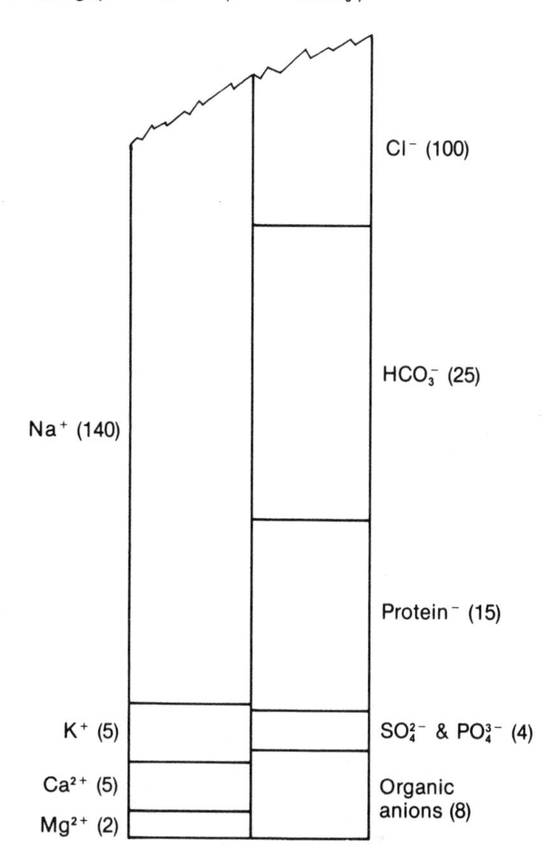

FIGURE 6–9. Scalar representation of the main ionic constituents of serum. (Values in parentheses are mEq/liter.)

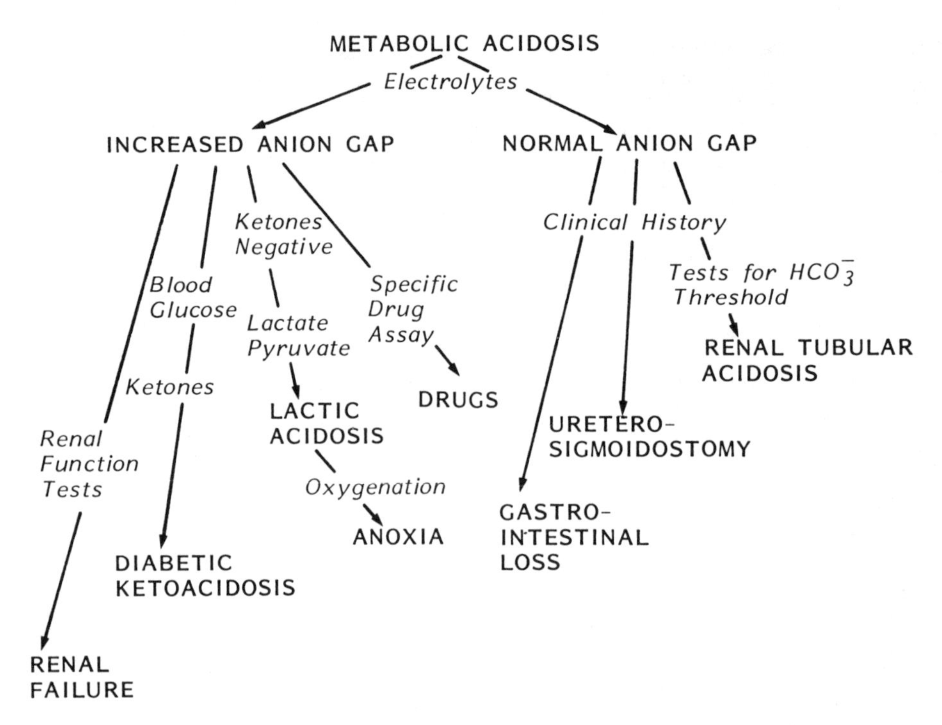

FIGURE 6–10. An approach to the differential diagnosis of metabolic acidosis.

requires the measurement of serum glucose and the demonstration of increased serum or urinary ketones (to exclude lactic acidosis). Infrequently, the ingestion of ethyl alcohol can lead to a ketoacidosis. In starvation, ketoacid production is increased and leads to ketosis, but buffering is sufficient to prevent ketoacidosis.

TABLE 6–7. Classification of the Causes of Lactic Acidosis

A. Increased lactate/pyruvate ratio
  1. Anoxia
    a. Acute respiratory failure
    b. Inadequate perfusion (shock)
    c. Anemia
  2. Alcohol
  3. Phenformin
  4. Tumors
  5. Idiopathic

B. Decreased lactate/pyruvate ratio
  1. Thiamine deficiency

C. Normal lactate/pyruvate ratio
  1. Diabetes mellitus
  2. Glycogen storage disease

Reference values (venous blood):
Lactate = 0.6–2.2 mmol/L (5–20 mg/dL)
Pyruvate = 0.04–0.10 mmol/L (0.3–0.9 mg/dL)
Lactate/pyruvate ratio = 6/1–10/1

3. ***Lactic acidosis*** can result from many causes, as listed in Table 6–7. Clinically, its presence may be presumed when a metabolic acidosis with an increased anion gap is found in the absence of ketosis or drug ingestion (with the exception of phenformin, a known cause of lactic acidosis). The diagnosis is confirmed by the measurement of plasma lactate and, if applicable, pyruvate.

Lactic acidoses fall into three *categories,* distinguished by the ***lactate/pyruvate ratio.*** This ratio is thought to be a reflection of the $NADH/NAD^+$ ratio, which, at least in part, is dependent on oxidative phosphorylation. All of these metabolic components are responsive to the availability of oxygen relative to the rate of glycolysis (i.e., generation of pyruvate).

The most common type of lactic acidosis is that characterized by an *increased* lactate/pyruvate ratio, resulting from impaired oxidative phosphorylation, due either to anoxia or to mitochondrial dysfunction. The mechanism whereby lactic acidosis may occur with a *normal* or *decreased* lactate/pyruvate ratio is not understood. In acute anoxic states (respiratory failure, cardiac arrest, etc.) lactic acidosis can develop within minutes and is life threatening. When caused by drugs, it develops in hours or days and has a mortality of 40% to 60%. In other disorders, lactic acidosis may exist as a chronic condition.

The specific procedures required to establish the various causes of lactic acidosis will not be covered here.

4. ***Drugs*** or other chemicals may cause metabolic acidosis. The mechanism common to all is the production of acidic metabolites. For example, acetylsalicylic acid → salicylic acid; methanol → formic acid; isopropyl alcohol → acetic acid; and ethylene glycol → oxalic acid. In many instances, either the drug or one of its metabolites also interferes with normal metabolic pathways, thereby retarding its own catabolism. Biochemical detection of the offending substance or one of its metabolites may confirm the cause, although the history is often adequate to establish a diagnosis.

Normal Anion Gap. If a metabolic acidosis is associated with a *normal* anion gap, it follows that the reduction in HCO_3^- is offset by an equivalent increase in Cl^-. This is a *hyperchloremic acidosis.*

A primary loss of HCO_3^- results from the loss of bicarbonate-rich fluid or from a defect in the renal reabsorptive mechanisms for HCO_3^-. To achieve electroneutrality Cl^- is retained by the kidney, leading to hyperchloremia, and so a normal anion gap is maintained.

Conversely, if the Cl^- concentration is increased by excessive saline administration, there is a secondary suppression of HCO_3^- reabsorption. The disorders may be classified into three groups, distinguishable primarily by the clinical history.

1. ***Gastrointestinal*** loss of bicarbonate-rich fluids is probably the commonest cause of hyperchloremic acidosis. The diagnosis is based on historical data. The three principal routes of HCO_3^- loss are
   a. *Vomiting,* secondary to intestinal obstruction, in which the vomitus contains duodenal contents—that is, pancreatic juice with high HCO_3^- content
   b. *Diarrhea* (especially in infants), in which the passage of intestinal contents is so rapid that pancreatic HCO_3^- is lost in the stool
   c. *Fistulas,* from the gut or pancreatic–biliary system, resulting in pancreatic HCO_3^- loss. This last category includes any tube drainage of intestinal contents.
2. ***Renal tubular acidosis*** (RTA) is an uncommon condition that may be inherited or acquired. In its most common form, cells of the proximal tubule are unable to reabsorb an appropriate amount of the filtered HCO_3^-, and in its stead Cl^- is reabsorbed, raising the plasma chloride concentration. In a less common form, HCO_3^- reabsorption (or regeneration) in the distal tubule is impaired. The diagnosis of RTA is suggested by a metabolic acidosis with a urine that is either alkaline or fails to acidify with an acid load. A more specific test is the demonstration of a low HCO_3^- threshold during the intravenous infusion of bicarbonate.
3. ***Elevated chloride*** levels will suppress renal HCO_3^- reabsorption. Such a disturbance in Cl^- may arise endogenously, as in a ureterosigmoidostomy, where the bowel mucosa in contact with urine selectively reabsorbs Cl^- (and Na^+), raising the plasma level of Cl^-. Exogenous causes are the ingestion or infusion of substances in which the ratio of Cl^- to Na^+ is greater than that normally present in plasma. Thus, sodium chloride administration over a prolonged period may result in mild acidosis.

Physiologic Effects

The effects of an increased H^+ concentration appear to be primarily *neuromuscular.* The usual manifestation is increased irritability with a hazard of cardiac arrythmias, which may progress to cardiac arrest. There is also depression of consciousness, which can progress to convulsions, coma, and death. Little is known about the effects of acidosis on intracellular metabolic processes in other tissues; nor do we know whether an extracellular acidosis is always accompanied by a proportionate drop in intracellular pH. One well-recognized metabolic effect is *insulin resistance* caused by ketoacids, which presents a problem in the treatment of hyperglycemic coma.

The *compensatory* response to metabolic acidosis is an increase in the respiratory rate. The increased H^+ concentration acts as a powerful stimulant of the respiratory center. The deep, rapid, gasping respiratory pattern (*Kussmaul breathing*) lowers pCO_2, thereby ameliorating the acidemia. Finally, it is important to remember that the symptoms displayed by the patient are a *combination* of the effects of the acidemia and the effects of the primary disorder causing the acidosis.

Worthy of note is the role played by the *blood–brain barrier,* which is freely permeable to many uncharged particles such as CO_2, but much less permeable to ions such as HCO_3^-. As the acidosis develops, equilibrium is established between blood and cerebrospinal fluid (CSF), so that HCO_3^-, CO_2, and H^+ concentrations are comparable in the two fluids. If the acidosis is corrected rapidly (1 to 2 hours) by the administration of bicarbonate, the increased respiratory drive may be diminished and pCO_2 will rise to normal levels. This soon equilibrates across the blood–brain barrier, thereby raising the pCO_2 in the CSF. However, the CSF bicarbonate may remain low since its equilibration with blood is slow, and so, for a time, the CSF acidosis is worsened.

METABOLIC ALKALOSIS

Etiology

The elevated HCO_3^- that characterizes metabolic alkalosis can result either from HCO_3^- administration or from improper renal handling of this metabolite. Plasma HCO_3^- concentration is controlled normally by renal regulatory mechanisms, which in turn are influenced by the anion content of plasma (Cl^- being most plentiful), as well as by hormones with mineralocorticoid activity and by plasma K^+. The theoretical possibility of impaired fixed acid production is not known to occur as a cause of metabolic alkalosis.

Differential Diagnosis

The most important diagnostic criteria are obtained from the clinical history, biochemical procedures often being only confirmatory. In virtually all forms of metabolic alkalosis, the serum Cl^- concentration is depressed; thus, the term *chloride depletion* refers to disorders in which the loss of Cl^- is primary, rather than to the observation that serum Cl^- is low. It is generally useful to assess the status of other serum electrolytes. A measurement of urinary Cl^- will yield some information on the appropriateness of the renal handling of this anion; 10 mmol/L is taken as the approximate boundary between high and low excretion. This test, however, rarely yields more information than that available from the history.

The causes of metabolic alkalosis may be *classified,* according to mechanism, into five categories, as shown in Figure 6–11.

1. ***Chloride depletion.*** If Cl^- concentration is diminished, less than normal amounts are available as a complementary anion for Na^+ and K^+ reabsorption in the proximal tubule.

FIGURE 6–11. An approach to the differential diagnosis of metabolic alkalosis.

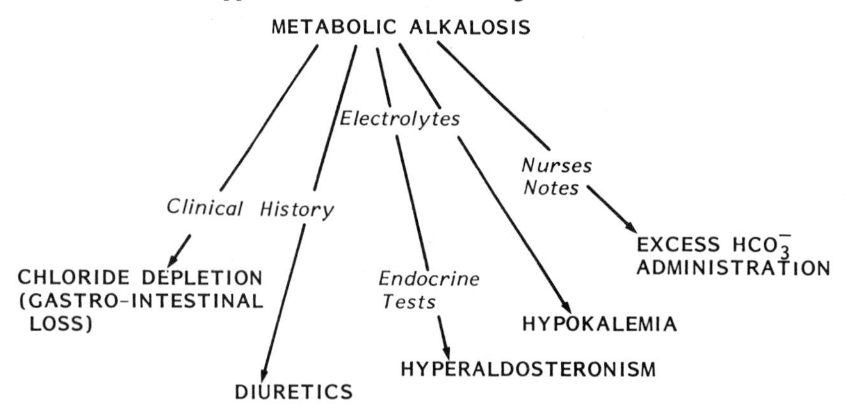

In order to maintain ionic balance, excessive amounts of HCO_3^- are reabsorbed, giving rise to a metabolic alkalosis. Such Cl^- depletion usually results from loss of fluid that has a Cl^-/Na^+ ratio greater than that normally present in plasma. In practice, this occurs with the loss of gastrointestinal fluids, most commonly gastric juice, in which the concentration of Cl^- may be twice as great as that of Na^+. Prolonged vomiting or the aspiration of gastric contents by gastric tube are the most frequent routes, but Cl^--losing diarrhea or villous adenoma of the colon may also be responsible. The acid content of lost gastrointestinal fluid is rapidly replaced by normal metabolic production of H^+ and does not appear to be responsible for the alkalosis, which can be readily reversed by the administration of chloride. This category of disorders is sometimes referred to as *chloride-responsive alkalosis.* The urinary Cl^- tends to be low (in urine, the ratio of Cl^- to Na^+ is normally greater than that in plasma).

2. ***Hypokalemia.*** Total body K^+ depletion results in an increased secretion of H^+ by the kidney. The reason for this is not completely understood, but the renal mechanisms for the secretion of H^+ and K^+ appear to be linked, so that a deficit in one ion results in an increased secretion of the other. The excessive secretion of H^+ then results in increased formation of HCO_3^- by the renal tubular cells.

   It is also postulated that a diminished level of extracellular K^+ causes an outward shift of intracellular K^+, and again, in order to maintain electroneutrality, other cations, including H^+, enter the cells, thereby facilitating the development of the alkalosis. Usually, only prolonged and severe hypokalemia will result in a significant alkalosis.
3. ***Hyperaldosteronism.*** Aldosterone is a potent stimulant of the renal mechanism that secretes K^+ and H^+ ions, depleting the body of the former and resulting in the excessive generation of HCO_3^- during excretion of the latter. Although excess aldosterone is the classic example, other hormones with mineralocorticoid activity, whether endogenous or exogenous, are included in this category. In particular, Cushing's syndrome due to ectopic ACTH production usually exhibits a hypokalemic metabolic alkalosis, which is more severe than that resulting from other causes of this syndrome.
4. ***Diuretics.*** Diuretics that impair the reabsorption of K^+ (as well as that of Na^+) may cause not only hypokalemia but also a metabolic alkalosis. The mechanism for this is probably similar to that described under hypokalemia.
5. Excess ***bicarbonate administration.*** Alkaline material ingested in an ordinary diet does not result in alkalosis because it is neutralized by the normal metabolic production of fixed acid. Nonetheless, prolonged antacid therapy may give rise to a metabolic alkalosis (milk–alkali syndrome). It is likely that the malabsorption of other ions such as Ca^{2+} plays a role in this. A more frequent and important cause of metabolic alkalosis is the rapid administration of excessive amounts of bicarbonate solution during the treatment of metabolic acidosis.

Physiologic Effects

The normal compensatory response to a metabolic alkalosis is a fall in respiratory drive, resulting from the decreased H^+ concentration at the respiratory center. This effect is usually not of great magnitude, but serves to elevate pCO_2 slightly and thus to ameliorate the rise in pH. Symptoms are generally unremarkable, being those of general malaise. In most circumstances the symptoms of the primary disorder are predominant, and the effects of electrolyte disturbances which usually accompany this condition should be considered.

RESPIRATORY ACIDOSIS

Etiology

Respiratory acidosis is an elevation of pCO_2 above the expected physiologic limit (hypercapnia). Usually this is 42 mm Hg, but it can be higher or lower depending on compensation that may have taken place in response to a metabolic disorder. Normally, the respiratory system, which is the sole route of CO_2 excretion, can deal easily with all the CO_2 that is produced metabolically. There is no known disorder characterized by *overproduction* of CO_2. Thus, all causes of respiratory acidosis have in common a defect in the *excretion* of CO_2.

Differential Diagnosis

Biochemical tests play only a limited role in the differentiation of causes of respiratory acidosis. Clinical and radiologic data and pulmonary function tests are most helpful. The causes of impaired respiratory function are numerous; a brief *classification* of the mechanisms follows:

1. ***Impaired respiratory drive.*** A number of central nervous system disorders may affect the respiratory center or its regulatory mechanisms, resulting in inadequate respiratory drive. Such disorders may be morphologic (vascular disorders, trauma, tumor), functional (epilepsy), or metabolic. The last category includes hypoglycemia, hypoxia, and drug ingestion, in which biochemical investigations may be important.
2. ***Impaired respiratory mechanics.*** Inability to move the chest wall to ventilate the lungs may result from disorders affecting the peripheral nerves (poliomyelitis), respiratory muscles (dystrophy), or ribs (trauma causing flail chest). Air in the pleural cavity (pneumothorax) will also limit inflation of the lungs.
3. ***Airway obstruction.*** Large airways may be obstructed by tumors, food, vomitus, or other foreign objects. Small airway obstruction is generally due to bronchospasm (asthma, chronic obstructive lung disease) or, less often, to the accumulation of mucus (cystic fibrosis). An important cause of small airway and alveolar obstruction is infection (pneumonia), which results in the production of an inflammatory exudate that may fill small air spaces.
4. ***Diffusion defect.*** Most frequently such defects arise because of alveolar destruction (emphysema), which reduces the surface area available for gas exchange. Thickening of the alveolar membrane may also result from diseases causing cellular proliferation. The accumulation of fluid in the alveoli (e.g., in congestive heart failure) also impairs diffusion, and by increasing surface tension results in alveolar collapse.
5. ***Abnormal ventilation/perfusion ratio.*** In destructive lung diseases, such as emphysema, portions of the capillary bed become isolated from alveolar ventilation, so that no gas exchange takes place. Although the defect in oxygenation is initially more severe, this abnormality does eventually result in CO_2 retention.
6. ***Circulatory impairment.*** Stagnation of blood peripherally (e.g., in limbs) will lead to a local 'respiratory,' as well as metabolic, acidosis. Cessation of the circulation proper (cardiac arrest) results in a rapid, catastrophic respiratory and metabolic acidosis.

Oxygenation

In general, all of the aforementioned causes are associated with impaired oxygenation of blood, which precedes the onset of hypercapnia. The reason for this is that the alveolar–arterial gas tension gradient is relatively much greater for CO_2 than for O_2. Therefore, in the presence of any defect, it is much easier to compensate for loss of CO_2 than for loss of O_2 exchange capacity.

Physiologic Effects

The normal *compensatory* response to a respiratory acidosis is the *renal* retention of HCO_3^-, which occurs over a few days. It is important to remember that it also takes some time for this to reverse, so that a patient whose compensated respiratory acidosis is corrected rapidly will be left, for a short time, with a metabolic alkalosis. The *symptoms* of a mild CO_2 excess are nonspecific and usually include increased neuromuscular irritability and the sensation of dyspnea. Marked elevations of pCO_2 (greater than 80 mm Hg) result in depression of the central nervous system, including the respiratory center (CO_2 narcosis). This eventually becomes a self-perpetuating process that terminates in death. The hypoxia that nearly always accompanies respiratory acidosis contributes significantly to this severe effect.

RESPIRATORY ALKALOSIS

Etiology

Respiratory alkalosis is a reduction of pCO_2 below the physiologically expected limit. Usually this is below 38 mm Hg, but it can be higher or lower, depending on compensation in response to a metabolic acid–base disorder. The production of CO_2 is a continuous metabolic process, and there are no known hypocapnic states caused by its impairment. Thus, respiratory alkalosis is always the result of excessive pulmonary excretion of CO_2, due to *hyperventilation.*

Differential Diagnosis

Differentiation of the various causes is not primarily a biochemical procedure, but rests mainly on the clinical history. Hydrogen ion is a potent stimulator of the respiratory center, but because such stimulation is a compensatory mechanism, it is not a cause of respiratory alkalosis. Likewise, vigorous exercise will cause a slight decrease in pCO_2 levels, but this is a normal physiologic phenomenon. The causes of respiratory alkalosis, i.e., of hyperventilation, may be *classified* as follows:

1. ***Pharmacologic stimulation.*** Many drugs act on the respiratory center, but probably the commonest to cause hyperventilation is salicylate. In this class of disorders, drug assays may have diagnostic value.
2. ***Psychogenic factors.*** Emotional states such as hysteria or anxiety, whether occurring in normal individuals as a response to stress or in psychotics as part of the disorder, often cause a feeling of stifling that leads to hyperventilation.
3. ***Hypoxia.*** Subnormal levels of pO_2 stimulate the respiratory center, increasing respiratory effort, which can lead to hypocapnia. As discussed in the previous section, there are quantitative differences between the alveolar–arterial gas tension gradients of O_2 and CO_2, so that mild impairment of pulmonary function may be associated with hypocapnia, while advanced disease is usually accompanied by hypercapnia.
4. ***Mechanical ventilation.*** Patients on artificial ventilators can be made hypocapnic if the respiratory rate or tidal volume is too great.

Physiologic Effects

Arterial CO_2 tension has a regulatory effect on *cerebral perfusion.* Subnormal pCO_2 levels cause constriction of small vessels in the brain, reducing the blood supply, which is perceived by the patient as light-headedness and, in the extreme case, may lead to syncope. *Peripherally,* the combination of low pCO_2 and alkalosis can cause a tingling sensation in the digits and around the mouth. This is due, in part, to the decreased level of ionized calcium, which results from increased protein binding at the high pH. Rarely, it may result in tetany.

The ***compensatory response*** to a respiratory alkalosis is an increase in renal HCO_3^- excretion, which lowers plasma levels. This adjustment takes several hours to days and is often incomplete, because the alkalosis is present for only a short period. As a rule, alkalosis is not a life-threatening condition. In metabolic alkalosis, the continuing, normal production of fixed acid has an ameliorating effect. The renal threshold for HCO_3^- (normally about 30 mmol/L) also has a limiting effect. However, it is very important to be mindful of the loss of other metabolites (especially electrolytes) in those disorders in which body fluid is being lost (e.g., vomiting). Therapy is usually aimed at correcting the primary problem, and if necessary replacing lost Cl^- (or K^+). Because it is produced endogenously, acid almost never has to be administered.

The *severity* of respiratory alkalosis is *limited* by the maximal respiratory rate that is physically possible. Usually, pCO_2 cannot be lowered much below 20 mm Hg. Furthermore, in psychogenic disorders, the vasoconstriction of cerebral vessels eventually results in syncope, which automatically reduces the respiratory rate to normal. Because in most instances the disorder is short-lived, no acid–base therapy is required. However, rebreathing exhaled air (breathing into a bag) may be used to correct symptoms and reduce the patient's discomfort and anxiety. (Remember that pO_2 falls when this is done.)

As a general principle, in the treatment of respiratory acid–base disorders, one must be mindful of the metabolic compensation that has usually taken place. Thus, rapid correction of the *respiratory* problem may leave the patient with a *metabolic* acid–base disorder.

THERAPEUTIC IMPLICATIONS

In practice, ***acidoses*** (metabolic or respiratory) are the life-threatening acid–base disorders. This appears to be because increased concentrations of H^+ can severely impair metabolic functions, and because the disease processes causing the acidoses are usually accompanied by other major metabolic disturbances. The first principle in the *treatment* of metabolic acidosis is, wherever possible, to treat the primary disorder in order to eliminate the excessive production of fixed acid. Accumulated acid may also be neutralized by the administration of appropriate buffers, usually bicarbonate, which also helps to restore the buffering capacity of extracellular fluids. Bicarbonate has the advantage over other buffers of combining with H^+ to form H_2O and CO_2 (which can be excreted by the respiratory system). It must be remembered, however, that this procedure does not remove other ions and metabolites which may also be present in excessive concentration, and which may require dialysis for their clearance.

Because assessment of the acid–base disorder is carried out on blood, it is important to be aware of the possibility of a severe ***intracellular*** disturbance. This can be difficult to judge, and thus the usual approach is to administer a *trial dose* of the alkali over a few hours and then to base further treatment on the patient's response. Allowance must be made for acid–base changes in the cerebrospinal fluid and the time (several hours) required for equilibration of metabolites across the blood–brain barrier. The result of treatment of the primary disorder must also be considered. For example, the revival of glucose metabolism during the treatment of ketoacidosis (by insulin administration) will greatly reduce the amount of ketoacid and cause potassium to shift into the cells.

In ***respiratory*** acidosis, the problem is an elevated pCO_2, which, except in very acute disorders, represents a readjustment of the steady state required to excrete metabolically generated CO_2. Although the acidemia can be corrected with bicarbonate administration, this does not correct the primary disorder or the hypercapnia. Doing so requires a reduction of pCO_2 levels by improved gas exchange.

CONCLUSIONS

Disorders of fluid, electrolyte, and hydrogen ion metabolism represent disturbances of chemical anatomy. The concept of *predictive values* for laboratory parameters is not readily applicable here, because these disorders are *definitions of status* rather than specific disease processes. For example, because hyponatremia is defined as a diminished concentration of serum Na^+, a serum sodium < 136 mmol/L is 100% specific and 100% sensitive for hyponatremia. Diagnosis therefore requires an attempt to determine predictive values for the various causes of hyponatremia.

When a disease process of this nature has caused a distortion of chemical anatomy, the disturbance usually represents a *new equilibrium state* in which metabolic processes continue to function. Therapy then, in effect, will cause a change in this new equilibrium, and the consequences of this physiologic 'insult' must be considered before treatment is undertaken.

It is essential to appreciate that each laboratory measurement is imprecise to some degree (*analytic error*), and all biochemical measurements for a given patient are subject to physiologic variation (*intraindividual variation*), which usually increases in disease. Consequently, the confidence that can be placed in the results of laboratory studies in both diagnosis and monitoring therapy is limited.

SUGGESTED READING

MORGAN B (ed): Electrolyte disorders. Clin Endocrinol Metabol, vol 14, July 1984

ROSE BD: Clinical Physiology of Acid–Base and Electrolyte Disorders, 2nd ed. New York, McGraw-Hill, 1984

SCHADE DS (ed): Metabolic acidosis. Clin Endocrinol Metabol, vol 12, July 1983

WEIL WB, BAILIE MD: Fluid and Electrolyte Metabolism in Infants and Children. New York, Grune & Stratton, 1978

7

Respiratory Disorders

Colin R. Woolf

When you can measure what you are talking about and express it in numbers, you know something about it; but, when you cannot measure it, when you cannot express it in numbers, your knowledge is of a meagre and unsatisfactory kind.
—Lord Kelvin.

In the investigation and management of respiratory disorders, arterial blood gas analysis is the link between the clinical biochemist and the practicing physician. Reliable measurements of blood pH, arterial carbon dioxide tension ($apCO_2$*), and arterial oxygen tension (apO_2*) should be available in all hospitals with more than 50 beds. A 24-hour blood gas measurement service is needed if seriously ill patients are treated, and the results should be reported within 30 minutes of the laboratory's receiving the specimen. Excellent quality control is necessary, as the reliability of the results must never be in question. The *arterial blood* sample should be free of air and the syringe should be placed on ice during transport to the laboratory. *Capillary blood* is satisfactory provided there is good vasodilatation and good blood flow through the sample site. Venous blood cannot be used for blood gas estimations.

Although abnormalities of blood gases may occur in all respiratory disorders, there are only a few underlying pathophysiologic mechanisms.

THE LUNGS: STRUCTURE AND FUNCTION

The respiratory system (Fig. 7–1) commences at the nose and mouth, which lead to the pharynx, then to the larynx (containing the vocal cords) and the *trachea,* which extends about 10 cm in length. The trachea terminates at the level of the fourth and fifth thoracic vertebrae, dividing into the right and left main *bronchi.* The bronchi divide into smaller and smaller air tubes, the smallest being the *bronchioles.* The terminal bronchiole opens into a lung unit, the *acinus,* or lobule, surrounded by a fibrous tissue septum. The terminal bronchiole divides into respiratory bronchioles, so named because air sacs or alveoli are found in their walls. The respiratory bronchioles finally enter collections of *alveoli.*

The right and left *pulmonary arteries* are branches of the main pulmonary artery, which arises from the right ventricle. The arteries divide

* The American Thoracic Society recommends the abbreviation Pa_{CO_2} for arterial blood carbon dioxide tension (i.e., partial pressure), and Pa_{O_2} for arterial blood oxygen tension (partial pressure). Because P also stands for plasma and Pa for pascal, we prefer the forms used in *The SI for the Health Professions,* a publication of the World Health Organization, Geneva, 1977.

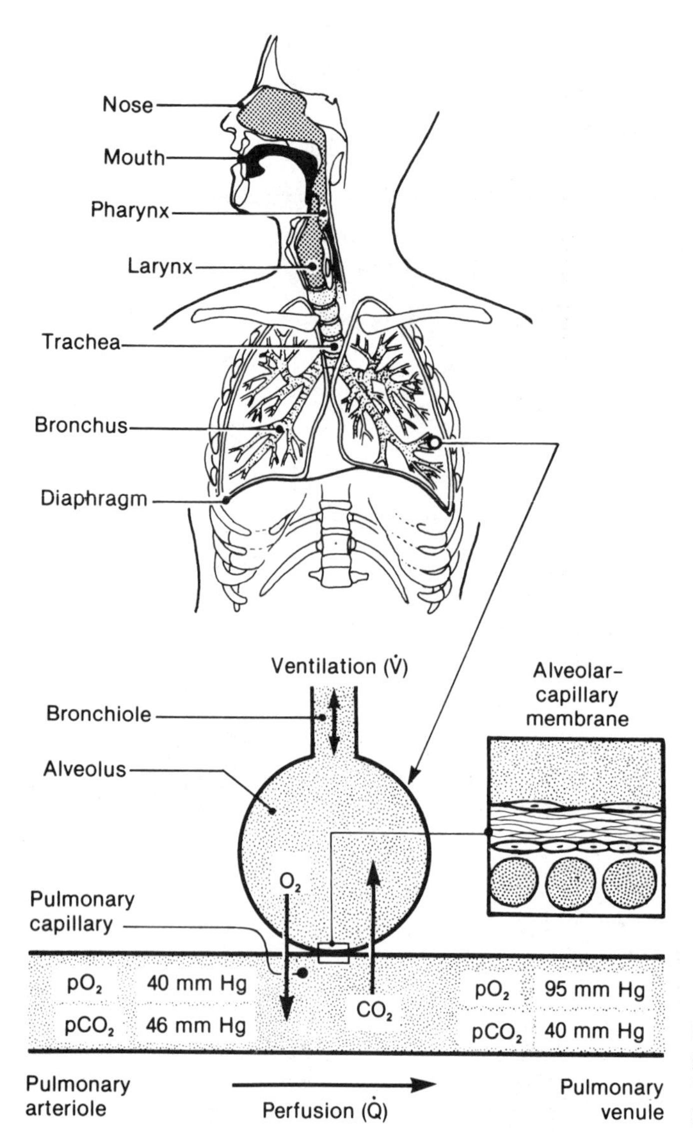

FIGURE 7–1. The respiratory system, with diagram of the relationship between the lung alveolus and the pulmonary capillary. Partial pressures are mm Hg.

into smaller and smaller branches, then into arterioles, and finally become the pulmonary *capillaries.* Pulmonary capillaries are in immediate contact with alveolar walls. The *alveolar capillary membrane* consists of a layer of flattened alveolar cells, a very little connective tissue, and the pulmonary capillary endothelial cells. This thin membrane constitutes only a slight barrier to oxygen passing from the alveolus to the pulmonary capillary blood or carbon dioxide passing from the pulmonary capillary blood into the alveolus. The pulmonary arteries contain venous blood with a low oxygen tension (pO_2 approximately 40 mm Hg) and high carbon dioxide pressure (pCO_2 approximately 46 mm Hg). After the blood has passed the alveoli and entered the pulmonary veins it has a pO_2 of about 95 mm Hg and a pCO_2 of about 40 mm Hg. This blood passes to the left side of the heart and is pumped from the left ventricle to the aorta and thence throughout the systemic arterial blood system. Arterial blood gas analysis is performed on blood removed from the radial, brachial, or femoral arteries.

During ***inspiration*** the diaphragm and the inspiratory intercostal muscles contract. Air is drawn in through the mouth or nose, through the trachea and bronchial tree, and into the alveoli. The inspired volume of a single breath, termed the *tidal volume,* is usually 500 to 600 mL. Part of this volume fills the nose, pharynx, larynx, trachea, and airways down to the terminal bronchioles; this gas is not in contact with pulmonary capillary blood and is termed the *anatomic dead space.* Its volume (in milliliters) is approximately twice the lean body weight of the individual (in kg); that is, a person weighing 75 kg will have an anatomic dead space of about 150 mL. Therefore, with a normal breath of 500 mL, the volume available for gas exchange will be about 350 mL.

The total volume of gas moved in or out of the lungs in one minute is the *minute ventilation* ($\dot{V}$) and is normally about 5 liters (3 to 4 $L/min/m^2$ body surface area). The amount of blood (cardiac output) perfusing the alveoli ($\dot{Q}$) is normally also 3 to 4 $L/min/m^2$. The relationship between ventilation and perfusion of the alveoli is expressed as the *ventilation–perfusion ratio.* This $\dot{V}/\dot{Q}$ ratio is the minute ventilation divided by the cardiac output and is normally 1. If the $\dot{V}/\dot{Q}$ ratio is >1.0, then ventilation is in excess of perfusion and some of the ventilation is being wasted. If the $\dot{V}/\dot{Q}$ ratio is <1.0, then there is more blood passing the alveoli than can be fully oxygenated; in this case blood with a low oxygen content passes to the pulmonary veins; this is termed intrapulmonary venoarterial admixture, or *shunting.* These types of mismatch between ventilation and perfusion may lead to serious abnormalities in the arterial blood gases.

ARTERIAL BLOOD CARBON DIOXIDE TENSION

The *reference range* of $apCO_2$ (Pa_{CO_2})* is at most 35 to 45 mm Hg (4.6 to 5.9 kPa [kilopascals]). Changes are easy to interpret:

1. If the $apCO_2$ is greater than 45 mm Hg there is *hypoventilation;* that is, the patient is not producing sufficient alveolar ventilation to get rid of the carbon dioxide produced by the body and so maintain a steady state of 35 to 45 mm Hg. The ***effective*** or ***alveolar ventilation*** is the amount of gas moved in or out of the alveoli each minute; it can be calculated by the following formula:

   $$\text{Alveolar ventilation} = \text{respiratory rate} \times (\text{tidal volume } \textit{minus} \text{ physiologic dead space})$$

   The *tidal volume* times the *respiratory rate* gives the ***minute ventilation,*** and from this must be subtracted the anatomic dead space. But air may enter alveoli that are not supplied by pulmonary capillary blood because of disease. This volume also is not available for gas exchange and should be added to the anatomic dead space to give the total ***'physiologic' dead space.*** Thus ***alveolar ventilation*** (normally 2.0 to 2.5 $L/min/m^2$) is the amount of gas that enters alveoli supplied by pulmonary capillary blood and is able to take part in gas exchange. If alveolar ventilation is low the $apCO_2$ will be high.
2. If the $apCO_2$ is less than 35 mm Hg there is *hyperventilation;* that is, the alveolar ventilation is high and carbon dioxide is breathed out faster than it is produced by the body.
3. If the $apCO_2$ is 35 to 45 mm Hg, but clinically the patient has a high minute ventilation (this is best measured, but can be approximated by looking at the patient's breathing pattern), then a portion of the gas breathed must be entering parts of the lung where there is no contact with pulmonary capillary blood (an increased *physiologic dead space* must be present). Thus, even a 'normal' $apCO_2$ may be indicative of a severe abnormality of pulmonary function if there is associated high minute ventilation. It is important, therefore, in the interpretation of an $apCO_2$ result, to have some knowledge of *minute ventilation.*

* See footnote at beginning of this chapter.

ARTERIAL BLOOD ACID–BASE RELATIONSHIPS SIMPLIFIED

The acidity of the blood is indicated by pH (reference range 7.36 to 7.45) or hydrogen ion concentration (44–36 nmol/L). If the pH is less than 7.36, *acidosis* is present; if above 7.45, then *alkalosis* is present.

The ***Henderson–Hasselbalch equation*** relates pH to blood bicarbonate and carbon dioxide tension. The original equation is

$$pH = pK + \log \frac{(HCO_3^-)}{(H_2CO_3)}$$

Since H_2CO_3 is in equilibrium with dissolved carbon dioxide, and dissolved carbon dioxide is in equilibrium with pCO_2, the equation can be simplified to

$$pH \text{ is proportional to } (\alpha) \frac{HCO_3^-}{pCO_2}$$

The kidney controls H^+ ion balance by retaining or excreting HCO_3^-, and this is closely related to and affected by the lungs, which increase or decrease CO_2 excretion by altering minute ventilation.

As a *clinical* concept the ultimate expression then becomes

$$pH \ \alpha \ \frac{\text{renal handling of } HCO_3^-}{\text{lung handling of } CO_2}$$

Using these simplified equations one can resolve any type of acid–base abnormality. It is essential to keep in mind that carbon dioxide tension can be affected very rapidly by changes in ventilation (within minutes), but bicarbonate is excreted or retained by the kidney only over many hours.

Respiratory Situations

Consider the *respiratory* examples given in Table 7–1:

1. There is ***acute hypoventilation:*** the $apCO_2$ rises, the denominator of the equation is increased, and pH falls. There is no time for bicarbonate to increase as a compensatory mechanism because of the slow action of the kidney. The patient now has an acute *respiratory acidosis.*
2. There is ***acute hyperventilation:*** the denominator of the equation is decreased and pH rises. There is no time for a compensatory fall in bicarbonate because of the slow action of the kidney. The patient now has an acute *respiratory alkalosis.*
3. There is ***chronic hypoventilation:*** the $apCO_2$ rises and there is, at first, a fall in pH and a respiratory acidosis. However, if the patient chronically underventilates and $apCO_2$ rises slowly or stabilizes, then the kidney has time to retain bicarbonate, the numerator of the equation rises, and the pH moves back toward 7.36. It is unusual for the pH to be restored to within the reference range, so that finally there is an almost but not completely compensated *respiratory acidosis.*
4. There is ***chronic hyperventilation:*** the $apCO_2$ is low, and the pH is high initially, but if chronic hyperventilation occurs there is time for bicarbonate to be excreted by the kidney. The numerator decreases and the pH falls toward 7.45; the final result is a compensated *respiratory alkalosis.*

Metabolic Situations

Metabolic conditions produce an effect opposite to that of respiratory conditions (Table 7–1):

5. ***Acute metabolic acidosis*** may occur: bicarbonate falls and there is a marked fall in pH. However, (6) the respiratory center in the medulla responds rapidly to an acid environment, and hyperventilation occurs. The $apCO_2$ quickly (within minutes) decreases and compensation results. If (7) the acidosis is very profound, even maximum hyperventilation cannot compensate and the patient then has a partially compensated metabolic acidosis. Comparable examples related to metabolic ***alkalosis*** are shown in Table 7–1 (8, 9, and 10).

Chronic metabolic changes will result in data similar to those of acute metabolic changes with respiratory compensation, because the ventilation changes occur acutely and maximally. However, if there is a chronic metabolic *acidosis,* compensation may initially be complete but subsequently fail, as respiratory hyperventilation may decrease in time owing to fatigue. Conversely, if there is a compensated metabolic *alkalosis* (e.g., from recurring vomiting), and sudden emotional or nervous stress induces hyperventilation, the alkalosis will suddenly become worse, at times sufficient to lower ionized calcium and induce tetany or convulsions. Treatment is to promote rebreathing of expired air until the $apCO_2$ rises again.

TABLE 7–1. Examples of Acid–Base Abnormalities

	$apCO_2$ (mm Hg)	HCO_3^- (mmol/L)	pH
Reference ranges	35–45	23–26	7.36–7.45
Respiratory situations			
1. Acute hypoventilation; acute respiratory acidosis	60	26	7.25
2. Acute hyperventilation; acute respiratory alkalosis	25	22	7.58
3. Chronic hypoventilation; compensated respiratory acidosis	60	33	7.35
4. Chronic hyperventilation; compensated respiratory alkalosis	25	17	7.45
Metabolic situations			
5. Acute metabolic acidosis; no respiratory compensation	40	12	7.10
6. Acute metabolic acidosis; respiratory compensation	22	12	7.36
7. Acute metabolic acidosis; partial respiratory compensation	18	6	7.12
8. Acute metabolic alkalosis; no respiratory compensation	36	35	7.60
9. Acute metabolic alkalosis; respiratory compensation	51	35	7.46
10. Acute metabolic alkalosis; partial respiratory compensation	64	48	7.50

A DEDUCTIVE APPROACH

When analyzing an acid–base situation from a set of blood gases look first at the $apCO_2$. If this is *high* there is hypoventilation. Look next at the pH. If this is only *slightly low* then the hypoventilation is chronic and compensation has occurred. The lower the pH the less the compensation and the more acute the hypoventilation.

If the $apCO_2$ is *low* there is hyperventilation. Look at the pH. If the pH is only *slightly elevated* then the hyperventilation is chronic and compensation has occurred. The higher the pH the more acute the hyperventilation.

If the $apCO_2$ is low, but the pH is also low, then there is a metabolic *acidosis.* If the $apCO_2$ is high and the pH is also high, there is a metabolic *alkalosis.*

From the *laboratory data* alone it may be very difficult to distinguish a compensated respiratory acidosis (Table 7–1, no. 3) from a compensated metabolic alkalosis (no. 9). In the former the pH is near the lower limit, and in the latter near the upper limit of the reference range, but exceptions to this rule occur. Similarly, it may be difficult to distinguish a compensated respiratory alkalosis (no. 4) from a compensated metabolic acidosis (no. 6). In the former the pH is slightly high and in the latter slightly low, but exceptions occur. The purpose of these examples is to show that some knowledge of the ***clinical situation*** is helpful and at times essential in the interpretation of an arterial blood gas result.

If a patient is known to have severe *lung disease* then a high $apCO_2$ is probably caused by hypoventilation from the primary pulmonary disease. If a patient is known to have good lungs but has been taking large quantities of bicarbonate, then the raised $apCO_2$ is probably a compensation for a metabolic alkalosis. Extremely complex situations may arise as the result of a mixture of metabolic and respiratory components, and this can be worked out only by having a detailed knowledge of the patient's clinical situation. Therefore, it is best for a laboratory to report only the results of blood gas measurements to the clinician, who will then interpret these results in the light of the patient's clinical condition. Interpretation of the results of arterial blood gas measurements, in the absence of a knowledge of clinical details, can result in major errors.

ARTERIAL BLOOD OXYGEN TENSION

The arterial blood oxygen tension (apO_2, Pa_{O_2}) decreases with age. In normal young people (under 25 years of age) the *reference range* is 95 to 100 mm Hg (12.5 to 13.2 kPa). An elderly person (over 70 years) might be normal with an apO_2 of only 85 mm Hg (10.6 kPa). The following formula (Sorbini et al, 1968) can be used to relate normal arterial blood oxygen tension to age:

$$apO_2 = 103.7 - (0.24 \times age)$$

LOW apO_2

An abnormally low arterial blood oxygen tension may occur under two pathophysiologic circumstances:

1. ***Hypoventilation***—If a patient hypoventilates, there will be a rise in $apCO_2$ and a rise in alveolar gas carbon dioxide tension. Carbon dioxide displaces oxygen in the alveolus, and there is a lower alveolar oxygen tension and consequently a lower apO_2.
2. ***Intrapulmonary venoarterial shunting***—If an alveolus is not ventilated but there is still pulmonary capillary blood passing it, this venous blood will not be exposed to oxygen and will pass to the arterial side of the pulmonary circulation unchanged. Blood with a low oxygen tension is mixed with blood with a normal oxygen tension coming from ventilated alveoli; the result is arterial blood with a lower than normal oxygen tension.

To decide whether a low apO_2 is due to one or the other or both of the above situations is simple, provided one knows both the $apCO_2$ and the inspired gas oxygen tension, from which the alveolar–arterial oxygen gradient can be calculated.

THE ALVEOLAR–ARTERIAL OXYGEN GRADIENT

To calculate the alveolar–arterial oxygen gradient, or difference, proceed as follows:

Step 1: Calculate the inspired gas oxygen tension (IpO_2) from the oxygen concentration (% O_2) of the inspired gas. This is done very easily by multiplying the percent oxygen by 7. The 7 is obtained by assuming a barometric pressure of 750 mm Hg and subtracting an assumed water vapor pressure at 37°C of 50 mm Hg, giving the figure of 700. Inspired gas oxygen (IpO_2) then becomes $(\%O_2/100) \times 700 = \%O_2 \times 7$. For example, room air is approximately 21% oxygen, so the IpO_2 would be $21 \times 7 = 147$ mm Hg. If the patient is receiving oxygen by mask and the inspired gas oxygen is 40%, then the IpO_2 will be 280 mm Hg.

Step 2: Divide the $apCO_2$ by 0.8 (the usual respiratory quotient) or multiply it by 1.25; the result will be the alveolar gas carbon dioxide tension ($ApCO_2$).

Step 3: Now, from the IpO_2 subtract the $ApCO_2$. This will give the alveolar oxygen tension (ApO_2). Thus, $ApO_2 = \%\ O_2 \times 7$ *minus* $apCO_2 \times 1.25$, which normally is $(21 \times 7) - (40 \times 1.25) = 97$ mm Hg.

Step 4: The difference between alveolar and arterial oxygen tensions ($ApO_2 - apO_2$) is known as the *alveolar–arterial oxygen gradient;* it should be less than 10 mm Hg when the patient is breathing room air.

If the alveolar–arterial oxygen gradient is ***<10 mm Hg*** and the $apCO_2$ is high, then the hypoxemia is entirely due to *hypoventilation.*

If the alveolar–arterial oxygen gradient is ***>10 mm Hg,*** then intrapulmonary venoarterial shunting must account for at least some of the hypoxemia. If the alveolar–arterial oxygen gradient is high and the $apCO_2$ is normal or low, then the hypoxemia is all due to *shunting.* If the alveolar–arterial oxygen gradient is high and there is also a high $apCO_2$, then the hypoxemia is due to a combination of intrapulmonary shunting and hypoventilation.

It should be noted that an important factor is the alveolar–arterial gradient, and the ability to make this calculation is essential if the cause of hypoxemia is to be determined.

Hypoxemia from intrapulmonary venoarterial ***shunting*** may also occur when alveoli have a decrease in ventilation in the presence of a relatively good pulmonary capillary perfusion. Under these circumstances there is a mismatch of ventilation ($\dot{V}$) and perfusion ($\dot{Q}$), resulting in a low $\dot{V}$ and a less affected $\dot{Q}$ (low $\dot{V}/\dot{Q}$ ratio). This causes similar but less marked shunting and hypoxemia than occurs when alveoli are completely unventilated. Under perfect conditions the normal $\dot{V}/\dot{Q}$ ratio is 1, because the ventilation of 3 to 4 $L/min/m^2$ matches the cardiac output of 3 to 4 $L/min/m^2$.

OXYGEN SATURATION

There is some difference of opinion as to whether the oxygen in blood should be expressed as oxygen saturation or oxygen tension. Oxygen *saturation* is the actual amount of oxygen carried in the hemoglobin of red cells, expressed as a percentage of the maximum oxygen that these cells can carry. A normal person has an oxygen saturation in the blood of 94% to 98%, but in older individuals it might fall as low as 90% and still be within reference limits.

The S-shaped (Sigmoid) oxygen dissociation curve for hemoglobin defines the relationship between blood oxygen ***tension*** and oxygen ***saturation*** (Fig. 7–2). If the oxygen tension is greater than 90 mm Hg, hemoglobin is usually fully saturated. For example, oxygen saturation will be 100% if the blood oxygen is at any level between 90 mm Hg and 600 mm Hg. Therefore, if a patient is breathing added oxygen, the most accurate way to estimate the effect of the oxygen therapy is to measure the apO_2. If a patient has a blood oxygen tension of less than 90 mm Hg, whether breathing air or added oxygen, then it makes little difference whether one is measuring oxygen tension or oxygen saturation, as changes in blood oxygen will be reflected accurately by either measurement.

Respiratory disorders may result in a shift of the oxygen dissociation curve due to a ***changed relationship*** between oxygen tension and oxygen saturation. The position of the curve is expressed as the oxygen tension required to cause 50% oxygen saturation of red cell hemoglobin (p50), normally 27 mm Hg apO_2. Acidosis, high $apCO_2$, high temperature, or increased red cell 2,3-diphosphoglycerate will all shift the oxygen dissociation curve to the right (high p50). Alkalosis, low $apCO_2$, low temperature, or decreased red cell 2,3-diphosphoglycerate shifts the curve to the left (low p50). If the oxygen dissociation curve is shifted to the right it will require a higher oxygen pressure to provide a satisfactory saturation of oxygen in the blood cells, and this oxygen must be taken up in the lungs. On the other hand, in the tissues this same situation results in a higher oxygen tension for a given hemoglobin saturation and oxygen is more easily released from the blood to the tissues. On the whole, a shift to the right has more advantages than disadvantages.

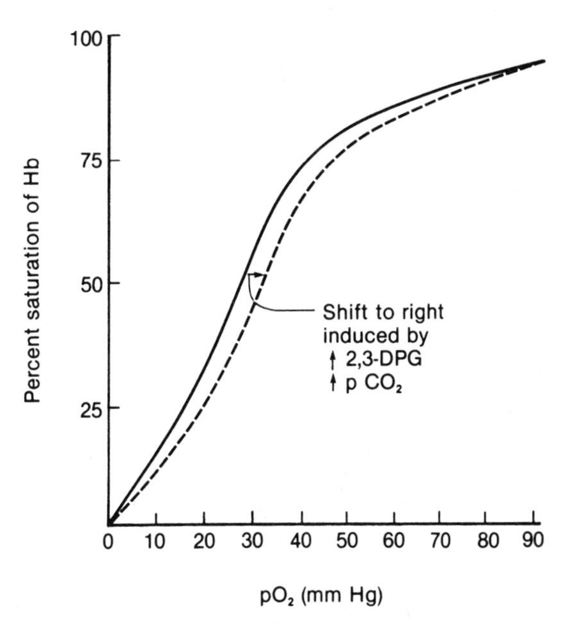

FIGURE 7–2. Oxygen saturation curve, showing the relationship between the partial pressure of oxygen (pO_2) in the blood and percent saturation of hemoglobin (Hb). Note the effect of increased 2,3-diphosphoglycerate (2,3-DPG) and pCO_2, making oxygen more easily available to the tissues.

Oxygen *saturation* has once more become popular, with the introduction of a noninvasive method of measuring it, using an oximeter that clips on the pinna of the ear (Saunders et al, 1976). Light shines through the ear from one side and is detected by photoelectric cells on the other. As the blood oxygen saturation falls (the ear becomes bluer), less light passes through the ear to be detected and recorded on a meter. This instrument allows a continuous readout and is particularly useful in measuring oxygen saturation during exercise, during changes in oxygen treatment (when the oxygen saturation is less than 100%) and for long-term monitoring such as making measurements during sleep.

ARTERIAL BLOOD GAS ABNORMALITIES IN RESPIRATORY FAILURE—CLINICAL SITUATIONS

As a general principle, when a clinical disorder causes ***hypoventilation,*** the $apCO_2$ is high and the blood oxygen is low. In situations causing ***intrapulmonary venoarterial shunting*** the arterial blood oxygen (apO_2) is low and the $apCO_2$ may also be low. This occurs when there are enough relatively normal alveoli that can be hyperventilated. With increased ventilation there is a progressive fall of carbon dioxide tension in well-functioning alveoli, and this compensates for the increased carbon dioxide in the venous blood that is shifted to the arterial side past alveoli that are underventilated. Because of the sigmoid oxygen dissociation curve, hyperventilation can raise oxygen saturation only to 100% and no further. This cannot compensate for the low oxygen tension and low oxygen saturation from the shunted blood. Thus, hyperventilation of normal alveoli can compensate for the high carbon dioxide, but not for the low oxygen, in the shunted blood.

However, if the disease condition *progresses* there will be fewer and fewer well-ventilated alveoli available, and finally there will be insufficient alveoli to compensate for the shunted blood carbon dioxide. Therefore, at late stages of a disease in which shunting of blood is the main problem, a rise in carbon dioxide tension will occur. There will then be hypoxemia due to hypoventilation in addition to hypoxemia due to venoarterial shunting, and the situation will become increasingly critical. All patients who start with moderate intrapulmonary venoarterial shunting as the main problem may at a late stage have both a high carbon dioxide tension and a low blood oxygen.

CAUSES AND MECHANISMS OF RESPIRATORY FAILURE

The following are brief descriptions of the pathophysiologic changes that occur with the various clinical causes of acute respiratory failure. The causes are arranged starting with problems in the brain and progressing through causes in the spinal cord, the peripheral nerves, the chest wall, and the lung itself.

Sedative Drug Overdose. A typical cause is a suicide attempt. Simple hypoventilation occurs, resulting from a decreased ability to breathe due to depression of the respiratory center. However, vomiting and aspiration of the stomach contents may complicate the situation, because this results in obstruction of alveoli, causing additional intrapulmonary venoarterial shunting.

Cerebral Edema. This may occur as the result of trauma or a brain operation. There is hypoventilation from depression of the respiratory center. However, multiple injuries may be present, including a crushed chest, and this may result in additional venoarterial shunting (see the following).

Spinal Cord and Peripheral Nerve Lesions. Spinal injury, especially at about the fifth cervical vertebra, and poliomyelitis, infective polyneuritis, and myasthenia gravis all result in weakness of the intercostal muscles, the diaphragm, or both. This leads to hypoventilation. As the tidal volume decreases there may also be a collapse of alveoli, resulting in additional venoarterial shunting.

Crushed Chest with Flail. This usually occurs as the result of trauma, often an automobile accident in which the chest is crushed against the steering wheel. If the ribs are broken on both sides of the sternum, or if ribs are broken in two places in the lateral chest, then with each inspiration the chest will go in (because of the negative intrapleural pressure) instead of expanding outwards; this is termed a ***flail chest.*** In addition to poor expansion of the lungs there is also hemorrhage into alveoli, leading to obstruction of alveoli and venoarterial shunting. When the flail is slight the major problem is the venoarterial shunting, but when there is severe injury then hypoventilation occurs as well.

Kyphoscoliosis. Patients with kyphoscoliosis have severe chest deformity owing to the curvature of the spine. The chest wall moves very poorly, and when the patient becomes older (usually between 30 and 40 years of age) there is less reserve and hypoventilation occurs. Also, the inability to take a deep breath may cause a decrease in surfactant in the alveoli (this keeps the alveoli expanded), resulting in collapse of alveoli and venoarterial shunting.

Primary Pulmonary Disease. There are two groups:

1. Acute respiratory failure may be due to primary pulmonary disease where there was *pre-*

viously normal pulmonary function. *Examples* are extensive pneumonia, vomiting and aspiration into the lungs, or fat embolism, resulting in inflammation at the alveolar level. In all these the emphasis is on venoarterial shunting with, in the terminal stages, increasing hypoventilation.
2. Acute respiratory failure superimposed upon *previously abnormal* lungs. The most common disorder is ***chronic obstructive lung disease*** from chronic bronchitis and/or emphysema. Patients with these disorders have obstruction to air flow. Where this is partial, hypoventilation occurs, but in some cases the small air tubes are completely closed, collapse of alveoli occurs, and there is venoarterial shunting. These patients remain in a reasonable balance with chronic respiratory failure, as shown by a high $apCO_2$, low apO_2, and only very slightly low pH (there has been plenty of time for the kidneys to retain bicarbonate to compensate for the chronic rise in $apCO_2$). When some *additional* problem such as a respiratory infection, pulmonary embolus, or pneumothorax occurs, or myocardial infarction causes left heart failure, then the patient goes into acute respiratory failure superimposed upon chronic respiratory failure. The $apCO_2$ is high, the apO_2 is low, and the pH is low but not as low as one might have expected if there had been an acute rise in $apCO_2$ from a previously normal level.

Asthma also causes an obstructive ventilatory defect. During an acute attack there is obstruction to small airways owing to spasm and plugging with mucus, and this results in venoarterial shunting. At a late stage, as more and more small airways become obstructed, the patient becomes exhausted from the 'high' work of breathing, and hypoventilation occurs in addition to the shunting.

In extensive ***pulmonary fibrosis*** the scarring destroys alveoli but often the pulmonary capillary flow remains good, and venoarterial shunting is the main defect. In the terminal stages hypoventilation will occur in addition, due to lung stiffness.

Pulmonary Embolism. This produces a complex situation. Clots block pulmonary vessels, resulting in poor blood flow past alveoli that still remain patent. This causes an increase in physiologic dead space, and the patients have to breathe hard, with a high minute ventilation, in order to maintain a normal $apCO_2$. In addition, the clot induces the release of substances that cause bronchospasm in other parts of the lung. This results in decreased ventilation of alveoli. Blood is shunted past these poorly ventilated areas, resulting in venoarterial shunting. Therefore, the picture is that of a high minute ventilation, normal or low $apCO_2$, and low apO_2.

Postoperative Respiratory Failure. There are two types:

1. There was *no previous lung disease.* Hypoventilation occurs because chest movements are restricted owing to pain from the surgical incision. Because the patient cannot take a deep breath, collapse of alveoli occurs and there is venoarterial shunting.
2. There was *prior lung disease,* often chronic obstructive lung disease, before the operation. In addition to restriction due to pain, and collapse of alveoli due to low tidal volume, there is often retention of sputum, causing further restriction and also obstruction and collapse of more alveoli. This produces an increasingly serious rise in $apCO_2$ and fall in apO_2.

ARTERIAL BLOOD GASES AND THE MANAGEMENT OF ACUTE RESPIRATORY FAILURE

Three principles apply to all forms of respiratory failure, and in particular to the complex case of acute respiratory failure superimposed on the respiratory impairment of chronic obstructive lung disease.

Principle 1: Oxygen is good. Oxygen is essential for life and the first objective of therapy is to maintain an adequate supply of oxygen to the tissues. Tissue oxygen supply depends on a combination of satisfactory hemoglobin level, satisfactory cardiac output, and satisfactory oxygen saturation. If one assumes a normal hemoglobin and a satisfactory cardiac output, then one should aim to have an apO_2 greater than

50 mm Hg. This will provide an oxygen saturation greater than 85%, which should be enough for most of the tissue needs. Therefore, as a first step, supply oxygen to the patient to maintain a blood oxygen greater than 50 mm Hg.

However, some patients if given too much oxygen will then hypoventilate, because some of the respiratory drive depends on a lower than normal oxygen tension in the blood. A good ***practical rule*** is as follows: If the patient's apO_2 is *greater* than the $apCO_2$, give a *high* concentration of oxygen (>50%), as such patients are unlikely to have their respiration suppressed by oxygen. If the apO_2 is *lower* than the $apCO_2$ then use a relatively *low* concentration of oxygen (24% to 35%), as these patients will tend to hypoventilate when their blood oxygen tension rises above 50 mm Hg. It is very satisfactory to maintain the apO_2 between 50 and 60 mm Hg, and once this has been established, do not be influenced into changing the oxygen supply because of any other changes in the blood gases (see Principle 2).

Principle 2: Acidosis is bad. The level of $apCO_2$ in itself will not cause death, but the rise in carbon dioxide tension causes a fall in blood pH and it is the acidosis which may be fatal. Acidosis causes ventricular arrhythmias, and these can lead to sudden death.

Once a suitable blood oxygen level is established, if the $apCO_2$ then rises, this need not be a matter of concern until the pH falls below 7.25. Certainly a pH below 7.2 indicates a need for further action. The quickest way to deal with this acidosis is to get rid of carbon dioxide, the pH will then rise to a safe level. This brings one to Principle 3.

Principle 3: The only way to get rid of carbon dioxide is to breathe it out. This can be done effectively by using either nasotracheal or orotracheal intubation with a cuffed tube and connecting it to an intermittent positive pressure machine, which will provide the patient with assisted ventilation (Jones, 1974). Another technique for increasing breathing is to use intravenous respiratory stimulants (Woolf, 1970). With these methods one can usually reduce the carbon dioxide tension and bring the pH to a safe level of above 7.25.

The above principles are important for the *life support* of the patient, but for *survival* it is essential to reverse the factor(s) that caused respiratory failure in the first place. Assisted ventilation or respiratory stimulants will keep the patient alive and allows the medical, or possibly surgical, treatment of the case to be effective.

Satisfactory treatment of respiratory failure requires an understanding of the basic pathophysiologic situations and a capacity to monitor therapy by repeated arterial blood gas estimations.

Many disturbances of ***blood chemistry*** occur in chest diseases; for example, disturbances in blood electrolytes; low levels of α_1-antitrypsin; abnormalities of hemoglobin, as in carbon monoxide poisoning and methemoglobinemia; changes in serum and urine osmolality; and a host of others. However, in respiratory medicine, the essential service that the clinical biochemist provides is the rapid availability and accurate measurement of arterial blood gases.

SUGGESTED READING

JONES NL: Physical therapy—present state of the art. Am Rev Respir Dis [Suppl] 110:132, 1974

SAUNDERS NA, POWELS ACP, REBUCK AS: Ear oximetry: Accuracy and practicability in the assessment of arterial oxygenation. Am Rev Respir Dis 113:745, 1976

SORBINI CA, GRASSI U, SOLINAS E, MUIESAN G: Arterial oxygen tension in relation to age in healthy subjects. Respiration 25:3, 1968

WOOLF CR: The use of respiratory stimulant drugs. Chest 58:49, 1970

WOOLF CR: The Clinical Core of Respiratory Medicine. Philadelphia, JB Lippincott, 1981

8

Disorders of the Kidney and Urinary Tract

Andrew DeWitt Baines

Kidney diseases occur in both sexes with equal frequency (prevalence 0.1%). Disorders of the lower urinary tract are most frequent in women of all ages (prevalence 4% to 7%) and older men (2% to 6%). Patients who have genitourinary disease may complain of symptoms directly related to the urinary tract, such as polyuria, discolored urine, pain on urination (dysuria), or of nonspecific symptoms such as back pain, fever, edema, and lassitude. They may have acidosis, anemia, hypertension, fractured bones, or tremors and even coma. The symptomatology of genitourinary disease is so diverse because the kidney is essential for:

a. Regulation of fluid, electrolyte, and acid base balance
b. Excretion of toxic waste products
c. Calcium and phosphorus metabolism
d. Production of erythropoietin, which stimulates red cell production
e. Release of renin to regulate blood pressure and aldosterone secretion

This chapter summarizes the biochemical consequences of renal disease.

PHYSIOLOGY AND BIOCHEMISTRY OF THE KIDNEY

STRUCTURE AND FUNCTION

Because the kidney's excretory functions depend upon its highly organized structure they are easily disrupted by inflammation, scarring, or loss of nephrons. Renal epithelial cells are polarized to transport substances from tubular lumen to blood or from blood to lumen. Groups of these cells are organized in an ordered sequence to form nephrons (Fig. 8–1).

The ***nephron*** (Fig. 8–1B) begins with a *glomerulus* designed to provide an ultrafiltrate of blood. Then follows the *proximal tubule* with its cells capable of actively reabsorbing sodium, other electrolytes, glucose, amino acids, and uric acid, while simultaneously secreting some organic acids and bases from blood to tubular fluid. *Henle's loop* comes next. In it, attenuated cells of the descending thin limb are permeable to water, salt, and urea, while cells in the ascending thin limb are relatively impermeable to water, although they remain permeable to solutes. Probably no active transport goes on in the *thin* limb; but in the *thick ascending limb* which lies downstream to the thin limb, chloride and sodium are actively reabsorbed through a water-impermeable wall, leaving a hypotonic tubular fluid behind. The ascending thick limb leads into the *distal convoluted tubule,* which contains several different cell types. Here and in the *collecting duct* that follows, transport of electrolytes and water permeability are sensitive to aldosterone and vasopressin. Ordered sequences of specialized cell types are maintained to the end of the nephron; cells of the medullary collecting ducts differ from those in the cortex.

The processing of glomerular filtrate to produce

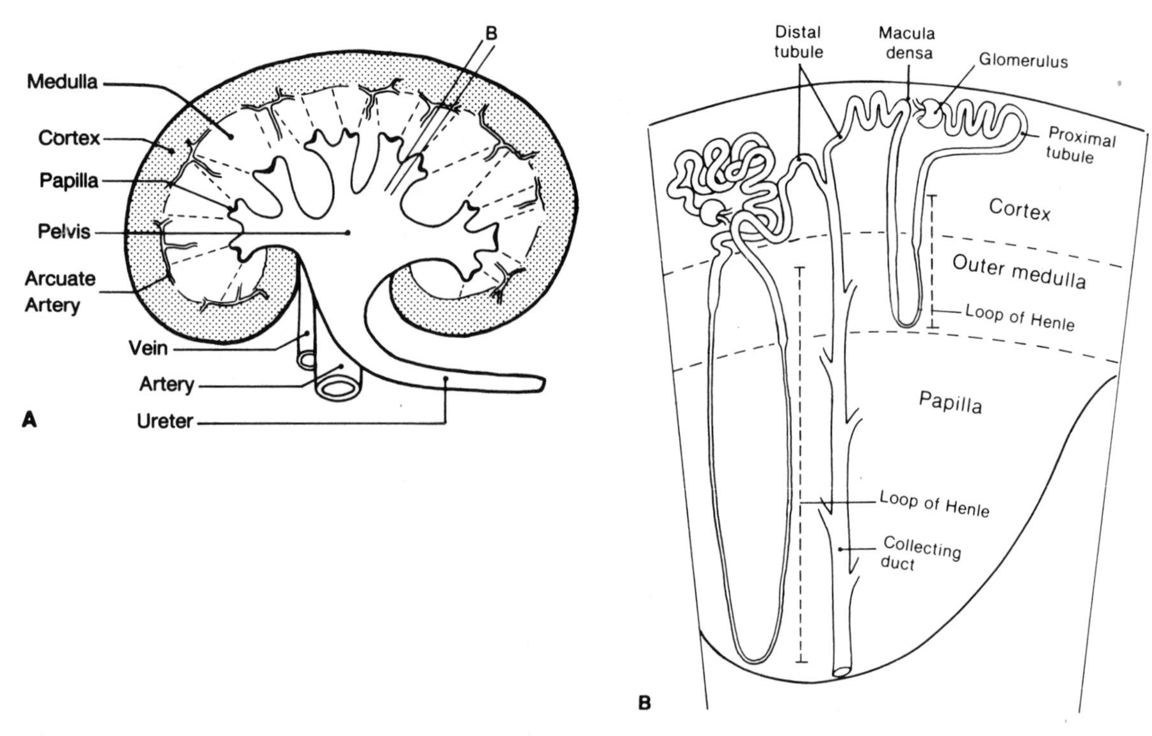

FIGURE 8–1. *A.* Cross section of the human kidney. *B.* Enlargement of section through the area indicated.

urine depends not only on the proper sequence of specialized cells but also on nephron conformation. The hairpin shape of Henle's loop is essential for concentrating urine. The junction between glomerular arterioles and distal tubule at the *macula densa* permits feedback regulation of the glomerular filtration rate and renin release in response to changing electrolyte flow in the distal tubule.

At a higher level of organization, the functional capacity of the kidney depends on the number of interacting nephrons; to achieve optimum performance, nephron function is integrated with peritubular blood flow.

EFFECTS OF DISEASE ON FUNCTION

Occasionally patients present with specific *genetic defects* at the cellular level. These produce defective transport of one or more substances such as amino acids, glucose, phosphate, or hydrogen ion. *Toxic* substances may also produce cellular transport defects. But these are rare causes of renal dysfunction. Common **disease processes** are much less specific in their effects. Infections, immunologic reactions, trauma, ischemia, and toxins may focus initially more on glomeruli than on tubules or may involve the interstitium primarily. However, any one of these processes, if it continues, will eventually destroy nephrons. Consequently, many of the features of chronic renal disease are reproduced experimentally by simple excision of renal tissue, a clear demonstration that functional capacity depends primarily on the number of functioning nephrons.

The ***diagnosis*** of renal disease requires the evaluation of functional capacity and elucidation of the pathogenic process. Methods of assessing function are described in the next section. Unraveling the intricacies of pathogenesis involves a consideration of the patient's history and physical condition and an examination of kidney structure through the use of biopsy and radiography.

ASSESSMENT OF RENAL FUNCTION

Renal Blood Flow and Glomerular Filtration Rate

One quarter of the cardiac output flows through the renal arteries, which distribute it to two million glomeruli. Glomerular filtration and the subsequent modifications of the filtrate by the kidney tubules depend upon an adequate flow of blood. Renal blood flow is determined by the arterial perfusion pressure and renal vascular resistance. The kidney, like many other tissues and organs, tends to autoregulate its blood flow unless there is some additional stimulus, such as increased sympathetic nervous activity. Glomerular filtration rate is also autoregulated.

Between 15% and 30% of the plasma water passing through the kidney is ***filtered*** by the glomeruli to provide the raw material which the tubules modify to create urine. The *rate of filtration* depends upon:

1. The surface area and permeability of the glomerular capillaries
2. Hydrostatic and oncotic pressure differences across capillary walls
3. The flow of blood through the glomeruli

Oncotic pressure* in afferent arteriolar blood does not vary widely. However, ***oncotic pressure*** may be responsible for the dependence of ***filtration rate*** on ***blood flow*** in the following way: Blood enters the glomerulus at a hydrostatic pressure of 45 mm Hg to 60 mm Hg. Hydrostatic pressure in the proximal tubule and Bowman's capsule is 10 mm Hg to 20 mm Hg. Protein in the blood entering the glomerulus exerts an oncotic pressure of 25 mm Hg across the capillary wall. This oncotic pressure acts to retain fluid in the capillary. As blood passes along the capillary, fluid is forced by hydrostatic pressure across the capillary wall into the capsular space. Protein does not leave with the fluid; therefore, the protein concentration and oncotic pressure rise until the latter equals or nearly equals the hydrostatic pressure. At this point filtration slows to a halt. The volume of the filtrate that is formed before oncotic pressure rises to equal net hydrostatic pressure is a function of blood flow through the capillary. Thus, the glomerular filtration rate (GFR) varies with glomerular blood flow if other factors are maintained constant.

In a healthy population the GFR correlates with height, weight, and body surface area. For purposes of comparison it is customary to express GFR in terms of the surface area of a 70-kg man, that is, 1.73 m^2. (There are tables for converting height and weight into surface area). A single protein-rich meal can increase GFR by 20% (or more) for several hours. A healthy person eating a low-protein diet (0.8 g/kg/d) may have a GFR of 80 mL/min. The same person on 1.6 g/kg/d could have a GFR of 150 mL/min.

Creatinine Clearance

Creatinine is generated spontaneously from creatine without enzymatic intervention. Consequently, the rate of production is proportional to lean body mass or muscle mass. Most creatinine is excreted in the urine, but some is metabolized or enters the intestinal lumen. Loss by extrarenal routes may increase as plasma creatinine levels rise with renal failure. Although creatinine does not penetrate the cells of the renal tubules, it appears to be distributed in a space equivalent to total body water.

Creatinine gains access to the urine primarily by glomerular filtration. Its concentration in filtrate equals that in serum water. The *quantity excreted,* therefore, is

$$\text{GFR (L/minute)} \times \text{serum creatinine } (\mu\text{mol/L})$$

Virtually none of the filtered creatinine is reabsorbed as it passes down the nephron. A small amount is added by secretion but it has been assumed that most of the creatinine excreted in the urine derives from glomerular filtration; therefore, filtered creatinine equals excreted creatinine, or $GFR \times S_c = U_c \times \dot{V}$ (where S_c and U_c are serum and urine creatinine concentration and $\dot{V}$ is urine flow in L/m). Rearranged, this yields the ***clearance equation*** for creatinine:

$$\frac{U_c \times \dot{V}}{S_c} = GFR$$

Substances which, like creatinine, are freely filtered and are neither reabsorbed nor secreted by

* ***Oncotic pressure*** is the osmotic force exerted by proteins across a membrane which is impermeable to proteins. In plasma most of the oncotic effect is contributed by albumin.

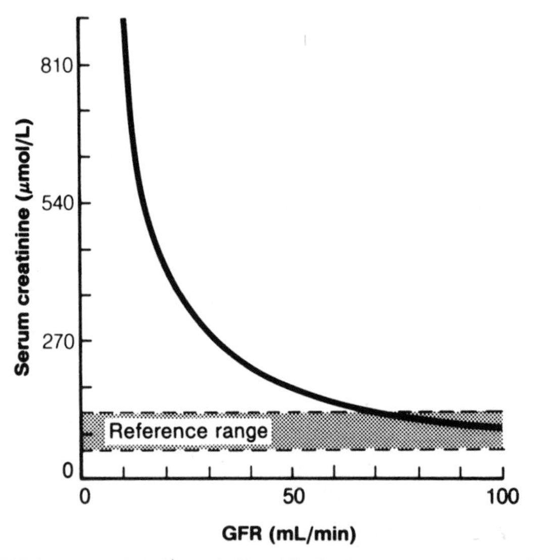

FIGURE 8–2. The relationship between serum creatinine and glomerular filtration rate (GFR).

the tubules, can be used to estimate the GFR; because urea is partly reabsorbed it cannot be used. *Inulin* is an example of an exogenous substance that is used to measure GFR. Creatinine clearance can be as much as 160% of the inulin clearance, because of creatinine secretion by renal tubules. There is considerable interindividual variation in creatinine/inulin clearance ratios. Certain drugs, notably cimetidine, inhibit tubular secretion of creatinine and lower creatinine clearance.

After a *sudden decrease* in the GFR, creatinine accumulates gradually in the body. During this time the S_c rises asymptotically to a new steady concentration. The rate of S_c increase, following a decrease in the GFR, is a function of the creatinine production rate, the volume of creatinine distribution in the body, and the new reduced GFR. A rapid rise in S_c, for example, 200 μmol/L (2.4 mg/dL) per day, requires a total or almost total loss of renal function, combined with high rates of creatinine production (as would be found in a muscular individual) and a contracted body fluid compartment. When the GFR falls suddenly from normal to 5% of normal, it may be days before a new stable S_c level is reached. GFR cannot be calculated accurately from creatinine clearance when S_c is changing.

Theoretically, one should be able to predict the GFR from the serum creatinine in a steady state. Assuming a constant rate of creatinine production, and excretion equal to production, it follows that as the GFR falls the S_c will rise. A fall in GFR to one-half of normal is associated with a rise in the serum creatinine from the normal 100 μmol/L to 200 μmol/L; a fall in GFR to one-fourth is associated with a rise to 400 μmol/L, and so on. This results in a hyperbolic relationship between GFR and creatinine levels (Fig. 8–2).

Factors influencing ***creatinine production*** also influence the relationship between serum creatinine and creatinine clearance. Muscle mass, which determines the production rate, increases with height and weight. Weight alone, however, is not an accurate indicator of muscle mass. The proportion of muscle to total body weight is less in females and in fat or sedentary individuals and decreases with age after the fourth decade. A young male excretes approximately 200 μmol of creatinine/kg/day; a young woman excretes 10% to 15% less. In an apparently healthy population the GFR decreases 10% per decade over the age of 40. A decline in creatinine production with age accounts for the relatively constant serum creatinine at all ages from 20 to 90 despite the gradual decline in renal function. Serum creatinine falls from a mean of 44 μmol/L in the first few days after birth to about 30 μmol/L from the age of 1 month to 2 years. It then begins to rise gradually, reaching 50 μmol/L at age 12 and 80 μmol/L at age 18.

Creatinine clearance per 1.73 m^2 surface area is low at birth (38 mL/min) but reaches adult levels (80 to 120 mL/min) by 6 months of age. Creatinine clearance increases by 20% during pregnancy. This increase, combined with an expanded extracellular fluid volume during pregnancy, causes the serum creatinine concentration to fall by approximately 20%.

Creatinine excretion is sometimes used as a reference for the excretion of other substances. It is only a rough guide, however, because the coefficient of variation (SD/mean × 100) for creatinine excretion in an individual is 10% to 15%. Variations are reduced slightly by controlling physical activity and diet. In practice the greatest source of variation comes from inaccurate urine collection, but ingestion of large amounts of cooked red meat will also raise serum and urine creatinine.

In some *analytic procedures* the creatinine measurement includes noncreatinine chromogens in the serum. Methods that measure only creatinine yield higher clearance values because the denominator (S_c) is lower than that measured with nonspecific methods. Some autoanalyzers use relatively specific procedures. Creatinine clearances determined by such analyzers are higher than those determined on analyzers that measure noncreatinine chromogens.

Blood or Serum Urea Nitrogen

The balance between protein catabolism and anabolism determines the urea production rate. Most of the urea is excreted into the urine by glomerular filtration, followed by a variable proportion of passive reabsorption. Urea reabsorption and water reabsorption are linked not only because the concentration gradient for passive diffusion is increased by removal of water from the tubular lumen but also because vasopressin (ADH) increases permeability of the collecting duct to urea. Thus *three factors* determine the blood or serum urea nitrogen concentration (BUN or SUN). These are: GFR, urea production rate, and water reabsorption or urine flow. BUN *rises* following increased protein intake or catabolism, or increased urine concentration, or decreased GFR. BUN *falls* with decreased protein intake, or increased anabolism, or increased renal water excretion (i.e., decreased urine concentration), or increased GFR. Processes that increase protein catabolism and urine concentration, without greatly reducing GFR, increase the ratio of serum urea/creatinine. Serum creatinine is a better index of GFR than BUN or SUN, but urea is more useful in monitoring the success of dialysis programs (see later).

Intravenous Pyelograms, Arteriograms, and Renograms

For diagnostic purposes estimation of the total GFR yields as much information as can be obtained from estimation of the total renal plasma flow (RPF) or blood flow; however, it is sometimes important to detect differences in renal blood flow between the two kidneys or within a single kidney. For this purpose, the most common test is the ***intravenous pyelogram*** (IVP). A radiopaque dye which is concentrated by the kidney is injected intravenously and radiographs of the kidney area are taken. Within 1 min to 3 min of the infusion the dye distributes diffusely through the renal cortex in what is called the *nephrogram* phase. Kidney tissue with a deficient blood supply picks up the dye less rapidly. After the nephrogram phase, the dye passes into the collecting system, which is outlined in the ensuing 5 min to 30 min.

Arteriograms are also employed to locate lesions in the renal vascular tree. Radiopaque dye for this purpose is infused directly into the aorta or renal arteries. Another technique, the ***renogram,*** employs radioactive compounds such as (1) ^{99m}technetium chelated with diethylenetriamine pentaacetic acid, which behaves like inulin; (2) ^{131}I-labeled orthoiodohippurate (Hippuran), which is filtered and secreted; or (3) chlormerodrin-^{197}Hg, which has an affinity for renal tubular cells. After intravenous injection of one or other of these compounds, the kidney area is scanned with scintillation detectors to reveal the shape and location of the kidney and to locate areas with a high or a low accumulation of radioactive material. Technetium and Hippuran scans can be used to estimate *relative rates of blood flow* to the kidneys. If blood flow is reduced, the radioactive material accumulates more slowly and is washed out less rapidly. Thus, the time course of radioactivity accumulation and washout in the affected kidney has a lower peak and is prolonged relative to the normal kidney.

Electrolyte Excretion

Sodium and Chloride. Excretion rates for sodium and chloride are extremely variable. An ability to vary the urinary excretion of salt over a 100-fold range is essential for the maintenance of constant extracellular fluid volume and osmolality.

Sodium excretion is regulated by

1. Glomerular filtration rate, which determines the quantity of sodium available for reabsorption by the tubules
2. Aldosterone, which regulates active reabsorption in the distal nephron
3. A group of poorly understood physical and humoral mechanisms lumped together as 'third' factors. This group includes physical variables such as perfusion pressure and filtration fraction (GFR/RPF), which act through their effects on hydrostatic or oncotic

pressure to influence fluid uptake from the proximal tubules into peritubular capillaries. The group also includes a variety of hormones whose physiologic role is not yet clear.

The following substances favor sodium loss, or *natriuresis:* prostaglandins, acetylcholine, kinins, and 'natriuretic hormones.' Under certain conditions angiotensin and catecholamines may induce natriuresis, but they usually act to decrease sodium excretion.

Chloride is actively reabsorbed in the ascending thick limb (driven indirectly by Na^+K^+ATPase); elsewhere it follows passively along with sodium reabsorption. Chloride and bicarbonate compete with each other to be reabsorbed with sodium. An increase in bicarbonate absorption is accompanied by a tendency for chloride absorption to decrease.

In the urine of a patient with *severe volume depletion* one expects to find <10 mmol sodium/L. Higher concentrations indicate an inappropriate response to volume depletion, which may result from the use of diuretics, or mineralocorticoid deficiency, or renal disease. Diuretics are the commonest cause of a continuing sodium loss in the face of volume depletion. Addison's disease, or adrenal cortical insufficiency, is a rare cause of inappropriate salt loss. So-called *renal salt wasting* is uncommon and tends to occur with diseases that involve the distal nephron, such as medullary cystic disease.

It is usually unnecessary to measure urinary electrolyte excretion; diagnosis can be made on the basis of history, physical examination, and serum electrolytes and creatinine. However, urinary 24-hour excretion rates or electrolyte concentrations can be valuable in the diagnosis of hyponatremia, acute oliguria, and metabolic alkalosis. For example, the commonest form of persistent *metabolic alkalosis* is associated with sodium chloride depletion. Measurement of *urine chloride* will help to decide whether the alkalosis will respond to simple replacement of the body's sodium chloride stores. Urine chloride should be <10 mmol/L in those who will respond to chloride replacement. Patients with chloride-resistant alkalosis may have primary hyperaldosteronism or an ectopic ACTH-producing tumor.

Potassium. Potassium is reabsorbed in the proximal tubule and loop of Henle. Secretion may occur in the distal tubule and collecting duct. Secretion is favored by the creation of a negative electrical potential in the tubule lumen as sodium is absorbed. Thus, distal sodium reabsorption encourages potassium secretion. The exchange is facilitated by aldosterone. However, *hydrogen ion* competes with potassium; hence, the secretion of hydrogen ion and of potassium may be reciprocally linked.

Potassium loss occurs with certain diuretics, with acidemia or alkalemia, with hyperadrenocorticism, with certain rare renal tubular disorders, and for unknown reasons in some patients with leukemia. A urine potassium of <10 mmol/L, when associated with systemic hypokalemia, implicates the gastrointestinal tract as the site of potassium loss.

Calcium and Phosphate. Absorption of these two compounds in the proximal tubule is linked to the transport of sodium and water. ***Parathyroid hormone*** (parathormone, PTH) decreases the proximal reabsorption of calcium, phosphate, sodium, and water. In the distal tubule PTH stimulates calcium reabsorption and inhibits phosphate reabsorption. The net effect of increased PTH activity is decreased calcium excretion and increased phosphate excretion. PTH also increases the urinary excretion of cyclic AMP. Phosphate absorption is almost complete at low plasma phosphate concentrations. There is a threshold for phosphate excretion and a maximum transport rate (T_m) as there is for glucose. PTH decreases the T_m for phosphate.

Abnormal renal excretion of calcium and phosphate may occur in association with renal stones or parathyroid dysfunction. This aspect of renal function is assessed by measuring 24-hour excretion rates when the patient is on a controlled dietary intake of calcium and phosphate.

Nephrogenous Cyclic AMP

Several hormones including parathyroid hormone, beta adrenergic agonists, and vasopressin (ADH) stimulate adenylate cyclase activity in the kidney and increase urinary cyclic AMP (cAMP) excretion. Nephrogenous cAMP (NcAMP) can be calculated from the urinary excretion minus the filtered load of cAMP.

$$\text{NcAMP} = \text{Urine cAMP} \times \text{V} - \text{Ccr} \times \text{plasma cAMP}$$

(where V = volume; C = clearance; and cr = creatinine)

This calculation is simplified by expressing the result as a fraction of Ccr (or GFR) thus:

$$\frac{\text{NcAMP}}{\text{Ccr}} = \frac{\text{Urine cAMP} \times \text{serum cr}}{\text{urine cr}} - \text{plasma cAMP}$$

The reference value is <30 nmol/L glomerular filtrate; high values reflect a biologic response to parathormone, which is the main regulator of NcAMP. (See also Chap. 17.)

Hydrogen Ion (H^+) and Bicarbonate Excretion

The kidney maintains H^+ *balance* by:

1. Secreting H^+ to regulate urine pH between 4.5 and 7.8
2. Excreting H^+ buffered by phosphate and organic acids as titratable acid
3. Excreting H^+ buffered as NH_4^+
4. Regulating bicarbonate excretion

Hydrogen ion secreted in the proximal nephron is buffered by *bicarbonate* in the tubular fluid. The carbonic acid formed in this process breaks down to CO_2 and water, and the CO_2 diffuses out of the lumen. The net effect is to lower tubular fluid pH to 6.5 and to cause bicarbonate reabsorption. Secreted H^+ that does not react with bicarbonate is buffered by phosphate, organic acids, and ammonia. In the *proximal* tubule carbonic anhydrase catalyzes the reaction $CO_2 + H_2O \leftrightharpoons H_2CO_3$. When this enzyme is inhibited by certain diuretics, secretion of H^+ and reabsorption of Na^+ with HCO_3^- is depressed. In the *distal* nephron hydrogen ion can be secreted against a larger concentration gradient. Thus, the pH of the filtrate can be lowered to 4.5. Secreted H^+ is buffered here primarily by phosphate and ammonia.

Renal H^+-excreting capacity can be tested by measuring (1) urine pH relative to plasma pH, (2) titratable acid minus bicarbonate excretion, (3) ammonium excretion, and (4) urine minus blood pCO_2.

pH. The pH of urine varies from 4.5 to 7.8 and should be appropriate to the homeostatic requirements of the individual. Thus, urine should be acid in patients with an acid load to excrete; when it is not, one suspects that the kidney may be contributing to the acidosis rather than compensating for it. The customary North American diet produces an excess 40 mmol to 80 mmol of H^+ that must be excreted daily by the kidney.

Titratable Acid Minus Bicarbonate. If the urine pH is greater than 6.0 some bicarbonate will be present. A known quantity of acid (A mmol) is added to convert urine bicarbonate to carbonic acid, which is removed (as CO_2) by applying a vacuum to the urine sample. At this point the urine contains its original titratable acid (TA) plus the A remaining after neutralizing the urine bicarbonate. Sodium hydroxide is then added to titrate this mixture of urine $TA + A - HCO_3^-$ back to the pH of plasma (7.4). To obtain the final answer the known quantity of A added initially is subtracted from the quantity of NaOH used in the final titration, thus: $TA - HCO_3^- = NaOH - A$.

The formation of urinary titratable acid depends upon H^+ secretion and the availability of buffers such as phosphate ($Na_2HPO_4 \rightarrow NaH_2PO_4$) and β-hydroxybutyrate. Bicarbonate excretion in the urine is subtracted because the loss of a bicarbonate ion is equivalent to the gain of a H^+ ion in so far as the body's acid–base balance is concerned.

Ammonium Excretion. Within renal tubular cells deamination and deamidation of glutamine yields ammonium (NH_4^+) and ammonia (NH_3) in a ratio of 100/1. Ammonia, being nonionized, diffuses across the cell membranes into tubular fluid and blood, therefore the NH_3 concentration is equal inside and outside the cell. Within the tubular lumen NH_3 is converted to NH_4^+ by reaction with secreted H^+ ions. The ratio NH_4^+/NH_3 increases tenfold for every 1 unit decrease in pH, therefore NH_3 is trapped effectively and is converted to NH_4^+ in the acidified tubular fluid or urine. Ammonium production is limited by enzymatic processes within the cell and by the supply of glutamine. Chronic acidosis induces enzymatic adaptations that may increase NH_4^+ production more than fourfold. Ammonium excretion requires secretion of H^+ ion by the tubular cell. Titration to pH 7.4 does not measure NH_4 because the pK of ammonium is 9. Ammonia in urine must be measured separately from titratable acid.

It should be remembered that the urinary pH is determined by H^+ secretion and the availability of buffers, primarily phosphate and ammonia. Hydrogen ion excretion may *increase* concomitantly with a *rise* in urine pH as more buffer is provided. This happens when the kidney adapts to chronic acidosis by increased synthesis of ammonia.

Urine Minus Blood pCO_2. Because there is no carbonic anhydrase in distal tubules and collecting ducts, H_2CO_3 formed by the reaction of secreted H^+ with luminal HCO_3^- does not dissociate to CO_2 and H_2O in the distal nephron. Delayed dehydration of H_2CO_3 raises pCO_2 in urine above that in blood. The difference between urine and blood pCO_2 is proportional to H^+ secretion in the distal nephron and is used as a measure of *distal* H^+ secretory capacity.

Test of H^+ Secretory Capacity. Renal capacity for H^+ excretion can be tested by feeding ammonium chloride 0.1 g/kg to the patient, then collecting urine every 2 hours for the next 6 hours. Urine pH should be <5.3 for *at least one* of these samples.

Effects of Disease on H^+ Excretion. The commonest cause of decreased H^+ excretion is generalized ***renal failure.*** With *acute* renal failure the lack of adequate glomerular filtration and urine formation is primarily responsible for reduced H^+ excretion. This effect may be reversible. Permanent defects appear as renal tissue is destroyed by progressive *chronic* diseases. As the GFR decreases, proximal bicarbonate reabsorption decreases; some distally secreted H^+ is then used up in reabsorbing the HCO_3^- overflow from the proximal tubule. In addition, as epithelial cells are lost, ammonia production decreases. As HCO_3^- excretion increases and NH_3 excretion decreases, net H^+ excretion falls below the rate needed to match metabolic acid production.

Rarely, acidosis is caused by one of the conditions grouped together as ***renal tubular acidosis*** (RTA). Three major forms of this disease are distal, proximal, and uremic or hyperkalemic RTA.

Classic ***distal RTA*** results from a hereditary defect in the ability to create an H^+ gradient in the distal tubule. As a result urine pH never falls below 5.5 to 6.0. Normally most of the filtered bicarbonate is absorbed in the proximal tubule, and only 10% to 15% is reabsorbed distally. With distal RTA, proximal bicarbonate absorption is normal; therefore, although there may be an inappropriate loss of bicarbonate it is rarely massive.

Most hydrogen ion is secreted in the proximal tubule, where it is used to reabsorb bicarbonate. In ***proximal RTA*** there is a hereditary or acquired defect in proximal H^+ secretion; consequently, the distal tubule is flooded with unabsorbed bicarbonate. The distal tubule, even in normal kidneys, has a limited capacity to secrete H^+. This capacity can easily be overwhelmed. Proximal RTA may lead to massive bicarbonate loss with defective titratable acid and ammonium excretion. However, when acidosis develops and plasma bicarbonate levels fall, the urine may become free of bicarbonate and its pH may again become acidic. Table 8–1 compares the findings in distal, proximal, and uremic RTA. (See also discussion under Chronic Renal Failure).

Excretion of Organic Compounds

Glucose. Urine always contains some glucose, but in a healthy person the concentration is usually less than 0.5 mmol/L (100 mg/L). Abnormally large quantities appear under two conditions: hyperglycemia with overflow or defective tubular reabsorption.

Overflow is the commonest cause of glucosuria. Quantities of glucose detectable by routine tests will be found in the urine when the plasma glucose rises above the threshold concentration of 10 to 11 mmol/L (180 to 200 mg/dL). This occurs most commonly in uncontrolled diabetes mellitus. As the plasma glucose rises, delivery of glucose to the proximal tubule approaches the maximum reabsorptive capacity and increasing amounts escape to appear in the urine.

Although in healthy young individuals the *threshold* and *maximum transport rate* for glucose are relatively constant, they are not invariable. Glucose reabsorption from the proximal tubule decreases along with water and salt reabsorption

TABLE 8–1. Findings in Renal Tubular Acidosis

	Type of Renal Tubular Acidosis		
	Distal	*Proximal*	*Uremic*
Acidemia	V	V	Yes
Net renal H^+ secretion at normal plasma HCO_3^-	N to ↓	↓↓↓	↓
Percent of filtered HCO_3^- excreted at normal plasma HCO_3^-	3%–10%	>15%	1%–30%
TA and NH_4^+ excretion at normal plasma HCO_3^-	↓	↓	↓
Urine acidification during acidosis	Impaired	V	Intact
Bicarbonaturia, filtered HCO_3^- excreted during acidosis	3%–10%	None	None
Serum K^+	N to ↓	N to ↓	N to ↑
GFR	N to ↓	N to ↓	↓↓↓

GFR = Glomerular filtration rate, N = normal, TA = titratable acid, V = varies, ↓ = decreased, ↑ = increased.

in patients undergoing a diuresis caused by fluid and solute overloading. Conversely, glucose reabsorption increases with extracellular fluid volume contraction. Furthermore, threshold and reabsorption may change relative to the GFR in some forms of chronic renal disease. Thus, one should never rely on the presence or degree of glucosuria to estimate the plasma glucose concentration.

A much less common form of glucosuria results from defects in ***tubular reabsorptive*** mechanisms. These may be caused by toxic or other forms of renal damage or by an inherited defect. Two forms of this defect exist: in one the threshold for glucosuria is lower than normal but the maximum rate of transport is normal; in the second form both threshold and T_m are below normal.

Renal glucosuria, which is significant glucose excretion when plasma glucose is less than 8 mmol/L, indicates either hereditary or acquired proximal tubular dysfunction. When it occurs alone, hereditary renal glucosuria is usually benign. Renal glucosuria may occur as part of the response to acute toxic or ischemic renal damage, but in these cases there is usually also aminoaciduria, phosphaturia, and the other stigmata of the Fanconi syndrome (Table 8–2). Hereditary forms of the Fanconi syndrome produce severe renal dysfunction.

Occasionally carbohydrates other than glucose are excreted in the urine. ***Galactosuria*** is the commonest of the underlying rare defects. Chemical tests for reducing substances will detect both galactose and glucose, but the widely used stick and tape tests for glucose depend upon a specific enzyme reaction and will not react with galactose. Therefore, when examining the urine of infants and children to detect metabolic defects one should use a nonspecific test for reducing substances.

TABLE 8–2. Causes of the Debré–deToni–Fanconi Syndrome (Aminoaciduria, Phosphaturia, Glucosuria, Hypouricemia, Acidosis)

Inherited
Cystinosis (cystine)
Galactosemia (gal-1-P)
Wilson's disease (copper)
Glycogen storage disease (glycogen)
Acquired
Nephrotic syndrome (protein?)
Myeloma (light chain)
Heavy metal poisoning (Pb, Cd, Au, Hg, U, Bi)
Antibiotics (streptomycin, kanamycin, neomycin)

Toxic substances are in parentheses.

Amino Acids. Specific or generalized defects in the tubular transport of amino acids also occur as hereditary or acquired defects. Abnormal quantities may appear in the urine because of a raised plasma concentration of substances which then overflow into the urine, or because of tubular defects which permit the substance to enter the urine at normal or below-normal plasma concentrations. These defects in renal function are uncommon and require specific measurements of amino acid excretion for diagnosis.

Phosphate. Phosphaturia may occur as part of the Fanconi syndrome, as a specific defect in hereditary sex-linked hypophosphatemic rickets, or as a response to hyperparathyroidism.

Urate. Urate enters the tubule by glomerular filtration and tubular secretion. It is also reabsorbed. Consequently, urate *clearance* may be greater or less than creatinine clearance depending upon the relative reabsorptive and secretory transport rates. A number of drugs inhibit one or another of the transport pathways. Although diuretics acutely increase urate excretion, ultimately extracellular fluid volume depletion, which accompanies the chronic use of diuretics, stimulates urate reabsorption, and raises serum concentration (normal <0.45 mmol/L). Probenecid and phenylbutazone are *uricosuric;* that is, they increase urate excretion. Salicylates decrease urate excretion at low doses and increase excretion at high doses.

The commonest cause of increased serum urate is reduced GFR. Urate concentration rises rapidly in the early stages of renal failure until it reaches 0.5 mmol/L (serum creatinine is 300 μmol/L at this point). With continued deterioration of renal function, urate concentration rises more slowly to a maximum near 0.65 mmol/L. The renal handling of urate is *assessed* by measuring 24-hour excretion. Measuring urate excretion is important in the diagnosis of gout and renal stones.

Urine Dilution and Concentration

Dilution. Some 60% to 70% of the glomerular filtrate is reabsorbed as an isotonic solution in the proximal tubule. The fluid then enters the descending limb of the loop of Henle where more water is lost. Next it enters the *ascending thick limb.* The walls of this structure have a low permeability to water and a high capacity to transport salt actively from the tubular lumen (Fig. 8–3A). Water reabsorption is negligible. Consequently, when the tubular fluid reaches the end of the ascending limb it has a greatly reduced salt content. In the absence of vasopressin (antidiuretic hormone, ADH) the cells of the *distal tubules* and *collecting ducts* are relatively impermeable to water (Fig. 8–3B). Active Na^+ transport removes more salt from the fluid, thereby reducing the concentration still further. The result is a urine containing as little as 1 mmol NaCl/L with an osmolality as low as 40 mosm/kg H_2O.

Excretion of ***dilute urine*** requires

1. Water-impermeable ascending thick limbs, distal tubules, and collecting ducts (glucocorticoids may assist in the maintenance of impermeable walls)
2. Absence of ADH
3. Active transport of salt in the ascending limb (inhibited by the diuretics furosemide and ethacrynic acid)
4. Active reabsorption of Na^+ in the distal tubule and collecting ducts (stimulated by aldosterone, inhibited by some diuretics)
5. Adequate delivery of filtrate from the glomerulus and the proximal tubule

Excretion of excess water has two components. *First,* glomerular filtration provides the volume of fluid upon which the tubules act. A reduction in GFR will affect the urine volume even if it has no effect upon the extent of urine dilution; thus, the capacity to excrete water decreases as the GFR is reduced. *Second,* concentration is lowered by active solute transport across the water-impermeable walls of the distal tubule and collecting duct. Capacity to dilute the urine is reduced by decreased delivery of tubular fluid to the diluting segment. This may happen to patients with heart failure as a result of reduced GFR and increased proximal tubular reabsorption.

Osmolar clearance is the hypothetic volume of urine (per unit of time) that would be required to excrete a solute load in isosmolar (same osmolality as plasma) form. In hyposmolar urine, the difference between urine volume and osmolar clearance is referred to as ***free water clearance*** (C_{H_2O}), and in hyperosmolar urine, the volume difference is

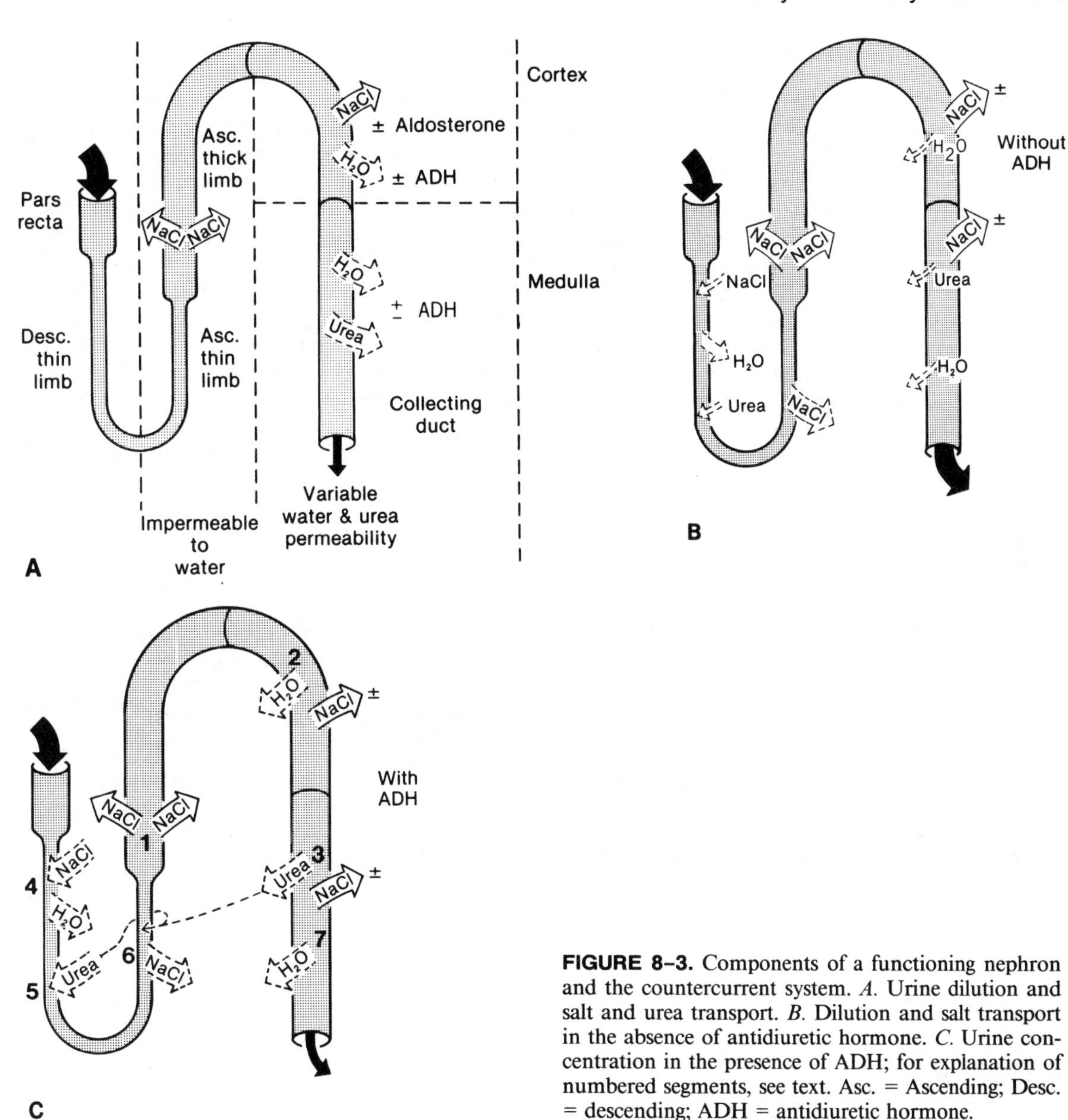

FIGURE 8–3. Components of a functioning nephron and the countercurrent system. *A.* Urine dilution and salt and urea transport. *B.* Dilution and salt transport in the absence of antidiuretic hormone. *C.* Urine concentration in the presence of ADH; for explanation of numbered segments, see text. Asc. = Ascending; Desc. = descending; ADH = antidiuretic hormone.

referred to as ***free water reabsorption*** (or negative free water clearance). Thus:

$$C_{osm} = \dot{V} \times \frac{U_{osm}}{P_{osm}}$$

$$C_{H_2O} = \dot{V} - C_{osm}$$

where $\dot{V}$ = urine flow rate, U = urine, and P = plasma.

Concentration. In the renal *medulla* there are three sets of tubules lying adjacent to each other in an interstitial compartment. Tubular fluid flows from the cortex toward the papilla in the descending limb of Henle, loops back, and flows toward the cortex in the ascending limb, and then again moves from cortex to papilla in the collecting ducts. This ***countercurrent flow*** in parallel tubules is an essential part of the concentrating mechanism, because it permits a small concentration difference set up across one segment of tubule wall to be multiplied along the length of the papilla. The resulting concentration in all the tubules and capillaries at the papillary tip is up to five times greater than in the cortex (Fig. 8–4).

The process of ***urine concentration*** is shown in Fig. 8–3C. Active transport of salt from the water-impermeable ascending thick limb (step 1 in Fig. 8–3C) has three effects: the medullary interstitium becomes hypertonic, salt diffuses into the descending thin limb, and fluid entering the distal tubule is made hypotonic. In the presence of ADH, water moves passively down a concentration gradient from the late distal tubule and cortical collecting duct to capillaries in the cortex (step 2). This segment of the nephron is not permeable to urea; therefore, urea concentration in the cortical collecting duct rises as water diffuses out of the duct lumen. Urea diffuses out of the medullary collecting duct into the interstitium (step 3). Water is drawn out of the descending thin limb by the concentrated urea and salt solution in the interstitium (step 4). The water is then carried off by blood vessels called the *vasa recta.* As water is removed the concentration of salt and urea in the descending thin limb rises further. Urea diffuses into the de-

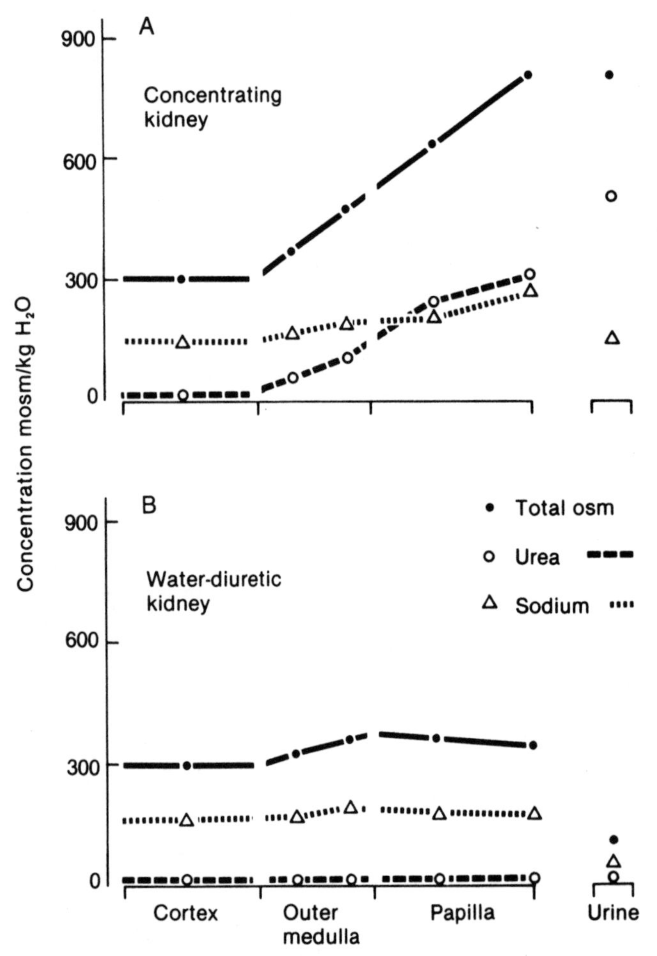

FIGURE 8–4. Solute concentrations in kidney tissue, from cortex to papilla, based on studies done on rats. *A.* During water deprivation salt and urea concentrations rise in the outer medulla and papilla. Total urine osmolality is roughly equal to twice the sodium plus urea concentrations. *B.* During water diuresis the salt concentration is highest in the outer medulla adjacent to the thick ascending limbs. The salt concentration is low in the papilla and the urine. The urea concentration is low throughout. (Atherton JC, Hai MA, and Thomas S: The time course of changes in renal tissue composition during water diuresis in the rat. J Physiol 197:429, 1968)

scending thin limb and probably into the ascending thin limb (step 5). Cycling of urea from the medullary collecting duct to the thin limbs and back raises the medullary concentration of urea.

As a result of these solute and water fluxes, solute concentration around the water-impermeable ascending thin limb is highest at the papillary tip and decreases to reach isotonicity at the corticomedullary border. This means that fluid entering the ascending limb at the tip of the papilla is more concentrated than the interstitial fluid surrounding the tubule in the region of the corticomedullary border. Consequently, as fluid moves up the ascending limb (step 6), salt diffuses out to the more dilute interstitium. This complex system creates a salt- and urea-rich medullary interstitium (step 7) which promotes passive water absorption from the collecting duct. In this way a small volume of concentrated urine is produced. The single active step in the process is the *metabolically driven salt transport* from the ascending limb. Salt is trapped in the medulla while water is absorbed in the cortical distal tubules and collecting duct.

A 'normal' urine concentration is that which is appropriate to maintain homeostasis. A healthy young person will excrete a urine of only 40 to 80 mosm/kg when faced with the need to excrete excess water, and will raise the concentration to 1200 to 1400 mosm/kg when deprived of water.

Excretion of a ***concentrated urine*** requires

1. Structurally intact loops of Henle and collecting ducts in normal numbers and physical relationships
2. Active electrolyte transport
3. An adequate supply of salt in the fluid entering the loop
4. The presence of ADH
5. Adequate supplies of urea
6. Appropriate rates of tubular fluid flow and blood flow

There is an optimum rate of urine flow into the descending limb. If flow decreases, concentrating ability is impaired by an inadequate salt supply; if tubular fluid or blood flow is too high, concentration is impaired by the washout of solutes.

Vasopressin release is controlled by atrial volume receptors and cerebral osmoreceptors. The threshold osmolality for stimulus of ADH release is lowered by reducing extracellular fluid volume. Thus, hypovolemic patients may continue to release ADH when their plasma osmolality is below normal. Inability to dilute the urine, for whatever reason, leads to hyponatremia and lowered plasma osmolality unless water intake is restricted.

Effects of Renal Disease on Dilution and Concentration. As normally functioning ***renal tissue*** is ***lost,*** concentrating ability decreases and the maximum volume of dilute urine is reduced. When GFR falls to 25% to 30% of normal, the kidney can no longer produce concentrated urine. Some ability to dilute the urine remains, but the individual usually excretes a urine with a concentration fixed close to that of plasma. At this stage regulation of body fluid balance must be accomplished primarily by variations in fluid and solute intake. The importance of the GFR in determining urine concentration can be seen by examining the response to severe prolonged exercise, which produces mild volume depletion with a paradoxically decreased urine concentration. The reduced concentrating capacity is attributed to a decreased GFR and reduced solute excretion.

Generalized renal disease is the most common cause of ***defective concentrating ability.*** In this condition the loss of nephrons and the medullary scarring interfere with the countercurrent system, and for unknown reasons the collecting ducts lose their sensitivity to ADH. Chronic hypercalcemia and hypokalemia may cause a reduced concentration capacity, possibly through an effect on collecting duct permeability. Hypokalemia also induces polydipsia, which contributes to the formation of a dilute urine. Reduced concentrating ability may also arise from deficient ADH release or insensitivity to its presence. When ADH release from the pituitary is impaired the condition is called *diabetes insipidus.* Nephrogenic diabetes insipidus is a rare inherited defect of the collecting ducts which renders them insensitive to ADH. Resistance to ADH may be produced by lithium, high doses of tetracyclines, some sulfonylureas, and methoxyflurane.

Tests of Urine Concentrating Capacity. The patient's ability to concentrate urine can be examined either by dehydrating the patient or by administering exogenous ADH. Combining these two approaches is useful in distinguishing diabetes

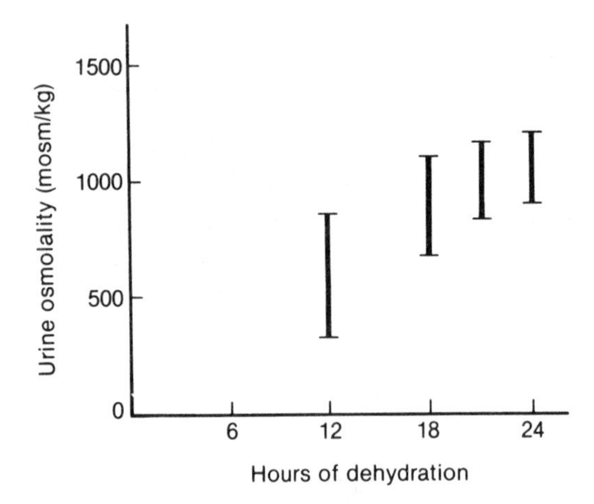

FIGURE 8–5. Changes in urine osmolality during dehydration. Range of values is shown between bars.

insipidus from psychogenic polydipsia or renal defects in concentrating ability. Patients with diabetes insipidus, unlike those with polydipsia, cannot concentrate urine after dehydration because they lack ADH, but should concentrate it after the administration of ADH. Patients with nephrogenic diabetes insipidus or generalized renal disease do not respond to ADH administration any better than to dehydration.

Careful attention should be given to the *preparation of patients* for these tests because in practice the responses are often not clear-cut. On the one hand, patients with long-standing polyuria caused by diabetes insipidus or excess water intake may respond poorly to exogenous ADH or dehydration. For example, urine osmolality in response to 500 mU of ADH averaged 1100 mosm/kg in eight normal subjects after 19 hours of hydropenia. After 48 hours of a high water intake (100 mL/kg over 24 hours) the maximum response to ADH was 850 mosm/kg. On the other hand, reduction of the GFR caused by volume depletion and hypotension will increase urine concentration in the total *absence* of ADH, probably by slowing the flow of tubular fluid through the medulla. Persons with reduced concentrating capacity rapidly *deplete* their *body fluids* when deprived of water. Hypovolemia, renal vasoconstriction, and a decreased GFR result. This sequence of events is responsible for a slightly increased urine concentration following water deprivation in patients whose primary defect is posterior pituitary malfunction.

Low-salt and low-protein diets also impair concentrating ability because they limit the excretion of salt and urea, which normally account for most of the solute in urine. A high intake of water and a low excretion of urea account in part for the low maximum concentration of urine in infants. Low protein intake may also contribute to the reduced maximum urine concentration in older people.

Dehydration Tests. These are done by stopping all fluid and food intake for 18 to 24 hours. The resulting urine concentrations are shown in Figure 8–5. Concentrating ability decreases with age and is reduced by inactivity or hospitalization. Once urine concentration rises above a specific gravity of 1.024 or osmolality of 960, even on a random urine sample, no further information will be gathered by prolonging the water deprivation. Healthy young volunteers deprived of all fluid until increments in urine osmolality were <30 mosm/hour reached this plateau within 18 hours. At this time values ranged from 870 to 1450 mosm/kg (mean, 1070). Hospitalized patients without renal disease, under the same conditions attained a mean of about 760 mosm/kg. In contrast, patients with diabetes insipidus could not concentrate urine above plasma osmolality. In fact, the mean maximum attained after 4 to 18 hours of dehydration was 170 mosm/kg, but after ADH administration this increased by at least 50%.

It must be noted that withholding water from a patient with neurohypophyseal insufficiency may precipitate severe volume depletion. These patients must be monitored carefully and the test stopped when a 4% reduction in body weight has occurred.

Urine Concentration. For many years this has been measured by determining the *specific gravity* (sp gr) of the urine. The method is simple, rapid, and inexpensive, but because it measures the density of the solution it will be influenced by the presence of heavy substances such as protein or the radiopaque dyes used in IVP examinations. Urine *freezing point depression* depends on urine osmolality and is a function of the number of solute particles independent of their relative mass. It is not subject to errors created by the presence of protein or x-ray contrast material. For most purposes, sp gr gives as much information as osmolality, but care must be taken to exclude the presence of abnormal solutes. The *refractive index* of urine also is a function of urine density. It can be measured rapidly on very small volumes of urine and is therefore a useful technique, especially for examining the urine of infants.

Tests of ***diluting capacity*** have little diagnostic value and are rarely done even in research units. Diluting capacity may be impaired by the loss of nephrons, by reduced or defective sodium reabsorption in the distal nephron, or by inappropriate ADH secretion.

Osmotic Diuresis. An increase in urine flow, or *diuresis,* can be induced by infusing large quantities of a solute that is filtered but not reabsorbed by the kidney. With very large solute excretion rates, such as those created by an intravenous dose of 100 g *mannitol,* urine flow rises to reach 8 to 10 mL/min within an hour and urine sodium concentration plateaus at 50 to 70 mmol/L. With still larger doses of mannitol 20% to 30% of filtered water and 15% to 20% of filtered sodium may be excreted in the urine.

When a non-reabsorbable solute is present in the proximal tubule, sodium reabsorption continues while the non-reabsorbable solute remains in the lumen. The proximal tubular epithelium is very water-permeable therefore the tubular fluid remains isotonic with peritubular plasma and as non-reabsorbable solute concentration rises, due to water reabsorption, sodium concentration in the lumen decreases. This increases the gradient for passive back diffusion of sodium from plasma to lumen. Consequently, net sodium reabsorption decreases which entails reduced water reabsorption. Non-reabsorbable solutes also impede water and solute reabsorption in Henle's loop by a mechanism that is not clearly understood.

The major action of ***osmotic diuretics*** appears to be on the loop of Henle and the countercurrent system. In nondiuretic states water is reabsorbed from the thin descending limb owing to the osmotic effect of salt and urea trapped in the interstitium by countercurrent flow (Fig. 8–3C). However, when the thin descending limb contains a high concentration of glucose or mannitol, water absorption by osmotic forces is impeded. This reduces water reabsorption from the descending limb, with a resultant increase in flow through the loop. Urea also may induce an osmotic diuresis; however, high plasma concentrations must be reached before the effect is significant. Unlike mannitol, urea is partially reabsorbed, which accounts for its weaker action as an osmotic diuretic.

Polyuria. The passage of urine in excess of 3L/day may result from a primary increase in fluid intake, or from a primary defect in concentrating ability which produces an increase in urine flow. *Primary polydipsia* may be a habit, a sign of functional mental disturbance, or the result of organic cerebral dysfunction. Large volumes of fluid are ingested, plasma osmolality is below normal, vasopressin secretion is suppressed, and urine flow is high. *Primary polyuria* may be caused by:

1. Reduced ADH secretion—disease of posterior pituitary (diabetes insipidus)
2. Disruption of the renal counter-current system—generalized renal disease
3. Decreased cellular response to ADH (nephrogenic diabetes insipidus)
   —congenital
   —acquired, chronic hypercalcemia, chronic hypokalemia
4. Osmotic diuresis—hyperglycemia

Certain renal diseases, such as pyelonephritis or polycystic disease, attack the medulla and papilla first and reduce concentrating capacity more rapidly than does glomerulonephritis. Sickle cell anemia is also associated with a defect in urine concentrating capacity.

Plasma osmolality and ADH are reduced by primary polydipsia and increased by primary polyuria. As discussed earlier, urine concentration should increase in response to dehydration in patients with primary polydipsia and in response to exogenous ADH in diabetes insipidus. Patients with generalized renal disease or nephrogenic diabetes insipidus respond poorly to both dehydration and exogenous ADH; they can be detected by finding reduced creatinine clearance and raised serum creatinine.

Frequency can occur without an increase in 24-hour urine volume. Frequency is the result of a reduced bladder capacity, or increased bladder irritability, which is most often caused by infection in the lower urinary tract. For healthy individuals, the volume of fluid intake determines urine flow (0.5L to 2.5L).

URINALYSIS

The chemical analysis of urine has been simplified by the use of dipsticks carrying small squares of paper impregnated with reagents. These sticks can be used to measure pH, protein, hemoglobin, glucose, ketones, bile pigments, nitrites, and leukocyte esterase. Not all of these tests are necessary for the routine assessment of *renal* function. Some sticks measure urine ionic strength, calling it specific gravity, but urea is not detected so the correlation with true specific gravity is only approximate.

Proteinuria

Normal Proteinuria. A 24-hour urine sample contains 200 mg to 300 mg of solid material, two thirds of which is carbohydrate, urea, and protein, the rest lipids and mineral salts. One expects to find between 20 and 120 mg protein/24 hours. Increased amounts of protein are found in the urine of people with many forms of renal disease. However, the transient proteinuria that frequently accompanies systemic illness, or extreme physiologic stress, results from hormonal effects on glomerular hemodynamics and permeability. *Transient* proteinuria does not indicate structural damage.

Some urinary proteins are derived from the *urinary tract.* Quantitatively the most important of these is Tamm–Horsfall glycoprotein, which comes from cells of the thick ascending limb and is excreted at the rate of 25 mg/24 hours. It is this protein that ordinarily provides the matrix for urinary casts (see casts, below). In addition, some urinary enzymes appear to originate in the urinary tract.

Proteins derived from plasma enter the urine by escaping through the *glomerulus.* The *glomerular filter* consists of a fenestrated capillary endothelium, an amorphous basement membrane, and the slits between the foot processes of epithelial cells (Fig. 8–6). This tripartite structure is a very effective barrier to large molecules. Molecules up to the size of inulin (M_r 2000) pass through without impediment. For larger molecules, permeability decreases with increasing size. Thus only 0.1% to 1% of plasma albumin escapes into the filtrate, and molecules larger than M_r 60,000 are virtually excluded from the urine. The main resistance to permeation appears to be in the basement membrane, although the slits between foot processes also play a role. A net negative charge in the glomerular filter hinders the penetration of negatively charged molecules.

The only *large* proteins ($M_r > 90{,}000$) found in any appreciable amount in the urine are IgG and IgA, which may come from the urinary tract. Among the *smaller* proteins (M_r 40,000 to 90,000), albumin is the most important, with an excretion of about 10 mg/24 hours. Other proteins of this size are also found, but because their plasma levels

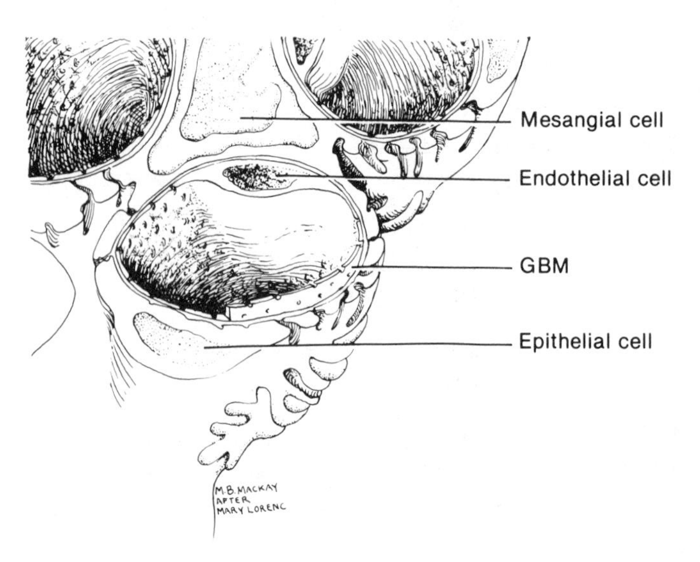

FIGURE 8–6. Details of a glomerular capillary: GBM = glomerular basement membrane. (Redrawn from Strauss MB, Welt LG [eds]: Diseases of the Kidney. Fig. 9. Boston, Little, Brown & Co, 1963)

are much lower than that of albumin, their urinary excretion is lower as well. There are only traces of proteins with a relative mass less than 40,000 in the plasma, but they constitute a significant fraction of the urinary proteins. Examples are α_2-microglobulin, β_2-microglobulin, and post-γ-protein.

Of the proteins that escape into the filtrate 80% to 90% are taken up by pinocytosis in the *proximal tubular cells* and are *catabolized.* Normal glomerular filtrate contains 0.1 g/L of protein. Approximately 14 g (144 L $\times$ 0.1 g) are filtered per day but less than 0.1 g is excreted, therefore 13.9 g are normally reabsorbed and metabolized in the proximal tubule. With massive proteinuria the kidney may become the major site for protein catabolism.

Abnormal Proteinuria. Increased protein excretion may occur for the following reasons:

1. Elevated plasma protein levels in which filtered load exceeds absorptive capacity
2. Increased permeability of the glomerular filter
3. Decreased reabsorption by the proximal tubules
4. Increased endogenous protein production in the urinary tract
5. Obstruction of the renal lymphatics, which leads to chyluria

Patients with ***myeloproliferative disorders*** excrete a unique class of proteins in their urine. These ***Bence Jones proteins*** precipitate at 50° to 60°C and tend to redissolve at 90° to 100°. Serologic analysis has revealed them to be monomers (M_r 29,000) or dimers (M_r 40,000 to 45,000) of the light chain of IgG. The concentration of these proteins rises in the plasma of patients with proliferative disorders of plasma cells and lymphocytes, and they are excreted in large amounts in the urine of about 50% of cases. Bence Jones proteins cannot be distinguished from other proteins by standard screening procedures; instead, electrophoretic and immunologic characterization must be carried out. Some of these proteins do not react with stick tests for protein.

Tubular proteinuria occurs when proximal tubules are defective. It is found with the Fanconi syndrome, cystinosis, chronic cadmium poisoning, Wilson's disease, myelomatosis, galactosemia, rejection of renal transplants, and other forms of acute tubular necrosis. The urine contains increased amounts of proteins with low relative mass (12,000 to 45,000) such as β_2-microglobulin, α_2-microglobulin, ribonuclease, lysozyme, and insulin.

Increased production of protein may occur with ***inflammatory diseases*** such as pyelonephritis and cystitis. The quantity excreted is small compared to that found in glomerular disease.

Proteinuria may be found in 1% to 1.5% of men and women aged 19 to 60 during an initial screening procedure but in 90% of these patients the proteinuria is *transient.* Some of the remaining patients excrete protein in all urine passed, while others excrete protein after standing up for some time but not during a period of recumbency. *Postural proteinuria* may become persistent with time, indicating the insidious onset of renal disease, but it is usually benign, as is transient proteinuria.

During ***severe exercise,*** urine protein concentration may increase 100-fold, because of reduced urine flow and increased protein excretion. High rates of protein excretion continue for several hours after cessation of the exercise. The proteins are largely of plasma origin but the condition is benign. When the urine is examined by electrophoresis it is clear that exercise proteinuria is not simply an exaggerated form of postural proteinuria. It may be caused by hemodynamic changes in the glomerulus or by altered permeability of the filtering barrier. Nonspecific proteinuria is also observed in patients with fever, stress, and heart failure.

The kidneys of a group of young men with ***persistent proteinuria,*** excreting less than 1.5 g protein/day, were biopsied; 8% had histologic evidence of renal disease, 45% had subtle nonspecific alterations of glomerular structure, and the rest were normal. These results are consistent with the experience of life insurance companies, which find that for those over age 30 the risk of extra mortality (the rate above that expected for age and sex) increases with proteinuria in excess of 500 mg/L.

Tests for Proteinuria. Qualitative tests for protein excretion are usually designed to give negative results with normal urine. These methods will detect gross proteinuria, but there is a danger of missing significant abnormalities when the urine is too dilute. More precise estimation of proteinuria requires 12 to 24 hours of urine collection and a quantitative determination of protein content.

TABLE 8–3. Overnight 12-Hour Excretion of Cells (±SE)

	Renal Tubular Cells	Leukocytes	Erythrocytes
Males	74,000 ± 2000	31,000 ± 2,000	32,000 ± 1800
Females	57,000 ± 2000	107,000 ± 13,000	23,000 ± 1300

Stick tests for protein are less sensitive for globulins and Bence Jones protein than for albumin. Turbidometric methods using sulfosalicylic acid are more sensitive and will detect any protein in concentrations greater than 0.15 g/L. Persistent proteinuria should be quantified either by measuring a timed 24-hour collection or by comparison with creatinine excretion during the day. Immunoelectrophoresis is necessary for the identification of Bence Jones proteins.

Hematuria

Blood may enter the urine at any point from glomerulus to urethra. The blood may be primarily in the form of intact red cells or may be hemolyzed either before or after it enters the urine. Red blood cell casts (see the following) or hemoglobin casts indicate an intrarenal source for the bleeding. The possibility that blood is coming from the lower urinary tract may be demonstrated by collecting two samples of urine sequentially during one episode of voiding the bladder. If bleeding comes from the urethra then the first sample will contain more blood than the second. If the bleeding is from the upper urinary tract both samples will contain equal amounts of blood. Dysplastic red cells, with blebs on their surface, are associated with glomerular hematuria.

Microscopic Urinalysis

The urine contains a number of formed elements which can be detected by examination with a light microscope: red blood cells, leukocytes, renal tubular cells, transitional epithelial cells from the bladder, squamous epithelial cells from the lower genitourinary tract, parasites (trichomonas, yeast), bacteria, crystals, and casts. This examination—which should be carried out by the *physician*—is an extension of the physical examination. Casts deteriorate after urine is passed, and 40% of leukocytes may disappear within one hour.

Cells. Everyone excretes red cells, leukocytes, and renal tubular cells; usual numbers are shown in Table 8–3. In an unstained urine sample it is difficult to distinguish leukocytes from renal tubular cells, and so the two are usually enumerated as one group. Stick tests for leukocyte esterase are specific and sensitive for leukocytes in urine. They can be used to screen for significant pyuria (negative predictive value > 95%).

Routine microscopic urinalysis is done on a 10 mL to 15 mL aliquot of freshly voided urine. The sample is centrifuged at 3000 to 5000 rpm for 3 to 5 min and the supernatant is discarded, leaving a button of sediment which is resuspended in 0.5 to 1 mL of residual urine in the tube. A drop is placed under a cover slip and is examined with the high-power objective of a light microscope to semiquantify cellular elements. A variety of techniques have been devised to improve quantification, involving the collection of timed urine samples, centrifugation at higher speed, and use of a counting chamber containing a known volume of fluid. However, they are not widely used because the simpler routine approach picks up most abnormalities.

Urine from *healthy* individuals examined in this way usually reveals no red cells and very few white cells. As will be seen, the excretion of red cells and white cells increases in a variety of diseases; other cells such as squamous epithelial cells from the lower urinary tract are not usually a sign of disease.

Casts. Red and white cells may enter the urine at any level from glomerulus to urethra. Casts, however, are formed only in the kidney tubules and are, therefore, much more specific indicators of renal disease. The usual site of formation is the distal nephron, and the shape of the casts often indicates their source. Some may be slightly convoluted or corkscrew in shape, but most are straight. Width and length are variable. Narrow casts are more common in mild or acute renal disease; broad casts are found in urine from patients whose chronic disease has produced dilated tubular lumens.

Glycoprotein from the kidney forms the basic substance of casts. Precipitation of this material is encouraged by the changes in pH and electrolyte concentration that occur in the distal nephron, and by the presence of other proteins. Therefore, excretion of casts increases whenever there is proteinuria. Casts containing glycoprotein and protein alone are clear, or *hyaline,* in appearance. Large numbers of these (100,000) are excreted daily by healthy people, but routine urinalysis does not detect more than an occasional hyaline cast. In the presence of proteinuria, hyaline cast excretion increases to the point at which the casts are easily detected with the microscope.

Erythrocytes, tubular epithelial cells, and leukocytes that are extruded into the lumen may be incorporated into casts. Such ***cellular casts*** are not normally formed or excreted in detectable numbers; their presence is evidence of an abnormal process in the kidney. ***Granular casts*** also provide evidence of kidney disease, the granules probably coming from degenerated cells. If red cells provide the material the result is a heme granular cast. White cell casts suggest inflammation. Casts from chronically diseased kidneys may have a waxy or ground-glass appearance.

One or two *heme granular casts* found after searching several urine samples is a significant observation in a patient suspected of having glomerular disease. It should be borne in mind, however, that excretion of cells and casts may increase with exercise, prolonged standing, and stress.

Crystals. These are frequently observed in routine urinalysis. Most of the oxalates, phosphates, and urates are of no significance, but in stone-forming patients crystals may provide a clue to the chemical nature of the stones.

Bacteria. Microorganisms are not seen in a normal urine microscopic examination. In heavily contaminated urine (more than 100,000 organisms/mL) bacteria may be observed. It is essential to have a clean fresh specimen of urine for urinalysis. Bacterial action will rapidly destroy cells and casts in urine that is allowed to stand unrefrigerated. Stick tests for *nitrite* are positive in about 50% of urines with significant bacteriuria.

DISEASES OF THE KIDNEY

APPROACH TO THE PATIENT WITH SUSPECTED URINARY TRACT DISEASE

Urinary tract disease is responsible for approximately 1.5% of all deaths. *Signs* and *symptoms* which suggest something amiss with the kidney or urinary tract are pain, nocturia, polyuria, frequency, thirst, proteinuria, hematuria, urinary casts, edema, acidosis, hypertension, azotemia, anemia, and bacteriuria. Few of these clues are produced exclusively by renal disease.

A *minimum* examination of *renal* function requires urinalysis, including urine pH, and qualitative measurement of protein, blood, and glucose; microscopic examination of the urine; serum electrolytes, hemoglobin, and creatinine. A more detailed examination will require measurement of calcium, phosphorus, urate, and in certain cases detailed microbiologic, hematologic, or immunologic evaluations.

Measurements of renal function are used to estimate the extent of kidney damage and to follow the progress of a disease. Therapy is based upon these estimates because tolerance for many drugs is related to their removal by the kidney. Renal function is often assessed by comparison with the responses of disease-free young adults to a physiologic stress. For example, a sodium concentration greater than 20 mmol/L in the urine of a hyponatremic volume-depleted patient indicates malfunctioning kidneys. This judgment can be made because we know that when healthy young adults are depleted of salt, urinary sodium falls below 20 mmol/L before they become hyponatremic.

Although we may define normal urine concentrating ability in terms of the response of *healthy young adults* to 24-hour water deprivation, we must be aware that this 'normal' does not apply to infants or older people, nor does it apply to those with a history of high water or low protein intake, and it may not apply to hospitalized or bedridden patients without renal disease. Table 8–4 gives some examples of the responses to various stresses expected from the kidneys of healthy young adults.

TABLE 8–4. Reference Ranges of Renal Function in Normal Young Adults

Sodium excretion
- Intake reduced to 10 mmol/day: Urinary excretion falls to 10 mmol/day after 5–7 days
- 9α Fluorohydrocortisone 0.5 mg bid for 3 days; Urine sodium < 10 mmol/day
- Response to salt load: Less than ⅓ excreted in 3 h; at least 24 h for complete excretion

Potassium excretion
- Urine output 10–20 mmol/day less than intake (remainder lost in feces)
- Intake reduced to 25–30 mmol/day; Urinary excretion falls to 25–30 mmol/day after 4–7 days

Water excretion and urine concentration
- Osmolality-40–1400 mosm/kg H_2O
- Specific gravity-1.001–1.040
- Response to water load: Maximum within 1–2 h; most excreted within 6 h; maximum* C_{H_2O} 8–30 mL/min/1.73 m^2
- Complete water deprivation: Maximum urine concentration attained after 24–36 h

Glomerular filtration rate
- Creatinine clearance: Decreases 10% per decade over age 40
- With fluid loss equivalent to 5%–7% of body weight, GFR falls 0–40%
- Saline loading: GFR increases 0–10%

Acid–base regulation
- Titratable acid 10–30 mmol/day
- Ammonium 30–50 mmol/day
- Maximum response to chronic acidosis:
  - Titratable acid 30–150 mmol/day
  - Ammonium 300–500 mmol/day
- Response to 1.9 mmol NH_4Cl/kg body wt orally as 1 dose: 5–8 h later, urine pH <5.3; titratable acid >25 mmol/3 h; ammonium >35 mmol/3 h
- Response to 140 mmol NH_4Cl/day for 5 days: Maximum titratable acid achieved after 3–4 days; maximum ammonium excretion after 4–5 days; daily acid excretion >120 mmol; plasma HCO_3^- and Cl^- do not change more than 6 mmol/L.

* $C_{H_2O} = \dot{V} - \dot{V}\frac{U_{osm}}{P_{osm}}$ where P = plasma, U = urine, and $\dot{V}$ = urine flow rate; bid = twice daily.

Our *concept of normal* must be qualified further because the kidney is part of many complex homeostatic mechanisms in the body, the cardiovascular and endocrine systems in particular. The expected renal response to hypoaldosteronism is natriuresis; the expected response to paroxysmal tachycardia is diuresis; the expected response to blood loss is a decreased GFR, and so on.

Structurally intact kidneys contribute to the *genesis* of some systemic disease states. This can be seen in patients with cardiac and liver failure. In these conditions the kidneys respond to abnormal systemic extracellular fluid distribution by retaining fluid and electrolytes. Signals to retain salt and water are sent to the kidney by reduced arterial pressure, increased ADH and aldosterone secretion, and increased sympathetic nervous activity.

GLOMERULONEPHRITIS

The classification of glomerulonephritis appears somewhat confused because of the rapid increase in knowledge in this area. The biochemical hallmark of this group of diseases is *proteinuria.* Some patients also exhibit microscopic or gross hematuria and red cell or heme casts in their urine. All patients are found to have histologic abnormalities in their glomeruli when biopsy specimens are examined with the electron microscope. In most cases abnormalities are visible with the light microscope as well.

The clinical presentation, the history, and histologic changes are used to distinguish the various forms of glomerulonephritis. Unfortunately, the borders between the sets of diagnostic criteria are not always clear. A recent international study of glomerulonephritis in children provides an example of the sort of information needed to rationalize the diagnostic process (Table 8–5). Unfortunately, this information is not available for all forms of glomerulonephritis.

Four broad categories of histologic change are associated with the various forms of glomerulonephritis (Fig. 8–7):

TABLE 8-5. Common Forms of Glomerulonephritis in Children (Results of an International Study, 1978)

		Characteristics of Patients (Frequency in %)			
Histologic Diagnosis	*Proportion of Patients (%)*	*Age ≤ 6 yr*	*Hematuria*	*Low Serum Complement*	*Raised Serum Creatinine*
Minimal-change	76.4	80	23	1.5	32.5
Membranoproliferative	7.5	2.6	59	74.3	50
Focal and segmental	6.9	50	48	3.7	40.6
Others	0.2	—	—	—	—

1. *Minimal change*—uniform thickening of the basement membrane and fusion of foot processes, seen only with the electron microscope; no evidence of immunologic processes
2. *Focal* (segmental) glomerulosclerosis—segmental sclerosis or hyaline deposition beneath endothelial cells of some glomeruli
3. *Membranous*—deposits of electron-dense material along the outer aspect of the greatly thickened basement membrane; thickening of the basement membrane may be visible with the light microscope
4. *Proliferative*—proliferation of one or more cell types in the glomerulus with occasional exudation of polymorphs, capillary wall thickening with subendothelial deposits

No single *biochemical* measurement can be used to make a diagnosis; however, there is a tendency for groups of abnormalities to be associated with certain histologic changes. For example, hypocomplementemia, hematuria, and raised serum creatinine are more common in children with membranoproliferative glomerulonephritis than in children with minimal-change disease. Edema and hypoalbuminemia are found more frequently in association with the minimal-change form than with other forms of glomerulonephritis in children.

Pathogenesis

Richard Bright in 1836 described the classic form of ***poststreptococcal glomerulonephritis.*** He observed that 2 to 3 weeks following an infection, frequently a pharyngitis (subsequently shown to be caused by certain types of streptococci), some patients develop acute renal failure, hematuria, proteinuria, fever, and malaise. After a course lasting days to weeks, more than 90% of the patients recover. However, some die in the acute phase and a few go on to chronic renal failure.

The latent period between the infection and the renal disorder suggests activation of the immune system. This hypothesis has been investigated by creating experimental models of glomerulonephritis in animals.

Circulating Immune Complexes. About 10 days after the injection of foreign proteins into an animal, antibody production increases to the point that soluble antigen–antibody complexes are formed. These complexes lodge in glomerular capillaries. Complement is activated (see Chap. 4), and the chain of inflammatory events that this sets off may contribute to further deposition of antigen–antibody complexes in the glomerulus. The response of a rabbit to a single injection of bovine albumin resembles the findings seen in human poststreptococcal glomerulonephritis. Repeated injections of foreign protein produce a variable response. Some rabbits develop what looks histologically like membranous glomerulonephritis; others show a more proliferative picture.

Antikidney Antibody. When antikidney serum from immunized guinea pigs is injected into a rabbit the antibodies adhere to the rabbit's glomeruli and initiate a series of events. Complement is activated, leading to the release of vasoactive compounds and chemotactic factors which attract polymorphonuclear leukocytes; proteolytic enzymes released by these polymorphs destroy segments of the glomerular capillary walls. The co-

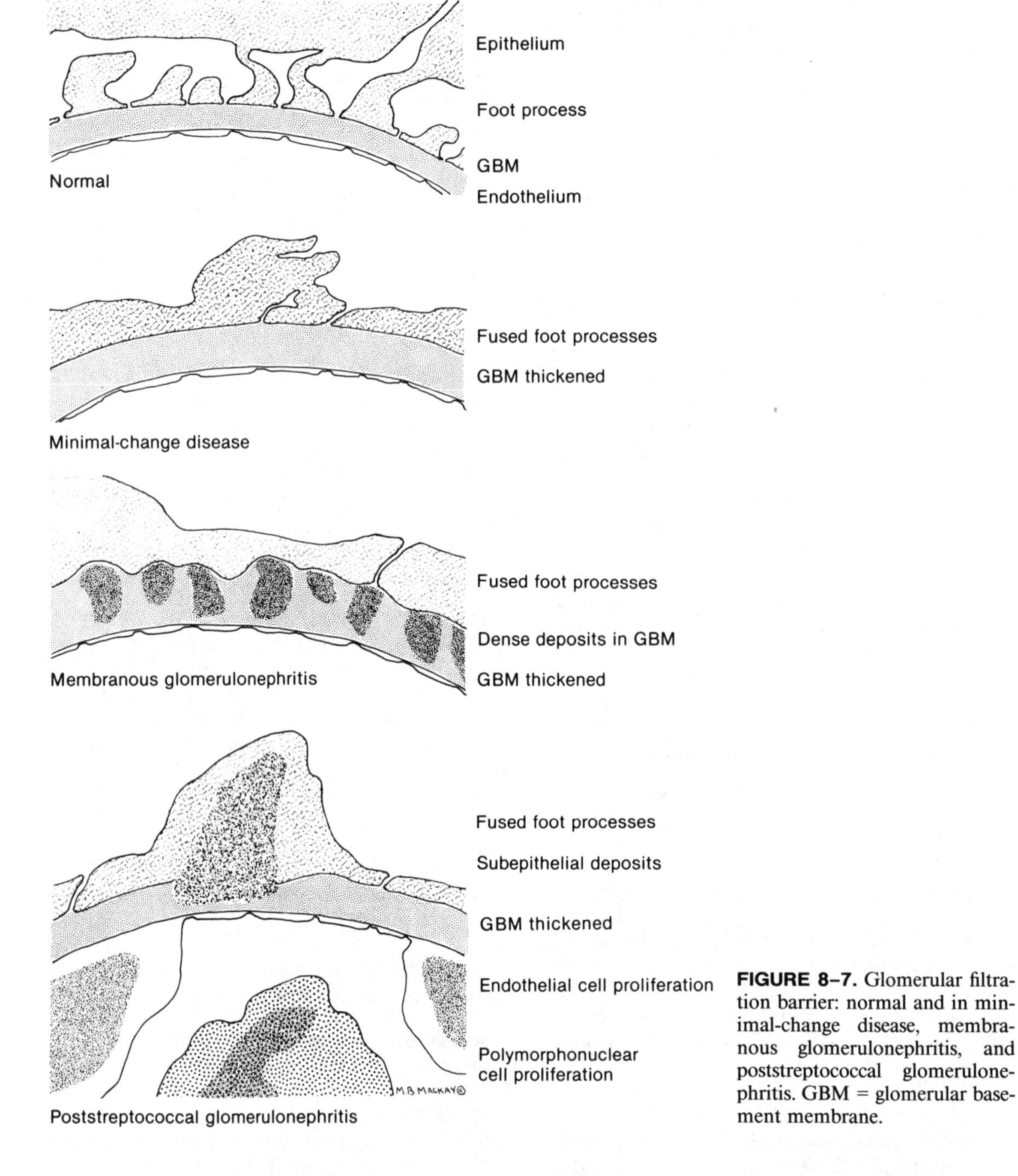

FIGURE 8–7. Glomerular filtration barrier: normal and in minimal-change disease, membranous glomerulonephritis, and poststreptococcal glomerulonephritis. GBM = glomerular basement membrane.

agulation and fibrinolytic systems are also activated. This inflammatory reaction is similar to that initiated by the injection of antiglomerular basement membrane antibodies.

Nature of Human Glomerulonephritis. Glomerulonephritis may occur as a primary renal disease, or as part of a systemic disorder such as *systemic lupus erythematosus* or *diabetes mellitus.* It is associated with deposition of *immune complexes* in 70% to 80% of the cases. Usually globulins are observed as granular deposits along the glomerular basement membrane (GBM) in biopsy specimens treated by immunofluorescent techniques. *Goodpasture's syndrome* is the name given to the rare condition in which antibasement membrane antibodies attack both the kidney and the lung. In this condition immunofluorescent stains show a smooth linear deposit along the glomerular capillary walls.

The *antigens* identified in immune-complex glomerulonephritis include some drugs, nephritogenic strains of streptococci, salmonella, and other bacteria, parasites, viruses, and endogenous antigens such as DNA or thyroglobulin. Relative immune deficiency, or inefficient removal of antigen/antibody complexes from the glomeruli, may predispose some individuals to develop glomerulonephritis. In roughly one third of the cases of glomerulonephritis there is little or no evidence of either immune-complex– or anti-GBM–initiated disease. *Minimal-lesion disease,* a common cause of proteinuria and the nephrotic syndrome in children, is an example. Nonetheless, this condition often responds to therapy that is presumed to act upon the immune system.

The ***diagnosis*** of glomerulonephritis relies upon a knowledge of several factors: historical details, including the age and sex of the patient; urinalysis; estimates of renal function; measurements of protein excretion; and finally renal biopsy. Table 8–6 shows some of the types of human glomerulonephritis. Note that the diagnosis is based largely on the histologic appearance of the glomeruli and that the same appearance can be associated with different clinical presentations. Laboratory findings for some immunologic tests are given to indicate

TABLE 8–6. Immunologic Findings in Some Forms of Glomerulonephritis

Syndrome	Disease	C_3	CIC	ASO	Ab
Nephritic	**Primary renal**				
	Post streptococcal	↓	+	+	−
	Membranoproliferative	↓	+	±	−
	Mesangial proliferative	N	±	±	−
	Systemic				
	Systemic lupus erythematosus	↓	+	−	Antinuclear
	Goodpasture's syndrome	N	−	−	Anti-GBM
	Disseminated vasculitis	↓-N	+	−	−
Nephrotic	**Primary renal**				
	Minimal lesion	N	±	−	−
	Membranous	N	±	−	−
	Membranoproliferative	↓	+	±	−
	Post streptococcal	↓	+	+	−
	Systemic				
	Systemic lupus erythematosus	↓-N	+	−	Antinuclear
	Leukemia and lymphoma	N	−	−	−
	Amyloidosis	N	−	−	↑ monoclonal light chains
	Diabetes mellitus	N	−	−	−
Asymptomatic urinary abnormalities	Berger's (IgA) disease	↑-N	±↑IgA	−	−
Chronic renal failure	Many different forms of primary renal and systemic glomerulonephritis				

C_3 = C_3 component of complement; CIC = circulatory immune complexes by one of several assays; ASO = positive antistreptolipin O- or anti DNAase B titer indicating prior streptococcal infection; nuclear Ab = antibodies to single or double stranded DNA of nuclear origin; GBM = glomerular basement membrane; N = normal; ↓ = decrease; + = positive; − = negative.

the variety of responses that occur. Prognosis and response to therapy are based on the age and sex of the patient, the renal biopsy, the estimates of renal and immune function and protein excretion.

Nephrotic Syndrome

All types of glomerulonephritis are associated with some increase in protein excretion but in certain cases the ***proteinuria*** can become ***massive.*** The proteinuria may occur by itself or in conjunction with hematuria. Protein appears to escape into the urine because of a change in the electrostatic charge of the GBM caused by a reduced proportion of negatively charged groups in the glycosaminoglycan sialic-rich glycoprotein component. The *severity or selectivity* of the lesion can be judged by measuring the ratio of small- to large-relative-mass proteins in the urine. Severe disease, with a poor prognosis, appears more frequently in association with nonselective proteinuria, in which the proportion of large proteins in the urine is greater.

By definition the ***nephrotic syndrome*** is said to exist when protein excretion exceeds 3.5 g/24 hours. The syndrome may accompany minimal-lesion disease or occur in the course of poststreptococcal glomerulonephritis and other types of nephritis. The natural history of the nephrotic syndrome is determined by the type of glomerulonephritis causing the proteinuria. A large proportion of the protein lost is albumin; hypoalbuminemia will result if synthesis cannot cope with the renal loss. Albumin synthesis by the liver increases in some patients and is unchanged or even decreases in others. Protein catabolism is greatly increased, much of it occurring in the kidney, as protein that escapes into the glomerular filtrate is taken up by the tubular cells and metabolized.

Pinocytosis of protein by tubular cells induces swelling and the formation of lipid deposits and vacuoles in these cells. Some cells are shed into the urine in which they appear as so-called oval fat bodies. Lipid in these cells is birefringent and when viewed with polarized light gives the appearance of Maltese crosses. Lipid droplets also appear in casts, which are excreted in large numbers by patients with proteinuria. The process that initiates the increased glomerular permeability may produce no other functional defect, at least in the initial stages of the disease.

Nephrotic Edema. Hypoalbuminemia interferes with the ***maintenance of plasma volume*** by its effect on fluid transfer across capillary walls. Normally, hydrostatic pressure forces fluid into the interstitium, while protein oncotic pressure tends to draw fluid back into the capillary lumen. However, when the albumin concentration falls the plasma has a diminished capacity to retain water, so that plasma volume falls and interstitial fluid rises. Baroreceptors responding to a fall in cardiac output and arterial pressure activate the sympathetic nervous system, causing vasoconstriction, reduced renal blood flow, and reduced GFR. The renin–angiotensin system is activated, causing secondary hyperaldosteronism. ADH is released in response to reduced stretch of receptors in the cardiac atria. In some cases, at the same time as protein permeability increases, glomerular permeability to water is reduced and the GFR falls.

In ***summary,*** renal excretion of salt and water decreases, owing to increased aldosterone and ADH levels, and in some cases to a reduced GFR. There are also hemodynamic and other unspecified alterations of the kidney which favor reabsorption of a greater proportion of filtered sodium and water in the distal nephron. It should be noted that these salt- and water-retaining factors may be operative only during the phase of edema formation; once sufficient fluid has been retained to raise the interstitial pressure a new balance point is established. At this point plasma volume, blood pressure, and hormonal secretion may all return to their previous normal ranges. But the patient is swollen with edema fluid, particularly in dependent portions of the body and in areas, such as about the eyes, where tissue is loose and easily expanded. The *hallmarks* of the nephrotic syndrome are proteinuria, hypoalbuminemia, edema, and hyperlipidemia.

Nephritic Syndrome

Many patients with glomerulonephritis do not develop the nephrotic syndrome. Instead they present with varying degrees of proteinuria, hematuria, and reduced GFR. In its mildest form the disease may produce nothing more than transient or constant mild proteinuria with or without hematuria. More severe disease is associated with a decreased GFR, and acute renal failure may occur. Depending on the nature of the initiating process the patient may recover or progress to chronic renal failure and uremia.

Nephritic Edema. Fluid retention and edema formation associated with the *nephritic* syndrome result from the low GFR and high fractional sodium and water reabsorption produced by glomerular disease. Fluid retention leads to hypertension with overflow from peripheral capillaries into the interstitium. Hypoalbuminemia is not a causal factor in this situation.

Persistent Asymptomatic Urinary Abnormalities

Persistent or recurrent hematuria is associated with deposition of IgA and IgG in the glomeruli of some patients. The commonest form of this condition, called *Berger's disease,* has usually a relatively benign prognosis. Similar asymptomatic changes may occur with certain hereditary and systemic diseases such as systemic lupus erythematosus.

A small proportion of the population (<0.1%) have persistent asymptomatic proteinuria with no other laboratory findings. At least half of these people have no structural changes in their glomeruli.

Cystic Diseases

These are a group of hereditary defects in kidney structure that may become manifest in childhood or adulthood. They lead usually to chronic renal failure. The commonest form is adult polycystic disease.

PYELONEPHRITIS AND INTERSTITIAL NEPHRITIS

Bacterial infection of the kidney in its acute form may produce edema, and infiltration with inflammatory cells. In the chronic form fibrosis with distortion of the tubules, atrophy, and obstruction are the predominant findings. A similar *histologic* picture, called interstitial nephritis, can be produced by radiation, immunologic attack on transplanted kidneys (i.e., graft rejection), toxic substances such as the analgesic phenacetin, metabolic disorders (hypercalcemia, hypokalemia, hyperuricemia, oxalate nephropathy), urinary tract obstruction, and sickle cell anemia.

Pyelonephritis may develop from hematogenous spread of bacteria, or more commonly from retrograde infection through the ureter. Urinary tract infection is common in the female population. Bacteriuria is found in 1% of young girls and 3% to 4% of teenage girls. At least 5% of women will have bacteriuria at some time. Very few develop pyelonephritis. Those who do usually have urinary tract obstruction or a defect which permits the reflux of urine from bladder to ureters.

ACUTE RENAL FAILURE

Prerenal Failure

Undoubtedly the commonest cause of decreases in urine flow and GFR is ***hypovolemia,*** usually associated with hypotension. Hemorrhage, burns, crush injuries, heart failure, diarrhea, vomiting, inadequate fluid intake, or bacterial shock may initiate the response. With decreased blood pressure intrarenal blood flow is altered and salt and water reabsorption increase. Systemic baroreceptors that respond to hypotension stimulate the release of ADH and aldosterone, which further enhance salt and water reabsorption. Consequently the kidney produces a small flow of concentrated urine, low in salt. Up to a point this essential adaptive response has only beneficial results. ***Diet*** is one of the factors that determines the point at which *oliguria* (low urine flow) becomes detrimental to health. A sick person eating no food can excrete daily hydrogen ion and protein waste products in 400 mL of urine. Carbohydrate does not add appreciably to the demands upon renal excretory capacity, but protein metabolites coming from either catabolism or the diet will accumulate in body fluids and cause toxic effects (*uremia*) unless excreted by the kidneys.

A person may sustain *blood loss* equivalent to 1% to 2% of body weight, or a loss of extracellular fluid equivalent to 4% to 5% of body weight, without large decreases in renal blood flow. However, if blood volume and pressure decrease further, the renal blood flow and GFR fall rapidly. *Vasoconstriction* induced by renal sympathetic nerves, circulating catecholamines, and the renin–angiotensin system plays an important role in the deterioration of function.

As discussed earlier, the kidney's capacity to *concentrate urine* is related to GFR because there

TABLE 8–7. Urine Sodium Concentration in Acute Renal Failure

	Percent of Patients	
Urine Sodium	*Oliguric*	*Nonoliguric*
>25 mmol/L	81	32
10–25 mmol/L	13	56
<10 mmol/L	6	12

(Brenner BM, Rector FC, Jr: The Kidney. Philadelphia, Saunders, 1976)

is an optimum rate of flow and solute delivery into the countercurrent system. At low flow rates the system does not function efficiently; thus, as the GFR falls to low levels so does the concentration of urine.

Plasma urea concentration rises in hypovolemia for two reasons, first because increased protein catabolism and urea production usually accompany the initiating disease process, and second because urea reabsorption from the tubular fluid increases at low urine flows. Creatinine concentrations rise also, but more slowly than urea levels.

Up to a point the decreased urine flow and rising blood urea produced by hypovolemia may be *reversed* by appropriate fluid and electrolyte therapy. If renal failure is reversed then as far as we know there are no structural or functional sequelae. The patient is said to have had *prerenal failure.* Unfortunately, in some cases volume repletion with fluid, electrolytes, and blood does not produce a return of renal function. Some clinicians believe that an osmotic diuretic such as mannitol, or a strong chemical diuretic such as furosemide, given at the appropriate time, can induce a return of renal function when volume expansion alone has failed. Should the condition prove to be *irreversible* the patient then has either *intrinsic acute renal failure* or *postrenal (obstructive) failure.* Note that volume expansion is the treatment for prerenal failure, but that continued volume expansion of a patient with intrinsic renal or postrenal failure may lead to congestive heart failure and death.

Intrinsic Acute Renal Failure

The progression from hypovolemia to prerenal failure to irreversible intrinsic renal failure, as just described, is probably responsible for many of the cases of acute renal failure associated with surgery, accidents, and blood loss during pregnancy. However, intrinsic renal failure may also result from toxic injury caused by chemicals or drugs, or from renal diseases such as glomerulonephritis. For up to 30% of cases the etiology is unknown.

The usual ***diagnostic criteria*** for acute renal failure are a urine flow less than 400 mL/24 hours and evidence of a decreased GFR. Below 400 mL a starving individual begins to accumulate toxic waste products and develops acid–base, fluid, and electrolyte abnormalities. Patients with prerenal failure tend to produce concentrated urine low in sodium, whereas those with intrinsic renal failure produce dilute urine with a sodium concentration greater than 25 mmol/L. As is often the case, classic descriptions do not apply in all instances. As shown in Table 8–7, the *urine sodium* concentration has a limited sensitivity and specificity for distinguishing between prerenal and intrinsic renal failure in the absence of other information about the patient's general condition and renal function. Up to 50% of patients with acute decreases in GFR and the systemic consequences of renal failure may never have oliguria while under medical care. These people are considered to have nonoliguric acute renal failure. *Fractional sodium excretion,* calculated as the ratio

$$\frac{\text{U/P sodium}}{\text{U/P creatinine}} \quad \text{where U = urine P = plasma}$$

may be a more sensitive index of renal tubular dysfunction. Values should be <1 in patients with prerenal failure.

Acute Tubular Necrosis. The general title of *acute tubular necrosis* (ATN) is currently applied to cases of intrinsic renal failure produced by toxins or by hypovolemia and shock. The former type is referred to as *toxic* ATN and the latter as *ischemic* or *vasomotor* ATN. The ultimate outcome of the disease process depends upon the age of the patient, the nature of the precipitating cause, and extrarenal involvement. Classically, the 40% to 60% of patients who survive have the following ***history.*** An initial period of oliguria, which lasts from several hours to weeks (mean, 10 to 14 days), is followed by a diuretic phase during which urine flow gradually rises to supranormal levels. There is then a recovery phase in which urine flow returns to normal levels and renal function improves over a period of weeks. Finally, there is a protracted convalescent period of 6 months to several years, during which the patient is able to function normally while renal function slowly improves and reaches a new stable state. Completely normal renal function may be regained by half the young people who survive the initial disease process.

Laboratory findings in acute tubular necrosis indicate that the urine frequently contains brown granular casts and tubular epithelial cells said to be characteristic of ATN, but in 20% to 30% of cases the urine sediment does not present specific distinguishing features. In most cases hyaline casts or heme granular casts, red cells, and leukocytes indicate the presence of nonspecific renal inflammation, urine stasis, proteinuria, hematuria, and cellular necrosis.

In the ***oliguric phase*** the blood urea nitrogen (BUN) rises daily by 7 to 11 mmol/L (20 to 30 mg/dL); however, with increased catabolism the rise may be as much as 35 mmol/L (100 mg/dL). Creatinine rises 40 to 90 μmol/L (0.5 to 1 mg/dL) per day unless there is destruction of muscle, in which case the daily increase may be up to 175 μmol/L (2 mg/dL). Metabolic acidosis is common; bicarbonate falls 1 to 2 mmol/L per day, as sulfate, phosphate, and other anions accumulate. Roughly 400 mL of water are released daily as cells break down and endogenous fat and protein are catabolized. This endogenous fluid must be included in the total body balance. Body weight should fall by 0.2 to 0.5 kg/day if the patient is properly hydrated. Overhydration, with consequent hypertension and circulatory failure, is a constant hazard. Hypernatremia or hyponatremia may develop, depending on the success or failure of clinical management of fluid and electrolytes.

Hypocalcemia and hyperphosphatemia occur early in the oliguric phase but tetany is rare, probably because of the concurrent acidosis, which increases ionized calcium concentration. Anemia develops as a result of decreased erythropoiesis, increased hemolysis, and a generalized bleeding tendency. Massive gastrointestinal hemorrhage is a common cause of death in patients with acute renal failure.

Hyperkalemia inevitably ensues unless diuresis or treatment by peritoneal or hemodialysis is instituted. The potassium comes from cells as they break down or as their contents are catabolized; thus, the rise in potassium concentration depends on the catabolic rate. Potassium increases usually by 0.5 mmol/L/day but under extreme conditions fatal hyperkalemia may develop in a matter of hours.

The retention of toxic products that are not normally measured probably accounts for many of the other manifestations of acute renal failure. These are anorexia, nausea, vomiting, confusion, stupor, convulsions, neuromuscular dysfunction, and decreased resistance to infection.

Studies of the ***pathogenesis and clinical course*** of ATN have involved experiments using animal models. These suggest that the initial oliguric phase is produced by renal vasoconstriction but that the continuation of the oliguria results from a combination of factors, including tubular obstruction by casts, compression of tubules by interstitial edema and inflammation, leaks in the tubule walls, and changes in the permeability and surface area of glomerular filtering barriers.

The conditions required for urine flow to return are not known. Usually when the ***diuretic phase*** begins the GFR is only 5% to 10% of normal. Urine flow may rise to 5 L/24 hours with little improvement in the GFR. Thus, while fluid and electrolyte balance return toward normal, accumulation of urea and protein waste products may continue. The magnitude of the diuresis is influenced by the extent of the fluid retention during oliguria. Present-day therapy strives to prevent overhydration; therefore the diuretic phase of ATN is no longer as dramatic as it was when fluid balance was ignored.

During the early diuretic phase the patient may occasionally become dehydrated. Usually fluid–electrolyte and acid–base management becomes somewhat easier at this point, but for the first few days renal function is still very poor and the consequences of impaired excretion may continue to worsen.

Return of renal function in the ***recovery*** and ***convalescent phases*** is brought about partly by maturation of newly formed cells and partly by hypertrophy of nephrons which escaped permanent damage. When recovery begins, with the patient on free fluid intake and the BUN falling, the GFR may be still only 20% of normal. The recov-

TABLE 8–8. Types of Renal Calculi and Their Causes

Type	% of Stones Containing	Causes
Calcium	90%	Idiopathic (Increased calcium absorption?) Hyperparathyroidism Renal tubular acidosis Excess calcium ingestion
Uric acid	5%–10%	Gout Idiopathic
Cystine	1%–2%	Cystinuria
Oxalate	65%	Rarely prime cause

ery phase may take up to 2 years and tends to be less complete in older patients. The duration of oliguria and the severity of the biochemical derangements in the acute phase do not correlate with ultimate recovery. Mortality also varies with the initiating cause.

Renal Failure in Glomerulonephritis and Interstitial Nephritis. Renal function may be reduced acutely by inflammatory processes in the glomerulus or tubules. Pyelonephritis and glomerulonephritis are common medical causes of acute renal failure, responsible for 20% to 25% of the cases. The course of acute renal failure in these conditions may resemble that in acute tubular necrosis. The underlying disorder can usually be distinguished by other features of the illness; however, renal biopsy is sometimes required to make a diagnosis before instituting specific therapy.

Postrenal (Obstructive) Renal Failure

Obstruction to urine flow from injury, stones, inflammation, and neoplasms should always be kept in mind as a possible cause for oliguria. Complete cessation of urine flow, or *anuria,* is most often attributable to lower urinary tract obstruction.

RENAL CALCULI

Stones form in the urinary tract when mineral salts precipitate about a mucoprotein core. Up to 60% of the material in a stone is protein; the rest is made up of varying proportions of Ca, P, Mg, NH_4, uric acid, occasionally cystine, and very rarely xanthine. Precipitation, which may begin in the terminal collecting ducts, is favored by reduced urine flow; concentrated urine; excess excretion of Ca, urate, cystine, or xanthine; absent or reduced excretion of protective substances; foreign bodies or other 'seeds' about which salts may precipitate; infection; and stasis of urine. Obstruction to urine flow in the ureters can predispose to infection and stone formation. Stones themselves may produce obstruction.

The types of renal calculi are shown in Table 8–8. Some 90% contain ***calcium.*** In most cases the problem is associated with increased Ca excretion, of unknown etiology, though possibly linked to increased Ca absorption in the intestine. This is called *idiopathic hypercalcuria* and the trait occurs more frequently in some families than in others. In 5% to 10% of patients hyperparathyroidism is responsible for the excess Ca excretion. In a few, renal tubular acidosis or excess Ca ingestion can be implicated.

Magnesium pyrophosphate and some polypeptides appear to prevent the precipitation of minerals in the urine of non–stone-forming people. People afflicted with stones may lack these substances in their urine.

Uric acid accounts for stones in 5% to 10% of patients. Uric acid solubility decreases greatly at a low urine pH. It has been suggested that a major defect in urate stone formers is their inability to raise the urine pH, possibly because of defective NH_3 formation. Urate hyperexcretion is associated with primary or secondary *gout* in some patients, but most have *idiopathic hyperuricosuria.* Only 10% to 25% of those with primary gout and 30% to 40% of those with secondary gout develop renal stones.

Cystine stones occur in 1% to 2% of calculi formers. Cystinuria is a recessive genetic trait which occurs in several forms. Although ***oxalate*** is found in two thirds of stones it is rarely the prime cause of the stone.

HYPERTENSION

(See also Chap. 9.)

The blood pressure in the renal artery influences salt and water excretion by a direct physical effect on nephron function. In ***normal kidneys*** a small rise in arterial pressure produces a large increase in salt excretion. If salt intake is increased salt and water are retained, which increases extracellular fluid volume, blood volume, cardiac output, and arterial pressure. The rise in arterial pressure increases salt excretion to bring intake and output back into balance. Doubling the salt intake produces an almost imperceptible increase in blood pressure (points A and B in Fig. 8–8) and in extracellular fluid volume. This is not true for people with ***advanced renal disease,*** for in their kidneys the natriuretic response to arterial pressure is blunted. For this reason they tend to operate with an expanded extracellular fluid volume and elevated blood pressure, even when they are eating a standard North American diet. As can be seen in Figure 8–8 (points A′ and B′) their blood pressure is much more responsive to changes in salt intake than is the case with normal individuals. These patients have ***volume-dependent hypertension.***

Not all cases of hypertension related to kidney dysfunction can be explained by the mechanism just described. In some cases salt reabsorption is stimulated by excess aldosterone secretion; in others, catecholamines and renal nerves may alter renal vascular tone and the relationship between arterial pressure and salt excretion. We have no explanation for the development of hypertension in the majority of people with normal renal function.

Hypertension may also result from overactivity of the *renin–angiotensin system.* Partial stenosis of one renal artery reduces blood flow to the kidney, which stimulates renin release. Renin acts upon angiotensinogen to form angiotensin, which is a potent vasoconstrictor and also stimulates aldosterone secretion. There are at least two forms of ***renovascular hypertension.*** In patients with *two kidneys,* only one of which is constricted, the hypertension is sustained by high renin secretion from the affected kidney. However, if only *one kidney* remains and its artery is constricted then renin may play a role in the onset of the hypertension, but in the chronic state extracellular fluid volume is increased and renin secretion is within normal limits.

Diagnosis of renal artery stenosis requires measurement of renal vein renin concentration and radiologic demonstration of the stenotic area. Both procedures are hazardous and should be undertaken only when surgical correction of the lesion

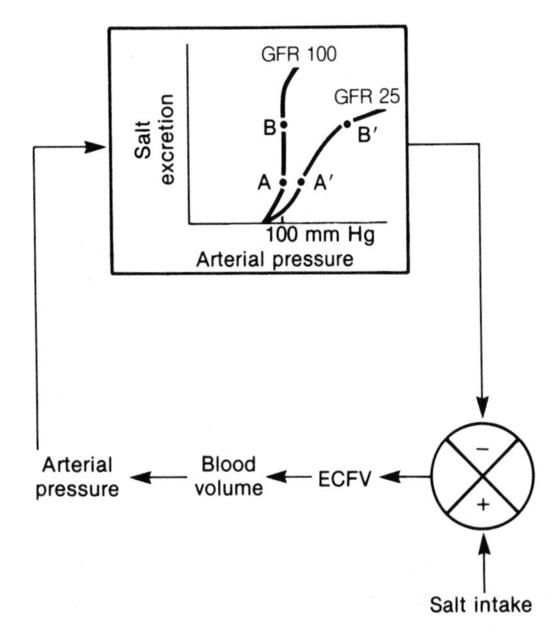

FIGURE 8–8. Regulation of arterial pressure by the kidney. Ingested salt increases extracellular fluid volume (ECFV), blood volume, and arterial pressure. A normal kidney (GFR = 100 mL/min) responds to small changes in arterial pressure; an increase in pressure immediately increases sodium excretion. When sodium intake is doubled an almost imperceptible rise in arterial pressure (points A and B) is required to maintain sodium balance. However, a diseased kidney (GFR = 25 mL/min) is less responsive to changes in arterial pressure; consequently higher arterial pressures are required to restore balance between intake and output when salt intake is increased (points A′ and B′).

is desirable and possible. *Plasma renin activity* (PRA) in a peripheral venous blood sample must be related to salt intake. This so-called *renin profile* is obtained by comparing the PRA with two 24-hour urine sodium excretion rates. The renin profile may be high, especially in young people, without anatomic evidence of renal artery stenosis, and it may be normal in patients with renal artery disease.

Some hypertensive patients, for unknown reasons, enter a phase in which the blood pressure rises rapidly to very high levels, with diastolic pressures greater than 140. They develop cerebral edema, edema of the optic nerve (papilledema), heart failure, and progressive renal failure. This type, ***malignant hypertension,*** is associated with raised plasma renin activity.

One of the consequences of prolonged hypertension is a degeneration of the arterial walls. This process, called *nephrosclerosis* when it occurs in the kidney, leads to a loss of glomeruli and renal functional capacity. It may at times be difficult to determine whether hypertension produced the renal disease in a patient or renal disease caused the hypertension.

Only 5% to 10% of hypertension can be attributed to primary renal disease or renal artery stenosis, but the kidney is a target organ for damage by sustained high blood pressure. *Routine examination* of a hypertensive patient requires urinalysis and serum creatinine to screen for renal disease; serum electrolytes to screen for primary or secondary hyperaldosteronism; blood glucose to screen for diabetes mellitus; serum urate because it is increased by mild renal failure and may be increased to pathologic levels by some antihypertensive drugs; and lipids to screen for hypercholesterolemia as a risk factor.

CHRONIC RENAL FAILURE

The end result of progressive renal damage is much the same no matter what the cause of the disease may have been. Subtle differences in the balance between glomerular and tubular function can be found in the early stages of deterioration when the extent of glomerular relative to tubular destruction may vary from one disease to another. However, these distinctions are of academic interest only, because the major deleterious effects of renal failure come from the ***loss of functioning nephrons.*** The remaining nephrons will hypertrophy. Recent evidence suggests that hypertension, and increased GFR per nephron, may by themselves produce glomerular sclerosis. The progression of chronic renal disease may be slowed by decreasing protein intake to reduce GFR, and by lowering blood pressure.

The GFR or ***creatinine clearance*** gives the best index of residual functioning renal tissue. Arbitrarily, we may classify renal disease as *mild* when creatinine clearance is greater than 50 mL/min, *moderate* when clearance is between 15 and 50 mL/min, and *severe* when clearance is 5 to 15 mL/min. Patients with a clearance of less than 5 mL/min will die unless treated by dialysis or transplantation.

If a person is to remain alive urine flow and waste product excretion must equal intake and metabolic production. To achieve this balance with reduced kidney tissue requires 'trade-offs.' The trade-offs involve accepting changes in body function or composition in order to facilitate the excretion of potentially toxic substances.

Substances such as *creatinine* and *urea* are excreted primarily by glomerular filtration. As nephrons are lost and the GFR falls these substances are retained in the extracellular fluid until plasma levels are raised and a new balance is attained, at which time excretion again equals production. Creatinine and urea are relatively nontoxic, so the retention probably produces few side effects.

Urate is excreted partly by filtration and partly by tubular secretion. With loss of functioning renal tissue, urate levels in the extracellular fluid tend to rise. However, some compensation for the loss of glomerular function is achieved by an increase in tubular secretion. Thus, urate levels do not rise as rapidly or to the same extent as creatinine and urea concentrations.

As nephrons are lost the capacity to excrete *phosphate* diminishes. Plasma phosphate levels rise; this leads to the deposition of calcium phosphate in bones and elsewhere, with a fall in plasma *calcium.* Parathyroid hormone (parathormone) is released in response to the hypocalcemia; in the early stages of renal failure this is sufficient to stimulate PO_4 excretion in the urine and to maintain balance. But the cost is an insidious development of abnormal bone metabolism due to the secondary hyperparathyroidism: In the normal kidney *vitamin D* is hydroxylated at the 25 position to form the active compound 1,25-dihydroxycholecalciferol. Synthesis of this compound, which stimulates Ca absorption in the intestine, is deficient in patients with chronic renal disease. Consequently, their Ca uptake decreases, thereby worsening their hypocalcemic state.

Another example of the trade-offs which are required to maintain balance occurs with *hydrogen ion* excretion. The loss of nephrons reduces the kidney's ability to form ammonia. This diminishes the H^+ excretory ability of the kidney, inducing a mild acidosis which is sufficient to stimulate the remaining renal cells to produce more ammonia.

A balance between production and excretion is regained but the cost is a systemic acidosis.

As nephrons are lost, so is their contribution to *sodium* excretion. If sodium intake is reduced in parallel with the loss of nephrons no adjustments need to be made in the function of the remaining nephrons. However, if sodium intake remains constant as kidney mass and GFR decrease, then the fractional reabsorption of sodium must decrease in order to maintain urinary excretion equal to intake. There is a price for this adaptation. As the GFR decreases sodium is retained, and the extracellular fluid volume and blood pressure rise until hormonal and physical changes bring renal sodium excretion back into line with intake. This new balance state is maintained with a sustained volume expansion and mild to moderate hypertension.

Potassium concentration in the extracellular fluid is well regulated until the terminal stages of renal failure. Potassium secretion in the distal tubule rises in response to the need to maintain secretion. Aldosterone secretion may play an important role in this process.

The ability to *concentrate urine,* which depends upon the integrated activity of many nephrons, is one of the first functions to show signs of failure as nephrons are lost. Mild polyuria is a problem for some patients in the early stages of chronic renal failure. Normally urine flow is low at night and high during the day. Patients with mild renal failure may lose this diurnal pattern of urine excretion; they then experience *nocturia.* Furthermore, as patients progress from mild to severe renal failure they lose the ability to withstand the ill effects of water deprivation. When they are subjected to water restriction their urine flow continues until volume depletion produces hypotension and a decreased GFR. With the reduced GFR comes further retention of toxic products which may cause nausea, vomiting, and diarrhea. Fluid deprivation is thereby aggravated and the patient's condition may become progressively worse.

The ability to excrete a *dilute urine* is retained after the power to concentrate urine has been lost. But the maximum volume of dilute urine is reduced. Thus, the patient is no longer able to excrete excess quantities of water rapidly and volume overload and hyponatremia may follow excessive water intake. The patient increasingly depends on the regulation of salt and water intake to maintain balance. Volume depletion or overload, hypotension or hypertension, hyponatremia or hypernatremia follow injudicious intake or unusual extrarenal losses.

Table 8–9 summarizes the ***common causes*** of chronic renal failure in adults.

Table 8–10 indicates that a simple measure of plasma or serum creatinine is not a sensitive index of reduced GFR when the patient has mild renal failure, and that urea is a somewhat less efficient index of renal disease.

TABLE 8–9. Etiology of Adult Chronic Renal Failure (CRF)

Etiology	Frequency in CRF (%)
Glomerulonephritis	28–44
Pyelonephritis, interstitial nephritis	16–28
Drug-induced, analgesic abuse	0.2–15
Etiology unknown	6–15
Cystic kidney disease	7–12
Renovascular hypertension	4–11
Hereditary diseases	1–5
Multisystem disease (*e.g.*, diabetes)	2–8
Other—tuberculosis, nephrocalcinosis, cortical necrosis	3–6

(Gurland HJ, Wing AJ, Brunner FP et al: Geographical distribution of renal diseases amongst patients accepted for dialysis and transplantation in Europe, pp 77–81. Proc VIIth Int Congress Nephrology, Montreal, 1978)

TABLE 8–10. Sensitivity and Specificity of Plasma Urea and Creatinine as Indices of Reduced GFR

1. GFR < 75% of normal, adjusted for age, sex, and body surface area

	Urea	Creatinine
Sensitivity %	67	69
Specificity %	91	96

2. GFR < 52% of normal, adjusted for age, sex, and body surface area

	Urea	Creatinine
Sensitivity %	88	96
Specificity %	84	89

GFR = Glomerular filtration rate.

Uremia

The final result of renal failure, whether it is acute or chronic, is a complex of signs and symptoms involving all organ systems and tissues. Before *dialysis* and *transplantation* were developed as treatments, uremia was the condition which heralded death. Most symptoms of the uremic syndrome are relieved by dialysis but a few are not. From this observation one concludes that a large part of the syndrome can be ascribed simply to a loss of the kidneys' ability to excrete waste products and to regulate fluids and electrolytes. Some of the waste products considered to be *toxic* are urea, ammonia, uric acid, hippuric acid, leucine, tyrosine, sulfates, phosphates, chlorides, potassium, H^+, organic acids, indican, guanidine, methylguanidine, guanidosuccinic acid, magnesium, phenols, aromatic oxyacids, and alkaloids. The *defects* they are assumed to be responsible for are stomatitis, esophagitis, gastritis, colitis, pancreatitis, pericarditis, pleuritis, dermatitis, bleeding, bruising, and meningitis. *Red cell* membrane function and the ATP-dependent ion pumps are inhibited; as a result, red cells are more fragile. *Nervous* function deteriorates. First there is impaired cerebration, then depression, and finally coma. Autonomic dysfunction is demonstrated by sweating, altered temperature control, diarrhea, and peripheral neuropathy.

In addition to these toxic effects are those resulting from defective *salt and water excretion*—edema, hypertension, congestive heart failure, hypernatremia, hyponatremia, hyperkalemia (with consequent neuromuscular dysfunction). Loss of *hydrogen ion* excreting capacity usually creates an acidosis, but an alkalosis can develop if intake of alkali is excessive.

The *anemia* that invariably accompanies renal failure is partly the result of toxic effects which depress the bone marrow, cause bleeding, and increase hemolysis. These defects clear with dialysis but the deficiency or inhibition of the action of erythropoietin does not. Hence dialyzed patients remain anemic.

Hypertension in the uremic state is largely caused by fluid retention but may, in some conditions, be aggravated by excessive renin release. Such hypertension cannot be controlled by dialysis alone.

KIDNEY TRANSPLANTATION

Kidney transplantation or dialysis can prolong the life of patients with chronic renal failure. Transplantation provides the best quality of life for most patients, but comparable survival rates and quality of life can be obtained with dialysis in some people.

An identical twin is the best donor for kidney transplantation; next best is a living related donor. The most common source, however, is a cadaver. Donor and recipient must have the same blood group. Best results are obtained when the histocompatibility genes also match; this may occur in a related donor but is unlikely with a cadaver donor. Four groups of genes code for the glycoproteins known as human leukocyte antigens (HLA). The group on chromosome 6 is called the major histocompatibility complex. There are four groups of HLA genes: A, B, C, and D (DR). Tissue typing identifies these HLA antigens.

A living donor's kidneys must be assessed by urinalysis, urine culture, creatinine clearance, and serum electrolytes before transplantation. The donor must be screened for cytomegalovirus, hepatitis B, and AIDS because these viruses can cause severe disease in the immunosuppressed recipient. Donors must also undergo thorough examination to ensure that they will not suffer from the loss of one kidney. Cadaver kidneys can be transplanted after being stored for more than 36 hours; some acute tubular necrosis, with serum creatinine up to 170 μmol/L and heme granular casts in the urine, is acceptable if the donor cadaver is young.

After transplantation the graft may be lost within minutes to hours as a result of donor-specific presensitization of the recipient. This is called *hyperacute rejection.* Two forms of *acute rejection* may occur during the first month. One form, which results from the development of antibodies against the donor interstitium and blood vessels, does not respond to treatment with steroids. The other, which results from antibodies against tubular antigens, does respond to steroid therapy. Acute rejection causes fever, pain, and swelling over the kidney and pyuria (leukocytes and casts in the urine). These symptoms are not usually associated with acute tubular necrosis, which can also occur following transplantation.

Chronic rejection occurs over a period of years as a result of interstitial infiltration and fibrosis.

DIALYSIS

Hemodialysis or peritoneal dialysis will sustain life when other therapeutic measures can no longer maintain fluid, electrolyte and acid-base balance. During *hemodialysis* the patient's blood passes through semipermeable tubules or across semipermeable membranes while a dialysis fluid circulates on the opposite side of the tubule or membrane. Fluid is removed from the patient's blood by ultrafiltration and small solutes are added or subtracted by diffusion. Filtration rate depends on the balance between hydrostatic pressure and protein oncotic pressure across the semipermeable membranes. Biochemical *monitoring* of the process requires measurement of electrolytes (including Ca, Mg and phosphorus), urea, creatinine, Hb, and hematocrit. Hypotension or hypertension can result from fluid imbalances. Symptoms of *dialysis disequilibrium* result from delayed transfer of urea from the brain when blood urea is decreased rapidly by dialysis. The brain becomes hyperosmotic to blood, resulting in cerebral edema, which causes altered levels of consciousness and convulsions. This condition can be prevented by dialyzing before blood urea levels are too high and by using shorter dialysis periods or a hyperosmolar dialysate.

During *peritoneal dialysis* the patient's own peritoneum is used as a semipermeable membrane. Dialysis fluid (1–2 L) is instilled through an indwelling catheter into the peritoneal cavity and is left there for up to 8 hours, then withdrawn. Diffusion, or clearance, occurs as during hemodialysis. Fluid is extracted by raising the effective osmotic pressure of the dialysis fluid with dextrose; this causes water to move from blood to peritoneal cavity. The quantity of water removed is regulated by changing the amount of dextrose added. Peritoneal dialysis may be used intermittently in hospital, or continuously with exchanges of 2 L every 4–8 hours during normal activity. The latter approach is called *continuous ambulatory peritoneal dialysis* (CAPD).

Management of the patient with end-stage renal failure requires careful adjustment of protein intake and frequency of dialysis. Plasma urea concentration is a good guide to therapy because it reflects protein metabolism. Many of the toxic and acidic substances that accumulate with renal failure are products of protein metabolism. If we assume that there is a single body compartment for the distribution of urea, then plasma urea depends on fluid volume, urea production (protein catabolism), and rate of removal. BUN is lowest immediately after dialysis and highest at the beginning of dialysis. In planning a dialysis program it is useful to aim for an *average* BUN over time and to adjust dialysis rates to reach this target.

SUGGESTED READING

BRENNER B, RECTOR Jr FC (eds): The Kidney. Philadelphia, WB Saunders, 1981

KLAHR S (ed): The Kidney and Body Fluids in Health and Disease. New York and London, Plenum Medical Book Co., 1983

LEVINE DZ (ed): Care of the Renal Patient. Philadelphia, WB Saunders, 1983

NEPHROLOGY FORUMS: In Kidney International, Official Journal of the International Society of Nephrology. New York, Springer–Verlag (ongoing series)

SELDIN DW, GIEBISCH G: The Kidney: Physiology and Pathophysiology. New York, Raven Press, 1985

9

Cardiovascular Disorders

Robert L. Patten

THE CARDIOVASCULAR SYSTEM

Disorders of the cardiovascular system are major causes of mortality in North America and Europe, accounting for approximately one half of all deaths. Most of this illness is caused by a degenerative disease of the arteries, *atherosclerosis,* which impairs blood flow to essential organs. *Ischemic heart disease,* caused by atherosclerosis of the coronary arteries, is responsible for over 50% of all deaths due to cardiovascular disease; vascular disease affecting the brain is responsible for another 20%. *Hypertension* contributes significantly to the development of these disorders as well as causing more specific lesions of the heart, kidney, and brain. Other disorders, such as congenital *malformations* of the heart and acquired abnormalities of the heart valves, also contribute to morbidity and mortality but occur less frequently. The clinical biochemistry laboratory is involved in the diagnosis and management of many of these disorders and in the identification of risk factors that are known to accelerate the atherosclerotic process.

STRUCTURE OF THE HEART

The heart is a muscular pump consisting of four chambers—the right and left atria and the right and left ventricles (Fig. 9–1). It is enclosed in a fibrous sac called the pericardium. Blood returns to the heart from the venous system by way of the superior and inferior venae cavae and enters the right atrium. During *diastole* it flows through the tricuspid valve into the right ventricle, and with ventricular systole into the pulmonary artery, and then to the capillaries of the lungs. It is collected by the pulmonary venous system and flows in the pulmonary vein to the left atrium, and then into the left ventricle through the mitral valve. With ventricular *systole* it is pumped through the aortic valve into the aorta, then through progressively smaller arteries throughout the body to the arterioles and then to the tissue capillaries, where the exchange of gases, nutrients, and wastes occurs. Representative pressures in the cardiovascular system during systole and diastole are given in Table 9–1.

STRUCTURE OF THE BLOOD VESSELS

The walls of the arteries (Fig. 9–2) consist of three layers: the innermost tunica intima, the tunica media, and the outer tunica adventitia. The *intima* includes a layer of endothelial cells, which forms the inner lining of the artery, and an elastic layer. The *media* is composed of smooth muscle cells interspersed with elastic material and an outer layer of elastic fibers. The *adventitia* consists mainly of elastic fibers, but there are also some collagen fibers. The size of these three layers varies with the size and location of the arteries.

The structure of *arterioles* is similar, but the layers are very thin and in the smallest arterioles the adventitia is almost nonexistent. *Capillaries* are small tubes, 7 μm to 9 μm in diameter, composed of endothelial cells surrounded by a fine basement

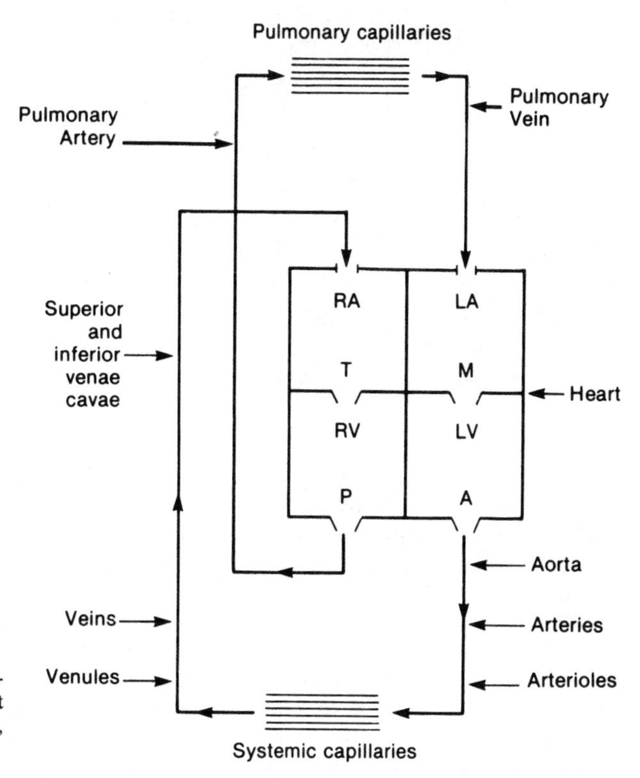

FIGURE 9–1. Schematic representation of the cardiovascular system. RA, LA = Right and left atria; RV, LV = right and left ventricles; T, M, P, A = tricuspid, mitral, pulmonary, and aortic valves.

membrane. *Veins* also consist of three layers of tissue and these vary considerably with the size and location of the vein. In general, veins have less elastic material, fewer smooth muscle cells, and more collagen fibers than arteries.

REGULATION OF CARDIAC FUNCTION

Contraction of the heart is normally initiated by specialized muscle cells located in the sinoatrial node in the right atrium. These cells regularly and automatically *depolarize,* which results in a wave of depolarization that spreads to the muscle of the atria. This wave is delayed briefly at the atrioventricular node before it proceeds down specialized fibers that are responsible for the conduction of the electrical impulse to the muscle of the ventricles.

In the *contraction* of cardiac muscle fibers, the intracellular thin filaments of actin slide past the thick filaments of myosin. This process requires adenosine triphosphate (ATP), which is derived mainly from the oxidation of fatty acids in the tricarboxylic acid cycle. Other substrates, such as glucose, pyruvate, and ketone bodies, are also utilized by the myocardium. Creatine phosphate in heart muscle serves as a reservoir of high-energy phosphate bonds for the generation of ATP.

TABLE 9-1. Pressures in the Cardiovascular System

	Systolic/Diastolic Pressure (mm Hg)
Right atrium	5/2
Right ventricle	20/2
Pulmonary artery	20/9
Pulmonary capillaries	10
Pulmonary vein	10
Left atrium	14/8
Left ventricle	120/8
Aorta	120/80
Brachial artery	120/80
Arterioles	50
Capillaries	30
Venae cavae	10

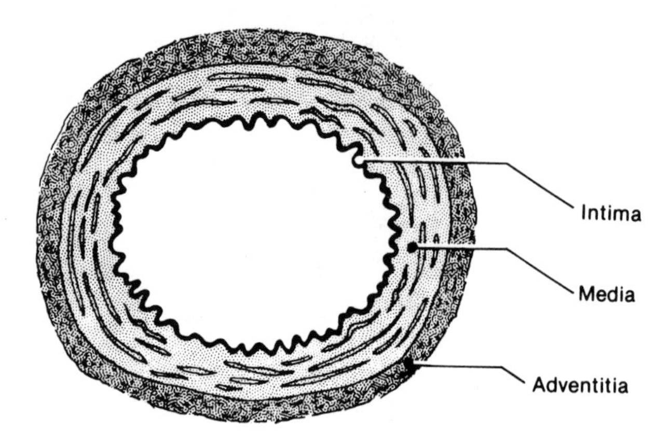

FIGURE 9–2. Component parts of the wall of a larger artery.

The volume of blood pumped by the heart each minute (***cardiac output***) varies according to physiologic needs, increasing, for example, during exercise. Changes in cardiac output are caused by changes in the heart rate and by changes in the volume of blood pumped with each beat (***stroke volume***). Normally a requirement for increased cardiac output is met by increases in both heart rate and stroke volume.

The *heart rate* is increased by stimuli from the sympathetic nervous system and by circulating catecholamines and is decreased by stimuli from the parasympathetic (vagus) nervous system. The *force* and *velocity of contraction* of the heart muscle are increased by sympathetic nervous system activity, and to a lesser extent by circulating catecholamines; this is called the *inotropic effect.* The force of contraction is also increased, within limits, by the amount of stretching of the myocardial fibers prior to contraction, which is proportional to the volume of blood present in the ventricles. An increase in the force of contraction results in an increase in the stroke volume.

REGULATION OF BLOOD PRESSURE

The level of blood pressure is determined by the rate of blood flow *into* the arterial system and the resistance to blood flow *out of* the arterial system. *Peripheral resistance* is a function of the degree of vasoconstriction in small muscular vessels, the arterioles, and the precapillary sphincters. The moment-to-moment regulation of blood pressure is mediated by the ***autonomic nervous system*** (primarily sympathetic) which effects changes in both cardiac output and peripheral resistance. The autonomic nervous system responds, through the hypothalamus, to stimuli from pressure receptors located in the carotid arteries and aorta and to stimuli from higher brain centers, the pathway through which emotional changes alter cardiovascular function.

The ***kidneys*** have a role in the regulation of blood pressure, but their effect is less rapid. If renal perfusion and glomerular filtration decrease, fluid and electrolytes are retained, increasing the extracellular fluid volume. This results in increased ventricular filling, increasing cardiac output. In addition, decreased pressure in the afferent arterioles of the kidneys releases the enzyme *renin* from the juxtaglomerular cells, resulting in increased *angiotensin II* production. Angiotensin II has a direct pressor effect, causing constriction of arteriolar smooth muscle. It also stimulates *aldosterone* production by the adrenal cortex and aldosterone increases salt and water retention by the kidneys. The increase in blood pressure resulting from these two actions of angiotensin II normally inhibits further release of renin.

In addition to these mechanisms there is increasing evidence that *prostaglandin* PGA_2, which is secreted by the kidney, acts to lower blood pressure. Similarly, *kalliden* and *bradykinin,* potent vasodilators produced from kininogen (an α_2-globulin) by the action of the enzyme kallikrein, appear to play a similar role.

EXAMINATION OF THE CARDIOVASCULAR SYSTEM

SYMPTOMS AND SIGNS

The major symptom of cardiac disease may be pain in the chest, shortness of breath, or edema. Symptoms of arterial disease are primarily due to an obstruction of blood flow and may depend on the location and severity and on the rapidity of development of the obstruction.

Cardiac disease may cause changes in the pulse rate and rhythm, changes in the blood pressure, and changes in the heart sounds. Damaged heart valves will produce abnormal sounds called *murmurs.* Abnormal sounds can be heard over the lungs when there is pulmonary congestion. There may be increased jugular venous pressure, peripheral edema, or an enlarged liver caused by right ventricular failure. Signs of peripheral arterial disease include diminished or absent pulses (usually in the legs), and bruits (abnormal sounds) over major arteries in which blood flow is partially obstructed.

INVESTIGATION OF THE CARDIOVASCULAR SYSTEM

The electrocardiogram reflects electrical events occurring in the myocardium. Changes are observed with disorders of cardiac rate, rhythm, and conduction and with disorders of the myocardium. The chest x-ray may reveal abnormalities of the heart or great vessels. Diagnostic ultrasound (echocardiography), phonocardiography, and myocardial 'imaging' with radioactive isotopes are other useful techniques for *noninvasive* diagnosis of cardiac disease.

Right- and left-sided cardiac catheterization and angiography (visualization of the coronary circulation) are used for the diagnosis and evaluation of many forms of cardiac disease. There is some risk with these *invasive* techniques, so their use should be restricted to those patients who may be candidates for cardiac surgery.

THE ROLE OF THE LABORATORY

The clinical biochemistry laboratory is involved in the differential diagnosis of cardiovascular disease, in monitoring therapy, in assessing the severity of complications, and in the evaluation of risk factors underlying the development of atherosclerosis. For example, certain *enzymes* are released from the damaged myocardium and the detection of increased levels of these enzymes in the blood is useful in the diagnosis of myocardial infarction. Abnormalities of serum *electrolytes* and *blood gases* often occur in the course of congestive heart failure or shock and their detection and correction are important in the management of patients with these conditions. The measurement of serum levels of certain *drugs,* such as digoxin, may be helpful in some patients. Elevations of serum *lipids* are associated with an increased risk of premature atherosclerosis and it is thought that the treatment of elevated lipids may delay this process.

COMPLICATIONS OF HEART DISEASE

CONGESTIVE HEART FAILURE

Congestive heart failure is a clinical syndrome resulting from increased pressure in the pulmonary or peripheral venous systems. It may be due to an inability of the normal myocardium to pump blood against an obstructive lesion or a high pressure gradient, or to the inability of a diseased myocardium to pump the volume of blood returned to it. Some *causes* of congestive heart failure include myocardial infarction, myocardiopathy, hypertension, and valvular heart disease. Frequently there is both a relatively chronic *underlying cause* of congestive heart failure and a *precipitating cause.* For example, a patient may have impaired function owing to a previous myocardial infarction, yet not develop clinical signs of heart failure because of compensatory mechanisms. In this situation, the additional stress of a relatively mild anemia or pneumonia may precipitate acute congestive heart failure.

Mechanisms of Congestive Heart Failure. As myocardial function becomes impaired, stroke volume decreases and venous pressures increase. Renal retention of water and electrolytes occurs, increasing plasma and extracellular fluid volume and further increasing venous pressure. These changes result in increased ventricular pressure at the end of diastole. Under these conditions the myocardium can increase the force of its contraction (the *Starling effect*) and can thus restore cardiac output, but only within limits.

Clinical Manifestations. If the aforementioned changes maintain an adequate cardiac output, the heart failure is said to be *compensated.* Patients may maintain this state for years with only mild signs and symptoms, such as shortness of breath on exertion or minimal ankle swelling, until myocardial function deteriorates further or tissue demands for blood increase.

The manifestations of heart failure depend on the ventricle involved. With failure of the left ventricle (***left-sided failure***) symptoms and signs are caused by increased pressure in the pulmonary capillaries, which causes stiffness of the lungs and transudation of fluid into the alveoli and pleural spaces. The major symptoms are shortness of breath, at first on exertion, but progressing to shortness of breath at rest. The heart may be enlarged and there is usually a fast heart rate. Sounds caused by fluid in the alveoli may be present and a pleural effusion may develop.

Clinical findings in ***right-sided heart failure*** are caused by increased pressure in the venous system. An early finding is the presence of edema of the ankles and legs. An enlarged liver owing to passive congestion may occur, and finally jaundice, 'cardiac' cirrhosis, and ascites may develop.

Biochemical Findings in Congestive Heart Failure. In most cases of mild congestive heart failure there are no specific biochemical abnormalities, although total body water and sodium will be increased.

As ***left-sided*** heart failure worsens, there is impairment of oxygen transport into the pulmonary capillaries due to fluid in the alveoli, resulting in a decrease in arterial oxygen tension (apO_2) and, because of compensatory hyperventilation, a decrease in $apCO_2$ and an increase in arterial pH. If the failure becomes severe enough to impair CO_2 transport, or if there is coexisting pulmonary disease, the $apCO_2$ may be elevated and the pH may be decreased.

In ***right-sided*** heart failure there may be elevations of serum aspartate aminotransferase (AST), alanine aminotransferase (ALT), and γ-glutamyl transferase (GGT), owing to liver congestion. The elevation of serum ***enzymes*** may be transitory if the heart failure is treated, and the pattern of the AST increase and subsequent return to normal may resemble that found following acute myocardial infarction. This can result in serious confusion, because myocardial infarction is a frequent cause of congestive heart failure. However, serum creatine kinase (CK) and the MB isoenzyme of CK will remain normal if the elevation of AST is caused by hepatic congestion. The serum bilirubin may also be elevated because of hepatic congestion. Serum proteins are not usually below reference limits, in spite of an increased blood volume, unless liver function has been severely impaired. Proteinuria and elevations of blood urea due to impairment of renal function are not uncommon.

Serum ***electrolyte*** disturbances are common in patients with congestive heart failure, but these are usually secondary to treatment. A mild *dilutional hyponatremia* may be found in untreated failure, but more severe forms of hyponatremia are associated with dietary restriction of sodium. Total body sodium is still elevated but not in proportion to total body water, a situation that develops because the ability of the kidneys to excrete excess water is limited in heart failure. Hyponatremia may be precipitated or aggravated by diuretic therapy if water intake is unrestricted.

An electrolyte disorder that is observed frequently with diuretic therapy is *hypochloremic alkalosis* with *hypokalemia.* In this condition there is a reduction in serum Cl^- that is out of proportion to the serum Na^+, with a decrease in serum K^+ and an increase in serum bicarbonate. This is due to a disproportionate loss of Cl^- and K^+ in the urine and a compensatory increase in bicarbonate retention by the kidney. Potassium depletion can be severe, because the total body K^+ may be significantly reduced before a change in the serum K^+ concentration is observed.

In very severe congestive heart failure, tissue hypoxemia may result in increased production of lactic acid, with the development of *lactic acidosis.*

SHOCK

Shock is a clinical syndrome in which the cardiac output is not sufficient to ensure adequate tissue perfusion. In most cases there is a reduction in blood pressure, a rapid heart rate, cold moist extremities, weakness, and impaired mental acuity.

Etiology of the Shock Syndrome. A decrease in cardiac output may be due to a deficient blood volume, an inability of the heart to pump blood, or pooling of the blood in the peripheral tissues. A deficient *blood volume* may result from

1. Hemorrhage (either internal or external)
2. Loss of plasma into burned areas
3. Loss of fluid and electrolytes through vomiting, diarrhea, diabetic ketoacidosis, or adrenal insufficiency

Inability of the heart to *pump blood* may be caused by

1. Myocardial damage from infarction or myocarditis
2. Severe tachycardia
3. Extrinsic lesions such as a large pericardial effusion or pulmonary embolus, restricting the flow of blood

Examples of conditions that result in shock due to extensive vasodilation and subsequent peripheral *pooling of blood* are endotoxic shock due to septicemia by gram-negative bacteria and anaphylactic shock.

Mechanisms in the Shock Syndrome. The initial response to a decrease in cardiac output is increased *sympathetic nervous system* (SNS) activity, which increases the heart rate and causes vasoconstriction in most organs and tissues. Vasoconstriction occurs primarily in the splanchnic, renal, and cutaneous blood vessels, so that blood flow is directed to the heart and the brain. If the condition is mild, cardiac output and blood pressure can be maintained by these compensatory mechanisms. However, if the condition worsens, arterial pressure begins to decrease in spite of intense vasoconstriction. At this point perfusion becomes seriously impaired and tissue hypoxia develops, resulting in cellular damage and cell death. The extent of damage depends on the severity of the hypoxia and its duration.

The ***complications*** of shock include the following:

1. Acidosis due to the formation of lactic acid from anaerobic glycolysis is a common finding.
2. Renal failure due to acute tubular necrosis often complicates severe shock.
3. Stasis of blood in small vessels results in disseminated intravascular coagulation.
4. Ischemia of the intestines permits access of bacteria to the circulation.
5. Severely impaired blood flow may result in infarction of the kidneys, intestine, heart, or brain.

In some forms of shock, such as *endotoxic* shock, cardiac output may be normal or increased. However, the perfusion of vital organs is reduced because of extensive vasodilation and pooling of blood in such areas as the skin.

Clinical Findings. Patients developing the shock syndrome may be restless but are usually apathetic. They become increasingly weaker and mentally obtunded as the disorder progresses. The skin is usually cold, pale, and moist, although patients with endotoxic shock may have a warm skin due to cutaneous vasodilation. The pulse is rapid and weak and the blood pressure is low. There may also be features specific to the disease responsible for the syndrome.

Biochemical Findings. There are no biochemical abnormalities that are specific for or diagnostic of the shock syndrome. The development of *acidosis,* due to the accumulation of lactic and pyruvic acids from tissue anoxia and of phosphate because of decreased renal function, is to be expected as shock progresses. Elevations of serum *potassium* are common because of the acidosis and impaired renal function. Other electrolyte abnormalities are found, but these depend in large part on the cause of shock, or on its treatment, and are not specific. Elevations of blood glucose are frequently observed, although hypoglycemia may occur terminally. Urea and creatinine levels are usually increased. Often the serum *enzymes* AST and lactate dehydrogenase (LDH) are increased, sometimes to high levels, because of tissue anoxia. The extent of these biochemical abnormalities depends on the severity and duration of the shock syndrome.

CARDIAC ARRHYTHMIAS

Cardiac arrhythmias consist of disorders of heart rate, rhythm, or conduction. Some may occur in the absence of any detectable cause, but most are *secondary* to organic heart disease, to endocrine disorders such as hyperthyroidism, to drug therapy (particularly with digitalis), or to electrolyte disturbances. Frequently more than one potential cause of the arrhythmia is present and it can be difficult to decide which is of primary and which is of secondary importance.

The effects of *electrolyte* disorders on the myocardium are complex and may be difficult to interpret in the clinical situation because multiple ion abnormalities are frequently present. Also, electrolyte abnormalities modify the effects of other disorders and drugs. For example, potassium depletion is more likely to cause an arrhythmia in patients with organic heart disease or in patients receiving digitalis therapy.

Most arrhythmias are detectable on clinical examination, but an electrocardiogram will be necessary for full interpretation of the abnormality. In the *assessment* of arrhythmias, serum electrolytes, including calcium and magnesium, should be measured. Serum digoxin levels may be useful, but must be interpreted with full knowledge of the patient's clinical and electrolyte status.

MAJOR DISORDERS OF THE CARDIOVASCULAR SYSTEM

ATHEROSCLEROSIS

Atherosclerosis is a disease in which specific lesions known as *atheromas* develop in the intima of large and medium-sized arteries. Three major types of atheromas are recognized:

1. The *fatty streak* is a localized accumulation of intimal smooth muscle cells and macrophages containing cholesterol and cholesteryl esters.
2. The *fibrous plaque* is a collection of extracellular lipid in the intima covered by a layer of lipid-containing smooth muscle cells, collagen, and elastic fibers.
3. The *complicated lesion* is basically a fibrous plaque that has undergone hemorrhage, calcification, or both.

The *cause* of atherosclerosis is not clear. It is thought by some to be a degenerative process which is accelerated by specific ***risk factors,*** of which hyperlipoproteinemia, diabetes mellitus, cigarette smoking, and hypertension have been clearly recognized. Similarly, the ***pathogenesis*** of this disease has not been established, although there are several hypotheses. It has been suggested that the initial lesions are due to mechanical trauma caused by shear stress from normal blood flow. The endothelial lining is thought to be damaged, following which blood platelets come in contact with the underlying connective tissue, releasing a substance that stimulates the focal proliferation of smooth muscle cells. This lesion probably heals by re-epithelialization and regression, but if the injury is repeated frequently, or is continuous, then further smooth muscle proliferation, accumulation of connective tissue, deposition of lipid, and ultimately deposition of calcium will result. Hypertension probably accelerates the process by increasing endothelial injury, and hyperlipoproteinemias act either by increasing lipid accumulation or by impairing the re-epithelialization of injured areas. (See also Arterial Disease, Chap. 2.)

The early lesion of atherosclerosis, the fatty streak, causes no clinical problems. As the disease progresses, however, the involved arteries become increasingly obstructed (see Fig. 2–13). Clinical manifestations will depend on the site and severity of the obstructing lesions and are caused by a gradual or sudden reduction of blood flow to the organ involved. Atherosclerosis is a disease that involves many arteries, commonly the aorta, the coronary arteries, the carotid and cerebral arteries, and the arteries of the legs.

The *detection* of atherosclerosis is by clinical examination. Usually, however, the process is well advanced before symptoms or signs are apparent. The biochemistry laboratory has little to offer in the diagnosis of atherosclerosis except for the identification of risk factors. *Noninvasive* techniques may be used to assess the severity of peripheral vascular disease, and arteriography to define the extent and location of atherosclerotic plaques.

ISCHEMIC HEART DISEASE

Ischemic heart disease is caused by an inadequate blood supply to the myocardium, which causes areas of local hypoxia, accumulation of metabolic products, and, if severe, necrosis of part of the cardiac muscle. It is a leading cause of death and disability in North America; approximately 30% of all deaths in males aged 45 to 55 are caused by ischemic heart disease. In most instances it is caused by atherosclerosis of the coronary arteries. Lesions of the coronary arteries due to atherosclerosis develop early in life, but remain asymptomatic and undetectable by clinical and laboratory examination until the process is advanced.

There are three major *syndromes* due to coronary artery disease: angina pectoris, 'unstable angina,' and myocardial infarction. The differentiation between these syndromes is usually easy to make, but may be difficult in some situations.

Angina Pectoris. Angina pectoris develops when coronary artery disease is severe enough to prevent the increase in coronary blood flow that is required by the myocardium to maintain an increased cardiac output. The patient develops oppressive discomfort or pain in the chest on exertion. The pain increases rapidly in severity, then decreases and disappears when the patient is at rest. Sublingual nitroglycerin tablets decrease the severity and duration of the pain and no damage to the myocardium appears to occur during these attacks.

Unstable Angina. Unstable angina (coronary insufficiency) has been variously defined, but can be considered as a syndrome between angina pectoris and myocardial infarction in severity. The pain occurs with minimal or no exertion and does not disappear quickly with rest. It may be a prelude to myocardial infarction. The electrocardiogram often reveals changes, but the serum cardiac enzymes are normal or only slightly increased.

Myocardial Infarction. Myocardial infarction is the result of a reduction in blood supply to the point at which it is insufficient to permit survival of a portion of the myocardial muscle. Some controversy exists over the exact mechanism of the decreased blood supply, which in many cases appears to be sudden, but there is frequently a thrombus in an area of the coronary arteries damaged by atherosclerosis. *Clinical* manifestations include chest pain, sweating, weakness, anxiety, and nausea. Shock, congestive heart failure, or conduction disturbances resulting in arrhythmias or sudden death are complications.

Myocardial infarction may or may not be preceded by angina or the coronary insufficiency syndrome. It often has its onset when the patient is at rest or sleeping. The pain, although similar in nature to anginal pain, is usually more severe and generally requires narcotics for relief. Subclinical myocardial infarctions, however, do occur.

On physical examination the patient is usually in pain and apprehensive and may be cold and perspiring. The heart rate is generally increased and there is often a lowering of blood pressure. The heart sounds may be normal or muffled or there may be extra heart sounds suggesting impending cardiac failure. The *diagnosis* is made from the history and physical examination and is usually confirmed by the electrocardiogram. In most patients, laboratory tests serve only to confirm what is already known, although in some, changes in the serum enzymes are essential to establishing the diagnosis.

Biochemical Confirmation of Ischemic Heart Disease. To assist a diagnosis of ischemic heart disease, three serum enzymes are commonly determined. These are aspartate aminotransferase (AST), creatine kinase (CK), and lactate dehydrogenase (LDH). Although as a rule only the total activities of these enzymes are measured, determinations of individual isoenzymes of CK and LDH have more diagnostic value, especially if related to the time course of the illness.

AST levels begin to rise within a few hours after the onset of myocardial damage, reach a peak of more than three times normal between 24 to 48 hours, and gradually subside to the reference range by the fourth or fifth day (see Fig. 3–3). However, many other conditions may also elevate AST levels: hepatitis, cholecystitis, renal infarction, tachycardia (if severe or prolonged), muscle diseases or injuries (including intramuscular injection), and heart failure (with associated hepatic congestion) are but a few.

CK activity follows approximately the same time pattern as AST, although the increase may occur slightly earlier and the peak may be higher (see Fig. 3–3). It usually returns to normal on the third day. Although increased CK activity is more specific for myocardial damage than increased AST activity, elevations of total CK can occur following intramuscular injections, in muscular dystrophies and injury, in hypothyroidism, in cerebrovascular accidents, in alcoholism, following seizures, and in severe tachycardias. CK activity may be elevated after myocardial damage even though AST activity remains normal.

LDH activity increases about 12 hours after myocardial infarction and peaks between the second and third day, returning to normal on the sixth or seventh day after the infarction (see Fig. 3–3). The activity of this enzyme may also be elevated in other conditions: following intramuscular injections, in hemolytic disease, in megaloblastic anemias, in liver disease, and in muscle injuries and disease.

Although the lack of specificity of these enzyme tests interferes with the interpretation of results, increases in specific *isoenzymes* of CK and LDH have a higher predictive value.

There are three major ***isoenzymes of CK,*** each composed of two subunits. The BB isoenzyme is found mainly in brain; the MB isoenzyme is found mainly in the myocardium and to a slight extent in skeletal muscle; and the MM isoenzyme is the predominant isoenzyme of skeletal muscle. The serum MB isoenzyme is elevated following myocardial infarction and in some rare skeletal muscle disorders. It may be slightly increased in the unstable angina syndrome.

Of the five ***isoenzymes of LDH*** there are two that predominate in serum, LD_2 normally exceeding LD_1. It is usual at some time during the first few days following a myocardial infarction to find LD_1 exceeding LD_2. The presence of both an elevated CK–MB isoenzyme and a 'flipped' LD_1/LD_2 ratio is considered virtually diagnostic of myocardial infarction.

Other enzyme determinations have been used in the diagnosis of myocardial infarction but these have not gained widespread acceptance. The presence of *myoglobin* in the serum or urine has been described following myocardial infarction, but experience with this test is limited.

Prognosis. The natural history of ischemic heart disease is unpredictable. Coronary atherosclerosis develops slowly and may be present for years without symptoms. The initial manifestations may be angina pectoris, unstable angina, myocardial infarction, or sudden death. Angina pectoris may remain stable for years or go on to unstable angina or myocardial infarction. The course of myocardial infarction is also unpredictable and depends partly on the extent of collateral circulation, on the amount of myocardial damage, and on the extent of atherosclerosis in other coronary vessels.

The *complications* of myocardial infarction include acute and chronic congestive heart failure, arrhythmias, ventricular aneurysms, shock, and rupture of papillary muscles. If any of these complications develop, the prognosis is worse.

HYPERTENSION

Hypertension, defined as a sustained elevation of arterial blood pressure to more than 140 mm Hg systolic or 90 mm Hg diastolic, is present in approximately 15% to 20% of the North American population. It occurs at all ages and is an important risk factor for the development of cardiovascular disease, cerebrovascular disease, and renal disease.

Etiology. The majority of patients with hypertension are considered to have 'essential' hypertension; probably fewer than 10% of hypertensive patients will have one of the specific causes listed in Table 9–2.

Pathogenesis. The pathogenesis of ***essential hypertension*** has not yet been elucidated. Some of the factors that may be involved in its development are

1. A genetic predisposition
2. Excessive intake of salt
3. Increased arterial wall sodium, resulting in increased vascular reactivity
4. Increased vascular reactivity from other causes
5. An overactive sympathetic nervous system
6. Inappropriate retention of sodium and water by the kidney
7. Inappropriate activity of the renin–angiotensin system

In this context it should be noted that the *plasma renin activity* (PRA) in most patients with essential hypertension is within the normal reference range—at a level that is high considering the patient's blood pressure levels. In some patients with essential hypertension, plasma renin levels are lower than normal and this group may represent a different form of the disease.

In ***other types of hypertension,*** the mechanisms are somewhat better understood. The majority of patients with chronic renal failure appear to have hypertension because of an expanded extracellular fluid volume, although some have increased renin and angiotensin levels. When due to renal arterial lesions, hypertension is for the most part due to increased or inappropriate renin secretion. Hypertension of renal origin is discussed in Chapter 8.

Clinical Findings. In mild, uncomplicated *essential* hypertension there are usually no symptoms or signs, the disease being recognized by routine checking of blood pressure during physical examination or a health screening program. Headache, epistaxis, dizziness, tinnitus, and faintness may occur. Nocturia may appear when hypertensive renal damage impairs the urinary concentrating mechanisms. Physical signs are limited to changes in the blood vessels of the optic fundi. Symptoms and signs become more obvious as the disease progresses. In *secondary* hypertension there may be signs and symptoms attributable to the disease causing the hypertension.

Investigation of Hypertension. Because of the large numbers of patients with hypertension and the small number of these who will have *secondary* hypertension, a thorough investigation for causes of secondary hypertension in all patients is impractical, from both an economic and logistic point of view.

When the presence of hypertension has been established, initial screening is done by a careful history and physical examination, with particular attention to symptoms or signs of any causes of secondary hypertension, to the presence of other risk factors for atherosclerosis, and to the presence of complications of hypertension.

If the history and physical examination are noncontributory then the following *screening tests,* which are considered by many physicians to be part of a routine investigation, should be done: routine urinalysis and hematology; serum creatinine (or blood urea nitrogen [BUN]), cholesterol, triglycerides, uric acid, and glucose; chest x-ray; and electrocardiogram.

In certain patients there is an increased probability of ***renovascular*** hypertension and a more detailed investigation of the kidneys should be undertaken. Such patients include

1. Those less than 30 years of age with a diastolic blood pressure over 110 mm Hg
2. All patients with a diastolic pressure over 130 mm Hg
3. Those who do not respond to treatment
4. Those with upper abdominal bruits

TABLE 9–2. Some Causes of Hypertension

Endocrine
Acromegaly
Carcinoid syndrome
Congenital adrenal hyperplasia
Cushing's syndrome
Estrogen therapy
Hyperthyroidism*
Hypothyroidism
Hypercalcemia
Pheochromocytoma
Primary hyperaldosteronism
Renal
Parenchymal renal diseases
Renal vascular insufficiency
Renin-producing tumors
Vascular
Aortic valvular insufficiency*
Arteriovenous fistula*
Beriberi*
Coarctation of the aorta
Paget's disease of bone*
Rigidity of arteries*
Miscellaneous
Anxiety
Hypervolemia
Increased intracranial pressure
Polycythemia
Porphyria
Toxemia of pregnancy

* Primarily systolic hypertension.

5. Those with a sudden onset of hypertension
6. Those with a history suggesting renal disease

In these patients a rapid-sequence intravenous pyelogram (IVP) is indicated and, if necessary, renal angiography with a determination of renal vein renin activity. Plasma renin activity will also help to distinguish between renovascular hypertension, primary hyperaldosteronism, and 'low-renin' essential hypertension.

If hypertension remains *untreated* many patients will develop cerebral vascular accidents, congestive heart failure, coronary artery disease, dissecting aneurysm, or chronic renal failure. A few patients will develop *accelerated hypertension* (high blood pressure with exudates and hemorrhages in the optic fundi), *malignant hypertension* (diastolic pressure above 140 with hemorrhages, exudates, and papilledema) or *hypertensive encephalopathy* (a rapid and severe elevation of blood pressure with headache and impaired consciousness).

CEREBROVASCULAR DISEASES

The most common cerebrovascular disorder is thrombosis in association with atherosclerosis in an artery supplying the brain, resulting in infarction. Emboli from a thrombus in the left atrium or ventricle, or from other sources, may also result in cerebral infarction. Intracerebral hemorrhage may occur from rupture of an artery supplying the brain, usually in association with hypertension. Rupture of an aneurysm causes hemorrhage into the subarachnoid space. The clinical features of a cerebral vascular accident, or *stroke,* vary with the nature, size, location, and duration of the lesion.

There are no specific biochemical abnormalities that contribute to the diagnosis or investigation of a cerebral vascular accident. Elevations of blood glucose may occur, but this is usually transient. With cerebral thrombosis or embolism the cerebrospinal fluid (CSF) is usually normal but its protein content may be moderately increased. Following a cerebral hemorrhage the CSF pressure is often elevated, the fluid may be bloody, and protein will be increased owing to the presence of plasma proteins.

Blood in the CSF may be the result of a cerebral hemorrhage, or may be due to damage to a small blood vessel during the spinal puncture. The CSF must be inspected after centrifugation. A yellow color (xanthochromia) indicates that the blood has been present in the CSF for some time and that it is not due to the spinal puncture itself.

PERIPHERAL VASCULAR DISEASE

Peripheral vascular disease affects predominantly the legs and is usually due to atherosclerosis of large and medium sized *arteries.* The main symptom of this slowly developing disease is pain in the legs on exertion, which disappears with rest (*intermittent claudication*). As the blood supply to the legs becomes progressively impaired, ulceration and gangrene can develop. A major artery may become suddenly occluded by an embolus, or by a thrombus developing in an area of atherosclerosis.

A nonatherosclerotic disorder involving the small arteries of the hands and feet—Buerger's disease—has been described.

CONGENITAL HEART DISEASE

Developmental abnormalities of the heart and great vessels occur in about 1% of newborns and vary in severity from very mild to those incompatible with life. Some are single isolated congenital abnormalities, others are very complex and involve several sites in the heart and major blood vessels. Congenital heart disease may result in obstruction to blood flow, in shunting of blood from the left to the right side of the heart, or in right-to-left shunting of blood. Systemic hypoxia results in many cases.

RHEUMATIC FEVER

Rheumatic fever is an inflammatory disease that may involve the heart, the central nervous system, the joints, and the skin. It follows an infection, usually a sore throat caused by a group A streptococcus and is thought to be caused by an autoimmune reaction. Involvement of the central nervous system (resulting in chorea), of the joints (resulting in arthritis or arthralgia), and of the skin, is usually self-limiting and without complications. Involvement of the pericardium may result in the development of a pericardial effusion and myocarditis, which can impair ventricular function enough to cause congestive heart failure. Involvement of a heart valve may result in permanent damage with consequent stenosis or insufficiency of the valve—usually the mitral or aortic valve.

The *diagnosis* of rheumatic fever depends on clinical evidence; a high erythrocyte sedimentation rate or elevated C-reactive protein levels may be helpful but these are nonspecific.

DISORDERS OF THE HEART VALVES

Disorders of the valves of the heart may result in an obstruction to blood flow due to stenosis or in blood flowing back through the valve (*valvular insufficiency*), or both. Most commonly the aortic and/or the mitral valves are involved. Valvular heart disease is usually congenital or a consequence of rheumatic fever. Infections of the heart valves by bacteria (bacterial endocarditis or syphilis) may also result in damage.

MYOCARDIAL DISORDERS

Disorders of the myocardium have a variety of causes and are usually classified as either myocardiopathy or myocarditis. *Myocardiopathy* may be caused by infiltration of the heart muscle during the course of such diseases as amyloidosis or sarcoidosis; it may be familial, or due to alcoholism, or result from unknown causes. *Myocarditis* is usually due to a viral infection of the heart, most frequently by coxsackie A and B strains, but may be due to toxins such as diphtheria toxins. The major clinical manifestation of myocardial disorders is congestive heart failure.

PERICARDITIS

Pericardial disease may be caused by

1. An infectious process, usually viral but occasionally bacterial or mycotic
2. Autoimmune disease such as rheumatic fever or systemic lupus erythematosus
3. Diverse conditions such as myocardial infarction, uremia, trauma, or irradiation

It may be acute and self-limited, or chronic. Pericardial effusions develop frequently during the course of pericarditis. These effusions may be small and asymptomatic, or severe enough to impair filling of the heart during diastole, causing the development of congestive heart failure.

DISORDERS OF THE VEINS

There are two common disorders involving the venous system. ***Varicose veins,*** a dilatation of the superficial veins of the legs, result from incompetence of the valves of the veins. The etiology is unknown, but pregnancy (causing increased pressure in the venous system) and heredity are known to have a role. ***Thrombosis*** (thrombophlebitis) usually occurs in the veins of the legs or pelvis. Although this may develop for unknown reasons, it often occurs as a result of immobilization, trauma, or surgical procedures. The most serious complication of venous thrombosis is *pulmonary embolism.*

SUGGESTED READING

BRENNER BM, STEIN JH (eds): Hypertension. New York, Churchill–Livingstone, 1981

COHN PF (ed): Diagnosis and Therapy of Coronary Artery Disease, 2nd ed. Boston, Martinus Nijhoff, 1985

GALEN RS: The enzyme diagnosis of myocardial infarction. Hum Pathol 6:141–155, 1975

HURST JW (ed): The Heart, 6th ed. New York, McGraw–Hill, 1986

LITTLE RC: Physiology of the Heart and Circulation, 3rd ed. Chicago, Yearbook Medical, 1985

SELZER A: Principles and Practices of Clinical Cardiology, 2nd ed. Philadelphia, WB Saunders, 1983

10

Porphyrins, the Heme Proteins, Bile Pigments, and Jaundice

J. Thomas Hindmarsh / Allan G. Gornall

PORPHYRINS

The porphyrins are tetrapyrrole ring compounds which, as metalloporphyrins, have been adapted by biologic systems for oxygen activation and oxygen transport. Disorders of porphyrin metabolism are not common, but diagnosis is often important to the patient's welfare. The metalloporphyrins form protein complexes, the most notable of which is hemoglobin. There are several important abnormalities of the heme proteins and the red blood cells. Heme is metabolized to bile pigments, and defects in the metabolism and excretion of bilirubin often result in jaundice. A synopsis of these topics is presented here. Pigment metabolism in hepatobiliary disorders is described also in Chapter 11.

PORPHYRIN AND HEME METABOLISM

Synthetic Pathways

Heme is an iron–porphyrin and forms the prosthetic group of two classes of heme proteins. The *major* portion is synthesized in red blood cells and remains incorporated in hemoglobin (Hb) for the life span of the cell (averaging about 120 days). A *significant* portion is synthesized in other tissues, forming part of the myoglobin in muscle, and of the cytochrome, catalase, and peroxidase enzymes in liver and other cells. These enzymes have a rapid turnover rate, varying from a few hours to a few days. Heme that is not immediately incorporated into a protein complex is soon metabolized to bile pigments.

The ***pathways*** to heme synthesis are shown in Figure 10–1. The conversion of glycine and succinyl CoA to 5-aminolevulinic acid (ALA) is effected by *ALA synthetase.* ALA leaves the mitochondrion and two molecules are coupled by *porphobilinogen synthetase* (ALA dehydratase) to form porphobilinogen (PBG). Four molecules of PBG are then deaminated by *PBG deaminase* (*UPG I synthetase*) which also assembles the pyrrole methanes into the tetrapyrrole ring structure of uroporphyrinogen (UPG). UPG I synthetase, when acting alone, forms UPG I isomers only (normally a minor pathway). These porphyrins have no known function and must be excreted. *UPG III cosynthetase,* acting in concert with UPG I synthetase, form UPG III (normally the major pathway). UPG cosynthetase effects a translocation of the aminomethyl group so that the fourth pyrrole can be assembled in reverse order. The acetate groups occupy positions 1, 3, 5, 7, in UPG I and 1, 3, 5, 8 in UPG III. Isomers of alternative hypothetic series do not occur in nature.

By decarboxylation of the 1,3,5,8-acetates, the octacarboxyl UPG III is converted successively to less polar hepta-, hexa-, penta-, and finally a quadracarboxyl coproporphyrinogen (CPG III). In the same manner UPG I may be converted to CPG I. All the porphyrinogens can be oxidized (−6H) to porphyrins (uroporphyrin I [UP I], coproporphyrin I [CP I], UP III, and CP III). CPG III reenters the mitochondrion and its conversion to protoporphyrinogen (PPG) appears to occur in two stages. Decarboxylation and oxidation convert the 2-propionic acid to a vinyl group, yielding harderoporphyrinogen (rare cases of harderoporphyria have been described); further action of the two enzymes converts the 4-propionic acid to a vinyl group, forming PPG. PPG therefore has a 1,3,5,8-methyl, 2,4-vinyl, 6,7-propionic acid structure. PPG is converted to protoporphyrin (PP) by *oxidases* that act on two of the pyrrole amino groups and convert the four methane to methene bridges (−6H). The uroporphyrins, with eight carboxyls, are water-soluble and readily excreted in the urine. The coproporphyrins and protoporphyrin, being much less polar, are preferentially excreted in the bile but appear in both feces and urine.

Finally, by action of the enzyme *heme synthetase* (ferrochelatase), Fe^{2+} is complexed with the four pyrrole nitrogens to form heme. Heme synthesis takes place in almost all cells under aerobic conditions, but mainly in bone marrow and liver. The adult bone marrow mass is about 4 liters and it is here that heme is normally taken up by the globin in developing red cells to form stable hemoglobin (Hb). Iron has a coordination number of six; the fifth position complexes to a histidine residue in globin while the sixth is available to function in oxygen transport.

Cofactors and Control of Heme and Globin Synthesis

The initial formation of succinyl CoA requires vitamin B_{12}, and ALA synthetase requires vitamin B_6. Heme synthesis is divided between the *mitochondria* and the *cytosol,* and requires that adequate Fe^{2+} be available and that protein be present for stabilization.

ALA synthetase is the rate-limiting enzyme by which heme synthesis is *regulated.* In erythroid cells heme appears to be a cofactor in what is probably a repressor–protein complex that blocks the synthesis of ALA synthetase, which has a half-life of just over 1 hour. In the liver, certain drugs can compete for the repressor protein, interfere with feedback control, and result in excess porphyrin synthesis (*porphyria*). Others, by inducing the apoprotein of cytochrome P450, deplete heme and thus result in secondary induction of ALA synthetase. During differentiation of erythroid cells, heme regulates not only ALA synthetase production but also the synthesis of *globin.* Excess heme will undergo rapid oxidation to hemin (Fe^{3+}), which may be the main factor in feedback regulation.

When ALA synthetase activity is increased *UPG I synthetase* probably exerts a secondary control over heme production. In the *acute* porphyrias UPG I synthetase activity is usually normal (porphyria variegata and hereditary coproporphyria) or decreased (acute intermittent porphyria). Consequently porphyrin precursors (PBG and ALA) accumulate in these disorders. In the *non-acute* porphyrias UPG I synthetase activity is also increased and the precursors do not accumulate in sufficient quantity to produce neurologic symptoms.

17-Oxosteroids play an ill-defined role in the control of heme synthesis. In acute intermittent porphyria a deficiency of hepatic 5α-reductase alters androgen and estrogen metabolism and may account for the higher incidence of this disorder in the 15 to 45 age group. The relative excess of 5-β metabolites probably induces the production of ALA synthetase.

Porphyria and Porphyrinuria

Under normal circumstances the porphyrin synthetic pathways are directed almost exclusively to the production of heme. Relatively small amounts of UPG I are formed, and only small amounts of UPG III or CPG III are available for conversion to uroporphyrins or coproporphyrins. As long as the pathway to heme is open, the liver is normal, and feedback control is effective, only *microgram* amounts of these porphyrin metabolites appear in the plasma, feces, or urine. Normally the *ratio* of urine CP III to CP I is about 3:1, and

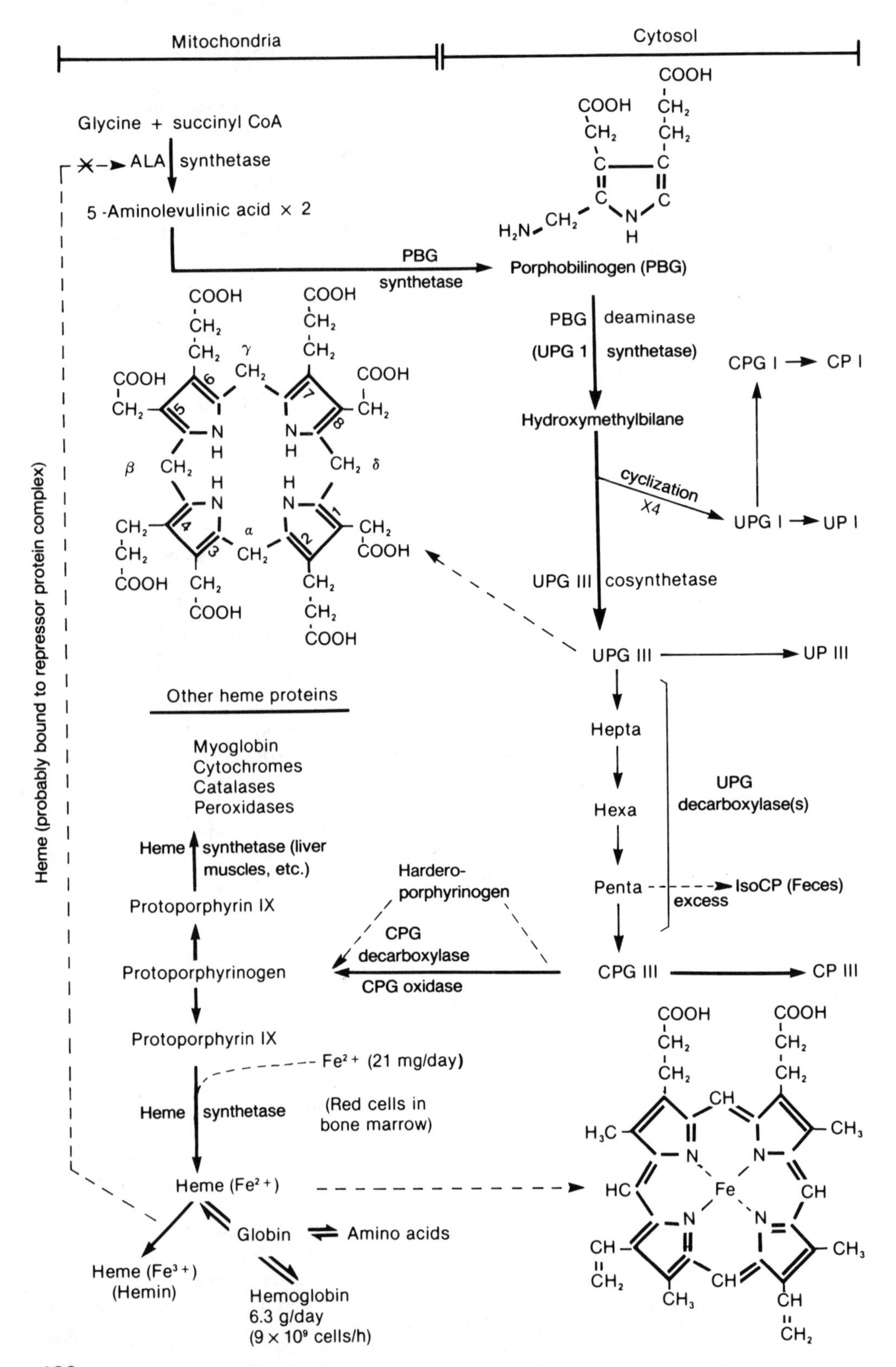

Mitochondria
Cytosol
Glycine + succinyl CoA
ALA synthetase
5-Aminolevulinic acid × 2
PBG synthetase
Porphobilinogen (PBG)
PBG deaminase (UPG 1 synthetase)
Hydroxymethylbilane
cyclization X4
UPG I → UP I
CPG I → CP I
UPG III cosynthetase
UPG III → UP III
Hepta
Hexa
Penta
UPG decarboxylase(s)
excess
IsoCP (Feces)
CPG III → CP III
Hardero-porphyrinogen
CPG decarboxylase
CPG oxidase
Protoporphyrinogen
Protoporphyrin IX
Heme synthetase (liver muscles, etc.)
Other heme proteins
Myoglobin
Cytochromes
Catalases
Peroxidases
Fe²⁺ (21 mg/day)
Heme synthetase
(Red cells in bone marrow)
Heme (Fe²⁺)
Heme (Fe³⁺) (Hemin)
Globin
Amino acids
Hemoglobin
6.3 g/day
(9 × 10⁹ cells/h)
Heme (probably bound to repressor protein complex)

FIGURE 10–1. Pathways of heme synthesis and feedback control. ALA = 5-aminolevulinic acid; CoA = coenzyme A; CP = coproporphyrin; CPG = coproporphyrinogen; PBG = porphobilinogen; UP = uroporphyrin; UPG = uroporphyrinogen.

the ratio of total CP to UP is about 6:1. New methods have made older reference values obsolete, but total urinary CP approximates 365 nmol/day (240 μg) and total UP 50 nmol/day (35 μg).

When a primary abnormality of porphyrin metabolism results in disease, it is called *porphyria*, which may be inherited or acquired. *Porphyrinuria* occurs in certain of these disorders, but also in other conditions such as liver disease, hemolytic anemias, and lead poisoning.

DISORDERS OF PORPHYRIN METABOLISM

A current classification of the *Porphyrias* is given in Table 10–1.

Porphyrias Producing Cutaneous Photosensitivity

Congenital Erythropoietic Porphyria. This very rare disease, inherited as an autosomal recessive, presents in infancy with skin lesions, attrib-

TABLE 10–1. The Porphyrias

Condition	Mode of Inheritance	Demonstrated or Suspected Enzyme Defect	Predominant Site(s) of Metabolic Expression
Non-Acute Porphyrias			
Porphyrias producing cutaneous photosensitivity			
Congenital erythropoietic porphyria	Autosomal recessive	Uroporphyrinogen III cosynthetase	Erythroid cells
Porphyria cutanea tarda	Autosomal dominant or sporadic	Uroporphyrinogen decarboxylase	Liver
Hepatoerythropoietic porphyria	Autosomal recessive	Uroporphyrinogen decarboxylase	Erythroid cells and liver
Toxic porphyria	Acquired	Variable	Liver
Protoporphyria	Autosomal dominant	Ferrochelatase	Erythroid cells and liver (?)
Acute Porphyrias			
Porphyrias producing neurologic symptoms			
Acute intermittent porphyria	Autosomal dominant	Uroporphyrinogen I synthetase (PBG deaminase)	Liver
Porphobilinogen synthetase deficiency	Autosomal recessive	Porphobilinogen synthetase (ALA dehydratase)	?
Porphyrias producing both neurologic and cutaneous manifestations			
Porphyria variegata	Autosomal dominant	Protoporphyrinogen oxidase	Liver
Hereditary coproporphyria	Autosomal dominant	Coproporphyrinogen oxidase	Liver

utable to photosensitivity, and with hemolytic anemia. The primary *biochemical lesion* is a deficiency of UPG III cosynthetase, which leads to decreased UPG III and heme, with increased ALA synthetase activity and UPG I. The child may compensate to produce normal amounts of heme, but UP I and CP I are markedly increased in red cells, other tissues, feces, and urine (their red-brown color stains the diapers). The photosensitivity leads to mutilating skin lesions, and the patient must always be protected from light as much as possible. *Splenectomy* may help by reducing hemolysis and hence the stimulus to porphyrin synthesis.

Porphyria Cutanea Tarda (PCT). This hepatic disorder is the commonest of the porphyrias. It probably involves an inherited trait, but appears clinically as an acquired, or sporadic, disease in adults. The lesion is a deficiency of UPG decarboxylase. *Symptoms* are mainly cutaneous and include blistering and fragility of the skin in light-exposed areas. Pigmentation and hirsutism may also occur. The *clinical course* tends to parallel dysfunction of the liver. The disease is often precipitated by over-indulgence in alcohol, occasionally by estrogen therapy, or a toxic chemical. Porphyrins are increased but as long as their biliary excretion is adequate there appears to be no problem. If liver disease develops, porphyria and porphyrinuria will occur.

The inheritance aspects of PCT are not yet fully elucidated. Considerable deficiency can be tolerated, because UPG decarboxylase is not a rate-limiting enzyme in heme production. In more severe forms (homozygotes?) both liver and red cell levels of UPG decarboxylase are reduced, usually to <20% of normal (hepato-erythropoietic porphyria?). In the sporadic type enzyme activity is normal in erythrocytes but is decreased in the liver to about 50% of normal. This is true whether the patient's disease is active or in remission and indicates that the reason for the low enzyme level is different from the factor which precipitates the disease. In the rare familial form of PCT red cell and liver enzyme concentrations are both approximately 50% of normal. The defect will also be present in some clinically normal relatives.

Iron has been implicated in the etiology of the disease, either as an inducer of ALA synthetase or as an inhibitor of UPG decarboxylase. Repeated phlebotomies (which remove iron) usually produce a remission and are a common form of treatment, although they have no demonstrable effect on liver UPG decarboxylase activity. Hepatic siderosis develops in many patients with PCT.

Diagnosis is made by the association of a typical clinical picture with elevation of *urine* uroporphyrins, usually >1000 nmol (830 μg)/day, and exceeding the urine coproporphyrins. Plasma UP levels are increased and can be used to monitor the effectiveness of therapy. The UPG decarboxylase defect results in increased carboxyporphyrins. The presence of *isocoproporphyrin* in the feces is characteristic of PCT. Excess pentacarboxyl CPG (2,4,6,7 propionate, 5 acetate) is decarboxylated, oxidized to dehydroiso CP (2 vinyl, 4,6,7 propionate, 5 acetate), and excreted in the bile. The vinyl is reduced by intestinal bacteria to an ethyl group, forming iso CP, which can readily be identified in an extract of feces using thin layer chromatography.

Hepatoerythropoietic Porphyria. Hepatoerythropoietic porphyria is a very rare disease and clinically resembles congenital erythropoietic porphyria.

Toxic Porphyrias (TP). The clinical and biochemical features of these disorders often resemble those of porphyria cutanea tarda (PCT). Toxins that cause TP include hexachlorobenzene and related compounds, the polychlorinated biphenyls, and chlorinated dibenzodioxins. The proportion of cases that result from exposure to these toxins generally exceeds the proportion of alcoholics who develop PCT. It seems likely that they cause a direct toxic inhibition of one or more steps in heme synthesis.

Protoporphyria. This disease usually presents in childhood with erythema and/or urticaria related to exposure to sunlight. The primary *biochemical lesion* is a deficiency of heme synthetase (ferrochelatase) in the bone marrow and liver. As a result of the imbalance between protoporphyrin synthesis and its conversion to heme, there is an excess of protoporphyrin in red cells and sometimes in the plasma and feces. *Clinically* the child shows some photosensitivity, associated with itching, edema, and erythema of the skin. Sunlight must be avoided. Hepatic cirrhosis may develop in later years. Urine porphyrin output is normal unless the cirrhosis has produced sufficient obstruction to divert significant amounts of coproporphyrin from the bile to the urine. The *diagnosis* is usually based on demonstrating the marked increase in protoporphyrin in the patient's red blood cells.

Porphyrias Producing Neurologic Symptoms

Acute Intermittent Porphyria (AIP). This is a serious though not very common disorder. It is more common in females and does not present usually until after puberty. It is not associated with cutaneous sensitivity. The primary *biochemical lesion* is a relative deficiency of UPG I synthetase (PBG deaminase) leading to decreased heme but increased ALA and PBG, and these precursors are excreted in the urine. The urine may appear normal, but will turn a wine color on standing or if acidified and heated to convert PBG to uroporphyrins. The enzyme deficiency shows autosomal dominant inheritance, but the degree of penetrance can be quite variable.

Patients with acute intermittent porphyria have recurrent attacks characterized by abdominal colic, nausea, vomiting, and constipation (easily mistaken for other acute disorders), often accompanied by neurologic and/or psychiatric features. Inappropriate ADH secretion may occur with attendant manifestations of fluid retention. *Latent* disease can be exacerbated by starvation, infections, and by many drugs (including barbiturates and sulfonamides) which increase ALA synthetase. Patients should in general avoid drugs and wear a 'Medic-Alert' bracelet. A high-carbohydrate diet, and intravenous heme (hemin, hematin) during acute attacks, help to relieve symptoms by reducing ALA synthetase activity. It is important to identify latent cases so that precipitating factors can be avoided. Death can occur during the acute attack, usually from respiratory paralysis (see also Chap. 15).

The *diagnosis* is made by equating the typical clinical picture (except for a rare PBG synthetase deficiency, AIP is the only acute porphyria that does not exhibit dermatologic lesions) and the characteristic biochemical features. During an *acute* attack urine PBG and ALA are increased and the Watson-Schwartz test (for PBG in urine) is positive. Between attacks quantification usually reveals moderate elevations of urine PBG and ALA, although in rare cases these are normal. UPG I synthetase levels in red cells are low in all patients, whether latent or active; however, there is an overlap between the normal and abnormal ranges.

Porphobilinogen Synthetase Deficiency. PBG synthetase (ALA dehydratase) deficiency is more commonly a result of lead poisoning. It is also a feature of hereditary tyrosinemia. It is now known, however, to occur as a rare inherited enzyme defect in patients who present with acute attacks, which include vomiting and polyneuropathy. It is characterized by an increase in ALA but not PBG.

Porphyria with Photosensitivity and Neurologic Symptoms

Porphyria Variegata. This disorder has highly variable manifestations. The major *symptom* is photosensitivity, which may be accompanied by skin pigmentation and hirsutism. Acute attacks occur, similar to those of acute intermittent porphyria, and may be precipitated by drugs or hormones. The *biochemical lesion* is probably a partial deficiency of protoporphyrinogen oxidase. The main abnormal finding during latency is excess fecal coproporphyrins and protoporphyrin. During attacks, PBG, ALA, and porphyrins may be present in the urine. *Diagnosis* is based on demonstrating excess protoporphyrin in the feces, the presence of PBG and ALA in the urine during attacks, and the combination of photosensitivity and abdominal and neurologic manifestations. Urine porphyrin excretion may be normal between attacks but fecal porphyrins are always elevated.

Hereditary Coproporphyria. This resembles acute intermittent porphyria in its neurologic manifestations and drug sensitivity. Cutaneous manifestations include bullous lesions and fragility in light-exposed areas. It is characterized by a relative excess of CP III in the feces and to a lesser extent in the urine. The *biochemical defect,* inherited as an autosomal dominant trait, is probably at the level of CPG oxidase, resulting in decreased heme, derepression of ALA synthetase, and excessive production of coproporphyrin and its precursors. During acute attacks urinary PBG and

ALA will be increased. The *diagnosis* depends on identifying the excess coproporphyrins.

Miscellaneous Related Disorders

Lead poisoning. Lead inhibits several steps in heme synthesis: most notably decreased are PBG synthetase in red cells, heme synthetase, and CPG oxidase, which results in porphyrinuria. Urine ALA is elevated and the patient develops a hypochromic anemia. Aggregates of ribosomes (basophilic stippling) can often be seen within the erythrocytes. Because lead also produces abdominal pain and neurologic disturbances the syndrome resembles acute intermittent porphyria. Diagnosis is aided by Pb determinations on blood and in urine after EDTA administration. A useful procedure in screening for lead poisoning is the measurement of red cell zinc protoporphyrin concentrations. Other metal poisonings may produce similar effects (see also Chap. 22).

Drugs. Certain drugs (e.g., barbiturates, sulfonamides, and dilantin) can accelerate heme synthesis by inducing ALA synthetase and can thus produce slight increases in urine porphyrins and their precursors.

Dubin–Johnson Syndrome. This rare abnormality of biliary excretion of conjugated bilirubin is characterized by an inversion of the ratio of CP I to CP III in the urine. Normally, CP III makes up >65% of total urinary coproporphyrin; in Dubin–Johnson syndrome it constitutes <20% of urine CP. This finding can be attributed partly to a diversion of CP I from the bile (where it is the predominant isomer) to the urine; but there also appears to be some suppression of CP III excretion by the kidney.

Laboratory Investigation of Porphyria

Normally up to 450 nmol/day (300 μg/d) of porphyrins are excreted in the urine. Of this 85% are copro- and 15% uroporphyrins, even though coproporphyrins are less polar. Fecal excretion of porphyrins is quite variable, with protoporphyrin (<5300 nmol/d; <3000 μg/d) and coproporphyrin (600 to 1700 nmol/d; 400 to 1100 μg/d) predominating, largely the I isomer. Normal fecal excretion of uroporphyrin is only 12 to 50 nmol/d (10 to 40 μg/d). In erythrocytes there are <960 nmol/L (800 μg/L) of porphyrin and most of this is protoporphyrin.

The biochemical differentiation of the disorders of porphyrin synthesis or metabolism are summarized in Table 10–2.

Most laboratories will be able to test for and distinguish between *urobilinogen* and *porphobilinogen,* which give the same reaction with Ehrlich's *p*-dimethylaminobenzaldehyde reagent. In the Watson–Schwartz test the red color due to urobilinogen is extracted with chloroform followed by butanol; that due to porphobilinogen is not. Drugs such as phenothiazines and chlorpromazine may cause false positives.

More definitive and quantitative tests should be available in regional laboratories. Notable among these are tests for *red cell* protoporphyrin, UPG I synthetase, and PBG synthetase; *fecal* coproporphyrin, isocoproporphyrin and protoporphyrin; and *urine* uroporphyrins, coproporphyrins, aminolevulinic acid, and porphobilinogen.

ERYTHROCYTES AND HEMOGLOBIN

THE RED BLOOD CELLS

Erythropoiesis

The tissues involved in oxygen transport can be considered as a system. Erythrocytes are the central component of this system and in a normal adult contain about 775 g of hemoglobin (Hb). The *fetus* begins to form heme proteins and red cells at an early stage, involving in succession the mesenchyme, liver, spleen, and bone marrow. By the third trimester the bone marrow has become the dominant site of erythrocyte production, and from birth onward it is the only site. *In utero* an adequate oxygen supply to the tissues is maintained by a relative excess of red cells carrying mainly fetal Hb (HbF; see Hemoglobins, below), which has a higher affinity for oxygen than adult Hb (HbA). *At birth* the normal infant has red cell and Hb levels above the average for adults. Levels of some of these components at various stages of life are shown in Table 10–3. Although levels of haptoglobin and other necessary components of the blood may be low at birth they are generated quickly in the neonatal period.

TABLE 10–2. Biochemical Features of the Porphyrias*

	Congenital Erythropoietic Porphyria	Protoporphyria	Acute Intermittent Porphyria		Porphyria Variegata		Hereditary Copro-porphyria		Porphyria Cutanea Tarda
			A	*B*	*A*	*B*	*A*	*B*	
Urine									
Uroporphyrin	++ I isomer	Normal	++	±	+	N			++
Coproporphyrin	+ I isomer	Normal	++	±	+	N	+	±	±
Porphobilinogen	Normal	Normal	++	+ Usually	+	N	+		–
δ-ALA	Normal	Normal	++	+ Usually	+	N	+		– Or slight +
Feces									
Uroporphyrin	+		±	N					±
Coproporphyrin	+	+	±	N	+	+	+	±	±
Protoporphyrin		++			++	+			
Isocoproporphyrin/Coproporphyrin ratio					<0.05				++ >0.05
RBC									
Uroporphyrin	++ I isomer								
Coproporphyrin	++ I isomer	+							
Protoporphyrin		++							
Enzymes			PBG deaminase ↓						
Plasma									
Uroporphyrin	++ I isomer								
Coproporphyrin	++ I isomer	±							
Protoporphyrin		+							
Other tests	Red cell fluorescence	Red cell fluorescence			Serum-peak fluorescence				Urine—carboxy porphyrin pattern Serum—peak fluorescence

* Reproduced by Permission of the Editors of Clinical Biochemistry.

A = during attacks; B = between attacks; + = moderate increase; ++ = marked increase; ↓ = decreased; ± = may or may not be increased; – = not found.

TABLE 10–3. Levels of Erythrocytes and Related Factors

Age	Red Blood Cells (10^{12}/L)	Hemoglobin (g/L)	Hematocrit (%PCV)	HbF (% of Total Hb)
Newborn (full-term)	4.8	165	53	75
range	(4–5.6)	(140–190)	(44–62)	(69–90)
Infant (6 months)	—		—	12
range				(5–20)
Child (1 year)	4.5	112	35	<5
Child (10 years)	4.7	129	37.5	<1
Adult male	5.5	155	47	<1
range	(4.5–6.5)	(140–170)		
Adult female	4.8	140	42	<1
range	(4.0–5.6)	(120–160)		

HbF = hemoglobin F (fetal hemoglobin); PCV = packed cell volume; single figures are *mean* values.

The main factor ***regulating*** erythropoiesis is the oxygen supply to the tissues. The mechanism involves the elaboration of erythropoietin (Ep; a glycoprotein) produced by the kidneys and the liver. The main source of Ep shifts from the liver to the kidneys at birth, but small amounts arise still from the liver. In states of chronic anemia, or advanced renal disease, the liver may again become the major source. In some situations 'toxic' factors can suppress the action of Ep on erythroid cells.

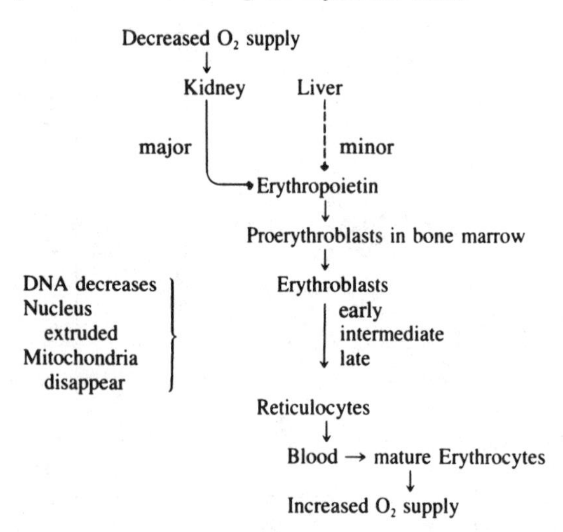

To meet an increased rate of red cell production, Hb is made more rapidly and heme is depleted, which removes feedback repression and porphyrin and globin synthesis increase. This requires adequate vitamins, iron for heme synthesis, and other nutritional factors to support accelerated cell division. When adequate numbers of erythrocytes containing normal amounts of Hb result in an increased oxygen supply to the tissues, erythropoiesis slows down.

Properties and Functions of Erythrocytes

Red cells approximate 7 × 2 nm in size and their main function is to provide a package that prolongs the life span of hemoglobin. In these cells O_2 is carried to the tissues and about 20% of excreted CO_2 is transported between the tissues and the lungs. Erythrocytes also contribute to the buffering capacity of the blood. During red cell development iron carried in the plasma by transferrin is taken up at the cell membrane by ferritin and is made available for heme synthesis. The mature erythrocyte maintains its ionic composition by transport mechanisms that depend on adenosine triphosphate (ATP). This source of energy is derived from glucose, which enters the cell by facilitated transport that does not require insulin.

Glucose metabolism in the red cell is summarized in Figure 10–2. More than 90% of the glucose is metabolized via the glycolytic pathway. A byproduct is reduced nicotinamide adenine dinucleotide (NADH), which serves to prevent the accumulation of methemoglobin. A byproduct of the pentose pathway is reduced nicotinamide adenine dinucleotide phosphate (NADPH), which serves to maintain glutathione in its reduced form. Glu-

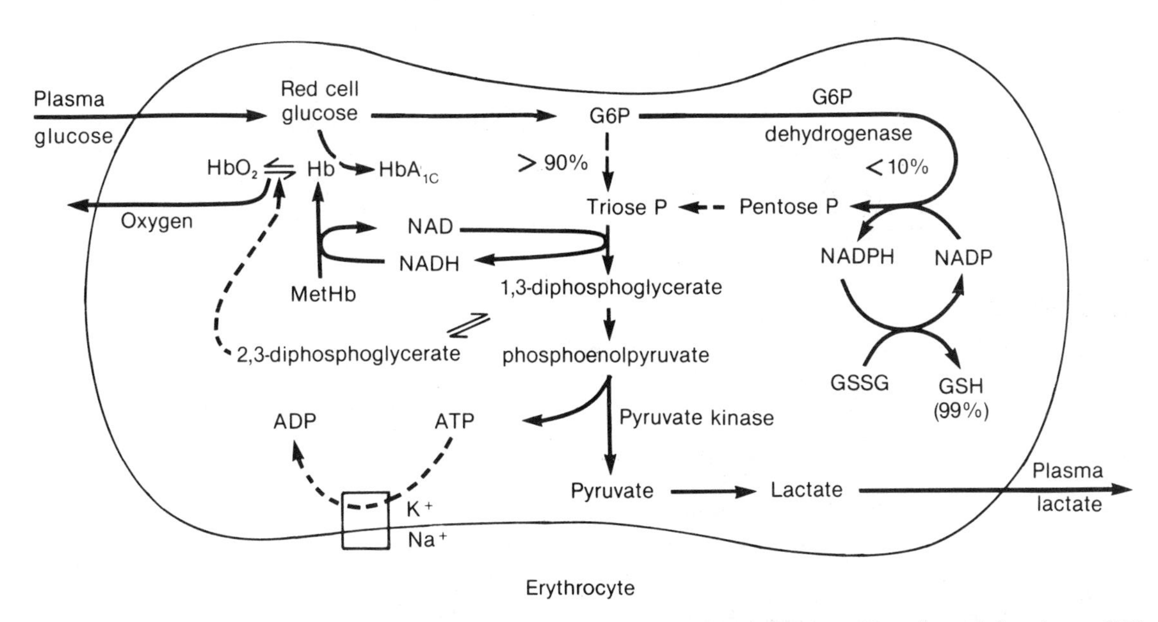

FIGURE 10–2. Glucose metabolism in the red cell. ATP = adenosine triphosphate; G6P = glucose 6-phosphate; GSH = reduced glutathione; GSSG = oxidized glutathione; Hb = hemoglobin; HbA_{1C} = glycosylated hemoglobin; MetHb = methemoglobin; NAD, NADH = nicotinamide adenine dinucleotide and its reduced form; NADP, NADPH = NAD phosphate and its reduced form; P = phosphate.

cose reacts nonenzymatically with Hb, forming first a *labile* aldimine intermediate and from this, more slowly, a *stable* ketoamine adduct at the N-terminal valine of the β-chain. The main product of this glycosylation is HbA1c, which accumulates during the life span of the erythrocyte (see p. 196).

The main organic phosphate in red cells is ***2,3-diphosphoglycerate*** (2,3-DPG), which can combine with the β-chain of Hb and decrease its affinity for oxygen. When the tissue demand for oxygen lowers the oxygen tension, the red cells produce more 2,3-DPG, which promotes oxygen release. Stored blood is very low in 2,3-DPG and when transfused does not function effectively in oxygen transport until 2,3-DPG is replenished, which may take several hours in a critically ill patient.

The *erythrocyte count* is expressed as cells per liter (or per cubic millimeter), and *hemoglobin* as grams per liter (or deciliter), or millimols per liter. Both are affected by the state of hydration. Also important clinically is the *hematocrit* (Hct = 42% to 52% in males, 37% to 47% in females), which is the percent packed cell volume (an index of blood viscosity). These values are required to calculate values such as the mean corpuscular volume (MCV = (Hct × 10)/(RBC per L ÷ 10^{12}); reference range = 80 to 100 fL) and the mean corpuscular hemoglobin concentration (MCHC = Hb × 100/Hct; reference range = 300 to 350 g/L red cells). Erythrocyte size and shape depend on many factors which determine the integrity of the cell membrane and hence the resistance to swelling and hemolysis.

Normal and Abnormal Red Cell Breakdown

Without either a nucleus or mitochondria, erythrocytes lack the capacity to regenerate necessary enzymes and must rely on glycolysis for energy. Normally, after about 120 days, most have become senescent and have been taken up by the reticuloendothelial system (RES). This normal process of ***extravascular hemolysis*** yields the products shown in Figure 10–3A. Major RES sites, in order of importance, are the spleen (about 50%), liver, and bone marrow; but cells of mesodermal origin in connective tissue spaces throughout the body can be involved. When the spleen is removed, the liver and bone marrow increase their phagocytic activity to about 75% of the former total rate.

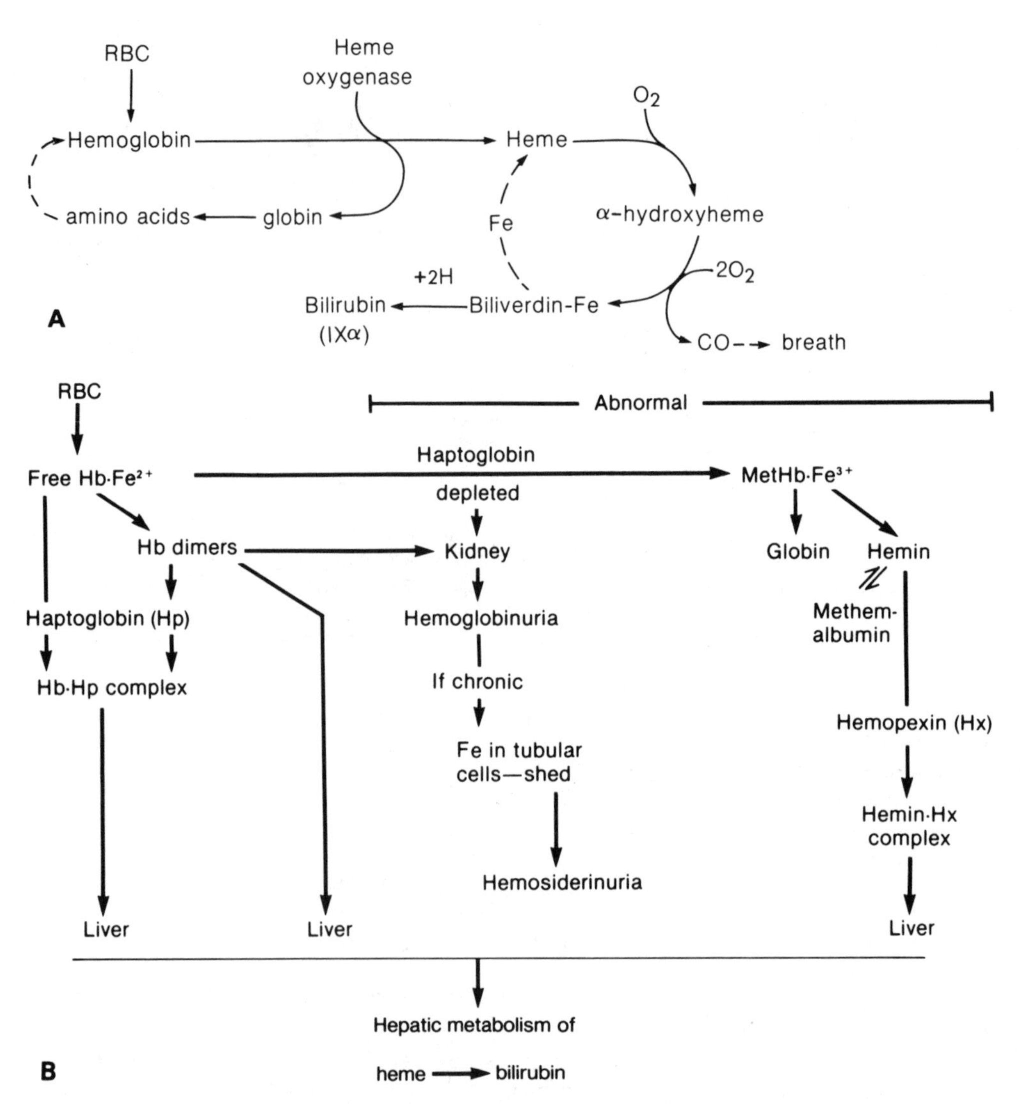

FIGURE 10–3. Products of hemoglobin metabolism. *A.* Following *extravascular* hemolysis. *B.* Following *intravascular* hemolysis. RBC = red blood cell; Hb = hemoglobin.

Under normal circumstances approximately 10% of erythrocytes break down *within* the vascular system, liberating about 0.7 g/day of Hb. Plasma hemoglobin remains below 40 mg/L and binds quickly by its α-chain to haptoglobin (a glycoprotein α_2-globulin, 1 to 3 g/L); this complex is then taken up by the liver. Normal demands require the regeneration of about 1 g/day of haptoglobin, and its level in the serum is inversely related to the rate of intravascular hemolysis. Levels are normally low in newborns, increase quickly to adult levels, and decline again with age. Low values may be found in <1% of normal adults and in 2% to 3% of patients over 60 (see also Chap. 18).

Under abnormal conditions, if erythrocytes are damaged (e.g., by antibody and complement, toxins, or mechanical trauma), more extensive ***intravascular hemolysis*** will occur (Fig. 10–3B). This may overwhelm the capacity of the liver to generate haptoglobin, resulting in its disappearance from the plasma. Free hemoglobin colors the plasma pink at 0.25 g/L and a portion dissociates into dimers which can be taken up by the liver. Above about 1.5 g/L, Hb is excreted by the kidney, and in chronic hemolytic states this will result in hemoglobinuria and hemosiderinuria.

Free hemoglobin is also rapidly oxidized to ***methemoglobin*** ($Hb \cdot Fe^{3+}$), which dissociates to liberate hemin. This hemin forms an equilibrium complex with albumin (methemalbumin) until removed by hemopexin (a β-glycoprotein produced in the liver). Methemoglobin in red cells remains very low (<1% of the total Hb) due to NADH dehydrogenase. Raised values are most often due to various drugs and can be lowered by intravenous methylene blue (which activates NADPH dehydrogenase). Less commonly, methemoglobinemia results from an inherited deficiency of dehydrogenase enzymes or a hemoglobinopathy (HbM). The resulting cyanosis may be well tolerated up to levels around 30% of total hemoglobin.

Sulfhemoglobin also has no respiratory function and is formed when a combination of toxic chemicals and sulfur or hydrogen sulfide is present. The reaction is irreversible and will lead to cyanosis at levels around 20% to 30% of total Hb.

Carboxyhemoglobin forms whenever carbon monoxide is inhaled, because of its high affinity for Hb. Levels above 20% of the total Hb cause symptoms, and at about 50%, unconsciousness. Smokers have levels of up to 10%, nonsmokers less than 2% of total Hb.

Anemia and Polycythemia

Normally, the bone marrow is capable of a six- to seven-fold increase in erythrocyte production; other things being equal, the lifespan of red cells could be as low as about 20 days without anemia. Tissue anoxia is the primary stimulus for compensatory increases, which are reflected in the proportion of reticulocytes found in the circulation.

Anemia is not a disease but is a manifestation of an underlying disorder. Only the major classes can be commented on here:

1. *Bone marrow failure.* Inadequate production of erythrocytes can be due to disorders of precursor cells, defective bone marrow, a deficiency of erythropoietin, endocrine disorders, inflammatory conditions, nutritional factors, or trace metal deficiency. A deficiency of *iron* leads to hypochromic (low MCHC), microcytic (low MCV) anemia. Deficiency of *vitamin* B_{12} or *folate* leads to a megaloblastic anemia. Lack of *vitamin* B_{12} (due to gastric atrophy and absence of intrinsic factor) leads to pernicious anemia. Many diseases, through chronic nutritional, metabolic or toxic effects, are associated with *secondary* anemia.
2. *Increased red cell destruction.* Hemolytic anemia may be the result of *intracellular* defects, either congenital (defective enzymes, abnormal hemoglobin) or impaired metabolism, affecting cell membrane integrity. The anemia may also be due to *extracellular* factors such as increased activity of the reticuloendothelial system and various toxic agents. Antibodies are a principle cause of *acquired* hemolytic anemia.
3. *Blood loss.* This may result from either acute or chronic hemorrhage.

Polycythemia, the presence of an increased red cell mass, occurs naturally under conditions of low oxygen availability. *Polycythemia rubra vera* is seen in older adults and is due to a loss of regulatory control of the bone marrow, with chronic production of excess red blood cells (Hb and Hct are near or exceed the upper reference limit). The prevalence is about 1:100,000 and the common and safest treatment is regular phlebotomies. Proliferation of white blood cells and platelets accompanies the disorder and there is a tendency to hyperuricemia and secondary gout. *Secondary* polycythemia results from any clinical disorder leading to tissue anoxia. A form of 'stress' polycythemia is also recognized, usually a relative erythrocytosis due to a decreased plasma volume.

HEMOGLOBINS

Fetal and Adult Hemoglobins

Normal ***adult hemoglobin*** (HbA) consists of two α and two β peptide chains, with a small proportion (HbA2, <3%) having α and δ chains and <1%

HbF. Each chain has a heme molecule bound to histidines in a hydrophobic cleft, stabilizing Fe^{2+} in an efficient oxygen transporting system. Excess heme is readily converted to hemin and either form may combine with a receptor–protein to repress the genome coding for ALA synthetase. Feedback control of heme synthesis is thus related to the need for hemoglobin and the ability to form enough to meet the oxygen requirements of the tissues.

Fetal hemoglobin (HbF) contains two α and two γ chains, the latter differing from β chains in only one amino acid. By two years of age only trace amounts of γ-chain Hb are found, except in the rare, benign condition of hereditary persistence of HbF, in which the homozygous individual produces no β or δ chains at all, while in heterozygotes about one third of the Hb is HbF.

Hemoglobin A1 (HbA1) is a mixture of glycosylated variants which normally make up about 6% of the total HbA. The only quantitatively significant form is HbA1c, which is produced readily in young cells, then remains fairly constant during the lifespan of the erythrocytes. Levels attained depend on the time-averaged plasma glucose levels over the preceding 2 to 3 months. The reference range for total HbA1 is about 5% to 8% of the blood Hb. It tends to be slightly higher during pregnancy, in obese people, in older age groups, and notably increased in diabetic patients (see Chap. 16).

The Hemoglobinopathies (see also Chap. 23)

Abnormal Hemoglobins. Many abnormal hemoglobins have resulted from random mutation throughout evolution; those that permit oxygen transport and are compatible with life have persisted. In the homozygous state they are usually associated with impaired function and shortened red cell survival. *Examples* are as follows:

HbS. Valine replaces glutamic acid in the β chain. HbS may result in sickle cell anemia. Heterozygotes, with <45% HbS, carry the sickle cell trait. Homozygotes have >70% HbS, which results in Hb polymerization, sickle cell formation, accelerated destruction, and anemia. The distorted red cells tend to plug capillaries, making it a painful and often fatal disease. Prenatal diagnosis on cells from amniotic fluid, using a molecular probe (e.g., DNA fragments) and restriction endonucleases to demonstrate fragment length polymorphism, is now possible.

HbC. Lysine replaces glutamic acid in the β chain. The HbS variant is present in about 10% of the black population in North America, and the HbC variant in 2% to 3%.

HbM—is a rare abnormal Hb in which the heme binding site is affected and is characterized by a tendency to methemoglobinemia (Fe^{3+}). Patients exhibit chronic cyanosis but are otherwise free of symptoms.

Rare examples of a growing number (>200) of other abnormal hemoglobins have been reported.

Thalassemia. The thalassemia syndromes are a group of inherited anemias resulting from an imbalance in the synthesis of one or more of the globin chains of Hb. They are characterized by a hypochromic microcytosis, often with a relatively normal total Hb, but an atypical appearance of the red cells in a blood smear.

Homozygotes (thalassemia *major*) for the α-chain defect do not live; those for the β-chain defect (*Cooley's anemia*) are seriously impaired and usually die in childhood with hepatosplenomegaly, hyperbilirubinemia, and anemia. The hemoglobin is mostly HbF, with some increase in HbA2. The repeated transfusions that are required lead to problems of excess iron build up.

Heterozygotes (thalassemia *minor*) for the α defect are often asymptomatic. The β defect is more common, associated with mild anemia and in rare instances hyperbilirubinemia. Usually HbA2, occasionally HbF, are increased and abnormal hemoglobins such as Hb Barts (γ4) and HbH (β4) are sometimes seen. Lacking α-chains the latter will not bind to haptoglobin. Definitive diagnosis may require radionuclide studies of globin synthesis *in vitro.* Thalassemia can now be diagnosed prenatally using a molecular probe and restriction endonuclease analysis.

Myoglobinuria

Muscle Hb consists of a single peptide chain (M_r 16,000) and one heme. It can pass the glomerular membrane and myoglobinuria should be suspected when the *urine* is red and the *serum* normal in color. The urine does not fluoresce, as it may with porphyria. Causes of myoglobinuria include crush injury or muscle infarction, certain toxins, and rare myopathies. It may occur spontaneously (paroxysmal myoglobinuria) and after extreme muscular activity. In excess it can cause acute renal failure.

BILIRUBIN AND JAUNDICE

BILIRUBIN METABOLISM AND EXCRETION

Bilirubin Formation and Transport

The conversion of heme to bilirubin proceeds as shown in Figure 10–3A. In the reticuloendothelial cells heme is removed from Hb as the result of competitive binding to microsomal heme oxygenase, which has a high specificity for the α-methene bridge. Approximately 7 g of Hb are degraded and replaced each day. Heme is oxidized first to form hydroxyheme, then on further oxidation the porphyrin ring is opened, with loss of the α-methene carbon as carbon monoxide. The resulting biliverdin-Fe first loses its iron to the metabolic pool and is then converted by biliverdin reductase (in the cytosol) to bilirubin IXα, yielding about 355 μmol (210 mg) per day.

An additional 120 μmol (70 mg) of bilirubin comes mainly from the relatively rapid turnover of heme enzymes and cytochromes in the liver (1 to 2 days), or from the rapid degradation of heme that has not been incorporated into Hb (1 to 2 hours), or that was incorporated into defective red cells which are quickly destroyed. Following ^{14}C-glycine administration these events are seen as early peaks of radioactivity excreted as bile pigment within the first 2 days. This rapid source of bilirubin assumes importance in conditions of ineffective erythropoiesis, in which an imbalance exists between heme and protein synthesis. It may occur in iron-deficiency anemia, pernicious anemia, or thalassemia, but has not been found to be due to excessive turnover of liver metalloporphyrins. It is known as ***'shunt' bilirubinemia*** and is usually of relatively mild degree. The bulk of the administered ^{14}C is recovered between 90 and 150 days as the red cells are destroyed.

Bilirubin is potentially toxic to plasma membranes, but is of clinical concern only in newborns. It has no biologic value and must be excreted. Because of intramolecular hydrogen bonding the molecule is nonpolar and hydrophobic (Fig. 10–4). Following its release from reticuloendothelial cells, bilirubin is ***transported*** in equilibrium with albumin, which has one high affinity binding site (K_D of about 10^{-8} M) and two less effective sites (K_D's of 10^{-5}–10^{-6} M). The *high affinity* capacity of plasma albumin for bilirubin is an important parameter in neonatal jaundice. Several drugs (e.g., sulfonamides and salicylates) can displace bilirubin and increase the risk of cytotoxic effects. Under normal circumstances only a small portion of these binding sites are occupied. In adults, with 40 to 50 grams of albumin per liter of plasma, it would take more than 500 μmol/L (30 mg/dL) of bilirubin to saturate the high-affinity sites. In infants, with their lower and qualitatively somewhat different albumin, particularly in the presence of competitive binding substances, bilirubin binding capacity and the risk of kernicterus caused by increased free bilirubin in the plasma, become matters of concern at total bilirubin levels greater than 170 μmol/L (10 mg/dL).

Metabolic Clearance and Excretion

In the plasma, bilirubin binds to a hydrophobic cleft in albumin and normally is not cleared by the kidney. The important site of *metabolic clearance* of bilirubin is the liver, where first a monoglucuronide (BMG) and then the more polar diglucuronide (BDG) are produced, and bile is the normal route of excretion. When for any reason conjugated bilirubin is returned to the circulation it is less firmly bound to albumin and is gradually cleared by the kidney. BMG exhibits relatively strong protein binding and has a low renal clearance (possibly 0.1 mL/min). BDG is cleared more rapidly (about 1 mL/min) and accounts for virtually all the bilirubin found in the urine. In babies where bilirubin has accumulated in the skin, treatment by exposure to light (440 nm to 470 nm) activates oxygen and converts significant amounts of bilirubin to conformational isomers called *photobilirubins* (Fig. 10–4). These pigments are hydrophilic, enter the plasma, bind less strongly to albumin, and can be excreted in the bile without being conjugated. Excessive radiation results in the accumulation of a photobilirubin-albumin adduct (delta B) which persists in the circulation for the lifespan of the albumin.

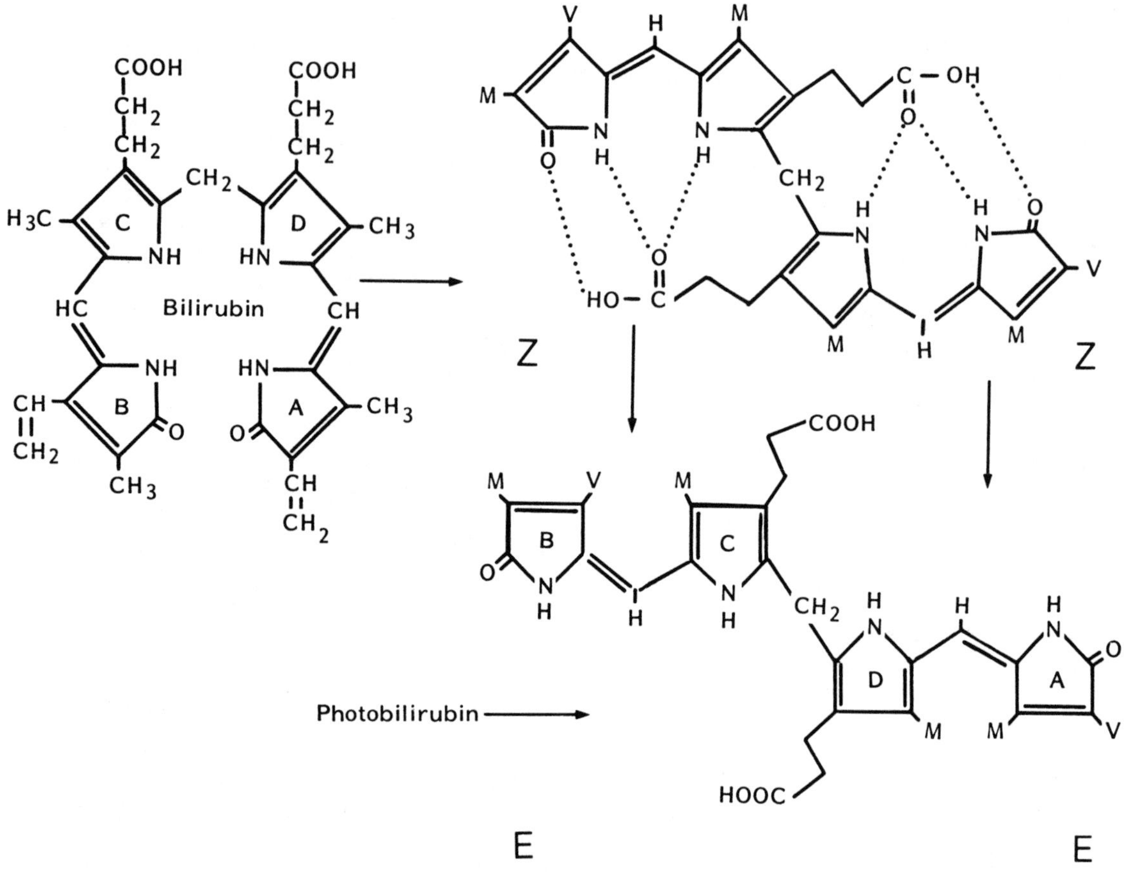

FIGURE 10–4. Bilirubin, showing conventional structure and hydrophobic ZZ configuration with hydrogen bonding. UV light (440 nm to 470 nm) causes changes that result in *photobilirubins* with a ZE or EZ and ultimately an EE spatial configuration. The loss of hydrogen bonding confers more hydrophilic properties, leading to easier clearance from the circulation.

The hepatic ***transfer*** of bilirubin from plasma to bile canaliculus is complex and is still not fully understood (Fig. 10–5). At the sinusoidal microvilli, possibly with the help of a membrane protein 'translocase,' a rapid interchange of bilirubin occurs between plasma albumin and cytosol ligandin (protein Y, glutathione reductase B) with about one third recycling back to complex with albumin. Ligandin, which constitutes around 5% of the cytoplasmic protein and binds a variety of substances, presumably facilitates the transfer of bilirubin to the smooth endoplasmic reticulum, where the enzyme uridine diphosphate glucuronyl transferase I (UDPGT I) forms bilirubin monoglucuronide (BMG). Another enzyme, UDPGT II, probably located closer to the canaliculus, forms the diglucuronide (BDG). *Hepatocyte* heme is converted to biliverdin in the microsomes (SER) and to bilirubin in the cytosol, where it binds to ligandin. Some escapes into the plasma but about two thirds is conjugated and is excreted directly into the bile.

The actual ***excretion*** of conjugated bilirubin into the canaliculus involves a carrier-mediated transport system that is shared with other organic anions (such as sulfobromophthalein) and is rate-limiting. Of the bilirubin found in bile about 80% is digluc-

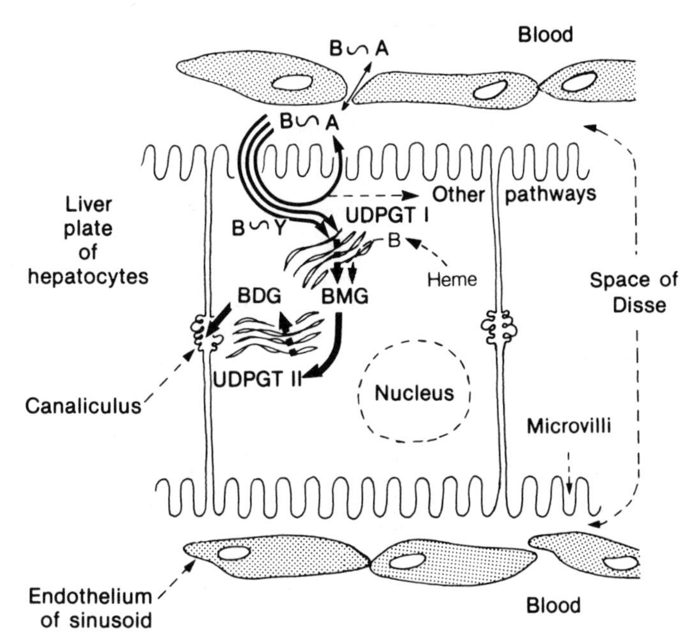

FIGURE 10–5. Normal excretion of bilirubin by the hepatocyte. B∼A = bilirubin bound to albumin; B∼Y = bilirubin bound to ligandin; UDPGT = uridine diphosphate glucuronyl transferase; BMG = bilirubin monoglucuronide; BDG = bilirubin diglucuronide. Some heme is metabolized in the hepatocytes themselves.

uronide, and 10% monoglucuronide; other conjugates and a small amount of free bilirubin make up the rest.

The uptake and conjugation of bilirubin are well preserved in the presence of liver cell injury. The 'transport-out' mechanism is sensitive to many forms of damage and accounts for the early cholestasis and bilirubinuria in acute hepatitis. A transient hyperbilirubinemia occurs when dyes used for gallbladder visualization saturate the transport system.

Conjugated bile pigments pass from the terminal hepatic ductules to the larger hepatic ducts, then via the common bile duct to the duodenum. The concentration of bilirubin in hepatic bile is about 170 ± 80 μmol/L (10 ± 5 mg/dL). In the gallbladder these pigments are concentrated several fold and free bilirubin has been implicated in the initiation of gallstones. On the average about 375 μmol (220 mg) of conjugated bilirubin reaches the duodenum each day.

Intestinal Metabolism and Urobilinogen Excretion

When bilirubin, mainly as the diglucuronide, reaches the *intestinal flora* of the lower ileum it undergoes extensive reduction, as follows:

Bilirubin diglucuronide
↓ +8H
Colon urobilinogens (colorless) { Mesobilirubinogens (10%)
↓ +4H
Stercobilinogens (90%)
↓ air, light −2H
Fecal { Urobilins (colored)

Variable portions of the conjugates (10% to 30%) are hydrolyzed in the intestine to *free urobilinogens,* which can be reabsorbed and carried in the portal circulation to the liver, where most (>90%) are actively taken up and reexcreted. A portion, however, reaches the general circulation and the kidney excretes urobilinogen by both glomerular filtration and tubular secretion. The latter process varies inversely with hydrogen ions in the luminal fluid. Variable portions of conjugated bilirubin may be cleaved into dipyrroles in the intestine, some of which are reabsorbed and excreted in the urine, where they contribute to normal urine color.

The conversion of bilirubin to urobilinogen is

subject to fluctuations, depending on bowel motility and bacterial flora. Infants that have not yet developed an intestinal flora, or patients whose flora has been suppressed temporarily by antibiotics, will excrete yellow bile pigment in the feces.

The normal daily excretion of urobilinogen in *feces* is 50 mg to 250 mg (average 110 mg/day); in *urine,* 0 to 4 mg (average 1.5 mg/day). In the absence of biliary obstruction, ***fecal urobilinogen*** reflects red cell destruction. In the presence of hepatobiliary disease, fecal urobilinogen reflects the degree of obstruction to bile flow. For the metabolism and excretion of bilirubin the liver has a large reserve capacity. For the excretion of portal blood urobilinogen the liver has a limited capacity—hence, this mechanism constitutes an endogenous overload test and ***urine urobilinogen*** is a rather sensitive index of impaired liver function.

Laboratory Indices

Urinalysis. When a patient's signs and symptoms suggest that the integrity of the liver may have been compromised, the urine should be tested qualitatively for bilirubin and urobilinogen (see Chap. 11). *Bilirubin* can be precipitated, filtered, and detected by mild oxidation to green biliverdin (Harrison's test), or reacted with commercial stick tests such as Bili-Labstix (or Multistix) or Chemstrip. *Urobilinogen* is best detected in fresh urine by Ehrlich's *p*-dimethylaminobenzaldehyde reagent, which reacts also with porphobilinogen (and *p*-aminosalicylic acid) to form a reddish brown color. Commercial Bili-Labstix (or Multistix) or Chemstrip may also be used. If the urine is not fresh it must be tested by Schlesinger's alcohol and zinc acetate procedure. This gives a greenish fluorescence with urobilin and a red fluorescence with excess porphyrins.

Bilirubinuria is a relatively early finding in cholestasis and reflects the degree of obstruction to bile flow. Tests for *urobilinogen* are designed to be negative for normal amounts, but positive when excretion exceeds 4 mg/day. Urine urobilinogen will be increased when there is excess hemolysis or when the liver is hypoxic or exposed to toxins of various kinds. It will also be increased in liver disease unless the accompanying cholestasis reduces bile flow to the point that little or no bilirubin is excreted to be converted to urobilinogen.

Quantitative measurements of urine or fecal urobilinogen may at times be of value but are seldom performed.

Serum Analyses. *Bilirubin* can be measured in several ways, most commonly by the van den Bergh reaction, which employs diazotized sulfanilic acid to form red azodipyrroles. In *aqueous* solution only conjugated bilirubin reacts rapidly (in 1 min), to give the so-called *direct reaction;* the upper reference value is 3.5 μmol/L (0.2 mg/dL). Unconjugated bilirubin reacts much more slowly unless its intramolecular hydrogen bonding is prevented by the addition of alcohol (or caffeine). In *alcoholic* solution the total bilirubin is measured; the upper reference value is 20 μmol/L (1.2 mg/dL). Subtracting the direct, conjugated bilirubin from the total gives the *indirect,* or *unconjugated, bilirubin.* Because of the high affinity of albumin for unconjugated bilirubin, the amount of free bilirubin is extremely low (<3 nmol/L; <0.2 μg/L).

In normal serum *total* bilirubin consists largely of unconjugated, with only small amounts of conjugated forms. Increased concentrations of the *conjugated* forms of bilirubin are a fairly sensitive index of impaired bile secretion (cholestasis) and lead to bilirubinuria. With continued retention of conjugated bilirubins a new complex, *covalently* bound to albumin, is formed called delta bilirubin (δB).

$$\text{BDG,BMG} \rightarrow \text{putative intermediate} + \text{albumin} \rightarrow \delta\text{B}$$

δB accumulates gradually and can make up 10% to 90% of the total bilirubin. It is not excreted by the kidney and remains in the plasma for the lifespan of the albumin. This accounts for the persistence of an icteric serum for a week or longer after bilirubinuria has cleared. δB gives the 'direct' van den Bergh reaction but it reacts more slowly with diazo reagent than the natural conjugates (BMG, BDG). It will be underestimated in total bilirubin analyses unless this is taken into account.

Transcutaneous measurement of bilirubin can be used to identify significant neonatal jaundice in full-term infants.

Bilirubin-Binding Capacity. The bilirubin binding or reserve capacity of plasma albumin can now be measured but is of clinical interest only in neonatal jaundice. Values below 40% of normal indicate a risk of kernicterus and may require special treatment. The test can be used to monitor the need for and benefits of transfusion (see also p. 204, Chaps. 11 and 21).

DISORDERS OF BILE PIGMENT METABOLISM LEADING TO JAUNDICE

Mechanisms of Jaundice

Jaundice may result from defects at any level of bilirubin metabolism: excess production, impaired delivery to the liver cells, defective uptake, impaired conjugation, defective excretion into the canaliculus, or obstructed drainage to the duodenum. The healthy liver has a reserve capacity for metabolic clearance of plasma bilirubin estimated at more than five times the daily production rate (BPR). Normally 70% of the bilirubin comes from old erythrocytes, 5% from ineffective erythropoiesis, and 25% from other sources, mainly hepatic. For a given influx of bilirubin into the plasma (≃BPR), the level of *unconjugated bilirubin* (UCB) relates to bilirubin clearance in a hyperbolic manner (Fig. 10–6). Plasma UCB levels increase relatively little until impairment of liver function is quite severe, but at this stage small decreases in bilirubin clearance are accompanied by marked increases in UCB.

Most cases of ***jaundice*** are due to hepatobiliary disease. The sclera of the eye becomes icteric at bilirubin concentrations around 34 μmol/L (2 mg/dL); the skin of Caucasians appears yellowish at levels above 50 μmol/L (3 mg/dL). Bilirubin excretion is a dynamic process in which any one of several steps may be impaired. ***Cholestasis*** is also a complex phenomenon, usually reflecting decreased hepatocyte function, which may have membrane effects or transport effects, and may be associated with altered bile acid metabolism, decreased nicotinamide adenine dinucleotide (NAD), increased lactate, changes in the endoplasmic reticulum, and impaired mitochondrial function. The effects of intra- or extrahepatic cholestasis on bilirubin excretion are illustrated in Figure 10–7.

When *conjugated bilirubin diglucuronide* cannot take the easy route into the bile, it is presumed to increase in the hepatocyte until it can diffuse into the interstitial space and pass by way of lymph channels and the thoracic duct to the blood. Inside

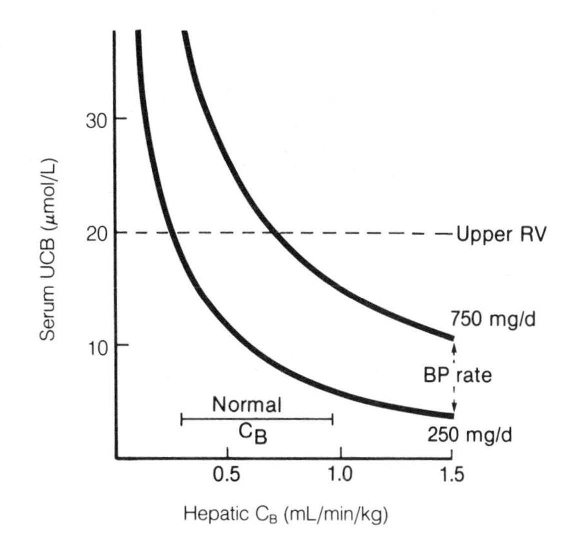

FIGURE 10–6. Relationship between serum levels of unconjugated bilirubin and bilirubin clearance C_B by the liver. BP = bilirubin production; RV = reference value; UCB = unconjugated bilirubin.

the cell it will depress UDPGT II and may be partly hydrolysed by glucuronidase to BMG. Increased levels of monoglucuronide in turn will suppress UDPGT I and some is regurgitated to the plasma. As binding sites on ligandin become saturated, more recycling of bilirubin back to albumin will increase the levels of *unconjugated bilirubin* in the plasma. More bilirubin is metabolized by other, less well known, pathways. When bile flow is impeded in the *biliary* system, bile pigments will 'regurgitate' back through the desmosome matrix to enter the paracellular spaces and then the plasma. In hemolytic jaundice analytic results may indicate up to 15% conjugated bilirubin in plasma; in hepatobiliary cholestasis conjugated bilirubin will exceed 50% of the total. Values between 15% and 50% suggest a mixed etiology and are usually found in cirrhosis of the liver.

Excess Production of Bilirubin

Increased production of bilirubin is usually due to a shortened half-life of red cells. When this is adequately compensated by accelerated erythropoiesis, anemia does not occur and the liver handles the increased load with relative ease. Generally, it requires the superimposed effect of a sudden loss of red cells, leading to anemia and tissue (liver)

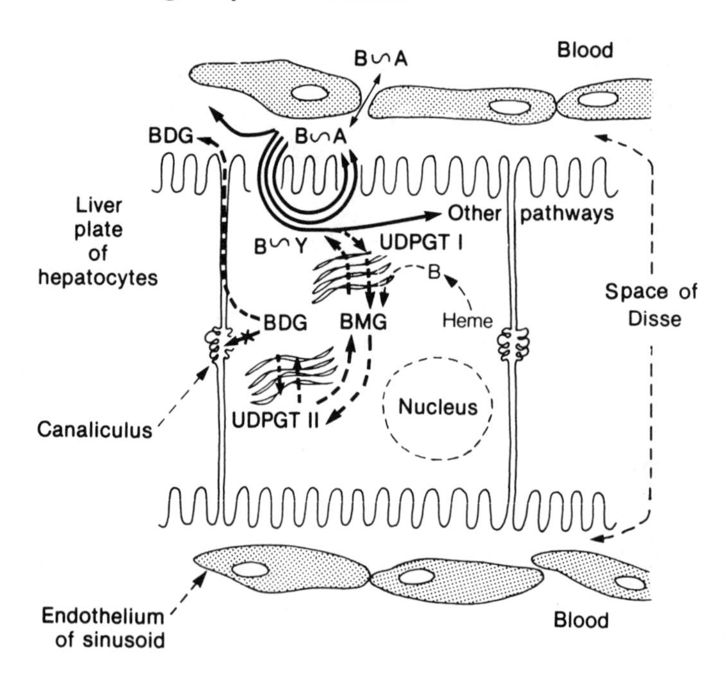

FIGURE 10–7. Effect of cholestasis on the excretion of bilirubin by the hepatocyte (see Fig. 10–5).

hypoxia, to produce significant unconjugated hyperbilirubinemia. The more common causes of *hemolytic anemia* are as follows:

Congenital Spherocytosis. This is probably a defect in the red cell membrane, inherited as an autosomal dominant trait. The cells lose their ability to regulate internal electrolytes early in their life span, become spheroidal, and are cleared by the reticuloendothelial cells especially in the spleen. Accelerated erythropoiesis, with 3% to 12% reticulocytes (normal $\simeq$ 1%), allows a delicately poised compensation. Infection, stress, toxic substances, or no obvious factor, may result in a crisis, during which red cell production ceases for perhaps a week (Fig. 10–8). Anemia develops rapidly and unconjugated bilirubinemia may be sufficient to produce a mild jaundice. During recovery erythropoiesis is greatly accelerated, with many reticulocytes present, until compensation is again achieved.

The disorder is usually diagnosed in early childhood but sometimes appears in the young adult as a sudden crisis with fever, abdominal pain, nausea, and vomiting. Treatment is surgical splenectomy, which removes the most active site of red cell breakdown.

Abnormal Hemoglobins. As mentioned previously, many abnormal *hemoglobins* have been identified, the most common being HbS and HbC. Thalassemia, also described earlier, is a defect in the synthesis of α or β chains of globin. In varying degrees these defects may affect the functional integrity and may shorten the life span of erythrocytes.

Red Cell Enzymes. Two enzymes that affect glycolysis are of special interest.

Glucose-6-phosphate dehydrogenase (GPD) is the enzyme that directs flow into the pentose pathway (see Fig. 10–2). This pathway generates NADPH, which is necessary to produce reduced glutathione, which serves as a sulfhydryl buffer against toxic peroxides. NADPH deficiency leads to early hemolysis. Although many millions of people, mainly in tropical countries, have a type or degree of the X-linked GPD defect, only a few are born with a deficiency in enzyme activity that leads to a non-spherocytic, hemolytic anemia. In many cases the defect is latent and hemolytic episodes may be drug induced. The enzyme in washed red cells is measured, after hemolysis, by the capacity to generate NADPH.

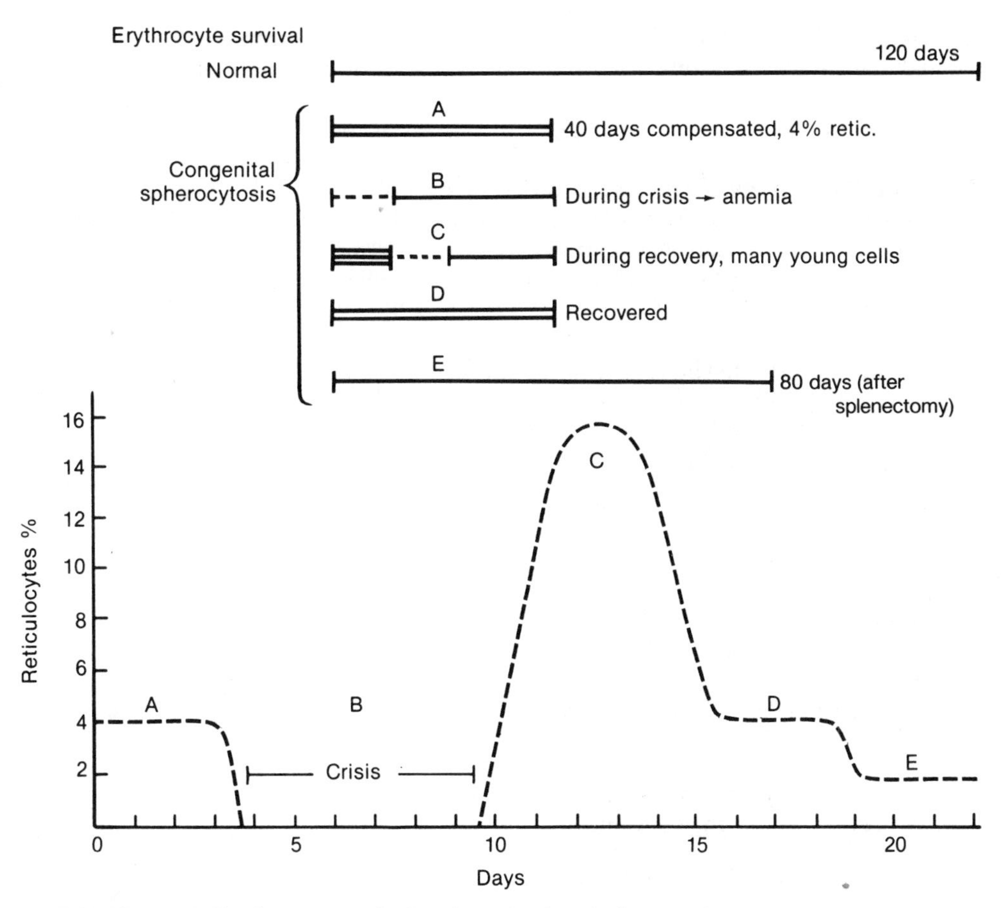

FIGURE 10–8. Erythrocyte survival and production during and following a hemolytic 'crisis' in congenital spherocytosis.

Pyruvate kinase (PK) deficiency is an autosomal recessive trait affecting far fewer individuals but with a significant incidence in Northern Europe and North America. Heterozygotes show about 50% of normal red cell activity, but only homozygotes, showing 5% to 25%, manifest clinical disease. Severe anemia requiring transfusions in infancy is rare. Latent disease in adults may become evident during an acute illness, but is generally asymptomatic. Red cells with PK deficiency have difficulty maintaining adequate amounts of ATP but the actual cause of their shorter life span has not been established.

Immunologic Causes. These can be *inherited;* for example, parental blood groups may result in ABO incompatibility, which is more common but generally milder than Rh incompatibility. The latter is serious when it develops but can now be prevented. Immune factors causing hemolysis can also be *acquired.* Many cases are idiopathic and some are part of another immunologic disorder. Some may be caused by a viral infection altering the surface of the erythrocyte so that antibodies to the red cell membrane are generated. Others are caused by drugs that can bind to red cells and can generate antibodies which in turn combine with the drug bound to the cell and cause its destruction by complement fixation and lysis.

Toxins and Other Factors. A number of toxic chemicals or drugs act directly on red cells and may cause rapid hemolysis. Severe trauma, burns, and radiation may also cause extensive red cell damage. Bacterial sepsis from a variety of infections can also cause hemolysis.

Shunt Bilirubinemia. As discussed earlier, this condition results from ineffective erythropoiesis and is one of the factors causing the mild jaundice in pernicious anemia, sickle cell anemia, and thalassemia.

Laboratory Studies. The expected findings in cases of excess bilirubin *production* are shown in Table 10–3. These include:

Plasma bilirubin, usually 35 to 85 μmol/L (2 to 5 mg/dL); conjugated bilirubin, normal;
Liver function, essentially normal (except when a toxic factor damages both red cells and hepatocytes);
Urine bilirubin, negative; urobilinogen, positive;
Decreased red cell survival, which can be shown with radionuclide–labeled cells;
A direct antiglobulin test (DAGT), which will usually identify red cell auto-antibodies (often called Coombs' test);
Special tests are required to identify specific Hb abnormalities or enzyme deficiencies.

Neonatal Jaundice (Hyperbilirubinemia)

Hemolytic anemia in newborns requires the replacement of red cell mass. An associated problem is the prevention of bilirubin toxicity.

During intrauterine life the fetus must be able to eliminate its bilirubin (probably bound to α-fetoprotein) by transport across the placenta and clearance by the mother. The *full-term* infant is born with a surplus of red cells (that have a reduced lifespan), an immature liver (which develops its full capacity to conjugate bilirubin during the first 2 to 4 weeks), and a significant enterohepatic circulation of bilirubin. It is normal for a mild hyperbilirubinemia to occur at 3 to 5 days of life, but this should not exceed about 150 μmol/L (10 mg/dL) and should disappear by the second week. There is a rough correlation between the development of hepatocyte ligandin and glucuronyl transferase activity and the fall in serum bilirubin (Fig. 10–9). In *premature* babies the induction of these enzymes takes longer; their neonatal jaundice may develop less rapidly but is more marked and is slower to clear.

In ***newborns*** a rise in serum bilirubin above 200 μmol/L (12 mg/dL) is caused for concern, presumably because of an increase in free bilirubin. The reserve bilirubin binding capacity of serum albumin is less in infants, and acidosis, fatty acids, and certain drugs may affect available binding sites. Values below 40% of normal adult values carry a risk of kernicterus. Exposure of the baby's skin to light at 440 nm to 470 nm will promote the formation of photobilirubin isomers which enter the plasma and can be cleared by the liver without being conjugated (Fig. 10–4). At levels > 305 μmol/L (18 mg/dL) exchange transfusion may be necessary to prevent damage to vital brain neurons

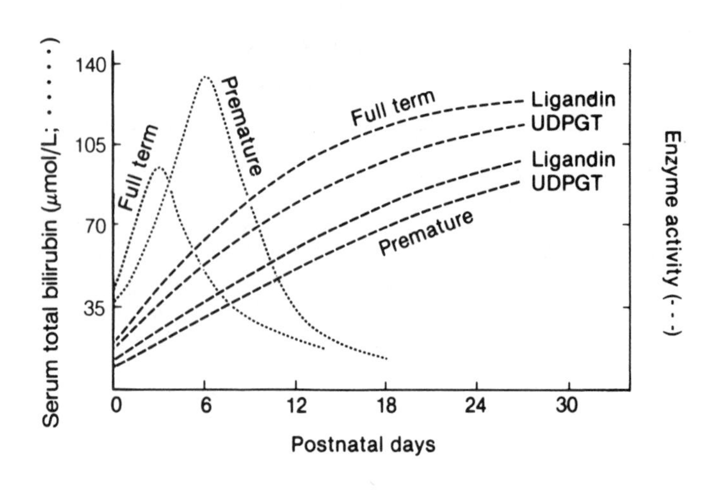

FIGURE 10–9. Correlation between serum bilirubin and hepatocyte ligandin and glucuronyl transferase in a newborn infant during the first few weeks of life. Adult levels are attained at about 3 months. UDPGT = uridine diphosphate glucuronyl transferase.

(*kernicterus*). In premature infants transfusion may be indicated when bilirubin exceeds 200 μmol/L (13 mg/dL).

Maternal factors may be a cause of severe neonatal jaundice. Some mothers pass on to their child a factor (an abnormal progesterone metabolite?) which delays for 2 to 3 weeks the induction of hepatic glucuronyl transferase (*Lucey–Driscoll syndrome*). If these children can be protected from bilirubin toxicity until the liver escapes from this inhibition the child will develop normally. Other mothers pass what may be a similar factor to their babies through their *breast milk*. Jaundiced babies recover quickly when changed to formula feeding if this has been the etiologic mechanism.

Impaired Delivery to the Liver

This may occur in a number of situations; most notable are cardiovascular disorders leading to hepatic congestion, and advanced fibrosis, in which the circulation makes much less effective contact with the hepatocytes in regenerated liver nodules. This leads to increased serum levels of unconjugated bilirubin in cirrhosis, a finding also seen following portacaval shunt operations for portal hypertension.

Defective Hepatic Uptake

The normal handling of bilirubin by the liver cell was shown in Figure 10–5. Plasma albumin and cytosol ligandin have similar affinities for bilirubin, and there is evidence that a membrane protein (bilitranslocase) facilitates its transfer between plasma and cells. About one third of the bilirubin taken up is recycled to the plasma. A small portion is probably converted by unknown pathways to other metabolites; the major portion is conjugated and excreted in the bile.

Impaired uptake occurs in newborn infants with an immature liver. It has also been demonstrated in about a third of patients with benign hyperbilirubinemia (*Gilbert's syndrome*), which has a prevalence of up to 5%. In this disorder the serum bilirubin fluctuates from 17 to 85 μmol/L (1–5 mg/dL), being influenced by factors affecting the general health of the individual. It is notably increased by 24 to 48 hours of fasting, which will help the diagnosis if subclinical hepatitis is excluded by enzyme tests (see Table 10–3). There may be a mild degree of increased hemolysis in some individuals, but evidence has increased for an abnormality of the glucuronyl transferase system in a majority of cases, with a higher proportion of BMG in the bile. The disorder is most clearly demonstrated by intravenous ^{14}C- or ^{3}H-bilirubin clearance studies, or by calculations based on the clearance of ^{51}Cr-tagged red cells—a procedure more likely to be available. Patients with Gilbert's syndrome may have a yellowish sclera, but the skin is rarely jaundiced. They lead normal lives and need only to be reassured and protected from medical or surgical intervention that they do not need. The disorder may represent a single gene defect for which the homozygous state is the Arias UDPGT type II syndrome (see Impaired Conjugation).

Secondary impairment of hepatic uptake, better described as increased recycling, occurs whenever cholestasis is present (Fig. 10–7). Bilirubin enters the cell and becomes conjugated, but the normal easy route to the bile is blocked. Considerable conjugated bilirubin finds its way via the interstitial space and lymph channels to the blood; but the net result is an increase in the levels of intermediate substrates, and saturation of ligandin binding sites. Uptake may be 'normal,' but only at higher plasma levels of the unconjugated pigment, with much more being recycled, and more finding alternate pathways.

Impaired Conjugation

Temporary deficiencies in the newborn have been discussed earlier. Permanent defects occur as relatively rare inherited deficiencies of glucuronyl transferase.

The ***Crigler–Najjar type I syndrome*** is due to the virtual absence of UDPGT I, probably inherited as a recessive trait. Homozygous offspring usually die in infancy, or endure lifelong severe jaundice; the serum unconjugated bilirubin may range from 250 to 750 μmol/L (15 to 45 mg/dL). Neither the monoglucuronide nor the diglucuronide of bilirubin is formed. Bilirubin uptake may be normal or increased; metabolism is mainly via alternate pathways and a large portion is recycled. The urine may be a deep golden yellow, but the color is not due to bilirubin. Mild deficiencies can be detected in relatives who carry only one defective gene (heterozygotes). They may have some degree of hyperbilirubinemia, even jaundice, but seem otherwise to be normal.

The ***Arias*** (or Crigler–Najjar ***type II***) ***syndrome,*** due to a deficiency of UDPGT II, is more benign. Bilirubin levels in serum range from 85 to 350 μmol/L (5 to 20 mg/dL), mostly as unconjugated bilirubin and monoglucuronide. Direct-reacting bilirubin is normal and bilirubinuria is rarely seen. Significant amounts of free bilirubin and the monoglucuronide are excreted in the bile. Histologic changes have been noted in the Golgi apparatus, and the deficient enzyme can be induced by phenobarbital with lowering of the serum bilirubin. The disorder may be transmitted as an autosomal dominant trait; some think inheritance is recessive, with homozygotes (1:40,000) showing constant jaundice and heterozygotes (1:100) making up part of the population with Gilbert's syndrome.

Defective Excretion into the Canaliculus

The excretion of conjugated bilirubin diglucuronide is an energy-dependent transport process that is shared with other organic anions (e.g., sulfobromophthalein [BSP]). The ***Dubin–Johnson syndrome*** is a relatively rare defect of this system, inherited as a recessive trait. In homozygotes the disorder may be noted early in childhood as a fluctuating hyperbilirubinemia with constant or intermittent bilirubinuria. Compensation is good and the disorder is relatively benign, although the patient may complain of fatigue or of vague gastrointestinal symptoms. In some cases the disorder is noted only under the stress of pregnancy or some illness, or as an incidental finding. The liver accumulates a dark pigment of uncertain composition, which led to the term 'black liver disease.' Heterozygotes carry the trait and may exhibit mild forms of the disorder, which can usually be demonstrated as a decreased transport maximum for BSP.

A characteristic feature of the disorder is an impaired ability to excrete BSP, resulting in a secondary rise in serum levels 60 to 90 min after intravenous injection, as the dye is returned to the circulation. Also useful in confirming the diagnosis is a ratio of urinary coproporphyrin III to total CP of <20%, as mentioned earlier (see also Table 10–4).

Another similar but milder, rare disorder, ***Rotor's syndrome,*** differs in the absence of liver pigment, absence of BSP reflux (although the T_m may be reduced), and a less significant but not absent effect on urinary CP I and CP III. An ill-defined condition called *hepatic storage disease* may also be associated with mild elevations of conjugated bilirubin in the serum.

In these conditions the elevated serum bilirubin and related bilirubinuria are often the only abnormalities noted. Usual liver function tests are normal and the importance of the diagnosis lies in reassurance that the condition is not serious.

Jaundice due to Hepatocellular Disease

Impairment of bile flow is one of the consequences of severe disease of the hepatocytes and leads to raised plasma levels of both conjugated and unconjugated bilirubin, usually <350 μmol/L (20 mg/dL). The final transport step into the canaliculus is probably rate-limiting, but ***cholestasis*** is a very complex phenomenon which reflects the widespread impairment of hepatocyte metabolism and function. A deficient energy supply for electrolyte regulation, or for normal bile acid synthesis, may be implicated directly in membrane injury or function. The development of cholestasis is accompanied by a sequence of changes which reflect increasing severity:

1. Elevated serum bile acids 2 hours after a meal
2. Elevated serum bile acids in the fasting state
3. Increased conjugated bilirubin in the serum, with dark urine due to bilirubin
4. Increased serum total bilirubin

Viral Hepatitis. This infection (see Chap. 11) occurs most frequently as a subclinical malaise without jaundice, although a transient mild hyperbilirubinemia and bilirubinuria may occur. In about 25% of cases jaundice is present and, in a few, complete *intrahepatic cholestasis* persists for periods ranging from 3 days to 3 weeks or longer. The jaundiced cases exhibit the pattern of bile pigment excretion shown in Figure 10–10. Both bilirubin and urobilinogen are present in the urine initially, but during the jaundice or cholestatic phase urobilinogen decreases or disappears. Onset of recovery is signalled by a return of appetite and increased urobilinogen in the urine. Bilirubin diglucuronide is rapidly excreted and bilirubinuria clears while the patient is still jaundiced. Serum bilirubin then falls slowly as δ-bilirubin is released and excreted. The final disappearance of urobilinogen during convalescence parallels the decline of elevated serum aminotransferase enzymes.

Chronic Hepatitis. Impaired bilirubin excretion occurs as a relatively late event in about a third of patients with *portal cirrhosis* but is a major feature in all cases of *biliary cirrhosis.* Urobilinogenuria is a relatively constant finding in the former and is usually absent in the latter.

Hepatobiliary Cholestasis

Included in this category are a variety of agents that appear to act specifically on the bile secretion mechanism, particularly as they affect bilirubin, to cause *cholestasis.*

All severe forms of cholestasis are associated with hyperbilirubinemia and bilirubinuria and usually with some secondary abnormality of liver function attributable to or responsible for the decrease in bile flow. Cholestasis associated with *hepatocellular* disease has already been discussed. Acute *idiopathic* cholestasis and *benign recurrent* cholestasis occur as primary, possibly genetic, disorders of biliary secretion with little or no evidence of hepatocellular dysfunction.

Drug administration is probably the most common cause of cholestasis. Drugs may have a direct toxic effect on the canalicular membrane of the hepatocyte, or may cause an inflammatory reaction in the portal tracts. Hepatitis from overindulgence in alcohol will elevate serum bile acids and serum bilirubin, and will be associated with a fatty liver and increased serum γ-glutamyl transferase (GGT), and not infrequently with jaundice.

Direct toxic effects of drugs or chemicals, such as chlorinated hydrocarbons and phosphorus, are related to the dose, occur in all exposed individuals, and appear after a brief latent period. Cholestasis may or may not be prominent.

Some types of jaundice in the newborn, in pregnancy, or in women taking anovulatory drugs have been attributed to the effects of steroid metabolites. Jaundice associated with drugs such as chlorpromazine, halothane, 17-alkyl steroids, other steroid hormones, and monoamine oxidase inhibitors, are unrelated to dose, occur in only a small percentage of cases, and exhibit a variable latent period. Different drugs show different histologic consequences, some mainly cholangiolitic with little hepatocellular damage. At least some of these effects are believed to be caused by a *hypersensitivity reaction.* Halothane and monoamine oxidase inhibitors tend to cause a more severe hepatocellular necrosis resembling hepatitis.

Immune mechanisms are a less common but important cause of cholestasis. They may result from a genetically sensitive immune system which perpetuates the response to an antigen derived from the hepatocyte canalicular membrane or biliary epithelium. A secondary *mitochondrial antibody* is regarded as a marker for autoimmune cholestasis.

Immune mechanisms are implicated in chronic progressive hepatitis, idiopathic recurrent cholestasis, recurrent jaundice of pregnancy, and cirrhosis of unknown etiology. Primary biliary cirrhosis involving inflammation, proliferation of ductal epithelium, and obstruction of the small bile ducts, appears to be an autoimmune disease. Certain forms of cholangitis of the larger bile ducts may have an immune basis, but are more commonly associated with an ascending infection of the biliary tree.

Metastases in the liver from cancer elsewhere frequently exert pressure on the small bile ducts. The associated obstruction will cause, or induce, increases in serum alkaline phosphatase, 5′-nucleotidase, and GGT, but not bilirubin, because the conjugated pigment is readily reexcreted by healthy portions of the liver. Intrahepatic *primary carcinoma* of the bile ducts is a rare cause of cholestasis, but when it occurs in the porta hepatis it can produce a syndrome very similar to biliary cirrhosis.

Bile acid metabolism is closely involved with bile flow. Some of the ***cholestatic syndromes*** can be caused by discrete disorders of bile production. Three specific pathogenic mechanisms have been identified: (1) The *monohydroxy bile acids,* lithocholic and 3β-OH-5-cholenoic acid, have been shown to cause direct injury to the canalicular membrane; (2) Ethinyl estradiol, methyltestosterone, and other *17-alkyl steroids* are believed to alter membrane permeability and decrease the fraction of bile secretion that is independent of bile acid; (3) A rare, possibly genetic 24-hydroxylase enzyme defect, leading to a severe and fatal cirrhosis, has been attributed to *coprostanic acid* accumulation.

(Text continues on p. 210)

TABLE 10–4. Diagnostic Aids in Disorders of Bilirubin Metabolism

Diagnostic Problem	Useful Tests	Usual Findings	Comment
Excess bilirubin production			
Congenital spherocytosis	Urine urobilinogen	Positive	No bilirubinuria; reflects excess bilirubin reaching intestine
	Serum bilirubin	35–85 μmol/L >85% unconjugated	
	Routine hematology—blood film	Hemoglobin normal or low, reticulocytes increased	Crisis may induce hypoxia
	Osmotic fragility	Increased	Defect in membrane?
	^{51}Cr red cell survival	Decreased half life (<20 d)	Measures severity
Acquired hemolytic anemia	Blood grouping	ABO, Rh incompatibility?	
	Direct antiglobulin test	Antibodies present	On red cells (or in serum)
Other causes			
Defective erythropoiesis	Hemoglobin electrophoresis	Hb S, Hb C, thalassemia, etc.	Keep possibility in mind
	Red cell enzyme assay	G_6P dehydrogenase?	Partial deficiency not uncommon
		Pyruvate kinase?	Rare
Megaloblastic anemia	Schilling test; serum B_{12}, folate	Impaired vitamin B_{12} absorption	Lack of intrinsic factor
Trauma	Myoglobinuria	May be present	Colors urine red, but not serum
Interstitial hemorrhage	Urine urobilinogen	Positive	Reflects excess bilirubin production
Impaired delivery to the liver			
Heart failure	Urine urobilinogen	Positive	Reflects hepatic congestion, hypoxia
	Serum AST	Elevated	Extent of cardiac necrosis and liver congestion
Advanced cirrhosis	Serum albumin	Low	Extent of functional impairment
	Prothrombin time	May be prolonged	Indicates advanced disease
Porta caval shunt	Blood ammonia	May be present	Impending coma
Defective uptake by liver			
Neonatal jaundice	Serum bilirubin	Elevated, unconjugated	Monitor need for therapy
	Serum bilirubin-binding capacity	May be decreased	Monitor need for, results of, therapy
	Hepatic ligandin in liver biopsy	May be low	Not a routine test

Benign hyperbilirubinemia (Gilbert's disease)	Serum bilirubin	17–85 μmol/L	Fluctuates
	Serum GGT	Normal	Sensitive test to exclude liver disease
	Serum bile acids	Normal	Sensitive test to exclude liver disease
	Fast 24–48 h, check for	2- to 3-fold increase in serum bilirubin	Simple clinical test
	^{14}C- or ^{3}H-bilirubin clearance	Low in ⅓ of cases	Establishes uptake or clearance defect
	^{51}Cr-tagged red cell clearance	Half-life normal (26–33 days)	Bilirubin up relative to red-cell destruction
Secondary uptake defects	See text		
Impaired conjugation of bilirubin			
Neonatal jaundice	Serum bilirubin	Elevated, unconjugated	Phototherapy at 170 μmol/L Exchange transfusion at 340 μmol/L
	Serum bilirubin-binding capacity	May be decreased	Better criterion to decide need for therapy
Lucey–Driscoll syndrome	UDPGT inducible by drugs	May lower serum bilirubin	Suggests partial deficiency (rare)
Mother's milk syndrome	Trial on formula feeding	Fall in serum bilirubin	Rare disorder
Crigler–Najjar syndrome	UDPGT I assay on liver biopsy	Enzyme absent	Permanent and often fatal
Arias syndrome	Transferase inducible by drugs	Lowers serum bilirubin by 50%	Partial deficiency
	UDPGT II assay on liver biopsy	Deficient enzyme, often <10%	Compatible with normal health
Defective excretion into canaliculus			
Dubin–Johnson syndrome	Urine bilirubin	Positive	Conjugated bilirubin
	Serum bilirubin	35–100 μmol/L	Fluctuates, compensated
	BSP excretion	Secondary rise at 60–90 min	Shares transport system with bilirubin
	Urine coproporphyrins	CP III < 20% of total coproporphyrin	Unexplained; normal >65%
Rotor's syndrome			Similar to Dubin–Johnson syndrome but a different disease
Hepatobiliary diseases	See Chapter 11		
Obstructed drainage of bile	See Chapter 11		

AST = Aspartate aminotransferase; BSP = sulfobromophthalein (Bromsulfalein); GGT = γ-glutamyltransferase; UDPGT = uridine diphosphate glucuronyl transferase; G_6D = glucose-6-phosphate.

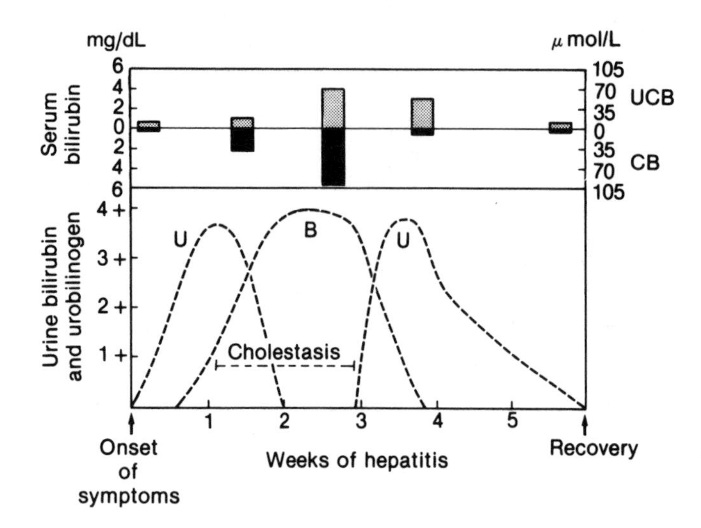

FIGURE 10-10. Bile pigments in hepatitis. Serum bilirubin, unconjugated (UCB) and conjugated (CB), and excretion patterns of urine bilirubin (B) and urobilinogen (U) during the course of viral hepatitis with severe cholestasis.

Extrahepatic Biliary Obstruction

In neonates, jaundice may be due to a congenital lack of development of a patent bile duct system; the capacity to excrete rose bengal is used as a measure of such *biliary atresia. Stricture* of the common bile duct is now a rare postoperative complication of biliary tract surgery.

Stones formed in the common bile duct, or entering it from the gallbladder, produce intermittent obstruction that is usually associated with acute pain, hyperbilirubinemia, and bilirubinuria. Cholecystectomy is commonly performed, but 10% of bile duct stones are discovered after gallbladder surgery. Mechanical removal of the retained stone (papillotomy) is usually advised.

Cancer of the larger bile ducts or, more commonly, of the head of the pancreas occluding the common bile duct, produces an unremitting jaundice calling for remedial surgery where possible. If the tumor is inoperable, a cholecystojejunostomy may restore the flow of bile.

In *extrahepatic* obstruction the serum total bilirubin is an index of the obstructive process and tends to plateau at about 500 μmol/L (25 to 30 mg/dL). The concentration of serum bile acids is usually high, and is generally associated with severe pruritus. Liver function is normal at first but suffers secondary damage in proportion to the degree of cholestasis and any biliary tract infection that may be present.

A full discussion of hepatic and biliary disorders is presented in Chapter 11.

SUGGESTED READING

BERK PD, JAVITT NB: Hyperbilirubinemia and cholestasis. Am J Med 64:311–326, 1978

CHOWDHURY NR, CHOWDHURY JR, WOLKOFF AW, ARIAS IM: Bile pigment metabolism. In Arias et al (eds): The Liver Annual, Chap. 14. Amsterdam, Elsevier, 1983

DREYFUS J-C, KAHN A: Red cell enzymopathies: Molecular mechanisms. Clin Biochem 17:331, 1984

GOLLAN JL, SCHMID R: Bilirubin update: Formation, transport, and metabolism. In Popper H, Schaffner F (eds): Progress in Liver Diseases, Vol 8, Chap. 15. New York, Grune and Stratton, 1982

HINDMARSH JT: Clinical disorders of porphyrin metabolism. Clin Biochem 16:209, 1983

STANBURY JB, WYNGAARDEN JB, FREDRICKSON DS (eds): The Metabolic Basis of Inherited Disease, 5th ed, Parts 8, 10. New York, McGraw–Hill, 1982

WEISS JS, GAUTAM A, LAUFF JJ ET AL: The clinical importance of a protein-bound fraction of serum bilirubin in patients with hyperbilirubinemia. N Engl J Med 309: 147, 1983

11

Hepatobiliary Disorders

Allan G. Gornall/
David M. Goldberg

LIVER STRUCTURE AND FUNCTION

ANATOMY AND BLOOD SUPPLY

The adult human liver weighs about 1500 g and lies in the right upper quadrant of the abdomen, with its lower edge behind the right costal (rib) margin. Embryologically, the liver develops as an extension of the gastrointestinal tract epithelium, from which are formed the bile duct system and parenchyma. The venous portal system resembles a tree, with its roots in the gut and its branches in the liver (Fig. 11–1).

The liver has a large right and smaller left lobe with the following connections: Beneath the liver is the *porta hepatis* where the blood supply, connective tissue, and nerves enter and from which the bile ducts and the major lymphatics leave. The hepatic veins and the minor lymphatics exit behind the liver.

The ***dual blood supply*** of the liver is unique. The hepatic artery arises from the celiac axis, which also sends arteries to the stomach and spleen. At the stage of the terminal hepatic arteriole, the blood pressure has fallen from 110 to about 35 mm Hg. The *arterial system* supplies 25% of the hepatic circulation at 95% oxygen saturation. The *portal vein* represents a confluence of the splenic vein (which includes the inferior mesenteric and drains the pancreas, spleen, and left descending colon) and the superior mesenteric vein (from the pancreas, small intestine, and right ascending colon) plus small contributories from the stomach and pylorus. There are *anastomotic* connections to the paraumbilical veins (anterior abdominal wall), the veins of the lower esophagus, and the hemorrhoidal veins. The *portal system* supplies 75% of the hepatic circulation at 70% oxygen saturation and a pressure of 10 mm Hg, falling to 8 mm Hg in the terminal portal venule. The high-pressure terminal arterioles empty into the low-pressure terminal venules in an intermittent pulsatile manner, controlled by junction sphincters that probably respond to changes in pressure or oxygen tension. The resulting blood flow through the liver sinusoids is about 25% of the cardiac output and varies from around 1.5 L/min (recumbent) to 1 L/min (erect), serving as a factor in regulating blood pressure and avoiding syncope.

A small fraction of the portal blood bypasses the parenchymal sinusoids and represents a 'physiologic shunt.' This may contribute to the normal excretion of urobilinogen in the urine and to postprandial transient elevations of serum bile acids. Such shunting can become much more extensive in liver disease.

Nerves enter the liver from a plexus around the celiac axis, derived from autonomic ganglia and the vagal and phrenic nerves. The role of these nerves is not fully understood, but they probably act mainly to control blood flow.

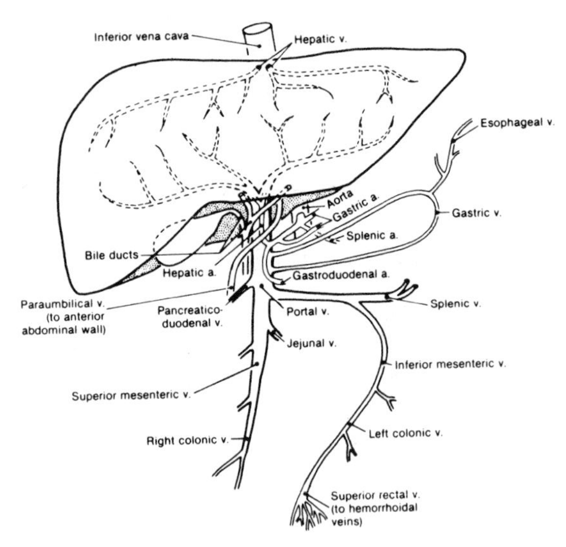

FIGURE 11–1. Circulatory connections of the human liver. a = artery; v = vein.

Intrahepatic Subdivisions

Segmental Structure. Within the liver the artery, portal vein, bile ducts, lymphatics, and connective tissue (a continuation of the liver capsule) subdivide like branches of a tree, in *portal tracts* that become progressively smaller. There are no anastomotic connections of significant size between segments of the liver supplied by major branches of the blood vessels. Because of this, surgical resection of anatomic portions of the liver can be achieved by following microvascular cleavage planes. The fine subdivisions of the connective tissue form the fibrous ectoskeleton of the liver.

Structural Unit of the Liver. As the branching vessels and ducts become smaller they supply in sequence an acinar agglomerate (about 16 units), a complex acinus (about 4 units), and finally a single ***acinus.*** This ultimate structural unit of the liver is the cluster of cells that surround the terminal trio of arteriole, portal venule, and bile ductule (Fig. 11–2A, B). It is within the acinus that the arteriole empties into the portal venule and the mixed blood percolates through the sinusoids toward the terminal hepatic venule (central vein) whence it drains through the hepatic veins to the vena cava.

The liver acinus is approximately 0.5 mm in diameter with about 15 sinusoids radiating from its inner core (zone 1) to its periphery (zone 3). Between these sinusoids are the *liver cell plates,* mostly one cell thick, and about 15 cells in length (Fig. 11–2B). The term 'lobule' should be considered obsolete, because it bears no useful relationship to the microcirculation of the liver.

THE LIVER CELLS

Parenchymal cells (hepatocytes) are specially adapted epithelial cells characterized by microvilli on their sinusoidal surfaces and a blind-end canaliculus between columns of cells (Fig. 11–2C). The sinusoidal microvilli make up about 75% of the surface area of the hepatocyte, the remaining 25% are divided about equally between the basolateral surfaces and the canalicular microvilli. Averaging about 20 μm in diameter, these cells have a prominent nucleus and nucleolus, an extensive smooth and rough endoplasmic reticulum and Golgi complex, about 400 mitochondria per cell, and many storage granules and lysosomes. They are linked by desmosomes, tight-junction fibrils located on either side of the canaliculus, consisting of a charged matrix that can regulate the passage of water and electrolytes between the paracellular spaces and the canaliculi. Hepatocytes, which show functional heterogeneity in different zones of the acinus, make up roughly 60% of the liver weight.

Endothelial cells lining the sinusoids of the liver are primitive reticular cells of mesodermal origin (Fig. 11–2C). They overlap irregularly and there are 0.1-μm to 1-μm pores that allow plasma, but not cells, to enter the *space of Disse* and come into direct contact with the hepatocyte microvilli. Very thin endothelial cells also line the *lymphatics* of the liver, which drain the interstitial spaces via the porta hepatis (major route) and along the hepatic veins (minor route) to the thoracic duct.

Kupffer cells are not part of the endothelial lining of the sinusoids, but occupy space within these tortuous channels, attaching themselves in various ways to the endothelial wall. Kupffer cells are probably derived from monocytes and can be regarded as *fixed macrophages,* functioning as guardian cells of the liver and the general circulation. They form a major part of the reticuloendothelial system (RES) and have important phagocytic, antitoxic, and possibly metabolic functions. They can be mobilized from cells in the bone marrow, and also appear to multiply within the liver.

The sinusoidal endothelial and Kupffer cells together make up about 25% of the liver mass.

Epithelial cells line the small bile ducts, which are formed by convergence of the terminal bile ductules (*cholangioles*) draining the several canaliculi of each acinus. These bile ducts are lined by cuboidal cells with microvilli on their luminal surface. The larger bile ducts are lined by columnar epithelium. In total they comprise about 10% of the liver mass.

Other cells normally make up a small portion of the liver substance. These cells include the lipocytes, which lie under the endothelial cells, and fibroblasts, plasma cells, and mast cells in the connective tissue spaces. Any inflammatory process in the liver will involve the so-called *mesenchymal reaction*—proliferation of phagocytic cells and fibroblasts, activation of mast cells, and attraction (infiltration) of polymorphonuclear cells, monocytes, and lymphocytes from the blood.

SECRETORY AND EXCRETORY FUNCTIONS

Formation of Bile

There are three main phases in the formation of bile—bile acid dependent and bile acid independent mechanisms in the hepatocytes and fluid transfer systems in the bile ductules. Of *primary* importance is a transport-osmotic mechanism that depends on an efficient uptake of bile acids from the portal circulation and their secretion into the canaliculi by a specific transport system. Lecithin is also secreted, and bile acid:phospholipid micelles are osmotically active in retaining fluid in the lumen. The *second* phase is a Na/K ATPase-dependent transfer of sodium from hepatocytes into the paracellular spaces, and thence through the desmosome matrix into the canaliculi. This process is linked to an active transport of HCO_3^- directly into the canaliculi and a passive movement of Cl^- and H_2O to maintain electroneutrality and osmolality. Myofibrils in the canalicular wall and microvilli propel fluid toward the bile ductules. The volume of canalicular bile has been estimated at 400 to 500 mL/day.

The *third* phase occurs in the bile ductules and ducts, where both secretion and reabsorption of fluid occur under the influence of secretin and probably other gastrointestinal hormones. The net output of hepatic bile is normally about 1 L/day and consists of 97% water, 1% bile acids, 1% electrolytes, and 1% everything else including cholesterol, phospholipids, pigments, porphyrins, and several important proteins, including IgA, which helps suppress the uptake of foreign antigens from the intestine.

During fasting and between meals a major portion of the bile is sequestered and may be concentrated up to tenfold in the gallbladder. Bile reaches the intestine in three main waves when the sphincter of Oddi relaxes and the gallbladder contracts after each meal. Approximately 500 to 600 mL of bile enter the duodenum each day. Pressure in the biliary tract is normally 12 to 15 mm Hg; resistance that raises pressure to >25 mm Hg will inhibit bile flow.

Bile Acid Formation and Metabolism

Bile acid synthesis is a major metabolic pathway for the removal of cholesterol from the body; the rate-limiting site of feedback control is the 7 α-hydroxylase step (Fig. 11–3). The primary products, cholic acid (3α,7α,12α-trihydroxy-5β-cholanic acid) and chenodeoxycholic acid (3α,7α-dihydroxy-5β-cholanic acid), have a pK between 5 and 6. They are conjugated with glycine or taurine to form glycocholates (pK = 4–5) and taurocholates (pK = 1–2), and are transported into the canaliculi mainly in this form. Bile acids can form esters, usually C-3 sulfates but also glucuronides, which render them more hydrophilic. Under normal conditions only the poorly soluble monohydroxy acids are esterified; during cholestasis esterification extends to all bile acids and facilitates their elimination in the urine.

At the pH of duodenal and jejunal contents ($\simeq$6.5) the conjugated bile acids are ionized and highly charged. This in addition to their strongly polar and nonpolar surfaces enables them to link hydrophilic and hydrophobic substances in stable ***micelles.*** The ratio of bile acids to phospholipid to cholesterol is about 10:3:1. Large aggregates incorporate bile pigments, fatty acids, and other lipids. Sodium and potassium salts of the free and conjugated bile acids are present, but the term *bile acids* is now used in preference to *bile salts.*

Bile acids in the *small intestine* aid the digestion

and absorption of fat and the absorption of a variety of other hydrophobic substances including the fat-soluble vitamins. Normally in the duodenum and jejunum there is very little metabolism of the bile acids. They encounter the anaerobic bacterial flora in the lower ileum where they are

1. Partly deconjugated to free bile acids
2. Partly dehydroxylated at the 7 position (creating the secondary deoxycholic and lithocholic bile acids)
3. Partly deesterified at the 3 position.
4. Partly oxidized to form 7-keto bile acids.

In bile acids *hydroxyl groups* are an asset. The trihydroxy acids are better choleretics, form more stable micelles, and are less toxic. Monohydroxy acids are relatively insoluble, will cause cholestasis, and are cytotoxic. Dihydroxy acids are intermediate. Very little monohydroxy acid is absorbed from the intestine. About 75% of the tri- and dihydroxy bile acids are ***reabsorbed*** in the lower

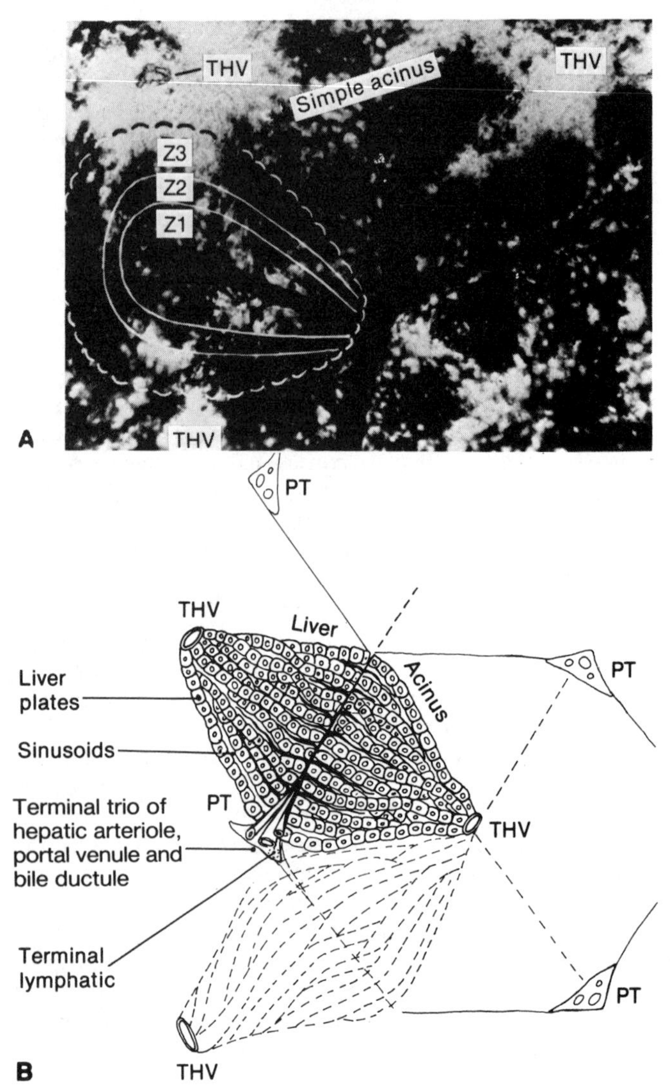

FIGURE 11–2. *A.* Blood supply and zones of the liver acinus (human liver injected with india ink). Small portal vein (center) has three terminal portal venules supplying the sinusoids of three simple acini. On the left are shown the three zones of an acinus. Zone 1 (**Z1**) is the region close to the terminal afferent vessels. **Z3** consists of the cells at the microcirculatory periphery of the acinus (these cells are more prone to ischemia). **Z2** is the intermediate zone. Incomplete perfusion has revealed the vulnerability of cells around the terminal hepatic veins (**THV**). (Photograph courtesy of Dr. AM Rappaport)

B. Representation of a liver acinus, showing cells arranged in plates one cell thick, and sinusoids between. Afferent vessels of the portal tract (**PT**) are the hepatic arteriole and portal venule; efferent channels are the bile ductule and the lymphatic. The blood leaves via the terminal hepatic veins (**THV**).

ileum (see Fig. 11–3); another 10% to 15%, mainly as free acid, in the colon. These bile acids return via the portal system to the liver, which has an efficient uptake mechanism for their extraction. In the liver cell the bile acids may be rehydroxylated, reconjugated, and reesterified, and are secreted to form new bile. Reabsorbed 7-ketolithocholic acid is converted to 7α-chenodeoxycholic acid, or 7β-ursodeoxycholic acid. Approximately 10% of the bile acid pool (about 500 mg) is lost each day in the feces and must be replaced by synthesis from cholesterol. Normally the liver extracts virtually all portal vein bile acids, so that plasma bile acid levels are low and show only a transient increase 60 to 120 min after each meal.

In *hepatobiliary disorders* with intra- or extrahepatic cholestasis, the accumulation of bile acids in serum depends on a new equilibrium between synthesis, feedback suppression, biliary excretion, intestinal reabsorption, hepatic reexcretion, esterification, escape into the plasma, and urinary excretion.

Formation of Gallstones

Stones may form when the quantity of cholesterol in the bile is excessive, or when qualitative changes in its constituents render the bile less efficient in maintaining cholesterol in micellar solution. Precipitation may tend to occur at night, when bile is concentrated in the gallbladder. Excess bilirubin from hemolytic disorders may form smaller 'pigment' stones for the same reason. Recent studies indicate that 80% of cholesterol stones may have been seeded by polymerized unconjugated bilirubin. At least four *theories* on the cause of gallstones have been postulated:

1. Increased cholesterol production may result from activation (or lack of feedback suppression) of β-hydroxy-β-methylglutaryl · coenzymeA (HMG · CoA) reductase, or suppression of 7α-hydroxylase.
2. Some patients excrete, or generate, abnormal amounts of unconjugated bilirubin in the bile.
3. The bile may have abnormal proportions or kinds of tri-, di-, and monohydroxy bile acids. Normal bile contains, in free and conjugated forms, about 40% cholic, 40% chenodeoxycholic, 19% deoxycholic, and perhaps 1% lithocholic acids.
4. Abnormal phospholipids (with atypical fatty acids) may have poor micelle-forming qualities.

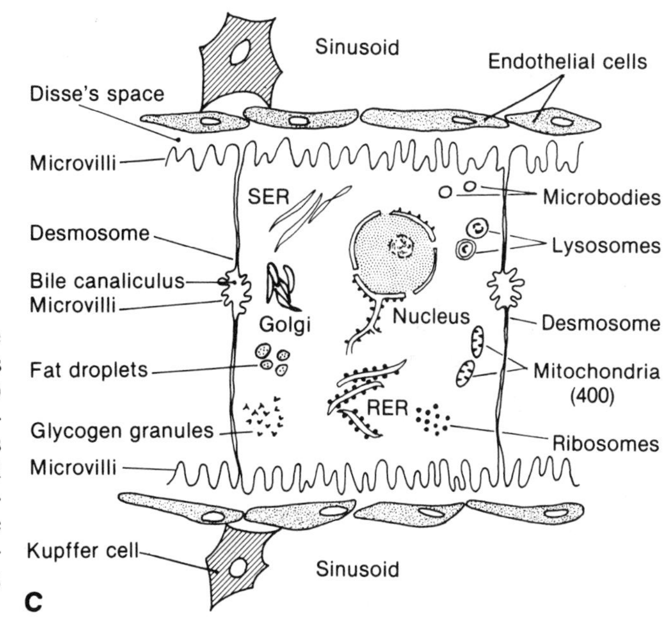

C. Schematic section through a liver plate shows endothelial cells lining, and Kupffer cells in, the sinusoids, tight junctions (desmosomes) linking the hepatocytes, and bile canaliculi between the cells. Picture blood in the sinusoids flowing toward you from behind the page, and bile in the canaliculi flowing in the opposite direction. Note the presence of microvilli on the sinusoidal surface of the cells and in the canaliculi. SER = smooth endoplasmic reticulum; RER = rough endoplasmic reticulum.

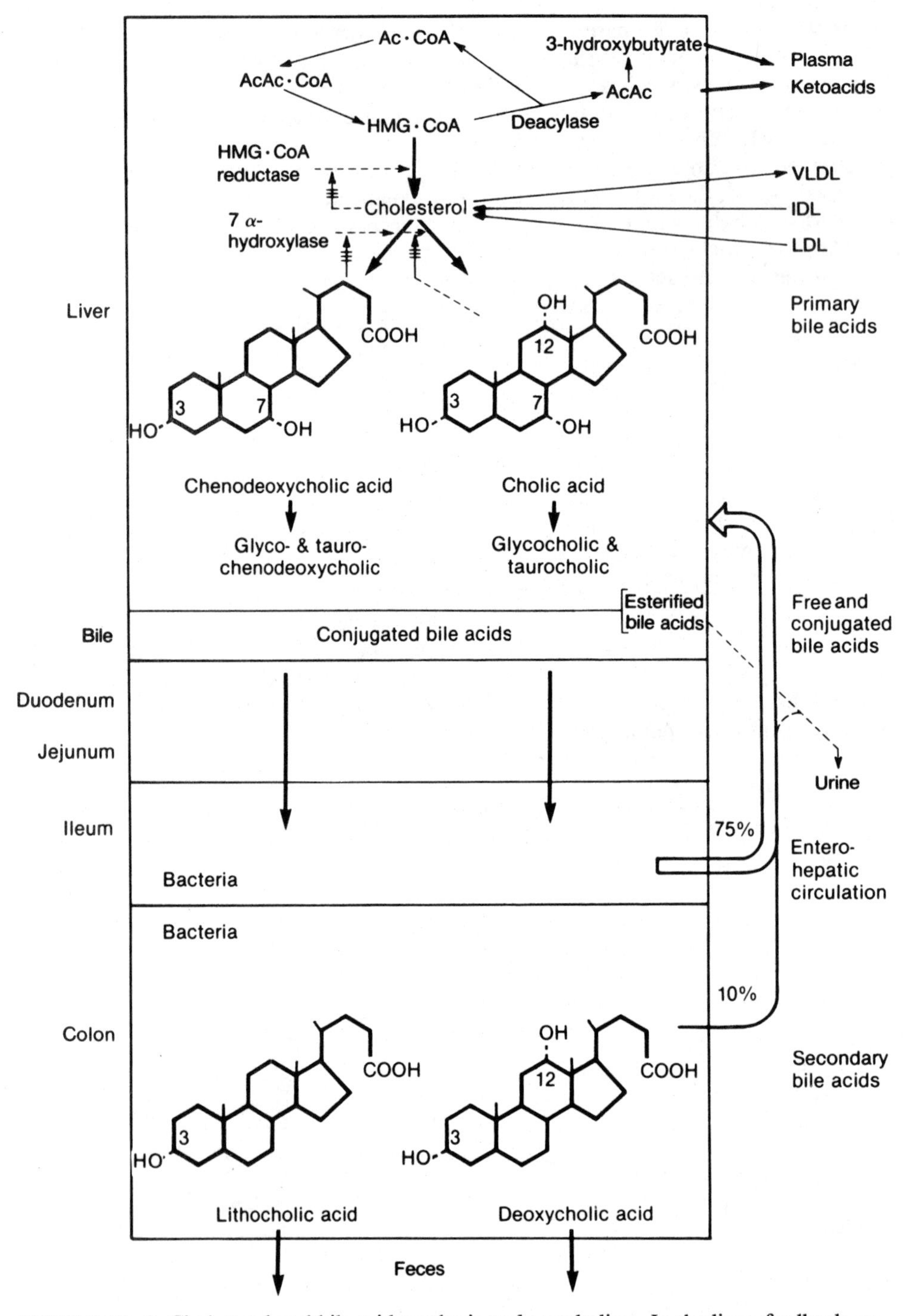

FIGURE 11–3. Cholesterol and bile acid synthesis and metabolism. In the liver, feedback regulation of HMG · CoA reductase by cholesterol, and of 7 α-hydroxylase by bile acids should be noted. The enterohepatic circulation, bacterial dehydroxylation, and excretion of bile acids are shown, as well as the excretion of the more hydrophilic esterified bile acids.

METABOLIC FUNCTIONS

Of the few hundred known functions performed by the liver it is possible to list here only those that have significance either in liver disease, in metabolic disturbances, or in relation to diseases of other organ systems.

Porphyrin and bile pigment metabolism and excretion are discussed in Chapter 10. The major metabolic pathways in the hepatocyte are shown in Figure 11–4.

Carbohydrates (See also Chap. 16)

Blood glucose is restored mainly from the liver, initially from glycogen, later by gluconeogenesis, mainly under the influence of glucagon. An inability to release glucose is seen in some glycogen storage diseases.

Other sugars are converted to glucose. Inability to do so is seen in such disorders as galactosemia and hereditary fructose intolerance.

Gluconeogenesis from amino acids is an important function of the liver.

Fat is formed from carbohydrate in the liver when nutrition is adequate and the demand for glucose is being met from dietary sources.

Lactate conversion to pyruvate and glucose is a major function of the liver.

Uridine diphosphoglucuronic acid (UDPG acid) formation is important in the metabolic clearance of hormones and as a detoxification mechanism. Low levels in newborns and possibly in Gilbert's syndrome may limit the capacity to clear bilirubin.

Proteins (See also Chap. 18)

The liver cell has special structural features that enable protein synthesis to occur at the endoplasmic reticular membrane with liberation of protein into the plasma. Complex proteins are formed in the Golgi region and probably released by reverse pinocytosis. Some proteins enter the bile by exocytosis.

Deamination of glutamate is the primary source of ammonia, which is then converted to urea via carbamyl phosphate. Most amino acids lose their amino group by transamination; the net product, however, is glutamic acid.

Transamination occurs in all body cells, but the activity of aminotransferases is high in liver cells.

Urea synthesis is almost confined to liver cells. Deficiencies of key enzymes in this process lead to specific diseases (see Ch. 18).

The liver is very important in *plasma protein* production, synthesizing

100% of albumin, 10 to 15 g/day (half-life 15 to 20 days)
about 90% of α_1-globulins, mainly consisting of α_1-antitrypsin, but also high-density lipoproteins, α_1-acid glycoprotein (orosomucoid), thyroxin-binding globulin, corticosteroid-binding globulin and testosterone-estrogen-binding globulin.
about 70% of α_2-globulins, including α_2-macroglobulin, haptoglobin, ceruloplasmin and very-low-density lipoproteins.
about 50% of β-globulins, including transferrin, hemopexin, and components of the complement system
100% of fibrinogen (rarely deficient)
100% of prothrombin, which is dependent on vitamin K absorption and hence on bile acids
almost all blood clotting factors, which have a short half-life
special proteins, including most of the apolipoproteins, erythropoietin, angiotensinogen, and somatomedin.

Normally, the liver synthesizes no γ-globulin; it does so only when immunologically competent cells infiltrate the liver as part of the mesenchymal reaction to injury.

Lipids (See also Chap. 20)

The liver plays a central role in fat metabolism:

Chylomicrons reach the circulation via the thoracic duct and are partially degraded by lipoprotein lipase; the liver is the major site of removal of these chylomicron remnants.

Acetyl CoA (AcCoA) conversion to fatty acids, triglyceride, and cholesterol is a major hepatic function.

Ketone body formation occurs almost exclusively in the liver. When the demand for gluconeogenesis depletes oxaloacetate, and AcCoA cannot be converted rapidly enough to citrate, acetoacetyl CoA (AcAc·CoA) and HMG·CoA accumulate and a deacylase in the liver liberates ketone bodies into the blood (Fig. 11–4).

Lipoprotein synthesis (very-low-density lipoproteins [VLDL]) occurs almost exclusively in the

(Text continues on p. 219)

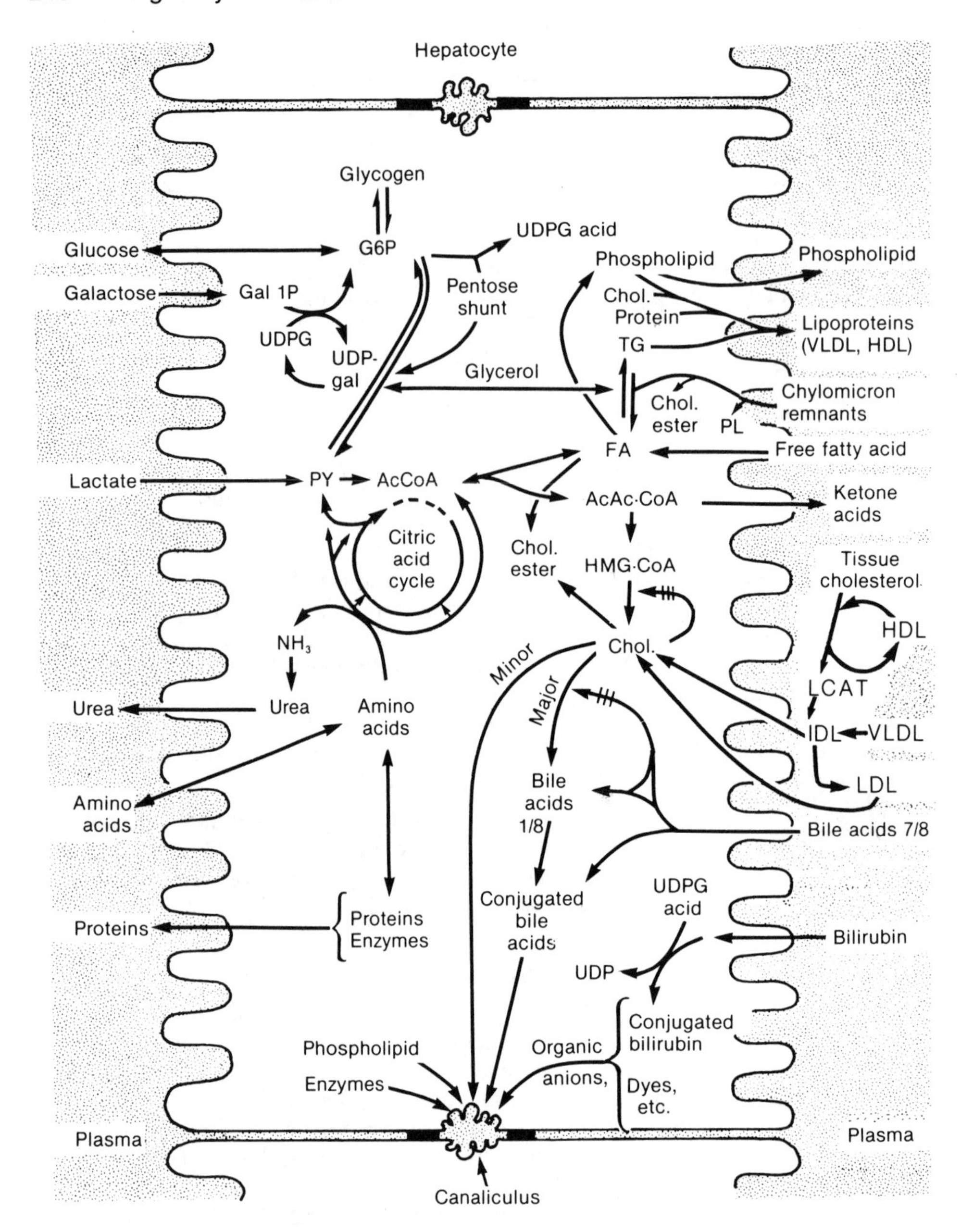

FIGURE 11–4. Major metabolic pathways in the hepatocyte. AcCoA = Acetyl coenzyme A; AcAc · CoA = acetoacetyl CoA; Chol. = cholesterol; FA = fatty acid; G6P = glucose 6-phosphate; Gal 1P = galactose 1-phosphate; HDL = high-density lipoprotein; IDL = intermediate-density lipoprotein; LDL = low-density lipoprotein; HMG · CoA = β-hydroxymethyl-glutaryl CoA; LCAT = lecithin: cholesterol acyltransferase; PL = phospholipid; PY = pyruvate; TG = triglyceride; UDPG = uridine diphosphoglucose; UDPG acid = uridine diphosphoglucuronic acid; UDP-gal = uridine diphosphogalactose; VLDL = very-low-density lipoprotein.

liver as a mechanism for transporting triglyceride to the tissues. High-density lipoprotein (HDL) and several of the apolipoproteins are also synthesized in the liver. The liver plays an important role in the degradation of chylomicron remnants, also of intermediate density lipoproteins (IDL) and low density lipoproteins (LDL).

Phospholipids are actively synthesized by the hepatocyte. Plasma phospholipid is normally about four times the cholesterol level and helps to keep this hydrophobic substance in micellar solution.

Cholesterol metabolism is largely centered in the liver (Fig. 11–3). Derived from local synthesis or from IDL-cholesterol or LDL-cholesterol, its major pathway of elimination from the body is by conversion to bile acids. This involves 7α and 12α hydroxylations, conversion of the 3 β-OH to 3α–OH, and removal of part of the side chain. Some cholesterol is excreted as such in the bile.

Other Substances

Vitamins. (See also Chap. 22.) The liver is a major site of metabolism and storage of vitamins A, B complex (especially B_6 and B_{12}), D, and K.

Hormones. (See also Chap. 14.) The liver is the source of somatomedin and angiotensinogen and a major site of metabolic clearance of most hormones. Liver disease can cause such side effects as fluid retention (from decreased clearance of aldosterone and vasopressin) and gynecomastia or postmenopausal bleeding (from decreased clearance of estrogen and altered metabolism of androgens).

Purines. (See also Chap. 19.) Xanthine oxidase is a liver enzyme but neither it nor uric acid levels are of value in the investigation of liver disease.

Enzymes. (See also Chap. 3.) Several liver enzymes have been found useful in the diagnosis of hepatobiliary disorders. Notable among them are AST (aspartate aminotransferase) and ALT (alanine aminotransferase), which escape into the plasma from damaged liver cells; ALP (alkaline phosphatase) and 5NT (5′-nucleotidase), which are induced or released when the canalicular membrane is damaged and biliary obstruction occurs; and GGT (γ-glutamyl transferase), which is increased in both hepatocellular and obstructive disorders.

Metals. (See also Chap. 22.) The liver synthesizes ferritin, plays a major role in iron storage, and is the main source of the iron-transport protein, transferrin. As the source also of ceruloplasmin, and metallothionein, the liver plays a key role in the transport, storage, and metabolism of copper and other metals.

DETOXIFICATION FUNCTIONS

The liver is interposed between the splanchnic circulation and the systemic blood. It serves to protect the body from potentially injurious substances (e.g. endotoxins) absorbed from the intestinal tract, as well as toxic byproducts of metabolism. There are many *mechanisms,* including: (1) the phagocytic (pinocytotic) function of reticuloendothelial (Kupffer) cells in the sinusoids, (2) the ability of liver cells to compete for molecules bound to plasma proteins (e.g. bile acids, bilirubin, thyroxin, cortisol), and (3) the capacity of the liver to metabolize, conjugate, and excrete foreign compounds.

Most important in detoxification is the microsomal *drug-metabolizing* system of the liver. This involves cytochrome P450 and its related hemoproteins, reduced nicotinamide adenine dinucleotide and its phosphate (NADH and NADPH), and reduced dinucleotide regenerating flavoprotein enzymes. The system is induced by many drugs and other foreign compounds and is responsible for most of the following detoxification mechanisms: Oxidation, hydroxylation, carboxylation, and demethylation.

Conjugation occurs mainly in the cytosol or smooth endoplasmic reticulum with such moieties as glycine, glucuronic acid, sulfuric acid, glutamine, acetate, cysteine, and glutathione.

HEMATOLOGIC FUNCTIONS

Blood Formation. Red cells are produced by the reticular cells during fetal life; the liver is one source of erythropoietin.

Blood Coagulation. The liver is the source of fibrinogen, prothrombin, and many other clotting factors.

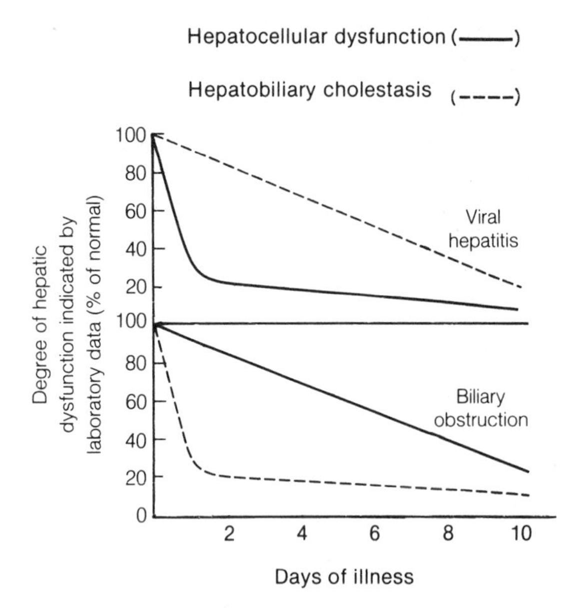

FIGURE 11–5. Simplified diagram showing the effect of time on laboratory indices of the degree (% change) of hepatocellular dysfunction and hepatobiliary cholestasis in viral hepatitis and in biliary tract obstruction.

Blood Volume Regulation. The splanchnic and hepatic vascular systems provide the 500 mL of blood required quickly in order to maintain circulation to the head in the change from a recumbent to an erect posture.

TESTS OF LIVER CELL INTEGRITY AND ORGAN FUNCTION

GENERAL CONSIDERATIONS

A number of factors have to be kept in mind when assessing the functional state of the liver:

The liver has a great *reserve* capacity for most of its functions and a remarkable ability to *regenerate* new cells. A liver cell can recover up to the point of mitochondrial damage. A liver acinus can regenerate as long as a layer of cells in zone 1 survives.

Metabolic, secretory, and excretory processes are *interdependent functions* of the liver, and no assessment of this organ is possible unless *time factors* are considered. Toxic injury to liver cells is often reflected in secondary cholestasis. Obstructive lesions of the bile ducts may lead to secondary hepatocellular dysfunction. Consequently, function studies obtained *early* in the course of liver disease or jaundice are more likely to discriminate between hepatocellular and obstructive disorders (Fig. 11–5).

Liver function cannot be evaluated out of context; that is, without knowledge of the *history* and *clinical findings*. Judged independently many tests have a relatively low sensitivity or specificity. Used appropriately, with all the relevant factors in mind, they can often bring a high level of probability to a clinical decision. A physician will make a proper diagnosis for 75% of patients with hepatobiliary disease from the history and physical examination alone. A few obvious laboratory tests should provide confirmation in these cases. Another 20% will be diagnosed correctly with the use of procedures appropriate to the problem at hand. Some cases will require all the diagnostic assistance available.

Several laboratory services are involved in assessing the functional state of the liver:

Radiology: The resolution achieved by the newer scanners has made *ultrasonography* a primary tool of the gastroenterologist. It usually permits a presumptive diagnosis and should precede all invasive procedures for the investigation of liver and biliary tract disorders. In distinguishing between the dilated ducts of extrahepatic obstruction and the nondilated ducts of intrahepatic cholestasis it has an accuracy >95%. It can help to demonstrate gallstones, poor gallbladder function, and cancer of the liver or biliary system. Occluding lesions can be localized in at least 70% of cases.

It may be necessary to use more complicated procedures such as ERCP (endoscopic retrograde cholangiopancreatography), PTC (percutaneous transhepatic cholangiography), CAT (computer assisted tomography), or EPT (endoscopic papillotomy, to remove a common duct stone). Liver scans for space-occupying lesions may employ ^{131}I rose bengal, taken up by hepatocytes, or ^{198}Au colloid or ^{99m}Tc sulfur colloid, taken up primarily by Kupffer cells.

Hematology: Red and white blood cell counts and differential smear; hemoglobin concentration and (where indicated) demonstration of an abnormal Hb; hematocrit; sedimentation rate; typing for ABO or Rh incompatibility; investigation of anemia. *Special* tests may include osmotic fragility; half-life of ^{51}Cr-tagged red cells.

Pathology: Examination of liver biopsy material

Immunology: Tests for hepatitis viral antigens and antibodies; tests for other infections (particularly of the newborn), the Paul–Bunnell test for heterophil antibodies (in infectious mononucleosis), the direct antiglobulin test for red cell autoantibodies (Coombs' test), and (rarely) agglutination tests for leptospirosis (Weil's syndrome)

Clinical Chemistry: These studies form the subject of the following section.

BIOCHEMICAL INDICES IN HEPATOBILIARY DISEASE

Laboratory tests involve examination of the urine, plasma, or serum and in rare instances the feces. They will be discussed here under general topic headings and later in relation to particular clinical problems.

The Bile Pigments

Urinalysis. ***Bilirubin*** in serum is normally *unconjugated,* hydrophobic, and tightly bound to albumin; it is normally undetectable in urine. If the urine gives a positive test for bilirubin it means that *conjugated* bilirubin is present in the serum, usually indicating a degree of bile secretory failure. The renal threshold is approximately 8 μmol/L (0.4 mg/dL), and urinary excretion reflects the level of less firmly bound plasma bilirubin diglucuronide. The clearance of bilirubin monoglucuronide is much lower (see Chap. 10). In acute viral hepatitis bilirubinuria occurs before jaundice is evident and disappears early in convalescence when the liver is able once again to clear conjugated pigment in the bile (See Fig. 10–10).

Urobilinogen has sometimes been measured semiquantitatively in freshly voided urine collected from 1400 to 1600 hours (2 PM to 4 PM), or quantitatively in a 24-hour collection. It has no apparent renal threshold but is subject to factors affecting its production in and absorption from the intestine (e.g., antibiotics) and its excretion or secretion by the kidney (e.g., hydrogen ion status). For most purposes simple *qualitative* testing will suffice and should be carried out. The quantity of urine urobilinogen is roughly proportional to red cell destruction and/or liver cell injury and inversely proportional to biliary obstruction. On standing it oxidizes fairly rapidly to urobilin and analytic steps must take this into account.

Plasma or Serum. ***Bilirubin:*** In normal serum *total* bilirubin consists largely of unconjugated, with very small amounts of conjugated forms, and has an upper reference value <20 μmol/L (<1.2 mg/dL). The *unconjugated* (free) bilirubin (B) establishes a reversible equilibrium with albumin involving mainly a high affinity ($K_D = 10^{-8}$ M) binding site. B + albumin ⇌ B∼A. Normally >99.9% is present as B∼A, only traces as free B. Measured amounts of *conjugated* bilirubin, giving the direct van den Bergh reaction (see Laboratory Indices, Chap. 10) are normally <3.5 μmol/L (<0.2 mg/dL).

Increased concentrations of the *conjugated* forms of bilirubin are a fairly sensitive index of impaired bile secretion (cholestasis). They bind less strongly to albumin and their presence is marked by bilirubinuria. When there is prolonged retention of conjugated bilirubin a new complex, covalently bound to albumin, is formed called delta bilirubin (δB). δB accumulates gradually to make up 10% to 90% of the total bilirubin, depending on the duration of jaundice. It is not excreted by the kidney and remains in the plasma for the lifespan of the albumin. This accounts for the persistence of an icteric serum for a week or longer after bilirubinuria has cleared.

Modest elevations of *unconjugated* bilirubin, 25 to 85 μmol/L (1.5 to 5 mg/dL), may occur in hemolytic disorders or shunt bilirubinemia, often reflecting the secondary anemia and a consequent relative hypoxia of the liver. Jaundice is infrequent and mild. Somewhat higher total bilirubin values, up to 300 μmol/L (17 mg/dL), may be found in acute hepatocellular disorders (e.g., hepatitis). The highest levels, as much as 500 μmol/L (30 mg/dL) or more, occur in obstructive (cholestatic) jaundice and in severe forms of neonatal jaundice.

In a *jaundiced* patient usually more than half the total bilirubin is made up of conjugated bilirubin. However, both intrahepatic and extrahepatic causes of bilirubin accumulation are associated also with significant increases in unconjugated bilirubin. Once jaundice is evident, determination of the total bilirubin provides a useful means of monitoring the severity of the disorder. Except in jaundiced newborns or when prehepatic hyperbilirubinemia is suspected, the separate measurement of unconjugated and conjugated bilirubin is seldom of value.

Bilirubin-binding capacity: The reserve binding capacity is normally in great excess and of no clinical concern; its evaluation may become important in the presence of neonatal jaundice as a means of avoiding *kernicterus* (bilirubin encephalopathy). The total binding capacity of albumin in adults is probably around 545 μmol/L (32 mg/dL). There are now several methods of measuring *bilirubin binding capacity* (BBC), but none has been shown to circumvent the theoretical constraints of a labile equilibrium, and thus give an accurate measure of the situation *in vivo.* The reserve BBC of newborn babies is <50% of adult values, but it normally increases appreciably in the first week. A baby whose BBC remains low (<40%) is probably at risk and should be treated. Factors that may reduce this capacity are serious neonatal illness, acidosis, heparin (in transfusions) which increases free fatty acids, and drugs such as salicylates, certain antibiotics, and diuretics. BBC measurements could be useful in monitoring the need for and effectiveness of an exchange transfusion, if they were more suited to routine analysis.

Fecal Urobilinogen. This mixture of stercobilinogen and mesobilirubinogen normally amounts to 85 to 420 μmol (50 to 250 mg) per day. The analysis requires a 3- or 4-day collection of feces, and so the test is rarely performed. Values above 500 μmol (300 mg)/day occur in hemolytic disorders; values below 8 μmol (5 mg)/day indicate total obstruction of bile flow. Antibiotic therapy that depresses the intestinal flora will render the measurement useless.

Bile Acids

In the presence of liver disease, even with no evidence of cholestasis, more bile acid remains in the circulation. Serum levels depend on the rates of intestinal absorption and the efficiency of hepatic re-excretion. The 2-hour *postprandial* value is a sensitive test of hepatobiliary dysfunction. *Fasting* levels become elevated when the cholestasis becomes more severe.

As a general rule serum bile acids are higher and pruritus is more severe when biliary cholestasis is associated with intact liver function and continued bile acid synthesis. Pruritus may be bothersome in viral hepatitis but is usually much more severe with idiopathic or drug cholestasis, and with biliary cirrhosis or obstructive jaundice due to carcinoma. Efforts thus far to prove that bile acids are directly responsible for the itching have failed.

Reference values should be established locally, but serum bile acid levels are usually <5μmol/L (<1.0 mg/dL) in the *fasting* state, and <8μmol/L (<1.6 mg/dL) 2 hours *postprandially.* A meal containing two eggs should induce bile flow and so provide an endogenous overload that the normal liver can handle in 90 to 120 min. Oral or intravenous radionuclide-labeled bile acids have been used as a test of the liver's capacity to extract bile acids from the blood.

Serum bile acid *trihydroxy/dihydroxy ratios* can be determined from high-performance liquid chromatography data and may have some value in differential diagnosis. Ratios >1 suggest primary obstructive jaundice, with decreased intestinal dehydroxylation and hepatocyte hydroxylation mechanisms still intact. Ratios <1 suggest primary hepatocellular disease, with a reduced capacity to rehydroxylate bile acids extracted from the portal blood. By measuring 3 h postprandial serum cholylglycine, cholic acid conjugates, and ALT, it has been possible in 75% of cases to distinguish chronic active hepatitis (in which they tend to be increased) from chronic persistent hepatitis (in which they do not).

Plasma and Serum Proteins (See also Chap. 18)

Albumin (reference range 35 to 48 g/L) is synthesized solely by the liver. When the need arises, for example, because of massive proteinuria, the liver may increase albumin synthesis to more than 20 g/day, but this compensatory mechanism is rather limited. Minor diurnal changes in protein concentrations reflect plasma volume fluctuations. In acute liver disease albumin levels fall slowly, because of the relatively long half-life. In chronic liver disease serum albumin is a useful index of serious deterioration of hepatic function.

α_1-Globulin is markedly reduced in neonatal jaundice when associated with α_1-antitrypsin deficiency, which leads to childhood cirrhosis and emphysema in early adult life.

α_2-Globulin will fall with intravascular hemolysis, owing to depletion of haptoglobin, but is increased in a number of inflammatory conditions, owing to the presence of 'acute-phase reactants' (see Malignant Disease of the Biliary Tree).

α_2-Globulins and β-globulins are characteristically increased in chronic cholestasis.

'Bridging' of the β- and γ-globulin peaks on electrophoresis is usually due to an increase in IgA, particularly in alcoholic liver disease (see Fig. 18–6).

γ-Globulin elevations in liver disease are attributed to an infiltration of the liver substance with immunologically competent cells (monocytes, lymphocytes, plasma cells). These cells reflect the extent of the mesenchymal reaction, and their presence may influence the biopsy diagnosis.

The ***immunoglobulins*** have proved useful in differential diagnosis. ***IgG*** (reference range 5 to 16 g/L) is more likely to be increased in chronic active hepatitis; ***IgA*** (reference range 0.5 to 3.5 g/L) is usually raised in portal cirrhosis, especially when it results from alcohol abuse; ***IgM*** (reference range 0.6 to 1.75 g/L) is increased in several disorders but almost invariably in biliary cirrhosis. Because of this high sensitivity, a normal IgM is strong evidence against primary biliary cirrhosis.

Prothrombin, measured as the clotting time after adding thromboplastin to plasma, is diminished when either liver damage is extensive or vitamin K deficiency has developed from prolonged cholestasis. In acute hepatobiliary disease with jaundice, if the prothrombin time is prolonged, the response to intramuscular vitamin K can aid the diagnosis: Clotting time will return quickly to control levels if the cause of jaundice is extrahepatic, but not if there is primary hepatic disease. In advanced chronic liver disease, even in the absence of jaundice, a prolonged prothrombin time means that liver damage is extensive; values more than 50% above controls suggest a poor prognosis.

Special tests include the detection of *ceruloplasmin* deficiency in Wilson's disease, the measurement of *α-fetoprotein* in suspected liver cell carcinoma, and the level of *lipoprotein X* in cholestatic jaundice.

Serum Enzymes

Because of the variety of methods and conditions employed, reference ranges for enzyme assays must be established locally.

Because serum enzymes are covered in detail in Chapter 3 only a summary is given here.

Aminotransferases: Increases in ***AST*** reflect damage to hepatocytes (also to the myocardium and skeletal muscle). Moderate elevations (3 to 20 times the upper reference value) are seen in anicteric, or subclinical, viral hepatitis; high values (>20 times the upper reference value) occur in acute viral or toxic hepatitis. ***ALT*** is more hepatospecific, reflecting membrane permeability changes and necrosis. The ***AST/ALT ratio*** is normally >1 because AST is present in higher concentration in the hepatocytes. Ratios >1 are usually found in chronic active hepatitis, cirrhosis, and hepatic metastases. The ratio is usually <1 in acute hepatitis, biliary obstruction, and chronic persistent hepatitis. The differences may be explained in part by the fact that ALT is cleared more slowly from the blood than AST. Emphasis on AST/ALT ratios presumes the use of methods capable of measuring these enzymes with considerable accuracy. In general, *peak activity* values give the best discrimination between hepatocellular and obstructive lesions.

ALP (alkaline phosphatase) is higher in childhood, late pregnancy, and old age. It is increased or induced by cholestasis, moderately raised in hepatocellular disease, and markedly increased in extrahepatic cholestasis. Other sources of ALP such as bone or the gastrointestinal tract must be excluded. Obstruction of small bile ducts by metastases in the liver will elevate the serum ALP.

GGT (γ-glutamyl transferase) is a sensitive but not a specific index of hepatobiliary disturbances. Values are high in toxic injury or cholestasis, the most marked elevations occurring in alcoholic cirrhosis. The increase in GGT in cholestasis can be useful because this enzyme, unlike ALP, is not increased in bone disease.

5NT (5′nucleotidase) is a canalicular and plasma membrane enzyme. It is not particularly sensitive but is fairly specific for cholestasis, and is a useful test in children as a substitute for ALP.

Serum Lipids

Cholesterol values in serum (reference range 4 to 7 mmol/L; 150 to 270 mg/dL) tend to increase with age and represent a balance between intake, synthesis, metabolism, and excretion; normally 70% exists as cholesteryl ester. In liver disease cholesterol may be normal or moderately increased, but the percentage of ester is usually decreased; in primary cholestasis the free cholesterol is often elevated. This is not an efficient diagnostic test, but may serve as an index of chronic biliary obstruction.

Triglyceride and free fatty acid estimations have no particular value in relation to hepatobiliary diseases.

Lipoprotein X correlates with a high cholesterol level in cholestasis and is associated with decreased levels of lecithin:cholesterol acyltransferase (LCAT).

LIVER FUNCTION TESTS

Many of the biochemical indices already mentioned are, in fact, tests of liver function. Injury leading to cytolysis and necrosis liberates various enzymes and impairs bile pigment excretion. Injury to mitochondria usually results in cell death. Injury to hepatobiliary membranes releases (or induces increases in) other enzymes. Inflammation leads to the mesenchymal reaction, with infiltration of immunologically competent cells and increased immunoglobulins. Chronic deterioration of liver function leads to decreased serum albumin and prothrombin.

Endogenous Overload Tests

The reexcretion of portal blood *urobilinogen* and of reabsorbed *bile acids* constitute endogenous overload tests of liver function. Both are sensitive indices of hepatocyte function; the latter is also a sensitive index of cholestasis.

Exogenous Overload Tests

The ***sulfobromophthalein (BSP) excretion test*** is based on the liver's ability to take up and excrete this dye. The usual intravenous dose of 5 mg/kg should be calculated from ideal body weight to avoid overdoses in obese patients. The dye binds firmly to albumin and, assuming 50 mL plasma/kg of body weight, a concentration of 10 mg/dL is taken to represent 100% retention. BSP is taken up rapidly by binding proteins in the liver cells, is conjugated mainly with glutathione, and is excreted into the canaliculi by a transport mechanism shared with conjugated bilirubin. Healthy young subjects show <5% retention after 30 min; older subjects, after 45 min. The test is quite sensitive but not specific for liver disease; it is affected by the hepatic blood flow, by toxic effects on the liver, and particularly by the presence of cholestasis. It is useless when serum bilirubin is elevated and has little value once liver disease is known to be present. The care with which the dye must be injected and blood sampling must be timed, coupled with rare sensitivity reactions, have limited its usefulness. A variation of the test that involves multiple sampling and calculation of the transport maximum has not added enough information to justify its use.

Indocyanine green clearance is as good as the BSP test, with fewer reactions, but is expensive and is therefore rarely used. The dose can range from 0.5 mg/kg to 5.0 mg/kg, depending on circumstances, and the concentration at 15 min reflects the functional reserve of the liver.

^{131}I Rose bengal excretion has proved useful in distinguishing neonatal hepatitis from biliary atresia, as discussed later.

Precoma Monitoring

A fall in blood urea and a rise in amino acids, coupled with an increase in ammonia, occurs in and may correlate with the encephalopathy of terminal liver disease. Occasionally, a fall in fibrinogen may result from disseminated intravascular coagulation.

Special Studies

The information provided by *ultrasonography,* supplemented by *biochemical* data, should indicate the nature of any further (hopefully definitive) diagnostic procedures that may be required.

Cholangiography, to visualize the biliary tree, can be carried out in different ways:

1. *Simple cholecystography* involves x-ray examination of the biliary tract following the oral or intravenous administration of a radiopaque compound; it is useless in the absence of bile flow.
2. When there is unexplained cholestasis (serum bilirubin >50 μmol/L; 3 mg/dL) the use of *endoscopic retrograde cholangiopancreatography* (ERCP) may help to identify the site of obstruction; it requires considerable skill and experience.

3. In obstructive jaundice *percutaneous transhepatic cholangiography* (PTC), which involves a needle search for dilated bile ducts, will sometimes locate mechanical obstruction of the biliary tree, and must then be followed by surgical intervention.
4. At an exploratory laparotomy an *operative cholangiogram* (via a catheter in the cystic duct) can be used to visualize the biliary tree within the liver.

Visualization of the arterial circulation of the liver, pancreas, spleen, and mesentery by *angiography* may be important if surgery is contemplated.

Needle liver biopsy is now a safe procedure in capable hands and can provide definitive information to an experienced pathologist. A prior check of the prothrombin time is mandatory.

Liver scanning after the intravenous injection of ^{99m}Tc sulfur colloid (taken up by Kupffer cells) may demonstrate tumors, metastases, or abscesses above 1.5 cm size. *Computer assisted tomography* (CAT) will sometimes demonstrate the cause of anatomic distortion of the liver.

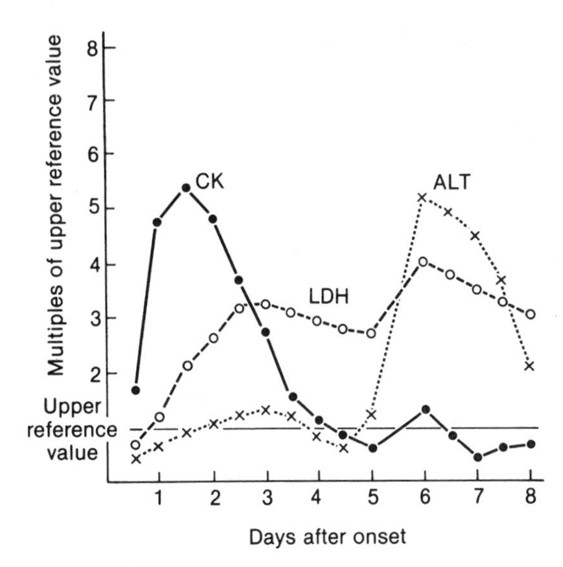

FIGURE 11–6. Serum levels of enzymes in congestive heart failure developing 5 days after an acute myocardial infarction. CK = creatine kinase; LDH = lactate dehydrogenase; ALT = alanine aminotransferase.

HEPATOBILIARY DISORDERS

ACUTE LIVER DISEASE: SECONDARY AND PRIMARY CAUSES

Many factors can cause acute damage to the liver and some are associated with jaundice. Liver disease may occur *secondary* to a disease of the cardiovascular system that results in heart failure or shock. ***Circulatory failure*** causes anoxia of liver cells, resulting in impaired function that is manifested by increased serum bilirubin, a reduced ability to clear BSP, and pronounced liberation of the aminotransferases ALT and AST, with the former predominating (Fig. 11–6). Enzymes associated with cholestasis are not much increased in these circumstances, although GGT may show quite pronounced elevations—partly owing to induction by drugs that have been used in the treatment of the primary illness. Correction of the circulatory failure is usually accompanied by full restoration of hepatic structure and function, although when anoxia is protracted, frequent, or relentless, it may progress to the chronic disease entity known as *cardiac cirrhosis.*

Primary acute liver disease is usually a consequence of infection by viruses or bacteria, or is caused by chemicals or drugs. The commonest agents are viruses responsible for the well-characterized disease entity known as *acute infectious (viral) hepatitis.* Two responsible viral agents have been identified, although they do not account for all cases of this disease. There are, thus, at least two forms of viral hepatitis; their natural histories show some characteristic differences, but the biochemical features are similar and are representative of those occurring in acute liver disease from whatever cause. These conditions will be described in some detail.

Viral Hepatitis

Etiology. The viral agent which causes the common form of this disease (***hepatitis A,*** often known as 'epidemic' hepatitis) is most often contracted through the fecal-oral route as a consequence of ingesting contaminated water or foodstuff such as milk or shellfish. Infection confers subsequent immunity to hepatitis A; development of the carrier state is rare, and fatalities are uncommon. The incubation period between contraction of the infection and development of symptoms is 15 to 49 days.

A viruslike particle, 27 nm in diameter, has been identified in the stools of patients with this form of hepatitis several days before symptoms develop. This hepatitis A *antigen* (HAAg) can now be quantified in feces in some centers. Antibodies to this agent (HAAb) appear in the serum in the course of the disease (Fig. 11–7A). The *first* to do so, shortly after the onset of icterus, is of the IgM class. It is diagnostic of the acute attack and disappears in 8 to 10 weeks. The *second* appears only during resolution of the disease and has little diagnostic value. It persists for at least 10 years, and probably indefinitely, in affected subjects, conferring immunity against a second infection.

A different form of the disease is due to an agent called ***hepatitis B*** virus. This form is usually contracted by the parenteral route. Oral infection seldom occurs, and the virus is not usually excreted in the feces. The incubation period is between 28 and 160 days, hence the former name 'long-incubation' viral hepatitis. Because the infection often followed administration of contaminated blood or other transfusion products, another synonym for the disease was 'serum' or 'posttransfusion' hepatitis. Infection can occur also as a consequence of other clinical procedures such as dialysis and organ transplantation. The use of contaminated needles or instruments represents another hazard, hence its association with drug addiction and tattooing.

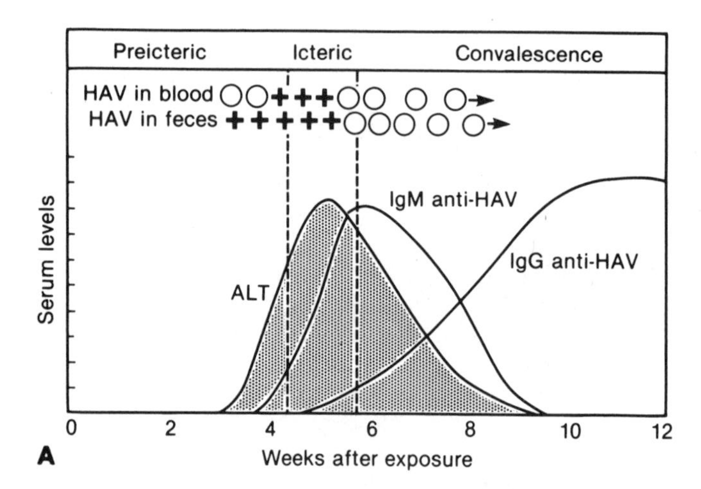

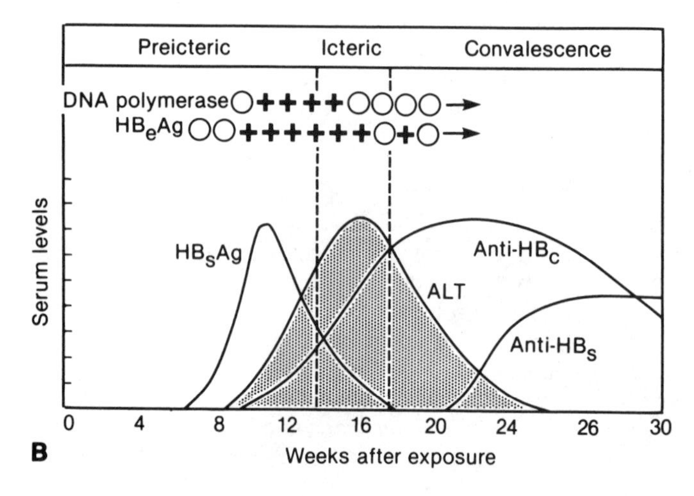

FIGURE 11–7. Serum levels of viral antigens and antibodies. *A.* In hepatitis A infection. *B.* In hepatitis B infection. ALT = alanine aminotransferase; anti-HB_c = hepatitis B core antibody; anti-HB_s = hepatitis B surface antibody; HAV = hepatitis A virus; HB_eAg = hepatitis B *e* antigen; HB_sAg = hepatitis B surface antigen. (Koff R: Viral Hepatitis. New York, Wiley, 1978. Redrawn)

Electron microscopy has shown the hepatitis B virus to be present in the liver of infected patients. The complete virion (Dane particle) consists of an inner core synthesized in the nuclei of affected hepatic cells and containing the *core antigen* (HB_cAg), the enzyme DNA polymerase, and double-stranded DNA. This complex passes to the cytoplasm, where a coat containing the *surface antigen* (HB_sAg) is acquired. The presence of Dane particles, which are 42 nm in diameter, can be revealed by electron microscopy of serum from infected patients; this may also show tubular particles 22 nm in diameter and 100 nm long, which consist entirely of excess HB_sAg. This surface antigen appears in the serum before the first symptoms develop, rises subsequently, and persists throughout the clinical course of the disease (Fig. 11–7B).

An *antibody* to the *core* antigen (HB_cAb) often appears in the serum at the onset of the acute illness and gradually falls to undetectable levels after 1 to 2 years. An *antibody* to the *surface* antigen (HB_sAb) makes its appearance during resolution of the disease and in the convalescent period, and may persist in those immune to infection. Its appearance is necessary to terminate the disease and is usually accompanied by disappearance of HB_sAg from the serum. If HB_sAb does not develop the disease will not resolve, and the patient will enter a carrier state typified by persistence of HB_sAg and HB_cAb. This is the outcome in 5% to 10% of reported cases in North America.

Chronic ***carriers*** of HB virus infection are a serious hazard to the healthy population, and under no circumstances should their blood products be used for transfusion purposes. Most blood donations are currently screened to exclude this possibility. A proportion of patients who are carriers will go on to develop chronic active liver disease. The *e antigen* system is believed to be an important determinant of this outcome. It is present in the serum early in the course of acute hepatitis B but has its main value in the assessment of chronic HB_sAg carriers. Those who remain positive for the *e* antigen are more likely to show abnormalities and develop chronic active liver disease. Those whose serum contains instead the *antibody* to *e* antigen tend to have normal biochemistry values and normal liver biopsies. The exact nature of *e* antigen has yet to be determined but it is probably part of the virion core.

HB_sAg ***subtypes*** have become recognized. All share a common determinant, *a*, and specific determinants *d* or *y*, and *w* or *r*, coded by the viral genome and not by the host. There are thus at least four subtypes: *adw*, *ayw*, *adr*, and *ayr*. This enables the epidemiology to be followed in any particular case of infection. Such studies have revealed that *ayw* is the most common form of antigen in viral disease spread by drug abuse.

Clinical and Biochemical Features. Both A and B forms of the disease begin with a generalized *prodromal illness* resembling influenza, together with symptoms of nausea, often with abdominal pain, vomiting, and diarrhea. The gastrointestinal symptoms precede the onset of jaundice by several days, and are frequently relieved when jaundice appears. The jaundice, if it develops, is variable in extent and duration. In *type A* hepatitis resolution is usually complete within 14 days, but can take longer when there is a phase of severe cholestasis. Epidemiologic studies have demonstrated that for every case developing icterus, two or three remain anicteric (*subclinical hepatitis*), and that in most of the latter cases the disease will go unrecognized and will be attributed to influenza or gastroenteritis. *Type B* hepatitis tends to be more severe, with a longer period of jaundice, and complete recovery may fail to occur in up to 10% of cases.

The urine is often dark owing to a mild hyperbilirubinemia, which is predominantly the conjugated form that is excreted by the kidney. Urinary urobilinogen is also increased, especially in the early stages of the disease, because the liver has a limited capacity to extract this constituent from the portal blood.

In approximately 10% of hepatitis patients exhibiting jaundice the ***cholestasis*** becomes a dominant feature of the illness. There is a marked diminution of bilirubin excretion via the bowel, as well as other biliary constituents such as bile acids, which will lead to pale stools with increased fat. In 40% to 50% of cases this is sufficiently marked to be classified as steatorrhea. The obstruction may last for periods varying from 3 days to 3 weeks or even longer. Urobilinogen will disappear from the urine during this phase of intrahepatic cholestasis (see Fig. 10–10). Its reappearance heralds the onset of recovery and is followed in turn by clearing of the bilirubinuria, return of the serum bilirubin to

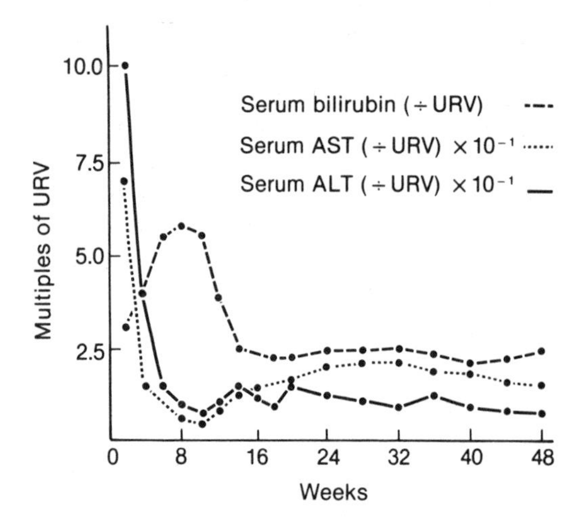

FIGURE 11–8. Serum levels of bilirubin, aspartate aminotransferase (AST), and alanine aminotransferase (ALT) in chronic hepatitis developing after an episode of acute hepatitis presumed to be of viral etiology. Note that enzyme values are ten times the multiples on the ordinate. URV = upper reference value.

normal or near normal concentrations, and ultimately clearing of the urobilinogenuria.

Rashes and excoriations are sometimes observed in the period of jaundice, which is associated with *pruritus.* Splenomegaly occurs occasionally, but *hepatomegaly* is frequent, the liver becoming enlarged owing to edema and the inflammatory process; even if not enlarged, it is usually tender to palpation.

Pathologic Features and Serum Enzymes. The pathologic features provide a rational explanation for the biochemical sequelae. Spotty necrosis of liver cells occurs throughout the liver acini, with most damage usually in zone 3. In more severe cases there is widespread necrosis with disruption of the liver architecture. This necrosis leads to a marked increase in parenchymal enzymes in blood serum and to impairment of liver function.

Aminotransferases. These enzymes have been widely used in the diagnosis of viral hepatitis. Dramatic increases ranging from 20 to 100 times the upper reference value occur in typical cases when appropriate assay methods are used; indeed, elevations within this range may be detected in anicteric subjects who have only mild symptoms. These enzyme determinations have therefore been of great value in screening contacts of hepatitis patients, and in epidemiologic studies following various outbreaks. Before the availability of direct viral tests, they were also helpful in the screening of blood donors. Enzyme activities *less* than 20 times the upper reference value, in an adult presenting within a few days after the onset of jaundice, make the diagnosis of viral hepatitis very unlikely (see Fig. 3–5).

After reaching a peak shortly after the onset of jaundice, the aminotransferases decline quite rapidly and approach reference values before the jaundice clears. These enzymes reflect the mass of cells undergoing necrosis and usually decline while the serum bilirubin is still increasing. They are useful guides to disease activity and can be used to *monitor* the need for *bed rest.* If the patient is ambulated before the inflammatory process has subsided, a sharp upturn in aminotransferase activity is often noted. Failure of resolution, leading to the condition of chronic hepatitis, is marked by continuing abnormality of the aminotransferases.

When enzyme elevations first occur, which often precedes the prodromal illness, the activity of *AST* exceeds that of *ALT,* because the former is present in the hepatocyte in higher concentration. However, the latter is cleared more slowly from the blood and accumulates to reach higher activities, so that in most cases *ALT* is the dominant aminotransferase when the patient first presents for medical attention. Typically, ALT activity remains high throughout the subsequent course of the illness and returns to normal values later than AST. In chronic persistent hepatitis, ALT usually remains the higher of the two aminotransferases. However, in cases which fail to resolve to the extent that postnecrotic cirrhosis develops, a cross-over in aminotransferase activity commonly occurs, with AST becoming dominant (Fig. 11–8). This is indicative of the continuing necrosis and should be regarded as an unfavorable prognostic sign.

Other enzymes. Many enzymes behave in a manner similar to the aminotransferases in acute viral hepatitis, but no test for other enzymes has any clear advantage over these well-established procedures. Some of the so-called *markers* of hepatic cell necrosis may, however, be elevated in toxic liver necrosis due to drugs or alcohol (e.g., isocitrate dehydrogenase and ornithine carbamoyltransferase) under circumstances in which the aminotransferases yield equivocal results. Isocitrate dehydrogenase has been reported to be a more reliable index of hepatitis due to infectious mononucleosis than the aminotransferases.

Venous stasis, inflammation, and edema within the liver and a general disruption of hepatocyte function in viral hepatitis lead to biliary stasis, with necrotic debris sometimes plugging the bile canaliculi. This intrahepatic cholestasis is usually reflected in modest increases in enzymes that are markers for this phenomenon, such as ALP, leucine aminopeptidase, 5NT, and GGT. The levels are generally below those seen in primary intrahepatic cholestasis and in extrahepatic obstruction.

Other Biochemical Changes. Destruction of liver cells causes the release of other constituents into the plasma. Notable among these are *iron,* released from liver stores, and *glucose,* which is released from the breakdown of glycogen. In early viral hepatitis, therefore, increased serum concentrations of both occur. In severe, prolonged forms of the disease, especially in fulminant hepatic failure, exhaustion of glycogen stores leads to hypoglycemia.

The ability of the infected hepatocyte to synthesize proteins is greatly impaired. Although serum *albumin* may diminish in acute liver disease, the levels rarely fall to pathologically low values. By contrast, blood clotting enzymes, which have a much shorter half-life, soon become depleted. Occasionally useful for diagnostic purposes, the *prothrombin* concentration (measured by the prothrombin time) shows a definite abnormality in most cases of acute hepatitis and is not restored to normal by parenteral administration of vitamin K. In subclinical anicteric cases, where there is a minimal disturbance of serum biochemistry, evidence of disturbed hepatic function can be provided by the measurement of serum *bile acids* (the 2-hour postprandial level being a more sensitive index than fasting levels), or by an exogenous overload of organic anions such as *BSP* or indocyanine green.

Inflammation around the portal tracts, with infiltration by lymphocytes and plasma cells, is a striking feature of viral hepatitis. The consequent activity of immunocompetent cells results in increased *γ-globulin* concentrations. Specifically, IgM is increased in the early stages of the disease in 80% of cases, while IgG is raised later in the illness in about 60%. Smooth-muscle antibody titers are elevated in approximately 80% of cases, but mitochondrial and nuclear antibodies are usually normal. Regeneration of hepatic parenchymal tissue is frequently accompanied by a transient elevation of *α-fetoprotein* levels in the serum.

Massive Hepatic Necrosis. Severe necrosis leading to fulminant hepatic failure is the most serious ***complication*** of viral hepatitis. It is associated with

a. Encephalopathy and a high blood ammonia concentration
b. Renal failure with an increasing concentration of creatinine in the serum
c. Severe lactic acidosis
d. A bleeding diathesis, which may be aggravated by disseminated intravascular coagulation
e. Hypoglycemia
f. Failure of urea production
g. Ultimately, cardiac and respiratory failure

The mortality rate is very high (80%), but some of those who survive seem to recover completely. Others develop postnecrotic cirrhosis.

Other Infectious Agents Causing Hepatitis

The commonest cause of *posttransfusion* hepatitis is now recognized to be a viral agent which is distinct from those responsible for type A and type B hepatitis. The ensuing illness is referred to as non-A, non-B viral hepatitis. Because this agent has not been identified, no tests are available to prove the diagnosis. Epidemiologic evidence suggests that there are at least three forms of non-A, non-B hepatitis. Another virus, *delta agent,* is an RNA particle that can reproduce only in the presence of hepatitis B virus. Patients harboring both viruses suffer a much more severe illness than those infected with type B virus alone. Commercial kits are available to test for delta agent. They are expensive, and a positive result is not likely to influence the management of the patient's illness.

A number of viruses have clearly been implicated in the hepatitis that develops occasionally during the *neonatal* period. These include *rubella, cytomegalovirus,* and *herpes simplex.* Confirmation of the diagnosis is usually obtained by sequential measurement of viral antibody titers, but occasionally the virus itself can be isolated from the placenta, or from a liver biopsy. The latter may also reveal the virus particles on electron microscopy.

In hepatitis due to these viral agents there is generally much less necrosis than in type A or B hepatitis. Consequently, the elevation of serum enzymes of parenchymal origin is rarely impressive, values generally being only 5 to 10 times the upper reference value. The inflammatory reaction may be quite severe and marked by giant-cell formation. Cholestasis, therefore, may be a pronounced histologic feature, but paradoxically it is usually accompanied by only modest increases in the activity of alkaline phosphatase and related enzymes. Because unconjugated hyperbilirubinemia is extremely common in the neonate, very high concentrations of total bilirubin are present in neonatal hepatitis.

A common cause of hepatitis is the *Epstein–Barr (EB) virus,* the presumptive causal agent of ***infectious mononucleosis.*** Hepatitis is a regular feature of this illness, even though the patient becomes icteric in only a minority of cases. The biochemical picture is indistinguishable from that of type A or B hepatitis, and the diagnosis is made on the basis of the clinical findings (generalized lymphadenopathy and splenomegaly), the characteristic blood picture (monocytosis with many aberrant cells), and positive tests for heterophil antibodies (Paul–Bunnell test).

Bacterial infections resulting in specific forms of hepatitis include *syphilis* and *brucellosis.* Leptospiral infection resulting in *Weil's disease* is a rare cause with a high mortality; hemorrhage and renal failure accompany the development of the usual features of acute hepatitis. Serologic tests will effectively make the diagnosis in all of these conditions. Overwhelming *sepsis,* either portal or systemic, may result in secondary liver failure, partly owing to circulatory changes, as well as to direct toxic effects of bacteria and their products on the liver cells.

Among the ***parasitic*** diseases, *malaria* is a globally important cause of hepatitis. In western societies, *toxoplasmosis* can cause hepatitis in the newborn with a picture similar to that already described for hepatitis due to viral agents in this age group. *Amebic* infestation can result in hepatocellular necrosis and the formation of large abscesses in the liver.

Toxic Hepatitis

Almost any drug, chemical, or industrial agent will, in sufficient quantity, result in damage to the liver. There are now more than 100 well-established hepatobiliary toxins, a fact which merits careful consideration when a patient's history is being taken. Some of the commoner hepatotoxic drugs, and the type of liver reaction which they cause, are listed in Table 11–1. Reactions to these agents often include ***generalized sensitivity,*** in which hepatic damage is accompanied by other manifestations of drug sensitivity such as skin rashes and lymphadenopathy.

Direct hepatotoxicity produces a histologic picture resembling viral hepatitis. The biochemical picture differs to some extent: immunologic abnormalities are usually absent, and because the liver damage occurs over a relatively short time span and affects all cells simultaneously, the pattern of enzymes in serum approximates that in liver tissue much more closely than in viral hepatitis; consequently AST is generally much higher than ALT.

The third reaction to drugs takes the form of a ***cholestatic syndrome,*** which is difficult to distinguish from any other form of intrahepatic cholestasis. In particular, because this may follow a sensitivity reaction, with features resembling the clinical symptoms of viral hepatitis, it may be difficult to distinguish drug-induced cholestasis from the latter condition. Once again, the lack of immunologic abnormalities in drug cholestasis is a help, as is the fact that eosinophilia is common in drug-induced reactions and hardly ever occurs in viral hepatitis.

Reye's syndrome is an acute illness occurring in children and adolescents. The liver becomes enlarged with fatty vacuoles. Damage to the renal tubules and the central nervous system is also prominent and death from coma and cerebral edema may occur in up to 50% of cases, although reduction of cerebral pressure dramatically improves survival. Onset is marked by severe vomiting and drowsiness, frequently following an upper respiratory infection. Salicylate therapy has been implicated in some cases, and one epidemic may

TABLE 11–1. Examples of Drug-Induced Liver Damage

Nature and Mechanism	Agent	Clinical Use
Hepatocellular damage		
Direct toxicity	Acetaminophen	Analgesic
	Chloroform	Anesthetic
Indirect toxicity	Salicylates	Analgesic
Sensitivity	Halothane	Anesthetic
	Rifampicin	Antituberculous
	p-Aminosalicylate	Antituberculous
	Iproniazide	Antidepressant
Canalicular damage		
Indirect toxicity	Methyltestosterone	Hormonal
	Oral contraceptives	Hormonal
Sensitivity	Amitriptyline	Antidepressant
	Chlorpropamide	Antidiabetic
	Tolbutamide	Antidiabetic
	Chlorpromazine	Tranquilizer
Fatty degeneration		
Indirect toxicity	Tetracycline (IV)	Antibiotic
	Dexamethasone	Hormonal
	Methotrexate	Antitumor

IV = intravenous.

have followed exposure to chemical pesticides. The dominant *biochemical* abnormalities include increased blood ammonia concentration and serum AST activity. A specific increase in the plasma concentration of basic amino acids is characteristic, and raised serum levels of mitochondrial enzymes reflect morphologic damage to that organelle. Jaundice is unusual. Survivors generally go on to complete recovery.

DISEASES OF THE BILIARY SYSTEM

Diseases of the biliary system fall into two categories: (1) these localized within the liver and resulting in intrahepatic cholestasis; (2) those external to the liver and causing extrahepatic cholestasis.

Intrahepatic Cholestasis

Distortion of the biliary architecture leading to cholestasis may occur in infiltrative diseases such as Hodgkin's lymphoma, leukemia, and some hepatic forms of secondary cancer. In this scenario, obstruction to bile-flow develops and the consequent regurgitation leads to high serum concentrations of bilirubin and other biliary constituents. More commonly, the condition is not associated with architectural disturbance and is best characterized as a *failure of bile secretion.* This may be accompanied by morphologic evidence of canalicular damage, but bile secretory failure may occur independently. Most such cases can now be attributed to drugs in the usual therapeutic doses (as aforementioned).

Secretory failure may occur in association with duct inflammation together with Zone 3 (centrilobular) canalicular bile retention; cholestasis due to chlorpromazine is a prototype of this form. Canalicular bile retention, however, may be the only feature, when the term 'pure cholestasis' is used; this form is represented by cholestasis due to birth control pills and anabolic steroids.

Pregnancy is another condition which can give rise to cholestasis, possibly involving a congenital or acquired sensitivity to estrogens or progestins. The recurrent cholestasis may occur with the first pregnancy, or the second, and becomes progressively more severe with succeeding pregnancies. It disappears shortly after delivery, but may recur if the patient uses steroid-containing contraceptive pills.

Benign recurrent intrahepatic cholestasis is a disorder characterized by acute attacks of jaundice and pruritus, often precipitated by a mild viral illness, with normalization of the liver and its function between attacks. It may occur at an average rate of once a year. The etiology of this condition is unknown but many cases occurring within a single family have been described.

Several ***dominant features*** characterize the many forms of intrahepatic cholestasis. Pruritus is intense, high levels of bile acids being present in the circulation. Hyperbilirubinemia may reach striking proportions and is predominantly of the conjugated variety resulting in marked bilirubinuria. Stools are pale or even colorless, often with marked steatorrhea, and urobilinogen is absent from the urine. For the most part, it is extremely difficult to distinguish *intrahepatic* from *extrahepatic* obstruction unless the causal agent can be identified. However, it is becoming recognized that the degrees of elevation of enzymes which are characteristic of biliary obstruction (ALP, leucine aminopeptidase, and 5NT) are much less pronounced, relative to the degree of hyperbilirubinemia, than one would expect in classic extrahepatic obstruction. The extent of aminotransferase elevation is also less than one sees in most cases of extrahepatic obstruction, and this readily distinguishes intrahepatic cholestasis from viral hepatitis, even when cholestasis is a prominent feature of the latter.

Extrahepatic Biliary Tract Diseases

These can be divided into nonmalignant and malignant disorders. In the former, etiologic factors are inflammatory and mechanical, both contributing to the obstruction in certain disease entities.

Acute cholecystitis (inflammation of the gallbladder) presents with fever, abdominal pain, and other features suggesting an acute abdominal emergency. Jaundice, if present, is slight. Elevation of biliary marker enzymes in the serum is usual, and increased activity of serum amylase may occur without apparent involvement of the pancreas. Aminotransferases are often raised, and in almost all instances ALT is the higher of the two.

Chronic cholecystitis gives rise to dyspepsia, especially after fatty meals. The condition has to be distinguished from chronic peptic ulcer. Serum biochemistry is usually normal. However, the chronically inflamed gallbladder is frequently the site of ***gallstone*** formation, and the passage of a stone into the common bile duct usually leads to acute extrahepatic obstruction.

Cholelithiasis can also result from chronic hemolytic conditions, when pigment stones may form as a result of the high concentrations of bilirubin in the bile. Pure cholesterol gallstones are prone to form in so-called lithogenic bile; the chemical basis for this phenomenon is probably a deficiency of the tri- and dihydroxy bile acids necessary to form stable micelles with cholesterol, which otherwise tends to precipitate. Recognition of this mechanism has led to the use of oral chenodeoxycholate in the treatment of gallstones, with a success rate in selected patients approaching 50%.

Gallstones, if they enter and obstruct the common bile duct (choledocholithiasis) will produce *intermittent* acute extrahepatic obstruction, usually with severe colic, with a parallel increase in the serum concentrations of bilirubin, bile acids, and biliary marker enzymes, the latter being increased to a greater extent than in almost any other condition. Serum aminotransferase activity increases in proportion to the degree of hyperbilirubinemia, being more marked than with idiopathic intrahepatic cholestasis, but much less than is seen in viral or toxic hepatitis. In the great majority of cases, ALT is higher than AST. As a rule gallstones *pass spontaneously* into the small intestine, but if a stone becomes impacted and obstruction is prolonged, a rise in serum cholesterol and in phospholipid concentration becomes evident. Unless surgery is performed the stools may become very pale, with complete disappearance of bile pigment, reflected in massive bilirubinuria and absence of urobilinogen from the urine. Steatorrhea may then develop and with it an impaired ability to absorb fat-soluble vitamins, including vitamin K. The resulting fall in prothrombin concentration, reflected in an increased prothrombin time, can, however, be fully corrected by parenteral administration of the vitamin, in contrast with the situation in primary parenchymal liver disease.

Strictures secondary to surgery involving the biliary tract may lead to extrahepatic obstruction with a biochemical picture similar to that seen in acute cholelithiasis, except that it is more insidious and unrelenting.

Cholangitis may develop as an ascending infection of the biliary tree proximal to obstruction due to either gallstones or stricture. It can develop also in patients who, owing to frequent passage of gallstones, have developed incompetence of the ampulla of Vater. Such patients show marked evidence of sepsis, including pyrexia and rigors. In contrast to patients with simple mechanical obstruction, whose γ-globulins and immunoglobulins remain normal, patients with ascending cholangitis frequently demonstrate abnormal concentrations of these constituents, especially IgG; however, organelle-specific antibody titers are usually normal. The serum activities of biliary marker enzymes are more prominently elevated than in simple mechanical obstruction, and the aminotransferases are likely to be elevated because secondary hepatocellular damage occurs more quickly.

Sclerosing cholangitis may occur as a complication of certain primary conditions—mainly inflammatory diseases of the gastrointestinal tract—of which ulcerative colitis is the most common; very rarely, it can occur as a primary disease entity in its own right. In both forms, episodes of obstructive jaundice lead to secondary biliary cirrhosis, and the dominant biochemical features are the high activities of biliary marker enzymes in serum, even when jaundice is mild.

Neonatal Hepatitis and Biliary Atresia

Bacterial and viral conditions causing hepatitis in the neonate have already been described. A number of rare *metabolic* conditions such as galactosemia, tyrosinemia, α_1-antitrypsin deficiency, and hereditary fructose intolerance may also give rise to acute liver disease in the newborn. A number of newborn infants, in whom all of these entities can be excluded, present with *primary* hepatic disease. Sometimes the jaundice is present from birth. In other cases it develops in previously anicteric infants. The histologic picture is similar to that described for the various forms of viral hepatitis in infancy (see Other Infectious Agents Causing Hepatitis), inflammatory infiltration and giant-cell formation being prominent; however, intrahepatic cholestasis and even obliteration of bile ducts is more dramatic in this condition which, for lack of better terminology, is called ***neonatal hepatitis.***

It is important to distinguish this condition, in which the extrahepatic ducts are entirely patent and which carries a relatively good prognosis, from one in which a similar picture may develop due to ***atresia*** of the extrahepatic or intrahepatic ***biliary system,*** and which until recently was invariably fatal. Surgical procedures have been devised that can restore biliary flow in some of these patients, although serious compromise of liver function may persist. The condition was formerly thought to be congenital, but the realization that the onset in a majority of cases occurs several weeks after birth has prompted speculation that the condition is really a toxic degeneration of the biliary tree. The etiology is attributed to some unidentified environmental or metabolic agent (in the latter regard, unusual bile acids have been identified in the serum and urine of a number of cases and it is speculated that these may be toxic for biliary epithelium).

It is also becoming appreciated that neonatal hepatitis and biliary atresia may represent opposite ends of a *spectrum.* It has proved impossible, on the basis of any single test or combination of tests on serum and urine, to distinguish clearly between the two conditions, yet this is essential to the proper selection of patients for remedial surgery. One test which has a useful role in this situation is the *^{131}I rose bengal test,* in which systemic administration of the labeled compound is followed by measurement of its appearance in the stool. Usually, 75% is excreted within 3 days; low values are strongly indicative of biliary atresia. The advent of liver transplantation can be expected to generate new knowledge on the biochemical aspects of liver graft rejection.

Malignant Disease of the Biliary Tree

Malignant obstruction of the biliary tree can occur as a consequence of cancer of the bile ducts, the gallbladder, or the ampulla of Vater. It can also arise as a consequence of external pressure by tumors such as those of the head of the pancreas, or tumor-infiltrated lymph glands, especially in the region of the *porta hepatis.* The clinical onset tends to be more insidious and less acute, but obstruction due to external pressure by cancer is generally relentless and progressive. Jaundice due to tumors within the biliary passages themselves may fluctuate as the tumor undergoes necrosis from time to time, with sloughing and therefore partial relief of the obstruction.

It is rarely possible to distinguish *malignant* from *nonmalignant* extrahepatic obstruction on the basis of biochemical tests alone. In both, there are increases in a group of serum proteins generally known as ***acute-phase reactants.*** These include α_1-antitrypsin, C-reactive protein, haptoglobin, and ceruloplasmin (the last has been reported to be especially valuable in distinguishing diseases of the biliary tree, in which it is elevated, from primary parenchymal liver disease, in which it is normal). Because these acute-phase reactants are predominantly α_2-globulins and glycoproteins, electrophoresis will show an increase in the α_2-globulin component.

These changes in serum proteins, as well as the nonspecific but quite useful *erythrocyte sedimentation rate,* are more dramatically abnormal in biliary obstruction due to tumor than with nonmalignant causes. Other clues may be provided by measuring the serum concentration of *cancer marker* proteins such as carcinoembryonic antigen (CEA). This is elevated almost as frequently in patients with pancreatic cancer as in cancer of the colon—the condition for which the test was first stated to be diagnostic.

Many attempts have been made to distinguish intrahepatic from extrahepatic obstruction. A test introduced for this purpose is the serum *lipoprotein X* assay. Unfortunately, it is elevated in any form of cholestasis, although higher values are found in extrahepatic obstruction.

Because the aminotransferases are elevated in biliary tract disease as well as in acute hepatic disease (although much more so in the latter), attempts have been made to find other enzyme tests which would be more specific for hepatic disease. So far this has been unsuccessful, because most serum enzymes showing any sensitivity in acute liver disease are also elevated in biliary tract disease. However, most functions of the liver continue relatively unimpaired for a short time after the onset of extrahepatic obstruction, even though rising back pressure within the biliary tree, ascending infection, and the accumulation of biliary waste products eventually lead to toxic necrosis of liver cells and recognizable deficiencies in hepatic function (see Fig. 11–5).

Secondary Cancer of the Liver

Metastatic infiltration of the liver by tumor cells can accompany malignant extrahepatic obstruction at the time of presentation, or can be a sequel to it. Under these circumstances, the patient will tend to be more severely ill, and the biochemical abnormalities more pronounced. Dominance of AST over ALT is more frequently seen in this condition, and the activities of ALP, 5NT, and GGT will be disproportionately elevated, relative to the degree of hyperbilirubinemia, than one would expect to find were hepatic metastases absent. Secondary cancer of the liver without icterus is a common occurrence, although jaundice may develop eventually as the condition progresses.

The *primary lesion* is usually to be found elsewhere in the gastrointestinal system, for example, in the body or tail of the pancreas, in the stomach, or in the large bowel. The form of presentation may be such that the differential diagnosis calls for the exclusion of chronic primary liver disease. The patient may present with a gastrointestinal disturbance accompanied by an enlarged liver (which may be nodular and tender, but not always so) before the primary malignant neoplasm is recognized. The diagnosis of hepatic metastases should be considered whenever high serum activities of ALP, 5NT, and GGT are encountered, with normal or minimally raised bilirubin concentration. A similar presentation may occur with primary biliary cirrhosis, but the cancer patient lacks the dramatic immunochemical abnormalities seen in biliary cirrhosis, and will sometimes have high concentrations of cancer-associated proteins such as CEA in the serum.

A *liver scan* may be helpful in revealing filling defects in the liver, but the picture cannot be distinguished with certainty from that seen in various forms of cirrhosis. Ultrasound will help to differentiate tumor nodules from fluid cysts. *Biopsy* can be diagnostic, but is prone to sampling error since an area of relatively normal tissue can be obtained. A thorough search will usually, but not always, reveal the primary site of the tumor. Cancerous infiltration may cause a form of intrahepatic cholestasis, and in this situation *biliary marker en-*

zymes in the serum will tend to be more elevated than in most forms of drug-induced cholestasis. Increased activities of serum aminotransferases are also seen in secondary cancer of the liver, with AST predominating, indicating steady necrosis of liver cells. Indeed, ALT is frequently within normal limits.

A different issue involves the ***monitoring*** of patients with known cancer who have had their primary lesion treated by surgery, radiotherapy, or chemotherapy. Metastases to the liver may occur several years after an apparently successful 'cure.' Again, cancer markers are of special value in monitoring this type of case, but the trio of enzymes (ALP, 5NT, GGT) will often detect the occurrence of metastatic liver disease before it becomes clinically obvious, and certainly before the patient develops jaundice.

Differential Diagnosis of Acute-Onset Jaundice

Unconjugated hyperbilirubinemia is established by the absence of bilirubinuria and an increase in serum bilirubin that is usually <85 μmol/L (<5 mg/dL). Higher levels occur only in the rare congenital disorders of bilirubin conjugation, especially in newborn infants.

The acute onset of ***prehepatic jaundice*** results in unconjugated hyperbilirubinemia, often with excess urobilinogen in stool and urine. It may result from circulatory impairment and anoxia due to external or interstitial hemorrhage, and the precipitating incident may be obvious. Overproduction of bilirubin also occurs in various hemolytic disorders and in defective erythropoiesis. These can usually be distinguished by routine hematologic investigations or by special tests, which may include survival of ^{51}Cr-tagged red cells and studies of hemoglobin or red cell enzymes. Abnormalities of most serum enzymes and proteins occur only rarely in these disorders; exceptions are the elevation of AST and LDH (which are present in erythrocytes in high concentration), and haptoglobin, which is reduced due to excessive intravascular hemolysis.

Hepatic ***uptake*** and ***conjugation defects*** can be demonstrated by appropriate tests, such as the clearance of intravenous ^{3}H-bilirubin or the measurement of glucuronyl transferase activity in liver biopsy specimens, but these tests are available in very few centers.

When ***conjugated hyperbilirubinemia*** is present urine levels of bilirubin will be high. Hepatocellular disease and intermittent obstruction (from choledocholithiasis) will show moderately high levels of serum bilirubin—usually <350 μmol/L (<20 mg/dL). Severe obstruction due to malignancy will show marked jaundice and very high levels, stabilizing around 500 μmol/L (25 to 30 mg/dL). The distinction between acute viral or toxic primary liver disease, intrahepatic cholestasis, and extrahepatic cholestasis is not difficult in the typical case, particularly when the patient is seen *early* after the onset of illness (Fig. 11–5). However, when presentation is *late,* or the course of the disease uncharacteristic, the full range of biochemical investigations may still leave doubt as to the diagnosis, which may be resolved only by radiology, biopsy, laparotomy, or (in the unfortunate few) autopsy. Table 11–2, summarizes those laboratory tests that may be helpful in making this *differential diagnosis* and their usual behavior in each disease entity. The most difficult problems are to decide whether cholestasis is intra- or extrahepatic, and in the latter case whether it is due to tumor or to stone.

CHRONIC LIVER DISEASE

Many chronic diseases of the liver terminate in a final common pathway of hepatic failure in which the clinical and biochemical features are nearly identical and the pathologic features so similar that unless the diagnosis has been established earlier it may be difficult to decide how the patient's ailment actually began.

Cirrhosis of the Liver

Cirrhosis is the consequence when the architecture of the liver is not restored to normal after injury. The essential morphologic features are: (1) increased fibrous tissue; (2) nodular parenchymal regeneration; (3) derangement of the microvasculature. This last feature contributes to portal hypertension and shunting of blood from the portal to the systemic circulation, two processes which account for some of the most constant and dramatic features seen in patients with the disease. Diffuse liver cell death is believed to be the precursor of cirrhosis, but evidence of necrosis may no longer be apparent. Less constant pathologic features are portal inflammation, focal or widespread cholestasis, and proliferation of bile ductules.

Cirrhosis of the liver may be *classified* according to etiology or according to morphologic criteria. Up to a point, the two are interrelated. Some of the factors that are known or believed to cause cirrhosis of the liver are listed in Table 11–3. However, the morphologic classification recognizes only two types based upon the average size of the liver nodules. In the congested or ischemic liver, and in alcoholic liver disease, the fibrous changes begin in Zone 3 around the terminal hepatic veins and spread gradually to link portal tracts and other ter-

TABLE 11–2. Tests to Distinguish Acute Diseases of the Liver and Biliary Tree (Tests that are generally normal in both or abnormal in both are not presented)

Test	Hepatocellular Disease (HCD)	Biliary Tract Disease (BTD)	Comment
Heme pigments			
Urobilinogen (U)	Variable	Absent	Increased in early and late phases of HCD
Coproporphyrin (U)	Variable	↑	Ratio of CPIII:total CP < 20% in Dubin–Johnson syndrome (normal > 65%)
Coproporphyrin (F)	Variable	↓	
Lipids			
Cholesterol (S)	Normal	Usually ↑	Measurement of free and esterified not very helpful
Triglyceride (S)	Normal	Usually ↑	Accompanied by raised VLDL in alcoholic liver disease
Lipoprotein X (S)	Usually normal	↑	↑ More dramatic in extrahepatic BTD
Proteins			
α_2-Globulins (S)	Variable	↑	β-Globulins also often ↑ in BTD
Immunoglobulins (S)	↑	Usually normal	Chiefly ↑ IgM and IgG in viral hepatitis; normal in toxic HCD
Hepatitis-associated antibodies & antigens (F or S)	Present in types A and B hepatitis	Absent	See Fig. 11–7 for specifics. Other serologic tests positive in HCD from other infectious agents.
Haptoglobin (S)	May be ↓	Usually ↑	Also low in hemolytic jaundice
C-Reactive protein (S)	May be ↓	Usually ↑	↑ in BTD is nonspecific inflammatory response
α_1-Antitrypsin (S)	May be ↓	Usually ↑	Absent in specific neonatal hepatitis owing to congenital defect
Prothrombin (P)	Usually ↓	May be ↓	Responds to vitamin K only in BTD
Other clotting factors	Often ↓	Usually normal	Most common cause of death in acute fulminant hepatitis
Ceruloplasmin (S)	May be ↓	Usually ↑	↑ Paralleled by ↑ in serum copper in BTD
Enzymes			
Alkaline phosphatase (S)	Modest ↑	Moderate to marked ↑	Similar for 5NT, leucine aminopeptidase, and GGT. Lesser ↑ when BTD is intrahepatic
Aminotransferases (S)	Very high	Moderate ↑	AST > ALT in malignant BTD and toxic HCD. Lesser ↑ in neonatal HCD and intrahepatic BTD
Pseudocholinesterase (S)	May be ↓	Normal	Check if operation is considered, to avoid prolonged apnea

TABLE 11–2. *Continued*

Test	Hepatocellular Disease (HCD)	Biliary Tract Disease (BTD)	Comment
Miscellaneous			
Iron (S)	May be ↑	Normal	From breakdown of liver iron stores
Copper (S)	Normal	May be ↑	Accompanied by increased ceruloplasmin
Glucose (S)	Often ↑	Normal	May be low in fulminant HCD
Bile acid tests			
Fasting bile acids (S)	Often ↑	Invariably ↑	Trihydroxy/dihydroxy ratio increased in BTD
2-h Postprandial bile acids (S)	Invariably ↑	Usually no further ↑	Food increases bile acids in enterohepatic circulation if biliary tree is patent
IV Cholyl-^{14}C-glycine excretion test	Abnormal	Usually normal	Measures disappearance of ^{14}C from the blood due to extraction by the liver
Other function tests			
Cortisone test	Jaundice ↓	Jaundice unaffected	Given as 50 mg tid
IV ^{131}I rose bengal test	Normal	Abnormal	Especially useful in neonatal BTD

↑ = Increased; ↓ = decreased; ALT = alanine aminotransferase; AST = aspartate aminotransferase; F = feces; GGT = γ-glutamyl transferase; 5NT = 5′-nucleotidase; P = plasma; S = serum; U = urine; VLDL = very-low-density lipoprotein.

minal hepatic veins (Fig. 11–9A). The nodules of viable parenchymal cells are, however, invariably small, and *micronodular cirrhosis* is widely used to describe the histologic appearance of the liver in this disease.

In *post hepatitic disease* and *chronic active hepatitis* the initial fibrosis is in Zone 1, and the nodules of viable parenchymal cells are much larger; thus *macronodular cirrhosis* is the term applied to the resulting microscopic appearance of the liver. Surviving cells in Zone 1 regenerate, but when numbers of acini are destroyed, structure and functional efficiency may be impaired, with more blood vessels bypassing the hepatocytes. The normal hepatic architecture is eventually destroyed by bands of fibrous tissue separating nodules of regenerating cells. Neoplastic transformation of such a regenerating nodule occasionally gives rise to *hepatoma.*

The clinical and biochemical features of cirrhosis will be described in relation to the principle etiologic forms. The ultimate picture is that of chronic liver failure, as described later.

Alcoholic Liver Disease. Chronic alcoholism, especially when associated with malnutrition, leads to a characteristic disease of the liver dominated

TABLE 11–3. Etiologic Forms of Hepatic Cirrhosis

Established associations

- Viral hepatitis
- Alcoholism
- Metabolic Diseases:
  - Glycogen storage disease
  - Sphingolipidoses
  - α_1-Antitrypsin deficiency
  - Wilson's disease
  - Hemochromatosis
  - Cystic fibrosis
  - Galactosemia
  - Hereditary fructose intolerance
- Biliary disease:
  - Primary biliary cirrhosis
  - Disease involving major bile ducts
- Venous outflow obstruction:
  - Budd–Chiari syndrome
  - Veno–occlusive disease
- Drugs and toxins

Proposed associations

- Autoimmunity
- Toxins
- Parasitic disease
- Malnutrition

Modified from MacSween RNM, Anthony PP, Scheuer PJ: Pathology of the Liver, p 266. Edinburgh, Churchill Livingstone, 1979

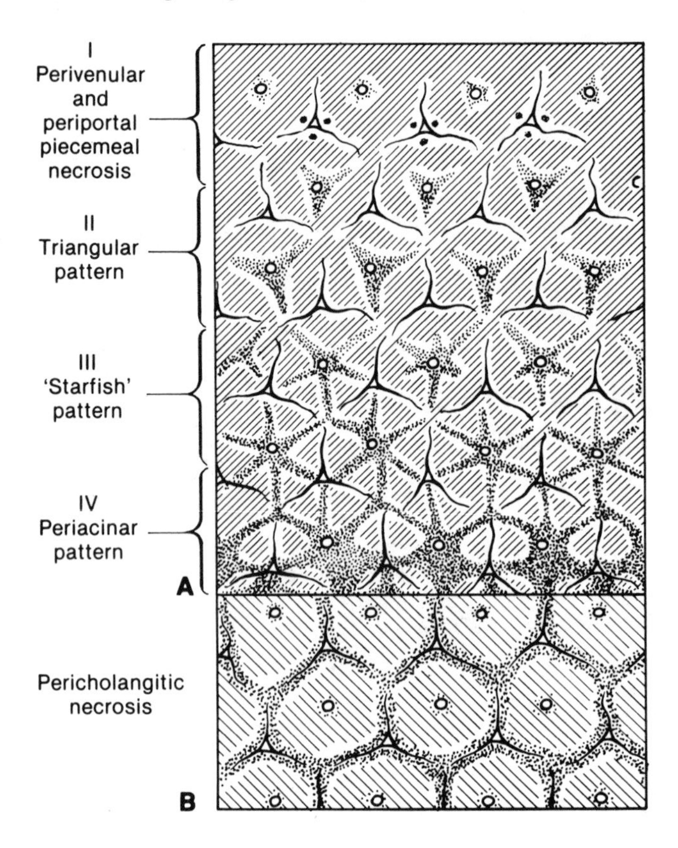

FIGURE 11–9. *A.* Schematic illustration of the development of fibrosis in so-called portal cirrhosis. Triangular structures are portal tracts; circular structures are terminal hepatic venules. Fibrosis appears first around the terminal hepatic veins, often with small patches of piecemeal necrosis near the portal tract. The fibrosis spreads in zone 3, exhibiting first a triangular pattern, then a 'starfish' pattern. In time the lesion circumscribes the liver acini (periacinar pattern), reducing the surviving portions to zone 1 or destroying numbers of acini entirely. *B.* In biliary cirrhosis the primary lesion begins in the small portal tracts (triangular structures), extends along the terminal bile ductules, and progresses until it outlines what was formerly called the hepatic lobule. There is some secondary necrosis around the terminal hepatic venules.

by fatty infiltration, degeneration of intracellular organelles, and the formation of hyaline bodies. Hepatomegaly is usually present. The incidence varies with different populations of alcoholics. The process develops no further in the majority of cases, and the biochemical picture is dominated by the pronounced increase in serum GGT activity, together with less pronounced and less frequent elevations in other serum enzymes, notably AST, ALT, and glutamate dehydrogenase. Type IV hyperlipoproteinemia (see Chap. 20) is frequently present. Jaundice is rare, but more subtle tests such as BSP retention and measurement of serum bile acid concentrations may reveal abnormal hepatic function.

In about 10% of alcoholics with fatty liver the condition progresses to frank ***alcoholic cirrhosis;*** these cases account for more than 80% of all persons with cirrhosis seen in North America. The manifestations of alcoholic cirrhosis are often insidious and typically develop after 10 or more years of alcohol excess. Weakness, fatigue and anorexia are first noticed and a firm enlarged liver is usually present. Ankle edema may occur, together with any of the manifestations of portal hypertension and hepatic failure illustrated in Figure 11–10. The prognosis is poor, with 50% or more failing to survive 5 years from diagnosis. Massive hemorrhage and hepatic coma are common terminal events.

During 'binge' drinking acute ***alcoholic hepatitis*** may develop, marked by necrosis of liver cells, increased activities of parenchymal enzymes in the serum, and the onset of jaundice. Clinical features resemble viral or toxic liver injuries, with fever in half the patients, and tender hepatomegaly as a common finding. Occasionally, cholestatic jaundice may develop during the course of alcoholic liver disease.

Prior to chronic liver failure, in the well-compensated patient, there may be few detectable *biochemical* abnormalities. Most useful are the reduction in serum albumin concentration and the increase in γ-globulin concentration with β–γ bridging on electrophoresis. Analysis of the immunoglobulins shows the dominant abnormality to be in the IgA fraction, and a proportion of patients demonstrate increased titers of smooth-muscle antibodies, which are found in patients with autoimmune or viral inflammations.

As the alcoholic fatty liver gives way to the typical histologic picture of micronodular cirrhosis, there is a reduction in serum *GGT* activity; thus, in the alcoholic patient, this can be used to *monitor* abstinence (when values return to normal) or the development of cirrhosis (when values fall but still remain considerably elevated). A feature of alcoholic cirrhosis worth emphasizing is that the *aminotransferases* are normal in a proportion of cases, but when they are elevated AST shows the dominant increase. Another serum enzyme, *adenosine deaminase,* is raised in a high proportion of patients with cirrhosis and shows some specificity for this condition.

Posthepatitic Cirrhosis. Relatively few patients with cirrhosis of the liver give a history of previous acute hepatitis, although it is a reasonable assumption that many have had subclinical forms of the disease. A proportion are asymptomatic *carriers* of HB_sAg, and some of these may have passed through the stage of chronic active hepatitis without its recognition. In this form of cirrhosis the liver is usually shrunken, and the size of the nodules separated by fibrous tissue is very much larger than in alcoholic cirrhosis, thus the term macronodular. Various industrial chemicals, drugs and bacterial infections may be causative factors, but it is rarely seen in those who abuse alcohol. The signs and symptoms resemble those of alcoholic cirrhosis, with portal hypertension and retention of salt and water quite prominent. Jaundice, however, is an earlier feature in this condition. The prognosis is even more bleak than in alcoholic cirrhosis. Hypergammaglobulinemia is more common in posthepatitic cirrhosis, with IgG elevations being as pronounced as those of IgA, or even more so. Otherwise, the two conditions are not easy to distinguish on laboratory criteria alone.

Idiopathic Cirrhosis. In cases of idiopathic cirrhosis there is no evidence of previous infection or alcoholic excess. Nutritional factors, environmental toxins, and trace metal contamination have all been suggested as contributory but evidence on these points is very scanty. The best guess is that the cirrhosis develops from a subclinical and unrecognized viral hepatitis.

Other Types. *Cardiac cirrhosis* has already been mentioned (see Acute Liver Disease). Various *infiltrative* diseases of the liver may lead to cirrhosis. Most of these are genetically determined rare conditions such as (1) glycogen storage disease; (2) the various sphingolipid disorders; (3) α_1-antitrypsin deficiency (a condition in which this protein is low or absent in serum because the liver is unable to glycosylate and secrete an abnormal gene product. This engorgement leads to necrosis of the hepatocytes); (4) Wilson's disease (in which copper deposition in the liver leads to necrosis and fibrosis). In most of these conditions, the biochemical abnormalities associated with *liver* pathology play a secondary role in diagnosis, compared with the severe and unique abnormalities attributable to the *primary* disease entity.

Primary Biliary Cirrhosis

The etiology of this disease is unknown although an autoimmune mechanism is suspected. ***Clinically,*** it presents first as pruritus which may be due to bile acid accumulation resulting from bile secretory failure. This can be the only manifestation for the first year or two of the illness. Mild, fluctuating jaundice may then become apparent and gradually more severe. A striking feature can be the increased serum *cholesterol* concentration, which often precedes the hyperbilirubinemia, and is usually much higher than one would expect to find in a case of extrahepatic obstruction with comparable icterus. The high levels of plasma cholesterol lead to cutaneous xanthomas.

Histologically, the disease attacks the small bile ductules, which initially show epithelial proliferation but then become damaged by invading masses of lymphocytes and plasma cells. Hepatocytes adjacent to the bile ductules are affected by the inflammatory process and undergo necrosis. The developing fibrosis follows the path of the terminal ductules (Fig. 11–9B), eventually outlining portions of several acini with a terminal hepatic vein in the center. Increasing periportal fibrosis, granuloma formation, and bile stasis eventually develop. The stasis causes *chronic intrahepatic obstruction* and may result in the absence of bile acids from the intestine, leading to malabsorption with steatorrhea. This in turn leads to defective absorption of fat-soluble vitamins and calcium, so that metabolic bone disease and bleeding due to low prothrombin concentrations (reflected in an increased prothrombin time) are common features of the illness.

The ***laboratory results*** are dominated by elevations of serum cholesterol and serum ALP and 5NT activities, even in the early stages and long before the development of jaundice. As obstruction develops, the serum bilirubin starts to increase but always lags well behind the elevated enzymes. Perhaps the most useful diagnostic features are elevation of the serum *IgM* concentration and the development of *mitochondrial antibodies* in the serum. The former reaches levels never seen in any other condition, while the latter are relatively specific for primary biliary cirrhosis, occasionally occurring in drug-induced jaundice and chronic active hepatitis, but virtually never in extrahepatic obstruction. Both immunochemical features are prominent at the early preicteric stage of the disease, as are dramatic elevations in serum bile acid concentrations. Large immune-complexes have been found in the serum of patients with primary biliary cirrhosis.

Secondary Biliary Cirrhosis

This can result from any chronic obstruction of the extrahepatic biliary ducts, including idiopathic sclerosing cholangitis. A few patients with ulcerative colitis exhibit cholangitis which can lead to secondary biliary cirrhosis. When associated with infection in the biliary tract, the disease progresses more quickly to hepatocellular damage. The histologic picture differs from that seen in primary biliary cirrhosis; moreover, the immunologic abnormalities are not encountered, and elevations of cholesterol and biliary tract enzymes commonly occur in parallel with rising bilirubin concentrations.

Chronic Active Hepatitis

This progressive disease manifests episodes of liver cell necrosis (marked by fairly high aminotransferase activities in the serum, with AST predominating), inflammation (marked by high concentrations of serum IgG and raised titers of smooth-muscle antibody and antinuclear factor), and ultimately fibrosis terminating in the final common pathway with cirrhosis and chronic liver failure. The disease is believed to be associated with a disordered immunity or to have an *autoimmune etiology.* The latter is marked by frequent involvement of other organs, resulting in skin lesions, joint inflammation, colitis, and thyroiditis; indeed, these manifestations may precede the discovery of liver involvement.

Approximately one third of the cases first come to attention during an acute illness indistinguishable from viral hepatitis; some of these patients, and others who present insidiously, have continuing HB_sAg titers detectable in their serum, together with HB_eAg. The ***clinical presentation*** of the remaining two thirds includes fatigue (with mild jaundice as an incidental finding), or manifestations of low-grade biliary tract disease as described above. Hepatomegaly and splenomegaly are common, and other manifestations of chronic liver disease may be presenting features, depending upon the duration and severity of the illness.

The ***laboratory findings*** will depend upon the activity of the disease and its response to therapy with steroids or immunosuppressants. Monitoring of aminotransferase activities is very useful in this situation. Biliary tract enzymes are generally only marginally raised, and the serum albumin concentration does not fall until the disease has become well established. In addition to the immunologic abnormalities already described, IgM may be moderately raised, and some patients may have positive titers of mitochondrial antibodies, making it imperative to distinguish this condition from primary biliary cirrhosis.

The disease has also to be distinguished from ***chronic persistent hepatitis,*** which does not have extrahepatic manifestations, does not require drug therapy, and carries a favorable prognosis. It can be distinguished *histologically* by the absence of 'piecemeal necrosis,' a characteristic degeneration of parenchymal cells at the periphery of many hepatic acini which is the hallmark of chronic active hepatitis. A useful *laboratory* distinction is that in chronic persistent hepatitis the increases in IgG and IgM tend to be parallel, whereas in chronic active hepatitis the IgG increase is dominant and usually quantitatively greater.

Hepatoma

Primary carcinoma of the liver is common in Africa and Asia, where it is frequently associated with chronic hepatitis B infection and occasionally with ingestion of fungal products such as aflatoxin. In Western countries, almost all cases occur in livers that are the site of alcoholic or postnecrotic cirrhosis, and onset of the disease is difficult to detect in such patients, although a recent increase in hepatomegaly or onset of pain should arouse suspicion. Biliary marker enzymes such as ALP and 5NT, as well as AST, are disproportionately high in relation to serum bilirubin concentration which is often normal. The most characteristic biochemical feature is an increase in α-fetoprotein (AFP), the principle circulating protein in early fetal life which disappears around the time of birth. It is virtually absent in healthy adults and is present in high concentrations in more than 70% of patients with hepatoma. Unfortunately, moderate elevations occur in approximately 30% of patients with various acute and chronic liver diseases, especially when regeneration is prominent. This limits its use as a diagnostic aid, and the course of the disease is usually so rapid that its role in monitoring progress is of little practical significance. The protein is of considerable utility in pregnancy evaluation (see Chapter 23).

Major Complications of Chronic Liver Failure

These are depicted diagrammatically in Figure 11-10 and can be summarized as follows:

Portal hypertension develops when there is acute congestion of the liver, but is important mainly in chronic disease associated with extensive fibrosis and disordered architecture of the liver. It leads to splenomegaly and dilatation of collateral veins in the anterior abdominal wall, lower esophagus, and rectum.

Ascites is an accumulation of fluid in the peritoneal cavity which occurs when portal hypertension, low plasma albumin, and continued intake of salt coexist. It is usually associated with a decreased blood volume on the arterial side, which results in excess secretion of aldosterone and vasopressin (antidiuretic hormone, ADH) and decreased free-water clearance by the kidney. Fluid accumulation can be arrested by restricting dietary sodium to a few millimoles per day. Clearance of the fluid may require treatment with aldosterone antagonists, cortisol, and diuretics. Catheter drainage may be necessary to relieve suffering, but results in the loss of about 25 g albumin/liter of fluid, an amount which the diseased liver will find it hard to replace.

Bleeding or frank ***hemorrhage*** is a particularly serious problem in chronic hepatobiliary disease. Prolonged obstruction will lead to vitamin K deficiency and consequent low prothrombin levels, but this is easily corrected. Advanced liver disease will result in impaired prothrombin synthesis; when the prothrombin time is prolonged for this reason, it is a grave prognostic sign. Vomiting of blood, or black tarry stools, due to massive hemorrhage from rupture of *esophageal varices,* is a life-threatening complication. It may justify portacaval anastomosis if other indices such as the serum albumin level indicate a reasonable prospect of benefit.

Patients with chronic liver failure generally have an ***anemia*** marked by the presence of 'target cells,' which is caused partly by impaired metabolism of vitamin B_{12} and folate, as well as other hematinics generated by the liver in normal health. Yet another factor is hypersplenism, which causes accelerated destruction of erythrocytes, leukocytes, and platelets. Depletion of the formed elements of the blood leads to hyperactivity of the bone marrow in an attempt to overcome these losses. *Disseminated intravascular coagulation* is sometimes precipitated by factors that are poorly understood. This can result in rapid depletion of fibrinogen, and other clotting proteins, with subsequent bleeding.

Malnutrition is frequently a factor in chronic liver disease. The dietary neglect of the alcoholic, the lack of appetite in alcoholic cirrhosis, and the cholestasis in biliary cirrhosis all contribute to vitamin deficiencies, weight loss, steatorrhea, protein depletion, and impaired carbohydrate metabolism.

Hepatic coma, or ***encephalopathy,*** results from effects on the central nervous system attributed to or produced by the action of toxic substances that can no longer be cleared effectively by the diseased liver. Early symptoms may include irritability, personality changes, and bizarre or manic behavior. Later signs are tremor, unsteadiness, delirium, and coma. In the end stages plasma ammonia levels show a rough correlation with severity.

Frank brain damage is rare but a disturbance in cerebral electrophysical activity seen in the electroencephalogram is common. This has been at-

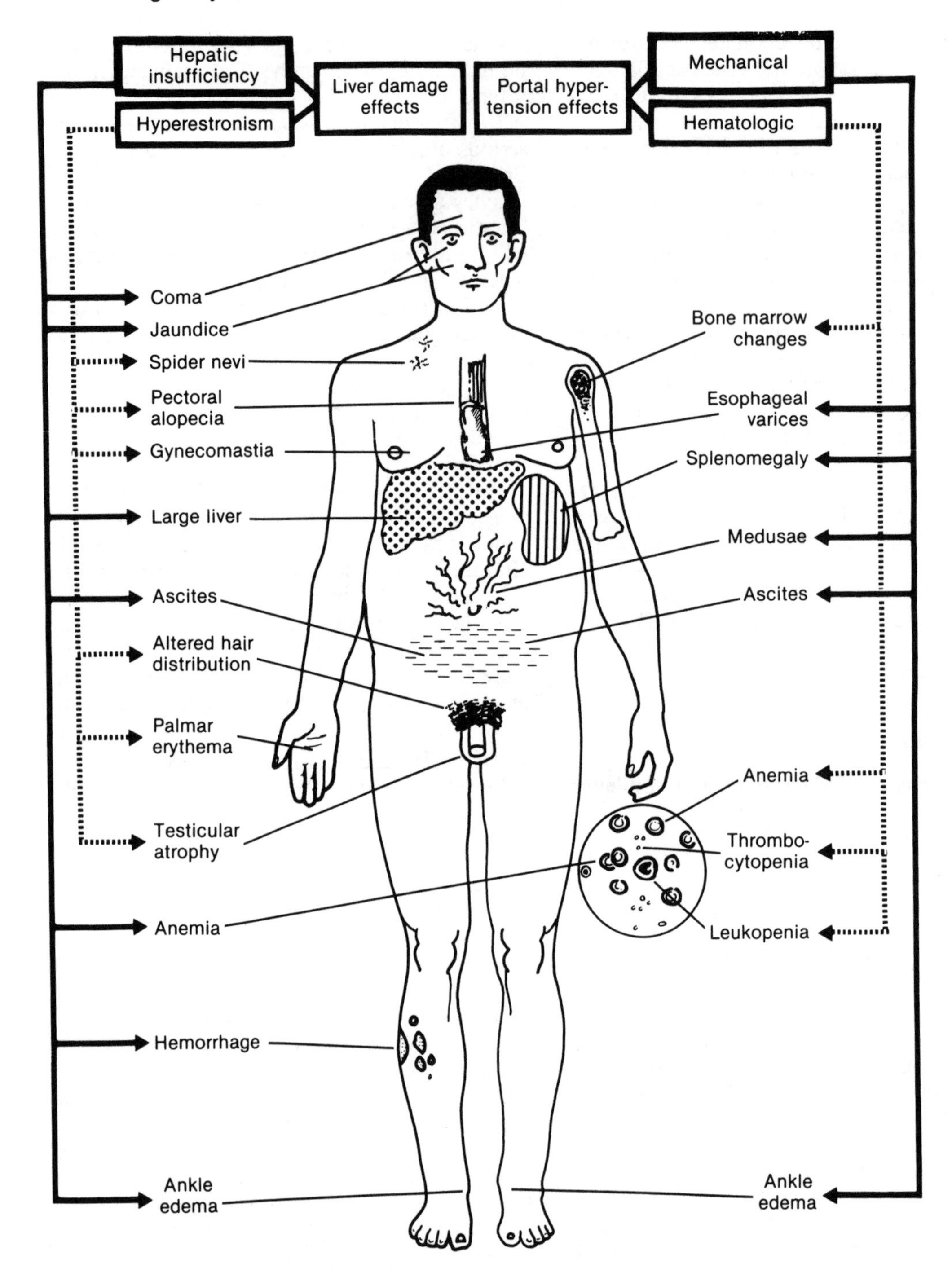

FIGURE 11–10. Clinical complications of chronic liver failure. (Adapted from an original painting by Frank H. Netter, M.D., for the Ciba Collection of Medical Illustrations; copyright Ciba-Geigy Corp.)

tributed to the presence of ***false neurotransmitters.*** Many of these are believed to be produced by bacterial action from protein in the gut and bypass the liver owing to portal–systemic anastomoses. In addition to the raised blood ammonia, elevated glutamine and γ-aminobutyric acid (an inhibitor of neurotransmission) have been reported. Phenylalanine and tyrosine, which normally generate the true neurotransmitters dopamine and norepinephrine, give rise in the presence of liver failure to competing amines such as octopamine and phenylethylamine. Finally, the accumulation of free fatty acids in the serum, in particular the short-chain fatty acids butyrate, valerate, and octanoate, interfere with neurotransmission and represent another possible mechanism for the neurologic disturbances seen in this condition. Hypoglycemia is yet another factor which has to be taken into account. The false neurotransmitter theory has received some support from reports of the beneficial use of levodopa in the treatment of acute hepatic encephalopathy.

Endocrine changes also occur. The failure of the liver to inactivate ADH and aldosterone causes retention of water and salt, contributing to the ascites described previously. Because of the liver's inability to conjugate estrogens, and the impaired synthesis of steroid-binding globulin, the half-life of active estrogens in the serum is greatly prolonged. This gives rise to gynecomastia, testicular atrophy, and loss of pectoral and pubic hair in the male. In both sexes, a peculiar flushing of the palms and the presence of tiny superficial arteriovenous anastomoses (*spider nevi*) have also been attributed to a relative excess of free estrogen.

The *syndrome* of liver failure may occur with acute massive hepatic necrosis, or as the terminal phase of cirrhosis. The number of manifestations present in any one case increases as the disease advances until, in the terminal stage, all may be present and all are severe. Death in such cases may result from unremitting hepatic coma, hemorrhage, or disseminated intravascular coagulation, which is especially prone to precipitate terminal renal failure.

HEPATOBILIARY DIAGNOSTIC TESTS

General Considerations

Although several groups of tests have been proposed as liver 'profiles,' few have proved of value. Even the better tests of liver cell integrity and organ function have relatively poor sensitivity and/or specificity. Consequently, one should select two to four of the best tests and avoid those in which the average of sensitivity and specificity is less than 80% (see Chap. 1), recognizing that misclassification will increase if more tests of lower efficiency are used. Laboratory tests that do not increase the

TABLE 11–4. Serum Tests of Value in Differential Diagnosis of Chronic Liver Disease and Liver Cancer

Test	Alcoholic Cirrhosis	Primary Biliary Cirrhosis	Chronic Acute Hepatitis	Liver Cancer
Cholesterol	Normal or ↓	Usually ↑	Normal	Normal
Alkaline phosphatase	Moderately ↑	Very high	Moderately ↑	Very high
5′-Nucleotidase	Moderately ↑	Very high	Moderately ↑	Very high
Aspartate aminotransferase	Slightly ↑	Moderately ↑	High	High
Alanine aminotransferase	Normal or marginally ↑	Moderately ↑	Moderately ↑	Slightly ↑
IgA	High	Usually normal	Usually normal	Usually normal
IgM	Slightly ↑	Very high	Slightly ↑	Usually normal
IgG	Moderately ↑	Slightly ↑	High	Usually normal
Antinuclear factor	Absent	Absent	Frequently present	Absent
Mitochondrial antibodies	Absent	Usually present	May be present	Absent
Smooth-muscle antibodies	May be present	May be present	Often present	Absent
α-Fetoprotein	May be ↑	Normal	Usually normal	↑ In hepatoma
HB_sAg and HB_eAg	Usually absent	Absent	May be present	Absent

↑ = Increased; ↓ = decreased; HB_sAg = hepatitis B surface antigen; HB_eAg = hepatitis Be antigen.

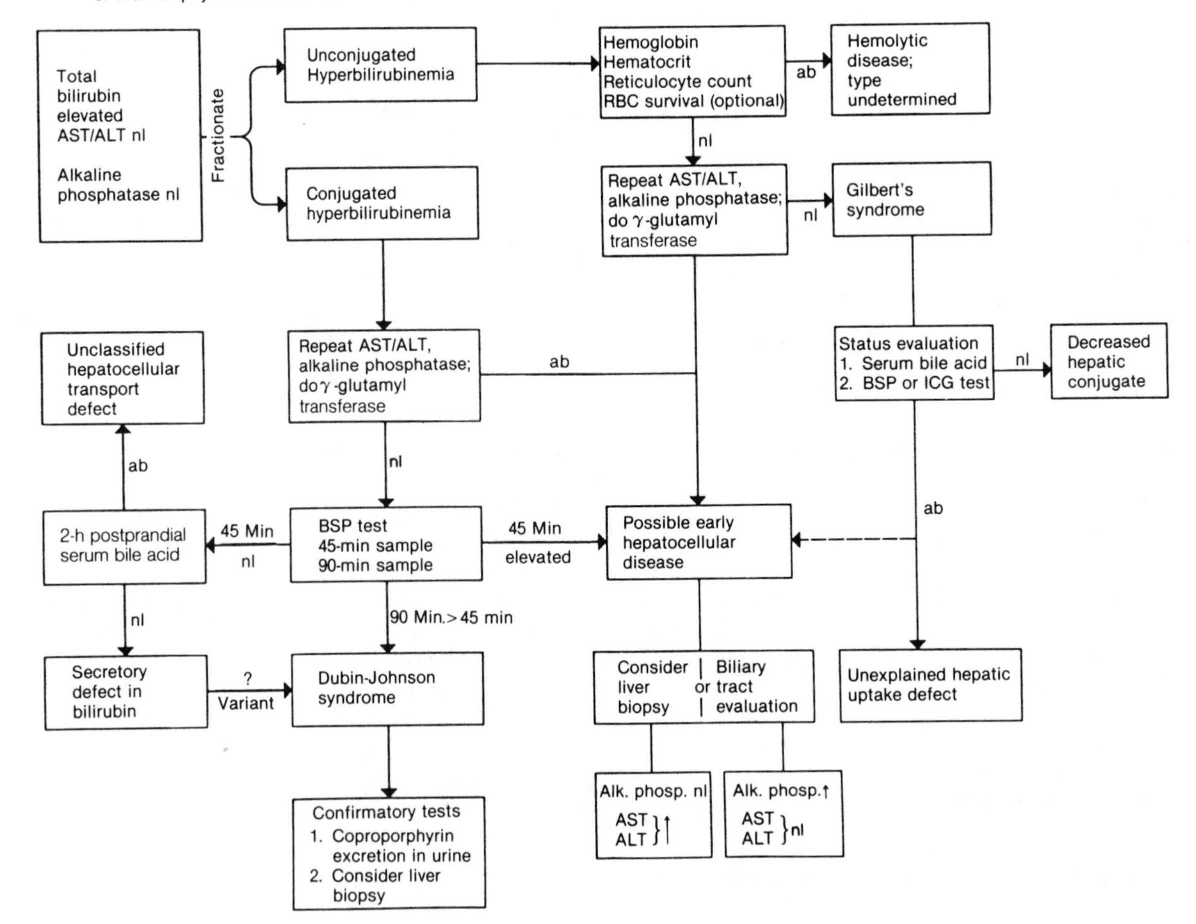

FIGURE 11–11. Algorithm for the investigation of an elevated serum bilirubin. ab = abnormal; nl = normal; ALT = alanine aminotransferase; AST = aspartate aminotransferase; BSP = sulfobromophthalein (Bromsulphalein); ICG = indocyanine green. (Javitt NB et al: Material Diagnostica—A Manual of Liver and Kidney Tests for the Medical Profession. Puerto Rico, Searle and Co., 1977.) (Reproduced, with modifications, by permission of Dr. NB Javitt and Searle Diagnostics)

information gain, or that are of borderline efficiency, will not be included in this discussion. To achieve a high predictive value liver function tests must be weighed in conjunction with all clinical factors that affect disease probability (history, physical findings, the time course of the present illness, and hematology studies).

The basic relationships between *biochemical* and *morphologic* damage to liver cells are only partly known and are to some extent interdependent. Viral infection of liver cells causes widespread damage to the endoplasmic reticulum; drug-induced injury may center on the mitochondria. Either may compromise the functional integrity of

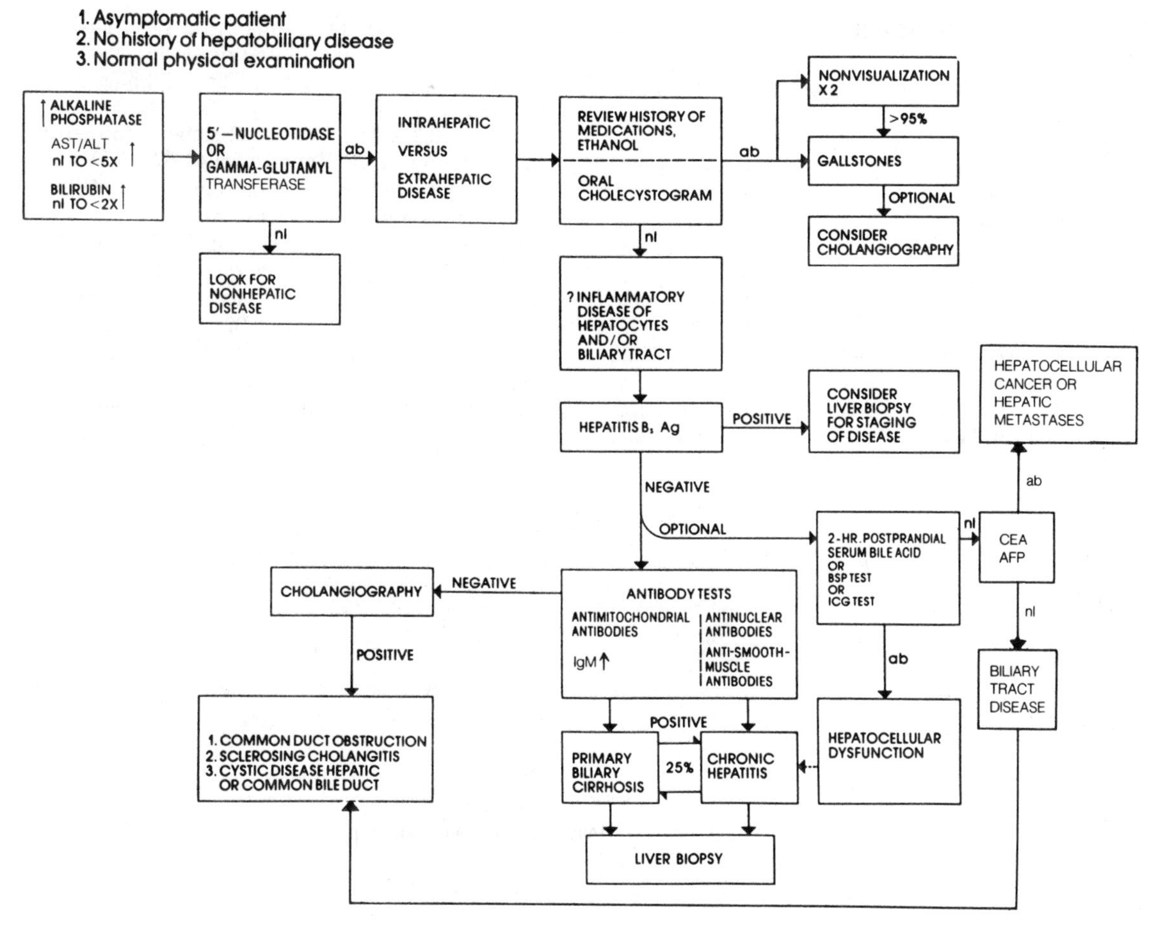

FIGURE 11–12. Algorithm for the investigation of a raised serum alkaline phosphatase. ab = abnormal; nl = normal; Ag = antigen; ALT = alanine aminotransferase; AST = aspartate aminotransferase; BSP = sulfobromophthalein (Bromsulphalein); ICG = indocyanine green; CEA = carcinoembryonic antigen; AFP = alphafetoprotein (Reproduced with modifications by permission of Dr. NB Javitt and Searle Diagnostics)

the cell or the canalicular membrane and lead to cholestasis. Intra- or extrahepatic factors that interfere with bile flow will produce membrane changes followed by cytotoxic effects leading to hepatocellular damage.

Recommended Diagnostic Tests

Laboratory tests that are likely to be of assistance with each of the many problems that occur clinically in hepatobiliary disorders are listed in Table 11–2 for *acute* and Table 11–4 for *chronic* disorders. Usual findings are presented along with relevant comments.

A logical approach to solving two common problems is illustrated in Figure 11–11—an algorithm for the investigation of elevated serum *bilirubin* levels in an asymptomatic patient—and in Figure 11–12—an algorithm for the investigation of elevated serum *alkaline phosphatase* activity in an asymptomatic patient.

SUGGESTED READING

ARIAS IM, FRENKEL M, WILSON JHP: The Liver Annual, Vol 3. Amsterdam, Elsevier, 1983

DEMERS LM, SHAW LM: Evaluation of Liver Function. A Multifaceted Approach to Clinical Diagnosis. Baltimore, Urban and Schwarzenberg, 1978

ELLIS G, GOLDBERG DM, SPOONER RJ, WARD AM: Serum enzyme tests in diseases of the liver and biliary tree. Am J Clin Pathol 70:248–258, 1978

FISHER AM, ROY CC: Pediatric Liver Disease. New York, Plenum Press, 1983

HOFMANN AF: The enterohepatic circulation of bile acids. In Sleisenger MH, Fordtran JS (eds): Gastrointestinal Disease: Pathophysiology. Diagnosis, Management, 3rd ed., pp 115–131. Philadelphia, WB Saunders, 1983

POPPER H, SCHAFFNER F: Progress in Liver Diseases, Vol 8. New York, Grune & Stratton, 1986

SCHIFF L: Diseases of the Liver, 5th ed. Philadelphia, Lippincott, 1982

SCOTT JONES R, MEYERS WC: Regulation of hepatic biliary secretion. Annu Rev Physiol 41:67, 1979

SHERLOCK S: Diseases of the Liver and Biliary System, 7th ed. Oxford, Blackwell Scientific, 1985

SHERLOCK S, THOMAS HC: Treatment of chronic hepatitis due to hepatitis B virus. Lancet ii:1343–1346, 1985

THOMPSON RPH, MARIGOLD J: Hepato Biliary Diseases. In Goldberg DM (ed): Clinical Biochemistry Reviews, Vol 3, pp 407–427. New York, Wiley, 1982

WARD AM, ELLIS G, GOLDBERG DM: Serum immunoglobulin concentrations and autoantibody titers in diseases of the liver and biliary tree. Am J Clin Pathol 70: 352–358, 1978

12

Gastrointestinal and Pancreatic Disorders

W. H. Chris Walker

THE GASTROINTESTINAL TRACT

Disorders of the gastrointestinal tract are among the most frequent reasons for seeking medical attention, usually because of peptic ulcer, diarrhea, abdominal pain, or constipation. In addition, acute manifestations such as vomiting, diarrhea, hematemesis, and melena can lead to profound and sometimes life-threatening changes in body fluid compartments.

In recent years our understanding of the endocrinology, immunology, and microbiology of the alimentary canal, and of bile acid metabolism, have increased considerably. Fiberoptic technology now permits direct visualization and biopsy of the stomach and duodenum, and cannulation of the pancreatic and common bile ducts. Small intestinal biopsies are readily obtained for morphologic and biochemical studies. Immunoassay allows the measurement of hormones; gas chromatography permits the ready estimation of hydrogen in expired air. We may expect continuing advances in investigative techniques, involving perhaps telemetry and specific electrodes. Direct measurements with better clinical sensitivity and specificity will probably supplant current methods of analyzing blood, urine, and feces.

STRUCTURE AND FUNCTIONS

General Structure

The alimentary canal (Fig. 12–1) extends from the mouth to the anus. Food enters the pharynx and thence the *esophagus,* passing through the diaphragm in the midline to enter the *stomach,* which is suspended by the lesser omentum and terminates in a muscular sphincter called the pylorus. From here food enters the *duodenum,* which lies behind the peritoneum close to the head of the pancreas. The common bile duct and the pancreatic duct enter the duodenum near its midpoint and allow bile and pancreatic juice to enter its lumen. Food and digestive juices pass into the *small intestine,* which is termed the *jejunum* in its first half, then becomes the *ileum,* to enter the *large bowel* at the *cecum.* Intestinal contents then pass into the *colon,* termed in sequence *ascending, transverse, descending,* and *sigmoid.* The terminal segment of bowel, lying within the floor of the pelvis, is called the *rectum.* It ends at the skin as a muscular sphincter, the *anus.*

Essential Functions

The alimentary canal serves as a reaction vessel for the emulsification and digestion of food. In the process, autonomic signals are activated, hormones are secreted into the portal circulation, and digestive enzymes are released into the lumen, together with bile and large volumes of fluid. Bile acids, water, and electrolytes are conserved by distal reabsorption. Any impairment of this reabsorptive process can lead to rapid and severe depletion of body fluid.

Absorption of the products of digestion occurs by passive diffusion and by specific active transport mechanisms; defects of absorption may be gen-

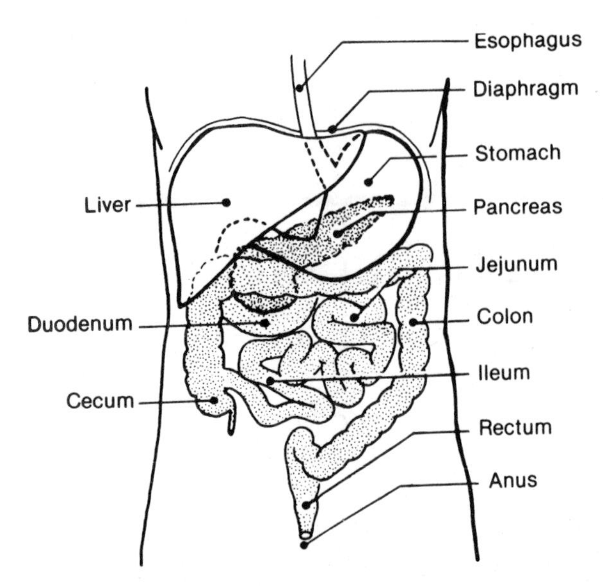

FIGURE 12–1. The alimentary tract.

eralized or selective. The alimentary canal also serves as the route for the elimination of unabsorbed food (mostly fiber) and of many end-products of metabolism.

Gastrointestinal Hormones

Several classes of specialized peptide- and amine-secreting cells are widely distributed throughout the intestinal mucosa. They belong to a family of cells characterized by their capability for *amine precursor uptake and decarboxylation* (APUD cells), and also known by their histologic staining reactions as enterochromaffin, argyrophil, or argentaffin cells. They respond to neural or luminal stimuli and release amine or polypeptide hormones into the portal circulation. After passage through the liver, these gastrointestinal peptides act like hypothalamic or pituitary hormones to influence the secretion of other hormones and to regulate cell metabolism. Most of their effects are on the gastrointestinal tract, where they control the secretion of digestive juices and regulate mucosal cell growth. They qualify as hormones because they are released in response to physiologic stimuli and have defined regulatory actions at physiologic concentrations in the blood. They are shown, along with systemic blood levels, in Table 12–1.

Gastrin (G–17), with 17 amino acids, has its major locus of activity in the C-terminal tetrapeptide. It is available in pharmacologic form as a pentapeptide (Pentagastrin). Postprandially, about three quarters of the molar concentration of gastrin in the blood is composed of a 34-amino acid molecule (G–34), 'big gastrin,' whose molar potency is less than half that of G–17 although it has a much longer half-life. A still larger molecule, 'big big gastrin,' reacts with gastrin antibody but has no biologic activity. It is the predominant form of gastrin in fasting plasma, but is only 1% of the immunoreactive gastrin in antral cells and tumor extracts.

Cholecystokinin has 33 amino acids and shares its N-terminal 6-amino acid sequence with gastrin. It is now known that pancreozymin is identical to cholecystokinin, and the latter name has replaced the former in deference to the priority convention.

The ***secretin*** family of peptides has no active fragments, the entire molecule being required for activity. *Secretin* has 27 amino acids, *vasoactive intestinal peptide* (VIP) has 28 and *gastric inhibitory peptide* (GIP) 43; nine amino acids are common to all three hormones. It is not known whether VIP is released in response to physiologic stimuli; even if it were, its rapid destruction by the liver would limit its action as a hormone. It is known to stimulate small intestinal secretion when produced in a high concentration by tumors. *Glucagon* has 29 amino acids, 14 in common with secretin. It is secreted into the blood by the α cells of the pancreatic islets. *Enteroglucagon* comprises a heterogeneous group of peptides with glucagon-like immunoreactivity arising from cells in the gastric fundus and the duodenum. The polypeptides *bombesin* (14 amino acids), *motilin* (22), and *pancreatic polypeptide* (36) have profound effects on the pancreas at pharmacologic levels but their physiologic significance is unknown. The physiologic stimulus of glucose entering the duodenum causes a release of GIP that is sufficient to stimulate insulin secretion. This explains, at least in part, the higher insulin response to oral than to intravenous glucose. GIP also inhibits 90% of gastrin-induced acid secretion. Secretin and cholecystokinin each potentiate the action of the other. *Somatostatin,* a cyclic 14-amino acid polypeptide found in cells of the hypothalamus and the intes-

TABLE 12-1. Gastrointestinal Peptide Hormones and Systemic Blood Levels

Hormone	Secretory Stimulus	Target Organ	Physiologic Action	Sensitivity# of Assay (ng/L)	Blood Levels (ng/L)		
					Fasting	*Post-prandial*	*Tumor*
Gastrin	Vagus, antral distension	Fundal parietal cells	Acid secretion; mucosal growth	5	20–70	100–400	200–100,000**
Cholecystokinin	Fat, protein in duodenum	Gallbladder Pancreatic acini	Contraction Enzyme secretion; cell growth	5	25	>100	—
Secretin	Acid in duodenum	Pancreatic ducts	Water and bicarbonate secretion	25	25	25	(200–1000)†
Gastric inhibitory peptide	Glucose in duodenum	Gastric fundus Pancreatic islets	Inhibits acid secretion Insulin secretion	100	70–500	1000	—
Vasoactive intestinal peptide	Not known	Small intestine	Not known	100	100	—	600–9000‡

Limit of detection

** Antral G cell hyperplasia causes fasting levels that overlap with tumor levels.

† Secondary to increased acid in duodenum caused by gastrin-secreting tumors.

‡ Tumor is often nonpancreatic.

tinal mucosa, competitively inhibits the action of secretin and inhibits the release of cholecystokinin.

THE STOMACH

STRUCTURE AND FUNCTIONS

The stomach (Fig. 12–2) consists of a thin-walled *body* leading to a more muscular *antrum* and terminating in the *pylorus,* which acts as a sphincter leading to the duodenum. The wall is composed of smooth muscle bounded externally by peritoneum and internally by submucosa and gastric mucosa, arranged in deep folds, or *rugae.* The stomach's function is to mix food with gastric secretions, break down solids and start the digestive process. The resultant fluid is called *chyme.* Contractions in the antrum, acting together with the pyloric sphincter, regulate the rate at which the chyme is released into the duodenum. Without this regulation, food would pass so rapidly through the small bowel that digestion and absorption would be severely impaired.

Gastric Secretions

The gastric mucosa (Fig. 12–2) has specialized secretory cells: the chief cells and mucous cells secrete pepsinogen I; the parietal cells, acid and intrinsic factor. The pyloric mucosa lacks parietal cells.

The control of ***hydrogen ion secretion*** is complex. Acetylcholine (from vagal fibers) and gastrin are the most potent stimuli; histamine is 10^{-3} times as active as gastrin on a molar basis. Antral G cells secrete gastrin in response to acetylcholine and to local distention. Acetylcholine, therefore, has a dual role in stimulating acid secretion, acting directly on the parietal cell and indirectly through gastrin release.

Pepsinogen I is secreted in response to local cholinergic reflexes triggered by acid receptors in the gastric mucosa. Pepsinogen I is present in plasma at a mean concentration of 60 μg/L, and is excreted in the urine. The urinary content varies widely and is not useful as a measure of gastric secretion. *Pepsinogen II* is secreted by pyloric glands in the proximal duodenum. It is present in the plasma at a concentration of about 10 μg/L and does not appear in the urine.

The *rate of acid secretion* varies widely throughout the day; dilution by neutral secretions containing sodium chloride leads to an inverse relationship between the concentrations of hydrogen

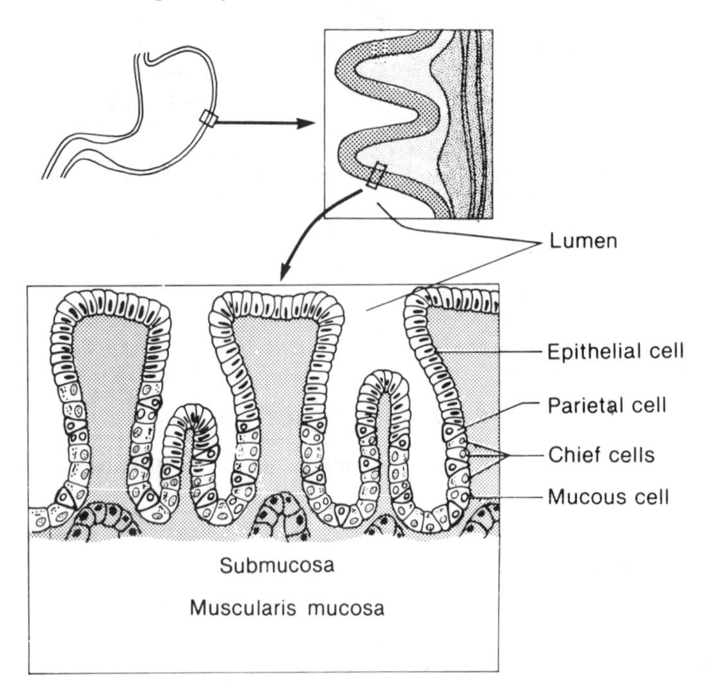

FIGURE 12–2. The stomach and gastric mucosa.

and sodium ions. Chloride concentration remains unchanged at about 150 mmol/L; potassium concentration lies within the range of 8 to 20 mmol/L. A 70–kg adult will secrete some 2500 mL of gastric fluid per day. Basal acid secretion is always <10 mmol H^+/hour; maximal secretion after pentagastrin stimulation seldom exceeds 30 mmol/hour. Vomiting or continuous gastric aspiration can lead to severe depletion of hydrogen, sodium, and chloride ions, with resultant hypochloremic metabolic alkalosis and extracellular volume depletion.

Intrinsic factor is secreted at a rate of about 2000 units/hour. Because 500 units are sufficient to permit absorption of the daily requirement of 1 μg of vitamin B_{12}, intrinsic factor is normally in 100-fold excess. Despite their common parietal cell origin, intrinsic factor secretion and acid secretion can become dissociated. In some cases of juvenile pernicious anemia, for example, intrinsic factor secretion is abolished while acid secretion persists.

DISORDERS OF THE STOMACH AND DUODENUM

Gastric and Duodenal Ulceration

Gastric ulceration is probably related to the action of pepsin, an enzyme inactivated above pH 5. The stomach is normally protected against self-digestion by the gastric mucosal barrier, which is composed of epithelial cell walls and tight junctions between adjacent cells. In *gastric* ulceration, acid secretion is usually not increased. Rarely, benign gastric ulceration has been reported in the absence of measurable acid, but in such situations gastric carcinoma must be excluded by histologic examination. Chronic *duodenal* ulcer is nearly always associated with a maximal acid output greater than 25 mmol/hour. The basal acid output is of no diagnostic help in duodenal ulcer.

Endoscopy now constitutes the most appropriate diagnostic approach to gastric and duodenal ulceration. Gastric ulcer must be distinguished from carcinoma; duodenal ulcer, from previous scarring. Up to half of the patients with dyspepsia and x-ray findings consistent with duodenal ulceration are found to have no ulcer when examined by endoscopy.

Gastrin-Secreting Tumor (Zollinger–Ellison Syndrome). Gastrin-secreting tumors usually result in distal duodenal or jejunal ulceration, diarrhea, recurrent stomal ulcers after surgery, and multiple ulcers.

The tumors occur mainly in the pancreas; 60% are malignant, 30% are benign but multiple, and only 10% are solitary benign tumors that offer a reasonable prospect of surgical resection. The treatment is directed at the acid hypersecretion, which is controlled by cimetidine, if necessary with addition of anticholinergic drugs and highly selective vagotomy. Total gastrectomy is used if medical therapy fails.

The tumor is sometimes associated with hyperparathyroidism and with a family history of endocrine tumors in the parathyroid and pituitary glands (*multiple endocrine adenopathy* type I, MEAI). Hyperparathyroidism alone probably does not produce an increase in gastrin or acid secretion, but removal of a parathyroid tumor, in a patient with an associated gastrin-secreting tumor, may result in a short-lived decrease in gastrin levels, owing to the removal of the hypercalcemic stimulus.

Diagnosis depends on clinical presentation, documentation of a basal acid output >15 mmol/hour, fasting gastrin concentration > 5 times the upper RV, or a paradoxical increase in gastrin in response to secretin provocation. The malignant potential of a tumor may be revealed by increased serum concentrations of the α-subunit of chorionic gonadotropin (hCG), and by a predominance of G–17 rather than G–34 immunoreactive serum gastrin.

LABORATORY INVESTIGATIONS

Gastric Acidity

The measurement of gastric acid before and after a provocative stimulus has lost much of its early popularity. Gastroscopy is now more helpful in peptic ulcer disease; intrinsic factor and vitamin B_{12} assays and the Schilling test establish a diagnosis of pernicious anemia. Measuring acid secretion may have relevance in the following cases:

1. In patients with duodenal ulcer who are to be treated surgically
2. After surgery when recurrent ulceration is suspected
3. Following vagal stimulation in assessing residual vagal innervation after vagotomy

Pentagastrin (6 μg/kg body weight, subcutaneously), has replaced histamine and its analogs for stimulating *maximal acid secretion.* Basal secretions are collected at 15-min intervals for 1 hour. Pentagastrin is then given and poststimulation secretions are collected at 15-min intervals for a further hour. Overnight collections add no useful information. Maximal acid secretion is closely related to the parietal cell mass.

Hydrogen ion (H^+) secretion is measured by titrating the gastric juice to pH 8; at this pH all buffering effects are eliminated. Two-step titration to pH 3 and then to pH 8 to yield 'free acid' and 'total acid' is no longer considered helpful. The amount of H^+ is derived from the concentration and the volume of each sample; the secretion rate is expressed as mmol of H^+/hour, either by totaling the four poststimulation samples (*maximal* acid output) or by addition of the two highest rates and multiplying by two (*peak* acid output).

The presence of bile, with its implicit admixture of alkaline duodenal juice, invalidates the acid measurement. If demonstration of achlorhydria is required, a simple glass electrode measurement of pH suffices; a pH of <3.5 indicates the presence of hydrochloric acid.

A *basal secretion* greater than 10 mmol H^+/hour, or 5 mmol H^+/hour following partial gastrectomy, justifies an estimation of the serum gastrin. *Maximal secretion* usually does not exceed 30 mmol H^+/hour, but the normal response has great variability and overlaps with duodenal ulcer patients who oversecrete.

The 'tubeless' dye test meal is now little used. It employs a basic dye–cation exchange resin which at low pH releases its dye for absorption and urinary excretion. It has fair sensitivity (92%) but low specificity (70%) and therefore falsely indicates many normal subjects to be achlorhydric.

Serum Gastrin

The *reference range* of fasting serum gastrin is usually 20 to 70 ng/L, depending on assay specificity. Great variation in gastrin assays with corresponding differences in reference ranges has led to the use of 5× the upper RV as a useful diagnostic threshold. An increase in serum gastrin occurs when there is a loss of the feedback control by which gastrin normally stimulates acid secretion, and acid entering the gastric antrum then inhibits gastrin release. Hypergastrinemia may be primary, due to hyperplasia of the antral G cells in chronic gastric outlet obstruction; it may be secondary to achlorhydria or an isolated retained antrum after gastrectomy; or it may be due to a gastrin-secreting non-β islet cell tumor of the pancreas or duodenum (Zollinger–Ellison syndrome).

Estimation of serum gastrin is indicated (1) in patients with peptic ulcer, especially those with multiple ulcers or ones in unusual sites, those with weight loss, hypercalcemia, or diarrhea, and those with a basal acid secretion greater than 10 mmol/hour; and (2) in patients with recurrent ulcer after gastric resection or vagotomy.

Increases up to five times the fasting gastrin concentrations are seen

1. With achlorhydria, especially in chronic atrophic gastritis with circulating parietal cell antibodies
2. After vagotomy
3. With gastric ulcer
4. In advanced renal failure
5. Rarely, with pheochromocytoma

Although *gastric* ulcers are usually associated with normal acid secretion, fasting serum gastrin concentrations may be increased two- to fivefold. In *duodenal* ulcer, acid secretion is increased but the fasting gastrin is usually normal, apparently because of impaired feedback inhibition by antral acid. Similar increases occur normally within 2 hours after a meal. When concentrations in excess of five times normal occur, or when no apparent reason exists for lesser increases, it is necessary to perform *provocative tests* to distinguish G-cell hyperplasia, isolated retained antrum, and gastrin-secreting tumor. Very high concentrations may occur in pernicious anemia.

Provocative Tests

1. ***Secretin.*** Karolinska secretin is given as a bolus, 2U/kg body weight, and serum gastrin is measured at 5, 10, 15, and 20 min. In G-cell hyperplasia and isolated retained antrum there is no change or a decrease in gastrin concentration. In 75% of patients with a gastrinoma, serum gastrin increases more than 200 ng/L above the basal concentration within 20 minutes.
2. ***Test meal.*** From 1 to 2 h after a protein meal there will be more than a twofold increase in the gastrin level in normal subjects and in those with G-cell hyperplasia. The patient with an isolated antrum or a gastrin-secreting tumor does not show this increase.

An *obsolete test,* calcium infusion, has an effect similar to secretin, but the sensitivity and specificity of this test are unacceptably low (many false-negative and false-positive results occur).

Assessment of Completeness of Vagotomy. The effective surgical treatment of a duodenal ulcer by vagotomy requires that the vagal innervation of parietal cells be completely interrupted. Ideally, a test during surgery would indicate complete section, but constraints of time and drug interference limit the value of existing *intraoperative* procedures. Tests administered 10 days or more after surgery still have relevance; the usual procedure is the ***Hollander test,*** in which insulin-induced hypoglycemia is used as a vagal stimulus and the resultant gastric acid is measured.

Subcutaneous insulin (0.15 units/kg) is given, and gastric juice is collected over 2 hours. The blood sugar must fall below 2 mmol/L (40 mg/dL); otherwise, a larger insulin dose is given. Standard *criteria of inadequate vagotomy* are an increase in acidity of more than 20 mmol/L above the basal concentration in any 15-min specimen up to 2 hours following the insulin dose, or an absolute concentration of 10 mmol/L if the basal secretion was zero. It has been shown that a positive response within 45 min of insulin dosage has a greater prognostic significance as a marker of recurrent ulceration. The test has a sensitivity of about 75% and a specificity of 80% and none of the many modifications of interpretative criteria improve these values.

Postgastrectomy and Postvagotomy Syndromes

Gastrectomy may lead to two clinical sequelae:

1. *Postgastrectomy hypoglycemia* is a result of the rapid transit of a glucose-containing meal. Accelerated duodenal absorption leads to a rapid rise in blood sugar, which serves as a maximal stimulus for the release of insulin. Increased insulin concentrations persist beyond the transient period of glucose absorption and lead to a rapid reduction in blood sugar, with hypoglycemic symptoms 1 to 2 hours after a meal.
2. The term *dumping syndrome* describes a feeling of abdominal discomfort, nausea, and lightheadedness occurring shortly after a meal,

which is considered to be caused by the rapid entry of hypertonic fluid into the duodenum. The osmotic transfer of water and electrolytes into the lumen results in hypovolemia and transient hypokalemia. The release of kinins and other vasoactive materials is also involved.

Postvagotomy syndrome is a state of postprandial diarrhea occurring after vagotomy. It may be caused by altered intestinal motility and appears to be associated with bile acid malabsorption.

THE PANCREAS

STRUCTURE AND FUNCTIONS

Structure. The pancreas (Fig. 12–3) is composed of specialized secretory cells of endocrine and of exocrine function. The *endocrine cells* are grouped into *islets of Langerhans* and are responsible for the secretion into the portal circulation of insulin (β cells) and glucagon (α cells). The *exocrine cells* are arranged in acini which lead into the pancreatic ductules and ultimately join to drain into the duodenum through a single pancreatic duct, which usually merges with the common bile duct. The pancreatic *acinar cells* are responsible for the secretion of digestive enzymes. The *duct epithelium* secretes water and electrolytes.

Functions. Exocrine cells constitute about 50% of the pancreatic cellular compartment. Little is known about *basal exocrine* pancreatic secretion, but the ingestion of food stimulates secretion through neural and hormonal influences. Secretin and cholecystokinin are released from the duodenum in response to an influx of gastric contents and are the dominant regulators of pancreatic secretion. Autonomic activity has a direct regulatory effect and stimulates the secretin and cholecystokinin release. The hypothalamus has the potential to suppress pancreatic secretory responses through the action of somatostatin.

When stimulated, pancreatic ***acinar cells*** secrete amylase, lipase, and several proteases. An adaptation of the relative proportions of these enzymes to long-term dietary changes does occur, but there is no clear evidence that the pancreas can modify its enzyme secretion to meet the specific needs of a single meal.

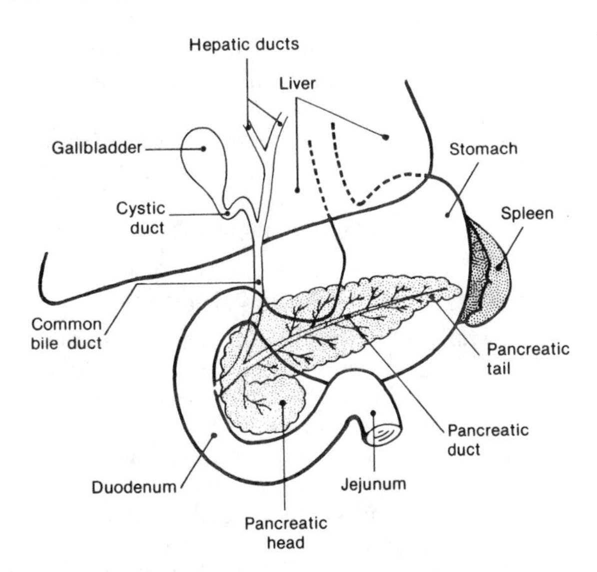

FIGURE 12–3. The biliary and pancreatic systems.

Ductal secretion of water and electrolytes is regulated independently of enzyme secretion. The normal pancreas secretes 1 to 2 liters of fluid/day containing about 135 mmol of sodium and 4 mmol of potassium per liter. The cation concentration is independent of flow rate. The bicarbonate concentration increases as flow is stimulated, and there is a corresponding fall in the chloride concentration.

PANCREATIC DISORDERS

Chronic Pancreatitis

This condition nearly always presents with episodic or continuous upper abdominal pain, and alcoholism is an important contributory factor. It is characterized by multifocal parenchymal damage that develops over months or years, sometimes with cysts and pseudocysts, and ultimately with pancreatic calcification, diabetes mellitus, and pancreatic exocrine deficiency. It must be distinguished from *acute recurrent* pancreatitis, which is often the result of biliary tract disease and is usually curable by appropriate surgery.

The early lesions are confined to foci within the parenchyma, leaving the larger ducts intact. Smaller ducts become plugged with eosinophilic protein deposits that consist of pancreatic enzymes and mucoprotein, together with lactoferrin, a protein not found in normal pancreatic juice. Destruction of the acinar epithelium and fibrous scarring occur, with chronic inflammatory cell infiltration. In the late stages there is replacement of the exocrine pancreas by fibrous tissue, isolation of islets, and the formation of cysts which may rupture into the adjacent parenchyma to form pseudocysts.

Laboratory Tests in Chronic Pancreatitis

Secretin Stimulation Test. A double-lumen catheter is introduced into the duodenum under fluoroscopy to permit the continuous aspiration of gastric juice through one lumen, while ten serial collections of duodenal juice are made at 10-min intervals. The first two serve as a baseline; secretin is then given (1 unit/kg body weight, IV) and the other samples are collected. The volume and bicarbonate concentration of all specimens are measured. In a normal individual the total poststimulation volume is at least 2 mL/kg/80 min, and the peak bicarbonate concentration is at least 90 mmol/L. Low volume is characteristic of obstruction; low bicarbonate, of chronic pancreatitis.

The following *problems* are associated with this test:

1. Secretin preparations vary in purity and their biologic activity is somewhat poorly defined.
2. The test is normal in 5% of subjects with pancreatic disease (sensitivity = 95%). It is abnormal without pancreatic disease in about one third of patients with vagotomy and pyloroplasty or gastrectomy.
3. Cholecystokinin contamination of the secretin may lead to gallbladder contraction and the inclusion of bile in the duodenal aspirate.
4. In cirrhosis, there may be a diminished bicarbonate concentration with an increased volume.
5. Wrong positioning of the tube may cause duodenal fluid to be discarded with the gastric juice, or gastric juice to enter the duodenum and neutralize the bicarbonate.

Cholecystokinin Stimulation. Cholecystokinin is restricted to experimental work in North America. No reference values are established for the secretion of pancreatic enzymes. The hormone causes contraction of the gallbladder and other alimentary smooth muscle, and abdominal pain is a recognized and sometimes severe side effect.

Lundh Test Meal. Following a standard meal of 18 g corn oil, 15 g casein hydrolysate, and 40 g glucose in 300 mL water, the contents of the duodenum are aspirated over a 2-hour period and the trypsin content is measured.

The following *points* should be noted:

1. Intact duodenal mucosal villi must be present to mediate the hormonal response; a diminished response is seen in villous atrophy.
2. Vagotomy or duodenal bypass surgery (Billroth II gastrectomy) will diminish the response.
3. The sensitivity in chronic pancreatitis is 90%, and in carcinoma, 80%.
4. The test does not help in the early diagnosis of pancreatic carcinoma, or in assessing the role of minor pancreatic malfunction in otherwise unexplained abdominal pain.

BT–PABA Excretion. An indirect test that correlates well with the Lundh and secretin tests measures 6 h urinary excretion of p-aminobenzoic acid (PABA), conjugated with glycine to form hippuric acid, after ingestion of 0.5 g N-benzoyl-L-tyrosyl-PABA (BT–PABA). This is a specific substrate for *chymotrypsin.* Intestinal, hepatic, and renal variables that influence absorption of PABA can be corrected for by adding 5 μCi ^{14}C–PABA to the BT–PABA. The test has a sensitivity for pancreatic disease of 75% and a specificity of 96%.

Lactoferrin, Trypsin, and Pancreatic Polypeptide. The concentration of *lactoferrin* in pure pancreatic juice from patients with chronic pancreatitis has been shown to be increased to 800 to 12,000 μg/L, while in patients with pancreatic carcinoma it falls within the reference range, 0 μg/L to 400 μg/L. Duodenal juice must be free of lactoferrin-containing saliva and gastric juice. Sensitivity and specificity have not been adequately defined. It is postulated that lactoferrin may complex with albumin to form precipitates which plug the small pancreatic ducts.

A radioimmunoassay for *trypsin* permits measurement of plasma immunoreactive trypsin and trypsinogen. Decreased concentrations occur in advanced pancreatic disease (sensitivity 60%) but less often in early disease (sensitivity 20%).

Measurement of *chymotrypsin* in feces reflects pancreatic secretion because intestinal bacteria do not produce this enzyme. *Pancreatic polypeptide* (serum reference range 0 to 600 ng/L) appears to be more closely related to pancreatic exocrine rather than endocrine function.

Further Investigation of Chronic Pancreatic Disorders

With a diagnosis of pancreatic exocrine malfunction established, further tests are required to distinguish chronic relapsing pancreatitis, pancreatic carcinoma, and pancreatic pseudocyst.

The following is an assessment of *other procedures:*

1. *Cytologic study* on pancreatic juice demands skill and experience and lacks general usefulness.
2. *Radiography* will demonstrate calcium-containing gallstones and calcification of the pancreas. A barium meal with hypotonic contrast medium will reveal deformation of the duodenal mucosal outline.
3. *Ultrasound* is reliable in the diagnosis of pseudocyst. *Isotope scanning* has many false positives, with a specificity of only 65%.
4. Guided *needle biopsy* allows the nonoperative diagnosis of pancreatic cancer.
5. *Endoscopic retrograde cholangiopancreatography* (ERCP) involves cannulation of the pancreatic duct under visual control through an endoscope; the retrograde injection of radiopaque dye then demonstrates the pancreatic ducts. This is a recent and valuable adjunct to diagnosis, but skill in cannulation and experience in interpretation are needed. There are rare adverse reactions. This procedure frequently distinguishes between pancreatic carcinoma and chronic pancreatitis.

It is unfortunate that laboratory testing can still do little to meet two critical clinical requirements: (1) the definition of pancreatic malignancy at a stage sufficiently early to permit curative surgical intervention, and (2) the demonstration of mild chronic pancreatic disease as the cause of intermittent upper abdominal pain of uncertain origin.

Acute Pancreatitis

This condition presents with the sudden onset of upper abdominal pain and tenderness, accompanied by tachycardia and hypotension. Gallstones are present in about half the cases, alcoholism in about one quarter. Less frequent associations are hypertriglyceridemia, hyperparathyroidism, chronic renal failure, upper abdominal surgery, afferent loop distention following gastrectomy, and severe malnutrition. Carcinoma of the pancreas may cause obstruction of the pancreatic duct, and malignancy should be suspected in any patient over 40 years of age presenting for the first time with acute pancreatitis and without obvious predisposing factors.

A gallstone or other pathology obstructing the common orifice of the biliary and pancreatic ducts may cause the disorder. Increased pressure in the pancreatic ducts probably leads to regurgitation of pancreatic juice and bile into the acini, damaging the parenchymal cells. The powerful digestive enzymes then enter the circulation and are usually inactivated by binding to plasma α_1-antitrypsin and α_2-macroglobulin. When these mechanisms are exhausted the enzymes activate complement and phospholipase C, leading to local destruction of cell membranes with the release of more digestive enzymes and their precursors. Tryptic digestion activates proteolytic enzyme precursors and further cell damage results. The mechanism of alcoholic predisposition is unknown; associated hyperlipemia may be a factor. Hypocalcemia may be caused by the formation of insoluble calcium soaps with free fatty acids released by lipase activity, or by hypoalbuminemia, or by hormonal changes leading to parathyroid hormone resistance.

Laboratory Diagnosis of Acute Pancreatitis

Amylase: Amylase (see also Chap. 3) occurs in the pancreas, salivary glands, small intestine, and kidney. It is freely filtered at the glomerulus and reabsorbed by the tubules. Serum amylase in health is predominantly of the *salivary* type, and this fraction may be increased up to three times normal in acute inflammation of the parotid gland and in mumps. In acute pancreatitis or common duct stone, and after the administration of morphine, the increased amylase in both serum and urine is of the *pancreatic* type.

Pancreatic *isoamylase* quantification may assist in establishing a diagnosis of acute pancreatitis. It appears to be more specific than total amylase, which is increased in many acute abdominal disorders.

An increase in amylase/creatinine clearance ratio was believed to distinguish acute pancreatitis from other causes of increased amylase such as biliary obstruction, perforated ulcer, small bowel infarction, diabetic ketoacidosis, burns, and pancreatic carcinoma. The clearance ratio is now known to have too low a specificity to be useful in diagnosis and the test is no longer used.

Serum amylase concentration may be used *to follow the course* of acute pancreatitis, but the magnitude of the increase does not always reflect the severity of the attack and such increases may be transient (see Fig. 3–7). Increased plasma triglycerides inhibit amylase activity. The specificity of serum amylase alone for acute pancreatitis is poor, and the sensitivity is no better than 90%, so that false-negative results will occur in 10% of cases of acute pancreatitis.

Urinary amylase concentration depends on renal function as well as on pancreatic pathology. It lags behind serum changes and sometimes remains increased up to 7 days after serum levels have returned to normal (see Fig. 3–7).

Reference ranges for serum amylase are 70 to 300 U/L, and for urinary amylase, 100 to 7000 U/24 hours.

Serum amylase may be bound to plasma proteins in the absence of pancreatic disease, a condition termed *macroamylasemia.* The serum concentration is increased due to the lack of glomerular filtration and the urine concentration is low. Confusion in the interpretation of amylase measurements may occur in both health and disease.

Lipase: Serum lipase (see also Chap. 3) is frequently increased in acute pancreatitis. Newer rapid assay techniques for lipase may offer added diagnostic information over that provided by amylase assay, but in general lipase concentrations parallel amylase.

THE SMALL AND LARGE INTESTINES

STRUCTURE AND FUNCTIONS

The Small Intestine

The small intestine (Fig. 12–4) is composed of a tube some 3 m long and 4 cm in diameter. It starts as the *duodenum,* lying behind the peritoneum and then becomes the *jejunum* and finally the *ileum,* both lying suspended within a double layer of peritoneum called the *mesentery.* The lining mucosa is arranged in circular folds which

FIGURE 12–4. The structure of the small intestine, showing mucosal villi and microvilli.

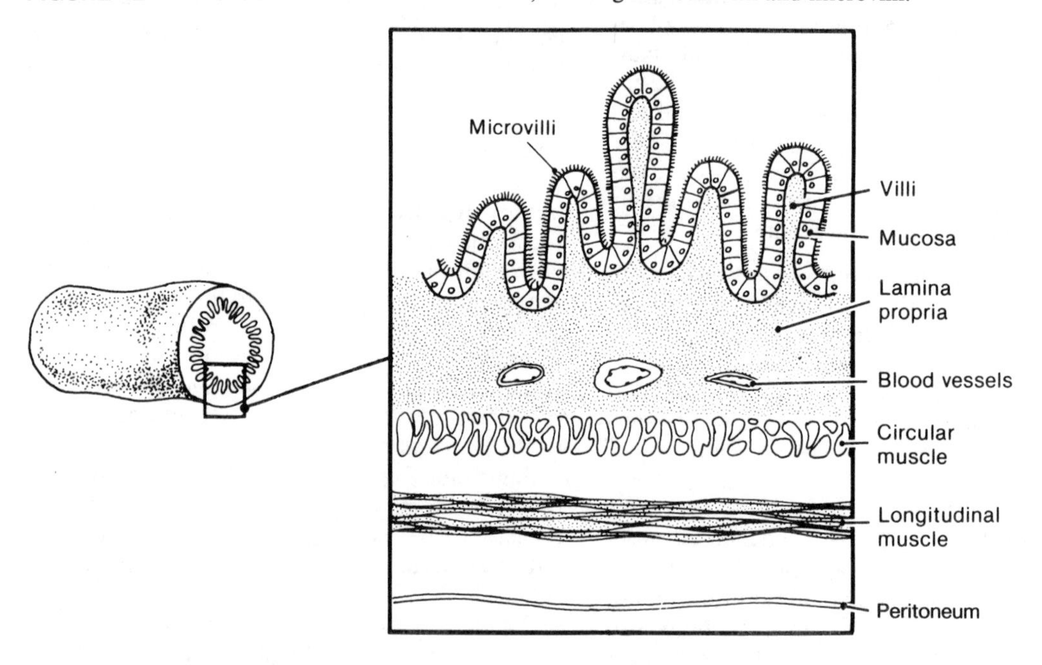

greatly increase its surface area. The surface is further folded into villi 1 mm long which are covered with columnar epithelial cells. The luminal border of the cells is arranged in microvilli 1 μm long, yielding a final absorptive surface area of some 500 m^2. Cells originate in the crypts between villi, migrate to the tips of the villi in about 2 days, and are then shed into the lumen.

The small intestine performs the following ***functions:***

1. ***Motility.*** Stationary contractions provide a *mixing* function; progressive contractions (*peristalsis*), a *propulsive* function. The 'barium meal and follow-through' remains the standard method for evaluating small bowel motility. The barium normally enters the cecum in 1 to 2 hours.
2. ***Secretion.*** Endocrine secretions of the small intestine mucosal cells pass into the blood. Exocrine secretions pass into the lumen; they include water and electrolytes (Table 12–2), protein, and the enzyme enterokinase. Intestinal mucosal cells have a very rapid turnover and about 250 g of cell debris is shed into the lumen each day.
3. ***Digestion.*** The lumen provides a 'reaction vessel' for digestion of fat, protein, and carbohydrates. Disaccharides are further hydrolyzed at the mucosal brush border and oligopeptides are digested within mucosal cells.
4. ***Absorption.*** Fat, protein, and carbohydrate are absorbed mainly by the proximal jejunum, vitamin B_{12} and bile acids by the ileum. Nearly all the secreted water and electrolytes are reabsorbed before reaching the colon.

Impairment of any one of these functions will lead to malabsorption, which may initially be selective but often becomes generalized owing to (1) changes in the luminal water content, (2) bacterial overgrowth, and (3) a decrease in the time that the luminal contents remain within the small bowel.

Immunology and the Alimentary Canal

Lymphoid tissue exists widely throughout the intestinal mucosa in the epithelium and in the submucosa (lamina propria). In the ileum, lymphoid tissue is aggregated as Peyer's patches. Like other mucosal surfaces, the intestinal mucosa is capable of generating a local immune response that is independent of systemic immunity. The predominant immunoglobulin in intestinal plasma cells is ***IgA.*** This local antibody production in response to ingested antigenic stimuli is an important protective mechanism against bacterial and viral insults. After passage through the epithelium, IgA is secreted into the intestine as *secretory* IgA, an 11S dimer (M_r 390,000) consisting of two 7S IgA monomers associated with a polypeptide synthesized in mucosal cells, termed secretory component or transport piece, and with a small protein, the J (junction) chain. In this form, secretory IgA is more resistant to intraluminal proteolysis.

Microbiology of the Alimentary Canal

Normally, bacteria in the upper intestinal tract are few in number and are mostly of dietary origin until the mid-ileum. In the normal colon there is about 1.5 kg of bacteria (wet weight). More than 99% of the organisms are nonsporulating anaerobic bacilli of the genera *Bacteroides, Bifidobacterium,*

TABLE 12-2. Fluid Transfer in the Alimentary Tract

Source	Volume L/day	Ion Concentrations (mmol/L)				
		H^+	Na^+	K^+	Cl^-	HCO_3^-
Food	1.5					
Saliva	1.5	—	30	20	30	20
Stomach	2.5	70	80	8–20	150	—
Bile	0.5	—	135	4	110	25
Pancreas	1.5	—	135	4	100–60	40–80
Small intestine	12	—	130	15	110	35
Total	19.5*					

* Most of this fluid is reabsorbed in the small intestine, leaving less than 1.5 L to be absorbed in the colon.

and *Eubacterium*. The flora in the rectosigmoid is similar to that in feces, and differs from that at the ileocecal junction.

Bacteria influence metabolism by glycosidase activity and by fermentation. *Urea* diffusing from the blood into the colon is hydrolyzed to ammonia, which is reabsorbed and converted by the healthy liver back to urea. Some 100 mmol of urea, about one third of the body urea pool, is hydrolyzed each day. The glycoside cathartics *senna* and *cascara* are hydrolyzed by bacteria to release their active aglycones, while hemicellulose is hydrolyzed within the colon to the extent of about 20% to yield small organic molecules with a large osmotic effect.

When ***bacterial overgrowth*** occurs in the small intestine, there is competition with the normal absorptive mechanisms for substrates. Bile acid losses may be severe owing to deconjugation and reduced solubility; carbohydrates are fermented to organic acids; and, especially when a blind loop is present, tryptophan is degraded by bacteria to indoles with a resultant increased urinary excretion of indoxyl sulfate (normal 20 to 80 mg/day).

Fat Absorption

The average western daily diet contains 100 g to 150 g triglyceride and about 5 g phospholipid, mostly lecithin. A further 20 g to 40 g of lipid enters the upper intestine daily from desquamated cells and secretions, including 10 g to 20 g lecithin in bile. A healthy adult excretes less than 6 g fat daily in the feces, half being from dietary fat and the rest from cell walls of colonic mucosa and bacteria.

Lipid ingestion is followed by *emulsification, digestion, dispersion* of hydrolytic products, mucosal *absorption,* intracellular lipid *resynthesis,* and *chylomicron* formation and release into the lymphatics. Emulsification is aided by bile acids; digestion requires pancreatic enzymes; absorption and subsequent steps require intact mucosal function.

Emulsification of fat starts during its 1 to 4 h residence in the stomach, aided by dietary lecithin, oligopeptides and mechanical mixing. Up to 30% of triglyceride is hydrolyzed within the stomach by *salivary lipases* which have a broad pH optimum and act as true lipases at the lipid-aqueous interface. The resultant long-chain fatty acids are protonated at the low gastric pH and they partition with the lipid core of fat droplets. Short and medium-chain fatty acids of milk fat, being more hydrophilic, are absorbed by gastric and duodenal mucosa, enter portal venous blood unchanged, and pass to the liver bound to albumin.

Entry of acid into the duodenum causes the release of *secretin,* which stimulates pancreatic secretion of bicarbonate. Long-chain fatty acids become partially ionized to form 'acid soaps' and their absorption stimulates release of *cholecystokinin* with consequent gall bladder contraction and pancreatic enzyme secretion. The concentration of bile acids exceeds the critical micellar concentration (2 mmol/L) and bile acids aggregate with lipids, lecithins, monoglycerides, calcium fatty-acid soaps, cholesterol, and vitamins A, D and E to form *mixed micelles.*

Bile acids inhibit adsorption of pancreatic *lipase* and *phospholipase* A_2 to the lipid-aqueous interface, but pancreatic *colipase* reverses this inhibition and allows enzyme activity to occur. Pancreatic nonspecific *esterase,* by contrast, is activated by bile acids and hydrolyzes esters of cholesterol and vitamins A, D, and E at the surface of micelles, as well as short-chain triglycerides and lysolecithin in aqueous solution.

Absorption of the products of lipid digestion occurs in the proximal half of the small intestine. Bile acids remain in the lumen to form more mixed micelles and are absorbed eventually in the ileum. Long-chain fatty acids and 'acid soaps,' dispersed in the aqueous phase as liquid crystals and liposomes, have to diffuse in sequence through the bulk water, the unstirred water layer, mucin gel, and cell glycocalyx before entering the cell where they bind to a specific cytosol long-chain-fatty-acid-binding protein.

Intracellular Events. Microsomal enzymes of the mucosal epithelial cells catalyze resynthesis of triglycerides, which are coated with a layer of apolipoprotein, phospholipid, and free cholesterol. The resultant *chylomicrons* are released into the intestinal lymphatics (lacteals) by reverse pinocytosis (Fig. 12–5). The rare condition of *abetalipoproteinemia* results in defective fat absorption owing to the complete absence of apolipoprotein B. Mucosal cells become packed with lipid droplets that cannot be discharged into the lacteals.

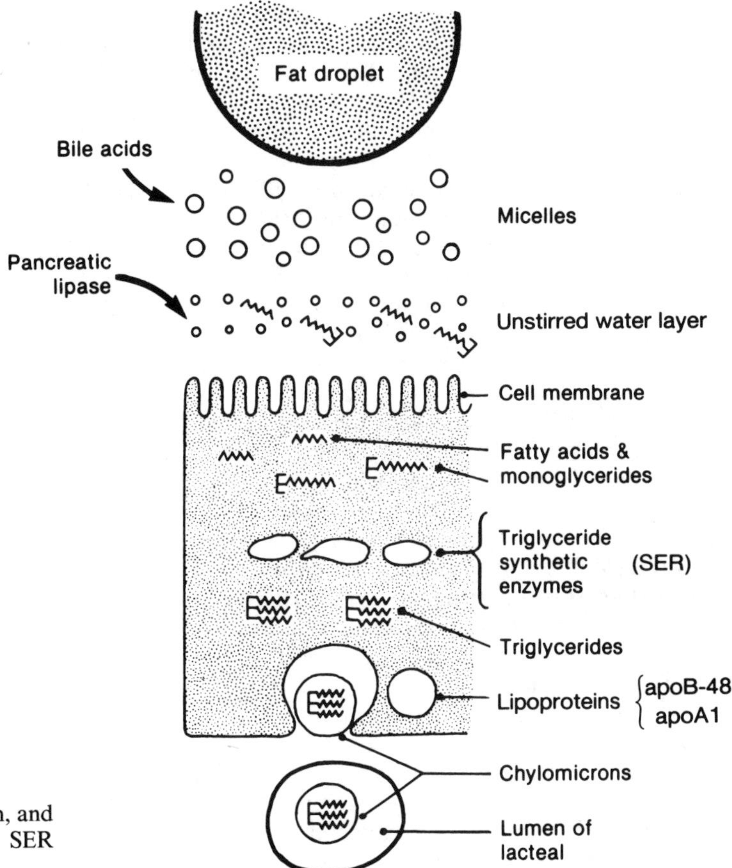

FIGURE 12–5. The digestion, absorption, and transport of fat in the small intestine. SER = smooth endoplasmic reticulum.

Carbohydrate Absorption

Absorbable carbohydrate in the diet consists mainly of starch, sucrose, and lactose. The α-1,4 linkages of starch are hydrolyzed by *amylase* within the upper jejunum to yield maltose, maltotriose, and α-limit dextrins with an average of eight glucose units and an α-1,6 branch linkage. Among carbohydrates *not* digested are cellulose with a β-1,4 linkage; raffinose and stachyose, present in legumes; and lactulose, used in tests of small intestinal transit time (see p. 265).

Lactase, maltase, and *sucrase* are synthesized within the mucosal cells at the apices of the jejunal villi, and become transferred to the brush border, where they exert their hydrolytic activity. *Isomaltase* is associated with sucrase structurally, but its function is to cleave the α-1,6 linkages of limit dextrin.

The active transport of sugars is mediated by specific cell-surface receptors for fructose and for glucose; the glucose receptor also transports galactose and, with lesser avidity, xylose. Glucose absorption is linked directly to sodium absorption.

The newborn child is deficient in pancreatic amylase, but has a full complement of surface hydrolytic enzymes. There is no evidence that these enzymes can be regulated by dietary manipulation.

Protein Absorption

Each day, 150 g to 300 g of endogenous protein from secretions and desquamated cells is digested along with 50 g to 100 g of dietary protein. Fecal nitrogen loss is normally less than 3 g per day.

Pancreatic proteolytic enzymes are secreted as inactive precursors (zymogens) and become active after cleavage by enterokinase in the mucosal cell brush border and by trypsin, itself formed from trypsinogen. Pepsin is secreted as pepsinogen and is activated by a pH less than 3. It is denatured at neutral pH.

Trypsin acts at peptide bonds involving arginine and lysine carboxyl groups; *pepsin* is specific for the amino groups of the basic amino acids; *chymotrypsin* cleaves at aromatic amino acids; and *elastase* at aliphatic amino acids. All these enzymes are endopeptidases, acting at two or more peptides distant from the ends of the polypeptides. *Aminopeptidase* and *carboxypeptidase* are exopeptidases acting respectively on the amino- and carboxyl-terminal peptide bonds.

Acting in concert within the intestinal lumen, these enzymes would eventually break proteins down to their constituent amino acids. This process would need several hundred hours, and only a fraction of this time is available for proteolysis. Instead, the jejunal mucosal cells have the ability to actively absorb not only free amino acids but small peptides of 2–6 amino acids as well. Further proteolysis occurs not at the brush border, but within the mucosal cells.

The active cellular uptake of some dipeptides is more rapid than that of their constituent free amino acids. The ability of patients with *cystinuria* to absorb dipeptides containing lysine, despite an inherited block in lysine uptake, accounts for the absence of amino acid deficiency states in these patients. Similarly, patients with *Hartnup disease* can absorb dipeptides containing phenylalanine, tyrosine, histidine, and tryptophan, but cannot absorb the free amino acids.

Folate, Minerals, and Water

Folate is ingested as a polyglutamate and is hydrolyzed within the lumen to monoglutamate prior to absorption in the jejunum. It is converted within the mucosal cell to 5-methyltetrahydrofolate before release into the portal blood. Reference ranges are 4 to 22 nmol/L (2 to 10 μg/L) in serum and 270 to 1000 nmol/L (120 to 440 μg/L) in red cells.

Iron is absorbed in the duodenum in the ferrous form. Its absorption is regulated by unknown intracellular mechanisms; ferritin is not involved. The release of dietary protein-bound iron is facilitated by gastric acid; gastroferrin, a mucopolysaccharide released by the stomach, may be involved in intraluminal transport.

Calcium absorption is regulated by a calcium-binding protein synthesized in duodenal and jejunal mucosal cells under the control of circulating 1,25-dihydroxycholecalciferol (Fig. 17–3).

Water absorption follows the active resorption of sodium, which is linked in the jejunum with the absorption of amino acids, bicarbonate, and glucose, and in the ileum with the absorption of bile acids. Bicarbonate is rapidly absorbed in the jejunum but is secreted in the ileum, where an exchange occurs with chloride. The ascending colon is the main colonic site of water absorption, with a capability of up to 1500 mL/day.

Overall fluid and electrolyte transfers in the upper intestinal tract are shown in Table 12–2.

When obstruction of the ureters or bladder is treated by transplantation of the ureters into the colon (***ureterosigmoidostomy***), the prolonged contact of urine with the colonic mucosa leads to profound metabolic changes. Sodium and chloride are reabsorbed in exchange for potassium and bicarbonate, which are secreted into the intestinal lumen. The resultant hyperchloremic acidosis with hypokalemia resembles the metabolic disorder that occurs in renal tubular acidosis. Osteomalacia may occur as a late complication. Ureteric transplants are now normally made into an ileal pouch that opens directly to the surface, but a similar biochemical change can occur if the outlet orifice of the pouch becomes obstructed.

Bile Acids (See also Chap. 11)

The primary bile acids, cholic and chenodeoxycholic acids, are synthesized in the liver and secreted into the bile as conjugates with glycine or taurine. Bile acids are not significantly absorbed or metabolized in the upper small intestine. When they encounter the anerobic bacteria in the lower ileum, they may be deconjugated, deesterified, and dehydroxylated at C-7 to form secondary deoxycholic and lithocholic acids. (see Fig. 11–3). Most free and conjugated bile acids are actively reabsorbed in the terminal ileum.

Bile acids have a half-life of 2 to 3 days. The total bile acid pool is 8–12 mmol and it recirculates up to eight times each day. Recycling may be interrupted by ileal pathology or resection, or by bacterial overgrowth in the small intestine with premature enzymatic deconjugation and dehydroxylation. *Neomycin,* given to inhibit intestinal bacteria, forms an insoluble complex with bile acids, as does the antiacid aluminum hydroxide. *Cholestyramine* is used therapeutically to bind luminal bile acids and increase cholesterol degradation by depleting the bile acid pool.

About 2 mmol of bile acid escapes reabsorption and is lost daily in the feces. Changes in this fraction can be quantified by the cholyl-^{14}C-glycine breath test. In *Crohn's disease* postprandial peak serum cholylglycine is consistently reduced. Other measures of bile acid malabsorption, such as reduced bile acid pool size, increased ratio of blood glycine- to taurine-bile acid conjugates, fecal bile acid loss, and postprandial total serum bile acids, all have insufficient sensitivity and specificity to be useful diagnostically.

Vitamin B_{12} (Cyanocobalamin)

The absorption of vitamin B_{12} and other cobalamins involves (1) formation of a salivary R-protein-cobalamin complex and its digestion by trypsin, (2) formation of an intrinsic factor–cobalamin complex (IF–B_{12}), (3) binding of this complex to mucosal cells of the distal ileum, and (4) the subsequent absorption of vitamin B_{12}.

Intrinsic factor is a glycoprotein (M_r 60,000) which is secreted by gastric parietal cells. In atrophic gastritis, especially with antibodies to parietal cells, the secretion of intrinsic factor becomes deficient. The IF–B_{12} complex may not reach the ileum if bacterial overgrowth occurs in the jejunum, or if there is infestation with the fish tapeworm, *Diphyllobothrium latum.* Trypsin deficiency, or ileal pathology as in regional enteritis (Crohn's disease), will also result in cobalamin malabsorption.

Normally, absorption of vitamin B_{12} amounts to about 1 μg/day and body stores are sufficient to last several years. Blood concentrations are therefore a poor indicator of current absorptive function, and the appropriate test for this purpose is the Schilling test, described later in this chapter (see also Chap. 22).

DISORDERS OF THE SMALL AND LARGE INTESTINE

Absorptive Defects

Malabsorption is a clinical term implying a selective or generalized failure to absorb nutrients due to gastric, biliary, pancreatic, or small bowel malfunction. The commonest causes are biliary obstruction, chronic pancreatitis, and malabsorption syndrome. Frequently associated clinical features are weight loss despite adequate food intake and bulky stools. Late changes reflect deficiencies of nutrients, minerals, and vitamins.

Severe, prolonged malabsorption will lead to malnutrition, osteoporosis, and osteomalacia due to protein, calcium, and vitamin D deficiency; edema due to low plasma albumin; hemorrhagic disorders due to vitamin K deficiency; and recurrent infections due to immunoglobulin deficiencies. On rare occasions the patient may present with bone pain due to osteomalacia, or with anemia due to iron, folate, or vitamin B_{12} deficiency.

Malabsorption syndrome (sprue) is a specific group of three jejunal diseases that show extensive mucosal changes (Fig. 12–6) leading to generalized malabsorption. Malabsorption syndrome comprises (1) celiac disease in children, (2) celiac sprue in adults (also called idiopathic steatorrhea or nontropical sprue), both of which respond to a gluten-free diet, but which may progress to refractory sprue, which does not respond, and (3) tropical sprue, which is unaffected by gluten but responds to broad-spectrum antibiotics.

Steatorrhea is a state of impaired fat absorption. Even when a primary defect of fat absorption occurs, as in obstructive jaundice or abetalipoproteinemia, the change in small intestinal contents and bacterial overgrowth leads to more generalized malabsorption.

The terms malabsorption, malabsorption syndrome, and steatorrhea are frequently used interchangeably, with consequent confusion.

Mechanisms of Malabsorption

Intraluminal. Inadequate lipolysis results from bile acid deficiency (due to biliary tract obstruction, ileal disease, or bacterial overgrowth), or from lipase deficiency (due to pancreatic disease, acid hypersecretion with destruction of lipase at a low pH, or postgastrectomy sequelae with inadequate stimulation of the pancreas). Pancreatic colipase deficiency may be the limiting factor, especially in pancreatic disease in children. *Pancreatic insufficiency* is usually due to chronic relapsing pancreatitis, pancreatic duct obstruction, or pancreatic carcinoma.

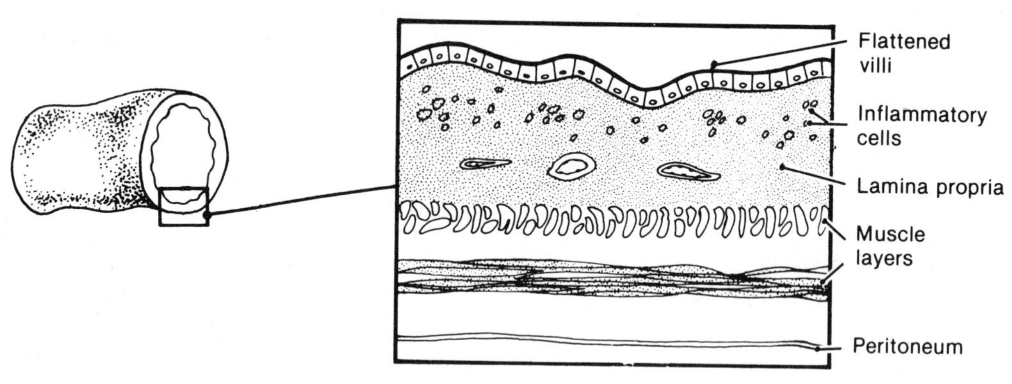

FIGURE 12–6. Mucosal changes as seen in celiac disease.

Defective motility of the small intestine, as in scleroderma, may lead to an intraluminal increase in bacteria sufficient to deconjugate bile acids and reduce their concentration below the critical micellar level.

Mucosal. Causes may be generalized, due to a loss of villous architecture (malabsorption syndrome, lymphoma, radiation, parasitic infestation, mast-cell disease), or selective (disaccharidase deficiency, pernicious anemia, abetalipoproteinemia, Hartnup disease, cystinuria).

Any disease causing ***structural changes*** locally in the intestine may result in malabsorption. Examples are regional enteritis (Crohn's disease), Whipple's disease, parasitic infestations (*Giardia lamblia, Strongyloides stercoralis*), lymphoma, enteroenteric fistula, intestinal strictures, systemic lupus erythematosus, and amyloidosis. Similar absorptive defects usually occur after total gastrectomy and sometimes after partial gastrectomy or even vagotomy, and after small intestinal resection, especially when a cul-de-sac or blind loop is formed.

Celiac sprue results from sensitivity to gluten. It appears to be caused by a change in cell mediated immunity to a peptide fragment of gluten, rather than by circulating antibodies or by abnormal digestion of gluten. There is a genetic linkage with HLA–B8.

Lymphatic. Whipple's disease, lymphoma, and intestinal lymphangiectasia lead to impaired lymphatic drainage.

Selective Carbohydrate Malabsorption

Pancreatic amylase is secreted in great excess and *starch* is digested normally even when there is severe fat malabsorption of pancreatic cause. Brush border ***disaccharidase deficiencies*** can be secondary to any disorder of the normal villous architecture. A primary deficiency of disaccharidases or of monosaccharide transport receptors is rare. *Lactase* is the slowest acting of the disaccharidases and its deficiency is often the most apparent. The accumulation of lactose within the lumen has a water-retaining osmotic effect, which is compounded by bacterial degradation of lactose to organic acids. Abdominal distention and watery diarrhea result, sometimes associated with vitamin and other nutritional deficiencies. Symptoms are related to ingestion of the target sugar; there is normal fat absorption and the microscopic appearance of the jejunum is normal. Diagnosis is based on carbohydrate tolerance tests, or breath hydrogen tests, followed by the demonstration of absent hydrolytic activity in a jejunal biopsy, or the failure to accumulate intracellular glucose after *in vitro* incubation with lactose.

Diarrhea

Feces normally have a volume of 100 to 200 mL/day and a mass of 200 g to 250 g. *Excessive volume* occurs with laxative abuse, steatorrhea, Zollinger–Ellison syndrome, Verner–Morrison syndrome, post gastrectomy syndrome, and bile acid malabsorption. Pancreatic disease generally leads to moderate diarrhea, with a volume up to 500 mL/day; idiopathic steatorrhea, 500 to 1000 mL/day; and intestinal hormonal disorders, 1 to 10 L/day. Diarrhea fluid has a potassium concentration of about 30 mmol/L and bicarbonate of 35 mmol/L, so that potassium depletion and metabolic acidosis may be severe, especially in infants.

Diarrhea due to ***bile acid malabsorption*** is often precipitated by meals, especially breakfast. After an overnight fast, bile acids in the gallbladder will approximate the body pool size. With reduced recirculation, the bile acids may be used only once or twice, instead of the normal eight times, and the concentration of bile acid entering the colon will be high. After bacterial deconjugation of chenodeoxycholate and dehydroxylation of cholic to deoxycholic acid, the resulting free dihydroxy bile acids are potent inhibitors of colonic water reabsorption. The liquid stool bulk is augmented by associated steatorrhea.

In extensive small bowel resection, steatorrhea predominates; in localized ileal resection, especially when the ileocecal valve has been removed, diarrhea due to excess bile acids is the major complaint. Both postvagotomy diarrhea and diarrhea associated with cystic fibrosis may be caused, at least partly, by the bile acid malabsorption that is associated with these conditions. Bacterial overgrowth in the small bowel causes watery diarrhea through premature bile acid deconjugation, dehydroxylation, and malabsorption. A *therapeutic test* for bile acid–induced diarrhea is the administration of 4 g of cholestyramine with each meal, which should result in a rapid reduction of stool volume and mass.

The diarrhea of *Zollinger–Ellison syndrome* is due to ***persistent acidity*** of the duodenal contents with a resultant increase in small intestinal motility, inactivation of pancreatic lipase, precipitation of bile acids, and steatorrhea.

Watery diarrhea with hypokalemia and achlorhydria is known as the ***WDHA syndrome,*** *Verner–Morrison syndrome,* or *pancreatic cholera.* Patients exhibit profuse diarrhea of tealike consistency with an output of up to 10 L/day. Potassium losses of up to 300 mmol/day may result. Skin flushing and tetany occur. Achlorhydria is found in about half of these patients. In some, tumors of the pancreas secreting vasoactive intestinal peptide (VIP) have been demonstrated, but it is by no means certain that VIP is the only causative agent. Many cases with gastric acid have gastrin-secreting tumors. Diarrhea associated with flushing, wheezing, and other symptoms occurs in *carcinoid syndrome.*

The ***laxative abuse*** syndrome occurs mainly in women and presents as a chronically disturbed bowel habit, with abdominal pain, weakness, and often depression. Constipation in childhood is frequently reported and patients often have associations with the medical profession. Hypokalemia dominates the metabolic picture and is commonly associated with hypocalcemia and tetany. Once the syndrome is diagnosed, the metabolic disorders resolve if the patient can be persuaded to abandon laxative usage. This syndrome must be carefully excluded if one is considering a diagnosis of WDHA syndrome.

INVESTIGATION OF INTESTINAL DISORDERS

Fecal Fat

The method of choice for fat analysis is the wet method of Van de Kamer. The patient is given a diet containing 100 g fat/day for 5 days. The fecal fat output, averaged over the 5 days, should be less than 20 mmol as fatty acid (5 g fat) daily. This is a highly specific indicator of fat malabsorption; there are no false-positive results if a fecal marker is used to ensure complete collection. It has poor sensitivity as an indicator of chronic pancreatic disease since 90% of the secretory capacity for lipase must be lost before the test becomes abnormal.

The following *comments* are relevant;

1. If the diet contains medium-chain triglyceride, the analysis loses validity, since these more polar triglycerides are not extracted and therefore not measured.
2. Fecal fat excretion is proportional to the dietary intake: a diet containing 200 g fat per day will result in a loss of up to 40 mmol (10 g).
3. A 50% reduction of an increased fat output after the administration of an oral pancreatic enzyme supplement indicates a pancreatic cause for the steatorrhea.
4. In patients with pancreatic disease, stainable fat droplets may be visible on microscopy. Sensitivity is no better than 85%.
5. Postprandial plasma turbidity and triglyceride concentration have sensitivities of 67% and 82% and specificities of 67% and 90% respectively and are unacceptable as screening tests.

Serum Carotene

Carotene is plentiful in vegetables, especially carrots, oranges, and tomatoes. A serum carotene concentration greater than 700 μg/L excludes steatorrhea with high sensitivity. A decreased value may occur without steatorrhea; specificity is therefore low. The serum level of vitamin A is dependent on hepatic synthesis of retinol-binding protein. This protein (M_r 21,000) has a short half-life (12 hours) which is a function of renal clearance. Serum levels of vitamin A do not correlate with hepatic stores and are not helpful in the diagnosis of malabsorption.

Vitamin D

Cholecalciferol (see also Chap. 22) is a hydrophobic lipid similar to cholesterol. Cholestasis and depletion of the bile acid pool impair its absorption much more than that of long-chain triglyceride. Dietary intake varies because of widespread food fortification, self-medication, and therapeutic supplements. Serum cholecalciferol or 25-hydroxycholecalciferol concentrations correlate poorly with lipid absorption and are not helpful in the diagnosis of malabsorption.

Xylose Absorption

Xylose is absorbed through the jejunal mucosa by facilitated diffusion. An overnight fast eliminates delayed gastric emptying and competition for absorption sites. D-Xylose is given, 25 g in 200 mL water (0.5 g/kg for children), with a further 200 mL water at 1 hour and at 2 hours. Urine is collected for 5 hours and should contain at least 2.5 g xylose.

This test presents certain *problems:*

1. Impairment of excretion occurs in patients with ascites and renal failure. When renal function or the adequacy of urine collection is in question, a plasma xylose concentration more than 2 mmol/L (300 mg/L) at 90 or 120 min indicates normal absorption. Plasma concentration is not as good an indicator of malabsorption as urine output.
2. The 25 g xylose test may cause diarrhea. The use of 5 g of xylose produces less adverse effects, and a plasma xylose concentration at 1 h of more than 1.3 mmol/L (200 mg/L) indicates normal absorption. Sensitivity is 70%; specificity 95%.

Breath Tests

If an oral bolus of ***^{14}C-lactose*** is absorbed normally, it is ultimately metabolized *in vivo* to $^{14}CO_2$ and water. If it is incompletely absorbed, bacterial fermentation in the lower intestine produces hydrogen and $^{14}CO_2$, both of which are absorbed by diffusion through the intestinal wall. Because the body pool of CO_2 turns over rapidly, the $^{14}CO_2$ appears in the expired breath, where it can be measured. Exhalation of $^{14}CO_2$ occurs whether or not absorption is complete; at best there may be differences in amount or a time delay between ingestion and exhalation. Metabolic factors may further confuse the interpretation of breath $^{14}CO_2$ measurement. $^{14}CO_2$ excretion is *low* in obesity, chronic respiratory disorders and parenchymal liver disease, and *high* in thyrotoxicosis and diabetes.

^{14}C-triolein ingestion leads to a 4 h cumulative excretion of $^{14}CO_2$ which correlates well with steatorrhea as defined by fecal fat output. Sensitivity is 100%; specificity 86%.

Cholyl-^{14}C-glycine (glycocholate) is absorbed from the ileum and resecreted intact in the bile (Fig. 12–7). Any that is not absorbed enters the colon, where the ^{14}C-glycine is cleaved from the molecule and absorbed either intact or after further bacterial decarboxylation to $^{14}CO_2$. Bile acid deficiency in the jejunum is caused either by abnormal bacterial deconjugation in the jejunum, with rapid passive absorption, or by defective active reabsorption of conjugated bile acids in the ileum and excessive loss of bile acids into the colon where bacterial deconjugation occurs. In either case, the bacterial breakdown of glycine results in the formation of $^{14}CO_2$. Exhaled $^{14}CO_2$ is a reliable marker of the extent of bile acid deconjugation, with a sensitivity in blind loop syndrome of close to 100%. The fecal excretion of ^{14}C during the subsequent 24-hour period is abnormally increased with ileal malabsorption but not with blind loop syndrome.

The measurement of ***exhaled hydrogen*** has a sound theoretic basis and is finding increasing use as a reliable diagnostic test. Molecular hydrogen is not produced in human metabolism and is a specific marker of unabsorbed substrate. Sampling is noninvasive and simple. Analysis requires only a low-cost gas chromatograph with a 'molecular sieve' column and a thermal conductivity detector. Substrates that have been used are disaccharides, for detecting a specific absorptive defect; xylose, for testing jejunal absorptive function; glucose, for identifying bacterial overgrowth; and lactulose, for

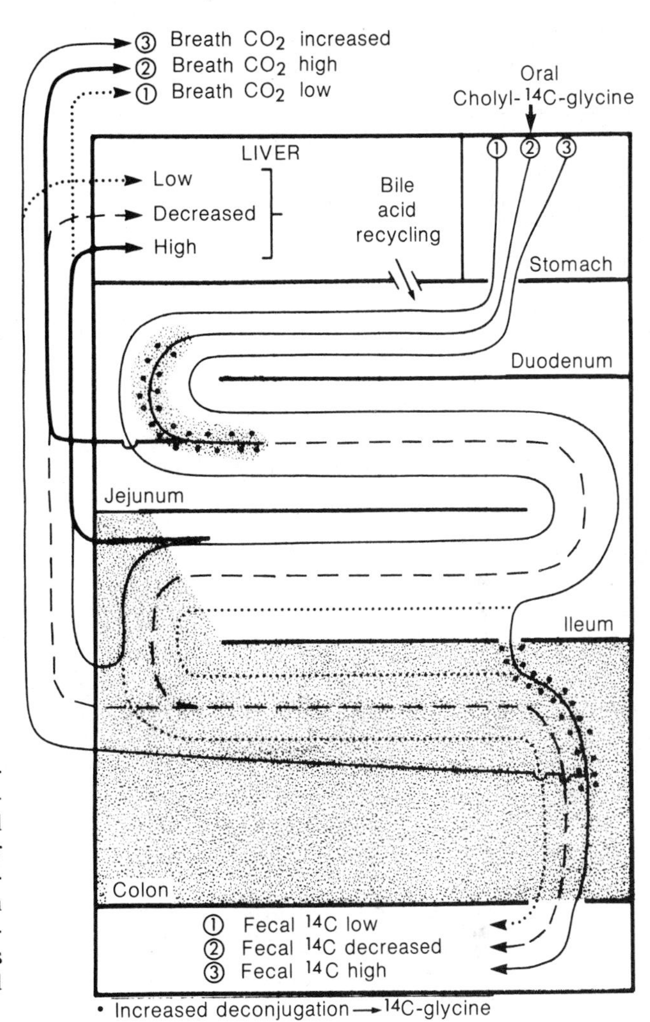

FIGURE 12–7. The cholyl-^{14}C-glycine test in malabsorption due to bile acid disorders. ① Normal—most cholyl-^{14}C-glycine recycled; breath $^{14}CO_2$ and fecal ^{14}C are low. ② Upper intestine blind loop or stasis. Bacterial *deconjugation* leads to increased ^{14}C-glycine absorption in the jejunum, increased breath $^{14}CO_2$ and decreased fecal ^{14}C. ③ Ileal malabsorption of bile acids (e.g., due to resection) increases *colonic* deconjugation, ^{14}C-glycine reabsorption and breath $^{14}CO_2$, but fecal ^{14}C is also increased.

measuring small intestinal transit time. Bacterial generation of hydrogen may be suppressed by antibiotic therapy, colonic preparation routines, or delayed gastric emptying. Reduction of colonic pH by fermentation of nonabsorbed carbohydrate also inhibits hydrogen production and the potentially malabsorbed carbohydrate should be removed from the diet for 48 h preceding the breath test.

Lactose Tolerance Test

Lactose (50 g) is given orally to a fasting subject and blood samples are taken for glucose estimation at 30, 60, 90, and 120 min. There is normally an increase of at least 1 mmol/L (20 mg/dL) of glucose above fasting levels. If a lesser response occurs, the test is followed by administering orally 25 g glucose and 25 g galactose to the fasting subject, with blood glucose measurements as before. Specific *lactase deficiency* has no effect on glucose or galactose absorption, while generalized intestinal disease causes defective absorption of both lactose and the mixed glucose and galactose load.

This test is sensitive for lactase deficiency, but flat curves (false-positive results) may occur in other malabsorptive states, resulting in a poor specificity. Table 12–3 compares the diagnostic efficiency of a variety of measures after lactose administration. The measurement of lactase activity in a jejunal biopsy is a definitive diagnostic procedure, providing the villous architecture is nor-

TABLE 12–3. Efficiency of Various Measures of Lactase Deficiency after 50 g Lactose Orally

Measure	Sensitivity (%)	Specificity (%)
Peak blood glucose	76	96
Peak blood galactose	96	96
$^{14}CO_2$ in breath	92	96
H_2 in breath	100	100

mal. Because lactase occurs in the brush border of cells only in the apical third of the villi, any villous atrophy must lead to a nonspecific loss of enzyme activity.

Jejunal Biopsy

Peroral biopsy of the small intestine mucosa may reveal evidence of partial or complete villous atrophy and of specific cellular defects. It is indicated in prolonged unexplained diarrhea, folate deficiency, unexplained osteomalacia, and with nonspecific abnormality in barium studies. In celiac sprue there is flattening of the mucosa, atrophy of villi, elongation of crypts, and inflammatory cell infiltration in the submucosa. Specific cellular abnormalities are seen in Whipple's disease, abetalipoproteinemia, amyloidosis, and mast cell disease. Biochemical tests on jejunal biopsies permit the quantification of disaccharidases and constitute a definitive approach to the diagnosis of disaccharidase deficiencies.

Tests for Protein Loss

Decreased serum protein levels may be due to defects in protein nutrition, impaired hepatic synthesis, or renal glomerular filtration. Serum proteins are also low in *protein-losing enteropathy.* This state can be demonstrated by intravenous infusion of 10 to 15 μCi ^{51}Cr-albumin and measurement of the excreted isotope in a 5-day collection of feces. In protein-losing enteropathy, losses of up to 60% occur (normal <1%). Clearance of endogenous α_1-*antitrypsin* is a nonisotopic alternative which reflects protein loss distal to the pylorus, because α_1-antitrypsin is destroyed at pH <3. Compared with ^{51}Cr excretion, it has a sensitivity of 93% and a specificity of 90%.

Other Tests

Oxalic acid is poorly absorbed in the normal individual but its absorption, probably from the colon, is increased in patients with steatorrhea. This may explain the increased incidence of nephrolithiasis in patients with bowel disorders. When patients are given 100 g spinach daily for 3 days, the average daily urinary oxalate on the second and third days is greater than 40 mg per day in all patients with proven steatorrhea. The test has a sensitivity of 100%, but false-positive results do occur and the specificity may be no better than 40%.

Tests of *intestinal permeability* may have value in screening for celiac sprue. A 5 h urinary excretion of cellobiose and mannitol in a mass ratio of less than 0.1, after ingestion of 5 g cellobiose and 2 g mannitol, has a sensitivity for celiac sprue of 96% and a specificity of 92%.

Reduced red cell folate and the presence of reticulin antibodies have poor sensitivity for celiac sprue (50%) but good specificity (95%).

Investigative Strategy for Malabsorption

An *algorithm* for the *diagnosis of malabsorption* is shown in Fig. 12–8.

Tests used to *monitor treatment* are:

1. Vitamin B_{12} absorption after intrinsic factor, pancreatic enzyme supplements, or a gluten-free diet
2. Xylose absorption after a gluten-free diet
3. Cholyl-^{14}C-glycine breath test after antibiotic therapy
4. Repeated jejunal biopsies
5. Serial red cell folate concentration

Occult Blood

The identification of hemoglobin degradation products in the feces in quantities less than those producing typical tarry stools requires special testing. The chemical tests depend on the peroxidase activity of iron-containing hemoglobin derivatives, which catalyzes the coupled reduction of hydrogen peroxide and oxidation of an organic substrate such as *o*-toluidine or guiac. Hemoccult slides using paper impregnated with guiac have suitable sensitivity. A *positive* result requires confirmation after the patient is placed on a meat-free and green-vegetable-free diet for 3 days. Bleeding from the gums or from hemorrhoids should be excluded by

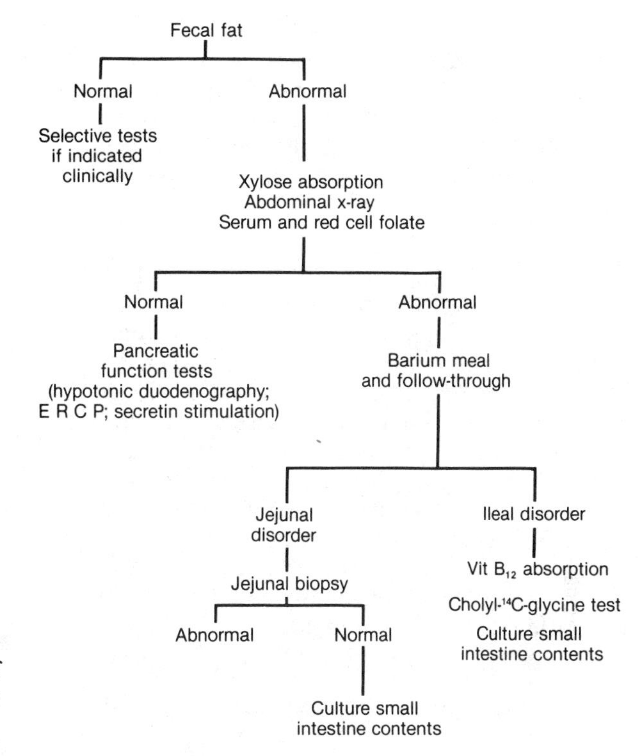

FIGURE 12–8. Algorithm for the investigation of malabsorption. ERCP = Endoscopic retrograde cholangiopancreatography.

clinical examination before attributing a positive result to intestinal blood loss. Hemoccult testing has a sensitivity of 78% and specificity of 99%. Some subjects may excrete a red pigment in the feces after ingesting beets, but this has no peroxidase activity.

The most sensitive, specific, and quantitative test of intestinal blood loss is the measurement of fecal radioactivity after the intravenous injection of the patient's own red cells tagged *in vitro* with ^{51}Cr.

Obsolete Tests of Intestinal Disorders

1. Fat content of dried stools: misleading
2. Fecal nitrogen: technically cumbersome; useful only in metabolic balance studies
3. Starch tolerance: the blood sugar response is difficult to interpret in the presence of possible pancreatic islet cell malfunction
4. Serum amylase, lipase: of no value in chronic pancreatic disease
5. ^{131}I-triolein and ^{131}I-fatty acid absorption: correlation with fecal fat has been poor
6. Fecal smell, appearance, and density: subjective; density reflects the content of gas more than of fat
7. Split and unsplit fat in feces: the absence of pancreatic lipase is masked by bacterial lipases, which will split unabsorbed triglyceride
8. Fecal trypsin: bacterial proteases mask the absence of pancreatic trypsin; fecal chymotrypsin may avoid this problem but is only semiquantitative
9. Amylase/creatinine clearance ratio: of no value in acute pancreatitis
10. ^{131}I-polyvinylpyrrolidone infusion: labeled preparation is unstable.

OTHER DISORDERS ASSOCIATED WITH THE ALIMENTARY TRACT

PERNICIOUS ANEMIA

Because this disorder requires lifelong therapy, diagnosis should be firmly established. The clinical features are accompanied by megaloblastic erythropoiesis, low serum cobalamin and normal serum folate levels, defective absorption of vitamin B_{12} but normal absorption of intrinsic factor–vitamin B_{12} complex, and the presence of anti-intrinsic factor antibodies in the serum. There is an absence of gastric acid secretion in adults but in children the loss of intrinsic factor secretion may precede the loss of acid secretion.

Schilling Test. ^{58}Co-Cyanocobalamin 1 μg (0.5–1.5 μCi) is given orally in 50 mL water, followed by 1000 μg of nonradioactive cyanocobalamin injected intramuscularly. A urinary excretion of more than 8% of the radioactive dose in the ensuing 24 hours indicates the absence of pernicious anemia and normal absorption. If the result is lower than normal the test can be repeated with intrinsic factor added to the oral B_{12}. The procedure then becomes a test of ileal absorption. In an improved version of the test ^{58}Co–B_{12} and ^{57}Co–B_{12}–IF are given simultaneously and the urine radionuclides are counted differentially.

Certain *problems* are associated with this test:

1. Excretion is reduced in patients who are vomiting or who have had a partial gastrectomy. Dehydration, renal disease, or incomplete urine collection will also give erroneous results.
2. In disease of the terminal ileum, the defective absorption may be inconstant. Correction of malabsorption by subsequent addition of intrinsic factor may merely reflect a transient improvement in ileal function, and not the absence of endogenous intrinsic factor.
3. Some patients with pernicious anemia excrete blocking antibodies into the gastric juice. Unless the intrinsic factor and labeled vitamin B_{12} are mixed in water before administration, the blocking antibody can prevent the normal action of intrinsic factor.
4. After prolonged vitamin B_{12} deficiency, the terminal ileum may become refractory to absorption even when intrinsic factor is added. The findings suggest terminal ileal disease, but absorption with intrinsic factor returns to normal after vitamin B_{12} therapy.
5. Other artifacts to be excluded are bacterial utilization of the vitamin and pancreatic insufficiency. These conditions will show a consistently normal absorption if the Schilling test is repeated following treatment with antibiotics or with pancreatic enzymes, respectively.

CARCINOID TUMORS

Tumor Metabolites

Tumors of ***argentaffin cells*** within the intestinal crypts, and of similar cells in the bronchus, ovary, biliary tract, or pancreas, occur with an incidence of about 1/year per 100,000 population. The tumor develops most frequently in the ileum with metastases to liver, mesentery, and lymph nodes. The cells are metabolically very active and may produce tryptophan metabolites, especially serotonin (5-hydroxytryptamine), but also kallikrein, prostaglandins, and histamine. Clinical manifestations include episodic flushes, cyanosis, chronic diarrhea, respiratory distress, and valvular disease of the right heart.

Tryptophan 5-hydroxylase and aromatic L-amino acid decarboxylase generate *serotonin,* which is stored in the cellular argentaffin granules. On release into the blood the serotonin is bound to platelets and is set free in the serum during clotting. Normally 1% to 2% of tryptophan is metabolized to serotonin; in *carcinoid tumor,* more than 60% may follow this pathway. Serotonin is metabolized (Fig. 12–9) in the liver by monoamine oxidase to 5-hydroxyindoleacetic acid (5HIAA), and this appears in the urine, usually in diagnostically significant amounts. Urine concentrations of tumor metabolites are listed in Table 12–4. Bananas, tomatoes, avocados, and plantain may result in misleading increases of urinary 5HIAA.

Bronchial *carcinoids* are often associated with severe symptoms of pyrexia, sweating, facial edema, vomiting, and explosive diarrhea. In other respects they are atypical: 5-hydroxytryptophan may predominate, the heart lesions are usually left-sided, metastases occur in bone, and other endocrine disorders such as Cushing's syndrome may coexist. Similar syndromes have been associated with primary tumors of the pancreas and thyroid, and with oat cell carcinoma of the lung.

Kinins and Carcinoid Flushing

Bradykinin is generated in the circulation during the flushes associated with carcinoid tumors. Kinins have a common nonapeptide sequence with additional N- or C-terminal residues. They arise from inactive precursors, kininogens, which are present in the plasma. An enzyme, *kallikrein,* is released from the tumor and cleaves bradykinin from its circulating precursor. The bradykinin increases capillary permeability, lowers blood pressure, and causes smooth muscle contraction. Patients with carcinoid tumors may have plasma bradykinin concentrations of 150 to 1200 μg/L (upper reference value, 100 μg/L). Radioimmunoassay of bradykinin is hampered by the absence of a tyrosyl residue for iodination. Flushing has been associated with the intake of food and, especially, of alcohol, and sometimes may be relieved by antihistamines.

TABLE 12-4. Urine Metabolites in Carcinoid Tumor

Solute	Tumor Range (mg/24 h)	Reference Range (mg/24 h)
Tryptophan	60–180	10–110
5-Hydroxytryptophan (5HTP)	Usually absent*	Absent
Serotonin	0.2–5.0	<0.2
5-Hydroxyindole acetic acid (5-HIAA)	15–1200	2–10
Histamine	0.02–7	<0.02

* Has been reported rarely as high as 50 mg/24 h.

CYSTIC FIBROSIS

Cystic fibrosis (fibrocystic disease; mucoviscidosis) is a generalized abnormality of secretions from liver, gallbladder, duodenum, pancreas, lungs, salivary glands, and genital tract. The sweat glands characteristically produce sweat with an increased chloride content (more than 50 mmol/L in children; more than 80 mmol/L in adults over 18 years). The defect is inherited as an autosomal recessive trait with a carrier prevalence of 1:20 and a disease prevalence of 1:2000 births. Respiratory problems occur in virtually all patients, leading to generalized obstructive emphysema. Gastrointestinal symptoms (especially steatorrhea) are found in 85% of patients; focal biliary cirrhosis is also common. Reduced concentrations of plasma *immunoreactive trypsin* occur with pancreatic dysfunction, but in neonates with cystic fibrosis abnormally *increased* concentrations may be present. The sensitivity and specificity of this test are not yet clearly established. Plasma pancreatic *isoamylase* may also be reduced, but the salivary component is often increased.

FIGURE 12-9. The synthesis and metabolism of serotonin.

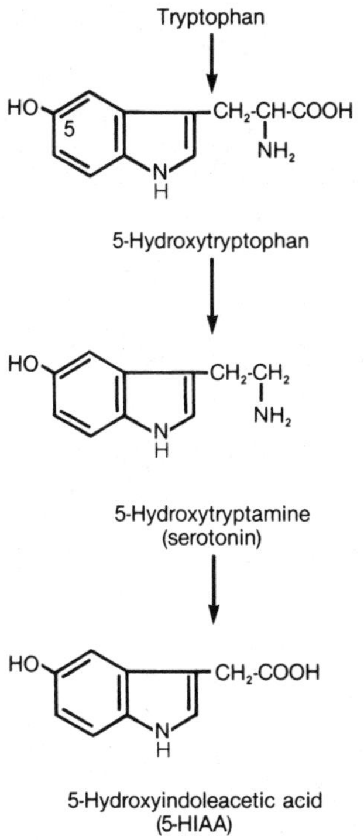

Tryptophan

HO–(5)–indole–CH_2-CH(NH_2)-COOH

5-Hydroxytryptophan

HO–indole–CH_2-CH_2-NH_2

5-Hydroxytryptamine (serotonin)

HO–indole–CH_2-COOH

5-Hydroxyindoleacetic acid (5-HIAA)

SUGGESTED READING

INVESTIGATION OF MALABSORPTION

BENINI L ET AL: Is the C14–triolein breath test useful in the assessment of malabsorption in clinical practice? Digestion 29:91–97, 1984

BO-LINN GW, FORDTRAN JS: Fecal fat concentration in patients with steatorrhea. Gastroenterology 87:319–322, 1984

HILL RE, HERCZ A, COREY ML, GILDAY DL, HAMILTON JR: Fecal clearance of alpha-1-antitrypsin: A reliable measure of enteric protein loss in children. J Pediatr 99: 416–418, 1981

MANOLIS A: The diagnostic potential of breath analysis. Clin Chem 29:5–15, 1983

ROSADO JL, SOLOMONS NW: Sensitivity and specificity

of the hydrogen breath analysis test for detecting malabsorption of physiological doses of lactose. Clin Chem 29:545–548, 1983

THEODOSSI A, GAZZARD BG: Have chemical tests a role in diagnosing malabsorption? Ann Clin Biochem 21: 153–165, 1984

PANCREAS

BARRY RE: Testing pancreatic exocrine function. J Royal Soc Med 78:342–343, 1985

KOOP H: Serum levels of pancreatic enzymes and their clinical significance. Clin Gastroenterol 13(2):739–761, 1984

LANKISCH PG, LEMBCKE B: Indirect pancreatic function tests: Chemical and radioisotope methods. Clin Gastroenterol 13(2):717–737, 1984

STEINBERG WM, GOLDSTEIN SS, DAVIS N ET AL: Diagnostic assays in acute pancreatitis. Ann Intern Med 102: 576–580, 1985

WILLIAMSON RNC: Early assessment of severity in acute pancreatitis. Gut 25:1331–1339, 1984

SMALL INTESTINE AND COLON

CAREY MC, SMALL DM, BLISS CM: Lipid digestion and absorption. Annu Rev Physiol 45:651–677, 1983

FARAH DA, CALDER I, BENSON L, MACKENZIE JK: Specific food intolerance: Its place as a cause of gastrointestinal symptoms. Gut 26:164–168, 1985

KAPADIA CR, DONALDSON RM JR: Disorders of cobalamin (vitamin B12) absorption and transport. Annu Rev Med 36:93–110, 1985

NEWCOMER AD, MCGILL DB: Clinical importance of lactase deficiency. N Engl J Med 310:42–43, 1984

PETERS TJ, BJARANSON I: Coeliac syndrome: Biochemical mechanisms and the missing peptidase hypothesis revisited. Gut 25:913–918, 1984

THOMSON ABR, PHILLIPS SF, BOWES KL ET AL: Pathophysiological basis of chronic diarrhea. Ann Royal Coll Phys Surg Can 16:670–680, 1983

STOMACH

GUTH PH: Pathogenesis of gastric mucosal injury. Annu Rev Med 33:183–196, 1982

HANSKY J: Gastrins and gastrinomas. Postgrad Med J 60:767–772, 1984

JENSEN RT: Zollinger–Ellison syndrome: Current concepts and management. Ann Intern Med 98:59–75, 1983

TAYLOR IL: Gastrointestinal hormones in the pathogenesis of peptic ulcer disease. Clin Gastroenterol 13(2): 355–382, 1984

13

Arthritic and Rheumatic Disorders

Joseph B. Houpt

DISEASE AND THE MUSCULOSKELETAL SYSTEM

There are over 100 different diseases that may affect the musculoskeletal system (joints, bones, and muscles) as well as internal organs and skin. Approximately 5 million persons in the United States suffer from *rheumatoid arthritis* (inflammatory disease of the joints) and an additional 12 million have some form of *osteoarthritis* (a degenerative disease of the joints). The American Rheumatism Association's *Classification of the Rheumatic Diseases and Rheumatism* is an approach to the understanding of the many diseases which affect the musculoskeletal system. Most of these diseases fall into one of four general *types:* (1) diseases usually affecting only the joints, for example, osteoarthritis; (2) systemic diseases affecting primarily the musculoskeletal system and also other internal organs, for example, rheumatoid arthritis; (3) systemic diseases manifested by pain in the muscles or joints without any structural change in the musculoskeletal system, for example, polymyalgia rheumatica; and (4) diffuse connective tissue disease involving primarily the internal organs of the body as well as the skin and musculoskeletal system, for example, systemic lupus erythematosus.

This chapter will deal with the rheumatic diseases and their laboratory diagnosis, recognizing that most of these disorders are characterized not by specific laboratory tests but by a constellation of symptoms and signs and by specific x-ray abnormalities of bones and joints. It does not include the muscular dystrophies (see Chap. 3).

The differential diagnosis of joint disease is very broad; however, a precise diagnosis and, therefore, a precise management plan can often be made.

PATTERNS OF PRESENTATION

1. Acute mono- or pauciarthritis (affecting one or a few joints)
2. Acute polyarthritis
3. Chronic mono- or pauciarthritis
4. Chronic polyarthritis
5. Non-articular 'rheumatism' (symptoms arising from periarticular tissues or muscles)
6. Arthritis associated with skin disease
7. Arthritis associated with gastrointestinal disease
8. Arthritis associated with pleuropulmonary disease
9. Arthritis associated with infection
10. Arthritis associated with hematologic and neoplastic disorders
11. Systemic disease associated with vasculitis
12. Arthritis associated with endocrine, biochemical or metabolic disorders
13. Hereditary diseases of connective tissue

SOME COMMON ARTHRITIC DISORDERS

RHEUMATOID ARTHRITIS

Rheumatoid arthritis occurs in approximately 3% of the adult population. It is primarily a *synovial* disease (i.e., affecting the joint lining) with secondary pathologic changes occurring in the synovial fluid, cartilage, periarticular tissues, and bone. The synovium becomes inflamed, causing pain and swelling in one or several peripheral joints. Fibroblasts, blood vessels, and chronic inflammatory cells proliferate, and the resulting granulation-tissue pannus extends over the surface of the articular cartilage, eroding and destroying it. The destructive changes may extend to articular and periarticular bone and to periarticular soft tissues, leading to joint deformities.

Clinically, the patient may have overwhelming fatigue, morning stiffness lasting many hours, fever, rash, and other signs of acute and/or chronic disease. Subcutaneous nodules may develop over any pressure area, but are found most often on the extensor surfaces of the forearm, just distal to the elbow. Microscopically they show an area of central necrosis, a surrounding zone of palisading fibroblasts with multinucleated giant cells, and an outer zone of chronic inflammatory cells. Many extraarticular manifestations occur, particularly in the eyes, spleen, lungs, heart, skin, and nerves. *Serologically,* the disease is characterized by the presence of rheumatoid factor (RF) in the blood and synovial fluid. However, this finding is not specific for rheumatoid arthritis; it is found in other collagen diseases and in a certain number of healthy aged individuals. ***RF*** is found also in persons with chronic stimulation of the immune system. This includes some patients with sarcoidosis; syphilis; viral infections; hepatitis and cirrhosis; pulmonary silicosis, tuberculosis, or interstitial fibrosis; leprosy; subacute bacterial endocarditis; macroglobulinemia; and trypanosomiasis.

Although many of the immunopathogenetic mechanisms underlying rheumatoid arthritis have been clarified by scrupulous laboratory 'dissection' over the past two decades, the ***etiology*** of this ubiquitous disease still remains a mystery. It is probable that the disease occurs in genetically susceptible individuals after exposure to an infectious—probably viral—agent. The response to the patient's own antigenic material (autoimmunity), as demonstrated by the finding of antibodies to IgG(RF), suggests an impairment of immune recognition, or 'tolerance.' The ability of synovial cells to produce antibody, the presence of immunoglobulin and complement in synovial lining cells, and the presence of large numbers of T lymphocytes in synovial tissue are all markers of *immune mechanisms* responsible for the perpetuation of joint inflammation. Biologically active products liberated following the interaction of antigen with antibody, complement components, and polymorphonuclear cells cause the release of lysosomal enzymes locally in the joint, which mediate many of the effects of inflammation and tissue damage.

SERONEGATIVE DISEASES

A group of diseases with some clinical resemblances, and having in common both the absence of rheumatoid factor and a high incidence of certain human leukocyte antigen (HLA) subtypes (see in Chap. 4) are called the 'seronegative' diseases. (This classification is open to modification in the light of new knowledge.) The model disorder is ***ankylosing spondylitis*** (Marie–Strümpell disease) and its variants, which affect primarily the sacroiliac joints and the small apophyseal joints of the spine as well as the peripheral joints. The spondylitic variants include the arthritis associated with *psoriasis* and the *enteric arthritides* (associated with Crohn's disease, ulcerative colitis, and Whipple's disease).

Reiter's syndrome may clinically resemble ankylosing spondylitis or psoriatic arthritis. It is associated with a nonspecific genitourinary infection thought to be transmitted venereally. The arthritis associated with *Yersinia enterocolitica* infection has a pattern comparable to the enteric arthritides. The association with certain HLA subtypes indicates that patients with a seronegative disease have a genetic predisposition to the disease and that the clinical expression is manifested only on exposure to a stimulating exogenous agent or event.

SYSTEMIC LUPUS ERYTHEMATOSUS (SLE)

This multisystem disease is characterized by the presence of a variety of circulating antibodies to nuclear and other tissue antigens. The course may be acute and fulminating, or slowly progressive over many years. It is a disease of unknown ***etiology*** occurring primarily in young women, with black persons appearing to be particularly at risk. The clinical expression of arthritis, skin rash, renal disease, and central nervous system, cardiac, and peripheral vascular involvement may be due to circulating *complexes of nuclear antigens* and their *antibodies.* These antibodies and nuclear antigens can be demonstrated in the basement membrane of glomerular capillaries and at the dermal–epidermal junction of the skin. Deposition of the antigen–antibody complexes is associated with complement fixation, leading to activation of the complement system and to the chemotaxis, increased vascular permeability, and enhanced phagocytosis related to inflammation. The reasons for formation of anti-DNA antibodies in SLE are not known; however, it has been proposed that they are synthesized when a latent virus is activated. Antinuclear antibodies do not necessarily have a pathologic effect, because they have been passively transferred to other humans without producing disease.

The ***clinical manifestations*** of this disease vary in severity from a mild skin rash to fulminating renal failure. Renal insufficiency, nephrotic syndrome, and hypertension are uncommon but may be the cause of death. The *renal lesion* is a glomerulonephritis that may be (1) diffuse proliferative, (2) focal proliferative, (3) mesangial, or (4) membranous. Kidney disease may present as mild or gross proteinuria with hematuria, red and white cell casts. A lupus-like syndrome may be induced by many drugs, especially the following: Antiarrhythmics (procainamide, quinidine); antibiotics (sulfonamides, penicillin, tetracycline); anticonvulsants; isoniazid; oral contraceptive agents; hydralazine; D-pencillamine; methyldopa, phenothiazines; and propylthiouracil.

SCLERODERMA

This uncommon disorder of connective tissue, also known as ***progressive systemic sclerosis,*** is manifested by fibrosis in the skin and in internal organs such as the esophagus and small bowel. There may be vascular abnormalities in the lungs and kidneys. Patients may present with Raynaud's phenomenon, then skin changes consisting of binding down of the skin of the hands, feet, face, and torso. A rheumatoid-like polyarthritis may be part of the disease. Telangiectases may be present. Changes in the arteries and arterioles, with proliferation of the internal elastic membrane, intimal deposition of proteoglycan, medial thickening, and adventitial deposition of connective tissue rich in collagen may exist prior to epidermal atrophy and fibrotic dermal thickening. Some patients may have Raynaud's phenomenon for many years before developing skin or internal organ involvement. The CREST variant of scleroderma (calcinosis, Raynaud's phenomenon, esophageal abnormality, sclerodactyly, and telangiectasia) is believed to have a better prognosis.

The abnormality in scleroderma may reside in the *skin fibroblast,* leading to increased collagen synthesis, thence increased collagen in the lower dermis and subcutis, thinning of the epidermis, loss of hair, and replacement of subcutaneous fat by collagen. Renal lesions consist of fibrinoid necrosis and damage to the intima of the intralobular arteries; the glomeruli are usually spared. Immunologic reactivity is not a characteristic of scleroderma, but a distinctive speckled nucleolar immunofluorescence is seen in many patients. The CREST variant is associated with an antibody to the centromere of the cytoplasm (see as follows).

POLYMYOSITIS—DERMATOMYOSITIS

This diffuse inflammatory disease involves primarily skeletal muscle. It is manifested *clinically* by weakness, mainly of the proximal muscles such as the shoulder and pelvic girdles. *Children* with this disease have a characteristic skin rash over the face, upper chest, and extensor surfaces of the joints. An asymmetric polyarthritis of small and large joints, mimicking rheumatoid arthritis, may occur. Esophageal muscle involvement can lead to swallowing difficulties. Subcutaneous and muscle calcification may occur, primarily in children, leading to difficulties in arm or leg movement or to muscle contractures. Some adults with this disease harbor an underlying malignancy. The clinical expression of polymyositis may occur as one feature of established connective tissue disease, such as systemic lupus erythematosus, rheumatoid arthritis, scleroderma, or mixed connective tissue disease.

Histologically, the major lesion is a focal necrosis of individual muscle fibers with infiltration of mononuclear inflammatory cells. Evidence of leakage of intracellular enzymes is found, with abnormally high levels of CK (and other muscle enzymes) in the blood, and high levels of creatine in the urine. Electromyography (EMG) is often helpful by showing the changes of inflammatory muscle disease.

POLYARTERITIS NODOSA

Polyarteritis (formerly, periarteritis) represents one entity in a spectrum of inflammatory diseases involving arteries and veins. It is a multisystem disease characterized by acute inflammation and fibrinoid necrosis of small and medium-sized vessels. The *etiology* of the vascular inflammation is unknown, although infectious and hypersensitivity mechanisms have been suggested because similar arterial lesions are seen in serum sickness, after allergic reactions to drugs, after bacterial infections, and in Australian antigenemia. A similar disease in Aleutian mink has a viral etiology.

The *clinical manifestations* of polyarteritis nodosa depend on the site and extent of arteries affected and may involve the brain, heart, kidneys, intestinal tract, or peripheral nerves. Arthritis occurs in approximately 50% of patients. Aneurysm formation or thrombosis may occur, and the inflammatory process may involve adjacent veins. In the 'healing' stage fibrotic obliteration of vessel lumina occurs, which may lead to local and distal vascular insufficiency.

Polyarteritis may occur as a disease on its own. Sometimes, however, it is difficult to distinguish this disease from other forms of vasculitis such as hypersensitivity angiitis, allergic granulomatous angiitis (with asthma and eosinophilia), or Wegener's granulomatosis (necrotizing granulomatosis of the respiratory tract, disseminated angiitis, and focal glomerulitis). Vasculitis also occurs in association with other rheumatic disorders such as rheumatic fever, rheumatoid arthritis, systemic lupus erythematosus, and giant-cell (temporal) arteritis. There are no characteristic diagnostic tests for this group of diseases, which are characterized by vascular inflammation, aside from *biopsy* of affected vessels. This procedure will demonstrate the vascular involvement, initially with edema of the *intima* and adjacent *media,* with subsequent infiltration of acute inflammatory cells, fibrinoid necrosis, and disruption of the *elastica.* The diagnosis may be suggested by the following histologic clues: the size of the vessels involved, the presence of vascular lesions of the same or different ages, and the presence of giant cells.

POLYMYALGIA RHEUMATICA

This syndrome, which occurs in patients over 50 years of age, is manifested by pain and profound stiffness of the shoulder and pelvic girdle muscles. Constitutional symptoms may include fever and weight loss. Some patients have arthritis, which is not a major feature but which may make it difficult to differentiate this syndrome from other arthritic disorders that occur in the elderly. Sudden blindness can occur, with involvement of the ophthalmic artery, if there is associated giant-cell or temporal arteritis. *Laboratory* studies show a mild normochromic, normocytic anemia, and an elevated erythrocyte sedimentation rate (>50 mm/hour, often around 100 mm/hour) is always present. An elevated serum globulin level is often present.

The symptom of muscle stiffness is exquisitely sensitive to small or moderate doses of oral prednisone. Patients feel remarkably improved within hours of starting steroid therapy.

OSTEOARTHRITIS

This common degenerative joint disease involves primarily the articular cartilage of central and peripheral synovial joints. Radiologic surveys of European and North American populations indicate that after age 30 an increasing proportion of the population shows evidence of this disorder. Although many elderly persons (>90%) may have radiologic evidence of osteoarthritis, relatively few (20%) are symptomatic. The degenerative process usually involves only a limited number of joints (hips, knees, proximal thumb joint, neck, or back) with slowly progressive pain and stiffness. There may, however, be severe widespread joint involvement, with joint destruction and serious disability.

The disease is believed to result from biochemical and biomechanical changes that occur, in which *reparative* mechanisms fail to keep up with normal *degradative* changes of articular cartilage. Abnormal physical stresses may affect the chondrocytes of cartilage; with release of enzymes, alteration of the surface of the cartilage may allow hyaluronidase from the joint fluid to enter, further damaging the cartilage matrix. The cartilage attempts to repair itself by chondrocyte proliferation and an increased synthesis of chondroitin sulfate. The articular cartilage becomes yellow and soft, with superficial pitting, fibrillation, and ulceration. With attempts at repair there is a proliferation of cartilage and bone, and the formation of osteophytic spurs. Gross deformity of the involved joints may occur. There are no diagnostic laboratory tests; the diagnosis is made on the basis of clinical findings and specific radiologic changes. *Synovial fluid analysis* rules out crystal-induced synovitis and other inflammatory arthritides.

CRYSTAL-ASSOCIATED ARTHRITIDES

Gouty Arthritis. This is the clinical expression of a metabolic disorder characterized by hyperuricemia and tissue deposition of monosodium urate (see Chap. 19). It commonly presents as an acute monoarthritis, most frequently in the first metatarsal phalangeal joint, but it may involve other peripheral joints, such as the ankle, knee, wrist, elbow, or an interphalangeal joint of the hand or foot. Precipitating factors may include local trauma, surgery, an illness such as myocardial infarction, starvation, diuretic therapy, or the use of uric-acid-lowering agents. *Synovial analysis* is helpful in differentiating this acute arthritis from other causes of acute inflammation of a peripheral joint, such as septic arthritis or other crystal-associated diseases.

Chondrocalcinosis (calcium pyrophosphate dihydrate [CPPD] crystal deposition disease). This is a calcification of the hyaline cartilage or fibrocartilage, usually with CPPD crystals. It is associated with a number of diseases, such as diabetes, hemochromatosis, ochronosis, osteoarthritis, and gout. The joint calcification may occur in many of the large peripheral and axial joints such as shoulders, wrists, hips, and knees, and may be present for many years without an associated arthritis. A crystal-induced synovitis caused by deposition of CPPD crystals may be seen in association with hyperparathyroidism.

The arthritis associated with chondrocalcinosis occurs primarily in persons over 60 years of age. It is called *pseudogout* and may be as acute, as agonizing, and as incapacitating as gout. It may occur as a single isolated event or may recur with increasing frequency, superimposed on a slowly progressive osteoarthritis. The *diagnosis* may be suggested by the finding of radiologic evidence of chondrocalcinosis and is confirmed by finding typical positively birefringent intra- or extracellular crystals in the synovial fluid with the polarizing microscope.

Other Crystal-Associated Arthritides. Sophisticated laboratory examinations of synovial fluid, synovium, and cartilage have found other crystals, such as hydroxyapatite and cholesterol. Other physical forms of calcium pyrophosphate have been found in experimental forms of arthritis. Newer research tools utilizing x-ray fluorescence and diffraction, and scanning and transmission electron microscopy, suggest that abnormal crystal deposition is associated with degenerative processes involving cartilage and bone.

SYSTEMIC AND JOINT DISORDERS ASSOCIATED WITH INFECTION

Many of the systemic arthritic disorders may be associated with as yet unrecognized infections in genetically susceptible individuals (e.g., rheumatoid arthritis and systemic lupus erythematosus). Ankylosing spondylitis has been linked to *Klebsiella* infection. Other disorders associated with clearly identified infectious agents are rheumatic fever (*streptococcus*) and the arthritis, tenosynovitis, and vasculitis associated with *Neisserian* infection. In some disorders an arthritis may be present, but culture of synovial fluid or blood is unrewarding and serologic tests to determine exposure to infectious agents must be carried out.

The presence of specific infection may be established by finding antibodies to agents such as the streptococcus, (i.e., Antistreptolysin 0 titer) or hepatitis virus. A newly described disorder, Lyme arthritis, is characterized by rash (erythema chronicum migrans), occasional carditis or meningoencephalitis, and often by subsequent arthritis. It is associated with a tick-borne spirochetal infection. Antibodies to this specific *spirochete* may be found in serum or synovial fluid. Erythema nodosum may be associated with systemic bacterial or fungal infections. Septic arthritis due to organisms within the joint may be diagnosed by appropriate culture of the synovial fluid. Some organisms such as *staphylococcus, streptococcus,* and *E coli* are relatively easy to culture. *N. gonorrhea* may be difficult to culture without appropriate laboratory precautions. Tuberculous infection of bones or joints may be difficult to diagnose without open biopsy and culture of synovium. Osteomyelitis or septic discitis (infection in disc of the spine arising from vertebral end plate) may require culture of needle aspirate or bone biopsy.

LABORATORY INVESTIGATIONS

HEMATOLOGY

Hemoglobin. An anemia of about 100 g/L, which is usually normochromic and normocytic, is commonly seen in active rheumatoid arthritis. Iron studies reveal the picture of chronic disease, namely, a low serum iron with a low iron-binding capacity. Iron-deficiency anemia may occur as the result of a chronic gastrointestinal blood loss secondary to the use of antiinflammatory agents. Autoimmune hemolytic anemia is seen in systemic lupus erythematosus. If the hemoglobin falls much below 100 g/L then a cause other than the anemia of chronic disease should be sought.

White Blood Cells. Leukocytosis may be seen with many of the inflammatory arthritides, particularly with the acute crystal-induced synovitis of acute gout. Leukopenia is seen in systemic lupus erythematosus and also in patients with rheumatoid arthritis and hypersplenism (Felty's syndrome).

Platelet Count. Thrombocytosis, with levels occasionally approaching 10^{12}/L may be seen in active rheumatoid arthritis. Thrombocytopenia is seen with the hypersplenism of Felty's syndrome and with systemic lupus erythematosus, and is occasionally drug-induced (e.g., by gold therapy for rheumatoid arthritis).

Erythrocyte Sedimentation Rate (ESR). The ESR and other tests for acute-phase reactants (plasma fibrinogen, C-reactive protein) are simple ways of establishing the presence or absence of inflammation and reflect changes in the intensity of the *inflammatory process.* The ESR correlates with plasma fibrinogen and is elevated in most cases of active rheumatoid arthritis, active rheumatic fever, and other inflammatory diseases of the joints. However, it is not necessarily elevated in active systemic lupus erythematosus, dermatomyositis, or ankylosing spondylitis. It is always elevated in polymyalgia rheumatica and is the only reliable test that one has to measure the progress of this disease. In acute crystal-induced synovitis the ESR may rise transiently during attacks. Normal rates may occur in a small number of patients with active rheumatoid arthritis and in active scleroderma. The ESR is usually normal in primary generalized osteoarthritis.

URINALYSIS

The urine is usually normal in uncomplicated rheumatoid arthritis. Proteinuria and urinary sediment abnormalities indicate other systemic diseases such as scleroderma, systemic lupus erythematosus, or polyarteritis.

EXAMINATION OF SERUM

Miscellaneous Biochemical and Serologic Studies. Protein abnormalities determined by *electrophoresis* and *immunoelectrophoresis* will often pick up nonspecific abnormalities in the rheumatic diseases. A polyclonal hypergammaglobulinemia is seen in rheumatoid arthritis and may occur in other collagen diseases. Multiple myeloma may present with a rheumatoid-like arthritis, but the specific monoclonal hyperglobulinemia would be noted on immunoelectrophoresis of serum or urine, and the bone marrow would show a characteristic number and type of plasma cells. Serum *calcium, phosphorus,* and *alkaline phosphatase* levels are helpful in delineating the osteopenia of metabolic bone disease.

Serologic tests such as the *antistreptolysin* titer should be carried out if recent streptococcal infection causing rheumatic fever is suspected. Arthralgia or arthritis may be part of the prodrome of hepatitis B infection, so that the determination of the presence of *hepatitis B antigen,* and subsequently *hepatitis B antibody,* may be required to diagnose an arthritis associated with jaundice. Serologic tests for *syphilis* may be falsely positive many years prior to the clinical expression of diseases such as systemic lupus erythematosus. Although the Wasserman, Venereal Disease Research Laboratory (VDRL), or Kolmer with Reiter protein (KRP) tests are positive, the more specific tests for the presence of *Treponema pallidum* infection, such as the *T. pallidum* immobilization (TPI) and fluorescent treponemal antibody absorption (FTA–ABS) tests, are negative. *Neuropathic* joint disease (Charcot's joints) may occur with the tertiary stage of syphilis. This destructive arthritis may occur as well with syringomyelia (upper limb) and diabetes (lower limb).

Rheumatoid Factors (RF)

Most adult patients who have rheumatoid arthritis possess rheumatoid factors (RF), a collective term for various *antibodies* reacting against *immunoglobulin G* (see also Injury Mediated by Immunologic Mechanisms, Chap. 4). The serum to be tested is mixed with IgG-coated particles (latex particles, red cells, etc.); if RF is present, agglutination occurs. The serum is diluted and if agglutination occurs with titers of 1/80 or greater it is considered positive for RF. The latex fixation test has 75% sensitivity and 75% specificity. The sensitized sheep cell agglutination test (SCAT) has only 50% sensitivity but 90% specificity.

Rheumatoid factor is now being measured by *rate nephelometric* techniques. Aggregated immune globulin (IgG) is the antigen used to detect RF (IgM—class antibody to IgG). Introduction of the antiserum containing a dye triggers a measurement sequence. The nephelometer measures the rate of increase in light-scatter by the antigen–antibody complexes; the reaction takes place in 17 to 20 seconds. Results agree well with those for the commonly used latex precipitation test. The new method is rapid and reproducible, has high sensitivity and specificity, and may provide a more reliable guide to diagnosis.

Rheumatoid factor is not specific for rheumatoid disease; it is found in a variety of *chronic inflammatory disorders* and in a small number of healthy subjects. Approximately 75% of patients with a diagnosis of rheumatoid arthritis exhibit a positive test, but there is no evidence that rheumatoid factor is involved in pathogenesis. Those with classic rheumatoid arthritis, with subcutaneous nodules and severe disease, and usually with vasculitis, generally have the highest titers, occasionally above 1/10,240 or 3000 I.U.

Some *healthy* individuals possess a large amount of anti-IgG. Transfusion of high-titer serum into human volunteers does not cause disease. It is possible that an unrecognized, prolonged immunologic stimulus is responsible for rheumatoid factor and for the disease rheumatoid arthritis itself. Newer methods for detecting lower-weight anti-IgG, as well as soluble immune complexes and cryoproteins, have been developed. A number of *promising techniques* for assessing other aspects of immune function in rheumatoid arthritis have become available. These include quantitative measurements of lymphocyte function based on mitogen stimulation or responsiveness to allogeneic cell suspensions, and qualitative methods for identifying T (thymus-dependent) and B (bone-marrow–dependent) lymphocytes by sheep red cell rosette formation or by cell membrane immunoglobulin receptors.

Lymphocyte Subpopulations

Advances in the development of antisera that recognize lymphocyte surface antigens have provided greater insight into the nature of T-cell subpopulations. Changes in the absolute number of T-lymphocyte subsets may be the basis of immune aberrations in certain diseases. Determination of the ratios of T-cell subsets is valuable for immunologic monitoring and treatment in recipients of renal allografts and for typing leukemias and lymphomas. In the rheumatic diseases the number and variety of lymphocyte subpopulations responsible for the complexity of control of the immune system is still under investigation. The development of cell-mediated or humoral immunity to 'self' constituents is a common feature of the rheumatic diseases; autoimmune mechanisms participate in maintaining the disease. Recent studies suggest that there may be hereditary alterations of the immune system that predispose individuals to the development of the various connective tissue diseases.

These tests are not recommended for the routine laboratory as there are no well defined abnormalities in the function of lymphocyte subpopulations that are specific for these disease entities.

Antinuclear Antibodies

Various immunologic methods are used to detect the presence of autoantibodies that react with cellular and nuclear components in patients with systemic rheumatic diseases. The latter are known collectively as *antinuclear antibodies* (ANA) (Table 13–1).

The LE cell phenomenon was the forerunner to tests examining ANA. When normal leukocytes and complement are incubated with serum from patients with systemic lupus erythematosus (SLE), a specific serum antibody to deoxyribonucleoprotein (DNP) reacts with the nuclear protein of some cells (DNA-histone complex), causing a change in staining characteristics which produces a pale and homogeneous appearance of the nuclear material. Other polymorphonuclear leukocytes are attracted to this globulin-coated material, and phagocytosis occurs. The characteristic LE cells can be seen with the light microscope. Although useful for many years the test is time consuming, difficult to perform and subject to interpretation.

The ***fluorescent antinuclear antibody test*** (FANA) has become the standard screening test for patients suspected of having connective tissue disease. The FANA has a sensitivity close to 99%, which virtually excludes SLE when the results are negative.

Dilutions of patient's serum are layered on cryostat sections of mammalian tissues (usually rat liver, kidney, heart, or stomach). After incubation and washing, the slide is 'stained' with a fluorescein-conjugated antiserum to human gammaglobulin. If ANA is present it will coat the nuclei of the cells and the tagged antiserum will then bind to the antibody. The fluorescent nuclei are visualized by *fluorescent microscopy*. Certain connective-tissue diseases have an increased frequency of one pattern of nuclear fluorescence over another (i.e., homogeneous, speckled, nucleolar, rim; see Table 13–1). Antinuclear antibodies may be found in normal people; the false-*positive* rate increases with age and is higher in women. Antibodies to *histones* are present in the serum of approximately 35% of patients with SLE, and in 96% of patients with SLE induced by drugs such as hydralazine and procainamide.

Many sera containing antinuclear antibodies react with DNA-free, *non-histone* nuclear antigens. These antibodies, directed against extractable nuclear antigens (ENA), are detected in a hemagglutination test in which there are two components. The antibody to *Sm antigen* is ribonuclease resistant and the antibody to ribonucleoprotein (nuclear RNP) is ribonuclease sensitive. Antibody to Sm antigen is highly specific for SLE. Antibody to *Nuclear RNP* is also found in SLE, but is most clearly associated with Mixed Connective Tissue Disease (which may be a form of SLE). Antibody

TABLE 13–1. Common Nuclear Antigens Seen in Rheumatic Diseases

Antigen	Pattern of Nuclear Fluorescence	Comments
DNP	Homogeneous (diffuse)	Antibody responsible for LE cell phenomenon and anti-DNP spot test. Found in SLE, RA, and other rheumatic diseases
DNA (double and single stranded)	Rim or peripheral (shaggy)	Found in active SLE often with nephritis
Histones (nuclear)	Rim or homogeneous	LE cell. High incidence in drug induced SLE
Non-histones (ENA)		
Sm (ribonuclease-resistant)	Speckled	Found mostly in SLE, occasionally in other rheumatic diseases
Nuclear RNP (ribonuclease-sensitive)	Speckled	Found especially in mixed connective tissue disease syndrome

DNA = deoxyribonucleic acid; DNP = deoxyribonucleoprotein; ENA = extractable (DNA-free) nuclear antigen; LE = lupus erythematosus; RA = rheumatoid arthritis; RNP = ribonucleoprotein; SLE = systemic lupus erythematosus; Sm = a non-histone protein.

to *SS-B antigen* is usually associated with Sjögrens syndrome. Antibodies to the above four non-histone nuclear protein antigens show speckled nuclear staining with immunofluorescence.

Antibody to chromosome centromere is detected on tissue culture cells. This antibody is found in some patients with primary biliary cirrhosis, in 10% of patients with systemic sclerosis, 30% of patients with Raynaud's disease, and over 95% of patients with the CREST syndrome. Absence of anticentromere antibody is strong evidence against CREST. Antibodies to other intracellular components (nucleoli, cytoplasm) have been found in patients with systemic rheumatic disorders.

Total ANA may include antibodies to single or double-stranded DNA, nucleohistone, extractable nuclear antigen (ENA), and double-stranded RNA. High levels of antibody to native (double-stranded) DNA are found in active SLE (usually with renal involvement). Determination of the presence of anti-DNA is carried out with a technique that measures the amount of radioactive labeled DNA left in the supernatant after the patient's serum is reacted with radioactively labeled DNA and the complex retained by filtration (Faar technique). Another method for the measurement of antibodies to double-stranded DNA uses immunofluorescence with an organism that is related to the trypanosomes. The hemoflagellate *crythidia luciliae* contains a kinetoplast which has a high concentration of native DNA with no other interfering antigens. This technique may provide a useful alternative to radioimmunoassay for detection of antibodies to native DNA. The method is specific and reliable and antibody is found only in sera of patients with SLE (48%) and mixed connective tissue disease (20%).

Complement (see also Chap. 4)

Normal complement activity requires the presence of nine major protein components reacting and interacting in a specific sequence. The complement cascade produces and amplifies immune responses in the presence of antigen–antibody complexes, releasing histamine and other potent mediators of chemotaxis, contributing to immune adherence and enhanced phagocytosis. *Increased* complement levels are seen in the presence of certain inflammatory states, infection, trauma, or acute illness (myocardial infarction). *Decreased* complement may be the result of (a) a specific deficiency of a complement component, (b) complement degradation caused by activation of the alternate pathway as may be seen in Gram-negative septicemias, (c) complement degradation caused by activation of proteolytic enzymes as a result of trauma, ischemia, or toxic necrosis, (d) complement depletion caused by the presence of high concentrations of immune complexes seen characteristically in active SLE, acute glomerulonephritis, membranoproliferative nephritis, serum sickness, acute vasculitis and severe rheumatoid arthritis.

A useful *routine complement screen* measures C3 and C4 levels by immunoassay and total hemolytic activity by CH50. This traditional method for determination of complement in serum measures the ability of the test specimen to lyse 50% of a standard suspension of sheep erythrocytes coated with rabbit antibody in a reaction which includes the entire classic reaction sequence. C3 is the most plentiful of the complement components; its serial measurement may be a predictor of disease activity. C4 is most sensitive to *in vivo* complement activation and serial measurements may reveal decreased levels in advance of clinical worsening prior to alteration of C3 or CH50.

Immune Complexes

Immunologic mechanisms of tissue injury are discussed in Chapter 4. One such mechanism, which is important in the systemic rheumatic diseases is mediated by *immune complexes.*

Foreign antigens (i.e., DNA, nucleoprotein, IgG, thyroglobulin, renal tubular antigen, and certain carcinomas of bowel, lung, or kidney) interact with antibody to produce immune complexes (ICs). The fate of ICs and the clinical manifestations of their presence are dependent on the nature of the antibody, the nature of the antigen, the nature of the complex itself, and the status of the reticuloendothelial system. With the same antigen and antibody a variety of ICs may form with different lattice sizes and structures, depending on a number of physical and chemical factors. *Immune complex disease* such as SLE occurs when a deposit of complexes accumulates on the basement membranes of blood vessels, glomeruli, or other tissues. The harmful effects of ICs are dependent on (a) their activation of the complement system, and (b) their interaction with cell-surface receptors, leading to the release of lysosomal enzymes.

Detection of Immune Complexes. The methods for detection of ICs depend on various biologic phenomena such as complement fixation and interaction with cell receptors. A detailed discussion of the methods and the significance of the various techniques is beyond the scope of this chapter; only a summary is given here.

Immunofluorescent microscopy is useful for demonstrating complexes in the tissues of patients with rheumatoid arthritis, systemic lupus erythematosus, and polyarteritis nodosa and other forms of vasculitis.

Electron microscopy has allowed the sites of deposition in the kidney to be localized precisely. Australia antigen and antibody (in patients with viral hepatitis) have been found in other tissues with this technique.

Analytic ultracentrifugation allows the separation and visualization of various components of circulating proteins.

Gel filtration and polyethylene glycol (PEG) *precipitation* with the first component of complement (C1q) has been used to detect complement components bound to ICs. This technique has been useful in the detection of abnormal proteins in patients with rheumatoid arthritis and systemic lupus erythematosus.

Cryoprecipitation can be used to develop precipitates rich in immunoglobulins and ICs. In the connective tissue diseases the cryoprecipitates contain immunoglobulins of various classes. They have been detected in serum and synovial fluid from patients with systemic lupus erythematosus and rheumatoid arthritis.

Immune complexes may lead to *complement consumption,* which can be detected by a group of techniques that assess the various components of the complement system (e.g., the C1q binding assay).

Interactions with cell-surface receptors for immunoglobulin (Fc) or for the third complement component (C3) are the basis for a number of test systems (platelet aggregation, human lymphoblastoid cell line with B-cell characteristics [Raji cells], and complement-coated erythrocytes).

Interaction with rheumatoid factors using binding, precipitation, or agglutination reactions may help to detect ICs.

Functional bioassays for ICs include the demonstration of cytotoxicity (in patients with inflammatory bowel diseases), and histamine release from lung preparations *in vitro* on exposure to serum containing ICs (biologically active rheumatoid factors).

TESTS FOR INFLAMMATORY MUSCLE DISEASE

Serum Enzymes. Elevated levels of creatine kinase (CK), aspartate aminotransferase (AST), or aldolase occur in the primary myopathies and in inflammatory myositis (polymyositis), probably by the escape of enzymes from damaged muscle or muscle with increased membrane permeability (see Chap. 3). CK is the most sensitive and most specific for muscle disease but can be elevated from a number of causes (myocardial infarction, intramuscular injection, strenuous exercise, subarachnoid hemorrhage, pulmonary infarction, hypothyroidism, malignant hyperthermia). Normal levels can occur in the presence of active muscle disease, especially if the patient is being treated with corticosteroids. Elevations of hepatic enzymes (AST, ALT, GGT, ALP) occur with insults to the liver that may accompany acute rheumatic fever, systemic lupus erythematosus, polyarteritis, or the use of drugs (e.g., aspirin in early stages of treatment of juvenile polyarthritis).

Urine Creatine. Creatine, synthesized mainly in tissues other than muscle (liver, kidney, pancreas), is rapidly taken up by muscle fibers and converted by the action of CK to creatine phosphate, an important energy reserve for muscle contraction. The anhydride, creatinine, arises from creatine and creatine phosphate and normally is excreted in constant amounts (8 mmol to 15 mmol [1 g to 2 g]/day, depending on muscle mass). The urine of normal *men* contains negligible amounts of creatine, up to 380 μmol (50 mg)/day. More may be found in the urine of *women* and growing *children.* Significant ***creatinuria*** may occur if the amount of normal muscle is being reduced, as in fasting, muscular dystrophy, poliomyelitis, and disuse atrophy, and in inflammatory destructive muscle disease such as polymyositis. Creatinuria is also found in catabolic states such as hyperthyroidism and with corticosteroid administration. Urine creatine excretion is also nonspecifically elevated in many other conditions affecting muscle. The measurement of urine creatine is useful, however, in diagnosis and follow-up of patients with polymyositis, because the elevations of serum enzymes (CK, AST, and aldolase) may be returned to normal by corticosteroid therapy despite the presence of continuing disease activity.

Corticosteroid-induced myopathy may occur in patients treated with significant doses (usually more than 15 mg prednisone/day) for any disease. It is a difficult entity to diagnose because it resembles the muscle weakness of a condition for which the steroid is given (rheumatoid arthritis, systemic lupus erythematosus, and polymyositis). The serum enzymes may be normal in the presence of proximal muscle weakness, which can be very profound. Significant creatinuria may occur (owing to enforced bed rest, the catabolic action of steroids, steroid-induced myopathy, or underlying inflammatory muscle disease). If the dose of corticosteroid is reduced and the CK and urine creatine rise, then a diagnosis of continuing *inflammatory muscle disease* can be made. If the CK does not rise, and the urine creatine falls, then a diagnosis of probable *steroid-induced myopathy* is made, enabling the physician to reduce the dose of corticosteroid still further.

Electromyography (EMG). Normal muscle is electrically silent at rest. Action potentials are produced when a muscle is voluntarily contracted. Skeletal muscle can be stimulated by the application of brief electrical pulses to the skin overlying its motor nerve. Deviations from normal are detected by (a) the occurrence of 'spontaneous' activity during relaxation (fibrillations, positive sharp waves, and fasciculations); (b) abnormalities in the amplitude, duration, and shape of single-motor-unit potentials; (c) a decrease in the number of motor units that can be recruited (i.e., with slowly progressive muscle contraction involvement of other motor units will be electrically present); (d) alteration in size, duration, or interpotential interval of action potentials recorded during graded single or successive voluntary or electrically induced muscular contraction; (e) the demonstration of special electrical or movement phenomena. The EMG is often very helpful in the diagnosis of *polymyositis* by showing electrical changes characteristic of inflammatory muscle disease.

EXAMINATION OF SYNOVIAL FLUID

The synovial space is a specialized compartment of connective tissue. The *synovial membrane,* or *synovium,* lines joints, bursae, and synovial tendon sheaths, covering ligaments within the joint and the joint capsule. It is a highly vascular structure, with a large surface area containing interconnecting plexuses of small vessels and capillaries. The absence of a basement membrane facilitates the diffusion of biologically active effectors of inflammation.

The synovial membrane has the ability to allow the selective passage of substances of various sizes. High-mass substances such as macroglobulins and the relatively lower mass haptoglobin (M_r 85,000) and prothrombin (M_r 62,700) are excluded from the normal fluid. Clotting factors are also lacking in normal synovial fluid.

The synovial lining cells (or underlying fibroblasts) are capable of manufacturing and secreting hyaluronic acid, which is a major determinant of the unique properties of the synovial fluid. Collagen is also synthesized by the synovium, and the synthesis of immunoglobulins can be demonstrated in pathologic states.

Normal synovial fluid is clear, pale yellow, or straw-colored, and usually does not clot. Only a small amount exists in the normal state; in pathologic conditions there may be rapid production of volumes approaching 100 mL in large joints such as the knee. Synovial fluid is basically an ultrafiltrate of plasma, but includes the hyaluronic acid produced by the synovial membrane.

Hyaluronic acid is a nonsulfated polysaccharide, composed of equimolar quantities of D-glucuronic acid and *N*-acetyl-D-glucosamine, which polymerize to form a long-chain asymmetric molecule of high relative mass. Studies of its molecular structure indicate that it is a random coil with a moderate degree of stiffness. It is this constituent which gives synovial fluid its high viscosity and affects the partition of diffusible solutes. The concentration of hyaluronate in *normal* synovial fluid is approximately 3 g/L. Small molecules such as glucose, uric acid, and bilirubin are found in approximately the same concentration in synovial fluid as in blood. The total protein in normal synovial fluid amounts to approximately 18 g/L and contains relatively more albumin and less globulin than serum.

Hyaluronate is *altered* in pathologic conditions, and this seems to alter the nature of the synovial fluid contents. Normal synovial fluid and that found in noninflammatory states has a high viscosity with a sticky feel when examined between the thumb and forefinger. In inflammatory conditions such as rheumatoid arthritis, the fluid may be watery, with a loss of the normal stickiness. Superoxide radicals generated by phagocytosis may contribute to the depolymerization of hyaluronic acid.

Enzymes of several types have been observed in normal and noninflammatory synovial fluids; many of them increase in amount in inflamed joints. In the synovial fluid many processes commence and interact to produce a self-sustaining active inflammation. They include complement, kinins, clotting factors, polymorphonuclear (PMN) phagocytosis and lysosomal release, PMN release of proteolytic enzymes, prostaglandins, and oxygen free-radicals. In rheumatoid arthritis a marked proliferative and invasive front replaces cartilage and subchondral bone through the action of neutral proteinases, cathepsins, elastase, and collagenase. Osteoclasts activated by prostaglandins, or soluble products of mononuclear cells, contribute to the mineral resorption of bone.

Laboratory Examination of Synovial Fluid

General. The synovial fluid obtained from diseased joints can provide useful diagnostic information, differentiating between noninflammatory and inflammatory, septic and crystal-induced arthritides. In obtaining synovial fluid, the following should be considered:

1. Fluid may be difficult to aspirate because of large amounts of fibrin or because of loculation of fluid within the fibrin, or within the joint. It may be difficult to differentiate between a thickened synovium and the continued presence of fluid within the joint.
2. The volume may alter during the day, and indeed effusions may disappear with bed rest.
3. If only a small amount of fluid is available, then this should be sent for Gram's staining and culture. Only a drop of synovial fluid is required for crystal analysis.
4. *Viscosity* can be assessed by noting the length of the 'string' as a drop falls from the end of the aspirating needle. With a pathologic watery fluid the string is short, if present at all; the 'string' in normal and noninflammatory states will be several centimeters in length.
5. Normal synovial fluid has a yellowish color, and print is easily read through the fluid. With inflammation the number of cells increases, making the fluid increasingly opaque. Opacity may also be due to fibrin or large numbers of crystals.

White Cell Count. Although changes in the total and differential white blood cell (WBC) count are not specific, they are most important in differentiating various kinds of synovial fluids. The standard WBC counting chamber is employed, but 0.3% saline must be used as diluting fluid (normal diluting fluid contains acetic acid, which will cause synovial fluid to clump). Usual findings are listed in Table 13–2.

Synovial fluid may contain various particles or inclusions. In a traumatic aspiration streaks of blood may be mixed with the synovial fluid. This is to be differentiated from a *hemarthrosis,* in which blood is the primary constituent, as may be seen in the effusions of hemophilia, posttraumatic arthritis, and villonodular synovitis.

Glucose. Synovial fluid glucose can be measured by standard techniques if the fluid is drawn into a 'fluoride' tube to prevent glucose destruction by synovial fluid leukocytes. A very low level suggests joint infection; levels approximating 50% of the blood level may be seen in active rheumatoid arthritis. Because of its lack of value in ordinary situations, glucose analysis is not carried out on every routine synovial analysis.

Lactate. Synovial fluid lactate can be measured if the fluid is collected in an appropriate 'fluoride' tube. It is normal in most cases of non-inflammatory synovitis, slightly raised in inflammatory synovitis. It is most helpful in the diagnosis of *septic arthritis* in which concentrations as high as 11 mmol/L may be found (using the enzymatic kit

TABLE 13–2. Examination of Synovial Fluid

Clinical State	White Cells/mm³		Polymorphs (%)	Mucin Clot Test
	Usual	*Occasional*		
Normal	<200		<25	Good
Noninflammatory effusion	<1000	Up to 500	<30	Good
Minimal synovial membrane inflammation (noninfectious)	<5000		<30	Fair
Severe inflammatory effusion (noninfectious)	15,000	50,000 + (e.g., in rheumatoid arthritis)	50–90	Poor
Effusion with acute bacterial infection	>50,000		>80	Poor

method). Septic arthritis secondary to *gonococcal* infection does not yield high synovial lactate levels. Synovial fluid lactate levels can be obtained rapidly, prior to microbiologic culture studies, thus leading to early diagnosis and treatment of septic arthritis.

Mucin. The hyaluronic acid content of synovial fluid can be measured by a variety of laboratory techniques, including a turbidimetric method, or digestion and analysis for hexosamine or uronic acid. A *qualitative mucin clot test* generally suffices as an estimate of the degree of depolymerization of synovial fluid hyaluronate. About 1 mL of the joint fluid being tested is added to about 5 mL of 2% acetic acid, mixed rapidly with a glass rod, and examined promptly. A normal, noninflammatory fluid forms a firm clot that does not fragment on being shaken. A softer mass with some shreds constitutes a 'fair' mucin, whereas a 'poor' mucin clot, which is seen in many inflammatory effusions, fragments easily and forms flakes, shreds, and cloudiness in the surrounding fluid. Little is known about how the changes in synovial fluid proteins that occur in inflammatory joint disease affect the result of this clot test.

Other Special Studies. Total hemolytic *complement* and C3 component (β_1 C) may be determined by standard techniques. In rheumatoid arthritis the serum complement is usually normal, while the synovial fluid level is less than 30% of the serum level. In active systemic lupus erythematosus both serum and synovial fluid complement levels may be low; in infectious arthritis, gout, and Reiter's syndrome they may be high. Other substances, such as antinuclear factors, enzymes, rheumatoid factor, and immunoglobulins, can be measured in synovial fluid, but these add little to the relatively simple studies which can be done on a routine basis.

The commonly used oral or parenteral *antibiotics* diffuse readily through the inflamed synovium and levels can be measured in patients with septic arthritis. If an organism is cultured the minimal inhibitory concentration of required antibiotic can be determined.

Microscopic Examination. Several drops of well-mixed synovial fluid (or centrifuged sediment) can be examined by phase contrast for fibrils and cartilage fragments or for the cytoplasmic inclusions seen in many types of joint inflammation. The specimen is then examined for extracellular and intracellular birefringent ***crystals.*** With compensated polarized light one can differentiate the crystals seen in gout from those seen in pseudogout: Urates show a strong negative birefringence. Urate crystals are usually needle-shaped, with short crystals seen in joint effusions within or outside of cells. (Those aspirated from tophi or from bursae may appear as very long needles.) Calcium pyrophosphate dihydrate crystals show a weakly positive birefringence. Cholesterol crystals are usually platelike with a notch in the corner, but can be needlelike. Other crystals identified in joint fluids may be corticosteroid esters used previously for intraarticular injection or oxalate crystals from inappropriately used collection tubes. Hydroxyapatite crystals may be associated with synovitis, but these are difficult to identify with the light microscope. The various crystals associated with degenerative and crystal-induced arthritides may be identified in the research laboratory by transmission and scanning electron microscopic procedures and x-ray diffraction techniques.

SUGGESTED READING

COHEN AS: Laboratory Diagnostic Procedures in the Rheumatic Diseases. Boston, Little, Brown & Co, 1975

DIETSCHY JM, COHEN AS (eds): The Science and Practice of Clinical Medicine, Vol 4. New York, Grune & Stratton, 1979

KATZ WA: Rheumatic Diseases. Philadelphia, JB Lippincott, 1977

KELLY WN, HARRIS ED, RUDDY S, SLEDGE CB: Textbook of Rheumatology. Philadelphia, WB Saunders, 1985

McCARTY DJ: Arthritis and Allied Conditions. Philadelphia, Lea & Febiger, 1985

MOSKOWITZ RW: Clinical Rheumatology. Philadelphia, Lea & Febiger, 1982

RODNAN GP, SCHUMACHER HR, ZVAIFLER NJ (eds): Primer on the Rheumatic Diseases, 8th ed. Atlanta, Arthritis Foundation, 1983

14

Endocrine Disorders

Allan G. Gornall /
Allan W. Luxton /
Bhagu R. Bhavnani

BASIC ENDOCRINOLOGY

Hormones are *regulators* of biologic processes, and most endocrine disorders manifest themselves as disturbances of homeostasis, producing a derangement of metabolism or function. Most glands have a large functional reserve, and usually more than 50% of a gland, often as much as 90%, can be destroyed before signs of a deficiency or a low reserve become evident. Diagnosis of most well-advanced endocrinopathies is relatively easy, but treatment by that time may be ineffective. *Early diagnosis* requires an ability to differentiate, for example, early Cushing's syndrome from simple obesity; Conn's syndrome from essential hypertension; hypothyroidism from chronic fatigue; diabetes insipidus from psychogenic polydipsia; and the various causes of amenorrhea, hirsutism, weakness, and loss of libido.

Hormone concentrations in the urine and the plasma are often within the reference range in the early stages of endocrine disease. For diagnosis it is necessary therefore to demonstrate a defect in the control mechanism, or an inadequate functional reserve. The disorder may exist as a primary target gland or tissue defect, a secondary pituitary problem, or a tertiary central nervous system/hypothalamic abnormality. The problem may be one of unregulated overproduction, hormone deficiency, defective release or transport, defective uptake by receptors, or an abnormal peripheral metabolism or metabolic clearance rate.

This chapter summarizes the endocrinology of the hypothalamic–pituitary axis, the thyroid, the adrenals, and the gonads. For a more complete perspective the reader should refer to Chapters 16 and 17, as well as portions of Chapters 12, 21, 23, 26, and 27.

A few aspects of basic endocrinology are emphasized in the following sections.

THE HYPOTHALAMUS–PITUITARY–TARGET GLAND AXIS

The components of the hypothalamus–pituitary–target gland axis (Fig. 14–1) are best understood for the thyroid, adrenals, and gonads. The same principles apply in lesser degree to growth hormone (somatotropin) and prolactin.

The *hypothalamus,* under the influence of higher neural centers, produces releasing hormones (*liberins*) and inhibiting hormones (*statins*) which pass down the axons, enter the hypophyseal portal blood system, and are transported to the anterior pituitary. These factors control the production and release, by specific cells of the *pituitary,* of *tropic hormones* which enter the circulation and are attracted to receptors on their target glands (see Fig. 14–3).

Most *target gland hormones* circulate in equilibrium with a carrier protein. It is the small amount of *free* hormone that acts on target tissues, exerts feedback control, and is subject to metabolic clearance and excretion. Some hormones (thy-

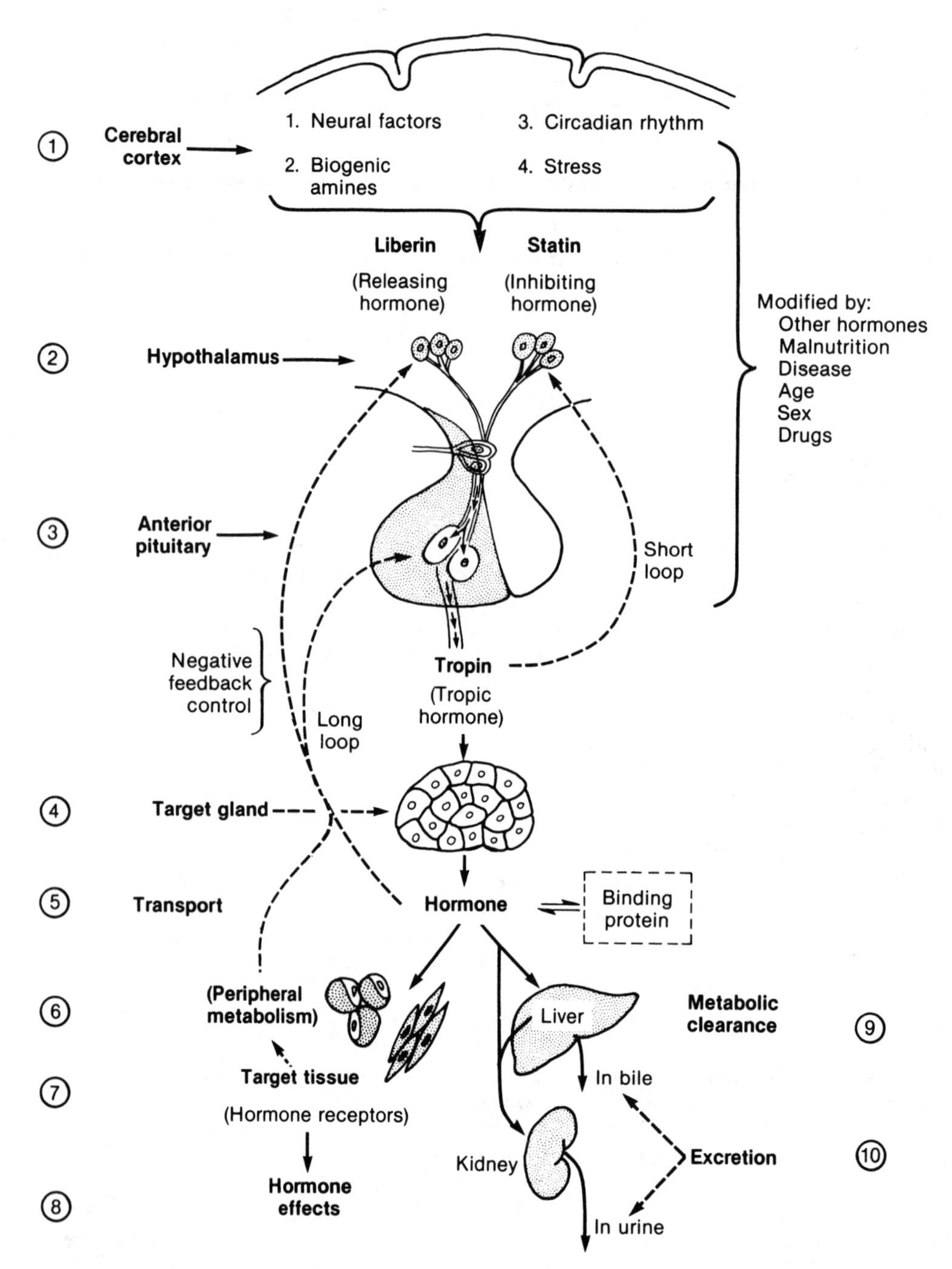

FIGURE 14–1. The relationships of the hypothalamus, pituitary, and target glands to hormone production, control, and metabolic clearance.

roxin, testosterone) undergo peripheral metabolism, which affects their binding to receptors and their hormonal activity. Some pituitary hormones (somatotropin, prolactin) act directly on target cells and are controlled by products of intermediary metabolism or by neural signals.

Peptide hormones are generated as *preprohormones,* because nascent chains must exceed 75 amino acids in order to enter the lumen of the rough endoplasmic reticulum; otherwise, biosynthesis would be terminated. Small peptides are derived by post-translational selective proteolytic deletion to form the inactive (or weakly active) *prohormones,* then the active *hormones.* Once circulated, the hormone may be degraded further to a carboxyl terminal fragment and an amino terminal fragment, the latter often having biologic activity.

A few additional statements are pertinent: Although each pituitary tropic hormone comes from a specific cell type and is under the primary control of a specific releasing and/or inhibiting factor, there are some crossover effects. *Feedback control* is usually negative and the most important mechanism is the 'long-loop' feedback of target gland hormone to the hypothalamus or pituitary. A 'short-loop' feedback effect of pituitary tropic hormone on the hypothalamus is recognized and an ultrashort feedback of hypothalamic hormones on hypothalamic neurones may exist. In general, stimuli that result in greater secretory activity of a gland, if prolonged, lead to hypertrophy of that gland. Daily (circadian), monthly, yearly, and even longer *rhythms* are known to exert a controlling influence on pituitary tropic hormone production. All of these cycles can be overridden by emotional, physical, and metabolic stresses.

REGULATION OF ENDOCRINE GLANDS

Secretory activity and homeostasis are influenced by many factors that vary with time and with the individual. For practical purposes the interaction and disorders of the system are often best understood by considering the hypothalamic–pituitary complex as a unit and feedback as the net result of all the factors that affect the level of circulating *free hormone.*

Taking the adrenals as a useful ***example:***

1. *Homeostasis* is maintained so long as normal levels of pituitary corticotropin (adrenocorticotropic hormone, ACTH) and target gland hormone (cortisol) are in balance.
2. *Partial ablation* (unilateral adrenalectomy) lowers cortisol output and feedback, causing ACTH to increase until the remaining adrenal gland undergoes hypertrophy and homeostasis is restored.
3. *Exogenous administration* of cortisol will depress ACTH and, if prolonged, will lead to atrophy of the adrenals. On withdrawal of the cortisol, the rate of recovery can be quite variable.
4. *Primary atrophy* of the adrenals leads to a decline in cortisol and a rise in ACTH until eventually all surviving cells are working at maximum capacity.
5. *Primary tumors* of the adrenal cortex are usually autonomous; the cortisol they produce suppresses ACTH, leading to atrophy of normal adrenal tissue.
6. *Bilateral hyperplasia* of the adrenals is usually secondary to either a pituitary adenoma secreting excess ACTH, or an ectopic source of this hormone.

MECHANISMS OF HORMONE ACTION

Three general mechanisms are known, and they appear to have a final common pathway involving increased nuclear transcription (Fig. 14–2).

Most protein and peptide hormones and biogenic amines bind to cell membrane receptors and activate one of two mechanisms for responding to the signal. Some (e.g., most releasing and tropic hormones, glucagon, parathormone, β-adrenergic agonists) act by stimulating adenylate cyclase to generate intracellular cyclic adenosine monophosphate (cAMP), which activates certain protein kinases. Others (e.g., growth hormone, angiotensin, oxytocin, prolactin, α_1-adrenergic agonists, and various growth factors) stimulate a membrane diesterase which splits a polyphosphoinositide to yield diacylglycerol (DG) and inositol triphosphate (IP_3). The latter increases intracellular calcium, while DG activates a Ca^{2+} dependent protein kinase. With either mechanism, the resulting phosphorylation of specific chromatin proteins in some way permits increased transcriptional activity,

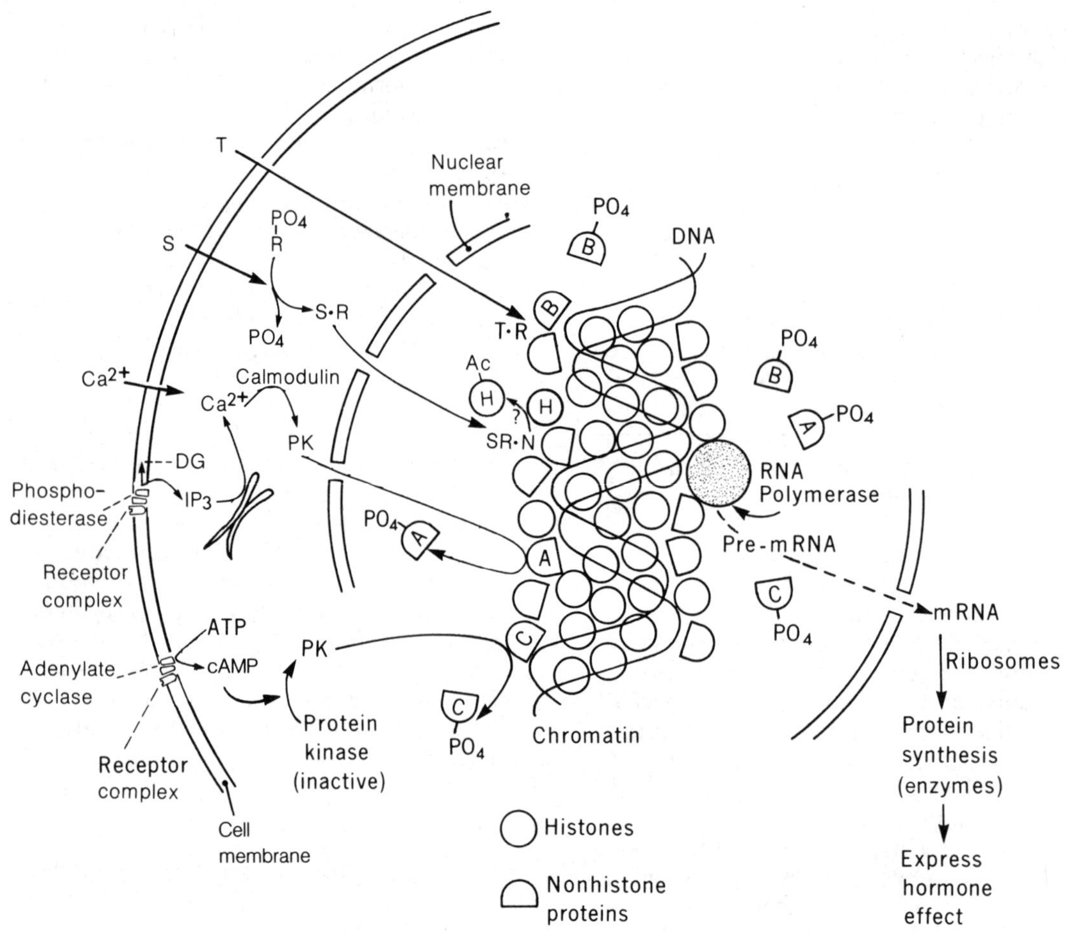

FIGURE 14–2. Postulated mechanisms of hormonal action. At least two types of membrane receptor complex recognize various peptide and other hormones; one acts through adenylate cyclase and the other through phosphodiesterase. Steroid hormones (S) are recognized by cytosolic receptors, and thyroid hormone (T) by nuclear receptors. A, B, and C represent different nonhistone proteins that appear to be phosphorylated, thus probably exposing DNA template and allowing RNA polymerase to function in the synthesis of messenger RNA. Some steroid receptor complexes may act by acetylating chromatin histones (H).

leading to synthesis of the protein or enzyme by which the function of the cell is expressed.

Steroid and thyroid hormones circulate mostly bound to plasma proteins, but the free hormone crosses the cell membrane rapidly, probably by facilitated transport. Steroids bind to a cytosol receptor protein; the resulting complex translocates to the nucleus where it interacts with a nuclear chromatin receptor and leads to increased transcription. Thyroid hormones bind directly to a receptor in the chromatin complex. In both cases covalent modification (e.g., phosphorylation) of nonhistone chromatin protein appears to be involved.

Some hormones may facilitate transport of Ca^{2+} through membrane channels. The increase in cytosolic Ca^{2+}, after binding to calmodulin, activates protein kinases which can phosphorylate cytosol or nuclear proteins.

The situation is undoubtedly more complex and not fully understood. Some hormones, for example, cause *rapid* effects—possibly allosterically, or by phosphorylation of an enzyme, a nucleotide or a phospholipid. Most have *delayed* effects that involve transcription and protein (enzyme) synthesis and occur after a latent period of about an hour.

Recent work on neurotransmitters, peptide hormones, and receptors has revived emphasis on *modulation,* whereby one substance alters the response of a cell to another substance. Catecholamines, for example, appear to modulate the sensitivity of somatotropes to somatoliberin, and vasopressin enhances the response of the corticotropes to corticoliberin.

THE HYPOTHALAMUS

STRUCTURE AND CONTROL OF THE HYPOTHALAMUS

The hypothalamus, located in the walls and floor of the third ventricle, is immediately above the pituitary gland and is connected to it by the pituitary stalk. The hypothalamus weighs only about 4 g but plays a vital role as a link between instinctive, psychic, neural, and humoral regulating influences. Neuronal connections to the hypothalamus are numerous and complex and are responsible for the mediation of circadian and cyclic rhythms in hormonal secretion, as well as for the integration of neuroendocrine and autonomic homeostatic responses.

The hypothalamus is composed of clusters of neurons referred to as *nuclei,* the most prominent of which are the supraoptic nucleus (SON) and the paraventricular nucleus (PVN). Axons of varying length extend from these nuclei down the pituitary stalk to terminate in dendritic processes on capillary walls in the posterior pituitary. Other regions of the hypothalamus have shorter axons that traverse the blood–brain barrier, cluster in the median eminence, and terminate on the walls of a rich capillary plexus from which portal vessels pass down the stalk to the anterior pituitary (see Fig. 14–3).

Various *biogenic amines* (dopamine, catecholamines, acetylcholine, serotonin, histamine) are released in the hypothalamus by neurons whose cell bodies lie in contiguous regions of the brain. The secretory activity of hypothalamic neurons is controlled in part by these biogenic amines. A cascade effect is achieved when *picograms* of biogenic amine induce *nanograms* of hypothalamic hormone, which induce *micrograms* of anterior pituitary hormone, which induce *milligrams* of target gland hormone.

Humoral factors in the systemic circulation reach specific receptors in the hypothalamus that respond to changes in electrolyte or metabolite concentrations, or to circulating peptide or steroid hormones.

HORMONES AND FUNCTIONS OF THE HYPOTHALAMUS

Control of the Anterior Pituitary

On the recommendation of the Commission on Biochemical Nomenclature the releasing factors (RF), or releasing hormones (RH), of the hypothalamus are designated with the suffix *-liberin,* the hypothalamic inhibitory factors (IF) with the suffix *-statin,* and pituitary hormones (stimulating hormones, SH) with the suffix *-tropin.* The factors or hormones thus far established are named according to the cells they influence. The six hypothalamic hormones shown in Figure 14–3 have been identified and structures are now known for:

SRH (somatoliberin; GRH, growth hormone releasing hormone) is a 44 amino acid (a.a.) peptide which acts on somatotropes to release somatotropin (STH; growth hormone, GH). Homologous 40 and 44 a.a. peptides have been isolated from pancreatic tumors causing acromegaly. SRH is active in humans, at a dose of 1 μg/kg.

SIF (somatostatin) exists in a 14 a.a. sequence (M_r 1600), and an N-terminal extended 28 a.a. sequence, both derived from a common prohormone. SIF-28 appears to be the more effective inhibitor of STH secretion and may also suppress TRH action on thyrotropes. These same hormones are found in a number of epithelial secreting cells, notably in the gastrointestinal tract and pancreas. SIF-28 strongly inhibits insulin secretion; both SIF-28 and SIF-14 suppress glucagon secretion. SIF-14 is more potent in the CNS.

TRH (thyroliberin) is the well known tripeptide pyro-Glu · His · Pro · NH_2. It is active in doses of 1 μg/kg, stimulating the thyrotropes to release TSH; it is capable also of liberating prolactin, but probably not as a physiologic control mechanism.

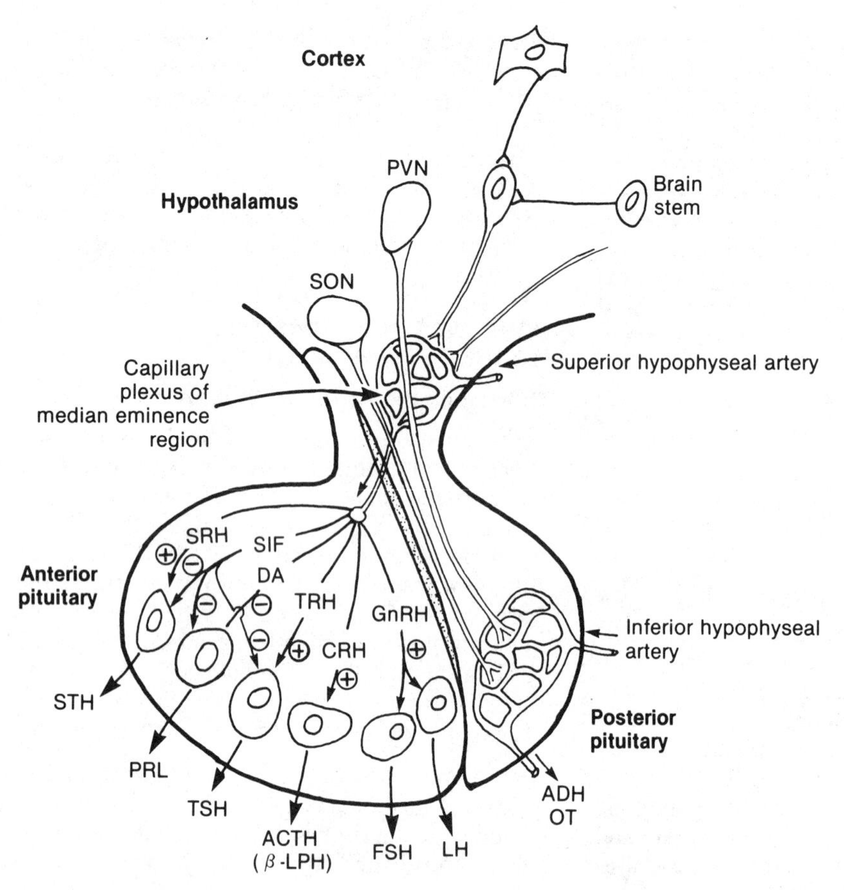

FIGURE 14–3. Structure of the pituitary gland and control of the different cells that produce the hormones of the anterior pituitary. ACTH = corticotropin, adrenocorticotropic hormone; ADH = vasopressin, antidiuretic hormone; CRH = corticoliberin, corticotropin-releasing hormone; DA = dopamine, prolactin-inhibiting factor (PIF); FSH = follitropin, follicle-stimulating hormone; GnRH = gonadoliberin, gonadotropin-releasing hormone; LH = lutropin, luteinizing hormone; β-LPH = β-lipotropin; OT = oxytocin; PRL = prolactin; PVN = paraventricular nuclei; SIF = somatostatin, somatotropin-inhibiting factor; SON = supraoptic nuclei; SRH = somatoliberin, somatotropin-releasing hormone; STH = somatotropin, growth hormone; TRH = thyroliberin, thyrotropin-releasing hormone; TSH = thyrotropin, thyroid-stimulating hormone. The significance of β-LPH as a by-product of CRH acting on corticotropes is uncertain. Shaded area is the 'intermediate' lobe.

CRH (corticoliberin) has a 41 a.a. sequence (M_r 5000) derived from a large prohormone. It acts on corticotropes at a dose of 1 μg/kg and has a relatively long half-life, but is less potent than insulin-induced hypoglycemia in stimulating ACTH release, suggesting that it is a multi-component system and that one or more synergistic cofactors exist.

GnRH (gonadoliberin; LHRH, luteinizing hormone releasing hormone) is a decapeptide controlling both lutropin (LH) and follitropin (FSH) through specific receptors coupled to a calcium/calmodulin messenger system. Gonadotropes are known to produce both LH and FSH coded by a single gene for a common large precursor protein. Whether two cell sub-types determine the proportion of each hormone is still uncertain. Pulsatile secretion may be regulated by β-endorphin, and feedback effects (e.g., by estrogen) may act on pulse frequency as well as gonadotrope sensitivity to GnRH.

DA (dopamine) is well established as prolactin inhibitory factor (PIF). The identity of a PRH (prolactin releasing hormone) remains uncertain, but may be a peptide homologous with vasoactive intestinal peptide (VIP), known to be produced in the hypothalamus.

Hypothalamic Receptor Functions

Hypothalamic ***chemoreceptors*** exist which are specific for androgen, estrogen, progesterone, cortisol, and possibly thyroxin. These receptors function in the long-loop feedback control of the anterior pituitary, while other receptors for the pituitary peptide hormones function in the short-loop feedback. Specific ***thermoreceptors*** in the neurons of the anterior hypothalamus are responsible for body temperature and caloric homeostasis. They perceive changes in body temperature and initiate appropriate responses such as shivering and sweating as well as metabolic rate adjustment, appetite, and behavioral changes.

The brain has ***glucoreceptors*** (glucose-sensitive neurons) which respond to hypoglycemia (both rate and degree of fall) by causing an autonomic discharge of norepinephrine and release of the hormones STH (GH) and ACTH. There is no evidence, however, of hypothalamic regulation of insulin or glucagon release. Closely related receptors control appetite, disorders of which range from anorexia to hyperphagia.

Osmoreceptors in the supraoptic nucleus respond to changes in the osmolality of the blood. At levels above 290 mosm/kg these neurons are stimulated to release arginine vasopressin (AVP, antidiuretic hormone, ADH) from the posterior pituitary, promoting water reabsorption in the collecting ducts of the kidney and a fall in electrolyte concentration. Conversely, a drop in osmolality will suppress vasopressin and result in diuresis.

Baroreceptors in the thorax (mainly in the right atrium) respond to a decrease in the effective circulating blood volume (falling arterial blood pressure) by stimulating the release of vasopressin, the signal being mediated via the vagus nerve. When volume depletion and hypoosmolality coexist, baroreceptors override osmoreceptors, resulting in vasopressin release and a state of hyponatremia that may be persistent in chronic wasting illnesses. Hypertonic stimulation of the hypothalamus prompts water-seeking behavior. States of inappropriate vasopressin secretion and drinking behavior can result from congenital, acquired, or psychic disorders involving the hypothalamus.

EVALUATION OF HYPOTHALAMIC FUNCTION

Until recently, test procedures to assess hypothalamic function have depended on the measurement of pituitary and target gland hormones (e.g., somatotropin and cortisol levels induced by insulin hypoglycemia). The diagnosis may still be one of exclusion, as in the cessation of menses, in which all other possibilities must be ruled out before the amenorrhea can be attributed to hypothalamic dysfunction. Immunoreactive assays for hypothalamic hormones have been reported, but their clinical value has yet to be established. Function tests using TRH and GnRH (LHRH) are routine; tests involving CRH and DA are coming into use; tests with SRH and SIF are still under investigation.

Table 14–1 gives the estimated relative frequency of various lesions as causes of endocrinopathies. Although these generalizations are not totally valid, such knowledge allows one to pursue investigations in a logical and cost-effective manner. The hypothalamus is not amenable to direct treatment, but because most hypothalamic defects are mediated via the pituitary and target glands, management of the disorder can be directed at these two areas.

Procedures that are unequivocal indicators of hypothalamic function include:

1. Documentation of *circadian rhythm,* as for cortisol, somatotropin, and prolactin. This requires frequent sampling for a minimum period of 24 hours, using indwelling catheters; for obvious reasons it is not a routine procedure.
2. *Stress testing,* as for cortisol and somatotropin. Giving insulin to induce hypoglycemia is the procedure of choice, with well-documented interpretative criteria.

TABLE 14–1. Approximate Frequency of Primary Lesion Sites in Common Endocrine Disorders*

	Site of Lesion				
Disorder	*Hypothalamus*	*Pituitary*	*Target Gland*	*Target Tissue*	*Ectopic Tropic Factor*
Thyroid					
Hypofunction	<0.1	2	98	—	—
Hyperfunction		<0.1	10	—	90
Adrenal					
Hypofunction	<1	80	20	—	—
Hyperfunction	5?	60	20	—	15
Gonadal					
Hypofunction—Women	95†	3	2	<0.1	—
—Men	60‡	20	20	<0.1	—
Somatotropin (Growth)					
Hypofunction	50	50	<1	<1	—
Hyperfunction	<1	100	—	—	—
Vasopressin (Water)					
Hypofunction (DI)§	90	10	<1	—	—
Hyperfunction (SIADH)‖	50	—	—	—	50

* Values expressed in percent (%)
† Main cause of amenorrhea
‡ Major cause of impotence
§ DI = diabetes insipidus
‖ SIADH = syndrome of inappropriate antidiuretic hormone

3. *Feedback blocking* procedures, as for cortisol. Blocking the production of the target gland hormone with metyrapone causes feedback to cease, and the hypothalamus is stimulated to function.

Additional tests that are used to evaluate hypothalamic–pituitary function probably reflect both endocrine levels, or even exclusively the pituitary level. Thus, responses to oral levodopa, intravenous arginine, or intramuscular vasopressin are not reliable indicators of primary hypothalamic disease. It may be possible to define a hypothalamic lesion more precisely when its hormones can be measured routinely.

DISORDERS OF THE HYPOTHALAMUS

Disease involving the hypothalamus can have widely differing manifestations, including growth, metabolic, endocrine, and psychic disturbances. The cause may be a tumor (craniopharyngioma, gangliocytoma, astrocytoma), a congenital malformation, an inflammatory lesion (encephalitis), or a degenerative process (arteriosclerosis). A rare familial hypothalamic disorder of hypogonadotropic hypogonadism, associated with one or more midline deformities (hair lip, cleft palate, loss of smell) is called *Kallmann's syndrome,* and appears to respond to GnRH. Other lesions are difficult to define pathologically but are probably associated with a defect in neuroendocrine or receptor function.

Defects recognized as caused by altered hypothalamic regulation of the *anterior* pituitary are:

1. Disturbances of sexual function—impotence, hypothalamic amenorrhea, and precocious puberty
2. Galactorrhea caused by hyperprolactinemia,

due to lack of DA (e.g., from traumatic stalk section), which is rare
3. Growth failure in emotionally deprived children due to depressed STH production
4. The multiple psychic and endocrine abnormalities of anorexia nervosa
5. The pituitary-target gland abnormalities seen in malnutrition and debilitating diseases

The most significant hypothalamic-*posterior* pituitary defect is diabetes insipidus. A final group of rare lesions in the hypothalamus itself may produce psychic disturbances, somnolence, impaired temperature regulation, and obesity or emaciation.

THE ANTERIOR PITUITARY

EMBRYOLOGY, ANATOMY, AND HISTOLOGY

The pituitary gland (Fig. 14–3) derives embryologically from two sources: (1) The *adenohypophysis* (*anterior pituitary*) originates as an outgrowth of ectodermal cells (Rathke's pouch) in the roof of the developing oral cavity. These cells migrate upward to form the anterior lobe and part of the pituitary stalk. (2) The *neurohypophysis* (*posterior pituitary*) is formed from neural cells in the floor of the developing third ventricle whose axons migrate downward. The pituitary separates from the oral cavity and becomes surrounded by bone in a depression called the *sella turcica.* Normally the sella measures about 12 mm to 16 mm in diameter, and radiologic evidence of enlargement (>18 mm) is an important diagnostic sign.

The pituitary gland weighs about 0.5 g, with the anterior pituitary accounting for 75%. The pituitary stalk passes through a covering called the *diaphragma sellae* to connect with the hypothalamus. The *optic chiasm* rests on the anterior portion of the diaphragma sellae and is susceptible to pressure from upward-expanding masses in the pituitary (suprasellar extension).

The *blood supply* to the pituitary is derived from the internal carotids via the superior and inferior hypophyseal arteries. The superior artery supplies a capillary network in the median eminence, merging with vessels from the hypothalamus to form a portal venous plexus which delivers blood and hypothalamic hormones to the cells of the anterior pituitary.

The adenohypophysis has developed toward a single cell type for each hormone secreted. There is some overlap in response to secretory stimuli, with one hypothalamic *liberin* or *statin* sometimes affecting more than one pituitary cell type. The old classification of acidophil, basophil, and chromophobe cells is now obsolete, as immunochemical techniques identify at least five cell types. These different cells exhibit resting, well-granulated (storage) forms, and large chromophobe (probably actively secreting) forms. Chromophobe adenomas have been found that secrete any one of the known hormones.

ANTERIOR PITUITARY HORMONES

The seven presently recognized pituitary hormones and details of their production and control are shown in Table 14–2 and Figure 14–3.

An important aspect of anterior pituitary hormone secretion is that it is intermittent, or pulsatile, yet under precise regulation by higher centers. The peripheral receptors for these hormones are dependent on such pulsed secretion, not only for their effective stimulation, but also to prevent down-regulation which would result from continuous exposure.

Somatotropin (STH, Growth Hormone, GH)

Active growth hormone (GH, STH) is a protein of 191 amino acids with two disulfide bridges and a striking homology to chorionic somatomammotropin (placental lactogen) and considerable homology to prolactin (PRL). The pituitary may contain some cells (somatolactotropes) that secrete both STH and PRL. Humans have more than one gene coded for STH and several forms of STH occur in the basal secretion. Physiologic or pharmacologic stimulation results in one major STH (hGH-B), M_r 22,000 or 22K, and one minor 20K hormone. It is the 22K form that predominates in plasma in acromegaly, that is present in the urine, and that accounts for the biologic activity of the hormone.

Somatotropin may have some direct effects on tissues to stimulate protein anabolism and lipolysis and to induce hyperglycemia. Its main action is to stimulate the hepatic production of a family of peptides, *somatomedins* (M_r 6000–8000), which mediate the action of STH on bone and cartilage

TABLE 14–2. The Anterior Pituitary Hormones

Preferred Name	Common Name	Peptide Length	Pituitary Cell Type	Relative Cell Frequency (%)	Estimated Hormone Production (μg/day)
Somatotropin (STH)	Growth hormone (GH)	191	Somatotrope	≃45	500–1000
Thyrotropin (TSH)	Thyroid-stimulating hormone	α: 96 β: 113	Thyrotrope	≃10	≃110
Corticotropin (ACTH)	Adrenocorticotropic hormone	39	Corticotrope (lipocorticotrope)	≃20	≃10
β-Lipotropin (β-LPH)	Same	91			?
Prolactin (PRL)	Same	198	Lactotrope	≃15	?
Follitropin (FSH)	Follicle-stimulating hormone	α: 96 β: 143	Gonadotrope	≃10	≃15
Lutropin (LH)	Luteinizing hormone	α: 96 β: 119	Gonadotrope		≃30

≃ = approximately

and act like STH in other tissues. Somatomedin-C is the same as insulin-like growth factor (IGF-I). STH is mildly diabetogenic and ketogenic and produces a positive calcium and nitrogen balance. In the fetus receptors for STH are immature and growth is probably determined by chorionic somatomammotropin. Growth in children depends on the regularity and amplitude of pulses of STH secreted by the pituitary. Plasma levels may not rise above adult values, so the question of tissue responsiveness must also be considered. Receptors for STH appear to be up-regulated by the hormone pulses.

Hypothalamic ***control*** of STH is mediated by somatostatin and somatoliberin. *Laron dwarfs,* who have elevated STH levels but lack IGF-I and IGF-II, provide evidence that negative feedback control may be effected by somatomedins. Higher control seems to be under the influence of one or more biogenic amines, and short-loop feedback suppression has been demonstrated. Basal levels of STH are <1 μg/L, but are subject to short bursts of secretory activity triggered by a variety of stimuli. These include decreasing plasma glucose, certain amino acids, fatty acids, the state of consciousness (sleep), physical stress (exercise), and psychic factors. Responses to these stimuli may be enhanced by catecholamines (central nervous system activity) and by estrogens.

Radioimmunoassay (RIA) for STH is well established, but because of the intermittent nature of the secretion, RIA is more useful under specified conditions: for example, 1 hour post sleep, following exercise, or a glucose load, or a provocative stimulus such as insulin hypoglycemia, L-dopa, or arginine infusion. In children of short stature the spectrum of somatotropin abnormalities can range from absolute deficiency to intermittent irregularities of hormone secretion, with reduced frequency and/or amplitude of STH pulses. *Reference ranges,* which allow for secretory pulses, in resting AM samples of serum or plasma are: Children: 0–8; men: 0–5; women: 0–9 μg/L (ng/mL).

Acromegaly: usually >10, with no suppression below 5 μg/L by glucose
Hypopituitarism: <2, with no increase induced by hypoglycemia

An assay for *somatomedin-C* (IGF-I) is also available and is potentially useful, because it is less subject to diurnal variation. It probably reflects biologically active growth hormone and has been used to identify children of short stature who may benefit from hGH therapy. Values are high in untreated acromegalics, usually low in STH deficiency, and always low in Laron dwarfs.

Thyrotropin (Thyroid-Stimulating Hormone, TSH)

Thyrotropin is a glycoprotein (M_r 28,000) consisting of an alpha (α) and a beta (β) chain; the α chain is virtually the same as that of FSH, LH, and human chorionic gonadotropin (hCG); all β chains show less homology. RIA specificity has been achieved with antibodies to the β chain. TSH regulates most phases of thyroid activity. Feedback inhibition by circulating free thyroxin (T_4) or triiodothyronine (T_3) occurs primarily at the anterior pituitary. Hypothalamic control is mediated by thyroliberin (thyrotropin releasing hormone, TRH), with no clear evidence that a thyrostatin exists. Responses to TRH are blunted in critically ill patients.

RIA for serum TSH is well established and provides a sensitive test for primary hypothyroidism, in which TSH levels are elevated. Increased TSH, even when T_4 and T_3 levels are normal, indicates a decreased thyroid reserve. *Euthyroid* patients show serum levels of <6 mU/L (<10 mU/L by some methods). The new monoclonal TSH assays are diagnostically useful in primary hyperthyroidism because they have the specificity and sensitivity to demonstrate the low concentrations that are characteristic of this condition. The reference range is 0.5 to 6.0 mU/L, for hyperthyroidism <0.5 mU/L.

Corticotropin (Adrenocorticotropic Hormone, ACTH)

ACTH is a peptide of 39 amino acids derived (along with β-lipotropin [β-LPH] discussed later) from a large precursor, proopiomelanocortin (POMC); see Figure 14-4. Biologic activity resides in the *N*-terminal 1–24 sequence, which is now produced synthetically and is not immunogenic. ACTH stimulates cortisol production by the adrenal cortex, and shows some influence on androgen and mineralocorticoid production as well. The half-life is approximately 10 min, and stimulation of cortisol production is rapid and transient. Hypothalamic ***control*** is mediated by corticoliberin (CRH), possibly modulated by other factors at the level of the corticotropes; negative feedback by cortisol occurs at the level of both the hypothalamus and the pituitary. A circadian rhythm reflects a major higher level of control, but this can be disturbed by CRH administration, by cortisol feedback and a variety of stresses.

Mean levels of plasma ACTH are highest in the early morning, with many small secretory spikes occurring throughout the day. ACTH can be measured by RIA but the hormone is unstable and the method is difficult and costly. An alternative assay of β-LPH, derived from the same POMC precursor, shows good correlation and may prove useful because it is more stable. An assay for β-endorphin appears to have similar advantages. The *reference range* for ACTH in resting nonstressed individuals at 0800 hours is 5–22 pmol/L (20–100 ng/L), but under stress the level can rise to well over 40 pmol/L. ACTH levels should be determined only in specific clinical situations, using special collection techniques, multiple samples, simultaneous cortisol determinations, and correlation with clinical

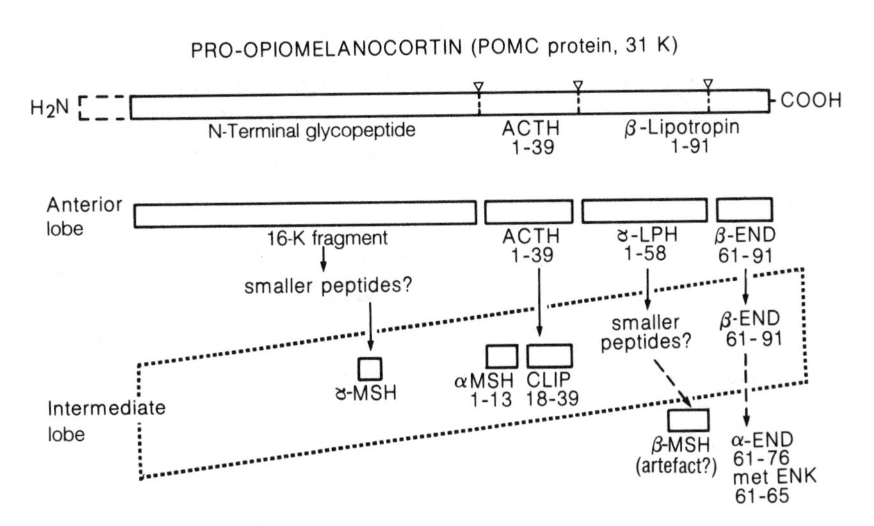

FIGURE 14-4. Hormones and by-products derived from proopiomelanocortin, a precursor protein synthesized by the corticotropes. Inverted triangles show Lys · Arg cleavage sites. ACTH = adrenocorticotropin; CLIP = corticotropin-like intermediate peptide; LPH = lipotropin; MSH = melanocyte-stimulating hormone; END = endorphin; ENK = enkephalin. Note different products of the anterior and intermediate lobes.

data. The assay procedure must be reliable and able to indicate levels less than 2 pmol/L to be of clinical value. Thus, for most purposes, the measurement of serum cortisol or its urinary metabolites is adequate and preferred.

Gonadotropins: Follitropin (Follicle-Stimulating Hormone, FSH) and Lutropin (Luteinizing Hormone, LH)

FSH and LH are glycoproteins consisting of a virtually identical α chain (also found in TSH and hCG) but different β chains, which confer specificity. The carbohydrate moiety comprises 20% to 25%. FSH stimulates ovarian follicle growth and estrogen production in women and testicular spermatogenesis in men. LH induces ovulation and thereafter maintains the corpus luteum and progesterone production in women, while it stimulates testosterone production by the Leydig cells of the testes in men.

Hypothalamic ***control*** of both FSH and LH appears to be by a common releasing hormone termed gonadoliberin (GnRH, LHRH), with *negative* feedback control at the hypothalamic level by estrogen in the female and testosterone in the male. In women, *positive* feedback by rising estrogen levels induces the LH surge that causes ovulation. The cyclic menstrual pattern recurs about every 28 days with a superimposed pulsatile (approximately hourly) secretion of gonadotropins. Control is at hypothalamic and higher neural levels and can be disrupted by stress. In men spermatogenesis requires sustained testicular gonadotropin stimulation over about 70 days involving pulsatile (approximately hourly) LH and FSH secretion. In addition to a negative feedback exerted by testosterone, testicular tubules appear to secrete a polypeptide termed *inhibin* which may suppress FSH (See Fig. 14–15).

Serum RIA measurements of FSH and LH based on antibodies to the whole molecule will suffer from nonspecificity at low levels. Proof of deficiency states may require the demonstration of nonresponse to stimulation. Results are reported in international units (IU/L, mIU/mL), because they are based on a bioassay reference standard. Typical reference values are shown in Table 14–3.

Prolactin (PRL)

Prolactin is a protein of 198 amino acids (M_r 21,000, or 21K) probably derived from precursor proteins of 40K or about 100K. Normally more than 80% of plasma PRL is the monomeric (21K) form. In cases of hyperprolactinemia the larger forms, which have low receptor affinity and weaker biologic activity, may make up 25%–75% of the total. The 100K form is increased in some women athletes. PRL acts directly on the mammary gland to stimulate milk production. In excess it increases adrenal androgens, decreases gonadoliberin, and blunts the response of the gonadotropes and gonadal cells to tropic factors. Thus, it induces amenorrhea in women and impotence in men. Prolactin levels increase in pregnancy, due to estrogens, and during lactation due to suckling. There is no target-gland hormone to exert a negative feedback; instead PRL is under tonic inhibitory control by the hypothalamus, self-regulating by a short-loop feedback. Prolactostatin (PIF) is now established as dopamine (DA). There is growing evidence for a *prolactoliberin,* but its identity remains in doubt. The control mechanisms result

TABLE 14–3. Serum Gonadotropin Reference Values

	Males		Females	
	FSH	*LH*	*FSH*	*LH*
	(IU/L; mIU/mL)		*(IU/L; mIU/mL)*	
Prepubertal	1–3	2–6	1–3	2–6
Adult	2–15	5–25		
Follicular phase*			3–15	5–30
Ovulatory spike*			10–50	50–150
Luteal phase*			3–15	5–40
Postmenopausal			30–200	30–200

FSH = follitropin; LH = lutropin

* Phases of menstrual cycle

in a pulsatile release, and a circadian rhythm having a nadir in late morning and a peak at mid-sleep. Other factors that increase PRL secretion are stress, serotonin, vasoactive intestinal peptide, and psychoactive drugs such as the phenothiazines. Dopamine agonists (L-dopa, bromocriptine, pergolide) suppress prolactin. Blockade of DA receptors (e.g., by metoclopramide) or lowering DA synthesis (e.g., by monoiodotyrosine) increases prolactin.

Excellent RIA methods are available for measuring prolactin. *Reference ranges* for serum PRL from non-stressed individuals between 1000 hours and 1600 hours (10 AM to 4 PM) are 0 μg/L to 15 μg/L (ng/mL) for men and 0 μg/L to 20 μg/L (ng/mL) for women.

β-Lipotropin (β-LPH)

β-LPH is a peptide of 91 amino acids derived from POMC and secreted by the corticotropes (Fig. 14–4). The intact peptide has no known biologic function, but it contains active sequences of β-MSH and β-endorphin (β-EP), and so may be involved in pigmentation and pain tolerance. In the *anterior pituitary* it is partly degraded to γ-LPH (a.a. 1–58) and β-EP (a.a. 61–91). In the *intermediate lobe,* which has negligible function in humans, γ-LPH can be deleted to yield β-MSH (melanocyte stimulating hormone); β-EP can give rise to α-EP (a.a. 61–76) and metenkephalin (a.a. 61–65); and ACTH can yield the deletion peptides α-MSH and CLIP (corticotropin-like intermediate peptide). The endorphins and enkephalins, particularly the pentapeptides, are present in many parts of the nervous, endocrine and other tissues. Some are considered to be endogenous opiates, in fact β-EP is the most potent naturally occurring analgesic known. Cushing's syndrome resulting from tumors of the intermediate lobe, or from ectopic tumors elsewhere, may release these unusual metabolites into the circulation. RIA methods for β-EP are available; it can substitute for ACTH assays and high concentrations may indicate that a more aggressive tumor is present.

EVALUATION OF ANTERIOR PITUITARY FUNCTION

The signs and symptoms of anterior pituitary disease may reflect space-occupying lesions (usually macroadenomas), hormone *excess* (from macro- or microadenomas or hyperplasia), or hormone *deficiency* caused by local or hypothalamic dysfunction. Investigation of these disorders has become easier with the tests and procedures now available. Successful treatment usually requires that the location and nature of the lesion be known.

The most ***useful investigations*** for anterior pituitary disorders are as follows:

1. Radiologic investigation of the hypothalamic/pituitary area, especially with high resolution computerized tomography (CT) scanning, can demonstrate space occupying masses responsible for hormonal or pressure effects
2. Measurement of basal concentrations of circulating anterior pituitary and target tissue hormones
3. Evaluation of circadian hormonal rhythms to demonstrate loss of control
4. Stimulation tests to demonstrate the adequacy of pituitary reserve, because low basal levels are often normal
5. Suppression tests to demonstrate a loss of control (i.e., autonomous function), for example, the glucose suppression test of STH for the diagnosis of acromegaly

The Triple-Bolus (Hormone) Test

This useful test of anterior pituitary function can be managed on a 1-day outpatient basis. The patient is given serial intravenous bolus injections of insulin (usually 0.15 U/kg), gonadoliberin (GnRH, 100 μg), and thyroliberin (TRH, 200 μg). Blood samples are collected at −15, 0, 20, 30, 45, 60, and 90 min, and the plasma from each sample is analyzed by RIA for STH, cortisol, LH, FSH, TSH, PRL, and glucose. After confirming an adequate hypoglycemia the following criteria for *normal responses* are commonly employed:

Insulin:	Glucose	fall to <2.2 mmol/L (40 mg/dL) and half the baseline level
	STH	increment of >8 μg/L
	Cortisol	increment of >200 nmol/L (70 μg/L)
TRH:	TSH	increment of >5 mU/L
	Prolactin	2-fold increment
GnRH:	FSH	increment of >3 IU/L
	LH	increment of >5 IU/L

The triple-bolus procedure varies somewhat between different institutions in terms of the timing of sampling, simultaneous versus sequential testing, and interpretative criteria. The commonest procedure is simultaneous hormone administration, but in certain cases giving each stimulus and taking samples is done sequentially (with a rest period in between) to avoid overlapping effects of the stimuli. The triple-bolus procedure is very costly for the laboratory tests alone and should be used only when clearly indicated, especially because results are often borderline and interpretation is difficult. In a recent modification of this test insulin is replaced by corticoliberin (CRH, 100 μg) and somatoliberin (SRH, 100 μg). Interpretative criteria are similar and the risks of hypoglycemia are avoided.

Selective Tests of Anterior Pituitary Function

The availability of hypothalamic hormones has made it possible to test pituitary function by selective stimulation. The advantages, however, are limited.

Somatoliberin (SRH; GRH) has potential value for both diagnosis and therapy. When given as a 100 μg bolus it acts selectively on the somatotropes in normal individuals to increase serum STH (GH) by >5 μg/L in 30 to 120 min. In acromegalic patients the response is variable, but in general higher and more prolonged. Growth-hormone-deficient patients will show a negative or minimal response when the somatotropes are not functioning. When growth failure is secondary to a hypothalamic disorder the response is definite but blunted, probably due to downregulation of receptors on the somatotropes.

Gonadoliberin (GnRH, LHRH) has been used successfully in the triple-bolus test for several years. It can be used selectively in the diagnosis of some gonadal disorders.

Thyroliberin (TRH) is used routinely for evaluating thyroid disorders; 200 μg is given intravenously, with serum levels of TSH measured at 20, 40, and 60 min, and (optionally) of T_3 at 180 min. In primary hypothyroidism TSH levels are elevated, although T_3 remains low. In secondary hypothyroidism, levels of TSH are usually low with little or no response to TRH. A sluggish but definite response should occur if the defect is at the level of the hypothalamus.

Corticoliberin (CRH) may earn a place as a useful diagnostic tool. It is given usually as a 100-μg bolus intravenously; maximum increments in serum ACTH and cortisol occur between 30 and 120 minutes later. CRH alone is not as potent a stimulus of ACTH secretion as is insulin hypoglycemia; in fact, a synergistic effect of vasopressin on the ACTH response to CRH has been noted. A *normal* response is approximately a 2-fold increase in ACTH by 30 min and a 2-fold increase in cortisol by 60 min.

DISORDERS OF THE ANTERIOR PITUITARY

Disorders of the anterior pituitary can present as endocrine hypofunction, endocrine hyperfunction, local problems, or any combination of them. Thus, a ***pituitary tumor*** secreting STH will cause acromegaly as a result of hyperfunction, but will cause hypothyroidism, hypoadrenalism, and hypogonadism if the tumor size impairs the secretion of other tropic hormones. If the tumor becomes large enough to erode and expand into structures surrounding the pituitary, it may cause local complications, which include blindness, nerve palsies, headache, and leakage of cerebrospinal fluid. The hormonal abnormalities and the resulting disorders are shown in Table 14–4.

The most *frequent* pituitary disorder is a benign hypersecreting adenoma, with a prolactinoma being by far the most common. The most *serious* pituitary disorder is ACTH deficiency causing life-threatening adrenal insufficiency. This may occur as an isolated defect, or in association with other deficiencies as the result of a pituitary tumor or other disorders. Tumors usually result in the loss of hormones in a specific sequence, with STH, LH, and FSH being more sensitive to pressure than TSH, and ACTH least sensitive.

Acute Anterior Pituitary Failure (Pituitary Apoplexy)

This is a rare but often fatal condition, usually resulting from acute hemorrhagic infarction of the pituitary, or as a result of associated disease (pituitary tumor, vascular problems, infectious or infiltrative disorders), or destructive processes such as surgery or radiation. The differential diagnosis has to exclude other conditions that can simulate the disorder. In those who survive the acute phase it is important to assess residual endocrine func-

TABLE 14-4. Disorders of the Anterior Pituitary

Hormone	Hypofunction	Hyperfunction
Somatotropin (STH)	Growth failure in children	Gigantism Acromegaly
Thyrotropin (TSH)	Hypothyroidism	Hyperthyroidism (very rare)
Corticotropin (ACTH)	Adrenal insufficiency (secondary)	Cushing's disease
Gonadotropins (LH and FSH)	Amenorrhea Infertility	Rare (no recognized syndrome)
Prolactin (PRL)	Postpartum failure of lactation	Galactorrhea Amenorrhea Infertility Impotence

tion. Because the pituitary is so well protected it is seldom affected by trauma to the head. Pituitary infarction as a complication of postpartum hemorrhage (*Sheehan's syndrome*) is now rare with good obstetric care. Transsphenoidal adenomectomy of pituitary tumors should decrease the incidence of pituitary apoplexy even further.

The *consequences* of a sudden loss of anterior pituitary function are generally as follows:

1. Days 1 and 2: a fall in FSH, LH, STH, PRL, and sex steroids occurs, and they remain low in almost all cases.
2. Days 4 to 14: owing to a decrease in ACTH, the debility of adrenal insufficiency develops in about 70% of cases, but usually no salt-losing state occurs.
3. Weeks 4 to 8: owing to a decrease in TSH, followed by the depletion of thyroid hormone reserves, cold intolerance and lethargy develop in 40% of cases.
4. Posterior pituitary function is usually spared but, depending on the extent of the lesion, the loss of ADH and consequent diabetes insipidus occurs in $<5\%$ of cases.

Patients usually present with severe headache, visual disturbances, hypotension, and shock or coma. *Emergency* treatment is required and hormonal confirmation of the diagnosis is deferred. The major factors contributing to death are the extent of the pituitary lesion (and the consequent effects on surrounding structures) and the lack of cortisol (from the ACTH deficiency). Later confirmation of the diagnosis requires the demonstration of low hormonal levels that do not respond to stimulatory tests.

Chronic Anterior Pituitary Insufficiency

Some of the conditions that can cause acute pituitary failure are, in fact, more likely to produce a gradual loss of anterior pituitary function. Rarely a congenital deficiency caused by the absence of cells or cell function occurs. In children, the presentation is usually one of growth failure, but if the condition develops in the adult a gradual loss of one or more target gland functions occurs, with the accompanying clinical features.

The *differential diagnosis* of chronic anterior pituitary deficiency, which presents as a syndrome of apathy, inertia, weakness, irritability, and hypothermia, requires the exclusion of severe malnutrition and primary disease of the thyroid, adrenals, or gonads. The initial demonstration of low or even low-normal thyroid, adrenal, and gonadal function must be followed by stimulatory tests (e.g., the triple-bolus test) to demonstrate the extent of hormonal reserve. *Treatment* depends on the causative lesion but usually pituitary function is permanently deficient and replacement target gland hormones (adrenal and thyroid, with or without gonadal hormones) are required.

Isolated Hypofunction of the Anterior Pituitary

STH Deficiency. In children who are slow to grow, or in adults with signs or symptoms of pituitary insufficiency, random STH assays are seldom useful. It is necessary to demonstrate that a provocative stimulus fails to induce the expected response. As a *screening test* in children one can measure serum STH 60 to 90 min after the onset of nocturnal sleep, 3 to 5 hours postprandially, or 20 min after exercise. Levels in normal individuals should be at least 7 μg/L. Values above this level exclude the diagnosis but low levels indicate only the need for more definitive investigation.

The *best test* still is a failure of STH concentrations to respond to the stimulus of hypoglycemia induced by regular insulin (0.05 to 0.15 U/kg IV). Blood samples are taken at 0, 30, 60, 90, and 120 min. Glucose levels must fall below 50% of the basal level and below 2.2 mmol/L (40 mg/dL) to achieve adequate stimulation. The STH level in normal individuals rises above 7 μg/L. An *alternative procedure* consists of giving 500 mg levadopa orally, with the same sampling times and reference values as for hypoglycemia. A new test is the response to synthetic somatoliberin (SRH-40 or SRH-44). In normal individuals a dose of 1 μg/kg (or a 100-μg bolus) will produce an increase of >5 μg/L in plasma STH, with a maximum between 30 to 120 min.

Growth failure can result from neurosecretory abnormalities ranging from an absolute deficiency of STH to a reduced number, or amplitude of pulses. It may be associated with a receptor defect in the somatotropes, the secretion of 'inactive' STH, or may be caused by excess somatostatin. Ectopic 'somatostatinomas' have also been reported. STH action is mediated by insulin-like growth factors, IGF-I (somatomedin-C), and IGF-II. Both these factors are very low in Laron dwarfs. IGF-I is low in the African pygmy. Somatomedin-C assays give an indication of circulating active STH. One of the commoner causes of STH deficiency is a craniopharyngioma, which is either suprasellar and interferes at the hypothalamic level, or invades down into the sella, causing pituitary destruction. Tumors require surgery; the hormone deficiency requires an ongoing series of injections of human STH given periodically prior to epiphyseal closure in order to obtain full development.

Other Deficiency States. These occur occasionally as idiopathic disorders, or in association with the 'empty sella' syndrome, or more frequently as a result of disease states causing pituitary destruction. Treatment is directed at the disease state if possible, with hormonal replacement for the target gland defect. The deficiencies include the following:

Loss of FSH and LH. In children puberty will be delayed or absent, increased growth of the arms and legs may be noted, along with a eunuchoid appearance and a tendency to osteoporosis (from the lack of anabolic steroids). In adults there will be a loss of libido, azoospermia or amenorrhea, some regression of sex characteristics, and aging of the skin.

Loss of TSH. Signs of hypothyroidism will develop, including lethargy, cold intolerance, bradycardia, and the dry and coarse skin termed *myxedema.*

Loss of ACTH. Evidence of adrenal deficiency will include weakness, a tendency to low blood pressure, and hypoglycemia. Under stress an adrenal crisis (nausea, vomiting, vascular collapse, and shock) may occur, which can be fatal unless promptly treated. Isolated ACTH deficiency can occur in the 'empty sella' syndrome.

Loss of PRL. This is not well documented as an isolated defect but postpartum deficiency leads to failure of lactation.

Loss of β-LPH. No known function, so no deficiency syndrome can be identified.

Hypersecreting Tumors of the Anterior Pituitary

These tumors are usually localized, small, slow-growing benign adenomas whose main effects are hormonal. As their size increases, local effects can occur in the structures surrounding the pituitary. Suprasellar extension and pressure on the optic nerve can cause visual loss, often one of the late presenting signs of larger tumors.

Somatotrope Adenoma. A pituitary tumor secreting excess STH is not uncommon. It causes *gigantism* in prepubertal children because proportional growth continues prior to epiphyseal closure. In adults there is enlargement of the distal parts of the body (*acromegaly*), with the bones and tissues of the face, hands, and feet particularly affected. A coarse facial appearance, generalized visceromegaly, deepening of the voice, hirsutism, and increased skin thickness all occur. Hypertension, impaired glucose tolerance, and diabetes are common. Galactorrhea is present in some and is probably due to an associated elevation of prolactin (possibly from somatolactotropes), which occurs in about one third of patients. In spite of the characteristic appearance of such patients, diagnosis is usually late and sellar enlargement is common, with headache and visual loss often the presenting complaints. Neurologic and musculoskeletal abnormalities also develop. An *ectopic* somatoliberin syndrome (causing acromegaly) has been described in association with tumors in the pancreas, gastrointestinal tract, adrenal gland or lung. A small subgroup of acromegalics may have a deficiency of *somatostatin* production.

Diagnosis is confirmed by demonstrating the failure of STH levels to suppress below 5 μg/L after the ingestion of 100 g glucose; in fact, a paradoxic rise sometimes occurs. Fasting serum inorganic phosphate is usually elevated, often >1.45 mmol/L (>4.5 mg/dL). In addition, intravenous TRH, which in normal individuals either lowers or has no effect on STH, in acromegalics usually causes a marked increase. Standard radiation therapy is generally ineffective and *treatment* now is transphenoidal adenomectomy or proton-beam radiation. A temporary alternative is the use of the dopamine agonist bromocriptine, which will suppress STH in some patients, notably those who also show an increase in PRL. Somatostatin-deficient acromegalics respond to treatment with synthetic analogues of this hormone. Untreated, the disease may burn out or progress until it causes death by pituitary apoplexy, heart disease, or other causes. Early treatment will reverse many of the abnormalities to some degree.

Lactotrope Adenoma. Since the discovery of prolactin, the common so-called nonsecreting chromophobe tumor of the past has turned out in most cases to be a PRL-secreting ***prolactinoma.*** In women presenting with *galactorrhea* and *amenorrhea* the probability of a prolactinoma is over 50%; with *amenorrhea* alone the probability is about 20%. While prolactin may have additional effects, no other specific clinical signs occur. The equivalent of amenorrhea in men appears to be a combination of impotence and reduced fertility. These gonadal effects of prolactin may be at the level of either the gonad, the pituitary, or the hypothalamus, or a combination of them.

Two syndromes have been defined. *Idiopathic hyperprolactinemia* is believed to be caused by a lack of sensitivity of the lactotropes to inhibition by dopamine (DA), probably from a decrease in receptor numbers or affinity. *Pathologic prolactinomas* are a primary pituitary disorder. They show a blunted increase in PRL when DA is suppressed by metoclopramine, indicating that they are less subject to hypothalamic control. DA agonists (e.g., bromocriptine) will suppress PRL in normal subjects and at higher doses in patients with prolactinomas. Microadenomas (<10 mm) appear to resist suppression by DA agonists more often than do the less common macroadenomas. Isolated hyperprolactinemia has been found in rare cases of 'empty sella' syndrome, but responds to therapy. Hyperprolactinemia caused by increased levels of the large precursor hormone, which has a low biologic activity, has been observed in women with galactorrhea but normal menses and fertility.

The *diagnosis* depends on RIA measurement of prolactin along with radiologic evaluation of pituitary structure. Prolactin levels greater than 200 μg/L are almost diagnostic of tumor and the sella is usually abnormal. Values between 50 and 200 μg/L are probably caused by a microadenoma and pose a dilemma both in diagnosis and *treatment.* Transsphenoidal adenomectomy is probably indicated if the sella is abnormal. Most elevated prolactin levels fall to normal on treatment with dopamine agonists (bromocriptine, pergolide) or following pituitary radiation. Galactorrhea ceases, menses resumes, and the tumor decreases in size. The natural course of the disease itself is relatively benign.

Other Adenomas. The ***thyrotrope adenoma*** (causing pituitary hyperthyroidism) is rare, about 1 in 250 pituitary tumors, and is discussed under thyroid disorders. The ***corticotrope adenoma*** causing Cushing's disease is more common, about 15% of pituitary tumors. Most are located in the anterior region, secrete mainly ACTH, and are amenable to surgical resection. A few arise in the intermediate lobe, and are harder to treat, and may give rise to more varied by-products. The subject is discussed further under adrenal disorders. The ***gonadotrope*** (glycoprotein producing) ***adenoma*** is extremely rare.

Although most nonsecreting chromophobe adenomas of the past were probably lactotrope adenomas, a small number of chromophobe or nonsecreting adenomas still exist. Also of interest are the relatively rare tumors that secrete more than one pituitary hormone (e.g., STH and TSH; STH, PRL, and TSH; STH and ACTH; PRL and

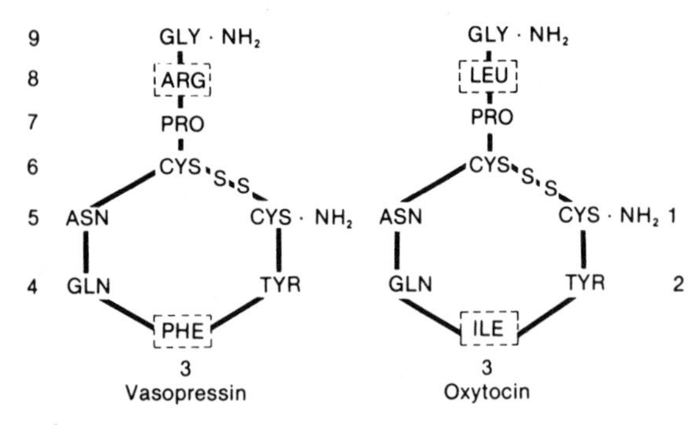

FIGURE 14–5. Structure of the hormones of the hypothalamic–posterior pituitary neurons. ARG = Arginine; ASN = asparagine; CYS = cysteine; CYS · NH_2 = cysteinamide; GLN = glutamine; GLY · NH_2 = glycinamide; ILE = isoleucine; LEU = leucine; PHE = phenylalanine; PRO = proline; TYR = tyrosine.

ACTH). These tumors may consist of one cell type, or two cell types (bimorphous). About *one* in *five* pituitary tumors is a hormonally inactive adenoma, or an oncocytoma.

THE POSTERIOR PITUITARY

STRUCTURE, HORMONES, AND FUNCTIONS

The posterior pituitary (neurohypophysis) consists of axons and dendrites of thousands of neurons whose cell bodies are found mainly in the supraoptic and paraventricular nuclei of the hypothalamus. Two hormones generated in these cell bodies are synthesized in precursor form, along with a 'carrier' protein *neurophysin* which is expressed by the same genome. ***Vasopressin*** (VP; antidiuretic hormone, ADH) is derived from propressophysin, and ***oxytocin*** (OT) from prooxyphysin. These complexes ($M_r \simeq 20{,}000$) migrate down the axons and are stored in the posterior pituitary. The active hormones (along with free neurophysin) are released on demand by exocytosis into the capillaries of the inferior hypophyseal blood vessels. Neurophysin has no known biologic function.

As shown in Figure 14–5, the hormones have seven amino acids in common; in the 3 and 8 positions, human vasopressin has phenylalanine and arginine (hence it is sometimes referred to as *arginine vasopressin,* AVP), and human oxytocin has isoleucine and leucine. This homology explains the degree of overlapping effects. ADH has no pressor activity at physiologic levels; it acts on the collecting duct system of the renal cortex and medulla as an antidiuretic, increasing water reabsorption and concentrating the urine. Recent evidence suggests that it also has neurotransmitter-like effects in the CNS (affecting learned behavior, body temperature, and blood pressure control) and can modulate the responses of other cells (e.g., corticotropes to corticoliberin). The main effects of oxytocin are to cause milk release, and to promote contraction of an estrogen-sensitized uterus at parturition. Synthetic injectable preparations of both hormones are available.

Vasopressin release and synthesis are under the primary ***control*** of osmoreceptors in the hypothalamus. It is only when blood volume decreases by more than 8%, or blood pressure falls by more than 5%, that baroreceptors in the great vessels of the thorax become dominant. Vasopressin is also stimulated by nausea, insulin hypoglycemia, pain, stress, exercise, sleep, and drugs such as nicotine, morphine, metyrapone, and barbiturates. It is inhibited by hypoosmolality, central volume expansion, cold exposure, and drugs such as alcohol, diphenylhydantoin (Dilantin), and glucocorticoids. Oxytocin concentrations in plasma are normally very low, and remain low during pregnancy; release and synthesis are under neural control, being stimulated by suckling during breast feeding and by cervical dilation during the expulsion phase of parturition. Both peptides appear to be metabolized mainly in the liver, by cleavage of the disulfide bond of cystine. The half-life of vasopressin is approximately 5 to 10 min; of oxytocin, 3 to 5 min.

EVALUATION OF POSTERIOR PITUITARY FUNCTION

Although radioimmunoassay (RIA) for ADH is not yet widely available, a definitive approach sometimes requires this procedure. More commonly, ADH levels are inferred from plasma osmolality. The usual *reference range* for osmolality, measured by freezing-point depression, is 285 to 295 mosm/kg of plasma H_2O. At levels below 285 mosm/kg, ADH values should be very low (1 to 2 ng/L). Above 285 mosm/kg, values rise quickly to a range of 5 to 10 ng/L at 295 to 300 mosm/kg.

Another approach is to relate urine osmolality to plasma or serum osmolality by a comparison of sodium levels in simultaneously collected blood and random urine specimens. A finding of low serum sodium along with high urine sodium values (>20 mmol/L) suggests a *hyper*vasopressin state.

A useful check on ADH *deficiency* (diabetes insipidus) is the ***water deprivation test.*** The patient is weighed carefully and the test is terminated when 3% to 5% of body weight is lost. No food or fluid is allowed and hourly urine specimens are collected for measurements of volume and concentration (osmolality or specific gravity). Once a constant urine osmolality ($\pm$30 mosm/kg) is reached, blood is obtained for the determination of plasma osmolality, and an injection of 5 U aqueous vasopressin is given subcutaneously. Hourly urine measurements are continued for 2 to 3 hours. Normal individuals will have a urine/plasma osmolality (U/P) ratio usually >2, those with psychogenic polydipsia >1; the increase in urine osmolality after vasopressin administration will be <5%. Patients with 'complete' diabetes insipidus (DI) will have a U/P ratio <1, those with partial deficiency usually < 2. Both will show a greater than normal response to administered vasopressin. Nephrogenic DI does not respond to vasopressin.

DISORDERS OF THE POSTERIOR PITUITARY

Hypovasopressin States (Diabetes Insipidus, DI)

Diabetes insipidus results from a failure to concentrate the urine, so that a large volume (5 to 15 L/day) of very dilute urine is excreted. There is a tendency to hypernatremia, but increased fluid intake and an obligatory loss of sodium moderates this development. The kidneys are unable to conserve free water, diuresis ensues, and plasma osmolality rises, which normally acts on hypothalamic receptors that provide thirst perception and increased fluid intake. In a rare disorder the receptors controlling *thirst* perception are also affected, so that water loss occurs without thirst, resulting in severe hypertonic dehydration, the syndrome of *chronic asymptomatic hypernatremia.*

Hypothalamic DI occurs most commonly following hypophyseal ablation in the management of cancer of the breast or gonads. It may be transient if enough functional (usually short-axon) neurons survive. Less commonly DI occurs as a result of infection, trauma, or degenerative or infiltrative lesions of the pituitary or hypothalamus. Idiopathic, possibly congenital, ***primary DI,*** manifesting in children or young adults, is relatively rare.

The expression of DI may be masked by a lack of anterior pituitary hormones, because STH, thyroxin, and cortisol affect free water clearance. Patients with such a lack will develop DI only when replacement therapy is instituted. Transient DI occurs from ingestion of alcohol, which probably inhibits ADH release.

Nephrogenic DI is caused by lack (or insensitivity) of renal receptors to ADH in the collecting duct system. It may be congenital and manifest itself in early infancy, more often in male children. It can be acquired later in life in association with hypokalemia, hypercalcemia, renal disease, or certain drugs.

Psychogenic polydipsia is a disorder that simulates DI, the polyuria resulting from compulsive water drinking. Urine volumes usually range from 4.0 to 6.5 L/day. It is distinguished from DI by responding to dehydration with an increase in urine osmolality. The marked fluid intake may deplete the renal medullary concentrating mechanism so that patients respond poorly to vasopressin. They may develop a stable hyponatremia due to a resetting of their osmoreceptors for ADH secretion.

Primary ADH deficiency is best *treated* with a nasal spray containing an analog of arginine vasopressin (DDAVP), or injectable preparations of the hormone. Drugs such as chlorpropamide, clofibrate, and carbamazepine stimulate the release and potentiate the effect of ADH, allowing control of some patients by oral medication alone. Children with nephrogenic DI are managed with diuretics (e.g., chlorothiazide) which lower the blood volume and decrease the glomerular filtration rate, so that sufficient reabsorption occurs in the tubules. Adults are managed by protein and sodium restriction to reduce the osmotic load. Psychogenic polydipsia is generally a psychiatric problem.

Hypervasopressin States (Inappropriate ADH Secretion)

These disorders can be divided broadly into a *hypovolemic* relative excess of ADH and a *hypervolemic* absolute excess.

A ***relative excess*** of ADH occurs in any disorder associated with hypovolemia from fluid and electrolyte loss. Baroreceptor function takes precedence and water retention occurs in spite of a low serum Na^+. ADH levels are appropriate to the low blood volume but inappropriate to serum osmolality. This situation may develop as a chronic state secondary to a variety of clinical disorders. It can also occur because of a malfunction of hypothalamic osmoreceptors or great vessel baroreceptors. ADH excess due to hypovolemia is characterized by a low *urine* Na^+ (<10 mmol/L). It is treated with intravenous saline, plasma, or blood as required.

An ***absolute excess*** of ADH can occur from the administration of too much vasopressin, or from ectopic ADH production. The *syndrome of inappropriate ADH secretion* (SIADH) is a state of dilutional hyponatremia with a *urine* Na^+ that usually exceeds 20 mmol/L. It is now recognized that 40% of patients with carcinoma of the lung and a smaller percentage of patients with a variety of other carcinomas have *ectopic production* of ADH, but only a few exhibit a serious electrolyte disturbance. Inappropriate ADH secretion occurs also in CNS disorders resulting from skull fractures, meningitis, and encephalitis. It can be induced by drugs such as chlorpropamide, clofibrate, vincristine, cyclophosphamide, morphine, or barbiturates. Pulmonary infections may impede blood return to the left atrium and activate the baroreceptors.

Clinical manifestations are related to sodium levels, values above 125 mmol/L being generally asymptomatic. Between 115 and 125 mmol Na^+/L *symptoms* are mild (anorexia, nausea, vomiting), but below 115 mmol/L they are usually severe (seizures, stupor, coma). The differential diagnosis must exclude thyroid, adrenal, and kidney disease.

Treatment of ectopic ADH production by removal of the tumor is occasionally successful. In all forms of SIADH the effect of the hormone on the kidneys can be blocked by lithium carbonate or demethylchlortetracycline (demeclocycline). Excess ADH secretion produces clinical symptoms only if fluid is given and free water is retained, so that treatment can usually be effected by severely restricting access to water. Patients often stabilize at serum Na^+ values of 125 mmol/L to 130 mmol/L and appear to reset their osmoreceptors to this lower osmolarity.

Oxytocin Disorders

Hyperoxytocin states are not known to exist. *Hypooxytocin* disorders may be suspected but are as yet ill-defined. Uterine contractions are sometimes weak during labor and appear to respond to oxytocin administration. Oxytocin is also used to contract the uterus and to reduce postpartum bleeding.

THE THYROID

Thyroid disorders are among the most frequent endocrine abnormalities encountered by the physician. The clinical expression of these disorders is so varied that the laboratory plays a central role in their assessment and an understanding of laboratory tests is essential to effective clinical management.

STRUCTURE AND FUNCTION

Origins, Anatomy, and Histology

The thyroid develops from the pharyngeal floor early in gestation and is functional by the end of the first trimester. Very little maternal thyrotropin (TSH) or thyroid hormone crosses the placenta, so that the fetus is dependent on its own resources, which are essential to further development. At birth the gland weighs less than 2 g, but it reaches 20 g in the adult, becoming relatively one of the larger endocrine organs.

The thyroid is located at the base of the neck, straddling the trachea. It consists of two main lobes ($4 \times 2 \times 2$ cm) joined by a thin isthmus, occasionally with a small central 'pyramidal' lobe extending upward. Two small *parathyroid* glands are closely attached to the posterior surface of each of the two main lobes.

The blood flow from bilateral superior and inferior thyroid arteries exceeds 100 mL/min and passes through a capillary network around each thyroid follicle. The gland is composed of clusters of these *follicles,* which are lined by a single layer of epithelial cells and contain stored thyroglobulin within their colloid. The outer surface of each thyroid cell is covered by a basement membrane; its apical inner surface has microvilli extending into the colloid. Cells of ultimobranchial origin, so called parafollicular, or C cells, that secrete calcitonin (calcium lowering hormone), are located both within, and in the interstitial spaces between the follicles.

Iodine Metabolism and Thyroid Hormone Synthesis and Storage

The essential features of thyroid hormone *synthesis* are shown in Figure 14–6. Circulating iodide from either dietary or endogenous sources is actively transported across the basal membrane into the thyroid cell, which can concentrate iodide to a ratio exceeding 30 times the plasma levels. This uptake of iodide is energy-dependent, is stimulated by TSH, and can be blocked by other monovalent anions, including thiocyanate or perchlorate.

Peroxidase enzymes convert iodide to an active oxidized state which can iodinate tyrosine. ***Thyroglobulin*** is secreted by the thyroid cell into the follicular lumen and forms the major component of both the follicular colloid and the normal thyroid mass. It is a 19S molecule (M_r 660,000), contains approximately 120 tyrosine residues, and is the repository of virtually all thyroid hormones and precursors within the gland. Peroxidase-facilitated iodination of thyroglobulin occurs at the cell-colloid interface, forming both monoiodotyrosine (MIT) and diiodotyrosine (DIT). The subsequent coupling of adjacent molecules of DIT to form tetraiodothyronine (*thyroxin, T_4*), and one each of MIT and DIT to form *triiodothyronine* (*T_3*), also

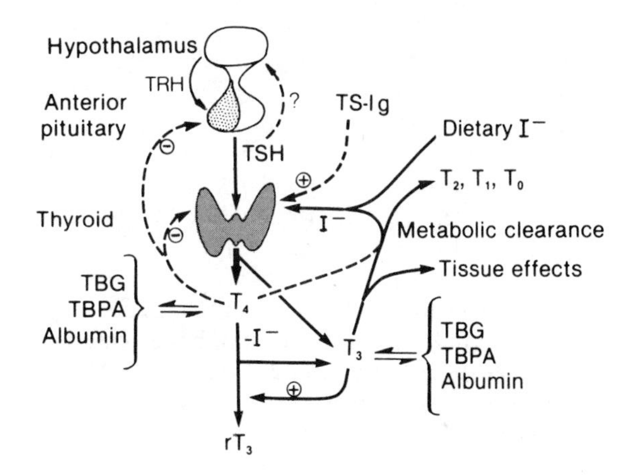

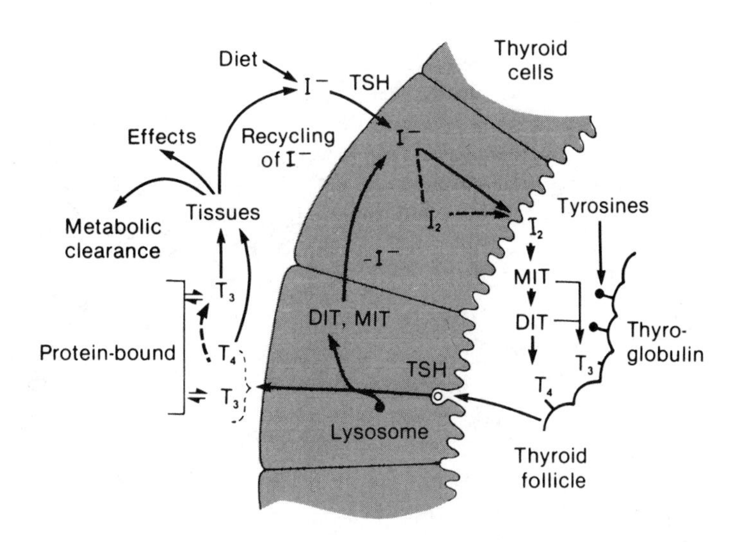

FIGURE 14–6. Synthesis, storage, secretion, metabolism, and control of hormones of the thyroid. I^- = iodide; DIT = diiodotyrosine; MIT = monoiodotyrosine; T_0 = thyronine; T_1 = monoiodothyronine; T_2 = diiodothyronine; T_3 = triiodothyronine; rT_3 = reverse T_3; T_4 = thyroxin; TBG = thyroxin-binding globulin; TBPA = thyroxin-binding prealbumin; TSH = thyrotropin; TS-Ig = thyroid-stimulating immunoglobulin. TRH = thyrotropin-releasing hormone (thyroliberin).

requires peroxidase. The thyronines remain as an integral part of thyroglobulin during this process and the thyroid hormones are *stored* as thyronines in the colloid. The processes of iodination (organification) and coupling are stimulated by TSH and inhibited by thiourea derivatives (e.g., propylthiouracil). A normal thyroid gland contains a 3-month supply of stored hormone (about 8 mg iodine). The structures of T_3 and T_4 are shown in Figure 14–7.

Secretion, Transport, and Metabolism of Thyroid Hormones

Under TSH stimulation, as required, small portions of colloid are taken into the thyroid cell by endocytosis; they are fused with lysosomes, and the thyroid hormones are released by proteolysis. Predominantly T_4, but some T_3, are passed into the circulation. Some DIT and MIT are also liberated, but they undergo intracellular deiodination and this iodine is salvaged for reuse. All steps of thyroid hormone synthesis and release are inhibited by iodine, given orally in large doses. Lithium, similarly, interferes with organification, coupling, and release, but not uptake.

In the *circulation* T_4 is 99.97% protein bound (0.03% free) while T_3 is 99.7% bound (0.3% free). Thyroxin-binding globulin (TBG, M_r 60,000) has a plasma concentration of 20 mg/L, a high affinity for T_4 (dissociation constant, $K_D = 2 \times 10^{-10}$ M), and accounts for 75% of T_4 binding. TBG has the capacity to bind 190 nmol (250 μg) of T_4/L and is normally about ⅓ saturated. Thyroxin-binding prealbumin (TBPA) is 20-fold more abundant, has a moderate affinity for T_4 ($K_D = 2 \times 10^{-8}$ M), but does not bind T_3. Albumin, at a much higher concentration, has a much lower affinity ($K_D = 2 \times 10^{-6}$ M) and normally binds only about 10% of plasma T_4. Both TBG and albumin bind T_3 about $^1/_{10}$ as efficiently as T_4. For clinical purposes in evaluating thyroid status only TBG (with rare exceptions described later) has to be considered. Certain drugs such as diphenylhydantoin (Dilantin), diazepam, salicylate, tolbutamide, and chlorpropamide are able to compete for T_4 binding sites on TBG.

Although T_4 may have some intrinsic biologic activity it is now regarded mainly as a prohormone, with T_3 the physiologically active product. It is important to note that while all the T_4 comes from the thyroid gland, about 80% of the T_3 is formed by 5′-deiodination in the peripheral tissue cells. In addition there is a slightly greater 5-deiodination forming reverse T_3 (rT_3), which is biologically inactive. The two thyronine monodeiodinases (5′MDI and 5MDI), by their capacity to generate active T_3 or inactive rT_3, serve as a peripheral regulating mechanism responding to a variety of physiologic and pathologic changes and determining the balance between hormone activation and inactivation. Drugs such as propylthiouracil, glucocorticoids, propranolol, and amiodarone, and

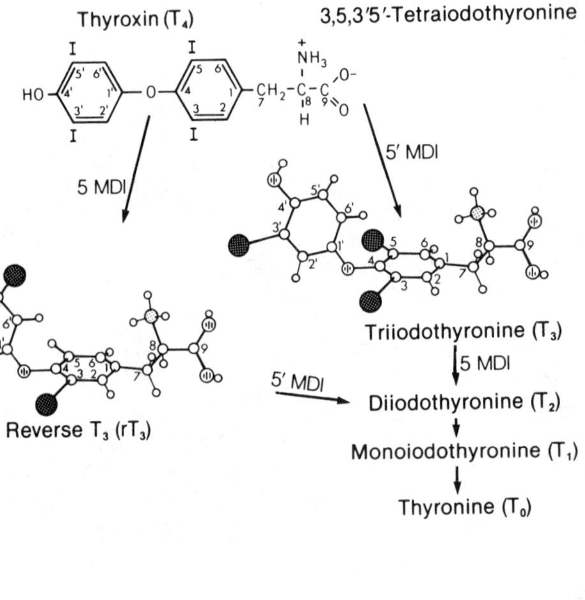

Thyroxin (T_4) 3,5,3′5′-Tetraiodothyronine
5 MDI
5′ MDI
Reverse T_3 (rT_3)
Triiodothyronine (T_3)
5 MDI
5′ MDI
Diiodothyronine (T_2)
Monoiodothyronine (T_1)
Thyronine (T_0)

FIGURE 14–7. Structures and metabolic pathways of thyroxin, triiodothyronine, and reverse triiodothyronine. MDI = monodeiodinase.

iodinated x-ray contrast agents such as iopanoic acid, suppress 5'MDI resulting in decreased serum T_3 and increased rT_3. The metabolism of T_3 and rT_3 leads in turn to diiodo- and monoiodothyronine and eventually to iodine-free thyronine (Fig. 14–7). In addition these intermediates may undergo deamination and decarboxylation, with excretion of free and conjugated metabolites in the bile and urine. Current estimates of normal values for thyroid hormone production and circulating concentrations are given in Table 14–5.

Mechanisms of Thyroid Hormone Action

Thyroid hormones are necessary for normal growth and development. The free hormones enter most body cells, probably by facilitated transport, where they exert a general stimulatory effect on many metabolic processes, and at sites in the nucleus, the mitochondria, and the plasma membrane. Receptors for T_3 probably occur in all three sites and the net result is a coordinated metabolic response directing the synthesis of structural and functional components of the cell. A fundamental effect involves the binding of T_3 to nuclear chromatin, leading to transcription of specific portions of the genome and the production of new proteins. A portion of the effect is probably directed to the activity of Na^+/K^+ dependent adenosine triphosphatase and the regulation of intracellular electrolytes. Many enzymes are known to be activated by thyroid hormone and the activity of microsomes in translating messenger RNA (mRNA) appears to be enhanced.

Thyroid hormones stimulate calorigenesis (although not by uncoupling of oxidative phosphorylation) and are required for the maintenance of body temperature, as shown by the cold and heat intolerance of hypo- and hyperthyroidism. Thyroid hormones are necessary for normal growth, yet in excess they switch protein anabolism to catabolism, resulting in a negative nitrogen balance, muscle wasting, and weakness. They affect virtually all aspects of carbohydrate metabolism and several areas of lipid and vitamin metabolism. They increase the number and function of both α- and β-adrenergic receptors, which probably accounts for the increased SNS activity in hyperthyroidism (tachycardia, sweating, and tremors) and the depressed SNS activity in hypothyroidism.

Mechanisms of Thyroid Hormone Regulation

Control of thyroid hormone secretion and action occurs at three levels:

a. The hypothalamic-pituitary-thyroid axis
b. Autoregulatory mechanisms within the thyroid
c. Peripheral hormone conversion

All three must be understood if laboratory thyroid monitoring is to be effective.

Hypothalamic TRH is necessary for the increased synthesis and release of TSH by the thyrotropes and TRH regulates the sensitivity of these cells to feedback control. Negative feedback by thyroid hormones occurs at the pituitary level; it correlates well with circulating concentrations of

TABLE 14–5. Estimated Production and Circulating Levels of Thyroid Hormones

	T_4	T_3	rT_3
Thyroid production (μg/day)	≃90 (100%)	≃7 (20%)	≃2 (4%)
Peripheral production from T_4 (μg/day)	0	≃28 (80%)	≃38 (96%)
Net production (μg/day)	≃25	≃35	≃40
Relative biologic potency	1	4	0
Half-life (days)	6–7	1–2	<1
Serum levels (mean)			
Total: nmol/L	103	2	0.38
μg/L	80	1.3	0.25
Free: pmol/L	31	6.1	1.2
ng/L	24	4.0	0.8

T_4 = thyroxin; T_3 = triiodothyronine; rT_3 = reverse T_3; ≃ = approximately.

free T_4, but is determined by the intracellular conversion of T_4 to T_3 by 5′monodeiodinase (5′MDI). In addition, the sensitivity of the thyrotropes to stimulation by TRH and inhibition by T_4/T_3, is influenced by other hormones, notably estrogens, dopamine and somatostatin.

Autoregulation within the thyroid results from the unique capacity of this gland to respond to its organic iodide content, which may reflect a dietary iodine deficiency. A low content of glandular iodide is associated with enhanced sensitivity to TSH, with the production of a higher proportion of T_3 relative to T_4.

Finally, a fine tuning occurs at the target cell level and may be selective within a particular tissue or organ depending on body demands. This modulation is effected by the activities of 5′MDI and 5MDI (see Fig. 14–7). In states such as malnutrition, liver disease, generalized debility, pregnancy, stress and steroid therapy, 5′MDI is inhibited, decreasing the production of T_3 and the metabolism of rT_3, thus decreasing metabolic activity within the tissue.

Thyroid Function in the Neonate, During Pregnancy and Aging

As stated earlier, the maturing fetus is dependent on its own hypothalamic-pituitary-thyroid axis. At term, the *normal infant* has slightly elevated serum TSH and free T_4 levels, but very low serum T_3 and high rT_3 values due to reduced 5′MDI activity. Rapid changes occur in the new temperature environment. Within 30 minutes of birth serum TSH values peak at about 100 mU/L, then fall by 48 h to around 15 mU/L, finally stabilizing at adult levels of <10 mU/L during childhood. Serum T_4 rises to a peak of 200 to 300 nmol/L at 24 h, then falls over several days to childhood values which are maintained about 30% above adult levels to the age of 5 years. Serum T_3 parallels serum T_4 but starts out very low and peaks at about 6 nmol/L in 24 h, while rT_3 falls to normal values by 1 week of age. These marked variations in the first few days of life must be considered when screening for thyroid disease.

During *pregnancy,* due in part to increased renal clearance of iodine, in part to increased TBG levels induced by estrogen, serum T_4 rises while free T_4 remains essentially normal. Thyroid function increases and goiter may become more apparent, but these changes revert to normal 6 to 8 weeks postpartum.

Thyroid function in the *aged* is not well defined; low serum T_3 levels have been reported, but may be caused by intercurrent disease. The prevalence of hypothyroidism increases from middle age onward. A significant number of institutionalized elderly people may be hypothyroid and may respond well to treatment.

TESTS OF THYROID FUNCTION

The *objectives* of thyroid testing are: (1) to define the level of thyroid function (hyperthyroid, euthyroid, or hypothyroid), (2) to identify the nature and location of the disorder, and (3) to monitor the effectiveness of treatment. Thyroid disorders are common and have clinical symptoms that overlap with many other complaints. Mild degrees of thyroid disease cannot be recognized clinically; the expression "clinically euthyroid, biochemically hypothyroid" (or "hyperthyroid") indicates the importance of reliable thyroid testing by the laboratory. Recent identification of individuals who are metabolically normal (or appropriate), but have abnormal biochemical indices of thyroid function, may require specially selected tests to avoid inappropriate treatment.

Serum Levels of Thyroid Hormones

When attempting to relate thyroid status to serum levels of the thyroid hormones one must keep at least *four variables* in mind:

1. Reference ranges are influenced by dietary factors and analytic methodology, and must be established locally.
2. The free hormone is the active form; thus, when TBG levels are abnormal, euthyroidism may exist despite increased or decreased total T_4 levels.
3. T_3 is 3 to 4 times more active than T_4, so that serum T_4 alone may not always indicate the patient's true status, as in T_3 toxicosis, in which T_4 is normal, and in T_3 euthyroidism, in which T_4 is low.
4. Although T_3 is more active than T_4 and is the biologically active hormone, serum T_4 is often a better index of thyroid status than T_3 in relation to the need for treatment.

Total Thyroxin (T_4). Competitive protein binding methods have been almost entirely replaced by radioimmunoassay (RIA) methods for the determination of serum T_4. Although T_4 is in all probability a prohormone, its measurement in serum is an excellent index of thyroid function and the first line of laboratory diagnosis in thyroid disease. Results are above the reference range in approximately 90% of hyperthyroid patients and below it in 85% of hypothyroid patients. Correct interpretation assumes normal levels of thyroxin-binding globulin (TBG), and the effect of altered TBG levels on T_4 values must always be kept in mind. The TBG level (normally 20 mg/L) is altered by a number of conditions (Table 14–6). Elevations of TBG increase the total T_4 but leave the amount of free T_4 unchanged; thus the patient is euthyroid although the T_4 test alone would suggest that the patient is hyperthyroid. A deficiency of TBG produces the reverse effects. This problem is resolved in most cases by either direct or indirect estimation of TBG.

Assessment of TBG (T_3 Resin Uptake and Effective Thyroxin Index). The direct immunologic assay of TBG is possible but not yet widely utilized. While RIA of TBG is now available, indirect methods for evaluating TBG abnormalities are far more commonly used. An individual with an increase in TBG will bind more free T_4, decrease feedback inhibition, and so increase T_4 production until homeostasis is restored. The result is an elevated total T_4, an increase in the number of both occupied and unoccupied TBG binding sites, and a normal *free* T_4. Conversely, a TBG deficiency state will show a low *total* T_4, a decreased number of occupied and unoccupied binding sites, and a normal *free* T_4. It is sometimes important to correct for changes in TBG, in fact some suggest that a serum T_4 assay should always be accompanied by an estimation of TBG.

The most widely used procedure to evaluate TBG status is the ***T_3 resin uptake*** (T_3RU, T_3U), which measures the empty binding sites that are present on TBG. Normal serum will bind a portion of labeled T_3 and the rest (25% to 35%) will be taken up by the resin. This quantity varies inversely with the TBG level. Thus, states of TBG *excess* have an elevated T_4 with increased numbers of empty binding sites and hence a low T_3 RU. TBG *deficiency* states have a low T_4, decreased numbers of empty binding sites on TBG, and thus a high T_3RU.

A number of mathematical formulae have been proposed in an attempt to use the T_3RU value to indicate the free or effective concentration of plasma T_4. Most widely used is the *T_3RU ratio.* The patient's T_3RU divided by the average 'normal' T_3RU gives a range from 0.85 to 1.15 for normal individuals. The actual T_4 result, multiplied by this ratio, gives a corrected T_4 value, sometimes referred to as the *effective thyroxin index* (ETI, ET_4) or *free T_4 index* (FTI). For example, a pregnant women might have the following results:

$$T_4 = 200 \text{ nmol/L} \, (N = 65\text{–}155 \text{ nmol/L})$$

$$T_3RU = 20\% \, (N = 25\%\text{–}35\%, \text{ aver. } 30\%)$$

$$T_3RU \text{ ratio} = 20/30 = 0.67 \, (N = 0.85\text{–}1.15)$$

$$ETI(FTI) = 200 \times 0.67 = 134 \, (N = 65\text{–}155)$$

TABLE 14–6. Factors Altering Thyroxin-Binding Globulin (TBG) Levels

	TBG Increased	TBG Decreased
Drugs	Estrogens (pregnancy, contraceptive medication, other exogenous hormone)	Androgens
	Phenothiazines	Anabolic steroids
		Glucocorticoid excess
		Diphenylhydantoin (phenytoin)
		Salicylates
		Phenylbutazone
Diseases	Chronic liver disease	Chronic liver disease
	Acute hepatocellular disease	Acromegaly
	Acute intermittent porphyria	Hypoproteinemia (nephrosis, malnutrition)
Other	Congenital excess	Congenital deficiency

This 'normalized' ETI (FTI) has the same numerical range as the total T_4 but no true dimensions or units.

Another commonly used device is to multiply the results of the serum T_4 by the T_3RU, the result being reported as the T_7 (or T_{12}). In the example above $T_4 \times T_3RU = 200 \times 20\% = 40$ (N = 20–47). These mathematical approximations are not measurements of free T_4. The correlation is good in normal sera, but often fails in abnormal situations in which it is most needed. Some drugs, for example, may interfere by inducing changes in TBG production, others by occupying hormone binding sites.

With an adequate history, an assessment of TBG is probably required in only 25% of patients, but in most areas the common practice is to order a T_3RU (or equivalent test) along with every T_4. T_3RU procedures can yield highly variable results, thus it is important for each laboratory to standardize its procedures and establish its own reference range.

Free Thyroxin (Free T_4). The reference method for serum free T_4, direct equilibrium dialysis, is a tedious procedure which precludes its routine use. Modifications using ultrafiltration may provide a reliable alternative. Recent commercial methods involve the use of competitive T_4 analogues, equilibrium kinetics, and encapsulated antibodies, and appear to give a reasonable approximation of serum free T_4. The ultimate aim is for direct estimation of serum free T_4 to replace the serum T_4 and T_3RU tests as a primary screen for thyroid disease. An accepted reference range for free T_4 is 10 to 30 pmol/L, but some methods give different results.

Triiodothyronine (T_3) and Free T_3. Methods for serum T_3 require high-affinity antibodies and special conditions to eliminate interference by other binding proteins. Measurement of serum free T_3, or its protein-bound reservoir of total T_3, does not necessarily correlate with intracellular biologically active T_3. Serum T_3 is influenced less than T_4 by changes in TBG concentration, but peripheral concentrations can be affected by non-thyroidal illness and by certain drugs, due to alterations of 5′-MDI activity. Reference means are given in Table 14–5, but *ranges* are best established locally. They are different in infants and in the elderly.

Serum T_3 is probably a superior indicator of *hyperthyroidism,* because thyroidal T_3 production is often increased disproportionally and early; some cases can be caused by a pure T_3 toxicosis, in which only T_3 is elevated. In early Graves' disease serum T_3 becomes abnormal before the serum T_4; during treatment the normal relationship between T_3 and T_4 may be altered. Thus, whereas serum T_4 and T_3RU may be acceptable as basic screening procedures, in problem cases, or for monitoring treatment, serum T_3 should also be measured.

By contrast, serum T_3 is an inferior indicator of *hypothyroidism.* Such patients tend to increase their peripheral production of T_3 to compensate for the deficiency of thyroidal T_4. A low T_3 (sick euthyroid) syndrome is discussed later. Serum T_3 is a useful test to monitor the adequacy of therapy for hypothyroidism, but not for the initial diagnosis.

The serum T_3 is low at birth, rises to somewhat above adult levels in early childhood, and then maintains adult values to age 30, after which it falls by about 0.08 nmol/L (0.05 μg/L) per decade. In patients who are elderly, or have cirrhosis, uremia, malnutrition, or cancer, peripheral conversion of T_4 to T_3 may be depressed. Conversely, in hypothyroidism due to iodine deficiency, T_3 may be synthesized in preference to T_4. Analysis of free T_3 has not proved to be helpful.

Reverse T_3 (rT_3). An RIA procedure for rT_3 is now available. The serum rT_3 comes almost entirely (96%) from peripheral deiodination of T_4, its concentration determined by the serum T_4 and the activities of 5 and 5′monodeiodinases. The test may have a place in defining the sick-euthyroid syndrome, and thus may help to exclude the diagnosis of hypothyroidism.

Serum TSH

Measurement of TSH by RIA is the most useful and sensitive test for *primary hypothyroidism.* When feedback suppression of the pituitary is reduced by a deficient production of thyroid hormones, the TSH rises in an attempt to increase thyroid hormone production (Fig. 14–8). This rise occurs while the patient is still asymptomatic and thus is an early and sensitive indicator of hypothyroidism. In a clearly hypothyroid patient, the finding of a *low* TSH indicates a deficient pituitary secretion of TSH as the cause.

TSH values are subnormal in all patients with hyperthyroidism (a result of feedback suppression), except for rare cases of a pituitary TSH-secreting adenoma. The nonspecificity of most routine TSH assays at low levels precludes their usefulness in the diagnosis of hyperthyroidism. Newer methods using monoclonal antibodies have the necessary specificity to distinguish the subnormal TSH levels of hyperthyroidism and are now a useful routine test for confirming this diagnosis. Values in hyperthyroidism are <0.5 mU/L, usually 0.3 or less. Serum TSH measurements are clearly useful in determining the response to exogenous TRH in the investigation of hypo- and hyperthyroidism and in differentiating between pituitary and hypothalamic causes of hypothyroidism.

The test may also be used to *monitor therapy,* because adequate thyroxin replacement in hypothyroidism is a dose that restores the serum TSH to low-normal levels. The serum TSH is also useful in following patients who have undergone thyroid surgery or developed hypothyroidism following radiation.

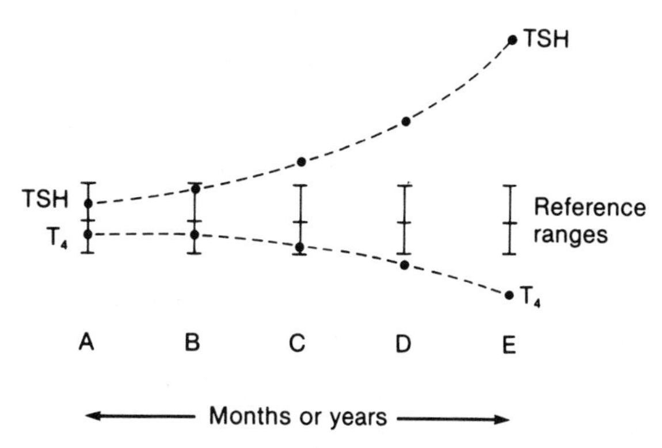

FIGURE 14–8. Changes in plasma TSH (thyrotropin) and T_4 (thyroxin) during the development of hypothyroidism. At **A** values are normal. At **B,** TSH is approaching the upper reference limit; goiter and antithyroid antibodies may be present. At **C,** T_4 is near the lower reference limit; TSH is raised and TRH (thyroliberin) will produce an exaggerated response. At **D,** TSH is distinctly abnormal and T_4 is now below the lower reference limit. At **E,** both TSH and T_4 are distinctly abnormal and signs and symptoms of hypothyroidism will be clearly evident.

Stimulation and Suppression Tests

TRH Stimulation. Blood is drawn for TSH assays before and 20, 30 and 60 min after an intravenous dose of 200 μg to 500 μg of TRH, although often 0 and 30 min specimens will suffice. In normal individuals the serum TSH peaks in 30 min, generally ranging between 5 mU/L and 30 mU/L. A normal response excludes the possibility of thyrotoxicosis. In *hyperthyroidism* no significant rise in TSH occurs, because the elevated thyroid hormones inhibit TSH production, but a subnormal response is not an absolute indication that treatment should be instituted. The TRH test has replaced the thyroid (T_3) suppression test in the investigation of borderline thyrotoxicosis. In primary *hypothyroidism* the TSH response to TRH is exaggerated and also diagnostic, but necessary only in subclinical or borderline cases. It is often useful in differentiating between pituitary and hypothalamic lesions as the cause of *secondary* hypothyroidism. No response suggests pituitary disease, whereas a sluggish response, often requiring sampling beyond 60 min, points to a hypothalamic defect.

T_3 Suppression (Werner Test). The radioactive iodine uptake or the serum T_4 is measured before and after a 7 to 10 day course of 75 μg to 100 μg of oral T_3 daily. Suppression is indicated by a fall in either index to <50% of the initial value; failure to do so indicates autonomy of thyroid function. Now that TRH is available this test is rarely used.

TSH Stimulation. This consists of measuring the RAIU before and after a 3-day course of intramuscular TSH injections (5 U daily) and is used in the investigation of pituitary hypothyroidism. It has largely been replaced by TRH stimulation but still has value in determining whether areas of nonfunctional thyroid tissue (as indicated by scanning techniques) are capable of function.

Radioactive Iodine Uptake (RAIU, ^{131}I Uptake)

Radioiodine uptake by the thyroid gland is used both to investigate and to treat hyperthyroidism.

Hyperthyroidism caused by thyroiditis (acute, subacute, or painless) or by exogenous thyroid hormone will have a RAIU of <1%, whereas hyperthyroidism from all other causes will have an elevated RAIU. The RAIU is also used as an aid in determining the dose of ^{131}I required for treatment purposes.

The test consists of an oral tracer dose of ^{131}I followed by measurement of the thyroid uptake at 24 hours. Reference values are area- and diet-dependent (iodine intake) but usually range from 5% to 30%. An intravenous test with uptake measured at 20 min is available which, in combination with the discharge effect of perchlorate administration, can be used to investigate thyroidal enzyme defects. The test is generally avoided in children and during pregnancy. The availability of ^{123}I (half-life of 12 hours) may widen the use of the RAIU test.

Thyroid Scanning by Radionuclides and Ultrasound

Radionuclide scans are indicated to evaluate thyroid nodules, locate ectopic thyroid tissue, investigate thyroid agenesis, and evaluate metastatic thyroid carcinoma. The procedure is similar to the RAIU except that a larger dose is administered and an image of iodine distribution within the thyroid tissue (or elsewhere in the body) can be obtained. Either ^{131}I, ^{123}I, or technetium (^{99m}Tc) may be used. Iodine is preferred as it undergoes both uptake and organic binding, while technetium reflects uptake only.

The common multinodular goiter is often a mixture of functioning and nonfunctioning thyroid nodules. *Ultrasonography* is now capable of distinguishing between cystic and solid nodules with about 90% accuracy. Cystic lesions are rarely cancerous but therapy may depend on examination of material obtained by needle aspiration. Solitary 'hot' nodules, with suppression of the rest of the gland, are infrequent and rarely malignant. Solitary, 'cold' solid nodules occur more often and 20% are malignant.

Other Thyroid Tests

Thyroid Autoantibodies One of the hallmarks of autoimmune thyroid disease is the detection in serum of antithyroglobulin antibodies (using tanned red cells) and antimicrosomal antibodies (by complement fixation). Markedly elevated levels are indicative of chronic lymphocytic (Hashimoto's) thyroiditis, but minimal to moderate elevations may occur in other thyroid disorders.

Thyroid-Stimulating Immunoglobulin. The sera of most patients with Graves' disease contains autoantibodies directed against thyroidal TSH cell-surface receptors. Initially termed long-acting thyroid stimulators (LATS), their antibody nature now allows them to be classified as thyroid-stimulating immunoglobulins (TSIg). TSIg assays based on either TSH binding inhibition, or on cAMP generation, now give a sensitivity of 95% in Graves' disease. Although TSIg is the etiologic agent, the test has been available in very few laboratories but there is now a reliable and commercially available assay.

Thyroglobulin. Circulating serum thyroglobulin levels can be measured by RIA. This is useful in monitoring metastatic carcinoma of the thyroid (elevated levels) and in the diagnosis of factitious thyrotoxicosis (absent levels), but only in the absence of circulating thyroid autoantibodies which interfere with the assay.

DISORDERS OF THE THYROID

Hyperthyroidism (Thyrotoxicosis)

This clinical syndrome is due to excessive levels of circulating T_4, T_3, or both, with the signs and symptoms being an exaggeration of the physiologic actions of thyroid hormone. The metabolic effects due to protein catabolism include weight loss in spite of an increased appetite, muscular weakness, hair loss, and skin and nail changes. Increased sympathetic activity causes nervousness, irritability, tremor, palpitations, heat intolerance, sweating, frequent bowel movements, and menstrual irregularities. On *physical examination* one finds tachycardia; fine tremor of the hands; warm, smooth, and moist skin; and hyperreflexia. The thyroid is usually enlarged.

In older patients particularly, hyperthyroidism may manifest itself by aggravating an underlying disorder. The possibility of thyroid disease should be kept in mind when congestive heart failure, worsening of angina, atrial fibrillation, unusual muscle weakness, unexplained weight loss, or personality change occurs.

Biochemical abnormalities in hyperthyroidism are many and varied. Urinary excretion of hydroxyproline is increased as a result of rapid collagen turnover. Hypercalcemia occurs in 27% (total Ca^{2+}) to 47% (ionized Ca^{2+}) of patients. Nitrogen balance is negative. Lipid levels (notably cholesterol) are reduced, and mobilization of free fatty acids is increased. Catecholamine production is normal.

Graves' Disease. Hyperthyroidism can have different etiologies (see Other Causes), but by far the commonest is Graves' disease. The full-blown entity includes hyperthyroidism associated with a smooth, nontender, diffusely enlarged thyroid, an infiltrative ophthalmopathy, and occasionally an infiltrative dermopathy (pretibial myxedema). Most commonly the disease presents as hyperthyroidism alone, with minimal eye involvement, but these two aspects can occur in any conceivable time sequence.

Graves' disease is an autoimmune disorder characterized by the production of thyroid-stimulating IgG antibodies (TSIg) directed at TSH receptors. They cause thyroid hyperactivity with consequent suppression of TSH. The basic 'lesion' is believed to be a suppressor T-lymphocyte defect which allows helper T-lymphocyte stimulation of β-lymphocyte production of TSIg. The disease occurs more frequently in young women, and is often precipitated by stress. The *diagnosis* is based on the clinical presentation along with laboratory evaluation that includes serum T_4, T_3, TSH, and T_3RU, along with a RAIU and scan to exclude other causes of hyperthyroidism. In problem cases the definitive tests of serum free T_4, TRH stimulation and TSIg determinations may be necessary.

Symptomatic *treatment* of hyperthyroidism consists of decreasing sympathetic activity by using β-adrenergic blocking agents (beta-blockers) such as propranolol. This can be started immediately as it does not interfere significantly with the usual investigations. The disease remits in approximately one third of patients in less than a year. Thus, one choice of treatment is a 1 year trial of blocking T_4 production with propylthiouracil or related drugs. If this is unsuccessful, partial ablation of the thyroid by either radioactive iodine or surgery is indicated. As a result of either of these latter treatments, hypothyroidism will ensue in most patients, given enough time, but can be managed by replacement therapy. The ophthalmopathy, if progressive, may be improved by steroids and surgery but can be very difficult to treat.

Other causes of Hyperthyroidism. Apart from Graves' disease, only the following are of significant frequency:

Subacute thyroiditis
Toxic adenoma
Toxic nodular goiter

Uncommon causes in order of frequency include:

Exogenous thyroid hormones
Acute exogenous excess iodine
Trophoblastic tumors
Stroma ovarii
Functioning thyroid carcinoma
Thyrotrope adenoma

These disorders can be *differentiated* easily from Graves' disease by RAIU and thyroid scanning. Toxic adenoma will show a single (and toxic nodular goiter at least one) hot spot on the scan, with reduced uptake in the rest of the gland, while subacute thyroiditis gives a RAIU of $<1\%$. The only other conditions that can give a RAIU of $<1\%$ are exogenous thyroid hormone administration or excess iodide intake. The remaining causes are quite rare and can be suspected from a good clinical history and confirmed by appropriate investigations.

Familial Dysalbuminemic Hyperthyroxinemia

Several families have been identified who have marked elevations of total serum T_4, along with normal TBG levels, normal T_3RU (normal numbers of empty TBG binding sites) and thus a markedly elevated ETI. The patients are euthyroid, serum free T_4 and serum T_3 are usually normal. Electrophoresis has revealed that in contrast to normal individuals, whose albumin binds 10% of circulating T_4, these patients have 50% or more of their T_4 bound to albumin. This appears to be caused by a genetic abnormality of albumin with autosomal dominant inheritance. These patients should not be treated because they are euthyroid. Prevalence may be higher than expected because an ability to identify the condition has led to the reporting of more and more cases.

Hypothyroidism

Inadequate production of thyroid hormones may develop at any time of life but the prevalence increases markedly in late middle age. Occurrence *in utero* may result in cretinism, while development later in life results in a slowing of metabolic processes. The *symptoms* include fatigue, decreased mental and physical performance, cold intolerance, change in personality, hoarseness, constipation, muscle cramps, paresthesias, dry skin, and facial and peripheral edema. On *physical examination* one finds bradycardia, dry scaly skin, and a delayed relaxation phase of the reflexes. The thyroid may be impalpable, normal, or enlarged.

Tissue infiltration by mucopolysaccharides produces facial puffiness, coarse features, enlarged tongue, and thickened vocal cords. Mild weight gain, reduced cardiac performance, increased lipid levels, dry yellow-tinted skin (from carotene accumulation), reduced body hair, anemia, and menorrhagia may all occur. The clinical presentation can range from virtually asymptomatic, to such extremes as hypothermic coma (myxedema coma) or overt psychosis (myxedema madness). Approximately 2% of senile dementia cases may be undiagnosed hypothyroidism.

The early symptoms of hypothyroidism are variable and nonspecific and its onset is usually so insidious that many patients are grossly hypothyroid for years prior to the *diagnosis.* Early diagnosis requires that clinical suspicion be high and laboratory investigation using both serum T_4 and TSH analyses be carried out routinely. The stages in the development of hypothyroidism are illustrated in Figure 14–8, showing the sensitivity of serum TSH as an indicator of hypothyroidism.

Several biochemical abnormalities occur in hypothyroidism, but are seldom helpful diagnostically and are occasionally confusing. Serum lipid levels (notably cholesterol) are elevated. Creatine phosphokinase and aspartate aminotransferase are increased. Other hormone levels are affected; in particular, serum prolactin levels may be elevated.

Etiology and Diagnosis of Adult Hypothyroidism. *Common* causes of hypothyroidism include:

Hashimoto's lymphocytic thyroiditis
Idiopathic atrophy (or aging)
Iatrogenic factors (surgery, ^{131}I, x-irradiation)
Drugs (propylthiouracil, iodide, lithium).

Uncommon causes include:

TSH deficiency (pituitary/hypothalamic damage)
Iodine deficiency
Intrathyroidal enzyme defects
Congenital dysgenesis

When the patient presents with a clinical picture suggestive of *hypothyroidism,* the diagnosis requires only the demonstration of a low serum T_4 and a high TSH. Thyroid autoantibodies (antithyroglobulin and antimicrosomal) are helpful in defining the etiology. The serum T_3 and free T_3 are normal in about half of patients with hypothyroidism and are not useful in diagnosis. In problem cases a free T_4 concentration, TRH stimulation, and a serum rT_3 may be helpful.

Approximately 98% of hypothyroidism is caused by *primary* thyroid failure and can be diagnosed by the raised serum TSH. The remaining 2% are caused by defects at the hypothalamic–pituitary level, resulting in deficient TSH production. A patient with a low serum T_4 and a low serum TSH requires further evaluation. This should include a serum T_3RU (or other TBG assessment), serum free T_4, serum T_3, and a TRH stimulation test to prove or disprove the diagnosis of pituitary hypothyroidism. Other causes of hypothyroidism can be resolved by the history (including drug intake) and appropriate investigations. Iodine deficiency is rare in North America. Treatment consists of daily oral L-thyroxin at a dose (usually 0.15 mg daily) which suppresses serum TSH to the low-normal range.

Neonatal Hypothyroidism. The incidence is about one in 4000 births, with about half caused by enzyme defects and half a result of thyroid dysgenesis. Neonatal hypothyroidism is a preventable cause of mental retardation and developmental abnormalities, and justifies the screening of newborn infants using serum TSH or T_4. The prognosis depends on the severity and duration of the deficiency before diagnosis is made and treatment is instituted.

Sick Euthyroid (Low T_3) Syndrome

Most of the T_3 (80%) and nearly all of the rT_3 (96%) found in the plasma arises by peripheral deiodination of T_4. Although there are two monodeiodinases, 5′MDI appears to be the major regulator because it effects both the production of T_3 and the degradation of rT_3 to T_2. Thus, changes in 5′MDI activity usually result in reciprocal changes in serum T_3 and rT_3 concentrations. The activity of 5′MDI is affected not only by drugs but also by systemic illness and the stresses of surgery, anesthesia, and severe malnutrition.

In patients with a *moderate* systemic illness, as a result of minor decreases in TBG concentration and possibly the presence of inhibitors of hormone binding, the serum T_4 is usually low normal while free T_4 is normal. Because the activity of 5′MDI is depressed, both T_3 and free T_3 levels are markedly decreased while rT_3 is increased. Calculated ETI will be misleadingly low, suggesting hypothyroidism, because of interference with the T_3RU test in such patients. In *severely* ill patients the serum T_4 concentration may also be depressed, with very low serum T_3 and free T_3 values; even serum free T_4 may be low. Again rT_3 will be raised.

It is noteworthy that serum TSH is *not increased* in the sick euthyroid syndrome. Systemic illness appears to have little effect on 5′MDI activity in the thyrotropes. Thus intracellular free T_4, reflecting serum free T_4, is appropriately converted to free T_3. The pituitary is 'unaware' of the peripheral hypothyroid state, which allows the body to maintain a condition of reduced metabolic demands. Whether these patients should be classed as euthyroid or hypothyroid is debatable. They are peripherally hypometabolic, yet for their clinical state may be metabolically appropriate.

The problem with these patients is that routine thyroid investigations may indicate hypothyroidism and because the TSH is not increased a lesion of the pituitary or hypothalamus is suspected. It is here that the rT_3 assay, which is not generally available, would be useful. In true hypothyroidism rT_3 is low, in the sick euthyroid syndrome it is usually elevated. The TRH test may help, but may fail to produce a normal TSH response in very ill patients. Treatment with thyroid hormone replacement is inappropriate and the situation is generally managed by clinical acumen.

Thyroiditis

Inflammatory disease of the thyroid is relatively common and occurs in two main forms:

1. ***Subacute thyroiditis*** is a self-limiting disorder (probably a viral infection) accompanied by flu-like symptoms, with pain, fever, and an elevated sedimentation rate and white count. There is usually a transient hyperthyroidism with an elevated serum T_4 (due to rupture of follicles) and a suppressed TSH, but a decreased RAIU ($<1\%$). This last feature is diagnostic and important, because the disorder may present painlessly as hyperthyroidism only. The disorder usually resolves in about 3 months but a transient hypothyroid phase may occur. Treatment is with aspirin, or with prednisone in severe cases.
2. Chronic (***Hashimoto's***) ***thyroiditis*** is the commonest cause of goitrous hypothyroidism in the adult. The inflammatory process has an autoimmune basis and gradually destroys the thyroid. Occasionally there is transient hyperthyroidism. Diagnosis is based on the clinical examination, tests of thyroid status, and the demonstration of usually high titers of autoantibodies to thyroglobulin and thyroid microsomal antigen.

Thyroid Nodules and Cancer of the Thyroid

Thyroid nodules occur in about 4% of the population, but thyroid carcinoma is seen clinically in only 0.004% of individuals; thus, most nodules are benign and, in addition, thyroid function is usually normal. Carcinoma is more likely if there is a history of previous neck irradiation, if the patient is under 40, if there are associated findings, and if the nodule is solitary, is changing rapidly, is cold on thyroid scan, is solid on ultrasonography, or fails to shrink on suppressive doses of L-thyroxin.

Most physicians when faced with a solitary nodule, or multiple nodules that are unchanging, do a serum T_4 to confirm euthyroidism, a RAIU and scan, and ultrasonography. Solitary cold solid le-

TABLE 14–7. Commonly Accepted Thyroid Test Reference Ranges

Test	Range
Thyroxin (T_4)*	
RIA methods	65–155 nmol/L (50–120 μg/L)
T_3 resin uptake (T_3RU)	25%–35%
T_3 resin uptake ratio	0.85–1.15
ETI or FTI ($T_4 \times T_3RU$ ratio)*	65–155 nmol/L (50–120 μg/L)
T_7, or T_{12} ($T_4 \times T_3RU/100$)	20–47
Thyroxin-binding globulin (TBG)	12–28 mg/L
Free T_4	10–30 pmol/L (8–24 ng/L)
Triiodothyronine (T_3)	1.2–3.4 nmol/L (0.8–2.2 μg/L)
Reverse T_3(rT_3)	0.23–0.54 nmol/L (0.15–0.35 μg/L)
Free T_3	3–9 pmol/L (2.5–7.5 ng/L)
^{131}I uptake (RAIU) by thyroid at 24 h	5%–30% (varies with local iodine intake)
Thyrotropin (TSH)	<6 mU/L (<10 mU/L by some methods)
Ultrasensitive method	0.5–6 mU/L

RIA = radioimmunoassay; RAIU = radioactive iodine uptake.

* The effective thyroxin index (ETI) or free thyroxin index (FTI) is a T_4 value corrected mathematically for the effect of plasma TBG levels and has the same normal range and units.

sions should be removed surgically. Other lesions may call for a needle biopsy, a trial of suppressive therapy, surgery, or simply close follow-up. A hot nodule is, for practical purposes, not cancer.

LABORATORY INVESTIGATION OF THYROID DISORDERS

The clinical pictures in thyroid disease are nonspecific and a definitive diagnosis rests with the laboratory data. In addition, treatment must be monitored by laboratory indices.

The usually accepted *reference ranges* for thyroid tests are given in Table 14–7. One must be aware, however, that these values vary among different laboratories (because analytic methods vary) and are also dependent on the diet and the geographic area in some instances.

Table 14–8 shows the expected results of the various thyroid indices in a variety of thyroid disorders. This does not mean that an extensive thyroid investigation is required or should be undertaken for every patient.

Table 14–9 lists the thyroid tests in a logical diagnostic approach to the various disorders of this gland (hypothyroidism, hyperthyroidism, and thyroid nodules). If the initial screen supports the diagnosis, it must be followed by confirmatory procedures. Thus *hypothyroidism* must be confirmed with a TSH assay, and *hyperthyroidism* with a RAIU scan. The RAIU scan is also important for the differential diagnosis and in decisions regarding the treatment of hyperthyroidism. A needle biopsy of *thyroid nodules* may help resolve the suspicion of malignancy. These tests serve medical diagnostic needs in over 90% of patients with thyroid disorders.

Some clinicians order thyroid antibodies in all patients with thyroid disorders in an attempt to establish an autoimmune etiology, which then implies other possible disease associations. In a small number of patients the results of screening and confirmatory procedures may not be conclusive, and definitive assessment (free T_4, TRH stimulation, surgery) may be required. In a few patients, with unusual aspects in either their presentation or biochemical data, it may be necessary to use every test available to establish the diagnosis. Those that are most helpful in these problem cases are shown in the lower half of Table 14–9. Treatment of the disorder may alter thyroid hormone metabolism and thus the laboratory results. A different set of tests may be appropriate for monitoring and follow-up than for diagnosis.

Concerning the *rational use of the laboratory,* two important points can be made. Although recommended as routine screening tests for thyroid disorders, it is not necessary to do a serum T_3RU

TABLE 14–8. Thyroid Indices in Various Conditions

	Euthyroid					Hypothyroid		Hyperthyroid				
	Normal	*TBG Excess*	*TBG Deficiency*	*Sick Euthyroid Syndrome*	*Familial Dysalbuminemic Hyperthyroxinemia*	*Primary (Thyroid)*	*Secondary (Pituitary)*	*Graves' Disease*	*Subacute Thyroiditis*	*Toxic Nodule*	*T_3 Toxicosis*	*Secondary (Pituitary)*
T_4	N	↑	↓	LN/↓	↑	↓	↓	↑	↑	↑	N	↑
T_3RU	N	↓	↑	↓/LN	N	↓	↓	↑	↑	↑	N	↑
ETI (FTI)	N	N	N	↓/LN	↑	↓	↓	↑	↑	↑	N	↑
TBG	N	↑	↓	LN/↓	N	N	N	N	N	N	N	N
T_3	N	N/↑	N/↓	↓	N	↓/N	↓/N	↑	↑	↑	↑	↑
rT_3	N	N/↑	N/↓	↑	N	↓	↓	↑	↑	↑	N	↑
TSH	N	N	N	N	N	↑	LN	↓	↓	↓	↓	↑
RAIU	N	N	N	LN	N	↓	↓	↑/N	↓	↑/N	N/↑	↑/N
Free T_4	N	N	N	LN/↓	N	↓	↓	↑	↑	↑	N	↑
Free T_3	N	N	N	↓	N	↓/N	↓/N	↑	↑	↑	↑	↑

N = Normal; ↑ = increased; ↓ = decreased; LN = low-normal; ETI = effective thyroxin index; FTI = free thyroxin index; RAIU = radioactive iodine uptake; T_3 = triiodothyronine; rT_3 = reverse T_3, T_3RU = T_3 resin uptake; T_4 = thyroxin; TBG = thyroxin-binding globulin; TSH = thyrotropin (thyroid-stimulating hormone).

TABLE 14–9. Laboratory Investigation of Thyroid Disorders

	Hypothyroidism	Hyperthyroidism	Thyroid Nodule
Tests used for initial screen	T_4 / T_3RU or T_4 / TSH	T_4 (T_3) / T_3RU or T_4 / TSH	T_4, T_3RU, TSH RAIU scan Ultrasound
Confirmation and diagnosis	TSH	TSH, RAIU scan	Needle biopsy
Definitive test	Free T_4, TSH	Free T_4, TSH	Surgery
Tests of value in problem cases	Free T_4 TRH stimulation TBG rT_3 Free T_3	Free T_4 rT_3, free T_3 TRH stimulation TSH TBG TSIg	
Tests for etiology	Thyroid antibodies	Thyroid antibodies	Thyroid antibodies
Monitoring treatment and followup	TSH T_3	T_4 and T_3 RU T_3 (TSH)	TSH T_3 (Thyroglobulin)

T_4 = thyroxin; T_3 = triiodothyronine; T_3RU = T_3 resin uptake; RAIU = radioactive iodine uptake; TSH = thyrotropin (thyroid stimulating hormone); TRH = thyroliberin (thyrotropin releasing hormone); TBG = thyroxin binding globulin; TSIg = thyroid stimulating immunoglobulin.

with every serum T_4, for example, with those that are requested during treatment and follow-up. Similarly, it is seldom necessary to do a TSH test to exclude hypothyroidism when the serum T_4 is >90 nmol/L (>70 μg/L). In fact, many clinicians when investigating hypothyroidism have discontinued ordering the T_3 resin uptake and now order: T_4 and TSH if T_4 <90 nmol/L. Laboratories may adopt this practice as a cost-effective approach.

THE ADRENAL MEDULLA

STRUCTURE AND FUNCTION

Origins and Structure

The sympathetic nervous system (SNS) and adrenal medulla have a common ectodermal origin in the neural crest. The primitive sympathogonia develop into sympathetic ganglion cells or into pheochromocytes. Preganglionic fibers of the SNS secrete acetylcholine (cholinergic), most postganglionic fibers secrete norepinephrine (adrenergic). About 90% of the pheochromocytes migrate at an early stage into the adrenal cortex, where they differentiate and mature by about age 3 years into secretory cells innervated by preganglionic fibers of the SNS. Most of the cells contain predominantly epinephrine (adrenalin); about 15% contain mainly norepinephrine (noradrenaline).

Catecholamine Synthesis and Metabolism

The biosynthetic pathway of the catecholamines is shown in Figure 14–9. Mitochondrial tyrosine hydroxylase is rate-limiting; the final *N*-methyltransferase is induced by glucocorticoids, which probably regulate epinephrine synthesis. The hormones of the adrenal medulla are stored as a complex with proteins (chromogranin A; dopamine-β-hydroxylase) and nucleotides (ATP) in 'chromaffin' granules, some of which on stimulation are released by a process of exocytosis.

Catecholamines include *dopamine* (DA, 1,2-dihydroxyphenylethylamine), found mainly in the central nervous system (CNS), *norepinephrine* (mainly in the SNS), and *epinephrine* (mainly in the adrenal medulla). They are stored as an inactive complex in equilibrium with a free fraction in the cytoplasm that exerts a negative feedback control over tyrosine hydroxylase. An increase in sympathetic nerve impulses (acetylcholine) causes release of the hormone(s), increased tyrosine hydroxylase activity, and (if prolonged) synthesis of the enzyme.

Released catecholamines have a short half-life. They are either taken up by sympathetic nerve

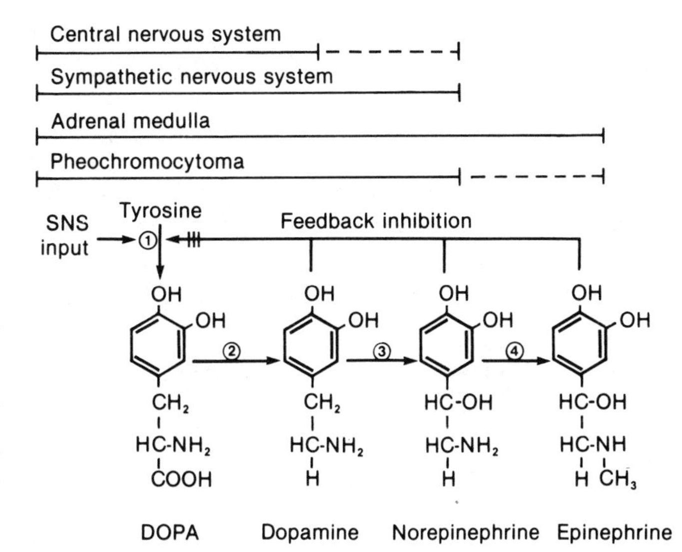

FIGURE 14–9. Pathways of catecholamine synthesis in the central nervous system, sympathetic nervous system, adrenal medulla, and a pheochromocytoma. The enzymes involved at the numbered sites are 1, tyrosine hydroxylase; 2, DOPA decarboxylase; 3, dopamine β-hydroxylase (in granule vesicles); 4, phenylethanolamine-*N*-methyltransferase (only in the adrenal medulla; induced by adrenal corticosteroids). DOPA = dihydroxyphenylalanine; SNS = sympathetic nervous system.

endings or metabolized by the liver and kidney and excreted; less than 1% of norepinephrine or epinephrine is excreted in free form in the urine. Normally, four fifths of urinary norepinephrine comes from the SNS, the small amounts of epinephrine, from the adrenal medulla. Some dopamine-β-hydroxylase and chromogranin A are released during exocytosis of portions of the chromaffin granules; their measurement may be an additional index of SNS activity.

The main ***metabolic pathways*** of the catecholamines are shown in Figure 14–10. They are determined by the activity of two enzyme systems, catecholamine orthomethyltransferase (COMT) in the liver and kidneys, and a more ubiquitous enzyme, monoamine oxidase (MAO). Some of the intermediates are conjugated and excreted as glucuronides or sulfates, but the excretion of unchanged metanephrines (although less than 5%) is a better index of SNS and adrenal medullary activity. The *deaminated* metabolites reflect to a considerable extent metabolism within the nerve endings, not hormone released in active form. The final products, 3-methoxy-4-hydroxymandelic acid (MHMA, vanillylmandelic acid, VMA) and 3-methoxy-4-hydroxyphenylethylene glycol (MHPG), reflect all sources and are less specific. Normally, normetanephrine (NMN) reflects the neuronal release of norepinephrine and metanephrine (MN) reflects the secretion of epinephrine from the adrenal medulla. MHPG is a major metabolite of CNS norepinephrine. Homovanillic acid (HVA) is the main metabolite of dopamine.

Effects of Catecholamines

In the brain these amines appear to function mainly as neurotransmitters. The hormones of the SNS and adrenal medulla act on specific receptors which are stimulated by some drugs (*agonists*) and suppressed by others (*blockers*). The diverse actions of the catecholamines are mediated by two types of α-receptors and two types of β-receptors. α_1-Receptors activate the phosphoinositol system and lead to contraction of smooth muscle in the blood vessels and genitourinary tract. They stimulate glycogenolysis in the liver. α_2-Receptors act through adenylate cyclase (via cAMP) to *relax* smooth muscle in the gastrointestinal tract and to *contract* smooth muscle in the vascular bed. β_1-Receptors act via cAMP to increase the rate and force of cardiac contraction. β_2-Receptors also act via cAMP to *relax* smooth muscle in the bronchi, blood vessels, gastrointestinal and genitourinary tracts. Insulin, renin, and norepinephrine release are increased by β_2- and are decreased by α_2-receptors. Lipolysis is increased by β_1- and is decreased by α_2-receptors. Adrenergic receptors exhibit high affinity and low affinity states modulated

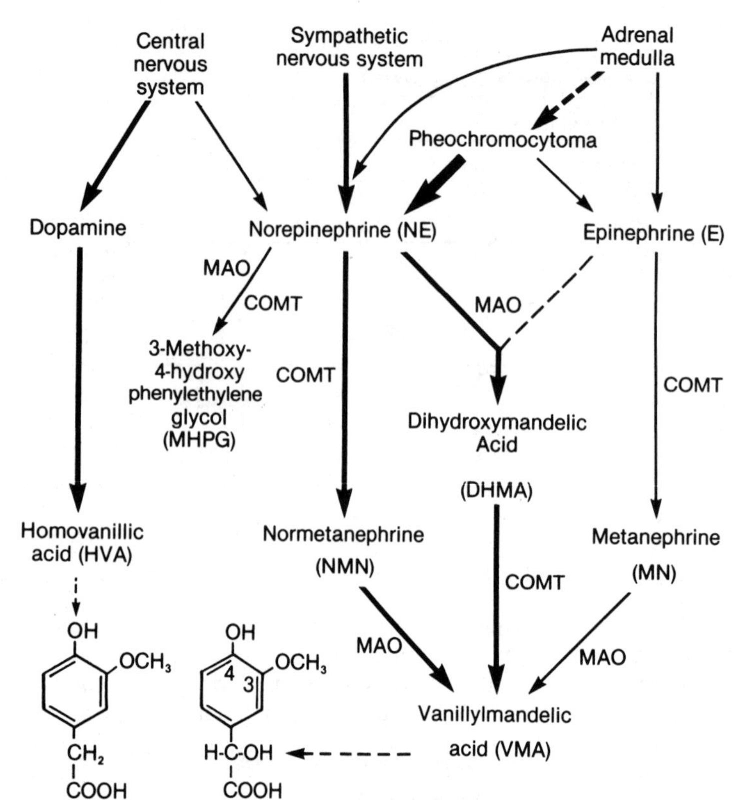

FIGURE 14–10. Metabolic pathways of the catecholamines. MAO = monoamine oxidase; COMT = catecholamine orthomethyltransferase.

through phosphorylation by GTP. The number of receptors is also subject to regulation by various factors, particularly by thyroid hormones. Prolonged use of β-blocking drugs (e.g., propranolol) leads to increased numbers of receptors and a *hypersensitivity withdrawal syndrome.*

Norepinephrine is primarily an α-receptor stimulant, whereas epinephrine stimulates both α- and β-receptors, although β effects tend to dominate. These hormones mobilize stored energy sources for essential muscle activity. By affecting renin release they enhance sodium reabsorption in the kidney.

The SNS has many connections to the CNS, and the entire sympathetic outflow is under central control. Together they regulate most routine circulatory, metabolic, and visceral functions. The adrenal medulla serves emergency requirements ("fight or flight"), to counter severe stress or shock, and to provide responses such as to hypoglycemia.

Two examples of drug *agonists* are methoxamine for α-receptors and isoproterenol for β-receptors; three drug *blockers* are phentolamine (Regitine) or phenoxybenzamine for α-receptors and propranolol for β-receptors. Drugs are now available that block β_1-receptors (in heart and adipose tissue) or β_2-receptors (broncho- and vasodilators), thus providing selective treatment for cardiac or respiratory problems. The adrenal medulla contains more than a fatal dose of catecholamines, and protection by blocking agents is a routine precaution during surgery for pheochromocytoma.

DISORDERS OF THE ADRENAL MEDULLA AND SNS

Pheochromocytoma

Pheochromocytomas are rare catecholamine-producing tumors; 90% occur in the adrenal medulla, 10% from chromaffin cell rests in the abdomen, thorax, or neck. The prevalence of this tumor is 0.1% to 0.2% of hypertensive patients. Undiagnosed or mistreated it can be fatal, but it is a curable disease, hence the importance of accurate diagnosis. The tumor loses its neural and feedback control and its close relationship to the adrenal cortex. The main product of biosynthesis is usually non-neurogenically controlled norepinephrine, rather than epinephrine, but various and unpredictable ratios of the two are secreted, with a corresponding variety of signs and symptoms.

An important ***clinical feature*** of the disorder is hypertension. Sustained hypertension may be found in about half the cases, with the remainder having paroxysmal hypertension lasting minutes to hours (but usually under 15 min), with intervals of normal blood pressure. In hypertensive patients, a symptomatic triad of sweating attacks, tachycardia, and headaches has a 94% specificity and 91% sensitivity for pheochromocytoma. Orthostatic *hypotension* is present in about 60% of cases and can be an important diagnostic clue. 'Crises' occur, with headache, sweating, anxiety, and abdominal or chest pain as common complaints. They may be induced by almost any stimulus, including stress, infection, or drugs. In most patients α-receptor effects dominate (high blood pressure, headaches, vasoconstriction); in others, β effects (flushing, sweating, palpitations). Metabolic effects may produce hyperglycemia, glucosuria, weight loss, and heat intolerance.

Pheochromocytomas have a long record of misdiagnosis, because the metabolic, cardiac, and gastrointestinal symptoms can mimic many other diseases. In most cases there is a single tumor in the adrenal medulla, usually less than 10 cm in diameter and weighing less than 100 g. About 10% are bilateral, 10% occur outside the adrenal, and fewer than 10% are malignant.

A familial form of pheochromocytoma with autosomal dominant inheritance is recognized, sometimes as part of a multiple endocrine adenopathy (MEA-II) syndrome which includes medullary carcinoma of the thyroid and parathyroid adenoma or hyperplasia. The coexistence of pheochromocytoma, thickened corneal nerves, alimentary tract ganglioneuromatosis, and (frequently) marfanoid habitus has been termed MEA type III. There is an increased prevalence of pheochromocytoma in patients with von Recklinghausen's neurofibromatosis. A condition of adrenal *medullary hyperplasia* can be associated with symptoms resembling pheochromocytoma. Urinary catecholamines are usually normal, but plasma concentrations may be increased during episodes of hypertension.

Diagnosis. A reliable diagnosis of pheochromocytoma requires that excess catecholamine secretion be demonstrated. This can be achieved by measuring the plasma catecholamines (NE + E) or their urinary metabolites, the metanephrines (MN), or vanillyl mandelic acid (VMA). All three tests have excellent specificity and, listed in decreasing order, show 95%, 95%, and 90% sensitivity, respectively.

When the clinical signs and symptoms point to pheochromocytoma and medications have been withdrawn for at least 48 hours, a *plasma* free catecholamine concentration >12 nmol/L (2000 ng/L) may be considered pathognomonic. A *clonidine suppression test* can be employed when values fall between 6 nmol/L and 12 nmol/L (1000 ng/L to 2000 ng/L). Clonidine in a dose of 0.3 mg will suppress neurogenically mediated catecholamine release seen in hypertension, but fails to suppress secretion by pheochromocytomas. Plasma free catecholamines will fall to <6 nmol/L in patients with essential hypertension (<3 nmol/L in normotensive subjects), but remain >6 nmol/L in pheochromocytoma (SENS 97%, SPEC 99%). In some patients the measurement of *urinary* MN or VMA may help secure the diagnosis. The upper RV for urinary MN in essential hypertensives is 10 μmol/day (1.8 mg/d) and in normotensives 7 μmol/d (1.3 mg/d). The upper RV for VMA is 55 μmol/d (11 mg/d) in hypertensives, and 35 μmol/d (7 mg/d) in normotensives. Dietary and drug interferences have been virtually eliminated by new and specific methods using gas or high performance liquid chromatography. Although in florid cases the diagnosis can be made on a random urine specimen (expressing the result per mmol of creatinine), a 24-h urine collection with the patient off all medications is recommended. Episodic hypertension may require blood or urine collection during the crisis itself to establish the diagnosis.

A *provocative* test is rarely indicated because of the risk and should be performed only when the diastolic blood pressure is <110 mmHg. The glucagon test (an IV bolus of 1 mg to 2 mg) has fewer side effects and is positive if there is a >3-fold increase in plasma catecholamines within 2 to 3 minutes.

Preoperative tumor *localization* is achieved usually by computerized tomography (CT) scanning, or by scintigraphy 2 to 3 days after giving ^{131}I-metaiodobenzylguanidine. Plasma catecholamine levels can also be useful in conjunction with vena caval sampling in the localization of extra-adrenal tumors. Preoperative *treatment* with both α- and β-blockers is required, with surgery usually resulting in complete cure. Inoperable tumors can be treated with a combination of α- and β-blockers along with α-methyltyrosine, a competitive inhibitor of tyrosine hydroxylase, which decreases tumor catecholamine production.

Other Disorders

Orthostatic hypotension can be induced by adrenergic blockers and may occur in adrenal insufficiency, or hypokalemia, and intermittently with pheochromocytoma. *Idiopathic* orthostatic hypotension is caused by a specific degeneration of preganglionic sympathetic neurons and is more common in middle-aged men. It usually presents as dizziness and can become incapacitating, with many other manifestations. Treatment with mineralocorticoids may help, but the prognosis depends on the extent of autonomic nervous system involvement.

Neuroblastomas are rare, usually malignant, tumors occurring in children. They secrete some catecholamines but rarely produce hypertension. Dopamine and homovanillic acid predominate in the urine. ***Ganglioneuromas*** are rare, benign tumors associated with excess catecholamines, hypertension, and diarrhea.

Congestive heart failure is a serious illness. Plasma norepinephrine concentrations offer a guide to prognosis that may be better than other indices of cardiac performance.

THE ADRENAL CORTEX

The human adrenal cortex produces three main classes of hormones: glucocorticoids, mineralocorticoids, and androgens. The first two of these are essential to life and deficiency states can be fatal. The diagnosis of disorders of the adrenal cortex depends on laboratory evaluation of adrenal cortical function, because the clinical features are nonspecific and relatively common.

STRUCTURE AND FUNCTION

Anatomy, Embryology, and Histology

Human adult adrenals weigh approximately 5 g each and fit like a cap on the upper pole of each kidney. A rich *blood supply,* mainly from the subarcuate arteries of the kidneys, reaches the adrenals through many small arterioles and then passes through sinusoidal capillaries between columns and clusters of cells toward the center of the gland. Collecting venules merge to form a central vein in the medulla which returns to the circulation usually by a single adrenal vein. The right adrenal vein is very short (1 mm to 5 mm) and passes directly into the inferior vena cava, while the left adrenal vein is 2 cm to 4 cm in length and empties into the left renal vein. *Nerves* enter the adrenal but pass directly to the medulla; the cortex appears to be entirely under humoral control.

There are remarkable differences between the ***fetal cortex*** and the mature human adrenal cortex. By the twelfth week of pregnancy the adrenals have formed from the migration of two successive bilateral clusters of cells, derived from the coelomic mesoderm, which surround the cells forming the adrenal medulla. The *inner* cortical layer is dominant during fetal life, secreting hormones from the third month and beginning to degenerate a few days before the onset of labor. Control appears to be first by hCG, later by ACTH. The main secretory product is the weak androgen dehydroepiandrosterone sulfate (DHEA-S), which is hydroxylated by the fetal liver to 16-OH-DHEA-S (see Chap 23, Fig. 23–2) and then desulfated and aromatized to estriol by the placenta, finally being excreted as estriol glucuronide by the mother. (Measuring maternal urinary estriol is thus one means of monitoring fetoplacental viability.)

The *functions* of the fetal cortex are not fully understood. The androgens may contribute to the rapid growth of the fetus, and may protect a male fetus from the high levels of maternal estrogen; they may also serve to delay estrogen-induced sensitization of the uterus to oxytocin until near term.

After 30 weeks the *outer* 'definitive' layer of the fetal cortex increases in size while the inner layer involutes. These cells produce cortisol under ACTH stimulation (also aldosterone) and assume a dominant role after birth. By age 3 years the three layers of the ***adult cortex*** have been established. The outer *zona glomerulosa* (ZG) consists of parenchymal cells in irregular clusters. These cells lack 17-hydroxylase and possess a special oxidase that diverts a precursor steroid to aldosterone. The *zona fasciculata* (ZF) consists of cords of polyhedral cells with microvilli on the capillary side, rich in lipid and ascorbic acid. The inner and narrower *zona reticularis* (ZR) contains irregular masses of more compact cells richer in protein and DNA. The major locus of steroid biosynthesis (glucocorticoids and androgens) appears to be in the border zone between the ZF and ZR. Adrenal cells are able to regenerate in local layers as required.

The steroid-secreting cells of the adrenal cortex are characterized by an extensive smooth endoplasmic reticulum (SER), a few parallel arrangements of rough endoplasmic reticulum (RER), abundant pleomorphic mitochondria, and a prominent Golgi complex. The total lipid content averages about 15% wet weight and one third of this is cholesteryl ester.

Structure and Mechanism of Action of Adrenal Cortical Hormones

Although steroids are most easily drawn as planar structures, their functions relate to the properties depicted in a three-dimensional molecule. The structures of the more important adrenal steroids are shown in Figure 14–11. Corticosteroids are to a degree lipid soluble and pass freely through cell membranes; their biologic activity depends on their binding to specific cytoplasmic receptor proteins. This complex translocates to receptors in the nuclear chromatin where transcription of mRNA is stimulated. The mechanism of action of these hormones is shown in Figure 14–2; it results in the production of enzymes that express the function of the target cells. Steroids may exert more rapid allosteric effects on cytosol enzymes as well.

Biosynthesis of Adrenal Hormones

Although the adrenal cortex synthesizes cholesterol from acetate, under normal conditions the plasma LDL-cholesterol serves as the precursor of about 75% of adrenal steroids. On a demand for increased synthesis more adrenal cholesterol is used. The main biosynthetic pathways are shown in Figure 14–12.

The most important steps involve oxidases leading either to hydroxylation or dehydrogenation. ***Hydroxylation*** proceeds via a common pathway involving reduced nicotinamide adenine dinucleotide phosphate (NADPH), and flavoprotein, a nonheme iron protein, cytochrome P_{450}, and oxygen. The overall reaction can be written

$$SH_2 + 2e + 2H^+ \text{ (NADPH)} + O_2 \rightarrow SHOH + H_2O$$

where S = steroid and e = electron.

The enzyme complexes for hydroxylation at C-17 and C-21 and dehydrogenation of the C-3 hydroxyl occur in the SER. The enzyme complexes for hydroxylation at carbons 20, 22, 11, 18, and 19 are located in the mitochondria. The mitochondria are therefore involved in the early and final stages of cortisol and aldosterone production, but Δ^5-pregnenolone must find its way to the SER to be converted to progesterone and for hydroxylation to occur at carbons 17 or 21.

The 'gate' from cholesterol to Δ^5-pregnenolone is controlled by ACTH, is rate limiting, but never completely closed. In the SER, Δ^5-pregnenolone has about an equal chance of undergoing either (1) 17-hydroxylation, followed by cleavage of the side chain to form dehydroepiandrosterone, or (2) C-3 dehydrogenation and shift of the double bond to form progesterone. Progesterone is preferentially hydroxylated in the sequence 17, 21, and 11 to form *cortisol.* Some is hydroxylated at carbons 21 and 11 to form *corticosterone,* then in the ZG at C-18 may be coupled with a second hydroxylation and dehydration to form *aldosterone.*

A major portion of *DHEA-S* is secreted and metabolized to androsterone and etiocholanolone, which are conjugated and excreted to form the bulk of urinary 17-ketosteroids (17-KS). Small amounts of DHEA-S and 17-hydroxyprogesterone are converted to Δ^4-androstenedione, which can be reduced to the 17-hydroxy form, testosterone. Successive 19-hydroxylation, ring A dehydrogenation, and loss of the C-19 carbon lead to estrone or es-

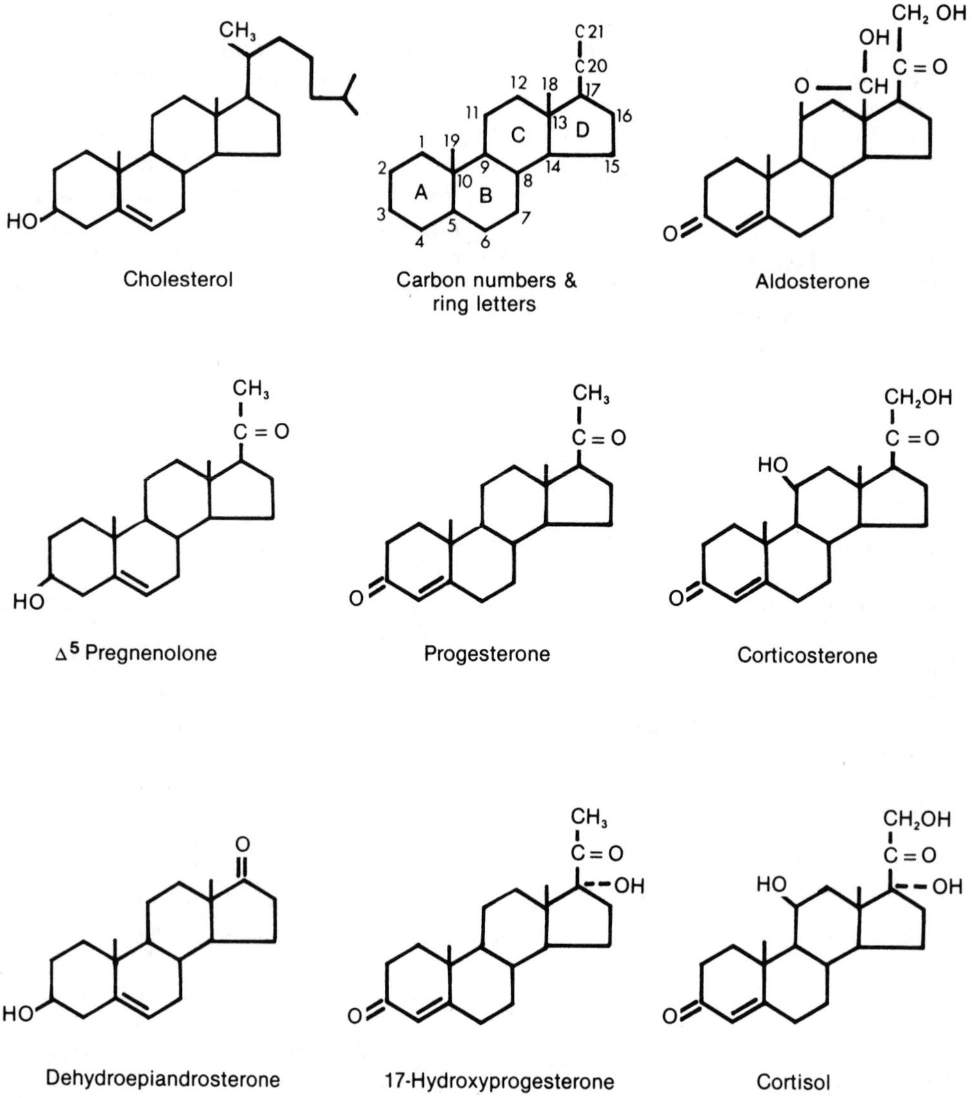

Cholesterol

Carbon numbers & ring letters

Aldosterone

Δ^5 Pregnenolone

Progesterone

Corticosterone

Dehydroepiandrosterone

17-Hydroxyprogesterone

Cortisol

FIGURE 14–11. Structures of the major adrenal steroids.

tradiol. Small amounts of glucocorticoid may lose the side chain to form 11-oxy-Δ^4-androstenedione and some is metabolized to 6β-hydroxysteroid.

Control of Adrenal Steroid Biosynthesis

Cortisol. As shown in Figure 14–13, adrenal cortisol production is regulated through CRH and ACTH in three ways:

1. Circadian rhythm (sleep/wake cycle)
2. Negative feedback by cortisol (and perhaps ACTH)
3. Stress override.

Hypothalamic secretion of corticoliberin (CRH) is regulated by higher neural stimulation (serotonin, acetylcholine) and inhibition (norepineph-

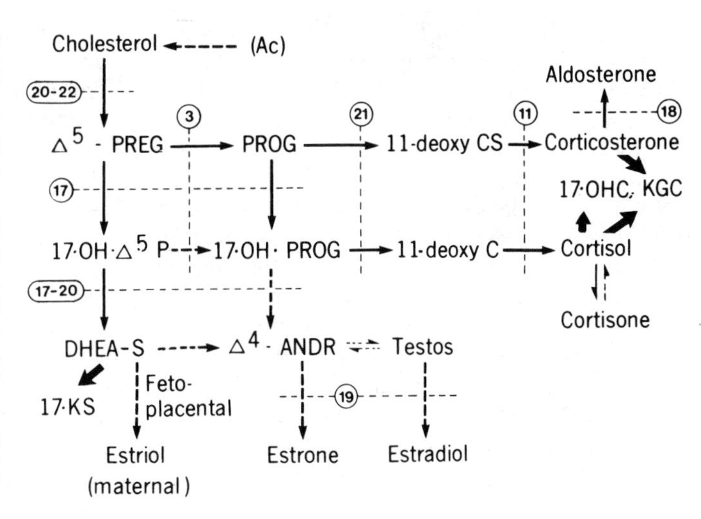

FIGURE 14–12. Biosynthetic pathways of the adrenal steroids with the enzymes involved: 20,22-hydroxylases, 20,22-desmolase; 3-hydroxydehydrogenase, Δ^{5-4}-isomerase; 17-hydroxylase; 17,20-desmolase; 21-hydroxylase; 11-hydroxylase; 18-hydroxylase; 19-hydroxylase. (Ac) = acetate; Δ^4-Andr = Δ^4-androstenedione; 11-deoxy-C = 11-deoxycortisol; 11 - deoxy - CS = 11 - deoxycorticosterone; DHEA-S = dehydroepiandrosterone(sulfate); KGC = ketogenic corticoids, 17-ketogenic corticosteroids; 17-KS = 17-ketosteroids; 17-OHC = 17-hydroxycorticoids, 17-hydroxycorticosteroids; 17-OH-Δ^5-P = 17-hydroxy-Δ^5 pregnenolone; 17-OH-Prog = 17-hydroxyprogesterone; Δ^5-Preg = Δ^5-pregnenolone; Prog = progesterone; Testos = testosterone. Major pathways, →; minor pathways, - - →.

rine). There is a circadian (diurnal) rhythm with a peak of CRH-ACTH-cortisol in the early morning (0600 h to 0800 h) and a nadir during the night (2300 h to 0400 h). Hormone release occurs in short bursts that decrease in frequency during the day and virtually cease just prior to or after sleep in the evening.

Circulating *free* cortisol exerts a negative feedback effect that serves to maintain homeostasis. It acts on the anterior pituitary to decrease corticotrope responsiveness to CRH. It acts on the hypothalamus (and possibly on higher neural centers) to depress the secretion of CRH. A short-loop feedback by ACTH may be exerted on the same target areas. *Stress,* working through the higher neural centers, can override these feedback regulatory mechanisms.

ACTH binds to membrane receptors on the adrenal cortical cells, activating adenylate cyclase. Within minutes cortisol is released and increased synthesis begins shortly after. Because ACTH controls the first step in the biosynthetic pathway, it increases precursor intermediates in all pathways and increases androgen and aldosterone production as well. If increased levels of ACTH are sustained, however, aldosterone levels will decline.

Aldosterone. Synthesis and release of aldosterone are controlled by two *major* factors: (1) the

FIGURE 14–13. Regulation of cortisol production. CRH = corticoliberin, ACTH = corticotropin.

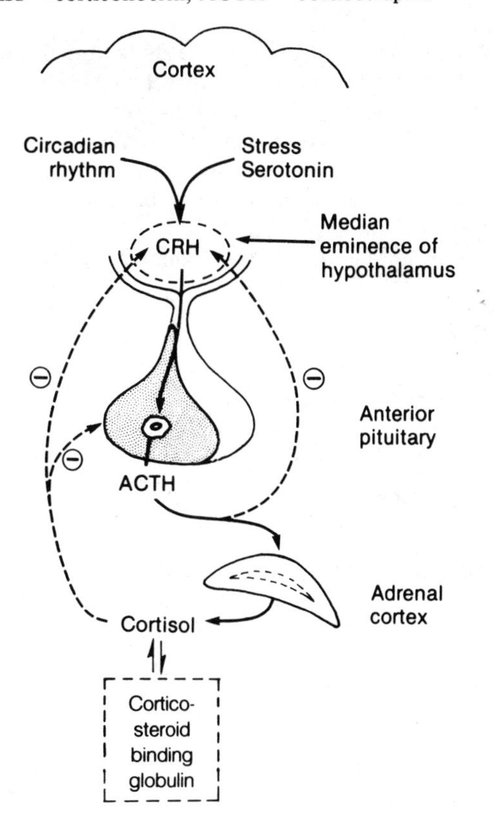

renin–angiotensin (R · AT) system and (2) electrolyte (Na/K) status, and two *minor* factors, ACTH and factor (x). Most important is the R · AT system, which responds to changes in the effective blood volume. When body fluid is depleted and/or pressure in the afferent arteriole of the kidney falls, renin is released from the juxtaglomerular cells. Renin is a proteolytic enzyme that acts on circulating angiotensinogen (an α_2-globulin) to split off the decapeptide angiotensin I (AT-I). During passage through the lung, an angiotensin converting enzyme (ACE) cleaves two amino acids to yield the octapeptide angiotensin II (AT-II), which is a potent vasoconstrictor and which acts through the phosphoinositide system to stimulate aldosterone synthesis and release (Fig. 14–14).

AT-II controls the 'gate' leading from corticosterone to an 18-hydroxy intermediate which can either form the stable 18–20 hemiketal of 18-hydroxycorticosterone or be acted on by a second oxidase (β-methyloxidase Type II) following by dehydration to form the stable 11–18 hemiacetal of aldosterone (Fig. 14–11).

Regulation by electrolyte status is also important, with K^+ having a more direct role than Na^+. Hyperkalemia *increases* and hypokalemia *suppresses* aldosterone secretion. However, in a moderately Na^+-depleted state ACTH and AT-II will no longer enhance aldosterone production, but further sodium depletion can.

The minor factors, ACTH (which generates precursors) and factor (x), appear to be permissive and are rarely rate limiting. The latter may be a complex of factors that modulate the terminal β-methyloxidase. These include probable tonic suppression by dopamine, possibly a pituitary aldosterone-stimulating glycoprotein, or a peptide derived from βLPH, any of which could be subject to hypothalamic control. The role of *atrial natriuretic factor,* a peptide that suppresses aldosterone secretion by ZG cells, remains to be clarified.

Aldosterone increases Na^+ retention which, along with the obligatory retention of water, restores effective blood volume. This, in turn, suppresses renin release.

Secretion, Metabolism, and Excretion of Adrenal Cortical Hormones

Generally accepted ranges of adrenal hormone production are shown in Table 14–10. Once secreted, cortical hormones circulate largely bound to protein. *Cortisol* binds to a specific, high-affinity, corticosteroid-binding globulin (CBG, transcortin). Small amounts bind to low-affinity albumin. The concentration of plasma CBG (an α_2-globulin) is fairly constant; it can bind cortisol to levels of 700 nmol/L (250 μg/L), and is normally close to saturation. CBG is increased by estrogen (e.g., in pregnancy), and decreased by proteinuria or congenital absence, with corresponding changes in plasma cortisol levels.

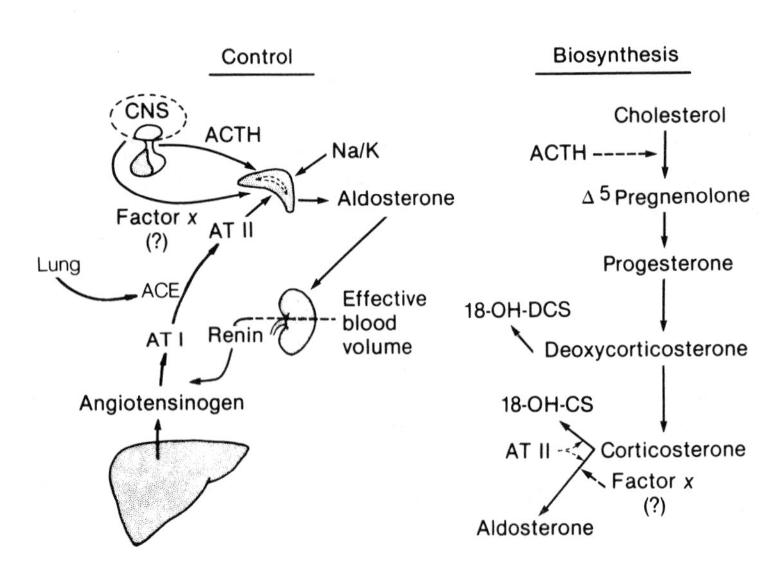

FIGURE 14–14. Control and biosynthesis of aldosterone. ACTH = Corticotropin; AT I = angiotensin I (10 amino acids); AT II = angiotensin II (8 amino acids); CNS = central nervous system; ACE = angiotensin converting enzyme; 18-OH-CS = 18-hydroxycorticosterone; 18-OH-DCS = 18-hydroxydeoxycorticosterone.

TABLE 14–10. Adrenal Steroid Production and Excretion

Steroid	Daily Production	Main Metabolite(s)	Measured as
Cortisol (cortisone)	10–30 mg	Tetrahydro-C3-conjugates	Urinary
Corticosterone	2–4 mg	(glucuronides and sulfates)	17–OHC or KGC
Aldosterone	50–200 μg	18-Glucuronide Tetrahydro-C3-conjugate	Aldosterone
Dehydroepiandrosterone (sulfate)	20–30 mg	DHEA–S Androsterone Etiocholanolone	17–KS
Δ^4-Androstenedione			
Males	6–8 mg	C3-conjugates	17–KS
Females	3–4 mg		
Testosterone	<1 mg		
Estrogens	10–30μg		
Progesterone	<1 mg		

For convenience and from custom we have used the following condensed abbreviations: 17-OHC = 17-Hydroxycorticoids, (17-hydroxycorticosteroids); KGC = ketogenic corticoids (17-ketogenic corticosteroids); 17-KS = 17-ketosteroids; DHEA–S = dehydroepiandrosterone (sulfate).

Under physiologic conditions approximately 83% of cortisol is bound to CBG, 12% to albumin, and 5% is free and metabolically active. Free cortisol is exposed to an active 11-dehydrogenase which establishes about a 1:1 equilibrium of cortisol to cortisone. All steroids are acted upon by reductases that add four hydrogens to saturate the Δ^4 double bond and form a C-3 hydroxyl. The resulting *tetrahydro* derivatives of cortisol, cortisone, corticosterone and 11-deoxycortisol are conjugated as glucuronides or sulfates and excreted efficiently in the urine.

Aldosterone is metabolized in ring A to form tetrahydro glucuronides, as aforementioned, but also forms a unique glucuronide on the C-18 hydroxyl. This acid-labile conjugate normally accounts for 7% to 15% of the total urinary aldosterone and is the fraction that is usually measured. Under physiologic conditions, approximately 30% of aldosterone is bound to CBG, 42% to albumin, and 28% is free. As a result of its weaker binding affinity, about 75% of circulating aldosterone is metabolically inactivated in a single pass through the liver.

Adrenal *androgens* bind to sex-hormone binding globulin (SHBG; testosterone–estrogen binding globulin, TEBG). Some DHEA·S is excreted as such in the urine, but most is converted to androsterone and etiocholanolone, which are conjugated and excreted to form the major portion of the urinary 17-ketosteroids. Some of the less polar androgen and estrogen metabolites are excreted in the bile as well. Although 24-hour urinary excretion values provide a more consistent index of adrenal activity, they are being requested with decreasing frequency, partly because the accurate collection of urine is more difficult to monitor than taking a blood sample. Plasma or serum values are useful when obtained under well established conditions.

Physiologic Effects of Adrenal Steroids

Cortisol (cortisone) counterbalances the effects of insulin in maintaining the homeostasis of carbohydrate, fat, and protein metabolism. It plays a role in regulating free water clearance by the kidney, calcium absorption in the intestine, blood cell production by the bone marrow, and neuronal interactions in the brain. Supraphysiologic levels are anti-inflammatory and immunosuppressive, stabilizing lysosomal membranes, inhibiting leukocyte diapedesis, suppressing antibody production, and inhibiting cell mediated immunity.

Aldosterone is known to induce citrate synthetase and Na^+/K^+ ATPase and exerts its main effects in the collecting ducts of the kidney, increasing Na^+ reabsorption in exchange for K^+ and, when hypokalemia develops, for H^+ ions. It affects ion transport systems also in the epithelial cells of the salivary glands, intestinal tract, and elsewhere. It probably influences the response of smooth-muscle cells to pressor agents.

Dehydroepiandrosterone (sulfate) is a weak androgen, and the female adrenal produces somewhat less than the male. The adrenals normally produce very small amounts of testosterone and estrogens. Although much less potent than gonadal hormones, adrenal androgens do influence facial, axillary, and pubic hair and, in excess, can cause hirsutism, acne, increased muscle mass, and masculinization in female children.

EVALUATION OF ADRENAL CORTICAL FUNCTION

The early stages of adrenal disease rarely produce easily definable clinical or biochemical abnormalities. Steroid 24-h secretion rates are rarely measured because they involve dilution techniques that require injection of a radionuclide. Routinely plasma and/or urinary excretion levels of the hormones are measured. *Plasma* is easy to obtain but steroid values are affected by episodic secretion, diurnal variation, metabolic and renal clearance, and the influence of such factors as drugs and stress. A representative value is difficult to obtain and the reference range is wide. *Urine* analyses require careful and complete collection of a 24-hour specimen. These problems require that abnormal results in screening procedures be confirmed by definitive evaluation of adrenal function using tests of the hypothalamic-pituitary-adrenal axis. The objective is to determine not so much what the gland is doing as what it is capable of doing when the axis is stimulated or suppressed.

Routine Tests

The common tests for evaluating adrenal cortical function are given in Table 14–11, with reference ranges and brief comments.

Serum cortisol, measured by RIA, is commonly performed at 0800 hours and later in the day (1600 h to 2000 h) to establish both the actual level and the presence or absence of diurnal variation. More precise studies require sampling every 2 to 4 hours for 24 hours. The determination of the 24-hour *urinary free cortisol* by RIA is one of the most reliable indicators of adrenal status. Cortisol secretion can be approximated by measurement of the 24-hour urinary excretion of cortisol metabolites, either as the 17-hydroxycorticosteroids (17-OHC) which accounts for 30% of secreted cortisol, or as the 17-ketogenic steroids (KGC). Measurement of urinary 6 β-hydroxycortisol, a major unconjugated metabolite of cortisol, may also prove useful.

Serum aldosterone, measured by RIA under standardized conditions of posture and salt intake, is used in conjunction with plasma renin activity. *Urinary aldosterone* determinations are based usually on the 18-glucuronide conjugate of aldosterone.

Serum dehydroepiandrosterone-sulfate (DHEA-S) is the major circulating adrenal androgen and can be measured by RIA. A more complete profile will require androstenedione and testosterone assays as well. *Urinary androgen* excretion is measured by the 24-hour urinary 17-ketosteroids (17KS), which consist of the conjugates of androsterone, etiocholanolone, and DHEA, with minor contributions from metabolites of testosterone and cortisol.

Plasma ACTH can be measured by RIA but the hormone is unstable, is present in very low concentrations, is secreted episodically and with diurnal variations, and is increased by stress. Thus, several samples must be analysed and the results interpreted with these factors in mind. The test is most useful with stimulation and suppression tests, in which it often provides a clear answer in the differential diagnosis of hypercortisolism. β-LPH, although not directly implicated in adrenal function, is secreted by the pituitary in equimolar amounts with ACTH (Fig. 14–4). β-Endorphin (β-EP), a fragmentation product of β-LPH, is also secreted in amounts that parallel ACTH. Both β-LPH and β-EP are more stable, have a longer half-life, and may become useful substitutes for ACTH assays.

Special Tests

Plasma Renin Activity (PRA). This test is often necessary to aid the distinction between primary aldosteronism (which is rare), aldosteronism secondary to renal ischemia (also infrequent), and aldosteronism secondary to a variety of hypovolemic states (which are common). The actual measurement is a RIA of either angiotensin I or angiotensin II, produced under standard conditions by the action of renin on its substrate angiotensinogen. Reference values can be interpreted only in relation to the patient's sodium status and other factors

TABLE 14–11. Laboratory Tests Used to Evaluate Adrenal Function

Steroid	Reference Range	Method	Comments
Plasma (or serum)			
Cortisol	0800: 200–660 nmol/L (70–240 μg/L) 1600: 140–500 nmol/L (50–180 μg/L) 2000: 55–390 nmol/L (20–140 μg/L) 2400: 55–275 nmol/L (20–100 μg/L) Nadir: <140 nmol/L (<50 μg/L)	RIA	Subject to major diurnal variations and frequent minor fluctuations; check at 0800 and 2000 for loss of diurnal rhythm
17-Hydroxyprogesterone	Men: < 6 nmol/L (<2 μg/L) Women: < 9 nmol/L (<3 μg/L)	RIA	Confirm 21-hydroxylase deficiency
Aldosterone Recumbent Standing	 0800: 80–300 pmol/L (30–110 ng/L) 1200: 140–700 pmol/L (50–250 ng/L) (On controlled 7 g salt intake)	RIA	For differential diagnosis of hypokalemia and hypertension in conjunction with renin assay
DHEA-S	Men: 4–10 μmol/L (1.5–3.9 mg/L) Women: 2–9 μmol/L (0.8–3.5 mg/L)	RIA	Major adrenal androgen
Δ^4-androstenedione (15–50 y)	Men: 3–8 nmol/L (0.9–2.4 μg/L) Women: 2–9 nmol/L (0.6–2.7 μg/L)	RIA	Both adrenal and gonadal origin
Testosterone (15–50 y)	Men: 11–35 nmol/L (3.0–10 μg/L) Women: 0.7–2.8 nmol/L (0.2–0.8 μg/L)	RIA	Major gonadal androgen
ACTH	0800: 5–22 pmol/L (20–100 ng/L) 2000: 10 pmol/L (45 ng/L)	RIA (heparinized plasma, collected in plastic and kept on ice)	For differential diagnosis of primary and secondary adrenal disorders
PRA (plasma renin activity)	High salt intake, supine 0.3–1.1 ng/L/s (1–4 ng/mL/h) Low salt intake, ambulatory 1.7–6.7 ng/L/s (6–24 ng/mL/h)	RIA (EDTA plasma kept on ice)	For differential diagnosis of hypertension
Urine (24-hour specimen)			
17-Ketosteroids (17–KS) androgen metabolites	Prepubertal, 10–20 μmol/d, gradual increase to adult values Men 35–75 μmol/d (10–22 mg/d) Women 20–55 μmol/d (6–16 mg/d) Gradual decrease after age 50	Zimmerman (m-dinitrobenzene in KOH) Red color specific for 17-keto group	Investigation of hirsutism
17-Hydroxycorticoids (17–OHC); metabolites of cortisol, cortisone, deoxycortisol	Men 5–30 μmol/d (2–11 mg/d) Women 5–25 μmol/d (2–9 mg/d)	Porter–Silber (phenylhydrazine in H_2SO_4) Yellow color specific for dihydroxyacetone side chain	Measure of glucocorticoid production in adrenal disorders
17-Ketogenic corticoids (KGC); broader spectrum of corticosteroid metabolites, including pregnanetriol	Men 20–60 μmol/d (6–17 mg/d) Women 15–50 μmol/d (5–15 mg/d)	Corticosteroids oxidized by Na bismuthate and measured as 17–KS	Raised in congenital adrenal hyperplasia, where 17–OHC are low
Free cortisol	55–300 nmol/d (20–110 μg/d)	RIA	Sensitive index of increased cortisol production
Aldosterone; 18-glucuronide metabolite	15–55 nmol/d (5–20 μg/d) (On controlled diet of approx. 7 g salt/d)	RIA, preferably after partial purification	For diagnosis of hypokalemia and hypertension
Pregnanetriol	Age 0–6 < 1.5 μmol/d (<0.5 mg/d) 6–16 < 4.5 μmol/d (<1.5 mg/d) Adult < 6.0 μmol/d (<2.0 mg/d)	Gas–liquid chromatography	Confirm 21-hydroxylase deficiency

ACTH = Corticotropin (adrenocorticotropic hormone); DHEA–S = dehydroepiandrosterone (sulfate); RIA = radioimmunoassay.

such as erect or recumbent posture. Units of expression and ranges are not yet standardized and must be established locally. A 6-hour integrated plasma *aldosterone/PRA ratio* is a sensitive discriminator between controls (<100) and primary aldosteronism (>100), but requires that blood be collected with a constant withdrawal pump.

ACTH Stimulation Test. This is indicated only in checking for possible primary or secondary adrenal insufficiency or decreased reserve.

An effective *screening* procedure consists of an intravenous injection of 0.25 mg synthetic ACTH (Cosyntropin, Cortrosyn), with measurements of serum cortisol at 0, 30, 60, and 90 min. A normal response is an increment of at least 200 nmol/L (70 μg/L) to a minimum level of at least 500 nmol/L (180 μg/L).

A *long* ACTH test should be carried out when the screening test indicates possible adrenal insufficiency. Urine and blood collections are made on 2 control days, and then on the subsequent 3 days, during which synthetic ACTH is given intravenously each day, or a long-acting preparation (Synacthen) can be given intramuscularly. Normally serum cortisol, urinary free cortisol, and 17-OHC should all rise at least 2.5-fold over the control values, with the maximum exceeding the upper reference limit. Patients with primary adrenal insufficiency show a minimal response; those with secondary deficiency (pituitary disease) show a stepwise increase in steroid production over the 3 days.

Metyrapone Stimulation Test. Metyrapone (Metopirone) is an inhibitor of 11β-hydroxylase; by blocking cortisol production and feedback inhibition it acts as an *endogenous* ACTH test. It can be used to check pituitary function in secondary adrenal insufficiency, and may be used in the differential diagnosis of Cushing's syndrome. The test is not without risk in primary adrenal insufficiency.

A short *overnight test* consists of 3 g of metyrapone taken orally at 2300 h with food. A fall in the 0800 h serum cortisol proves the drug was taken; a normal response is a rise of plasma ACTH to at least 22 pmol/L (100 ng/L) and a rise of 11-deoxycortisol by at least 200 nmol/L (70 μg/L).

In the *standard* metyrapone test blood and urine specimens are collected for 2 control days, then metyrapone is given orally (750 mg every 4 h for 6 doses) and specimens are collected during that and the following day. A fall in serum cortisol proves only that the drug was ingested. Normal responses consist of a rise in 0800 h plasma ACTH to at least 22 pmol/L (100 ng/L), in plasma 11-deoxycortisol by at least 200 nmol/L (70 μg/L), and an increase in urinary 17-OHC of at least 2.5-fold, the maximum exceeding the upper reference limit.

In primary adrenal insufficiency no increases will occur. In secondary deficiency neither adrenal steroids nor ACTH will rise unless some pituitary reserve exists, in which case a delayed and/or muted response may be noted. With hypercortisolism due to a pituitary adenoma the adrenals are hyperresponsive, whereas with Cushing's syndrome due to an adrenal tumor or the ectopic ACTH syndrome the response will be minimal or absent.

Dexamethasone Suppression Test. The cortisol analog dexamethasone (Δ1–10,9α-fluoro,16-methyl cortisol) suppresses ACTH secretion. It is not measured by RIA procedures for cortisol and can be used to demonstrate the integrity of feedback control of the corticotropes.

A single-dose *screening* test is useful in distinguishing 'normal' obesity from early Cushing's syndrome. At 2300 hours (11 PM) the patient is given 1 mg dexamethasone by mouth and the serum cortisol is determined at 0800 hours the next day. Normal values will be <140 nmol/L (<50 μg/L); levels >275 nmol/L (>100 μg/L) are strongly suggestive of Cushing's syndrome.

A positive screening test should be followed up with a *definitive* test of hypothalamic-pituitary-adrenal axis suppressibility. Dexamethasone, 0.5 mg every 6 hours, is given orally for 2 successive days while collecting 24-hour urine specimens and blood samples. In *normal* subjects, on the second day, plasma ACTH will be <5 pmol/L (<20 ng/L), serum cortisol <140 nmol/L (<50 μg/L), urinary free cortisol <80 nmol/d (<30 μg/d), and urinary 17-OHC <10 μmol/d (<4 mg/d). Failure to suppress normally is confirmatory evidence of Cushing's syndrome (hypercortisolism), and now the *etiology* must be determined. The most reliable procedure is to continue with a *higher dose* dexamethasone suppression test, usually carried out in conjunction with the low dose procedure when suspicion of hypercortisolism is high. The entire procedure consists of 2 control days, 2 days on low-dose dexamethasone (0.5 mg every 6 hours) and 2 days on high-dose dexamethasone (2.0 mg every 6 hours) while collecting 24-hour urine specimens and blood samples. Hypercortisolism, caused by pituitary adenoma (Cushing's disease), will usually show about 50% suppression of the various indices measured because these tumors are under a partial degree of feedback control. Hypercortisolism caused by ectopic ACTH production, or an adrenal tumor, fails to be suppressed (see Table 14–13).

Interpretation of this test will at times have to be tempered by the clinician's impression of the rapidity of onset, the severity of the disease, and the degree of autonomy of the pituitary adenoma or adrenal tumor. On occasion a pituitary tumor will be suppressed only after 3 or 4 days of the 2-mg dose, or only after the dose is increased to 4 mg every 6 hours. An adrenal adenoma may, on rare occasions, also be suppressed at this dosage.

A variant of the overnight dexamethasone screening test is now being used in the investigation of *depression.* Approximately half of the patients who have endogenous depression demonstrate resistance of the hypothalamic-pituitary-adrenal axis to suppression. Oral dexamethasone (1 mg) is given at 2300 hours and blood samples are taken for cortisol the following day at 0800, 1600, and 2300 hours. Values >140 nmol/L (>50 μg/L) indicate abnormal resistance of the axis to glucocorticoid suppression. *Chronic alcoholism* can present with a pseudo-Cushing's syndrome and a similar resistance to suppression.

Another variant is used to discriminate between primary hyperaldosteronism (PA) and *glucocorticoid suppressible hyperaldosteronism* (GSA). Oral dexamethasone (1 mg) is given at 2300 hours and another 0.5 mg at 0600 hours. Blood is taken supine at 0800 hours. In PA, plasma aldosterone concentrations are usually >300 pmol/L (>110 ng/L); in GSA usually <150 pmol/L (<55 ng/L).

CRH Stimulation. The CRH stimulation test has only recently become available but may be useful in the investigation of adrenal hypo- and hyperfunction. An intravenous injection of 200 μg of synthetic CRH is given, with blood samples taken at 0, 30, 60, 90, and 120 minutes for ACTH and cortisol assays. Hypercortisolism caused by a pituitary adenoma (Cushing's disease) results in responsiveness of both ACTH and cortisol. Cushing's syndrome caused by adrenal tumors or ectopic ACTH production gives no response. Secondary adrenal insufficiency also produces no response (see Table 14–13).

Other Tests. Other stimuli that have been used to stress the hypothalamic-pituitary-adrenal axis include vasopressin and insulin hypoglycemia. The insulin hypoglycemia procedure (described under the Triple Bolus Test) may be particularly useful in differentiating between hypercortisolism due to a pituitary adenoma, in which no response occurs, as opposed to alcoholic pseudo-Cushing's syndrome, or depression, in which there is a *rise* in ACTH and cortisol after hypoglycemia.

Adrenal Radionuclide Scan (Scintigraphy). During steroid synthesis the adrenals take up LDL-cholesterol from the plasma. If a cholesterol analog (e.g., ^{131}I-6β-iodomethyl-19-norcholesterol, NP-59) is given intravenously, the lumbar region can be scanned after 3, 5, and 7 days. In pituitary Cushing's with adrenal hyperplasia, both sides will show radioactivity. In adrenal Cushing's the tumor will be revealed, because the normal gland will be inactive due to ACTH suppression. To demonstrate an *aldosterone*-producing adenoma the patient is given dexamethasone (4 mg/d for 7 days) to suppress the normal cortex, prior to 1 mCi of NP-59. Imaging of the adrenals appears in 5 days in normal individuals, a day or two earlier with an adenoma or bilateral hyperplasia.

DISORDERS OF THE ADRENAL CORTEX

Disorders of the adrenal cortex are relatively rare. In advanced stages they are easy to recognize, but the conditions can be life-threatening and early diagnosis is important. Suspicion of an adrenal disorder, because of obesity, hypertension, weakness, or hirsutism, is common. A knowledge of effective screening procedures with high specificity is necessary to rule out adrenal disease. The main clinical disorders and their underlying hormone abnormalities, their prevalence, and their clinical features are shown in Table 14–12.

Hypercortisolism (Cushing's Syndrome)

Pathogenesis. Cushing's syndrome results from prolonged exposure to excess cortisol and a varied mix of other adrenal hormones. Prolonged corticosteroid therapy is the most common cause and is cured by drug withdrawal, if feasible. Cushing's syndrome may be classified as:

	Incidence
1. ACTH dependent	
pituitary tumor (Cushing's disease)	60%
ectopic 'ACTH'-producing tumor	20%
hypothalamic dysregulation	?
2. ACTH independent	
adrenal tumor (adenoma or carcinoma)	15%
bilateral multinodular adrenal hyperplasia	<5%
iatrogenic (exogenous glucocorticoid)	—

The ACTH-dependent causes are associated with bilateral adrenal hyperplasia caused by the high levels of ACTH or ACTH-like peptides. 'Pituitary' Cushing's occurs more often in adult women; 'ectopic' Cushing's syndrome is seen more often in older men.

Most *pituitary* tumors are small (<1 cm in diameter); a few may be large enough to affect the size or shape of the sella turcica. *Ectopic* ACTH-like peptides causing Cushing's syndrome occur in 2% of patients with lung cancer; many more tumors produce ACTH-precursor peptides without causing adrenal hyperplasia. Fewer than half of ectopic Cushing's syndrome cases arise from tumors in one of a dozen other locations.

Adrenal adenomas and carcinomas occur with equal frequency in both sexes and are associated with low ACTH levels and atrophy of normal adrenal tissue. 'Adrenal' Cushing's in children is almost always due to carcinoma and is usually fatal. Bilateral multinodular adrenal hyperplasia is rare and its pathogenesis is unclear because of the occurrence of nodules in 5% of normal adrenals.

Other disorders involving the glucocorticoids include:

Cyclic Cushing's syndrome, in which hypercortisolism is caused by periodic fluctuations in ACTH secretion, varying from 5- to 50-day cycles or more.

TABLE 14–12. Major Disorders of the Adrenal Cortex

Clinical Disorder	Hormone Abnormality	Estimated Prevalence	Characteristic Features
Glucocorticoid excess (Cushing's syndrome)	↑ Cortisol ↑ Aldosterone (?) ↑ Androgen (?)	Women 1:10,000 Men 1:30,000	Obesity, hypertension, muscle wasting, hirsutism, weakness, amenorrhea, acne, bruising, psychoses, osteoporosis, hyperglycemia
Aldosterone excess (Conn's syndrome)	↑ Aldosterone ↓ Renin (PRA)	1:2,000	Hypertension, headache, hypokalemia, weakness
Androgen excess (Congenital adrenal hyperplasia)	↑ 17-KS, ↑ DHEA ↓ Cortisol ↓ 17-OHC, ↑ KGC ↑ 17-hydroxyprogesterone ↑ Pregnanetriol ↑ Testosterone	1:15,000 homozygous 1:75 heterozygous	Hirsutism, amenorrhea, abnormal genitalia, short stature
Adrenocortical deficiency			
Primary adrenal destruction (Addison's disease)	↓ Cortisol ↓ Aldosterone ↑ ACTH	1:50,000	Abdominal distress, dehydration, weakness, hyponatremia, hyperkalemia, postural hypotension, pigmentation, weight loss, ↑ blood urea nitrogen
Secondary adrenal atrophy (ACTH deficiency)	↓ Cortisol ↓ ACTH	1:10,000	Weakness, weight loss, anorexia

ACTH = adrenocorticotropic hormone; 17-KS = 17-ketosteroids; KGC = ketogenic corticoids (17-ketogenic corticosteroids); 17-OHC = 17-hydroxycorticoids (17-hydroxycorticosteroids); PRA = plasma renin activity; DHEA = dehydroepiandrosterone.

Pseudo Cushing's syndrome, which occurs in some patients with chronic *alcoholism.*

Almost half of all patients with endogenous *depression* demonstrate increased cortisol secretion, loss of diurnal rhythm, and resistance to dexamethasone suppression. The defect is presumed to be at the hypothalamic level or higher.

Primary cortisol insensitivity (end-organ resistance) is a rare autosomal defect of the glucocorticoid receptors in target tissues. The patients have hypercortisolism without Cushing's syndrome.

Familial transcortin (CBG) deficiency, with very low plasma cortisol values but normal urinary cortisol excretion, has also been reported.

Pathochemistry. Excess cortisol is common to all forms of Cushing's syndrome and accounts for the obesity and mental disturbances. Androgen excess is variable and when present accounts for the hirsutism, acne, and amenorrhea in female patients. A relative decrease in testosterone, due to gonadotrope suppression, may explain the loss of libido and fertility in men. Mineralocorticoid effects are variable but may result in sodium retention and potassium loss leading to hypokalemic alkalosis, especially with ectopic ACTH production. The extent to which aldosterone or deoxycorticosterone contribute to the electrolyte changes and the development of hypertension is not yet clear. The level and duration of excess ACTH and associated melanocyte-stimulating peptides, is usually not sufficient to cause increased pigmentation.

Metabolic disturbances reflect the catabolic and anti-insulin effects of cortisol. There is protein and collagen wasting, leading to muscle weakness, thin skin, cutaneous striae, easy bruising, and poor wound healing. Accelerated gluconeogenesis produces hyperglycemia and glucosuria but not ketosis. Osteoporosis is a common feature and is caused by catabolic effects on bone and decreased calcium absorption from the intestine. Weight-gain results from cortisol effects on adipocytes in specific areas causing the characteristic plethoric 'moon' face, trunkal adiposity, and 'buffalo hump' in advanced cases.

Hypertension is often present and can be caused by a variety of mechanisms. Mental disturbances can range from mild sleep disorders to severe depression and frank psychosis. Superficial fungal infections occur because of decreased cell-mediated immunity. Hematologic abnormalities include mild neutrophilic leukocytosis, mild erythrocytosis, lymphopenia, and eosinopenia.

Clinical Features. Cushing's syndrome is a complex and highly unpredictable disease which can be fatal. In order of frequency, ranging from an average of 90% down to 40%, the clinical features of the disorder are obesity, hypertension, weakness, carbohydrate intolerance, hirsutism and acne, menstrual disturbances, plethoric facies, cutaneous striae and bruising, osteoporosis, and psychiatric disturbances.

No single manifestation is always present; even obesity and hypertension can be absent in cases that present with hirsutism, hypokalemic alkalosis, or osteoporosis. Several features of the disease are common to other disorders, but Cushing's syndrome can be excluded or diagnosed definitively only by the use of appropriate laboratory tests.

Laboratory Investigation. The diagnosis of Cushing's syndrome requires the demonstration of: (a) increased cortisol secretion, (b) loss of diurnal rhythm, and (c) failure of normal suppression. The *first objective* is to determine whether the patient has an overproduction of cortisol. This will involve analysis of both urine and plasma (or serum). A 24-hour urinary corticoid (17-OHC or KGC) determination and a fasting plasma cortisol have a moderate predictive value when negative. A normal urinary free cortisol level, and a normal diurnal fall in plasma cortisol, have a high predictive value and virtually exclude the presence of Cushing's syndrome. This exclusion is confirmed if the plasma cortisol suppresses normally in an overnight dexamethasone test. Abnormal results have a lower predictive value because stress and drugs can produce misleading data. Patients suffering from depression or alcoholism may fail to suppress on dexamethasone, but may be identified by the clinical situation and a normal cortisol response to insulin hypoglycemia.

A reliable out-patient *screening protocol* calls for: (1) a 24-hour urine collection for free cortisol, (2) blood samples at 0800 hours and in the late afternoon (1600 h to 2000 h) on 2 successive days for plasma cortisol, and (3) oral dexamethasone (1 mg) at 2300 hours (11 PM) on the second day with a final blood sample taken next morning at 0800 hours. If urinary free cortisol exceeds 300 nmol (110 μg) per day, if AM and PM plasma cortisol values reveal a lack of diurnal variation, and the value at 0800 hours following dexamethasone is above 140 nmol/L (50 μg/L), then a diagnosis of Cushing's syndrome is highly probable.

The *second objective,* once hypercortisolism has been established, is to determine whether the cause is a hypothalamic-pituitary lesion, an adrenal tumor, or an ectopic tumor. A definitive diagnosis is important because the treatment of each disorder is different. The endocrinologist may recognize differences between these clinical syndromes in typical cases, but will always require laboratory and radiologic confirmation.

Diagnosis is achieved by an evaluation of the dynamics of the hypothalamic-pituitary-adrenal axis using appropriate stimulation and suppression tests. *Most useful* are the plasma ACTH level, the high dose dexamethasone suppression test, metyrapone blockade, the insulin hypoglycemia test, and computerized transaxial tomography of the adrenals and/or pituitary. The expected results for these procedures for the usual clinical differential diagnosis are shown in Table 14–13. In practice the results are not always as clear cut as this table would suggest. For example:

1. Some pituitary tumors will be suppressed by dexamethasone only with longer administration or larger doses. Adrenal adenomas that still respond to ACTH, and in rare instances an ectopic tumor secreting 'ACTH,' may also be suppressed.
2. Although low ACTH values occur with adrenal lesions and high values are the rule in ectopic Cushing's syndrome, normal levels are usual in pituitary Cushing's because of the feedback effect of increased cortisol production. Intermediate plasma ACTH concentrations (25 to 50 pmol/L) may occur in either pituitary or ectopic Cushing's. Other procedures must then be used to resolve the dilemma. Venous sampling from the inferior petrosal sinus may identify a pituitary source of the ACTH.

In most cases these procedures will have achieved the *final* objective of localizing the tumor or lesion prior to treatment. There are some situations in which the CRH stimulation test, adrenal radionuclide scintigraphy, or adrenal arteriography can be helpful. Ultrasound will often locate larger tumors. In ectopic Cushing's about half the tumors are visible in a chest x-ray. Special procedures may be necessary to discover the neoplasm in other sites.

Treatment. Hypercortisolism caused by an anterior *pituitary* adenoma (Cushing's disease) can be cured by transphenoidal adenomectomy in about 80% of cases. Radiation of the pituitary is

TABLE 14–13. Response to Various Diagnostic Procedures in the Differential Diagnosis of Hypercortisolism

Diagnostic Procedures*	Pituitary Tumor (Cushing's Disease)	Ectopic ACTH Production	Adrenal Tumor (Adenoma, Carcinoma, Multinodular Hyperplasia)	Alcoholic pseudo-Cushing's; Depression
Plasma ACTH	Normal/high	Normal, high, or very high	Low	Normal/high
High dose dexamethasone	Suppression	No suppression	No suppression	Suppression
Metyrapone blockade	Response	No response	No response	Response
CRH stimulation	Response	No response	No response	Undefined as yet
Insulin hypoglycemia	No response	No response	No response	Response
CT Scan of adrenals	Bilateral hyperplasia	Bilateral hyperplasia	Unilateral or bilateral tumors	Normal
CT Scan of pituitary	Normal/abnormal	Normal	Normal	Normal

CT = Computer tomography
* See section on Laboratory Investigation for definition of responses.

curative in only 20% of cases. Intermediate lobe tumors are rare and may be harder to treat. The bilateral adrenal hyperplasia will rarely be treated by adrenalectomy and replacement steroids, because the lowering of feedback inhibition may allow the pituitary tumor to grow unchecked and cause Nelson's syndrome (hyperpigmentation). Cyproheptadine, a serotonin antagonist, will lower ACTH production temporarily in over half of the cases. Drugs such as metyrapone, aminoglutethimide and o,p-DDD block steroidogenesis but have undesirable side effects.

Hypercortisolism caused by a unilateral *adrenal* adenoma can be cured by surgery, because the remaining adrenal tissue will usually regain normal function with time. Adrenal carcinoma is often diagnosed at an advanced stage; temporary relief can be achieved by drugs that inhibit steroidogenesis, but the prognosis is poor.

Hypercortisolism caused by *ectopic* ACTH production is often associated with advanced metastatic malignancy with a poor prognosis. Inhibition of steroidogenesis can provide a temporary benefit.

Hyperaldosteronism (Conn's Syndrome)

Pathogenesis. Hyperaldosteronism is usually *secondary* and is a consequence, not a cause, of disease. Any disorder leading to a hypovolemic state (diarrhea, vomiting, sweating, hemorrhage, edema, ascites, burns) will activate the renin–angiotensin system, increase plasma-renin activity (PRA) and result in increased secretion of aldosterone (and vasopressin), a mechanism designed to restore blood volume. Only the underlying disorder requires treatment.

Primary aldosterone excess may be associated with either bilateral hyperplasia (pseudo-Conn's syndrome) or adenoma (Conn's syndrome) of zona glomerulosa cells of the adrenal cortex, leading to hypertension and hypokalemia. The adenomas are small and may take years to develop and to produce symptoms that are attributable, presumably, to the chronic inappropriate excess of aldosterone, which also suppresses PRA. Bilateral hyperplasia, also called idiopathic hyperaldosteronism, may be caused by a pituitary or hypothalamic factor acting on the zona glomerulosa; in rare instances it appears to have progressed to adenoma formation.

A special example of *secondary* hyperaldosteronism occurs with *renal* arterial or arteriolar lesions that reduce afferent blood flow to the kidney. This disorder may be distinguished from primary aldosteronism (Conn's syndrome) only by the high plasma renin activity.

Other causes of hyperaldosteronism include the following:

Glucocorticoid suppressible hyperaldosteronism (GSA) is now recognized as a rare autosomal dominant defect of the zona glomerulosa (ZG). Precursors arising in the zona fasciculata (controlled by ACTH) become available to ZG cell β-methyloxidase and give rise to aldosterone (AO). Plasma AO concentrations are elevated, but PRA is suppressed. Patients have the classic signs of primary aldosteronism but can be differentiated by the suppression of plasma AO when 2 mg of oral dexamethasone are given daily for 10 days.

Bartter's syndrome is a rare form of hyperaldosteronism associated with hypokalemia and a tendency to *low* blood pressure. Because of the low serum K^+ aldosterone production may approach normal. There appears to be a generalized defect in Na^+ transport, most notable as increased renal Na^+ and Cl^- excretion. Higher electrolyte levels in the distal tubule result in hypertrophy and hyperplasia of the juxtaglomerular cells. This causes increased plasma PRA and increased AO secretion, resulting in hypokalemia. The blood pressure tends to be low because of decreased arteriolar sensitivity to angiotensin II and norepinephrine. The disorder usually presents in childhood with weakness, polyuria, salt craving, and growth retardation.

Pathochemistry. Aldosterone secretion is controlled primarily by the renin–angiotensin system and plasma K^+, but the effect may be modulated by sodium status, ACTH levels, dopamine, and other factors. Primary aldosteronism is characterized by a level of secretion that is inappropriate for both the renin and potassium levels. Aldosterone induces Na^+ reabsorption in the renal collecting ducts in exchange for K^+ and, when hypokalemia develops, for H^+ ions. Persistent autonomous aldosterone secretion results in hypokalemic alkalosis, muscular weakness, and sometimes polyuria caused by tubular nephropathy. Potassium depletion may obscure the diagnosis by suppressing aldosterone production. Hypertension is attributed to the chronic relative excess of aldosterone, but the mechanism is still not clear.

Clinical Features. Patients with primary aldosteronism usually seek medical assistance because of frequent headaches (caused by hypertension) and/or fatigue and weakness (caused by hypokalemia). They are likely to appear perfectly normal, and so the condition must be considered a possibility in all who have essential hypertension (EHT) and are discovered to have hypokalemia. Because EHT (diastolic >100 mmHg) has a prevalence of about 10% in an adult population, and many have been given diuretic drugs that can produce hypokalemia, the *exclusion* of primary aldosteronism is a common problem. Fewer than 1% of hypertensive patients are likely to have an aldosterone-producing adenoma. The incidence is much higher when unprovoked hypokalemia is present. All such patients must be fully investigated, because prompt treatment usually results in a complete cure.

Laboratory Investigation. In a patient with hypertension and unexplained hypokalemia the *first objective* is to assess the electrolyte status. Patients with primary aldosteronism usually exhibit serum levels of $Na^+ > 140$ mmol/L, $K^+ < 3.5$ mmol/L, $HCO_3^- > 28$ mmol/L, and a blood pH > 7.42. It may then be necessary to establish that:

a. Renal potassium wasting is present and is augmented by salt loading
b. There is increased aldosterone production that is not suppressed by volume expansion
c. Plasma PRA is suppressed and fails to increase on volume depletion

Patients on diuretics presenting with hypokalemia should be taken off drugs and have their serum K^+ checked periodically in response to potassium supplementation. A 24-hour urine K^+ > 40 mmol in the presence of hypokalemia (<3.5 mmol/L) is highly suggestive and calls for further investigation. These analyses are then repeated after oral salt loading with 5 g to 10 g added NaCl per day for 5 days. In primary aldosteronism this will aggravate both the hypokalemia and the K^+ wasting.

Aldosterone data must be obtained under conditions of known salt intake, with attention to posture, preferably when hypokalemia has been corrected and the patient has been off natriuretic drugs for a few weeks. Aldosterone excretion (18-oxo-conjugate in a 24-hour urine specimen) usually exceeds the upper reference value of 60 nmol/day (20 μg/d). Plasma aldosterone is usually above the reference range of 90 to 360 pmol/L (30 to 110 ng/L) when the patient's salt intake is 7 g/day and the blood is taken while the patient is recumbent. A simple *volume expansion* procedure consists of a 3-day high sodium (>100 mmol) diet along with 200 μg of fluorocortisol thrice daily. Plasma aldosterone concentration will be suppressed to <90 pmol/L (<33 ng/L) in normal subjects, and urine aldosterone to <30 nmol/day (<11 μg/d). Plasma values should exceed 300 pmol/L (>110 ng/L) in primary aldosteronism.

Plasma renin activity (PRA) is suppressed in primary aldosteronism, but is low also in about 25% of people with essential hypertension. They can be differentiated by a *volume depletion* procedure consisting of a 3-day low-sodium (10 mmol), high-potassium (100 mmol) diet. After an overnight fast without water the patient is given a 40-mg oral dose of furosemide and ambulates for 4 hours, after which blood is taken for PRA measurement. Normal subjects will have elevated values; low values suggest primary aldosteronism.

A new procedure, the *captopril test,* appears to be safe, convenient, and cost-effective. A single 25-mg dose of this angiotensin-converting-enzyme inhibitor will block AT-II production. It is injected intravenously at 0800 h and blood is sampled 2 hours later. Plasma aldosterone decreases in normal subjects and in essential hypertensives, but remains elevated in primary aldosteronism.

For all these tests the protocol should be dictated by the laboratory, to conform to standards on which the reference ranges were established.

The *final objective* is localization and characterization of the lesion (unilateral tumor or bilateral hyperplasia). Computerized tomography (CT) of the abdomen and scanning of the lumbar region for ^{131}I-iodocholesterol uptake in dexamethasone suppressed adrenals, are most helpful. Retrograde venography, with adrenal vein sampling for PRA and aldosterone assays, may sometimes be required.

Treatment. Adenomectomy or unilateral adrenalectomy is curative for solitary adenomas (Conn's syndrome) with return to normal of the blood pressure, electrolytes, and adrenal function. Surgery is contraindicated in bilateral multinodular hyperplasia because it does not improve the hypertension and means lifelong steroid supplementation. In these cases specific blockade of aldosterone action by treatment with oral spironolactone (Aldactone) is usually effective in lowering blood pressure and restoring normokalemia. In fact a trial of spironolactone with marked improvement in blood pressure and serum K^+ is good presumptive evidence for primary aldosteronism.

Hyperaldosteronism and hypertension, caused by renal artery stenosis, may be corrected by vascular surgery or nephrectomy, provided treatment is effected before the contralateral kidney has been damaged.

Adrenocortical Insufficiency (Addison's Disease)

Pathogenesis. The deficiency may be caused by failure of the adrenal glands (Addison's disease), or secondary to lack of pituitary ACTH, or production of renin by the kidney. Addison's disease is a rare *primary* lesion of the adrenals which can present acutely, from hemorrhage while on anticoagulant therapy, or from gram-negative sepsis with endotoxic shock (e.g., Waterhouse–Friderichsen syndrome caused by meningococcemia). More often it develops insidiously with gradual destruction of the gland, usually on an immune basis, but occasionally because of granulomatosis from tuberculosis or sarcoidosis, or because of metastatic invasion. Autoimmune *adrenalitis* can now be diagnosed from the presence of circulating adrenal antibodies and may occur in association with other autoimmune disorders (e.g., Hashimoto's thyroiditis, pernicious anemia, and diabetes mellitus).

Secondary adrenal insufficiency also may develop acutely, following withdrawal of prolonged steroid therapy, from trauma to or surgical ablation of the pituitary, or (rarely) postpartum infarction of the pituitary (*Sheehan's syndrome*). It may develop chronically as an idiopathic atrophy of the pituitary (*Simmond's disease*), or because of a tumor, infiltrative or other destructive lesions within the sella. The disorder generally involves the entire pituitary and hypofunction of other glands usually precedes loss of adrenal function; however, solitary ACTH deficiency can occur.

Hypoaldosteronism is defined by a low urinary AO and high serum K^+. *Primary hypoaldosteronism* occurs as a rare genetic deficiency of the terminal methyloxidase that converts 18-hydroxycorticosterone (18-OHCS) to aldosterone (AO). The ratio of 18-OHCS/AO is very high (>150). *Pseudohypoaldosteronism* is a rare congenital insensitivity of the collecting ducts to aldosterone. Both PRA and AO are high but the ratio of 18-OHCS/AO is normal (2–12). Isolated *hyperreninemic* aldosterone deficiency occurs in some elderly people and frequently in critically ill patients. A *hyporeninemic* hypoaldosteronism is seen occasionally in long-standing diabetics with mild renal insufficiency; their serum K^+ is high.

Pathochemistry. In *primary* adrenal deficiency both cortisol and aldosterone are deficient and there is an inadequate response to stress. Lack of aldosterone results in decreased renal exchange of Na^+ for K^+, leading to salt wasting with hyponatremia and hyperkalemia. Dehydration and hypovolemia lead to hypotension, and to increased vasopressin (ADH) secretion, which accentuates the hyponatremia; this leads to poor renal perfusion and increased blood urea and creatinine. Lack of cortisol results in impaired gluconeogenesis and the unopposed action of insulin results in hypoglycemia. The loss of cortisol feedback inhibition results in high ACTH and β-LPH levels, which lead to the usual skin pigmentation of Addison's disease. The loss of aldosterone effects on blood volume results in high renin concentrations.

In *secondary* adrenocortical insufficiency, resulting from a loss of pituitary ACTH, there is cortisol deficiency but aldosterone production is fairly well maintained and salt loss is not a serious problem. The absence of an array of pituitary hormones gives a broader spectrum of clinical problems ranging from hypogonadism to hypothyroidism.

The low levels of ACTH and β-LPH result in hypopigmentation.

Clinical Features. Chronic adrenocortical deficiency (Addison's disease) does not become apparent until nearly 90% of the gland is destroyed. It usually has an insidious onset of fatigue, weakness, weight loss, and dehydration. Gastrointestinal complaints include pain, anorexia, nausea, vomiting, and diarrhea. Low blood pressure, postural hypotension, and hypoglycemia are significant findings, along with lethargy, depression, and inability to concentrate. In mild cases pregnancy can occur, sometimes resulting in an adrenal crisis postpartum. Pigmentation of the skin, mucous membranes, and areola is probably caused by β-LPH rather than ACTH. Localized vitiligo is caused by autoimmune destruction of melanin-producing cells in the area.

Adrenal insufficiency *secondary* to pituitary hypofunction differs in three ways. Hypopigmentation is usual because ACTH and β-LPH are very low. A functioning renin–aldosterone system prevents serious salt and water loss, thus dehydration and hypotension are unusual, but the patient may be unable to achieve an adequate adrenal response to stress. The lack of other pituitary hormones is usually clinically evident.

Acute adrenal insufficiency is rare and is usually caused by adrenal hemorrhage. Initial complaints include nausea, vomiting, fever and abdominal pain, progressing rapidly to hypotension, shock, and vascular collapse. Hyponatremia, dehydration, hyperkalemia, and hypoglycemia are often present and severe. Pigmentation is unusual at this stage.

Laboratory Investigation. In *acute* adrenal crises, the prime concern is to bring the situation under control and to assess the fluid and electrolyte disturbance. Although *chronic* insufficiency (classic Addison's disease) can be diagnosed by the clinical presentation, along with unequivocal low serum cortisol and high ACTH concentrations, less obvious cases and exclusion of this diagnosis require an ACTH test. These patients may have low-normal serum cortisol and urinary corticoids with high-normal plasma ACTH values. In *severe* deficiency serum Na^+, HCO_3^-, and glucose will be low, and K^+ and BUN will be elevated. A low serum cortisol is virtually diagnostic. In other forms of shock cortisol values are likely to be raised. In *milder* cases, electrolyte data are nonspecific and are not helpful. Low serum DHEA · S concentrations are suggestive but not definitive. The initial investigation in such cases is a *short ACTH stimulation test.* A normal cortisol response rules out primary adrenal insufficiency; no response confirms the diagnosis.

An equivocal response of serum cortisol levels could be caused by either primary or secondary adrenal insufficiency and calls for a *long ACTH stimulation test.* Failure of the serum cortisol, the urinary free cortisol, or urinary cortisol metabolites (17-OHC, KGC) to rise indicates *primary* adrenal disease. A delayed but stepwise increase probably means *secondary* adrenal insufficiency. The status and functional reserve of the pituitary can best be assessed by a *CRH stimulation test.* The endogenous ACTH test (metyrapone blockade) and the insulin hypoglycemia test have been in common use but these tests are potentially dangerous in this situation.

Treatment. An adrenal crisis requires immediate supportive therapy consisting of fluid and electrolyte replacement along with dexamethasone. A short ACTH test can be done during treatment because dexamethasone does not interfere with cortisol determinations. Return of adrenal function is unlikely, so lifelong replacement therapy is required. Oral cortisone (37.5 mg) or prednisone (7.5 mg) daily in divided doses, with a supplement of 9α-fluorocortisol (0.1 mg) if necessary, gives routine support. The patient must wear a 'Medic-alert' bracelet and take extra hormone in times of stress.

Congenital Adrenal Hyperplasia, CAH (Adrenogenital Syndrome)

Pathogenesis. These genetic defects in adrenal hormone synthesis are transmitted as autosomal recessive traits of varying prevalence. The disorders are summarized in Table 14–14. Almost all involve enzymes that affect cortisol (and aldosterone) production. This reduces feedback inhibition and increases ACTH levels causing adrenal hyperplasia. The most frequent clinical problem is excess androgen production; the next is salt loss. The commonest disorder is *21-hydroxylase deficiency,* which is characteristically associated with the HLA–B locus on chromosome six.

TABLE 14–14. Congenital Adrenal Disorders

Enzyme Defect (% frequency)	Circulating Hormones: *Deficiency*	Circulating Hormones: *Excess*	Increased Urinary Metabolites	Clinical Consequences
21-Hydroxylase (95%)	Cortisol Aldosterone (in 50%)	Progesterone 17-Hydroxyprogesterone Androgens	Pregnanediol Pregnanetriol 17-Ketosteroids	Hirsutism, amenorrhea, masculinization, hyponatremia, hyperkalemia, dehydration
11-Hydroxylase (2%?)	Cortisol Aldosterone Corticosterone	Androgens 11-Deoxycorticosterone (DCS) 11-Deoxycortisol (DC)	17-Ketosteroids Tetrahydro–DCS Tetrahydro–DC	Masculinization, hypertension
17-Hydroxylase (rare)	Androgens Corticosteroids Estrogens	11-Deoxycorticosterone (DCS) Corticosterone (CS) Progesterone	Tetrahydro–DCS Tetrahydro–CS Pregnanediol	Female immaturity, hypertension, hypokalemic alkalosis
18-Hydroxycorticosterone β-methyloxidase (rare)	Aldosterone	18-Hydroxycorticosterone	Tetrahydro-18-hydroxy-CS	Salt loss, hypotension
3β-ol Dehydrogenase (rare)	Adrenal corticoids Aldosterone	Dehydroepiandrosterone	17-Ketosteroids Δ^5-3β–Steroids	Salt loss, male hypospadias, female mild virilization
Desmolase (rare)	All adrenal hormones	Adrenal lipid (cholesterol)	None	Adrenal insufficiency

Pathochemistry. The enzyme defect can occur at any stage in the steroid biosynthetic pathways (Fig. 14–12) and may be partial, or virtually complete. Hormone precursors increase in amount and will follow any other synthetic pathway that is available. With a rise in precursor substrate the effects of the enzyme deficiency may be compensated, resulting in low-normal cortisol production, but at the expense of abnormal amounts of other hormones, particularly androgens.

A deficiency of 21-hydroxylase may be limited to the zona fasciculata and the substrate 17-hydroxyprogesterone, or include the zona glomerulosa and the substrate progesterone. The latter, which occurs in about 50% of cases, results in aldosterone deficiency and salt wasting. Excess 17-hydroxy precursors are partly metabolized and excreted (pregnanetriol); the rest follow the androgen pathways, leading to elevated DHEA-S and testosterone, and result in virilization. The much *less common* deficiency of 11-hydroxylase results in excess deoxycorticosterone and 11-deoxycortisol, weak mineralocorticoids which result in salt retention and hypertension. The *rare* syndrome of aldosterone deficiency due to the lack of 18-hydroxycorticosterone methyloxidase is relatively benign except in infants. *Other rare forms* of congenital adrenal deficiency are included in Table 14–14 and will not be discussed further.

Clinical Features. The number of heterozygotes for 21-hydroxylase deficiency varies in different populations but the average prevalence is about 1:75. The prevalence of homozygotes is about 1:15,000, but if one sibling is affected there is a 25% risk for the next child. The disorder ranges in severity from mild hirsutism and oligomenorrhea in young women to a masculinized newborn infant with a life-threatening salt-losing problem. In female infants there may be moderate clitoral enlargement or varying degrees of labial fusion. Diagnosis of mild defects in males is difficult.

Excess androgens also cause hirsutism, acne, increased muscle mass, and accelerated growth followed by early closure of the epiphyses and short stature. At puberty girls fail to menstruate. The block in cortisol production lowers tolerance to trauma and other forms of stress; a relatively mild illness can cause severe prostration. When aldosterone synthesis is also impaired there may be salt wasting, volume depletion, and hypotension. The differential diagnosis must exclude ovarian and adrenal tumors and exogenous hormones.

Laboratory Investigation. The diagnosis of CAH should be considered in all infants presenting with failure to thrive, severe prostration, salt-wasting crises, or precocious development. Plasma DHEA and testosterone concentrations will be elevated because of the 21-hydroxylase defect, and androgen metabolites cause an increase in urinary 17-KS. Cortisol precursors increase urinary KGC, but not 17-OHC. Low plasma and urinary cortisol values are suggestive, whereas increased levels of urinary pregnanetriol or plasma 17-hydroxyprogesterone (17-OHP) are usually diagnostic. In *milder cases,* in which the condition may be suspected in hirsute or amenorrheic young women, the same tests will usually establish that a deficiency of 21-hydroxylase exists. The status of the aldosterone pathway may need to be assessed. About half will have high renin, low aldosterone levels; the rest may have high aldosterone *and* renin levels because of competitive blocking of renal aldosterone receptors by progesterone and 17-OHP.

In *severe cases* in newborn infants with ambiguous genitalia the first requirement is to determine the fluid and electrolyte status. About half the infants will be salt-losers, and the hyponatremia, hyperkalemia, and dehydration will require appropriate management. The diagnosis can usually be made by measuring urinary pregnanetriol and/or plasma 17-OHP and cortisol. A buccal smear or karyotype will establish the infant's sex. Later the anatomic defects can be defined and corrected.

The diagnosis of the 11-hydroxylase deficiency is more difficult but follows the same principles and need be considered only when hypertension is present.

Treatment. The acute situation in newborn salt-losers is managed with intravenous saline, mineralocorticoids, and glucocorticoids. Lifelong maintenance therapy consists in replacement of the end-product(s) of the affected pathway(s). Cortisone or prednisone in divided doses, plus oral 9α-fluorocortisol if necessary, will usually be effective.

In children, in whom side effects on growth and sexual development are important, treatment should be *monitored.* Urinary 17-KS or pregnanetriol, and plasma 17-OHP, or androgens, have all been advocated for this purpose. Bone age will allow a prediction of growth status.

Most patients will do well with treatment but should wear a Medic-alert bracelet. They lack a normal response to stress or trauma and must receive extra prednisone at such times. Apart from stress crises the prognosis is generally good and normal development and fertility can be anticipated.

THE GONADS

Gonadal disorders present a wide spectrum of clinical abnormalities, which manifest at various stages of sexual maturation. An understanding of these disorders requires a knowledge of the embryologic and biochemical aspects of sexual differentiation, the complex hypothalamic-pituitary-gonadal interrelationships, and the role of the sex steroid hormones. Some aspects of the gonadal disorders are presented in this section.

GONADAL DEVELOPMENT AND DIFFERENTIATION

Normal human cells contain 46 chromosomes: 22 pairs of autosomes and a pair of sex chromosomes, XX and XY, representing female and male respectively. One member of each pair is derived from each parent.

The sex-determining genes are present on the X and Y chromosomes and these influence the primitive bipotential gonad to differentiate either as a testis or as an ovary. When a Y-containing sperm fertilizes an ovum, a genetic male develops; with an X-containing sperm, a genetic female develops. Thus, the *genotype* (genetic composition) is determined at the time of fertilization, whereas the *phenotype* (bodily appearance) is dependent on the subsequent sexual differentiation and maturation under the control of sex hormones.

The normal process of sex differentiation is a dynamic one; more easily understood when considered in four stages:

1. *Sex determination;* the fertilization of an ovum by either a Y- or an X-bearing sperm
2. *Gonadal differentiation;* the stage in which the bipotential undifferentiated gonad becomes either an ovary or a testicle, depending on the outcome of the first stage
3. Differentiation of the internal *genital duct system* (müllerian or wolffian)
4. Differentiation of the *external genitalia.*

The ***genital system*** arises in the human embryo during the fourth week of gestation, when a thickened area of coelomic epithelium, the germinal epithelium, develops on the urogenital ridge. The cells of the germinal epithelium and the underlying mesenchyme proliferate and give rise to the gonadal ridges on the medial side of each mesonephros. The complete differentiation of normal gonads depends on the arrival at the genital ridges of a sufficient number of viable *primordial germ cells* (primitive sex cells) from the yolk sac. By about 42 days 300 to 1300 germ cells are present in the primitive gonad. These cells will develop later into either oogonia or spermatogonia. If the germ cells fail to arrive, gonads do not develop. At about the sixth week, the primordial germ cells migrate into the underlying mesenchyme and become incorporated into the finger-like epithelial cords, called primary sex cords, which had developed a little earlier. At this stage the gonad is said to be bipotential (indifferent) and consists of an outer area, the *cortex* (a potential ovary), and an inner area, the *medulla* (a potential testis). If a testis develops, the cortex regresses and similarly if an ovary develops, the medulla regresses.

Development of the Testis

In the bipotential gonad both the wolffian (mesonephric) and the müllerian (paramesonephric) duct systems are present. However, between days 43 to 49, the medulla under the influence of a testis-organizing factor (histocompatibility Y, H-Y antigen), is transformed into an embryonic testis. The H-Y antigen is regulated by the Y-chromosome and is thought to be secreted by the primitive Sertoli cells of the gonadal blastoma. The embryonic testis secretes a glycoprotein, M_r 125,000 (müllerian inhibiting factor or müllerian regression factor), which causes the müllerian system to regress. At about 60 days, *Leydig cells* appear and the embryonal testes become fetal testes. Under the influence of human chorionic gonadotropin (hCG, synthesized by the syncytiotrophoblast) the Leydig cells produce *testosterone,* which promotes the differentiation of the wolffian duct system into the male genital tract (epididymis, vas deferens, and seminal vesicles). Also, under the influence of testosterone or its metabolite 5α-dihydrotestosterone, differentiation of the primitive external genitalia into the definitive external genitalia of the male occurs. At about 28 weeks the fetal testes have descended from the dorsal abdominal wall to the inguinal rings. Shortly before birth the testes descend into the scrotum.

Development of the Ovary

In the absence of the Y chromosome, gonadal development proceeds at a slow rate; transformation from the bipotential gonad into an embryonic ovary occurs around days 50 to 55. Primordial germ cells incorporated into the epithelial cords (cortical cords) are called oogonia; between the 11th to 12th week a large number of these oogonia enter meiotic prophase and are transformed into oocytes. After the completion of prophase, meiosis-I is interrupted and is not completed until years later, at ovulation. At about 16 weeks, the cortical cords break up and form primordial follicles consisting of an oocyte surrounded by a single layer of flat follicular (granulosa) cells. Between the 20th to 25th week, the number of primordial follicles reaches a maximum of about 7×10^6; this number decreases to about 2×10^6 at birth. However, only about 400 oocytes are ever ovulated; the rest degenerate.

During the formation of the ovary, the female genital tract (uterus, fallopian tubes, and the upper vagina) develops from the müllerian duct system, while the wolffian duct system regresses. The removal of the fetal ovary from a female embryo has no effect on fetal sexual development; thus, in the absence of testes (Y chromosome), or fetal ovaries, the müllerian duct system will develop, while the wolffian duct system will regress.

REGULATION OF NORMAL GONADAL FUNCTIONS

The Testes

The mature testis has two well-defined functions, one gametogenic and the other endocrine. Accordingly, it consists of two distinct entities, the *seminiferous tubules,* which produce the sperm, and the *Leydig cells* (interstitial cells), which secrete testosterone. Both of these functions are under the control of complex neuroendocrine interactions. A simplified mechanism is shown in Figure 14–15. Under the influence of gonadoliberin (gonadotropin releasing hormone, GnRH, LHRH), lutropin (luteinizing hormone, LH) released from the anterior pituitary stimulates the Leydig cells to secrete testosterone. The circulating testosterone exerts a negative feedback on the secretion of LH. This effect of testosterone appears to be mainly on the hypothalamus, whereby the secretion of GnRH is inhibited. A direct effect of testosterone on the secretion of LH at the level of the pituitary has also been described. Thus, when Leydig cells are damaged, or after castration, the plasma levels of testosterone decrease while the levels of LH increase. On the other hand, when testosterone levels increase LH levels decrease. There are also reports that high levels of LH may inhibit GnRH secretion by a 'short feedback loop', and the existence of an 'ultra short feedback loop', whereby GnRH may inhibit its own secretion, has been reported.

Follitropin (follicle-stimulating hormone, FSH) acts on the germinal epithelium to promote spermatogenesis. The regulation of FSH has not yet been established; the most favored hypothesis at present is that the seminiferous tubules (or more specifically the Sertoli cells) secrete a nonandrogenic, nonestrogenic, water-soluble substance designated *inhibin-F* (folliculostatin; FSH-suppressing substance) which selectively regulates FSH by a negative feedback mechanism. Whether the hypothalamus secretes a single releasing factor for the two gonadotropins or whether there are two distinct hypothalamic releasing hormones, one each for LH and FSH, has not been settled.

The decapeptide isolated from the hypothalamus and synthetic GnRH are capable of stimulating the secretion of both LH and FSH. GnRH binds to gonadotrope plasma membrane receptors, ionic Ca^{2+} is mobilized, and gonadotropins are released. Receptor numbers appear to be regulated by circulating sex steroids in a negative feedback mechanism. GnRH receptors have been identified also in both the ovary and the testis, and so this hormone may have a broader role in reproductive physiology than is presently understood. Secretion of GnRH is episodic, which results in a pulsatile release of LH and FSH into the circulation. Serum assays for these hormones require several samples

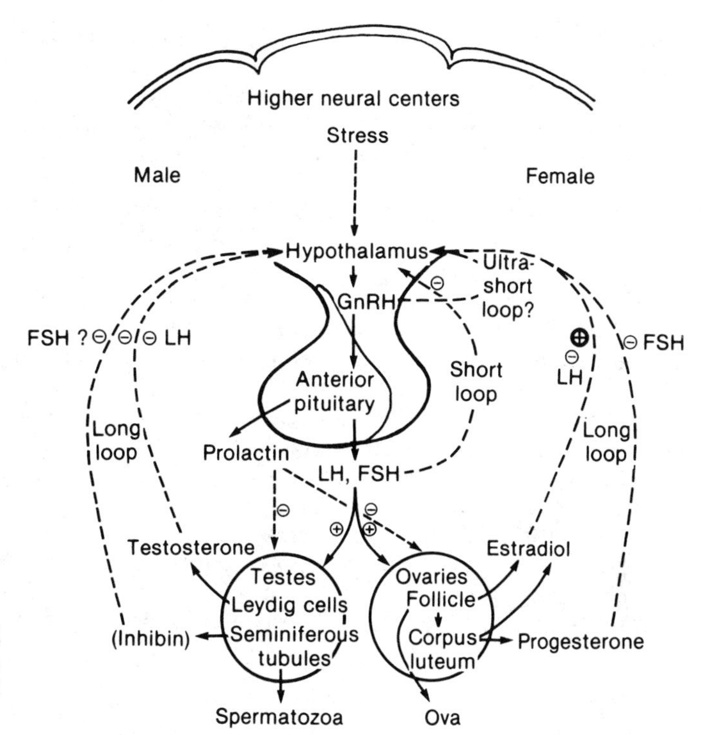

FIGURE 14–15. Secretions and control mechanisms of the hormones of the gonads. GnRH = gonadoliberin; FSH = follitropin; LH = lutropin; ⊕ = stimulation; ⊖ = inhibition.

at timed intervals to achieve a representative mean value.

LH binds to specific receptors on the Leydig cell membrane; adenylate cyclase is activated and the resulting cAMP releases a catalytic protein kinase which induces enzymes that increase steroidogenesis (Fig. 14–16). The mechanism of FSH action may be similar but is not as well established. Both LH and FSH are required for spermatogenesis.

The main steroids secreted by the testis are *testosterone, androstenedione,* and *dehydroepiandrosterone.* The major portion of the biologically active androgen 5α-dihydrotestosterone is formed by peripheral reduction of testosterone. It is clear that Leydig cells are the main site of testicular androgen biosynthesis. In some species the cells of the seminiferous tubules may also be a source of androgen that influences directly the differentiation of the germinal cells. The major steroid *biosynthetic pathway* for the production of testosterone is probably via the Δ^4 (4-ene-3-ketonic) intermediates such as progesterone and 17α-hydroxypro-

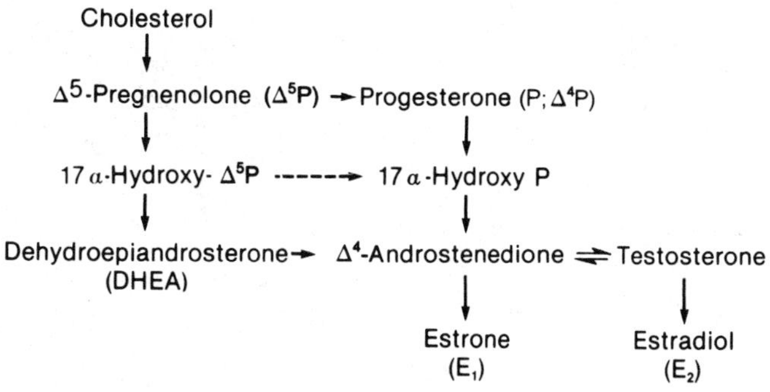

FIGURE 14–16. Biosynthetic pathways of the testicular androgens and ovarian estrogens. For estriol see Fig. 23–2. Gonadal hormone synthesis in the testes and corpus luteum is mainly by the Δ^4 pathway; in the adrenal cortex and ovarian follicle, mainly by the Δ^5 pathway. 17-Hydroxy-P = 17-Hydroxyprogesterone; 17-Hydroxy-Δ^5-P = 17-hydroxy-Δ^5-pregnenolone.

Cholesterol
↓
Δ^5-Pregnenolone (Δ^5P) → Progesterone (P; Δ^4P)
↓ ↓
17α-Hydroxy-Δ^5P - - - → 17α-Hydroxy P
↓ ↓
Dehydroepiandrosterone (DHEA) → Δ^4-Androstenedione ⇌ Testosterone
↓ ↓
Estrone (E_1) Estradiol (E_2)

gesterone as shown in Figure 14–17. The contribution of alternative pathways for the biosynthesis of testosterone in normal testis is minimal. The enzymes 17α-hydroxylase, the C_{17}–C_{20} desmolase, and the 17 β-hydroxydehydrogenase involved in the Δ^4 pathway are present in the testicular microsomal fraction and are NADPH dependent.

In normal men the *plasma testosterone* levels range from 11 to 35 nmol/L (3 to 10 μg/L) and over 95% of this testosterone is from the testis. Testosterone in the plasma, like other biologically active steroids, exists in both free and protein-bound states. Between 97% and 99% of plasma testosterone is bound either to plasma albumin (high capacity and low affinity) or to a high affinity β-globulin which is specific for gonadal steroids and is referred to as testosterone–estrogen binding globulin (TEBG; sex-hormone binding globulin, SHBG). It is the free (unbound) form of testosterone which can enter the tissues to exert a biologic effect or be metabolized. The protein-bound form serves as a reservoir and as a means of transporting the hormone. It has also been shown that TEBG can enter various cells of the genital system, but the significance of this finding is not yet apparent.

Circulating testosterone enters most body cells and may act directly in some of them (e.g., bone, muscle). In specific androgen target tissues (e.g., prostate) a cytosolic 5α-reductase converts testosterone to 5*α-dihydrotestosterone* which then binds to a specific cytosol receptor protein and this complex is transferred to the cell nucleus. An interaction with the chromatin leads to the production of messenger RNAs for specific proteins by which the hormonal effects of androgens are expressed.

Experimental and clinical observations have indicated that *prolactin* exerts an effect on the male reproductive system. Specific RIA measurements have indicated that the plasma levels of prolactin in males are about half those observed in females. The levels in males decrease after castration and are restored following treatment with gonadal ste-

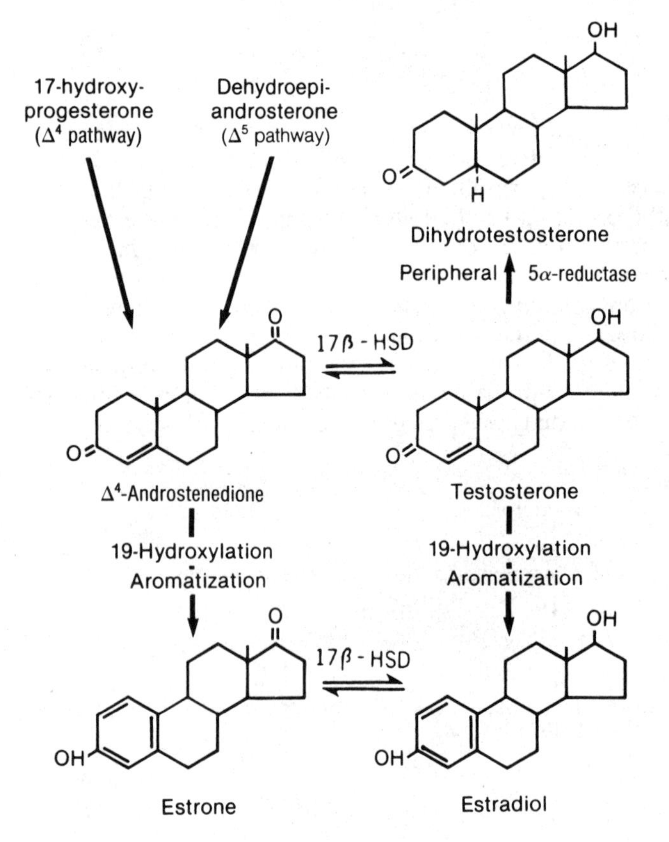

FIGURE 14–17. Structure and biosynthetic pathways of the major gonadal steroids. 17β-HSD = 17β-hydroxysteroid-dehydrogenase.

roids. Because prolactin receptors have been detected on the Leydig cells, it is presumed that prolactin can act directly on these cells. It has also been suggested that prolactin potentiates the effect of LH on Leydig cells. Men with prolactin producing tumors are generally impotent and lowering of the prolactin levels by treatment with bromocriptine, or by surgical removal of the pituitary tumor, restores sexual function. The mechanism by which prolactin influences sexual function is not clear, but it may act to downregulate gonadotropin receptors on Leydig cell membranes.

The Ovary

The normal ovary performs two well defined functions: (1) it stores and harbors ova and releases these into the genital tract for fertilization and (2) it secretes estradiol and progesterone which condition the whole individual and in particular, the genital tract. In mammals, the ripening and release of ova occur in a regular cycle which is controlled by tropic hormones. In primates this cycle is known as the menstrual cycle, which is divided into three phases: the *follicular* phase (proliferative), *midcycle* (ovulatory) and the *luteal* phase (secretory).

The *hormonal changes* that occur in a normal human ***menstrual cycle*** are summarized in Figure 14–18. The beginning of the cycle is initiated by a rise in FSH, which occurs in response to a decline in the secretion of 17β-estradiol (estradiol) by the corpus luteum of the preceding cycle (negative feedback). Under the influence of FSH, the *follicles* begin to grow and become responsive to the steroidogenic action of LH, producing increasing amounts of estradiol. At first, during the late follicular phase, estrogens rise slowly, then rapidly reach a peak on the day preceding a surge of LH. The intensity and the duration of the estradiol surge are critical. The mid-cycle surge of LH (positive estrogen feedback), is accompanied by a smaller surge of FSH. Approximately 24 hours after the surge of gonadotropins, the follicle ruptures and the egg is released (ovulation).

Following ovulation, the follicular cells hypertrophy and 'luteinize' to form the *corpus luteum* and the luteal phase of the menstrual cycle commences. The corpus luteum secretes both progesterone and estradiol, which reach peak concentrations by day 8 or 9 after ovulation. A serum progesterone value >16 nmol/L (5μg/L) is reliable evidence of ovulation. During the luteal phase, LH secretion is suppressed by the negative feedback of progesterone and estradiol. Following the peak of progesterone and estradiol secretion the corpus luteum begins to regress (day 10 to 12 after ovulation)

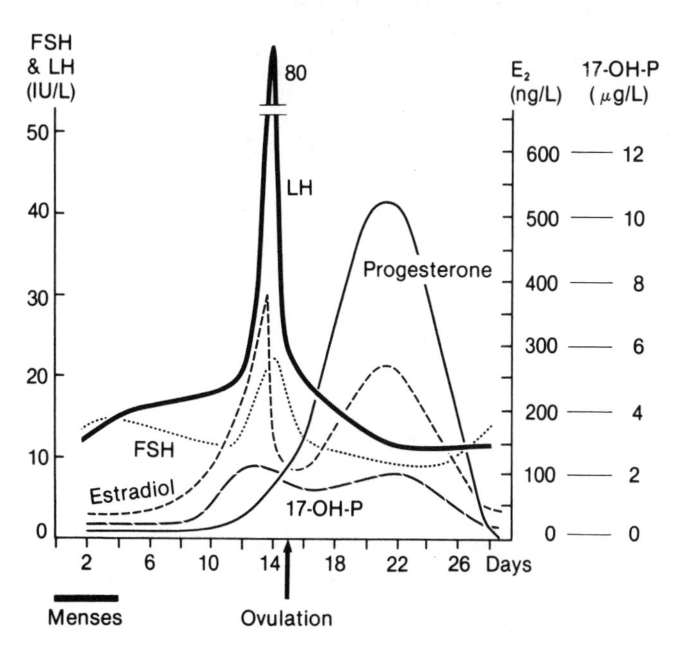

FIGURE 14–18. Hormonal changes in plasma during a normal menstrual cycle. FSH = follitropin; LH = lutropin; E_2 = estradiol; 17-OH-P = 17-hydroxyprogesterone.

and is followed by menstruation. The secretion of progesterone and estradiol decline to basal levels, resulting in the rise of FSH and initiation of a new cycle of follicular growth and maturation.

The *regulation* of this complex cycle involves the hypothalamus, the pituitary, and the ovarian axis. Estradiol plays a key role in this regulation. Both positive and negative feedback mechanisms are involved and these are summarized in Figure 14–15. The gonadotropins produced under the aforementioned control mechanisms are released in a pulsatile manner giving rise to recurring peaks in serum concentrations. Because GnRH stimulates the secretion of both FSH and LH, the divergence in the pattern or the rate of their secretion observed *in vivo* may be modulated by differential effects of gonadal steroids at the pituitary level.

The ovary produces three classes of steroids—*progestins, androgens,* and *estrogens.* The biosynthetic pathways leading to the formation of estrogen in the ovary are summarized in Figure 14–16. Two pathways for estrogen biosynthesis have been established: (1) the Δ^5-3β-hydroxy pathway (via pregnenolone) and (2) the Δ^4-3-ketone pathway (via progesterone). In the graafian follicle the theca cells utilize the Δ^5 pathway, while the granulosa cells utilize the Δ^4 pathway for estrogen production. The corpus luteum produces estrogen via the Δ^4 pathway. The production of estradiol involves both the theca and the granulosa cells. Under the influence of LH the theca cells synthesize androgens (androstenedione, testosterone) which diffuse into the follicular fluid and enter granulosa cells. These androgens, under the influence of FSH, are aromatized to estradiol. Inhibin-F activity has been demonstrated in follicular fluid and the granulosa cells are probably its source; this factor may have a direct feedback effect on pituitary FSH secretion.

Estradiol formed in the ovary is secreted into the circulation where it binds to plasma proteins. Approximately 38% of plasma estradiol is bound to testosterone–estrogen binding globulin (TEBG), 60% to albumin and 2% to 3% is free. Compared to its binding affinity for testosterone, TEBG has about a 3-fold greater affinity for 5α-dihydrotestosterone and approximately one-third as great for estradiol. TEBG synthesis occurs in the liver and is stimulated by estrogens and inhibited by androgens. Adult women have twice the plasma TEBG concentration of men. The metabolic clearance rate (MCR) of sex steroids is inversely related to the relative binding to TEBG and therefore is generally lower in women than in men.

Large amounts of *progesterone* (Table 14–15) are secreted during the luteal phase of the menstrual cycle, primarily by the corpus luteum. Progesterone is cleared rapidly from the blood (MCR-2100 to 2500 L/day). Although progesterone binds to corticosteroid binding globulin (CBG) with high affinity, under physiologic conditions the CBG

TABLE 14–15. Commonly Accepted Reference Values for the Gonadal Steroids

Hormone	Male	Female	Units
Serum			
Testosterone			
Prepubertal	<1.0 (<0.3)	<1.0 (<0.3)	nmol/L (μg/L)
Adult	11–35 (3–10)	0.7–2.8 (0.2–0.8)	nmol/L (μg/L)
Androstenedione			
Adult	3–8 (0.9–2.4)	2–9 (0.6–2.7)	nmol/L (μg/L)
Estradiol (estrogen)			
Prepubertal	<75 (<20)	<75 (<20)	pmol/L (ng/L)
Adult	<180 (<50)		pmol/L (ng/L)
Follicular		90–270 (25–75)	pmol/L (ng/L)
Ovulatory		750–2200 (200–600)	pmol/L (ng/L)
Luteal		375–1100 (100–300)	pmol/L (ng/L)
Postmenopausal		<180 (<50)	pmol/L (ng/L)
Progesterone			
Follicular	<3 (<1)	<6 (<2)	nmol/L (μg/L)
Luteal		8–80 (2.5–25)	nmol/L (μg/L)

binding sites are occupied by cortisol, which has a higher affinity for CBG and is present in much higher concentrations in the blood.

Estradiol and progesterone exert their effects on target cells by combining with their specific cytosolic receptors. The steroid-receptor complex translocates to the nucleus and stimulates the synthesis of new protein (enzyme) through which the biologic activity of each is expressed.

There are no significant changes in the plasma levels of *prolactin* during the menstrual cycle. Normally prolactin concentrations rise during sleep, reaching a peak between 0300 to 0500 hours. Prolactin concentrations, presumably because of high levels of circulating estrogens, are 10 to 20 times higher during pregnancy. The levels decline postpartum unless the infant is breast fed, in which case they continue to be elevated. Prolactin may play a role in normal reproductive function by influencing ovarian steroid secretion and by modulating the number of LH receptors.

Current concepts regard the *ovary* as the principle regulator of the menstrual cycle. Neuronal elements in the region of the arcuate nucleus discharge a bolus of GnRH into the hypophyseal circulation about once per hour. This is required for the functional integrity of the gonadotropes, but the effect is permissive. Release of gonadotropins is *controlled* by ovarian estrogens. The length of the menstrual cycle is determined by follicular development and the lifespan of the corpus luteum. In clinical disorders in which the secretion of GnRH is affected by a lesion in the arcuate nucleus, administration of GnRH in hourly pulses (using an IV pump) will maintain ovarian function and menstrual cycles. Recent attention has focused on the catecholestrogens, particularly 2-hydroxyestrone, as possible mediators of the *positive* feedback effect of ovarian estrogens.

Puberty

Puberty is the period of transition during which a child is transformed gradually into a young adult. This involves the maturation of the secondary sex characteristics, the occurrence of the adolescent growth spurt, the initiation of gametogenesis, and the attainment of fertility. The first sign of secondary sexual development in the male is testicular enlargement, while in the female it is the appearance of breast budding (thelarche) and pubic hair (pubarche). The mean chronologic age for these events in males and females is 11.6 years and 10.8 years, respectively. This stage of puberty and the progression to adulthood is generally completed in males by 15 years and in females by 14 years. The average age of menarche (menstrual periods) in industrialized Western countries is at present 13 years. The mean age of voice change in boys is about 13.5 years.

Although the specific *mechanism* for the onset of puberty is complex and not fully understood, it is generally believed that pubertal changes result directly or indirectly from the maturation of the hypothalamic-pituitary-gonadal axis and the subsequent increased secretion of sex steroid hormones. Current data suggest that the following sequential changes occur and result in the initiation of puberty:

1. During early childhood the hypothalamus is highly sensitive to negative feedback inhibition by sex hormones (low set point of the hypothalamic 'gonadostat') resulting in low levels of gonadotropins.
2. During the late prepubertal period the sensitivity of the hypothalamic gonadostat to negative feedback by sex hormones decreases (higher set point) resulting in increased secretion of GnRH, FSH, LH, and gonadal sex hormones.
3. The onset of puberty is associated with a further decrease in hypothalamic sensitivity to the negative feedback mechanisms, which results in the production of adult gonadotropin and steroid hormone levels. This phase is also accompanied by the characteristic sleep-associated increases in episodic (pulsatile) secretion of LH and the development of the secondary sex characteristics.

During the mid- to late-pubertal stage the positive feedback mechanisms and the capacity to exhibit an estrogen induced LH surge also mature. Once these events have been initiated and completed normally, *spermatogenesis* will occur in men and *ovulation* in women.

Menopause

Menopause is the stage in a woman's life when menstruation ceases, the mean age being 48 years. The period of 1 to 2 years preceding the menopause is termed the *climacteric,* which is characterized by endocrine, somatic, and psychologic changes. The menstrual cycles become irregular and ovulation is either infrequent or ceases completely. Due to the decline in estrogen production the negative feedback regulation of the hypothalamic-pituitary-ovarian axis is affected and results in an increased production of FSH and LH. Menopause results from the final depletion of follicles in the ovary.

A number of *symptoms* have been associated with the climacteric and menopause such as 'hot flashes', sweating, tiredness, increased irritability (vasomotor and emotional symptoms), with atrophy of the breast, vagina, vulva, and uterus, and loss of skin turgor.

Approximately 25% of postmenopausal women develop *osteoporosis.* This and other problems of the menopause can be controlled or eliminated by the administration of exogenous estrogens; however, the benefits of this therapy have to be weighed against an increased risk of carcinoma of the endometrium.

Estrogens in postmenopausal women are low and are formed by peripheral aromatization of mainly adrenal androgens. Evidence has accumulated that these estrogens contribute to a higher incidence of adenocarcinoma of the endometrium. Normal human endometrium contains 17β-hydroxysteroid dehydrogenase, and arylsulfotransferase, which convert estradiol to less potent estrone and estradiol sulfate, respectively. These enzymes are stimulated by progesterone. Thus, a capacity to suppress the biologic potency of estradiol is lost postmenopausally, exposing the tissues to 'unopposed' estrogen action. A resulting hyperplasia of the endometrium may lead to carcinoma.

EVALUATING THE FUNCTIONAL CAPACITY OF THE GONADS

Bioassay procedures for the determination of sex-steroid and gonadotropin concentrations are now largely obsolete. Specific *radioimmunoassay* (RIA) procedures are available for most of the biologically active steroid hormones. These methods are very sensitive (10 ng/L) and consequently require only small quantities of plasma or serum for reliable measurements. The methods can be modified for the determination of urinary steroids. In all of these assays the problem of cross reactivity between hormones for a particular antiserum has to be minimized by technical manipulations prior to RIA.

Although fairly specific RIA procedures are available for the gonadotropins, the problem of *cross reactivity* is still important. The glycoprotein hormones have two subunits, α and β. The α subunits of FSH, LH, hCG (and TSH) are nearly identical but the β subunits are different. Consequently antisera generated using the β subunits are more specific. RIA procedures for βhCG, βLH and βFSH are now available. RIA methods tend to give values for the gonadotropins which are relatively higher than the values based on their biologic activity. Unfortunately the peptide hormones measured by RIA procedures may not always reflect the amount of biologically active hormone. Clinical disorders, such as hypogonadism in a male, have been described with an immunologically active, but biologically inactive LH. In this situation, an *in vitro bioassay* may be necessary to resolve the problem.

Because many steroids exhibit diurnal variations, and the gonadotropins (LH and FSH) are secreted in a pulsatile manner, *single blood samples* (one point in time) for the determination of hormonal status are rarely adequate. Several samples should be analyzed to get an accurate picture. RIA procedures for plasma GnRH are of doubtful value because this decapeptide is metabolized in the pituitary and very little reaches the circulation. No practical method for determining inhibin-F is yet available. The measurements of *urinary* estrogens, pregnanediol, 17-ketosteroids (17-KS), 17-ketogenic corticosteroids, and 17-hydroxycorticosteroids reflect excretion over a 24-hour period and the classic colorimetric methods are still employed in many laboratories. It is important to note that the bulk of urinary 17-KS is of *adrenal* origin, so its diagnostic value in *gonadal* disorders is limited. Significant amounts of 17-KS are sometimes produced by the gonads and in such cases a dexamethasone suppression test may be able to differentiate between the two sources. Normal adrenal 17-KS production will decrease following dexamethasone administration.

Useful RIA procedures for the *evaluation of gonadal function* include LH, FSH, hCG, βhCG, prolactin, progesterone, testosterone, 5α-dihydrotestosterone, androstenedione, estradiol, and estrone. Approximate levels of serum gonadotropins and gonadal steroids under various conditions are given in Tables 14–3 and 14–15. Their diagnostic value in various clinical disorders will be described later.

Evaluation of Steroid Hormone Action at the Target Organ

Vaginal Epithelium. The vaginal mucosa is one of the target tissues of the ovarian sex hormones and characteristic changes occur during the normal menstrual cycle. Adequate amounts of *estrogen* cause proliferation, maturation, and desquamation of vaginal epithelium. When stimulated by estrogen, the nuclei of these cells become pyknotic (condensed by degenerative changes to a dense mass of chromatin) and the cytoplasm becomes eosinophilic. The cells are described as 'cornified.' The proportion of cornified cells in a vaginal smear is a rough indication of the amount of estrogen acting on the vaginal epithelium (karyopyknotic index). Immediately following menses, the smear consists of precornified cells with vesicular nuclei and basophilic cytoplasm.

Progesterone formed after ovulation acts on the epithelium (previously stimulated by estrogen) to bring about a progressive desquamation, clustering (clumping) and folding of cells. This effect is so pronounced that it is possible, by taking serial samples, to determine the day of ovulation. The effects induced by progesterone are not specific; androgens and adrenal steroids at high concentration can bring about similar changes. In the absence of hormone stimulation, the vaginal mucosa remains atrophic, its epithelium does not mature. The smears are composed of round parabasal and intermediate cells with a large vesicular nucleus, plus a number of inflammatory cells, neutrophils, lymphocytes and histiocytes. The usefulness of the *vaginal smear* (in amenorrheic patients) is primarily as an indicator of the degree of ovarian function present. The smear patterns can range from highly proliferative to completely atrophic.

Cervical Mucus. The secretion of cervical mucus is regulated by the ovarian hormones and characteristic changes are observed during the menstrual cycle. *Estrogens* stimulate the secretion of copious amounts of watery mucus, while progesterone inhibits this secretion. The physical and chemical properties also show changes during the menstrual cycle. During days 1 to 7 very little mucus is secreted and it is sticky and viscous. During days 8 to 21 the secretion increases about 10-fold, reaching a maximum at the time of ovulation. The viscosity decreases and the mucus is glossy, transparent, highly elastic and can be drawn into a long thread (Spinn-barkeit). Because of these characteristics, the mucus at this stage is permeable to spermatozoa. The dried mucus (on a glass slide) shows a characteristic ferning, which is at a maximum just prior to ovulation. Depending on the day of ovulation the mucus remains transparent and elastic up to days 20 to 21 after which, owing to the effect of *progesterone,* the secretion decreases and the ferning disappears. The mucus becomes thready with very little elasticity.

Sex Chromatin Analysis

Genetic sex can be determined by observing the presence or absence of an extra bit of chromatin at the periphery of the cell nucleus (nuclear membrane). This mass of chromatin, first observed by Barr and Bertram, is called the *sex chromatin* (or *Barr body*). The sex chromatin can be seen in many tissues, but in practice a ***buccal smear,*** appropriately fixed and stained, is used. In a normal human female, approximately 25% of these mucosal cells contain the Barr body. The *number* of Barr bodies present per cell is one less than the number of X chromosomes; thus, individuals with an abnormal 3X chromosomes will have two sex chromatin masses. Normal females are referred to as 46XX, chromatin positive (1 Barr body) and normal males as 46XY, chromatin negative.

The interphase nuclei of cells from males contain a *Y body* for each Y chromosome present. The long arm of the Y chromosome has an affinity for fluorescent stains and male sex can be determined in this way.

Chromosomal Analysis

Although sex chromatin and Y body analyses provide a simple and rapid means of determining genetic sex, definitive diagnosis should always be based on chromosomal analysis. Almost any tissue grown in culture can be used for the analysis, although white blood cells are most convenient. The cells are cultured with phytohemagglutinin (an agent which stimulates mitosis), then treated with colchicine, which arrests mitosis at the stage of metaphase. On exposure to a hypotonic solution the cells swell; the chromosomes will duplicate themselves and condense. Each chromosome consists of two chromatids joined by a centromere. Techniques are now available to stain chromosomes with various agents to produce specific banding patterns, which help identify each chromosome pair. Several cells are analysed and by these procedures even small abnormalities can be detected. The total number of chromosomes per cell are counted and any structural alterations noted. The resulting figure is referred to as a *karyotype* and is expressed as 46XX for a normal female and 46XY for a normal male.

Semen Analysis

It is estimated that the male partner is implicated in 40% to 50% of infertility problems. At least two semen analyses should be performed and the examination should include semen volume, sperm concentration, sperm morphology, % motility, and % progressive (forward) motility. The specimen should be collected by masturbation (preferably after abstinence for 48 to 72 hours) and analysed within 2 hours. The usual ejaculate volume is 2 mL to 5 mL, and liquefaction of the seminal fluid should occur at room temperature within 20 minutes. Although the criteria of male fertility have not been fully established, the following standards are considered as representative of a normal *fertile male.* There should be at least 20×10^6 sperm/mL; at least 50% of these should be mobile (at 37°), the rate of forward progression >5 (rated 0 to 10 at 37°), and at least 60% should have normal morphology. A new method, available in only a few laboratories, is based on determining the penetration of zona-free hamster eggs by *in vitro* capacitated sperm.

Basal Body Temperature

Disorders of ovulation account for approximately 10% to 15% of all infertility problems and may be caused by anovulation. Indirect evidence of ovulation can be obtained by the use of basal body temperature (BBT) charts. The temperature (oral or rectal) should be taken immediately upon awakening and before any activity. The BBT (36 to 36.4°C) begins to rise simultaneously with the LH mid-cycle surge, but a significant increase over 36.7° is not noted until approximately 2 days after the LH peak and coincides with a rise in peripheral levels of progesterone to >13 nmol/L (>4 μg/L). The temperature rise is due to progesterone and should be sustained for an average of 14 days, falling at the time of the subsequent menstrual period. BBT charts are useful in detecting ovulation but do not indicate the exact time of ovulation.

Diagnostic Tests for Gonadal Disorders

Stimulation or suppression of one or other endocrine gland is frequently used to establish the origin of a particular hormone and to determine the functional status and capacity of the gland in question. These procedures are referred to as *dynamic hormone tests,* some of which are now described.

Clomiphene Stimulation Test. Clomiphene citrate (Clomid) has been used to induce ovulation in anovulatory women and can also serve as a test of the hypothalamic-pituitary-gonadal axis. The mode of action is not clear, but it is established that clomiphene acts at the level of the hypothalamus or pituitary or both, probably by blocking the estrogen (or androgen) sensitive receptors and thereby abolishing their inhibitory action on gonadotropin release.

For purposes of testing in women a daily dose of 50 mg to 100 mg of clomiphene is given orally for 5 days, on days 5 to 9 following menses. Serum gonadotropin and estradiol levels should rise during the period of administration. In the *normal female,* an ovulatory surge of gonadotropins is observed approximately 7 days after completion of the course of clomiphene, and, in the absence of fertilization, is followed by a menstrual period approximately 14 days after ovulation. However, ovulation and a subsequent menstrual period are not essential to establish that the hypothalamus-pituitary axis is responsive. Generally, a 2-fold rise in gonadotropins, especially serum LH, is considered a positive response.

In men, administration of 200 mg of clomiphene daily for 7 days should result in a significant increase in the levels of serum FSH and LH beginning on the fourth day. The minimum normal response has been defined as an increase over control levels of 30% for LH and 22% for FSH.

Gonadoliberin (Gonadotropin Releasing Hormone, GnRH, LHRH) Stimulation Test. The test is used to assess the status of gonadotropin secretion by the pituitary gland. For routine testing, 100 μg to 150 μg of synthetic LHRH is administered intravenously over 30 seconds. Venous blood samples for the determination of LH and FSH are obtained at −30, −15, 0, +15, +30, +45, +60, +120, and +180 minutes. Responses vary widely among individuals, although in general following LHRH a rise in serum gonadotropins occurs rapidly, with a peak of LH occurring at about 30 minutes and of FSH at about 45 minutes. Thus, if the gonadotropins do not rise following the administration of clomiphene, and if a significant increase is observed following LHRH administration, then a diagnosis of hypogonadism owing to hypothalamic failure or unresponsiveness can be made. If neither FSH or LH levels rise following LHRH then pituitary failure is indicated.

Progestin-Induced Withdrawal Bleeding Test. This test is used to evaluate the functioning of the hypothalamic-pituitary-ovarian axis in amenorrheic patients. The patient is given either an oral dose of medroxyprogesterone acetate (Provera) 5 mg to 10 mg daily for 5 days or an intramuscular injection of 50 mg to 100 mg of progesterone in oil. Within 10 days after the conclusion of progestin medication, the patient will show withdrawal bleeding or will not bleed. Any quantity of uterine bleeding demonstrates the presence of sufficient estrogen to cause endometrial proliferation and at least minimal functioning of the hypothalamic-pituitary-ovarian axis. However, a negative test (no withdrawal bleeding) does not necessarily indicate that the axis is nonfunctional.

Dexamethasone Suppression Test. In patients with amenorrhea associated with hirsutism, the dexamethasone suppression test may indicate whether the source of excess androgens is mainly adrenal or ovarian. Dexamethasone will suppress ACTH-dependent adrenal steroidogenesis by a negative feedback effect on ACTH secretion. Suppression is carried out by oral administration of 2.0 mg of dexamethasone every 6 hours for 4 to 5 days. On the control day before and the last day of suppression 24-hour urine samples are collected for 17-ketosteroid determination. Serum testosterone is also measured during the test.

Diagnostic Work-Up of a Patient with Amenorrhea

In order to determine and treat the probable cause of amenorrhea, the patient must undergo a series of laboratory and therapeutic studies. Initial laboratory tests are for pregnancy, as well as adrenal, ovarian, and thyroid function. At the same time sex chromatin, karyotype, serum prolactin, and gonadotropin data should be obtained. The results from the prolactin and gonadotropin assays are of major importance in the further evaluation of the patient with amenorrhea. Sequential tests can be used to evaluate the functional status of the pituitary-hypothalamus, ovary and the uterus (target organ). Some of these tests have been described in earlier sections of this chapter.

DISORDERS OF THE TESTIS

Hypogonadism

Hypogonadism in the male may present as delayed puberty, postpubertal gonadal failure, ambiguous genitalia, gynecomastia, or merely as infertility. This disorder is classified in three ways:

1. *Primary* hypogonadism (primary testicular insufficiency) which is characterized by a hypergonadotropic state.
2. *Secondary* hypogonadism (secondary testicular insufficiency), the distinguishing feature of which is a hypogonadotropic state.
3. *Acquired* hypogonadism which may result from systemic infections, irradiation, surgery, or may involve psychogenic factors.

Primary Hypergonadotropic Hypogonadism

Klinefelter's Syndrome. Males with at least one Y chromosome and at least two X chromosomes have what is commonly referred to as Klinefelter's syndrome. The prevalence is about 1 in 500 males and is the most common form of male hypogonadism. It is characterized by varying degrees of seminiferous tubule dysgenesis, decreased

Leydig cell function, sterility (due to azoospermia), and a somewhat greater incidence of mental retardation. The chromosome complement most often associated with the classic form of Klinefelter's syndrome is 47 XXY. The abnormality is thought to result principally from nondysjunction during meiotic division in the formation of one of the gametes.

In classic Klinefelter's syndrome the external genitalia are usually clearly differentiated and the individuals are phenotypically male, but are chromatin positive. In the adult, the testes are small and firm. Gynecomastia is frequently present. Testicular histology reveals the most characteristic features of this syndrome, that is, the atrophy and hyalinization of the seminiferous tubules and hyperplasia of the Leydig cells. Plasma and urinary FSH and LH values are elevated while the concentration of testosterone ranges from 2 to 30 nmol/L (0.5 to 8.6 μg/L), overlapping in part the range for normal men (11 to 35 nmol/L; 3 to 10 μg/L). In rare cases, the level of LH may be normal. The 'free' testosterone levels are also below normal in some men with Klinefelter's.

Variants (*mosaics*) of the classic 47XXY condition are XY/XXY, XXYY and XXXY. The most constant feature of these variants is the increased incidence of mental retardation and somatic anomalies.

Sertoli-Cell-Only Syndrome (Germinal Cell Aplasia; del Castillo Syndrome). This is a relatively uncommon disorder of unknown etiology; the classic features being a male with normal secondary sexual characteristics, but sterile. The testes are normal or slightly reduced in size, with normal consistency and azoospermic. The FSH levels are elevated but LH and testosterone are normal. Testicular biopsy reveals the absence of germ cells (sperm precursors), while the Sertoli cells appear to be normal.

Reifenstein's Syndrome. This is a hereditary disorder in which the patient resembles Klinefelter's syndrome clinically and biochemically, but has hypospadias (urethral orifice on underside of penis) and a normal male karyotype. The mode of inheritance is either an X-linked recessive or a sex-linked autosomal dominant characteristic.

Functional Prepubertal Castrate Syndrome (Anorchia; Testicular Agenesis). These individuals have normally differentiated infantile external genitalia and there is no recognizable testicular tissue. Because these patients have a male phenotype, the destruction of the testicular tissue must have taken place between the seventh to fourteenth weeks of fetal life. Plasma testosterone in these individuals is low whereas FSH and LH are both elevated. These patients are infertile but the administration of androgens will result in the development of normal secondary sexual characteristics.

Male Turner's Syndrome. Most of these individuals have a normal 46XY karyotype although a number of chromosomal abnormalities have been detected. They have decreased testicular function along with various congenital somatic and visceral abnormalities that include short stature, webbed neck, cryptorchidism, and ocular and cardiovascular anomalies. Testosterone levels are low while the gonadotropin levels are elevated. The features distinguishing this syndrome from other hypergonadotropic syndromes are the visceral and somatic anomalies.

Secondary Hypogonadotropic Hypogonadism

Secondary hypogonadism is also called secondary testicular insufficiency, or primary hypothalamic-pituitary defect. A rare familial disorder, associated with the lack of a sense of smell and at times with harelip and cleft palate, is known as Kallman's syndrome. The endocrine defect responds to pulsatile GnRH treatment.

Hypogonadotropic Eunuchoidism. Because puberty in normal males is completed by the age of 15 to 16 years, boys who do not mature by this time are simply classified as 'late bloomers.' If everything else is normal these individuals eventually attain full maturity and develop into normal fertile men. If puberty is delayed beyond 20 years, the condition is classified as *hypogonadotropic eunuchoidism.* The reason for delayed puberty is not known. Because of a lack of gonadotropin stimulation the testes remain small and contain immature seminiferous tubules and poorly differentiated Leydig cells. Testosterone levels are low and complete development of the secondary sex characteristics does not occur.

In its classic form, the individual is usually tall and thin and has long tapering extremities, no facial or body hair (juvenile appearance) and scanty pubic hair. The plasma testosterone levels are in the range of a normal female while FSH and LH are well below the normal range. Treatment with hCG and human menopausal gonadotropin (hMG, Pergonal), separately or in combination, can restore the secondary sex characteristics and spermatogenesis. If sexual maturity fails following treatment and withdrawal of gonadotropins, permanent therapy with androgens may be required. Administration of LHRH to these individuals has resulted in restoration of spermatogenesis and secondary sexual development. These findings indicate that the primary defect in this disorder is probably at the level of the hypothalamus.

Fertile Eunuch (Isolated LH Deficiency). In these individuals the testes are of normal size but body features are eunuchoidal. Plasma testosterone and LH levels are below normal while FSH levels are normal. Testicular biopsy reveals mainly normal tubules, while the Leydig cells are absent or sparse. Spermatogenesis is evident in the biopsy, indicating that adequate androgen and LH are produced to allow normal sperm production, but the levels are not sufficient for full maturation of the secondary sexual characteristics. The ejaculate may contain viable sperm or can be azoospermic. Treatment with hCG or androgens will result in maturation, but this therapy is required permanently.

Acquired Hypogonadism

Orchitis. About 15% to 25% of males with epidemic parotitis (mumps) can develop acute seminiferous tubule failure. Plasma FSH levels may be elevated and there is oligospermia or azoospermia. Production of testosterone by the Leydig cells is normal, although in severe cases complete testicular failure may result. Histologically the testis resembles that present in Klinefelter's syndrome. If the infection occurs before puberty, the testes will recover completely. In rare cases gonorrhea and leprosy can cause tubular destruction resulting in infertility.

Irradiation. Exposure to x-rays, radioactive material and neutrons can damage the germinal cells to various degrees depending upon the dose. As in the case of orchitis, the levels of FSH increase while LH and testosterone stay in the normal range.

Chemotherapeutic Drugs and Toxic Chemicals. The antineoplastic agent cyclophosphamide may selectively damage the germinal cells of the testes, while Leydig cell function is usually not affected. These changes occur with the first few months of treatment and are generally reversible if therapy was for less than 18 months. Longer exposure can result in permanent sterility. Other antineoplastic drugs have been found to damage the testicular tubules. In some of these patients gynecomastia, associated with decreased Leydig cell testosterone production, has been noted. Prolonged exposure to the pesticide 1,2-dibromo-3-chloropropane (DBCP) has resulted in damage to the tubular germinal cells.

Adult Leydig Cell Failure (Male Climacteric). Although the question regarding the existence of a male climacteric analogous to a female climacteric has not been resolved, there are a number of cases where in otherwise normal men, Leydig cell function has decreased. These men may suffer from hot flushes, excessive sweating, increased irritability, episodes of depression, and a general decrease in libido and sexual drive. Some of these symptoms are characteristic of postmenopausal women.

Circulating levels of testosterone tend to decline progressively with advancing age and may be responsible for declining sexual function in aging men. Various aspects of sexuality, including activity level, sexual thought, enjoyment, nocturnal penile turgesence, and orgasmic function usually decline in men from age 60 to 90 years. There is often a decrease in plasma testosterone levels, an increase in TEBG, and thus a decrease in 'free' testosterone. Both FSH and LH may be increased, but no age-related changes in estradiol or prolactin have been noted. Other studies have found no change in total or free testosterone levels in selected aged men where there was a favorable psychic stimulus and continued sexual activity.

Other Male Gonadal Disorders

Testicular Neoplasms. The prevalence of testicular malignancies in the total male population is only 0.002%, however, in males between the ages of 25 to 35 years, testicular tumors are one of the most common forms of malignancy. There are two main groups of testicular tumors:

Germinal Cell Tumors. Approximately 95% of testicular tumors arise from the germ cells. They may be classified as seminoma, embryonal (embryoma, choriocarcinoma, teratocarcinoma), or combined seminoma and embryonal. In individuals with these tumors hCG is secreted (due to chorionic elements) and its presence should be demonstrated by a specific RIA utilizing hCG β-subunit antisera. As a result of the excess hCG these patients often have gynecomastia, which may also be caused by an increased production of estrogens by the trophoblastic tissue present in these tumors.

Non-Germinal Cell Tumors. Interstitial or Leydig cell tumors of the testis can be associated with the production of either androgens or estrogens or both; the excess androgens in prepubertal boys cause virilization and sexual precocity, while excess estrogens produce feminization. The urinary 17-ketosteroids and androgens are elevated.

Testicular Feminization (Target-Organ Insensitivity to Androgens). The term 'testicular feminization' has been used to characterize a rare hereditary disorder in which the individuals are *genotypically* male, with normal 46XY chromosomal constitution and chromatin negative nuclei, but are *phenotypically* female. At puberty normal breast development occurs, but pubic and axillary hair are generally absent and menstruation does not occur. The vagina is shallow and ends in a blind pouch; other internal genitalia are absent or rudimentary but intra-abdominal testes are present. Transmission of this disorder is by means of an X-linked recessive gene thought to be responsible for the intracellular receptor protein for androgens. The prevalence of the defect is approximately 1/50,000. These 'females' seek medical advice on account of primary amenorrhea, sterility, or an inguinal 'hernia.'

Testosterone levels in the plasma, urine, and spermatic vein are in the normal adult range. The 17-ketosteroid excretion is normal or slightly elevated. Estrogen levels are in the lower normal range for females and the concentration of 17β-estradiol in spermatic vein blood is elevated above the range for normal men. The serum FSH levels are usually normal while LH levels are elevated. Normal levels of testosterone in these patients indicates that this genetic disorder is not caused by some biosynthetic defect in the testes.

The clinical features result from an insensitivity of target tissues to androgenic hormones. Thus the crucial steps in male fetal sexual differentiation, which require androgens, do not occur and the development is totally female. In normal androgen target organs testosterone is converted to 5α-dihydrotestosterone which binds to a specific cytosol receptor protein, is transferred to the nucleus and induces the production of specific proteins by means of which the hormonal effects of androgen are expressed. Present evidence indicates that the pathogenic defect in patients with *complete* testicular feminization is an absence of the cytosol receptor protein for testosterone and 5α-dihydrotestosterone. As a result target cell tissues cannot respond to androgens; they can, however, respond to estrogens and the unopposed action results in feminization.

Because the patient is phenotypically and psychologically female, and the incidence of testicular malignancy or neoplasia in this disorder is high, the testes are removed surgically, once pubertal feminization has been completed, and the patient is maintained as a female on estrogen replacement therapy.

When there is only a *partial* deficiency of the androgen-receptor protein the genitalia are primarily female, but some masculinization will have occurred (clitoromegaly and partially fused labia). This is 'incomplete' as opposed to 'complete' male pseudohermaphroditism.

5α-Reductase Deficiency. The primary defect in this rare disorder is a deficiency of the enzyme 5α-reductase, and thus target tissue cells are unable to convert testosterone to 5 α-dihydrotestosterone. Because testosterone is required for male differentiation of the wolffian ducts, and 5α-dihydrotestosterone is required for masculinization of the urogenital sinus and the genitalia, individuals with 5α-reductase deficiency are genetic males with normal internal ducts but ambiguous external genitalia. Testosterone levels are normal; consequently pubertal changes (skeletal growth and maturation, deepening of the voice, muscular development) occur fairly normally. The 5α-dihydrotestosterone changes (prostate maturation, beard growth) do not occur.

Gynecomastia. Benign glandular enlargement (hypertrophy) of the male breast is called gynecomastia. Its etiology is not known in the majority of cases and some of the clinical gonadal disorders associated with it have been already described. In patients in whom gynecomastia is associated with *galactorrhea* (milk production), prolactin levels are elevated, most often indicating a pituitary tumor (prolactinoma). Gynecomastia can also occur after drug ingestion (marijuana, phenothiazines, alcohol, spironolactone, cimetidine, metoclopramide, alkylating agents, antitumor agents, and psychotropic drugs), and from a number of other causes such as hypo- and hyperthyroidism, malignancies of lung, liver and adrenal, and in cirrhosis and renal failure.

DISORDERS OF THE OVARY

Ovarian Hypofunction

Turner's Syndrome (45XO, Gonadal Dysgenesis). A child may be born with a total absence of gonads (gonadal *agenesis*), but more often gonadal function is present for some time in fetal life and subsequently becomes impaired (gonadal *dysgenesis*). In 1938, Turner described girls who had short stature (<5 feet), multiple skeletal abnormalities, webbed neck, and sexual infantilism. Patients with this syndrome lack normal gonads but have bilateral fibrous 'streak gonads' that are devoid of any germ cells. The frequency of Turner's syndrome with a 45XO chromosome pattern and its variants is about 1 in 3000 to 6000 phenotypic females and is one of the most common causes of primary amenorrhea.

Cytogenetically, these patients have only one normal X chromosome and 80% are chromatin negative. The second X chromosome is either completely or partially absent. At birth, the external genitalia are usually unambiguously female. Their internal genital organs (müllerian derivatives: uterus, cervix, fallopian tubes, upper vagina) are small (infantile) yet well differentiated. The streak ovaries do not produce estrogens and secondary sexual characteristics fail to develop at puberty. These girls do not menstruate spontaneously because they do not have the cyclic production of estrogens and progesterone. The feedback mechanism regulating the hypothalamic-pituitary-gonadal axis does not function normally in these patients and so the serum FSH and LH levels are elevated (hypergonadotropic hypogonadism). The urinary 17-KS tend to be reduced and some patients may have poor adrenal function.

Almost all individuals with gonadal dysgenesis require sex steroid replacement to achieve maturation of secondary sexual characteristics. If the epiphyses remain open some additional growth usually occurs when sex steroids are administered. Because of possible premature epiphyseal closure, estrogens should be withheld until the expected time for puberty.

Polycystic Ovarian Disease (Stein–Leventhal Syndrome). The principal features of this syndrome are secondary amenorrhea, bilateral enlargement of the ovaries, infertility, obesity, and hirsutism. Relatively few women fulfill all the criteria of the Stein–Leventhal syndrome. In contrast, *polycystic ovarian disease* in its broadest definition, (theca cell hyperplasia, hyperthecosis, LH-dependent ovarian hypersecretion of androgens) is a common disorder, which is widely recognized as one of the major factors in a large proportion of cases of infertility associated with anovulation.

Women who develop polycystic ovarian disease appear to grow and develop normally and pass through puberty uneventfully. Menstrual periods may commence in a normal fashion or irregular cycles may occur right from the start. In a typical patient the interval between periods becomes longer and these irregular and anovulatory cycles are followed soon afterward by either oligomenorrhea or amenorrhea and hirsutism.

The *histology* of the polycystic ovary is rather variable; in the Stein–Leventhal ovary there are large numbers of follicular cysts and relatively few atretic cysts, whereas in hyperthecosis there are few follicular cysts, large numbers of atretic cysts, and marked stromal hyperplasia with clumps of luteinized theca cells deep in the stroma. The capsule is thickened and the corpus luteum is absent. The nature of the endometrium of women with polycystic ovarian disease is also variable; it can be hyperplastic, atrophic, or (most frequently) proliferative. The incidence of carcinoma of the endometrium is higher in women with polycystic ovarian disease.

Biochemical investigations of polycystic ovarian disease have been quite extensive, yet the underlying endocrine abnormality has not been precisely defined. It is clear, however, that there is a disturbance of the normal hypothalamic-pituitary-ovarian axis resulting in failure of ovulation. *Endocrine* investigations reveal that in contrast to the normal menstrual cycle, the levels of estrogens and gonadotropins do not fluctuate (steady-state) in polycystic ovarian disease. Patients with this syndrome have higher mean concentrations of plasma LH (above 25 IU/L) while the levels of FSH are low or low normal. The elevated LH levels are also seen in primary ovarian failure, therefore are not diagnostic of polycystic ovarian disease. The serum FSH values, however, aid in the diagnosis of the aforementioned two conditions. They are either low or low normal in polycystic ovarian disease, whereas in ovarian failure LH and FSH levels are both elevated. Thus, the LH/FSH ratio in polycystic ovarian disease is abnormally high.

There is an exaggerated secretion of LH and a normal FSH response when GnRH is administered, but the explanation for this disparity is not yet clear. The negative feedback effect of circulating estradiol may be greater on FSH than on LH, or FSH release may be relatively insensitive to GnRH stimulation. The concentration of inhibin-F in follicular fluid obtained from patients with polycystic ovarian disease is significantly higher than in follicular fluid from normal cycling women. Thus, inhibin-F may selectively suppress the release of FSH from the pituitary in response to GnRH.

Another *characteristic biochemical feature* of polycystic ovarian disease is the excessive ovarian secretion of androstenedione and/or testosterone (3- to 4-fold higher). Because of variability in the levels of androgens, multiple plasma samples taken at various times during the day, and at various times in the menstrual cycle, need to be analyzed before an accurate diagnosis of polycystic ovarian disease can be made. Measurement of the *free testosterone index* (plasma TEBG binding capacity), or TEBG levels, are necessary in women suspected of having polycystic ovarian disease but in whom the plasma testosterone concentrations are not increased. The TEBG binding capacity is frequently low or low normal in polycystic ovarian disease. Thus, an increased level of the free biologically active fraction of testosterone can occur, even though the total testosterone is normal. When levels of *total* plasma testosterone exceed 7 nmol/L (2 μg/L), an adrenal or ovarian tumor is more likely to be present.

The urinary 17-KS in women with polycystic ovarian disease are normal or slightly elevated. If the excretion of 17-KS is less than 70 μmol/day (20 mg/d), adrenal disease is highly unlikely. If the values are higher, then dynamic tests using ACTH and dexamethasone, and hCG and estrogen-progestin are utilized to determine the extent of the involvement of the adrenal and the ovary, respectively.

The plasma *estrone* (E_1) concentrations can be elevated due to the peripheral conversion of androstenedione, while the levels of *estradiol* (E_2) are usually comparable to levels observed during the follicular phase of the normal cycle, however, the E_1/E_2 *ratio* is elevated. When the clinical assessment of ovarian status is uncertain, a diagnostic laparoscopy may be indicated.

Various hypotheses have been put forth to explain the *mechanism* underlying polycystic ovarian disease. The process may be triggered by elevated levels of LH which down-regulate the LH receptors on granulosa cells and which also cause hypertrophy of the theca cells. These changes would result in increased production of androstenedione and testosterone by the theca cells, which could stimulate the secretion of inhibin-F from the granulosa cells, which in turn would inhibit the secretion of FSH by the pituitary. Decreased LH receptors on the granulosa cells and low levels of circulating FSH result in poor follicular maturation. The increased circulating androgens are metabolized peripherally to estrogens; the latter augment pituitary sensitivity to endogenous GnRH, with the relative output favoring LH over FSH, thus perpetuating the cycle. The initiating lesion resulting in higher levels of LH remains to be determined. The whole process could also be triggered by an increased synthesis of adrenal androgens.

There are two specific aims in the *treatment* of these patients; they are the renewal of menstrual (ovulatory) cycles and the reversal of hirsutism. Treatment with oral contraceptives, glucocorticoids, clomiphene, and wedge resection of the ovaries have been used with some success. Somewhat better results are now being reported with pulsatile GnRH treatment.

Premature Menopause (Irreversible Ovarian Failure). Premature cessation of menses in women under 40 years, who previously had normal cycles, is termed premature menopause. These patients present because of amenorrhea (secondary) and vasomotor symptoms (hot flashes, sweating) that are indistinguishable from those of the postmenopausal woman. The *pathogenesis* probably represents a late manifestation of gonadal dysgenesis. The initial number of primordial follicles in these patients were either insufficient or more rapidly depleted, leading to hypogonadism. The *laboratory* findings are similar to those seen in patients with gonadal dysgenesis in that the pituitary gonadotropins are elevated (postmenopausal levels) and the estrogens are low. The ovary is unresponsive to stimulation and there is a permanent loss of ovulation.

Ovarian Tumors. A number of primary or secondary functional ovarian tumors have been reported and these may alter normal ovarian function. Because these tumors originate in the various cell types of the ovary (granulosa, theca, and hilus cells) they are capable of producing excess amounts of steroid hormones, particularly *estrogens* and *androgens.* These tumors can produce virilization, feminization, and precocious puberty. Several rare types of androgen-producing ovarian tumors, for example, arrhenoblastoma and tumors derived from lipid and hilar cells, may be responsible for amenorrhea (secondary) with hirsutism or virilization. They are most common in the third and fourth decades of life. Some rare tumors are known to influence ovarian function by producing an *ectopic gonadotropin* which is biologically and immunologically similar to hCG. There are still other types of ovarian tumors which secrete substances such as *serotonin* and *thyroxin,* which can result in clinical manifestations of carcinoid syndrome and thyrotoxicosis respectively.

Hyperprolactinemia. Association of secondary amenorrhea (with or without galactorrhea) and hyperprolactinemia is fairly common. Generally, elevated prolactin levels are associated with low plasma gonadotropins and estradiol levels, and no withdrawal bleeding is noted following the administration of progestins. However, some patients in spite of hyperprolactinemia have sufficient estrogen and will have withdrawal bleeding.

The *mechanism of amenorrhea* in patients with hyperprolactinemia is not completely understood. There are probably suppressive effects at several levels in the hypothalamic-pituitary-ovarian axis. Prolactin has been reported to inhibit estrogen synthesis in rat granulosa cells, and to stimulate selectively adrenal androgen production. Whether such effects play a role in the development of amenorrhea in these patients remains to be determined. Hypersecretion of prolactin can be *induced* by a number of conditions; the three most common are pituitary adenoma, hypothyroidism, and postpill (oral contraceptives) or postpartum amenorrhea. *Treatment* of patients with hyperprolactinemic amenorrhea depends on the cause of the syndrome. If due to hypothyroidism, the amenorrhea can be reversed by appropriate replacement therapy. Suppression of prolactin can be achieved by the administration of bromocriptine, and in the case of pituitary tumors producing either neurologic or endocrine disturbances, surgical or radiation treatment may be required.

Ovarian Hyperfunction

Precocious Puberty (Sexual Precocity). Sexual precocity in a female child is defined as the onset of sexual maturation before the age of nine. Accelerated somatic growth is often the first change in precocious puberty. This is followed by breast development and growth of pubic hair. Menarche usually occurs several months later, although occasionally it may be the very first sign. Sexual precocity occurs spontaneously due to hypersecretion of hormones from the hypothalamus, adrenal, or gonad, or as a result of target-organ hypersensitivity. In addition, it may arise from exogenous sex hormones, or from ectopic production of hormones. Precocious puberty has been divided into *true* (complete) sexual precocity, in which the sex hormones are secreted by the maturing ovary, and precocious *pseudopuberty* in which the source of the sex hormones is other than normal maturing ovaries (e.g., congenital adrenal hyperplasia).

Diagnostic procedures are used to determine, if possible, the etiologic factors causing the accelerated sexual maturation. However, approximately 90% of such cases are of unknown etiology (cryptogenic precocity), whereas the remainder usually have either a central nervous system or ovarian tumor origin. Plasma FSH, estrogen and 17-KS levels correspond to the patient's somatic or developmental age, rather than to her chronologic age. With feminizing adrenal tumors the excretion of 17-KS and estrogens is elevated above normal adult levels. The presence of pregnanediol in a child's urine indicates the possibility of an ovarian luteoma. Similarly a teratomatous choriocarcinoma is suspected if the presence of hCG is demonstrated and pregnancy has been ruled out.

Menstruation, breast development and growth can be inhibited in some patients by the administration of a potent progestational agent such as Depo-Provera, which inhibits the secretion of gonadotropins.

SUGGESTED READING

BONDY PK, ROSENBERGLE (eds): Duncan's Diseases of Metabolism, 8th ed. Philadelphia, WB Saunders, 1980

BRAVO EL, GIFFORD RW: Pheochromocytoma: Diagnosis, localization and management. N Engl J Med 311: 1298, 1984

CHIUMELLO G, SPERLING MA (eds): Recent Progress in Pediatric Endocrinology, Vol 4. New York, Raven Press, 1983

COHEN MP, FOÁ PP (eds): Special Topics in Endocrinology and Metabolism (Series), Vol 6. New York, Alan R Liss, 1984

DE GROOT LJ (ed): Endocrinology (3 vol). New York, Grune and Stratton, 1979

DONALD RA: Endocrine Disorders. A Guide to Diagnosis. New York, Marcel Dekker Inc, 1983

EZRIN C, GODDEN JO, VOLPÉ R (eds): Systematic Endocrinology, 2nd ed. Hagerstown, Harper & Row, 1979

KANNAN CR: Clinician's Approach to Endocrine Problems: 45 Case Studies. Chicago, Year Book Medical Publishers, 1983

KRIEGER DT: Cushing's Syndrome. Monographs on Endocrinology, Vol 22. New York, Springer–Verlag, 1982

KRIEGER DT, HUGHES JC (ed): Neuroendocrinology. Massachusetts, Sinauer Associates, 1980

NEW MI (ed): Congenital Adrenal Hyperplasia. Annals of New York Academy of Sciences, Vol 458, New York, 1985

RABIN D, MCKENNA TJ: Clinical Endocrinology and Metabolism. New York, Grune and Stratton, 1982

SCHWARTZ TB, RYAN WG (eds): Year Book of Endocrinology (Series). Chicago, Year Book Med Publishers, 1981–1985

WERNER SC, INGBAR SH: The Thyroid, 4th ed. Hagerstown, Harper & Row, 1978

WHITE DA, MIDDLETON B, BAXTER M: Hormones and Metabolic Control. London, Edward Arnold, 1984

WILLIAMS RH (ed): Textbook of Endocrinology, 6th ed. Philadelphia, WB Saunders, 1981

15

Neurologic and Psychiatric Disorders

John R. Wherrett / Stephen J. Kish

DIAGNOSTIC APPROACH

The nervous system mediates and integrates the interactions of an organism with its external environment and regulates its internal environment. It is a pervasive organ system with direct connections to special sensory organs and receptors and to all other tissues and organs through the brain stem, spinal cord, and nerves (Fig. 15–1). The brain is the substratum for human 'higher cerebral functions' of consciousness, emotions, language, memory, and cognition.

Individual functions such as vision, speech, or memory involve localized anatomic networks of neurons that have highly complex connections with networks subserving other functions. Lesions affecting localized networks or connecting systems result in activation or inhibition of specific functions. Current practice delegates the management of lesions of the nervous system *with* manifest pathology to neurology and neurosurgery, whereas disorders *without* manifest pathology, affecting the higher cerebral functions of thought and emotions, are delegated to psychiatry. There is growing evidence that in the major forms of chronic psychiatric illness the pathogenesis involves a 'lesion' at the cellular or molecular level. On the other hand, primary *organic* lesions of the brain are often accompanied by disorders of behavior that may be predominant, and thus management falls mainly to the psychiatrist.

For ***diagnosis of organic disorders*** of the nervous system, the neurologist applies a knowledge of neuroanatomy and neurophysiology to interpret the history of disordered nervous function and the findings of the clinical examination, and is able to postulate the site and distribution of lesions (see Fig. 1–2). A consideration of the evolution and pattern of the lesions, leads to a short list of etiologic hypotheses ranked in order of priority. At this point, various imaging and electrophysiologic techniques may be used to confirm the location and configuration of lesions, and various other laboratory procedures may be selected to test and confirm the postulations about etiology. The formulation and testing of diagnostic hypotheses begins as soon as the clinician encounters the patient. The knowledge, skill, and experience of the neurologist will determine the number and validity of hypotheses postulated. The same professional attributes will dictate the choice of biochemical tests needed to confirm a diagnostic hypothesis.

The psychiatrist initially employs a different strategy that depends on the recognition of patterns (syndromes) of symptoms from the history, and signs derived from a detailed assessment of behavior and of the higher cerebral functions. Psychiatrists, however, are thinking increasingly of mental disturbances as disturbances in neuroanatomic systems, and they are working to devise techniques to test the functional and biochemical integrity of these systems. Neurologists and psychologists play a role in conducting these tests.

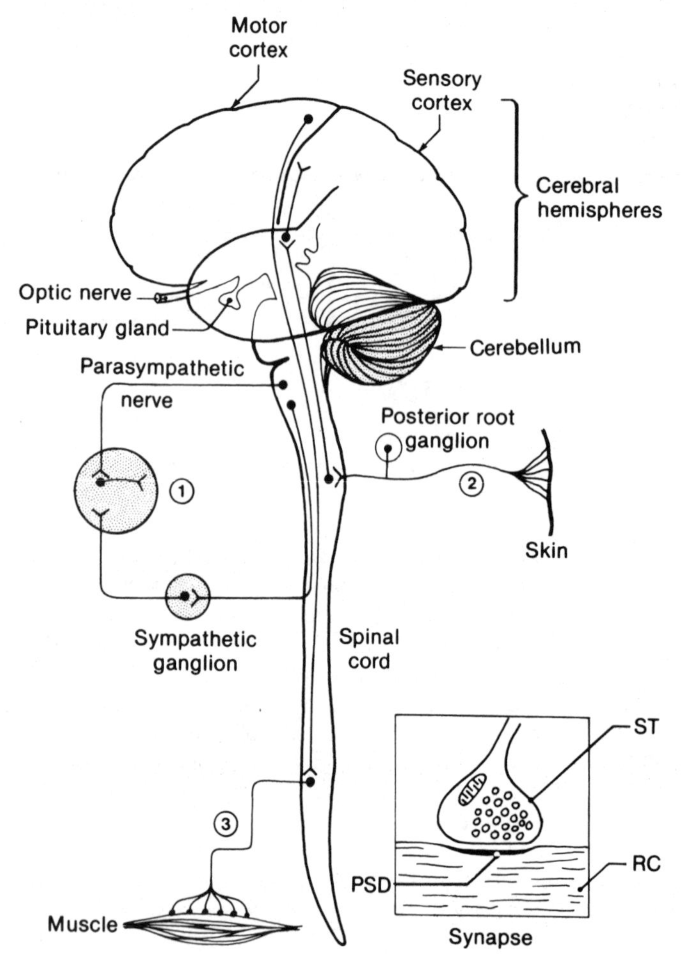

FIGURE 15-1. Diagram of the nervous system. 1. Autonomic fibers to organs and vessels. 2. Sensory nerve fiber connecting with a spinal tract fiber to the sensory cortex. 3. Lower motor neuron fiber going to a muscle and synapsing with an upper motor neuron from the cerebral cortex. **Inset** shows an enlargement of a synapse. ST = synaptic terminal containing synaptic vesicles; PSD = postsynaptic density; RC = receiving cell.

Psychotherapeutic drugs have selective actions often directed to specific neuroanatomic and biochemical systems. Technologies for 'imaging' specific brain functions, such as *positron emission tomography* and *nuclear magnetic resonance,* are becoming available. These techniques, by displaying the anatomic extent and activity of biochemical processes, hold great promise for understanding the pathophysiology of neurologic and psychiatric disorders.

To apply biochemistry in the diagnosis and management of neurologic and psychiatric disorders, we must understand the unique isolation and protection of the nervous system. This is embodied in the concept of the ***blood–brain barrier*** which denotes special structural features such as the tight junctions between vascular endothelial cells in the brain and physiologic mechanisms such as transport systems regulating entry of metabolites from the blood into brain parenchyma. The brain is highly dependent on normal functioning of other body organs, but at the same time complex mechanisms exist to protect the brain from large fluctuations in systemic metabolism. An implication of the existence of cells as highly specialized as neurons, which are capable of signaling through a variety of mechanisms over short as well as long distances, is that their functioning will be unusually susceptible to subtle changes in metabolism and structure. These subtle alterations may manifest clinically as changes in behavior caused by mental confusion, as epileptic seizures that reflect abnormal synchronization of neuronal discharges, or as a slow deterioration of function such as loss of strength and sensation in the limbs in a 'glove and stocking' distribution resulting from a 'dying back' of the longest axons in a toxic polyneuropathy.

INHERITED NEUROLOGIC DISORDERS

As environmental factors have come under greater control, genetic predisposition has assumed increasing importance as a cause of disease. A current catalog of inherited diseases lists about 3300 disorders that have been identified by clinical or laboratory criteria. If disorders of muscle and of the special senses, and many vascular disorders affecting the nervous system, are excluded, about 20% of the listed disorders involve exclusively or mainly the nervous system. Of these *about half* are inherited as *autosomal dominant* traits; that is, clinical manifestations appear in individuals whose diploid cells bear only one copy of the mutant gene and are heterozygous for the trait. The products of the mutant genes and the possible consequent metabolic disturbance are unknown in the vast majority of these disorders. When a gene is heterozygous for a disease-producing allele, so that only half of the gene product is abnormal, the molecular and biochemical disorder may be incomplete and subtle. If both copies of a gene are mutant, a more complete functional abnormality is likely to result and is likely to be more readily detected biochemically. Therefore, we may expect that biochemical abnormalities will have been detected in a greater number of *recessively* inherited diseases.

Recent advances in molecular biology permit direct characterization of genes affected by mutations that result in disease. From a relatively pure gene product one may proceed to prepare complementary DNA that can be used to detect, isolate, and characterize the gene and its mutations. Bacterial endonucleases will cut duplex DNA sequences at specific sites. High resolution gel electrophoresis, with labelled DNA 'probes' to detect specific base sequences, permits the direct detection of mutations in genes for diagnosis. A second approach is to detect inherited variations in DNA base sequence (restriction fragment length polymorphisms) which can be linked to inherited disease traits using classic genetic analysis. These DNA markers, if linked closely enough to the disease gene, provide a relatively simple biochemical test for diagnosis. Furthermore, discovery of DNA-linkage markers may localize disease genes to particular chromosome segments and conceivably lead to isolation of the gene even before its normal function is known. Already the number of human genes isolated has exceeded 200 and progress is accelerating.

It is apparent that an increasing proportion of disorders of the nervous system comprises a very large number of uncommon, even rare, variants for which specific biochemical tests of metabolites, protein products of genes, or of the genomic DNA *per se* are or will be available shortly. This potential capability will pose problems for health care systems attempting to provide genetic services for the large number of disorders. The approach in the past has been to develop formal or informal networks in which there is a division of responsibility among collaborating centers operating at regional, national, or international levels.

Some biochemically defined genetic disorders affecting the nervous system, such as *phenylketonuria* or *cretinism,* are diagnosed through formal biochemical screening programs; however, most patients present to individual clinicians as neurologic syndromes for diagnosis through standard clinical methods. It is inappropriate here to attempt to catalog all biochemically defined genetic diseases affecting the nervous system. Table 15–1 presents a ***classification*** based on the clinical syndrome, or phenotype, with which the patient presents that will allow the clinician to narrow the range of possibilities and to focus the investigation. The table represents only a small portion of the inherited syndromes which come under clinical scrutiny. Well-known and relatively common genetic diseases are excluded because the biochemical phenotype is unknown. In theory, the biochemical abnormality in all inherited diseases will be discovered eventually. The table is also only representative of biochemically defined disorders in each clinical category. As new biochemical phenotypes are elucidated, the 'checklist' for clinical phenotypes can be extended. For example, in patients that present with spinocerebellar ataxia, at least 22 different biochemical phenotypes have been detected. Included in the table are some dis-

TABLE 15–1. Clinical Classification of Genetic Disorders Having Primary Neurologic Symptoms

Disorder	Nature of Biochemical Abnormality
Neonatal Disorders (Disordered alertness, eye movements, motor function, breathing, and autonomic function; seizures)	
Aminoacidopathies	Defective amino acid metabolism or transport; defective catabolism of branched-chain amino acids
Hyperammonemias	Disorders of arginine–urea cycle and dibasic amino acid metabolism
Galactosemia	Galactose-1-phosphate uridyltransferase deficiency
Lactic acidosis	Pyruvate dehydrogenase or carboxylase deficiency
Pyridoxine dependency	
Trichopoliodystrophy (Menke's kinky hair disease)	Defective intestinal absorption of copper
Disorders of early infancy (Progressively delayed maturation; specific neurologic signs; ocular and systemic abnormalities)	
Lysosomal storage disorders—sphingolipidoses, mucopolysaccharidoses, mucolipidoses	Deficiencies of lysosomal acid hydrolases acting on heteromacromolecules; abnormal subcellular localization of lysosomal acid hydrolases; deficiency of activator protein
Sudanophilic leukodystrophies	Unknown defects in myelin metabolism
Spongy degeneration	Unknown
Alexander's disease	Unknown
Subacute necrotizing encephalomyelopathy	Possible abnormality of thiamine metabolism
Cerebrohepatorenal disease (Zellweger's disease)	Absent peroxisomes
Oculocerebrorenal disease	Unknown
Disorders of late infancy and early childhood (Delay or regression of intellectual and motor development; seizures; specific neurologic and systemic signs)	
Aminoacidopathies—phenylketonuria, Hartnup disease, etc	Defective amino acid metabolism or transport.
Lysosomal hydrolase deficiencies	
Metachromatic leukodystrophy	Sulfatide sulfatase deficiency
Late infantile Gaucher's disease	Glucocerebrosidase deficiency
Late infantile Niemann–Pick disease	Sphingomyelinase deficiency
Late infantile GM_1 gangliosidosis	Ganglioside β-galactosidase deficiency
Mucopolysaccharidoses	Deficient enzymes of mucopolysaccharide catabolism
Fucosidosis	α-Fucosidase deficiency
Mannosidosis	α-Mannosidase deficiency
Aspartylglucosaminuria	Deficiency of N-aspartyl-β-glucosaminidase
Other storage disorders	
Mucolipidosis	Sialidase deficiency in some
Lipopigment storage (Jansky–Bielschowsky syndrome)	Abnormal dolichol metabolism
Ataxia–telangiectasia	Disorder of DNA repair
Disorders of late childhood, adolescence, and later life	
Progressive cerebellar ataxias	
Abetalipoproteinemia	Absence of β-lipoprotein; vitamin E malabsorption
Friedreich's ataxia and spinocerebellar degeneration	Possible malate or glutamate dehydrogenase deficiency
Extrapyramidal syndromes (disorders of movement)	
Hepatolenticular degeneration (Wilson's disease)	Copper storage, low serum copper and ceruloplasmin, elevated urine copper
Lesch–Nyhan disease	Hypoxanthine-guanine phosphoribosyl transferase deficiency
Hallervorden–Spatz disease	Unknown
Pseudohyperparathyroidism	Calcification of basal ganglia; hypocalcemia
Neurovisceral lipidosis with ophthalmoplegia	Accumulation of bis(monoacylglycero)phosphate, and other lipids
Syndromes with myoclonus	
Lipopigment storage (Batten–Spielmeyer–Vogt, Sjögren, Kuf diseases)	Abnormal dolichol metabolism
Lafora body myoclonic epilepsy	Storage of polysaccharide
Cherry-red-spot myoclonus disease	Sialidase deficiency

TABLE 15–1. *Continued*

Disorder	Nature of Biochemical Abnormality
Progressive encephalopathy	
Adrenoleukodystrophy	Deficient β-oxidation of very-long-chain fatty acids in peroxisomes
Sudanophilic leukodystrophy	Unknown defect in myelin metabolism
Cerebrotendinous xanthomatosis	Cholestanol storage in tendon and brain
Cerebrovascular syndromes	
Homocystinuria	Cystathionine β-synthase deficiency
Fabry's disease	α-Galactosidase deficiency
Peripheral neuropathy	
Acute intermittent porphyria	UPG I synthetase deficiency
Fabry's disease	α-Galactosidase deficiency
Tangier disease	Analphalipoproteinemia
Heredopathia atactica polyneuritiformis (Refsum's disease)	Phytanic acid α-hydroxylase deficiency

orders of interest in which only histologic changes are known. The extent of phenotypic variation and genetic heterogeneity will not be apparent, but is discussed in the following sections.

In the account that follows, two genetic diseases whose primary clinical manifestations are neurologic have been selected for more detailed discussion to exemplify several principles of applied biochemistry.

ACUTE INTERMITTENT PORPHYRIA

Clinical Features. Acute intermittent porphyria (see also Chap. 10) is a dominantly transmitted disorder with various symptoms indicative of diffuse involvement of the nervous system. Symptoms occur only after puberty, more frequently in women, and only intermittently. Most individuals carrying the gene never experience symptoms, even under circumstances known to precipitate attacks. Symptomatic carriers experience periodic attacks of abdominal pain not unlike colic. Associated with these episodes but occurring also in their absence are mental confusion and delirium, accompanied by convulsive seizures that are indicative of a disturbance of the cerebral hemispheres, and an acute polyneuropathy which can progress to generalized paralysis and breathing failure. The confusional states and seizures in some cases have been attributed to brain edema, resulting from the inappropriate release of antidiuretic hormone because of a lesion in the hypothalamic–pituitary system. The name *porphyria* refers to the purple color which appears in urine excreted during an acute attack if it is left standing in the light. A most important feature of this disorder is the observation that acute attacks may be precipitated by a large number of drugs.

Pathogenesis. The known *enzymatic* and *metabolic consequences* of the presence of a *single mutant allele* account for many of the foregoing features, permit precise diagnosis, and provide a rationale for therapy of acute attacks. The trait is associated with a 50% decrease in the activity of uroporphyrinogen (UPG) I synthetase, an enzyme in the synthetic pathway for heme (see Fig. 10–1), now known to be coded for on chromosome 11. During attacks, two intermediates in the pathway, aminolevulinic acid (ALA) and porphobilinogen (PBG), are excreted in great excess. Knowledge of the regulation of this pathway is necessary to an understanding of the pathogenesis of acute intermittent porphyria.

Control of porphyrin synthesis occurs at the step involving ALA synthetase, which is the rate-limiting enzyme in the pathway. The synthesis of this enzyme may be inhibited by a heme repressor–protein complex, thereby providing negative feedback control of the pathway by the end-product. A variety of drugs and chemicals, and some steroid metabolites, can compete for the repressor protein to derepress the enzyme and lead to the increased formation of porphyrins. Some (such as phenobarbital) may also act indirectly to induce synthesis of the microsomal heme protein, cytochrome P450, in the liver. Many of these compounds can precipitate neurologic symptoms in porphyrics.

The key biochemical observation in patients with acute intermittent porphyria was a very high activity of *ALA synthetase* in liver biopsies. Because ALA and PBG are the only intermediates which are increased during attacks of acute intermittent porphyria, it was suggested that the pathway could be partially blocked at the step where PBG is converted to UPG I or UPG III. Assay of *UPG I synthetase* in liver, red cells, and fibroblasts from porphyrics confirmed that activity is reduced to approximately 50% of that found in unaffected individuals. This is compatible with a single 'dose' of the porphyria gene and the dominant mode of expression. It also accounts for other features of the metabolic disturbance. A partial block in heme synthesis limits the capacity of porphyric cells to respond to a sudden demand for heme such as that created by barbiturates during the induction of cytochrome P450 in the liver. In this circumstance, the heme in the porphyric liver becomes depleted, and there is derepression of the synthesis of ALA synthetase with increased formation of ALA and PBG, which accumulate because further metabolism is partially blocked.

Laboratory Approach. By combining assays of porphyrin precursors and porphyrins in urine and stool with assays of enzymes, precise diagnosis of symptomatic individuals and asymptomatic carriers is now possible. Before the affected enzymes were identified, the diagnosis could be confirmed chemically only by measurement of metabolites which commonly are not in excess in asymptomatic carriers. The availability of enzyme tests to detect asymptomatic carriers of the gene allows *counselling* of these individuals to avoid fasting and the use of agents which are known to precipitate attacks. Only about 10% of those carrying the genes ever develop neurologic symptoms and these are readily controlled in most instances. Knowledge of the metabolic derangement led to the use of heme (hemin, hematin) by parenteral administration during acute attacks. This results in repression of ALA synthetase and decreased production of metabolites and is associated with clinical improvement.

OTHER PORPHYRIAS

Mutations affecting single alleles of genes coding for other enzymes in the synthetic pathways for heme may also be associated with acute neurologic syndromes precipitated by factors identical to those acting in acute intermittent porphyria. In ***hereditary coproporphyria*** there is a 50% reduction in the activity of coproporphyrinogen III oxidase, and in ***variegate porphyria*** protoporphyrinogen oxidase is reduced. The reduced activity of these later enzyme steps results in accumulation of porphyrins, in addition to their precursors. Excessive porphyrins in the skin account for the appearance of dermal lesions induced by exposure to light (photosensitivity). Identification of the affected gene does not explain the pathogenesis of the neurologic disorder in acute attacks. Mechanisms postulated are a decreased formation of heme, which might be critical to some neuronal function, or the possible effects of excess ALA acting on GABA receptors in neurons.

METACHROMATIC LEUKODYSTROPHY

This is an example of a *recessively* transmitted lysosomal enzyme deficiency that presents as a progressive deterioration of nervous system function. It illustrates a biochemical lesion affecting primarily a function specific to the nervous system, the heterogeneity of clinical and biochemical phenotypes, the application of biochemical tests to prenatal diagnosis, and the special problems in the treatment of lysosomal enzymopathies affecting mainly the nervous system.

Clinical Pathology. A striking abnormality was discovered years ago by pathologists on examining the brains of patients who had died with progressive disorders of the nervous system: A diffuse disappearance of the glistening white myelin was readily apparent in gross specimens. Since 1912, these disorders have been referred to as ***Schilder's disease*** and have been found to comprise forms of *multiple sclerosis,* inflammatory disorders of white matter (*leukoencephalitis*), and inherited disorders of myelin metabolism (*leukodystrophy*). Among the hereditary forms were some which could be recognized by specific histologic features. One of these features was an unexpected staining reaction of the products of myelin breakdown, called *metachromasia,* in which a red-brown hue is noted (instead of purple) when a basic dye, such as cresol violet, is used. Leukodystrophy with metachromatic products of myelin breakdown occurred in four clinical variants (Table 15–2). Studies of the composition of myelin, the chemical basis of metachromasia and the histochemical properties of the products of demyelination indicated that the appearance of metachromasia was associated with the accumulation of an acidic glycolipid. This was identified as sulfatide (galactosyl-3-sulfate ceramide) which is uniquely concentrated in normal myelin membrane.

The relative *accumulation* of *sulfatide* could result from a defect in catabolism, such as cleavage of sulfate from the lipid, a presumed first step. In the absence of an assay for enzymes active toward the natural sulfatide, activities of enzymes called arylsulfatases, which act on 'artificial' substrates (catechol and nitrophenyl sulfates), were measured in tissues from patients with metachromatic leukodystrophy. Deficient *arylsulfatase* activity was found in the various forms of metachromatic leukodystrophy, and the degree of deficiency correlated with the dose of the mutant gene (10% or less of normal activity in clinically affected homozygotes and about 50% in obligate heterozygote relatives). It was later shown that sulfatase activity toward the natural sulfatide and other lipid substrates such as sulfogalactosyl-diglyceride is deficient and that the normal enzyme is identical to arylsulfatase A. *In vivo* the enzyme requires a noncatalytic protein activator. Deficiency of this activator has been found in rare variants of metachromatic leukodystrophy in which *in vitro* assay of sulfatase gave normal activities.

This account illustrates several features of the biochemistry of inherited deficiencies in the metabolism of complex lipids. Although the storage (accumulation) of complex lipid within cells is only relative in some forms of metachromatic leukodystrophy, these disorders are examples of storage disease resulting from a deficiency of *lysosomal hydrolase.* A characteristic property of lysosomal hydrolases acting on complex macromolecules is that they manifest specificity for both a particular chemical bond and for the molecule containing the bond. Thus, the single sulfatase, arylsulfatase A, has variable affinity for several glycolipid, steroid, and synthetic or artificial substrates. The ability to use artificial substrates has permitted the development of simple and sensitive enzyme assays for diagnosis and carrier detection.

The broad substrate specificities of lysosomal acid hydrolases usually results in the accumulation of more than one substrate; for example, both galactosylceramide sulfate and lactosylceramide sulfate accumulate in the more common forms of metachromatic leukodystrophy with arylsulfatase A deficiency, and glycosaminoglycan sulfates and sterol sulfates accumulate in the variant with dysmorphic skeletal changes and multiple arylsulfatase deficiencies. A molecular explanation for the different clinical phenotypes with similar arylsulfatase deficiencies is not yet available; unless enzymatic function *in vivo* is affected in ways that are not reflected in *in vitro* assays. Human mutants have been detected in which a deficiency of enzyme activity occurs only with artificial substrates and is not accompanied, even later in life, by a clinical disorder. Here the mutation appears to affect the specificity of the enzyme for various sulfate substrates. Conversely, the clinical phenotype accompanied by storage of sulfatide has been observed in association with normal activities of aryl and lipid sulfatases. In these patients there is a deficiency of a noncatalytic protein required for activation of the enzyme in the lysosome. In the *in vitro* enzyme assays which yield normal activities, detergents are thought to fulfil the function of this activating protein. As indicated in Table 15–2, heterogeneity of disorders with sulfatase deficiency extends to conditions which affect mainly nonneurologic systems and which are without demyelination. The new techniques using DNA probes for molecular diagnosis will help to resolve many of these problems.

Laboratory Approach. The *diagnosis* of metachromatic leukodystrophy will be suggested by clinical, electrophysiologic, and histochemical evidence that indicates progressive demyelination and is confirmed by enzyme assay. Arylsulfatase A will be deficient in plasma, white cells, urine, and cultured fibroblasts. Fibroblasts cultured from amniotic fluid may be used to determine the biochemical phenotype of the fetus *in utero.*

Although the deficient sulfatide catabolism in metachromatic leukodystrophy results in degeneration of myelin-forming cells, and a consequent disorder of neurologic function, as is often the case in storage disorders, other cell types may be affected without degeneration or functional disturbance. In metachromatic leukodystrophy the renal tubular cells also accumulate metachromatic lipid but without impairment of renal function. This can

TABLE 15–2. Inherited Disorders with Deficiency of Lysosomal Sulfatases

Disorder	Clinical Features	Deficient Enzyme
Metachromatic leukodystrophy		
Late infantile form	Onset, age 1–4 yr, of psychomotor regression, pain in limbs, absent tendon reflexes	Arylsulfatase A, cerebroside sulfate sulfatase
Juvenile form	Onset, 4–21 yr, of gait disturbance, incoordination, dementia	Arylsulfatase A, cerebroside sulfate sulfatase
Adult form	Onset after age 21 of psychosis, dementia followed by motor deterioration	Arylsulfatase A, cerebroside sulfate sulfatase
With multiple sulfatase deficiencies	Retarded early development, progressive psychomotor deterioration, mild dysostosis, hepatosplenomegaly	Arylsulfatases A, B, C; steroid sulfatases
Mucopolysaccharide storage disease		
MPS II (Hunter)	Dysostosis, mental retardation, heart disease (mild and severe forms)	Iduronate sulfatase
MPS III A (Sanfilippo)	Mild dysostosis, profound mental retardation	Heparan N-sulfatase
MPS IV (Morquio)	Distinctive, severe dysostosis; corneal clouding; aortic regurgitation	Hexosamine 6-sulfatase
MPS VI (Maroteaux–Lamy)	Dysostosis, corneal clouding, valvular disease, leukocyte inclusions, normal intellect	Arylsulfatase B, N-acetylgalactosamine 4-sulfatase

be detected by microscopic examination of the urine sediment. Increased sulfatide excretion can be demonstrated by thin-layer or high-pressure liquid chromatography of the lipids in the urine sediment and is necessary for the diagnosis of the variant of metachromatic leukodystrophy with 'activator' deficiency.

At present, there is no *treatment* which can reverse the demyelination process. Various strategies for enzyme replacement have been developed but many technical problems remain; for example, the introduction of an active enzyme into myelin-forming cells in the brain.

NEUROLOGIC AND PSYCHIATRIC DISORDERS SECONDARY TO DISORDERS IN OTHER ORGAN SYSTEMS

Patients may present with neurologic or psychiatric symptoms which are a reflection of disordered nervous system function that is secondary to disorders primarily affecting other body organs. Of course, many disease processes affect the nervous system and other organs concurrently.

Because the nervous system regulates the function of other organs, lesions of the brain may impair function of organs whose integrity in turn is necessary for normal neurologic function. Chronic insufficiency of non-neural organ function, in many instances prolonged by modern life-support systems, invariably leads to neurologic complications. Thus, it is important to recognize the special susceptibility of the nervous system to systemic biochemical disturbance, although it is poorly understood in some circumstances.

Seizures, encephalopathy, and coma are the common neurologic syndromes accompanying systemic biochemical disturbance. Clinical ***seizures*** reflect abnormal paroxysmal synchronous electrical discharges within neuronal networks in focal or generalized convulsive activity, or in a variety of other transient disturbances of nervous activity both excessive and deficient. A generalized *convulsion,* or tonic-clonic seizure, which is accompanied by sudden loss of consciousness and intense muscular activity, imposes a sudden metabolic stress of its own. The sudden tonic spasm of the body musculature, to be followed by alternating relaxation and spasm, interrupts breathing and leads to intense anerobic glycolysis in muscle. A sudden lactic acidosis develops, which dissipates in 30 to 60 minutes as the patient recovers. ***Encephalopathy*** is the term used to describe clinical states in which there is evidence of diffuse partial impairment of cerebral function. The commonest forms are those caused by trauma, toxins, drugs, and systemic biochemical (metabolic) disturbances. *Metabolic encephalopathy* appears as alterations of arousal and attentiveness, confused and inappropriate behavior, and hallucinations. A state of excessive arousal, overactive and disorganized mental processes, and hallucinations is called a *delerium* and is characteristic of encephalopathy resulting from toxins such as bromides and sudden abstinence from drugs and alcohol. Increasing severity of encephalopathy leads to complete failure of arousal mechanisms, which is ***coma*** and a true emergency. If recovery does not occur rapidly, life support systems are required to maintain the patient until brain function recovers. The following are metabolic encephalopathies caused by disorders of energy substrate supply, and of water and electrolytes, and by failure of other organs.

Hypoglycemia. The principal substrate for brain oxidative metabolism is glucose. Reduction of blood glucose to 1.5 mmol/L to 2 mmol/L is accompanied by confusion or delerium with lower levels leading to convulsions and coma. Coma, if prolonged beyond an hour, often results in irreversible damage to the cerebral cortex. Although the brain is uniquely dependent on an immediate supply of glucose in normal circumstances, it is capable under conditions of prolonged fasting, of adaptation to the use of ketone bodies derived from fat metabolism as the major substrates for oxidation. The chains of metabolic events that lead to coma and to brain damage, or to restoration of nervous function on replenishment of glucose, are only partly understood; they reflect more than a simple lack of substrate for oxidative phosphorylation. Causes of hypoglycemia are discussed in Chapter 16.

Hyponatremia and Water Intoxication. Serum hypoosmolarity results from sodium dilution or depletion. *Overhydration* results from: (1) the inability of the kidney to excrete a water load consequent to renal disease, (2) administration or inappropriate secretion of antidiuretic hormone, (3) excessive oral intake of fluids by mentally disturbed individuals, or (4) excessive parenteral fluid administration. Symptoms begin to appear when the serum Na^+ falls below 125 mmol/L, depending on the rate of development and presence of osmotically active solutes in the serum. The earliest symptoms are headache, apathy, asthenia, and nausea, followed by confusion and delirium. Further deterioration, leading to coma, is usually accompanied by marked neuromuscular irritability with focal and tonic–clonic seizures.

Excessive *sodium depletion* may result from prolonged sweating without replacement, excessive use of diuretics in renal disease, adrenal insufficiency, and myxedema. The depletion may lead to encephalopathy similar to that seen with mild water intoxication. Animal studies have shown that dilutional or depletional states lead to cerebral edema and a decreased sodium content in the brain, followed by decreased potassium. The syndrome of inappropriate secretion of antidiuretic hormone is a special form of hyponatremia discussed in Chapters 6 and 14.

Hypernatremia and Hyperosmolarity. The common cause of hypernatremic encephalopathy is the severe dehydration that occurs with prolonged diarrhea in infants. Isolated instances of accidental salt poisoning have also occurred in infants. With a serum Na^+ above 160 mmol/L, the patient becomes lethargic, with increased neuromuscular irritability; seizures may occur and coma may supervene. The occurrence of persistent focal neurologic deficits is characteristic and is caused by intracranial hemorrhages that result from mechanical changes in the cranium produced by osmotically induced shrinkage of the brain. In adults, water deficit occurs when stuporous patients are unable to communicate thirst, or when patients have received excess solute parenterally. Hypernatremia has occurred with lesions of the hypothalamus, which are thought to interfere with thirst perception and/or the secretion of antidiuretic hormone.

Calcium and Magnesium Imbalances. The cardinal symptom of *hypocalcemia* is tetany, characterized by acroparesthesias and progressive muscle spasms proceeding to painful cramps of skeletal and smooth muscle. Tonic–clonic seizures are common even in the absence of tetany. In long-standing cases and in some inherited forms of hypocalcemia there may be calcification of basal ganglia with development of the extrapyramidal syndromes of choreoathetosis and parkinsonism. *Hypercalcemia* may present with purely psychiatric symptoms or progress to a subacute encephalopathy. Proximal muscle weakness may suggest a primary myopathy.

Magnesium deficiency results in a syndrome of confusion, neuromuscular irritability, and seizures similar to that of hypocalcemia. It occurs as a result of dietary deficiency conditioned by excessive loss from the gut, infection, or alcoholism. The magnesium ion acts to depress neuromuscular and central transmission. *Hypermagnesemia* sufficient to cause hypotension, loss of tendon reflexes, stupor, and coma occurs rarely following the administration of magnesium in laxatives and as magnesium sulfate, given in the treatment of eclampsia.

Hepatic Encephalopathy. Acute hepatic failure from any cause leads to profound coma. Chronic hepatic insufficiency, accompanied by shunting of portal venous blood directly into the systemic venous circulation, is associated with a striking disorder of recurring coma that may be further complicated by a form of cerebral degeneration, or a form of progressive spastic paraplegia. The syndrome of ***recurring coma*** presents often with gradual and subtle onset of confusion and is accompanied by hyperventilation associated with respiratory alkalosis, a characteristic tremor with sudden lapses in posture of the limbs, called *asterixis,* and prominent motor signs. The coma may be precipitated by a number of factors involved in liver disease and its management, including intestinal hemorrhage, diuretics, sedative drugs, uremia, infection, and bowel stasis, and appears to be associated with increased brain susceptibility to mild metabolic stress. A variety of substances, exogenous or endogenous, that have escaped metabolism or detoxification by the damaged liver have been incriminated. The syndrome cannot be explained adequately by the toxic effect of any single metabolite. Blood ammonia is usually elevated and probably plays a role by interfering with cerebral energy metabolism and synaptic and membrane ion transport. Another factor creating the neurologic disturbance may be incompletely metabolized amino acids and related analogues acting as 'false neurotransmitters.'

Acquired hepatocerebral degeneration is a chronic irreversible disorder complicating advanced liver disease, characterized by tremor, choreoathetosis, ataxia, and dementia. Although the syndrome may closely resemble inherited hepatocerebral degeneration (Wilson's disease), copper metabolism is normal. A gradually progressing spastic ***paraparesis*** has occurred among patients having either a surgical or a spontaneous portacaval shunt, who are particularly intolerant to dietary protein. The pathologic changes are similar to those in acquired hepatocerebral degeneration, with degeneration of neurons in the motor cortex resulting in demyelination of lateral columns of the spinal cord.

Renal Failure. A variety of neurologic disorders develop consequent to renal failure and its treatment. Suddenly developing or untreated advanced uremia induces a progressive encephalopathy with acidosis, a twitching of muscles called myoclonus, and convulsive seizures. The encephalopathy is reversed by hemodialysis, indicating that it is secondary to retained small-molecular toxins or metabolites. Chronic uremia from any cause leads commonly to a sensorimotor polyneuropathy characterized by 'burning' feet, 'restless' legs, and cramps. In contrast to the encephalopathy, the polyneuropathy is only partially reversed by hemodialysis but can be fully reversed by transplantation. During hemodialysis, patients often experience a variety of nervous symptoms such as headache, irritability and confusion, and rarely convulsions and coma, referred to as 'dialysis dysequilibrium.' This is attributed to a sudden movement of water into the brain, as a result of removal of osmotic solute from the blood, inducing a form of water intoxication. A dementing disorder with involuntary movements and myoclonus has occurred in patients on long-term hemodialysis and appears to result from excessive exposure to aluminum in dialysis fluids and orally administered gels.

Cardiac and Pulmonary Failure, Cerebral Hypoxia, and Ischemia. Modern methods of resuscitation, as well as the artificial maintenance of circulation during surgery, have greatly increased the incidence of brain damage secondary to transient reduction of oxygen and blood supply to the brain. *Hypoxia* is distinguished from *ischemia,* because circulatory failure not only restricts the supply of oxygen but also the supply of substrates for metabolism and removal of products of oxidation and metabolism. Hypoxia in isolation, however, such as that occurring in acute carbon-monoxide poisoning, may initially affect the heart to precipitate hypotension and arrhythmias with accompanying cerebral ischemia. Consciousness becomes impaired when the apO_2 falls rapidly below 30 mm Hg and is lost within seconds of total cerebral ischemia. Complex interlocking cascades of physiologic and biochemical responses are set in motion during these events. In *acute ischemia,* high-energy phosphates, glucose, and glycogen are rapidly exhausted with increasing accumulation of lactic acid, release of free fatty acids, alterations in cyclic nucleotides and neurotransmitters, and disruption of protein synthesis. Persisting hypoxia and ischemia are accompanied by characteristic histologic changes of neurons (ischemic cell change) in selective regions of vulnerability. Current studies are seeking to identify irreversible steps in the pathophysiologic process, as well as factors promoting permanent damage either during ischemia or recovery. Special changes accompany chronic *pulmonary* insufficiency in which there is mild hypoxemia with chronic carbon-dioxide retention. In addition to encephalopathy, these patients may have signs of increased intracranial pressure resulting partly from marked cerebral vasodilatation.

DISORDERS OF CHEMICAL NEUROTRANSMISSION

In the past two decades, a great deal has been learned about biochemical mechanisms underlying neurotransmission in the nervous system. This has led to an increasing number of correlations with neurologic and psychiatric disorders. Such studies have indicated a selective involvement of transmitter systems in certain disorders and have provided the rationale for treatment with drugs and neurotransmitter precursors which influence the metabolism of neurotransmitters. An outline of the principles of chemical neurotransmission, with examples of selective involvement of transmitter systems in disease, follows.

PRINCIPLES OF CHEMICAL NEUROTRANSMISSION

Complex patterns of signaling in neuronal networks comprise the prime activity of the nervous system. The transmission and mediation of excitability between the electrically conducting neurons occurs by means of small molecules called *chemical neurotransmitters,* which act at specialized cell contacts called *synapses* (Fig. 15–1). The propagation of an impulse in the form of electrical depolarization along a nerve fiber and into the nerve terminal triggers the release of a quantum of chemical neurotransmitter (stimulus–release coupling). The transmitter diffuses across a narrow synaptic cleft between two neurons. It then encounters specific receptors in a specialized area of the membrane of the postsynaptic neuron. The binding of the transmitter to the receptor in turn triggers a change in the ion permeability of the membrane. This results in an adjustment of the potential of the postsynaptic membrane and subsequently of the receiving neuron toward either increased or decreased excitability, depending on whether the synapse is excitatory or inhibitory. The termination of transmitter action occurs either by enzymatic inactivation of the transmitter or by its uptake into the presynaptic terminal, the postsynaptic cell, or adjacent glial cells.

The mechanism of chemical neurotransmission appears to be subject to *regulation* at a number of sites, including the presynaptic nerve endings (transmitter synthesis, storage, and release), postsynaptic receptors (number and affinity), and neurotransmitter inactivation processes (catabolic enzymes, uptake).

The sensitivity of the receptor may be modified (modulated) by other molecules or ions acting on the receptor. The complexity of neuronal mechanisms is indicated not only in the 'wiring diagram' of neuronal networks but also by the presence of hundreds of synapses on an individual neuron, which set its excitability and employ various transmitters. Individual neurons release exclusively one or more specific neurotransmitters. Cytoskeletal elements within neurons may alter the geometry of dendritic spines bearing synaptic receptors to regulate further the excitability of the neuron.

CHOLINERGIC SYSTEMS

Cholinergic neurons synthesize, store, and release *acetylcholine.* Cholinergic pathways in the central nervous system involve extensive arborization in the cerebral cortex of fibers originating in the *nucleus basalis* of Meynert, a collection of cells in the basal forebrain. More is known about pathways in the peripheral nervous system. Examples are motor neurons in the anterior horn of the spinal cord, which send axons to skeletal muscle fibers, and postganglionic parasympathetic fibers, which supply smooth muscles and glands. The former are examples of cholinergic fibers ending on cholinoceptive cells with the 'nicotinic' type of acetylcholine receptors, while the latter are examples of those ending on cells with the 'muscarinic' type. Activation of *nicotinic* receptors results in rapid, brief, excitatory responses which can also be evoked by nicotine and blocked by curare. *Muscarinic* receptors are associated with slow, prolonged, excitatory or inhibitory responses evoked by muscarine and blocked by atropine. Acetylcholine action at postsynaptic receptors is terminated by the action of the enzyme acetylcholinesterase present in the postsynaptic complex.

Alzheimer's Disease. Alzheimer's disease is an example of a disorder of a central nervous system cholinergic pathway. It is characterized clinically by progressive deterioration of cognitive functions, particularly memory and language, and histopathologically by an excessive accumulation of filamentous proteins in neurons (neurofibrillary tangles), and by excessive numbers of lesions in the neocortex consisting of an amyloid-like material surrounded by degenerated nerve endings (senile plaques). Alzheimer's disease is associated with a marked deficiency in the activity of the acetylcholine synthesizing enzyme (choline acetyltransferase) in the neocortex and hippocampus. This reduced cholinergic enzyme activity may reflect degeneration of the cholinergic neurons with their cell bodies in the nucleus basalis. Pharmacologic efforts designed to increase brain cholinergic activity in Alzheimer patients, by supplementation of the diet with cholinergic precursors, have not resulted in any consistent improvement in cognitive function.

Myasthenia Gravis. Myasthenia gravis is an example of a disorder of the cholinergic neuromuscular junction. The characteristic clinical feature is weakness aggravated by repeated muscle contractions. The weakness, when severe, leads to paralysis of breathing. Synaptic transmission is impaired as a result of damage to the postsynaptic apparatus, caused by an abnormal autoimmune reaction against the acetylcholine receptor. Antibodies to the receptor can be demonstrated in the plasma. Treatment utilizes measures for suppressing immune mechanisms and, more recently, by plasma exchange for the removal of antibody. In a related disorder, the *Lambert–Eaton myasthenic syndrome,* muscle weakness results from decreased release of acetylcholine, associated with binding of an abnormal antibody to the presynaptic apparatus. Tumors may trigger the formation of the autoantibody in both of these disorders.

AMINERGIC SYSTEMS

Neuronal systems utilizing amine derivatives of amino acids have been known in the autonomic nervous system for a long time. It is only in comparatively recent years that systems in the central nervous system utilizing dopamine, noradrenaline, adrenaline, and serotonin have been defined.

Dopamine. Dopaminergic systems elucidated in the brain include neurons with their cell bodies in the substantia nigra of the brain stem projecting to the neostriatal component of the basal ganglia (nigrostriatal system) and neurons with cell bodies in the ventral tegmentum which innervate many limbic structures (mesolimbic system) and neocortical areas (mesocortical system).

Parkinson's Disease. Parkinson's disease is a disorder of later life characterized clinically by muscle rigidity, slowing of movements, and tremor, and histopathologically by degeneration of the pigmented neuronal systems of the brain stem, namely the nigrostriatal dopaminergic system and the locus ceruleus noradrenergic system. Neurochemical analysis of autopsied Parkinsonian brain has revealed a profound *reduction* of dopamine in the basal ganglia, the degree of this dopamine depletion correlates well with the severity of the observed symptoms. In addition, drugs that are potent dopamine blockers (major antipsychotic drugs) may also induce Parkinsonian symptoms.

The discovery of the reduction of brain dopamine in Parkinson's disease has led to the treatment of this disorder with the *dopamine precursor,* L-DOPA, which alleviates the slowing of movement and the rigidity.

The observation that the major antipsychotic drugs (such as the phenothiazines and the butyrophenones) are potent antagonists to the action of dopamine had led to the hypothesis that some forms of ***schizophrenia*** might be a consequence of *excessive* dopaminergic activity in brain. Some support for this hypothesis is derived from recent studies of autopsied schizophrenic brain demonstrating an increased concentration of *dopamine receptors* in the basal ganglia. The important question, however, of whether the increased number of brain dopamine receptors is related to the disorder itself, or simply a consequence of prolonged treatment with antipsychotic drugs, remains to be resolved.

Noradrenaline and Adrenaline. A major noradrenergic system arises from the locus ceruleus in the brain stem projecting widely to innervate the cerebral and cerebellar hemispheres. Other noradrenergic neurons, also with their cell bodies in the brain stem, give rise to descending fiber systems that project to the hypothalamus and basal ganglia. Central adrenergic tracts have also been described.

The observation that acute administration of some tricyclic antidepressants enhanced the availability of noradrenaline at the synapse (through presynaptic uptake blockade) led to the original 'catecholamine hypothesis' of affective disorders. This hypothesis states that ***depression*** is associated with a relative noradrenergic *deficiency,* whereas ***mania*** is associated with noradrenergic *overactivity.* Support for this hypothesis is derived from studies demonstrating that some depressed patients excrete in their urine a reduced amount of a major metabolite of noradrenaline, 3-methoxy-4-hydroxyphenylethylene glycol (MHPG), whereas the opposite holds true for some manic patients. The significance of these findings, however, is now uncertain because the proportion of renally excreted MHPG that is of central nervous system origin is unknown. Moreover, no evidence for a noradrenergic abnormality has been obtained in neurochemical studies of autopsied brains of patients dying with affective disorders.

Much evidence now suggests that the psychotic symptoms observed in one type of schizophrenic disorder, namely ***paranoid schizophrenia,*** may be related to an abnormality in the brain noradrenergic system. Studies of autopsied brains of patients dying with paranoid schizophrenia have revealed a marked increase in noradrenaline concentration in limbic areas. Consistent with these observations, elevated levels of noradrenaline have been observed in *cerebrospinal fluid* of paranoid schizophrenic patients.

Serotonin. The serotonin (5-hydroxytryptamine) system arises from clusters of cell bodies in the *raphe nuclei* of the brain stem projecting widely to the telencephalon, diencephalon, and spinal cord. Although the physiologic function of the serotoninergic system is unknown, the projections to the forebrain have been postulated to play an important role in the control of the sleep/wake cycle.

Recent evidence suggests that brain serotonergic activity may be reduced in some patients with ***depression.*** In support of this hypothesis, studies of autopsied brains of suicide and depressed patients have disclosed reduced concentrations of serotonin and 5-hydroxyindoleacetic acid (5-HIAA, the major metabolite of serotonin) in certain raphe nuclei. The finding of *reduced* CSF concentration of 5-HIAA in a subgroup of depressed patients is in agreement with this observation. The demonstration in electrophysiologic experiments of increased sensitivity of central neurons to iontophoretically applied serotonin, following either chronic antidepressant drug administration or repeated electroconvulsive shock, is also consistent with the aforementioned findings.

Patients who have suffered temporary brain hypoxia may develop a characteristic and disabling disorder of voluntary movement called ***action myoclonus.*** Some of these patients are found to have low concentrations of 5-HIAA in cerebrospinal fluid and respond to treatment with the serotonic precursor, 5-hydroxytryptophan.

AMINO ACID AND OTHER TRANSMITTER SYSTEMS

Considerable evidence supports neurotransmitter roles in the central nervous system for several amino acids such as the inhibitory amino acids: γ-aminobutyric acid (GABA), glycine, and taurine, and the excitatory compounds: glutamate and aspartate.

Reduced concentration of *GABA* in basal ganglia of patients with ***Huntington's disease*** and in the cerebral cortex of ***dialysis encephalopathy*** have suggested an involvement of the GABAergic neuronal system in these two disorders. Recent postmortem studies of patients with ***schizophrenia*** (which probably comprises a heterogenous group of psychotic disorders) has disclosed a reduction in GABA concentration in certain limbic brain areas.

Excessive concentration of *glycine* occurs in brain and cerebrospinal fluid of infants with ***glycine encephalopathy,*** an autosomal recessive disorder characterized by intractable seizures, lethargy, spasticity, severe mental retardation, and early death.

Increased concentration of the excitatory neurotransmitter *glutamic acid* occurs in biopsied temporal lobe epileptogenic cortex of some ***epileptic*** patients, suggesting that excessive brain glutamatergic activity may play an important role in some forms of epilepsy.

An explosion of knowledge has occurred about *small peptides* which appear to have synaptic actions either as transmitters or modulators. These compounds are localized to specific groups of neurons, act in various ways on neurons, may be released from neurons on stimulation, and may have specific receptors. A list that is not exhaustive would include thyroliberin (TRH), the endogenous opioid peptides (endorphins and enkephalins), angiotensin II, oxytocin and vasopressin, gonadoliberin (LHRH), neuronal substance P, neurotensin, and somatostatin. Some of these peptides have potent effects on behavior and many occur in cells outside the nervous system. These peptides can be measured by radioimmunoassay so that their concentrations in body fluids may now be correlated with clinical syndromes. Still other important physiologic substances, such as cyclic nucleotides, prostaglandins, and histamine are implicated in synaptic function.

Although few of these small peptides have been measured in neurologic conditions, preliminary evidence indicates the involvement of the endogenous opioid peptides in subacute necrotizing encephalomyelopathy, or ***Leigh's syndrome.*** This neurodegenerative disorder is characterized in part by depressed respiration and coma, symptoms resembling morphine intoxication. In autopsied brain tissue the cortical levels of *enkephalin* were >100-fold the control values. Significantly, the opiate antagonist naloxone is able (transiently) to restore consciousness in this disorder.

BIOCHEMICAL CHANGES IN THE CEREBROSPINAL FLUID

ANATOMY AND PHYSIOLOGY

A potential source of important biochemical information about the brain is the cerebrospinal fluid (CSF), which surrounds the central nervous system. In order to appreciate the value and limitations of biochemical analysis of cerebrospinal fluid, some understanding of the anatomy and physiology of the CSF is necessary.

The CSF circulates through the ventricular system and over the external surface of the brain and spinal cord in the subarachnoid space. The brain and spinal cord have a gelatinous consistency and lack internal support. The function of the CSF is primarily a mechanical one, to float the brain and cord and thereby buffer translational and rotational forces, including those of gravity, acting on these organs. However, other reasons, particularly its mode of circulation and special composition, suggest that the spinal fluid has additional physiologic functions.

Although there are several potential sites where CSF may be *produced,* most of it is formed by the choroid plexuses of the third and fourth ventricles. Estimates are that about 0.5% of the total CSF volume (140 mL in a 70-kg person) is replaced each minute, which amounts to a total formation of 0.5 L to 1 L/day. Formation occurs through the active secretion of solutes which are passively accompanied by water. The *circulation* of CSF through the ventricles and out through the foramina of the fourth ventricle is aided by the pumping effect of the systolic surge of blood into the brain. In the subarachnoid space of the posterior fossa the emerging fluid is thoroughly mixed in the to-and-fro motion of fluid through the foramen magnum, caused by rhythmic changes in blood circulation and in breathing. The fluid then flows up over the hemispheres and down around the spinal cord.

Normally, spinal fluid is removed from the subarachnoid space by transfer in bulk through the arachnoid villae. It is not certain whether this occurs through one-way valves or by transcellular vesicular transport mechanisms.

The CSF is thought to be in continuity with the *extracellular space* of the brain through gaps in the ependymal cell lining of ventricles and the pia-glial layer over the hemisphere. The central nervous system lacks a lymphatic drainage system and it seems that this function is provided by what has been called the 'sink action' of the spinal fluid. Thus, the ventricular and subarachnoid spinal fluids are seen as expanded lacunar extensions of the extracellular space which are continuously irrigated with fresh CSF. This system creates a gradient of metabolites released from cells or blood vessels in the substance of the central nervous system, which favors their removal from the brain by simple passive diffusion into the spinal fluid sink.

BIOCHEMICAL INFORMATION OBTAINED FROM SPINAL FLUID

In *theory,* the CSF is a clinical chemist's dream and should reflect a great variety of changes in cellular metabolites. In *practice* this is largely unrealized because the fluid can be sampled only by special procedures, such as *lumbar puncture,* which must be conducted by physicians. Lumbar puncture is commonly followed by unpleasant aftereffects and is associated with serious risks when the intracranial pressure is high. Furthermore, the concentration of metabolites in the CSF is low and also varies depending on the location of the fluid in the ventricular–subarachnoid system when it was sampled. Therefore, in line with the diagnostic approach to neurologic disorders described earlier, spinal fluid is sampled only to obtain data important to the testing of serious diagnostic hypotheses.

Cells and Cellular Constituents. Normal CSF contains a few mononuclear white cells, but no red cells. The presence of abnormal numbers and types of cells in CSF, and of microorganisms, can provide crucial information for the diagnosis of infections and neoplastic disorders. Immunologic assays may be used to detect either antigens released from microorganisms or antibodies to specific organisms produced by the patient. Monoclonal antibodies help to characterize subsets of lymphocytes in CSF which reflect an inflammatory response in the central nervous system.

The white cells should be removed by centrifugation before the analysis of soluble metabolites. There is a rich plexus of veins in the epidural space and if they are damaged during lumbar puncture, blood may contaminate the sample of spinal fluid. If the CSF is found to be bloody on removal, it is important to determine whether the blood was introduced by the needling procedure. If it was, the fluid will usually clear as it continues to drip from the needle. Therefore, *serial samples* are taken when fluid initially appears to be bloody and clearing is estimated visually. Bloody samples of CSF should also be centrifuged. The finding of a *xanthochromic supernatant* indicates that hemolysis has begun and that the entry of blood into the CSF preceded the needling procedure. Clinically important xanthochromia can be detected by visual comparison against a water blank in appropriate cuvettes. Blood that has leaked into the CSF begins to hemolyze in about 2 hours. The analysis of blood pigments by spectroscopy may be helpful in timing the onset of hemorrhage into the ventricular–subarachnoid fluid. *Oxyhemoglobin* is released when cells are hemolyzed, presumably through the action of macrophages, and after about 8 hours begins to be converted to *bilirubin.* Oxyhemoglobin will have disappeared after 10 days.

Glucose. In fasting normal subjects, the CSF glucose concentration is 60% to 70% of the plasma glucose concentration. Changes in blood sugar are reflected in the CSF but there is a delay of 90 to 120 min before a new steady-state equilibrium is reached. In clinical situations in which a low CSF glucose is suspected, a knowledge of the blood glucose may be helpful in the interpretation of spinal-fluid values. Glucose is transported into the spinal fluid from the blood by a process of facilitated diffusion which utilizes a bidirectional carrier system. The concentration of CSF glucose is regulated by the limiting rate of the carrier system, by a small rate of utilization in tissues adjacent to the spinal fluid, and by bulk reabsorption of the spinal fluid.

The CSF glucose concentration is a critical factor in the diagnosis of *bacterial meningitis,* in which it will be found to be low, in contrast to *viral meningitis,* in which the concentration is usually un-

TABLE 15-3. Normal Cerebrospinal Fluid

Pressure	4–25 cm water
Color	Clear and colorless
Cell count	<4 leukocytes/μL (lymphocytes 79%, monocytes 17%, others 4%)
Total protein	150–590 mg/L
Albumin	60–270 mg/L
IgG	5–50 mg/L
IgG/Albumin ratio	¼–⅙
Glucose	2.8–4.4 mmol/L (50–80 mg/dL)

affected. The glucose may also be low in *fungal meningitis, meningeal carcinomatosis,* and *sarcoidosis* of the central nervous system. Interference with the glucose carrier mechanisms and excessive glucose consumption by infiltrating abnormal and inflammatory cells accounts for the decreased concentrations. In the spinal fluid of patients with *fungal meningitis,* ethanol, the end product of glycolysis in fungi, can be demonstrated.

Proteins and Immunoglobulins. The ***proteins*** in the spinal fluid represent an ultrafiltrate of plasma in which smaller proteins are favored selectively and to which there are minor additions and subtractions of other proteins through various mechanisms. Standard procedures in the clinical laboratory for *total protein* assays employ nonspecific turbidimetric and nephelometric methods. There is a gradient in protein concentration between the ventricles and the lumbar subarachnoid space, the latter being the usual source of CSF submitted to the clinical chemist. Protein concentration in the ventricles is about one third that in the lumbar subarachnoid space. Some proteins are derived from cells within the blood–brain barrier and some from preexisting proteins through the action of peptidases and sialidases. Proteins specific to brain cells have now been identified in spinal fluid.

Characterization of the proteins in the CSF has been hindered by the fact that only small quantities can be obtained from an individual on any occasion and repeated puncturing is rarely justified. However, the use of modern techniques for the separation and characterization of proteins is giving much valuable information. The techniques include various forms of electrophoresis, microimmunoprecipitation techniques, and, more recently, radioimmunoassay. The Lange colloidal gold test for detecting an altered ratio of globulins to albumin has been largely abandoned in favor of more direct methods of protein estimation.

Common *patterns of protein abnormality* in the CSF reflect:

1. An abnormal composition of serum proteins, particularly the presence of gammopathies
2. Impairment of the blood–brain barrier, so that an increased concentration of protein occurs and more closely approximates the composition of plasma proteins
3. The presence of abnormal quantities of fractions produced within the blood–brain barrier, usually immunoglobulins

Normal protein concentrations in the CSF are given in Table 15–3. Fractions quantified by immunoprecipitation or immunoassay techniques do not add up to totals determined by chemical procedures.

An increase in ***immunoglobulins*** (mainly IgG), either absolute or relative to albumin, is found in acute infections of the brain or its covering meninges during the convalescent phase. The elevation of immunoglobulins is a striking characteristic of some chronic inflammatory diseases of the brain, namely, neurosyphilis, multiple sclerosis, subacute sclerosing panencephalitis, and progressive rubella panencephalitis. In these disorders, the excess immunoglobulin is produced by lymphocytes and plasma cells within the blood–brain barrier.

MULTIPLE SCLEROSIS

This chronic disorder, which commonly begins in young adults, is characterized by the occurrence of focal inflammatory lesions in the myelinated regions of the brain and spinal cord which are disseminated in regard to both incidence and location. This process (or processes) results in a wide spectrum of syndromes of neurologic deficits with highly variable chronic remitting or progressive course. In the early stages, these syndromes may be difficult to diagnose. The lesions or 'plaques' contain inflammatory cells which produce the *immunoglobulins* found in the CSF. In patients with widespread and active lesions, elevations of cell count and of total protein are found. When the disease is less active only the concentration of IgG or the ratio of IgG to albumin may be elevated. In some patients neither total protein nor IgG concentration, absolute or relative, is increased, but a qualitative alteration of immunoglobulins in which the γ-globulins separate into several discrete bands, called an *oligoclonal pattern,* may be recognized in high-resolution electrophorograms. Radioimmunoassay of the immunoglobulins has shown that IgG, IgA, and IgM are all increased. A protein specific to myelin, called *myelin basic protein,* has been isolated in pure form and has been used to develop a radioimmunoassay for the protein in spinal fluid. Determinations of the myelin basic protein concentration in the CSF of patients with multiple sclerosis and other demyelinating diseases have revealed increases in association with active disease.

SUGGESTED READING

COOPER JR, BLOOM FE, ROTH RH: The Biochemical Basis of Neuropharmacology, 4th ed. New York, Oxford University Press, 1982

SIEGEL GJ, ALBERS RW, AGRANOFF BW, KATZMAN R (eds): Basic Neurochemistry, 3rd ed. Boston, Little, Brown, 1981

Part Three

Metabolic Diseases

16

Disorders of Carbohydrate Metabolism

John A. Kellen

CARBOHYDRATES AND CLINICAL CHEMISTRY

The metabolic pathways of carbohydrates are interwoven with those of practically all the other major classes of foodstuffs ingested. It is obvious that defects in carbohydrate metabolism may afflict one or more of these other biochemical pathways. It is often difficult or impossible to isolate and define the single primary alteration.

Within the framework of this chapter, only relatively frequent disorders can be dealt with in detail. The processes involved in the digestion, absorption, storage, release, and utilization of carbohydrates are complex, and will be found in standard textbooks of biochemistry.

The rare carbohydrate deficiencies, esoteric meliturias, and storage defects certainly deserve attention and may have unforeseeable genetic implications, but the brunt of the clinician's interest will be concentrated on the common conditions. Determinations of glucosuria and blood sugar are still the tests most frequently performed in hospital laboratories. Justified or not, they are requested in the hope of a fast answer to a wide variety of problems.

LABORATORY TESTS

Blood Sugar

The reserve of carbohydrates in a healthy adult is only approximately 370 g, stored as liver and muscle glycogen. When needed, these stores can be mobilized and depleted rapidly, and so our dependence on regular intake is very evident. Glucose in the plasma amounts to less than 3 g, and blood sugar levels are an imperfect reflection of physiologic extremes between fasting and feeding and are affected by exercise as well as the impact of many diseases. A precisely tuned homeostasis usually maintains fasting blood glucose levels within rather narrow limits of 3.6 to 6.0 mmol/L (65 to 108 mg/dL). Indeed after the correction of fasting levels for age and a possible effect of drugs (for interference by the latter see Table 16–1), values significantly below or above the reference range signal an abnormality of possibly serious consequences. The fasting blood glucose determination actually is becoming redundant; in almost all cases a 2-hour postprandial (pp; postcibal, pc) is now preferred.

In the past, glucose was measured as so-called *reducing substances* in whole blood. Measuring reducing substances caused a variable positive bias, and using whole blood caused a more predictable negative bias. Although the concentration of glucose is usually the same in the plasma water and in erythrocyte water, there is relatively less water per unit volume of erythrocytes than of plasma. Thus, reference ranges for glucose are lower in whole blood than in plasma.

The trend has been to use *enzyme assays* for glucose and to carry out all analyses on serum or

TABLE 16–1. Drugs and Factors That Affect Blood Glucose

Drugs that may *decrease* blood glucose (by a pharmacologic or toxic effect)	
Acetaminophen	Monoamine oxidase inhibitors
Acetylsalicylic acid (large doses)	Oxytetracycline
Anabolic steroids	Propranolol
Ethanol	Sulfonylureas
Fenfluramine	Biguanides
Guanethidine	
Drugs or factors that may *increase* blood glucose (by a pharmacologic or toxic effect)	
Acetazolamide	Glycerin
Arginine	Hypokalemia
L-Asparaginase	Levodopa
Caffeine (in diabetics)	Lithium carbonate
Clonidine	Marijuana
Dextrothyroxin	Nicotinic acid
Diazoxide	Oral contraceptives
Diphenylhydantoin	Phenothiazines (e.g., chlorpromazine)
Diuretics	Sympathomimetics
Glucagon	Theophylline
Glucocorticoids	

plasma. This has meant that reference ranges are not very different but, to distinguish between the two systems, results by the reducing methods are called 'sugar' and by the enzymatic methods 'glucose.' In certain clinical conditions (e.g., chronic renal failure) results for 'sugar' and for 'glucose' may differ considerably. *Plasma glucose* is probably more representative of the glucose concentration in body fluids in general than is whole blood glucose, and is to be preferred.

Urine Sugar

Functioning kidneys exert a regulatory effect on the great majority of solutes in our circulation. Normally, glucose is filtered by the glomeruli but returned to the blood via the proximal tubules, in an active process involving phosphorylation. There is a limit to the rate of glucose reabsorption, termed the *tubular maximum* for glucose (T_mG), which is approximately 1.8 mmol (325 mg)/min.

Generally, when plasma glucose levels are around 9 to 10 mmol/L (160 to 180 mg/dL) the reabsorption capacity is exceeded and glucosuria results. This level is commonly called the ***renal threshold*** and is estimated by following plasma and urine glucose values at 15- to 30-min intervals during a glucose tolerance test. In elderly individuals, with diminished cardiac output and renal arteriosclerosis, the glomerular filtration rate will be lower, so that glucosuria may not be seen even with plasma levels as high as 16 to 18 mmol/L (300 mg/dL).

As with blood, there is a problem in reporting *reducing substances* versus *glucose* in urine. The method used and its specificity must be clearly stated. Again, an increasing number of drugs is being found to influence the outcome of various tests for glucosuria, awareness of which will decrease the number of false-positive results. Table

TABLE 16–2. Substances That Affect the Measurement of Urinary Glucose

Agent	Method: Copper Reduction	Agent	Method: Glucose Oxidase
Cephaloridine	FP	Ascorbic acid	FN
Cephalothin	FP	Gentisic acid (from salicylates)	FN
Chloral hydrate	FP	Levodopa (metabolites)	FN
Glucuronides	FP	Mercurial diuretics	FN
Isoniazid	FP	5-Hydroxyindoleacetic acid	FP
Levodopa	FP	Methyldopa	FP
Metaxalone	FP		
Methyldopa	FP		
Nalidixic acid	FP		
Nitrofurantoin	FP		
Penicillin (massive doses)	FP		
Probenecid	FP		
Salicylates	FP		
Streptomycin	FP		
Sugars (galactose, lactose, fructose)	FP		

FP = False positive; FN = false negative.

16–2 summarizes the more common interfering agents.

In approximately 20% of apparently *healthy* individuals, random urine samples contain reducing substances. Infants and children up to the age of 14 excrete traces of almost every sugar (<100 mg/L); during the first 9 days of life, galactose and lactose can be found in much greater amounts. Increased or persisting melituria is characteristic of prematurity and can also be found sporadically in over 80% of women during pregnancy. The possibility of unusual sugars appearing in the urine can never be dismissed (Table 16–3).

The Glucose Tolerance Test (GTT)

The response of the blood glucose level to a load of glucose (oral or intravenous) can be evaluated as a measure of metabolic efficiency for glucose. Glucose *tolerance* implies that in general, after a specified, empirically established load, no glucosuria is observed, and in particular, the blood sugar levels at certain time intervals are found to follow a typical curve (Fig. 16–1). Ideally, the glucose load should be adjusted to the body weight (or body surface) of the patient (1.75 g/kg); in practice, a flavored standard load of 100 g glucose is given orally. Various large-scale studies have defined 'abnormal' levels (see, for example, Table 16–10); by convention, the 60-min peak should not exceed 9 mmol/L (160 mg/dL) and the blood glucose should return to the reference range within 180 min. A flat curve is a normal variant which occurs in nearly 20% of patients and is observed more often in young, slender individuals. It may, however, be one of the findings in hypothyroidism or malabsorption states.

The most common *nondiabetic* clinical conditions that may be associated with an *abnormal* oral GGT are the following:

1. Lack of dietary preparation (low carbohydrate intake on previous days)
2. Severe liver insufficiency
3. Chronic diseases with malnutrition (alcoholism, uremia)
4. Prolonged physical inactivity (bed rest > 72 h)
5. Acute stress (fever, trauma, major surgery, stroke, myocardial infarction)
6. Endocrine diseases (active acromegaly, Cush-

TABLE 16–3. Melituria and Its Causes

Sugar	Causative or Associated Condition
Inositol	Diabetes mellitus Nephropathies
Lactose	Insufficient intestinal lactase Lactation
Disaccharides	Insufficient intestinal invertase, maltase, isomaltase Pancreatic lesions Moncrieff's syndrome
Galactose	Alimentary (increased ingestion in infants) Congenital enzyme defect (incidence 1:180,000, USA; 1:70,000, Europe)
Fructose	Alimentary, increased ingestion Essential hereditary form (incidence 1:130,000) Hepatic cirrhosis
Mannose	In infants
Pentoses (xylose, arabinose, xylulose, ribose, ribulose)	Alimentary, increased ingestion Congenital (incidence 1:50,000) Hepatic cirrhosis Toxic (allergy, morphine)
Heptuloses	Alimentary, increased ingestion
Fucose	During lactation Infants

FIGURE 16–1. Representative glucose tolerance curves following the ingestion of 100 g glucose. Stippled area = normal curve. Typical responses are shown for liver disease and for mild and severe diabetes.

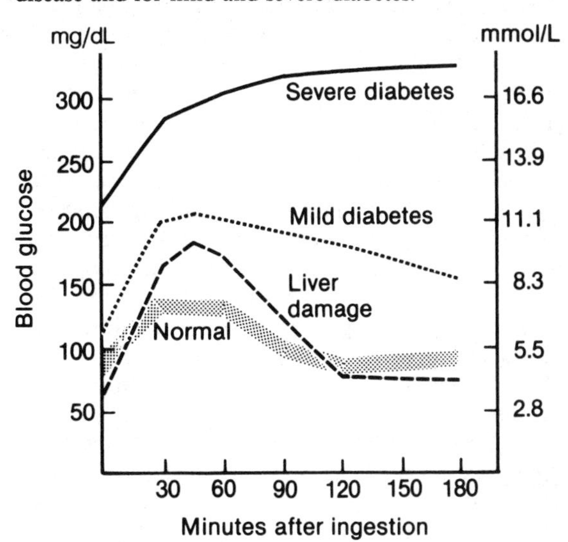

ing's syndrome, thyrotoxicosis, pheochromocytoma, insulinoma, glucagonoma)
7. Drug use (prolonged corticosteroid therapy; see also Table 16–1)
8. Overweight (>20% above average)
9. Gastrointestinal disorders

Glucose tolerance decreases significantly with *age.* This is true particularly after the age of 60, when more than half of a 'normal' population will have 'abnormal' GTTs. Almost half of *obese* subjects have abnormal GTT curves, and normal individuals will develop such intolerance if they become obese. The 'lag-type' GTT, in which undue hyperglycemia develops after 30 to 60 min, but disappears by 2 h, is attributed to a combination of insulin resistance and an early delay in the insulin response to the concomitant hyperglycemia. In obese patients such lag-type curves result solely from insulin resistance, attributed to a decreased concentration of insulin receptors.

A diagnostic refinement imparting greater sensitivity can be achieved by combining a *provocative agent* with the glucose load. Cortisone (given 8½ and 2½ hours before the test) is the one most often used, but glucagon, tolbutamide, epinephrine, and insulin have each been used and its concurrent effect on the GTT curve evaluated. Finally, the simultaneous determination of *plasma insulin* concentrations can add decisively to the information obtained with a GTT.

As already mentioned, gastrointestinal disorders may influence the absorption of an oral load and invalidate the results. An *intravenous tolerance test* provides a means of circumventing this problem, although the rapidly achieved blood glucose peak gives a different shape to the normal curve.

When all influencing factors have been taken into consideration and eliminated, GTTs are a *reasonably precise* indication of carbohydrate intolerance; they are time-consuming and certainly not suitable for large-scale screening. Although the GTT is the classic way of comparing plasma glucose and insulin responses between individuals, it has two serious drawbacks. Reproducibility of the GTT is poor, and it does not provide information on the ambient levels of plasma glucose and insulin, present in day-to-day life. Responses to a mixed test-meal would mimic more closely the real life situation.

In many cases, however, the determination of a postprandial blood glucose (2 h pp) after a carbohydrate-rich meal will serve as a guide to the necessity for further investigations. An approach to the judicious use of these simple tests in the laboratory diagnosis of carbohydrate intolerance is outlined in an algorithm (Fig. 16–2).

In a recommendation of the National Institutes of Health (US), the diagnostic criteria of impaired glucose tolerance should be based on the effect of a 100 g (or 1.75 g/kg ideal body weight where indicated) glucose load on the *sum* of the 1-hour and 2-hour glucose values, which should not surpass 16.9 mmol/L (300 mg/dL).

INSULIN

Synthesis and Secretion

The pancreas is a "dispersed gland of internal secretion." In a sea of cells producing digestive enzymes, internal secretions are formed in tiny islets, representing about 1.5% (or 1 to 2 g) of the total weight of the gland. Evidence that the β cells are indeed the *only site* of insulin synthesis, storage, and secretion is unequivocal. Synthesized as a large *preproinsulin,* the hormone is processed in the Golgi apparatus and is stored as β granules. The secreted products include insulin and C-peptide in equimolar amounts, along with small amounts of proinsulin and minor intermediate products. Electron microscopy has provided a detailed picture of the morphologic structure as it relates to active secretion. Insulinlike activity (ILA), determined by bioassay, also exists in the plasma but the substances appear to be of extrapancreatic or even tumor origin.

Insulin is a protein (the M_r of the monomeric form is 5734); usually the hormone exists in polymeric forms, depending on the pH and zinc content. Most tissues actively degrade insulin; its half-life in man is 7 to 15 min. Degradation is by cleavage, the enzyme involved being glutathione insulin transhydrogenase. Also, the gastrointestinal tract inactivates insulin rapidly and completely, much to the chagrin of diabetics who are forced to administer their exogenous insulin requirement parenterally. Genes carrying the genetic code for human insulin have been inserted into a strain of *Escherichia coli.* Once inside the bacteria, the genes are 'switched on' by the microorganism to tran-

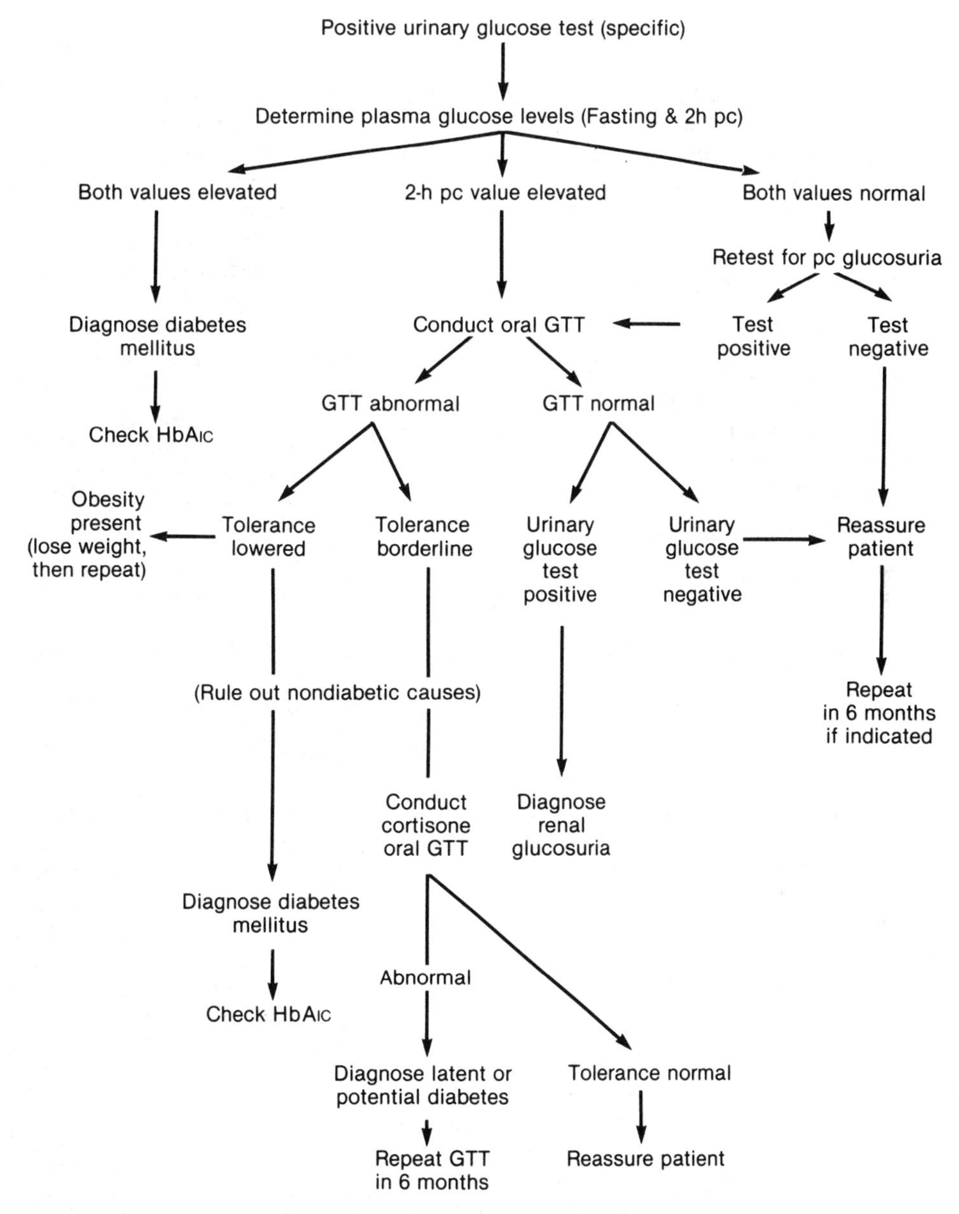

FIGURE 16–2. Algorithm for the investigation of a positive test for glucose in the urine. GTT = glucose tolerance test; HbA1c = glycosylated hemoglobin; pc = postcibal (postprandial).

scribe the code into insulin. This has been the first step toward an unlimited availability of human insulin for the diabetic who depends on this hormone for survival.

By the end of the third month of fetal life, the pancreas is capable of synthesizing insulin. A total of approximately 40 to 50 units of insulin are *produced* per day in a healthy adult; 20% of this amount is stored, the rest secreted. The stimulation of its release by glucose (or tolbutamide) occurs within 30 to 60 sec. Circadian variations in blood sugar and insulin levels have been noted in man.

A failure to respond adequately to a *glucose stimulus* may be one of the fundamental defects in human diabetes, as a lesion of the fetal islet cell. Insulin secretion gradually decreases as the adult ages. It is sometimes difficult to distinguish between physiologic exhaustion and disease.

Factors *stimulating insulin release* directly are:

a. Glucose
b. Leucine, arginine, histidine, phenylalanine
c. Sulfonylureas
d. ACTH, glucagon, somatotropin
e. α_2-adrenergic blockade
f. β_2-adrenergic stimulation
g. Vagal stimulation
h. Gut hormones: glucagon, secretin, cholecystokinin

Factors *inhibiting insulin release* directly are:

a. α_2-adrenergic stimulation
b. β_2-adrenergic blockade
c. Thiazide diuretics
d. Diphenylhydantoin
e. Diazoxide

Factors *decreasing the tissue response* to insulin are:

a. Growth hormone, placental lactogen, glucocorticoids, estrogens, progestins, oral contraceptive agents
b. Obesity
c. Inactivity
d. Poor physical condition
e. Low carbohydrate diet

The *autonomic nervous system* functions as one ***regulator*** of insulin secretion. β-Adrenergic stimulation enhances and α-adrenergic stimulation inhibits both the β-cell level of cyclic adenosine 5′-monophosphate (cAMP) and insulin secretion. The adrenergic influence on insulin secretion explains the decrease in glucose tolerance seen in patients with increased levels of catecholamines (from stress, pheochromocytoma).

The composition of the *diet* is also a factor in insulin secretion. Carbohydrates and proteins have a synergistic stimulating effect; prolonged caloric restriction results in decreased glucose tolerance and decreased insulin secretion in response to a glucose load. Ethyl alcohol, administered either intravenously or orally several hours prior to a glucose pulse, significantly increases insulin release. The total daily insulin requirement is not greatly affected when the proportions of carbohydrate, protein, and fat in the diet are changed under isocaloric conditions. The requirement, however, is higher immediately after a high carbohydrate meal. Consistency in the carbohydrate load per feeding is therefore important for people who take insulin, but not for those who do not.

The insulin released in the early phase of the response to a blood glucose increase most likely represents preformed, stored insulin; that released later is both preformed and newly synthesized. *Proinsulin,* the large, coiled precursor from which insulin is cleaved, constitutes 5% to 15% of the total insulin circulating in the blood and is biologically less effective.

Frequently, but not always, diabetics differ from normal persons by having a *delayed* and *sluggish* rise of insulin after stimulation. Such an insulin response is probably genetically determined, can be found in 15% to 20% of healthy adults and children, and may account for maturity-onset diabetes.

Effects on Metabolism

Insulin acts as an anabolic and anticatabolic hormone, influencing the rates of carbohydrate, lipid, protein, and electrolyte metabolism in an integrative and coordinating manner (Table 16–4). It decreases gluconeogenesis in the liver and possibly in the kidney; its antilipolytic effects are most readily observed in adipose tissue, whereas the glycogenic effects are most apparent in skeletal muscle. The metabolism of cells traditionally thought of as unresponsive to insulin (erythrocytes, brain, gonads, lymphoid organs, kidney, adrenal) may actually be influenced by insulin, according to recent experimental work.

Insulin can increase the activity of cAMP phosphodiesterase and inhibit adenylate cyclase, but the lowering of hormone-stimulated cAMP levels cannot explain all of the known effects of insulin. The *diverse effects* that insulin exerts can be summarized as follows:

1. Stimulation of membrane transport
2. Activation and inhibition of both membrane-bound and soluble enzymes
3. Stimulation of protein synthesis
4. Inhibition of protein degradation
5. Stimulation of messenger RNA synthesis
6. Stimulation of DNA synthesis

TABLE 16–4. Metabolic Responses of Tissues to Insulin

Tissue	Fasting; Low Insulin State (<10 mU/L serum)	Fed; High Insulin State (>10 mU/L serum)
Adipose	Failure to inhibit cAMP-induced lipolysis stimulated by glucagon, ACTH, STH, epinephrine Increased free fatty acids	Activation of lipoprotein lipase clears serum of endogenous and exogenous lipoproteins; increased glucose uptake and triglyceride synthesis
Muscle	Release of amino acids; leucine, valine, isoleucine (ketogenic); alanine, glycine, threonine, serine (glycogenic)	Amino acid uptake and protein synthesis; glucose uptake and glycogen synthesis
Hepatic	Gluconeogenesis from amino acids with energy from free fatty acids	Activation of glycogen synthetase; glycogenesis

ACTH = Adrenocorticotropic hormone, corticotropin; cAMP = cyclic adenosine 5′-monophosphate; STH = somatotropin, growth hormone.

Receptors

Insulin acts when combined with its specific receptors, localized on portions of the plasma membrane of target cells. The activated receptors alter adjacent enzyme systems and set in motion a cascade of events amplifying the original signal, resulting in a wide variety of effects: increased glycogen, protein, and fatty acid synthesis, or altered plasma membrane functions as evidenced by increased rates of entry of glucose, amino acids, potassium, and phosphorus. The receptor exhibits insulin-dependent tyrosine kinase activity and is capable of phosphorylating other cytoplasmic proteins, but beyond this point the mechanism of its actions remains a mystery. The binding of insulin to its receptor depends on the concentration of hormone, the number of receptors, and the affinity of the receptor for the hormone. Fasting results in a fall in plasma insulin and an increase in insulin binding to receptors. Brain cells have traditionally been considered as insulin independent; however, specific insulin receptors localized in the brain blood vessels have been identified. Their function is as yet unexplained.

In addition to the plasma membrane, subcellular organelles, such as the nucleus, may also have specific binding sites for insulin. It may be that insulin, freed from its internalized receptor complex, can mediate certain as yet ill-defined intracellular functions.

GLUCAGON

Pancreatic glucagon (M_r 3485) is secreted by the α cells of the islets of Langerhans. The liver and adipose tissue appear to be the main *target organs,* in which glucagon activates adenylate cyclase and increases intracellular levels of cAMP. In the liver, cAMP increases gluconeogenesis; in adipose tissue, it stimulates lipolysis. In an oversimplification, the action of glucagon can be considered as diametrically opposite to that of insulin; glucagon functions as a signal for the mobilization of stored and newly formed glucose from the liver and of free fatty acids and glycerol from fat tissue. It appears that the glycogenolytic and gluconeogenic effects of glucagon prevent insulin-induced hypoglycemia following the ingestion of a predominantly protein diet. A coordinated interaction between the two hormones brings about matched increases in production and utilization of glucose. Table 16–5 shows serum values of glucagon under specific conditions.

Whenever there is a reduction in functioning pancreatic tissue, a combined deficiency of both

TABLE 16–5. Reference Values of Serum Glucagon in Normal Subjects

Clinical State	Glucagon [ng/L; mean (±SD)]	Range (ng/L)
Fasting	108 (±10)	50–220
During glucose infusion	57 (±8)	
During carbohydrate-rich meal, maximum fall	54 (±9)	20–120
During arginine infusion, maximum rise	239 (±19)	100–400
During protein-rich meal, maximum rise	121 (±11)	50–180

glucagon and insulin may be present. The opposite, a glucagon-secreting islet-cell tumor (glucagonoma), appears to be rare. Recent studies suggest that most *diabetics* with a family incidence of the disease have at all times *excessive glucagon* levels relative to the actual glucose concentration. Absolute hyperglucagonemia may be present in diabetic patients who are in a state of poor control. In severely ketoacidotic patients, plasma glucagon may exceed 400 ng/L. Inappropriately high glucagon levels may be a secondary effect of a lack of insulin. Relative or absolute hyperglucagonemia has been identified in every form of endogenous hyperglycemia. Insulin deficiency in the absence of glucagon does not cause severe endogenous hyperglycemia.

Decades of intensive research centered around insulin have led to a relative neglect of other possible factors influencing carbohydrate homeostasis; glucagon is a striking example of this widespread attitude. It is difficult to evaluate the balance between insulin and glucagon, but an extreme view holds that a glucagon excess may be the principal cause of the overproduction of glucose in diabetes.

SOMATOSTATIN

Somatostatin, a tetradecapeptide originally isolated from the ovine hypothalamus but found also in secretory epithelial cells, prevents the secretion of growth hormone, thyrotropin, gastrin, insulin, and glucagon. Because it depresses plasma glucagon, somatostatin has been examined as a possible *therapeutic* agent in diabetic patients; it reduces fasting hyperglycemia and blunts the blood glucose rise after the ingestion of carbohydrate-containing meals, even in the absence of insulin administration.

The *infusion of somatostatin* in insulin-dependent diabetics results in a 75% to 100% reduction in the blood-glucose peak after oral glucose administration. Glucagon administration does not reverse this effect. In diabetes, somatostatin reduces postprandial hyperglycemia, primarily by decreasing or delaying carbohydrate absorption rather than by enhancing carbohydrate disposal. This effect may be mediated, in part, by a reduction in splanchnic blood flow. The manipulation of carbohydrate absorption does not represent a physiologic approach to the treatment and control of diabetes; therefore, the therapeutic potential of somatostatin should be viewed with caution.

A substance similar or identical to somatostatin has been found to be released by the peripheral nervous system and appears to be involved with the conduction of painful and thermal stimuli.

SOMATOMEDINS

A group of growth-promoting, tissue-specific peptides named somatomedins share an insulinlike effect (a 'positive pleiotypic response'), as opposed to the catabolic, adenylate cyclase-stimulating hormones. Somatomedins inhibit adenylate cyclase and may bind to the same cell membrane receptor as insulin, with a half-life of 12 hours. Their primary action is an effect on cartilage, stimulating protein synthesis. Total hypophysectomy is followed by a fall of somatomedin levels.

The original *bioassay* was based on measuring sulfate incorporation into cartilage (hence the name 'sulfation factor'); more recently, the effect on incorporation of thymidine into DNA has been determined.

PANCREATIC POLYPEPTIDE

A pancreatic polypeptide (PP) of 36 amino acid residues with hormonal properties has recently been described. Immunohistochemical studies show that PP is stored in an endocrine-type cell that is different from islet cell types A, B, and D; the PP cells are found both in the peripheral part of the islets and scattered throughout the exocrine parenchyma. Increased numbers of PP cells are found in the pancreas of animals with experimental diabetes, in the pancreas of patients with pancreatitis, and in the uninvolved pancreas of patients with islet tumors.

Pancreatic polypeptide can be *measured* reproducibly in serum and tissue extracts by means of a radioimmunoassay. The mean basal plasma level of PP is elevated in patients with diabetes mellitus, in particular in the juvenile-onset type of diabetes mellitus and in the more severe maturity-onset type (requiring treatment with insulin). Levels of plasma PP tend to increase with the clinical severity of diabetes as ascertained by fasting hyperglycemia. Basal plasma levels of PP increase with age. Despite extensive investigations its physiologic significance is still unclear.

GLUCOSE TOLERANCE FACTOR

Glucose tolerance factor (GTF), a biologically active organic chromium complex, is a dietary agent that potentiates the action of insulin in animals and man. It is water-soluble and relatively stable and its M_r is approximately 500. The earliest detectable sign of GTF deficiency is glucosuria, which progresses to fasting hyperglycemia, impaired growth, decreased longevity, elevated serum cholesterol, and corneal opacities. The quantitative dietary requirement for *chromium* in man has not yet been specified.

Chromium deficiency has been implicated as a contributing factor in impaired glucose tolerance in man; in one study, about 50% of diabetics showed significant improvement in their impaired glucose tolerance after chromium supplementation. It is possible that in advanced cases of diabetes, individuals may not respond to chromium supplementation because they have lost the ability to convert chromium to GTF. With increasing age, the body supply of chromium decreases progressively.

DISORDERS OF CARBOHYDRATE METABOLISM

DIABETES MELLITUS

General Considerations

Although it is one of the commonest diseases in man, there is no satisfactory *definition* of diabetes mellitus. It is not a single disease entity; rather it is a syndrome, perhaps a collection of several different diseases all sharing the common characteristic of glucose intolerance. A 3% overall prevalence is assumed to be a conservative estimate and some clinicians believe that the potential for developing diabetes exists in up to 20% of the North American population. Besides ranking as the fifth leading cause of death in the United States, diabetes cost the national economy an estimated $9.5 billion in 1982.

A wide variety of derangements in carbohydrate, fat, and protein metabolism is found, without a clear causal relationship to the vascular and neurologic manifestations. In view of the importance of the latter, it has been postulated by some that vascular disease is the primary defect; others consider it a complex disease of glucose metabolism. Although insulin deficiency is perhaps the most frequent feature, not all diabetics lack insulin.

At present, it is sufficient to *classify* diabetes with regard to its recognition by chemical tests and its potential for shortening the life expectancy of the patient. While clinical symptoms on presentation are well known and rather typical (such as polyuria, polydipsia, polyphagia, weight loss, pruritus, fatigue, blurred vision, and susceptibility to infections), 'asymptomatic' patients form at least 1% to 2% of the population. This group can be detected only by various provocative tests which challenge the capacity of the patient to handle a glucose load. On the other hand, in the early stages of clinical diabetes, urine glucose tests and random blood glucose levels may be normal.

In short, detection of diabetes and confirmation of the diagnosis, as well as monitoring of its course and the effects of treatment, lie mainly in the hands of the biochemical laboratory. Data on alterations of carbohydrate and lipid metabolism may be lifesaving in acute and chronic situations, but at present little can be done to prevent life-threatening vascular disease.

Etiology

According to traditional thinking, the clarification of its etiology should lead to ideal medical management or prevention of any disease. With diabetes, the difficulties are manifold: There is good evidence that diabetes is not one entity and data on factors associated with this disease are based on incidence and not on proved causal relationships. Among the factors considered to increase the *risk of developing* or the *probability of having* diabetes are the following:

1. Inheritance (see Table 16–8)
2. Obesity
3. Arteriosclerosis
4. Type IV (Fredrickson) hyperlipemia
5. Chronic pancreatitis
6. Cirrhosis of the liver
7. Endocrinopathies: thyrotoxicosis, pheochromocytoma, Cushing's syndrome, primary aldosteronism
8. Iatrogenic factors: Long-term treatment with corticosteroids, estrogen–progesterone com-

binations, diazoxides, or derivatives of nicotinic acid

This sequence does not necessarily indicate the importance of each factor, and the list is by no means complete.

Obesity, multiple pregnancies, and increasing age adversely influence *carbohydrate tolerance;* the age at menarche does not play a role. Pregnancy represents a mild stress effect on carbohydrate tolerance, probably through the action of a number of factors such as placental lactogen, corticosteroids, and other insulin antagonists. This must be recognized when one is attempting to distinguish physiologic from pathologic glucosuria during pregnancy.

Insufficient production of insulin was thought for many years to be the basic cause of diabetes. For only about 10% of patients, those known as *juvenile* diabetics, is this actually the case. It may be that in some individuals prone to develop diabetes, the predetermined potential for β-cell division is particularly low and becomes rapidly exhausted in response to increased insulin demands. The number of *functionally intact* β cells in the islets is still considered of decisive importance for the development, course, and outcome of diabetes mellitus. The total β-cell mass reflects the balance between the renewal and loss of these cells.

In the vast majority of sufferers from *non-insulin-dependent* diabetes, the pancreas does produce some insulin, even quantities that are above normal. In these cases, the problem obviously lies in a reduced sensitivity of peripheral tissues to the hormone; the cause of this 'insulin resistance' is not yet clear.

It had been assumed that hyperglycemia is nearly always associated with insulin deficiency; we now know that many patients with hyperglycemia have, in fact, normal or even supernormal concentrations of circulating insulin. In most obese patients with insulin resistance the first step in insulin action, the binding of the hormone to its ***receptor sites,*** is found to be decreased. This is caused by a reduction in the concentration of receptor sites on cell surfaces throughout the body, including those in liver and muscle. Recent data indicate that target cells, rather than being passive recipients of humoral stimuli, are continuously and actively regulating the number of receptor sites and hence their responsiveness to stimulation.

Insulin and glucagon receptors have been isolated. A knowledge of the structure and properties of these receptors should lead to substantial progress in our understanding of diabetes. Reduced binding of insulin to monocytes from prediabetics has been reported; for research purposes, this method can now be used as a *screening test* to identify individuals who are likely to develop diabetes.

Recently, a still unidentified substance produced in the pituitary gland has attracted attention as a possible key to the origin of diabetes mellitus. This agent may be a somatotropin molecule that has been acted on by an enzyme in the pituitary. When STH (GH) is treated with a protease called subtilisin, it is cleaved into three fragments each capable of producing diabetic symptoms.

For more than a century, diabetes has been associated with ***viral diseases.*** The agents implicated most often are mumps, rubella, coxsackie B_4, infectious hepatitis, influenza, and Epstein–Barr virus, the agent of infectious mononucleosis. Indirect evidence of the possible role of a viral infection in the etiology of diabetes has been produced by animal experiments showing islet damage and diabetes induced by the inoculation of a particular type of encephalomyocarditis virus. A viral origin of human diabetes cannot yet be settled, but some indirect evidence in favor of what is probably an infectious origin has come from observations of families in which two or more members appear to have acquired diabetes almost simultaneously.

The term ***insulitis*** has been introduced to describe a condition in the islets of Langerhans characterized by the presence of inflammatory cells. The most probable cause of insulitis is either a viral infection or an autoimmune reaction, or a combination of the two. The actual incidence of insulitis is still controversial and its importance in the pathogenesis of diabetes remains difficult to assess. It may be an important indication that an environmental aggressor, for example, a virus, can act either directly or through immune mechanisms to inflict severe and irreversible damage on the islet tissue. Diabetes caused by *chronic pancreatitis* appears to be increasing. This increase may relate to the known increase in alcohol consumption in developed countries. Malnutrition may be the chief etiologic factor in underdeveloped countries.

Classification

Classification schemes should be useful and should guide physicians in the selection of optimal treatment programmes. The accepted classification of diabetes mellitus, however, is based on the empirical observation of dependence on insulin. This is a clinical phenotype; successful treatment is the retrospective guideline for diagnostic grouping in most patients who have glucose intolerance, commonly termed *idiopathic diabetes.*

Many attempts have been made to introduce a working classification of diabetes, using criteria such as age at onset, family history, proneness to ketosis, and body weight. The National Diabetes Data Group (USA) accepted, in 1979, a scheme that is presented in Table 16–6. They recommended abandoning such terms as chemical, latent, borderline, subclinical, and asymptomatic diabetes mellitus, in favor of *impaired glucose tolerance.* Also, the proposed classification abandons the grouping into 'juvenile' and 'mature-onset' diabetes in favor of *insulin-dependent* (IDD) and *noninsulin dependent* (NIDD) diabetes.

Although the two main types, 1 (IDD) and 2 (NIDD) diabetes, differ in several clinical characteristics, there is a considerable overlap in age at onset, body weight, and family history. In principle, the IDD type is characterized by *insulinopenia,* but there may still be some residual β-cell function. If insulin treatment is withdrawn, these patients will readily become ketotic. The NIDD type is by far the most common; up to 80% of these patients are obese and rarely exhibit spontaneous ketosis. Their hyperglycemia is caused by a combination of insulin *resistance* and relative insulin *deficiency.* Insulin resistance exists, by definition, whenever normal concentrations of insulin fail to elicit a normal biologic response. The primary result is a decreased glucose uptake by peripheral tissues.

Secondary diabetes encompasses a heterogeneous group of patients in whom glucose intolerance is clearly associated with another disease. In pancreatic disease, the resulting diabetes is essentially type 1, whereas diabetes secondary to endocrine syndromes is generally type 2.

Gestational diabetes appears *de novo* during pregnancy and disappears when the pregnancy is terminated. In North America, 1 in 400 pregnancies involve a known diabetic; a further 2% to 3% of pregnant women develop gestational diabetes, the result of an impaired β-cell reserve, unmasked by the hormonal stress of pregnancy. This inherent vulnerability to diabetes is apparent from the fact that the incidence of NIDD is twice as great in women with gestational diabetes compared to

TABLE 16–6. Classification of Diabetes Mellitus

Idiopathic Diabetes	
Type 1 Insulin-dependent diabetes (IDD)	Thin, ketotic without insulin, short history, usually <age 30, female:male ~1, family history +
Type 2 Non-insulin-dependent diabetes (NIDD)	Usually obese, ketotic only with stress, long history, usually >age 30, female:male ~4:3, family history +++
Diabetes Secondary to	
Pancreatic disease	Pancreatectomy, pancreatitis, carcinoma, hemachromatosis
Hormonal excess	Cushing's syndrome, acromegaly, pheochromocytoma, primary aldosteronism, glucagonoma
Drugs	Diuretics, glucocorticoids, oral contraceptives, diphenylhydantoin (Dilantin), phenothiazines, tricyclic antidepressants
Insulin receptor unavailability	With and without circulating autoantibodies
Genetic syndromes	Hyperlipemias, myotonic dystrophy, lipoatrophy, leprechaunism, Friedreich's ataxia, Prader–Willi syndrome
Gestational Diabetes	Mild and self-limiting

~ = approximately; + = occasionally; +++ = frequently.

those manifesting normal glucose tolerance. The presence of obesity increases this risk.

Staging

Although the clinical and chemical manifestations of diabetes are sometimes dramatic and abrupt, one can safely assume that the disease syndrome evolves gradually in a sequence of stages. The terminology for these stages varies; most widely accepted is the one introduced by the American Diabetes Association (ADA), although some textbooks and papers also use the terminology of the British Diabetes Association. Table 16–7 presents the ADA staging, including comments concerning the biochemical data.

Glycosylated Hemoglobin (HbA1c)

The red blood cells do not require insulin for glucose uptake, and so intracellular levels reflect the extracellular glucose concentration. Glucose converts to L-amino,1-deoxyfructose which combines non-enzymatically with the N-terminal valine of the β-chain of hemoglobin to form glycosylated derivatives (HbA1). The only quantitatively significant isomer is HbA1c. A *labile intermediate* form of HbA1c responds to fluctuations in plasma glucose levels over the preceding 2 to 3 hours; the *stable* form of HbA1c represents the time-averaged glucose concentrations over the preceding 2 months.

The reference range for normal individuals, as a *percent of total Hb,* is 4% to 7% for HbA1c and 5% to 8% for total HbA1. Values will vary with different methods and should be established locally. The range increases by about 1% during pregnancy and by up to 2% in aged individuals, probably reflecting a less efficient glucose homeostasis.

A practical application of the degree of hemoglobin glycosylation is the use of HbA1c levels as a means of monitoring carbohydrate control in diabetic patients. In *stable* diabetics the level of HbA1 (or HbA1c) will reflect the extent to which the blood glucose has been maintained under control. The simpler HbA1 assay can be used because the labile fraction is not likely to be a significant factor. Values of 8% to 10% are commonly found and it is sufficient to monitor such patients at intervals of about 3 months. Young or *unstable* diabetics may have values ranging from 8% to 18% and the labile component may contribute as much as 10% to 30% to the result. Here the analytical procedure should be modified to ensure that the labile fraction is removed and that only stable HbA1c is determined. Monitoring of these patients should be at monthly intervals. HbA1c has a higher affinity for O_2 and can induce sufficient tissue hypoxia to cause a relative polycythemia in uncontrolled diabetics.

Periodic monitoring of HbA1c levels offers the physician and the diabetic patient the possibility of assessing metabolic control on an outpatient basis in a more objective manner than has previously been possible. The glucose tolerance test is best for diagnosis; HbA1c is best for following glucose regulation under treatment.

Diagnosis

Diagnostic tests for diabetes are indicated in the following clinical situations:

1. To confirm the disease in a patient with obvious clinical symptoms.
2. To confirm the diagnosis in a patient with equivocal symptoms.

TABLE 16–7. Stages of Diabetes Mellitus

Stage	Laboratory Diagnosis
I. Prediabetes	Insulin response to glucose delayed; glucose tolerance normal
II. Suspected diabetes (Synonym: latent diabetes)	Insulin response to glucose delayed; glucose tolerance abnormal during pregnancy, after stress or cortisone
III. Latent (chemical) diabetes (Synonym: asymptomatic diabetes)	Oral glucose tolerance abnormal; fasting blood glucose normal to mildly elevated
IV. Overt diabetes (Synonym: clinical diabetes)	Hyperglycemia and glucosuria in the absence of treatment

3. To establish the diagnosis in patients with conditions often associated with diabetes.
4. To detect gestational diabetes, and
5. To rule out diabetes and thus avoid the psychic trauma of patients being mislabeled.

The National Diabetes Data Group (USA) has stated that the diagnosis of diabetes in nonpregnant adults should be restricted:

a. To levels of fasting venous plasma glucose greater than 7.8 mmol/L (140 mg/dL) on more than one occasion and,
b. To those who, if the fasting glucose levels are less than 7.8 mmol/L, exhibit at any time during the GTT a value greater than 11.1 mmol/L (200 mg/dL).

It is seldom necessary to do a glucose tolerance test when the 2 h pp plasma glucose is clearly elevated (>8.0 mmol/L). In pregnancy a 100-g glucose load should be administered between the 24th and 28th week and a 2-h pp plasma glucose determined. Here an upper reference value of 6.6 mmol/L (120 mg/dL) is recommended.

Genetic Aspects

Diabetes occurs more frequently in both monozygotic than in both dizygotic twins. This is probably the best evidence for a *predisposition* to this disease. Geneticists prefer to use the vague term *'multifactorial inheritance'* (inherited by more than a single pair of genes and not in a strict statistical pattern) for the nature of genetic transmission in diabetes. Attempts to define a diabetic genotype have been inconclusive; a gene (or genes) for structurally altered insulin, and insulin release that is unresponsive to glucose have been considered.

For *genetic counseling,* empiric risk figures derived from family histories allow for an approximation of the probability of developing diabetes. Examples are given in Table 16–8. If a relative develops diabetes when under the age of 20, or if there is more than one relative afflicted, the risk factor is doubled (except for the age group over 60). Also, a positive family history is more common in patients with juvenile-onset diabetes than in those with maturity-onset disease. These facts should be taken into consideration when defining populations at risk for screening purposes.

TABLE 16–8. Risk of Developing Diabetes when One Parent, Sibling, or Child is Afflicted

Age of Nondiabetic (Years)	Risk (%)
0–19	<1
20–39	1
40–59	3
over 60	10

One of the factors determining susceptibility to insulin-dependent diabetes is the presence of a gene, or genes, closely linked to the human leukocyte antigen (HLA). HLA–DR4 seems to confer susceptibility to IDD in all ethnic groups, while HLA–DR3 demonstrates the association only in Caucasians and American blacks but not in African blacks and the Japanese. In these, however, HLA–DR8 is associated with IDD. HLA–DR2 has a negative association with IDD in all groups and may confer resistance to IDD. This has been attributed to 'linkage disequilibrium': that is, the two loci being very close together, by chance certain combinations of HLA and 'diabetogenic genes' occur at a higher-than-expected frequency in the population. The mode of action of such genes remains hypothetical; there is now speculation about interactions with the immune system. On the other hand, the HLA patterns in NIDD are no different from those in a control population.

Others believe that the pattern of inheritance in diabetes is mendelian *recessive* with incomplete penetration of the trait. One of the arguments for this view is based on the incidence of clinical diabetes as the offspring of diabetics approach 95 years of age. A logical counter-argument is the gradual decrease of glucose tolerance after 50 years of age. Most Caucasians who manage to escape strokes, coronaries, accidents, and malignant growths will finally end up with some degree of glucose intolerance.

In summary, the following 'high-risk' factors should be considered (in decreasing order):

a. Identical twin of a clinical diabetic
b. Offspring of two diabetic parents
c. Offspring of one diabetic parent with a positive family history for the other

In these individuals, tests for the detection of glucose intolerance will often show abnormalities, usually long before any clinical manifestation is detected. It remains to be seen whether *early detection* of this susceptibility, by the laboratory methods available, gives results beneficial to the patient. Too much zeal in predicting diabetes may be of more harm than help to the individual. There is little we can do about the serious vascular complications at present and the psychologic stress of being diagnosed as diabetic is considerable.

The *primary* inherited *defect* in diabetes is unknown, as well as the relative importance of genetic versus environmental factors. The clinical biochemist can assist in predicting future clinical diabetes with considerable accuracy, but eugenic elimination of diabetes, even if desirable, appears impossible. Diabetes has been with mankind for a long time and will remain unless genetic engineering can reach science-fiction efficiency.

Diabetes Mellitus and Autoimmunity

Circumstantial evidence is accumulating that insulin-dependent diabetes is associated with disorders of the *immune mechanism.* Serologic work has revealed a high prevalence of humoral, organ-specific autoantibodies in this disease; coexistent antibodies to thyroid and gastric parietal-cell cytoplasm, to intrinsic factor, and to adrenal cortical tissue have been found. Cell-mediated immunity to pancreatic elements other than β cells has been demonstrated by the leukocyte-migration technique.

Islet-cell antibodies (ICAbs) are now known to be common in newly diagnosed juvenile diabetics; they can be detected in only 0.5% of the general population and may precede the onset of diabetes by several years. Whether the ICAbs are causal in the genesis of diabetes or whether they should be regarded as a marker for the condition remains speculative. They have been found in over 50% of insulin-dependent and in 5% of non-insulin-dependent diabetic subjects. ICAbs were present in 85% of these patients immediately after the onset of symptoms and became less common as the duration of the disease increased. The antibodies were equally common in both sexes, and were directed against cytoplasmic components of islet cells, not against insulin itself. In multiple endocrinopathy syndromes the detection of ICAbs could predict the future occurrence of diabetes.

Insulin Antibodies. Even homologous insulin, when injected subcutaneously, produces an anti-insulin-antibody response that is virtually identical to that produced by highly purified pork insulin. The clinical importance of these antibodies is unclear, but it raises questions about other manifestations of insulin therapy such as insulin allergy and insulin lipodystrophy.

To mention the *cellular* arm of the immune response, phytohemagglutinin stimulation of lymphocytes is depressed in IDD. This suggests a biochemical abnormality in the T-lymphocytes.

Lipid Metabolism

One cannot ignore lipid metabolism in describing the complications of diabetes. The discovery of insulin has undoubtedly increased the average life span of diabetics, but life expectancy is still shorter than for the general population. The mortality rate for diabetics decreases from ten to four times normal from ages 15 to 50. Death from hyperglycemia and ketosis, infections, or gangrene has become relatively rare. Diabetics most commonly die of vascular diseases and their complications, thus joining the ranks of those with the prime cause of mortality in the Western population.

Diabetes is associated with disease of the small and large blood vessels. In morphology, diabetic atherosclerosis does not differ from the findings in nondiabetics. Microangiopathy, however, is probably specific to diabetes. ***Vascular disease*** appears to occur at a younger age in diabetics, but possibly because they tend to bring their problems to medical attention at an earlier date. The cause of the increased incidence of vascular disease in diabetes is unknown. Vascular disease may be the sole presenting feature, and there is no direct relationship between this complication and the severity of diabetes. About 50% of patients with diabetes develop the long-term complication of diabetic retinopathy.

One of the conclusions of the carefully controlled Framingham study was that hyperglycemia is an independent *risk factor* for atherosclerosis, at least as important as hypertension, hypercholesterolemia, or obesity. The effect of diabetes on lipid metabolism is difficult to understand. Insulin promotes the synthesis of VLDL–triglyceride (TG) in the liver, and the storage of TG in adipocytes. It stimulates lipoprotein lipase (and thus the peripheral clearance of TG) and hepatic HMG · CoA reductase (and thus the synthesis of cholesterol).

Abnormalities of ***serum lipids*** are statistically related to atherosclerotic vascular disease in diabetics (Table 16–9) but there is no explanation for the increased frequency of atherosclerosis in these patients. The low incidence of atherosclerosis found in diabetics in some parts of the world (e.g., Japan) points to the importance of genetic and environmental factors. Up to 30% of diabetics have elevated plasma *triglycerides,* either as the result of an increase in the splanchnic production of very-low-density lipoproteins or a defect in their removal from the circulation. In the untreated diabetic, there is a marked increase in the two major precursors of triglyceride synthesis—plasma free fatty acids and glucose. Triglyceride production from glucose is apparently the lesser in importance since hepatic lipogenesis is minimal in insulin deficiency.

A continuing excessive intake of dietary fat is frequent in diabetics with polyphagia and results in a rising *chylomicronemia.* Insulin is essential for the clearance of triglyceride-rich lipoproteins; irregularities in insulin dosage and resorption, dietary excesses, and the presence of insulin antibodies contribute to widely and rapidly fluctuating plasma triglyceride levels in diabetics. Single determinations are therefore of very limited value.

The electrophoretic *lipoprotein* pattern (see Chap. 20) is nonspecific. In addition to chylomicronemia, a type I or V pattern is frequently found. A type V pattern is characteristic of a profound dissociation between the rate of entry of triglycerides (both exogenous and endogenous in origin) into the circulation and the ability of the patient to clear glycerides. During treatment, or in ketoacidosis, a type IV pattern is often observed.

Diabetes in Children

Diabetes is less common in children than in adults and is very unusual indeed in babies. It occurs equally in boys and girls, and the commonest age at onset is about 10 years for girls and 13 years for boys. It is estimated that 4% of the diabetic population are children. Data on prevalence vary from 1 in 1200 to 1 in 6000 for children under age 17.

The *onset* of diabetes in children is usually more abrupt than in adults, although the diagnosis is often delayed. Polyuria is severe enough to cause enuresis; dehydration results from the combination of vomiting and osmotic diuresis due to hyperglycemia. The consequences of this can be catastrophic. Considerable effort is being expended on devising methods for the *diagnosis* of the various stages of diabetes in childhood; the glucose tolerance test may be used. There is a consensus that the Fajans–Conn criteria are most useful but the Rosenbloom system should also be noted (Table 16–10).

Almost all children who develop diabetes need *insulin.* When the diabetes is first diagnosed, large amounts may be necessary to control the metabolic disorder. The type of insulin to be used and the frequency of injection are matters which must be

TABLE 16–9. Lipid Levels in Diabetics (Age 30–59 Years) in Relation to Atherosclerosis

	Control	Diabetic without Atherosclerosis	Diabetic with Atherosclerosis
Triglyceride			
>1.7 mmol/L (>150 mg/dL)	19%	15%	50%
Mean triglyceride			
mmol/L	1.2	1.16	1.86
mg/dL	106	103	165
Cholesterol			
>6.5 mmol/L (>250 mg/dL)	10%	8%	23%
Mean cholesterol			
mmol/L	5.5	5.6	6.0
mg/dL	202	205	229

TABLE 16–10. Criteria for Interpretation of Oral Glucose Tolerance Curves: Diagnostic Reference Values in Children

	Fajans–Conn		Rosenbloom	
	Venous Plasma Glucose		Venous Plasma Glucose	
Time	*mg/dL*	*mmol/L*	*mg/dL*	*mmol/L*
0 min			100	5.6
60 min	185	10.3	160	8.9
90 min	160	8.9		
120 min	140	7.8	140	7.8
180 min			130	7.2
240 min			116	6.5

In the Fajans–Conn system, glucose intolerance is present if all of the three values are exceeded; in the Rosenbloom system, diabetes is present if two values after fasting exceed the above criteria, or if one value after 30 min is greater than 11.1 mmol/L (200 mg/dL).

decided for each individual. Chronic administration of excessive insulin may produce the *brittle* diabetic. This term refers to the difficulty in establishing and maintaining a relatively normal state of metabolic activity and a tendency toward development of ketoacidosis or hypoglycemia. As many as 20% of insulin-treated diabetic children may fall into this category. Adequate consideration should be given to the possibility that it is sometimes the patient and not the diabetes that is unstable.

The onset of *puberty* may be a particularly trying time for the diabetic child, because of the changing metabolic demands. The variability in the rate of energy expenditure is far greater in the child than in the adult. A vicious cycle of hypoglycemia, either symptomatic or asymptomatic, followed by reactive hyperglycemia and ketonuria becomes established and insulin insensitivity often develops.

The *objectives of management* of diabetes in children, based on laboratory data, are as follows:

1. Avoidance of ketoacidosis and prevention of gross abnormalities of serum lipid concentration
2. The maintenance of blood-glucose concentration within reference limits; however, achievement of normoglycemia and aglucosuria greatly increases the risk of serious hypoglycemia
3. Control of urinary frequency and nocturia by preventing excessive urinary glucose wastage. This is achieved by objective 2

The younger the child and the shorter the duration of the disease, the more variable will be the concentration of glucose in blood and urine. Because of this it is essential to treat the child, not the laboratory results.

Prognosis

The overall death rate of diabetics is well above that of controls for an age-matched population, mortality ratios for teenage patients approaching 10 and decreasing to about 2 at age 60. This excess mortality can be related to a higher prevalence of hypertension, obesity, coronary disease, and kidney disease in diabetics. It is highest in 'severe' diabetics with high insulin requirements, earlier age of onset, and recurrent glucosuria.

It would appear that little can be done to *prevent complications.* Clinicians generally agree that hyperglycemia must be controlled by diet and treatment, because it increases the proneness to infection and the pain of diabetic neuritis. Cataract formation and neuropathy may also be linked to abnormal glucose metabolism. The incidence of retinopathy and proteinuria shows a positive correlation with the duration of the diabetes.

With individualized knowledge of the life-style

and eating habits of the patient, and with the intelligent use of the laboratory to monitor glucose, lipids, and more recent indices (such as glycosylated hemoglobin [HbA1c]), every effort should be made to keep the diabetic adjusted, comfortable, and in a reasonably balanced metabolic state. For many diabetics, the achievement of physiologic levels of plasma glucose has become possible only with the advent of home glucose monitoring. Glucose analysers have improved in reliability, portability, ease and speed of testing, and in safeguards against unskilled use. It is still believed that careful control of the blood sugar can mitigate the early development of complications.

TABLE 16–11. Symptoms of Ketoacidosis

Symptoms	Remarks
Vomiting	In 75% of patients, all ages
Abdominal pain	In 40% of patients; can mimic surgical problems
Tachypnea	Precedes Kussmaul overbreathing (appears when serum HCO_3^- is reduced to 10 mmol/L)
Circulatory collapse	In 60% of patients, from dehydration and salt depletion
Weakness, paralysis, paresthesia	From changes in K^+ levels; may also develop during treatment

Diabetic Ketoacidosis

Diabetic ketoacidosis (DKA) is an acute, life-threatening medical *emergency,* requiring rapid diagnosis and institution of proper treatment. Since the advent of insulin therapy the importance of DKA as a cause of death has declined progressively; but it continues to be characterized by a mortality rate of 5% to 15%. DKA develops as a consequence of an absolute or relative deficiency of insulin, which may result from a failure of endogenous insulin secretion, from inadequate administration of exogenous insulin, or from an increased requirement for insulin. A variety of factors can contribute to the problem: Simple omission of the usual dose of insulin, infections, insulin resistance, emotional and physical trauma, pregnancy, surgical interventions, thyroid crises, and many similar stresses. Minimal residual function of the β-cells protects against DKA.

The *clinical symptoms* of DKA are summarized in Table 16–11. Decreased tissue utilization of glucose and *glucose overproduction* by the liver characterize the insulin-deficient state. Glucose is released from the liver at a rate of 150 mg/min to 200 mg/min (2 mg/kg/min) in normal fasting subjects; in DKA, the rate of glucose production may be increased to 400 mg/min to 600 mg/min. Hyperglycemia does not develop from a failure to metabolize ingested carbohydrate; it is a result of the overproduction of glucose from endogenous precursors.

Protein-derived amino acids are the only precursors readily available for *de novo* glucose synthesis; thus, the *catabolism* of body protein stores and the development of a negative nitrogen balance is implicit. A threat to the patient, however, is the osmotic diuresis produced by the glucose and urea, which leads to urinary losses of water and sodium, as well as to the development of serum *hyperosmolality.* Coincident with the increase in blood glucose, *ketone bodies* (acetoacetic acid and β-hydroxybutyric acid) accumulate progessively in the blood, reaching levels of 8 mmol/L to 15 mmol/L.

The metabolic site within the liver that is responsible for this activation of ***ketogenesis*** may reside in the *carnitine acyltransferase* reaction. This enzyme catalyzes the transfer of long-chain fatty acids across the mitochondrial membrane. Carnitine is an essential factor in fatty acid oxidation, and hence in ketoacid production. Because β-oxidation of fatty acids occurs solely within the mitochondria, this accelerated transfer leads to augmented acetyl CoA production. The increase in AcCoA exceeds the capacity for its oxidation via the Krebs cycle, resulting in the condensation of AcCoA molecules to form ketone acid precursors. The *mechanism* whereby insulin deficiency leads to increased carnitine acyltransferase activity is believed to involve augmented transfer of carnitine from extrahepatic sites to the liver. Abnormal plasma levels of free carnitine and acylcarnitines in diabetics return to normal promptly, in parallel with β-hydroxybutyrate, when insulin is administered.

Hyperketonemia in diabetes (even in patients with a mild lack of insulin) is also a consequence of decreased utilization of β-hydroxybutyrate and acetoacetate by muscle tissue. Contributing to the problem is the fact that a lack of insulin results in augmented lipolysis in adipose tissue, which leads to increased delivery of free fatty acids to the liver.

The ***diagnosis*** of DKA is firmly established by examining three biochemical indicators: (1) urinary glucose and ketones, (2) arterial blood pH and blood gases, and (3) serum ketones. All of the following *findings* must be present:

1. 4+ glucose and a strong nitroprusside reaction for ketones in the urine
2. A pH below 7.3 and pCO_2 of <40 mmHg in arterial blood
3. A positive nitroprusside reaction in undiluted serum

These measurements permit a rapid diagnosis before the laboratory reports of blood glucose and serum bicarbonate determinations become available. The nitroprusside reaction does not detect β-hydroxybutyrate and, in situations where DKA is accompanied by lactic acidosis, a weakly positive reaction may not reflect the true magnitude of ketonemia. An accumulation of ketone acids sufficient to cause metabolic acidosis may be seen in the *absence* of diabetes in poorly nourished alcoholic patients after repeated bouts of vomiting. More than 60% of patients with DKA demonstrate abnormal increases in salivary and serum amylase, and this increase is often accompanied by upper-abdominal pain suggestive of pancreatitis.

Despite efficient treatment in intensive care units, with careful *monitoring* by the laboratory and other special services, patients still die from ketoacidosis. Clinically, the *duration* and *degree* of acidosis are the main ***prognostic*** indicators, taking into account accompanying complications such as severe infections, vascular accidents, and diabetic glomerulosclerosis. By definition, full ketotic coma occurs with a blood sugar exceeding 45 mmol/L (800 mg/dL) and a serum ketone reaction that is positive at a dilution of 1:32. From subclinical mild ketoacidosis to this full-blown picture, the development may be slow or fulminant and the clinical symptoms insidious or dramatic. In the presence of chronic obstructive lung disease, *respiratory* acidemia may occur and obscure the acid-base disturbance resulting from ketosis. Rare cases of *ketoalkalosis* have also been reported, resulting from gastrointestinal fluid losses and alkali ingestion. The rational and efficient treatment of these conditions depends almost entirely on a rapid and knowledgeable response to information obtained from laboratory investigations.

Nonketotic Hyperglycemic Hyperosmolar Coma (NKHHC)

The hyperosmolar diabetic state, one of the acute complications of diabetes mellitus, is characterized by the *absence* of ketoacidosis and *extremely high* plasma glucose levels resulting in hyperosmolarity. NKHHC is considered an independent disease entity, mainly restricted to the infirm, the neglected, and undiagnosed diabetics. Such coma may be one end of a clinical spectrum for patients who suffer uncontrolled osmotic diuresis. Mortality rates have ranged from 40% to 60% in critically ill, dehydrated diabetic patients in the older age group. The recent improvement in prognosis, with a fall in the mortality rate to about 15% to 20%, can probably be attributed to a prompt recognition of the syndrome based on laboratory results.

The criteria for making a ***diagnosis*** of NKHHC are arbitrary. A blood glucose >45 mmol/L (800 mg/dL), or a serum osmolality >350 mmol/kg, is strong evidence for the presence of the hyperosmolar syndrome. An *approximation* of the osmolality can be obtained from the serum sodium, plasma glucose, and blood urea nitrogen by using the following formula:

$$\text{Serum osmolality (mosm/kg } H_2O) \simeq \underset{\text{mmol/L}}{2 \times Na^+} + \underset{\text{mmol/L}\ \left(=\frac{\text{mg/dL}}{18}\right)}{\text{plasma glucose}} + \underset{\text{mmol/L}\ \left(=\frac{\text{mg/dL}}{2.8}\right)}{\text{BUN}}$$

Azotemia with or without an elevated serum creatinine is common; a modest, metabolic *acidosis* is occasionally observed. Serum *sodium* may range from 120 to 160 mmol/L; concentrations as high as 188 mmol/L have been reported. Total body *potassium depletion* is an invariable concomitant caused by K^+ loss during diuresis, although most patients are normokalemic. Serum bicarbonate and chloride concentrations are likely to be normal with little or no *anion gap*. The plasma concentration of *free fatty acids* is found to be considerably less than is observed in diabetic ketoacidosis.

Our understanding of the *pathogenesis* of NKHHC is far from complete; controlled studies are ethically unfeasible because of the need for prompt treatment of this life-threatening syndrome. It has been suggested that an adequate insulin supply to the liver (which inhibits hepatic ketone synthesis) combined with an inadequate insulin supply to peripheral tissues is a causative factor. The *management* of NKHHC includes:

a. Recognition (from biochemical data)
b. Prompt correction of the hypovolemia by fluid replacement
c. Insulin administration

The ***differential diagnosis*** of ketoacidosis and NKHHC involves several considerations:

1. Acidosis also occurs—in both the diabetic and nondiabetic alike—in salicylate poisoning, uremia, loss of base (diarrhea, duodenal drainage), methanol poisoning, lactic acidosis, dehydration, and alcoholic ketoacidosis.
2. Coma may be associated with stress hyperglycemia and variable glucosuria in CNS disease, stroke, meningitis, and head trauma.
3. Hypotension and stress hyperglycemia may be associated with:
   a. Myocardial infarction
   b. Gram-negative septicemia
   c. Acute abdominal emergency or gastrointestinal bleeding
   d. Excessive sedation

LACTIC ACIDOSIS

Lactic acidosis is present when serum lactate levels are greater than 2 mmol/L (18 mg/dL) and blood pH is less than 7.35. Clinical lactic acidosis is apparent when the serum lactate levels are higher than 7.0 mmol/L (63 mg/dL). A clinical classification of the *causes* of elevated serum lactate is given in Table 16–12. The liver normally can metabolize lactate at rates far above those necessary to handle physiologic production. Hepatic insufficiency can predispose to lactic acidosis. An 'occult' form of diabetic ketoacidosis may occur concurrently with lactic acidosis.

The *clinical* picture is characterized by the sudden onset of stupor or coma with Kussmaul respiration. *Laboratory* findings include a decreased serum HCO_3^- (or total CO_2), and relatively normal values for BUN, ketones, chloride, and sodium, resulting in an 'unexplained' anion gap. Hypoglycemia may be present. Diagnostic features are a low pCO_2 and pH, lactic acid levels over 7.0 mmol/L (>63 mg/dL) and a lactate:pyruvate ratio over 10:1 (normally considerably under 10:1). *Arterial* lactate levels with an upper reference limit of 1.5 mmol/L (13.5 mg/dL) are a guide to the prognosis.

The mortality rate of untreated lactic acidosis approaches 100%, so that aggressive *therapy* is mandatory. The administration of insulin and glucose frequently produces a striking lactate fall; intravenous bicarbonate therapy is usually indicated as well. Hemodialysis may have to be considered.

TABLE 16–12. Cause of Elevated Blood Lactate Levels

Classification	Etiology	Lactate Concentration (mmol/L)
Without excess lactate	Infusion of $NaHCO_3$	up to 2.4
	Glycogen storage disease	3.3–14.4
With or without excess lactate	Hyperventilation	1.8–22
With excess lactate	Exercise	
	Epinephrine	
	Cardiovascular insufficiency	
	Shock	2–35
	Acute hypoxemia	3–4
	Leukemia	12.2–25
	Diabetes mellitus	10.2–30.8
	Phenformin	10–31
	Ethanol	up to 8

HYPOGLYCEMIA

General Considerations

Subnormal levels of blood glucose cause a symptom complex of hypoglycemia. Etiologic factors are intricate; glucose homeostasis is deranged as the result of an imbalance between processes that add glucose to the circulation (glycogenolysis, gluconeogenesis, and absorption of glucose) and processes that remove glucose from the blood (utilization by liver, brain, muscle, adipose tissue, and other tissues).

Clinical symptoms are dependent on the *rate* of decrease of blood glucose levels and are due mainly to activation of the autonomic nervous system.

Symptoms caused by this catecholamine release (sympathetic discharge, norepinephrinemia) include, in decreasing order of frequency:

Sweating
Trembling (shakiness)
Tachycardia
Nervousness and anxiety
Weakness
Hunger
Nausea and vomiting

Correction of the hypoglycemia rapidly relieves these symptoms. A decline in the blood glucose causes *inadequate delivery* of glucose to the brain. This is accompanied by a decrease in oxygen utilization by the brain, with the following order of symptoms:

Headache
Slow cerebration
Irritability
Personality changes (bizarre behavior)
Speech and gait disorders
Mental confusion
Convulsions, coma

Repeated severe hypoglycemic episodes cause permanent neurologic and mental damage; a correct diagnosis and clarification of the causes is therefore important, and this is usually dependent on laboratory information. Publicity in the press has led many people to believe (erroneously) that hypoglycemia is common and is the cause of a number of physical and psychic ills.

Classification

The current etiologic classification of hypoglycemia is clinical and is frequently trivial, because many underlying mechanisms are not fully understood.

Reactive Hypoglycemias

Type 1. *Reactive functional hypoglycemia,* which is relatively frequent in adults, is characterized by transient postprandial hypoglycemia (2 to 4 hours after a meal); the *fasting* blood sugar is normal. The majority of symptoms are due to norepinephrinemia. Excessive insulin levels in response to a normal rise in blood glucose are found only in infancy and a family history of diabetes is usually absent.

Type 2. *Late-reactive hypoglycemia* is secondary to diabetes mellitus. In addition to the aforementioned, the patient demonstrates a decreased glucose tolerance and frequently has a positive family history. Hypoglycemia occurs (on the average) 1 hour later than in type 1. The fasting blood sugar may be slightly elevated.

Type 3. *Alimentary hyperinsulinism* occurs after gastroenterostomy and subtotal gastrectomy and is associated with accelerated absorption of glucose and significant postprandial hyperglycemia. Symptoms occur soon after the ingestion of food. The oral glucose tolerance test shows an elevated peak (after 30 to 60 min), followed by hypoglycemia. Also, chronic excessive ethanol ingestion, together with inadequate food intake, often precipitates hypoglycemic episodes. The depletion of liver glycogen reserves contributes to the mechanism.

Organic Hypoglycemia

Insulinoma. A rare condition, this form of hyperinsulinism is due to pancreatic islet-cell disease. Hypoglycemic symptoms develop insidiously; in fact the patient often presents as a psychiatric problem. The islet tumors are sometimes associated with multiple endocrine adenomas of the parathyroids or other endocrine glands and with the Zollinger–Ellison syndrome. In these cases the hypoglycemic symptoms can be masked by the more pronounced effects of other types of hormone imbalance. Prolonged fasting (up to 48 hours) is most helpful as a diagnostic procedure (although rather unpopular with the patient).

Reliable confirmation of the diagnosis can be obtained by a determination of plasma insulin, or C-peptide, concentrations in conjunction with blood glucose levels.

Other Organic Causes. These include diffuse chronic liver disease, cholangitis, and hepatic cirrhosis. Gluconeogenesis is gradually impaired, glycogen availability decreases, and compensation during periods of fasting becomes inadequate. *Factitious hypoglycemia* from erroneous, iatrogenic, surreptitious, or suicidal administration of insulin or tolbutamide must always be considered a possibility.

Laboratory Tests

Fasting Plasma Glucose and Insulin Levels. A decreased plasma glucose associated with an elevated plasma insulin is an abnormal finding, seen in patients with insulinoma, idiopathic hypoglycemia of infancy, after administration of insulin, in leucine sensitivity, and in those receiving sulfonylurea.

Five-Hour Oral Glucose Tolerance Test. This test is valuable for confirming reactive functional hypoglycemia, alimentary hyperinsulinism, and late-reactive hypoglycemia in patients with early diabetes mellitus. It is not useful in detecting insulinoma.

Intravenous Tolbutamide Test for Insulinoma. At 2, 5, 15, 30, and 60 min after a dose of tolbutamide (1 g in saline intravenously), the plasma insulin is measured, and at 15, 30, 60, 90, 120, 150, and 180 min, the plasma glucose is measured. In patients with an insulinoma the response is usually 'supernormal,' with plasma insulin values exceeding 150 mU/L, although at early stages there may be a delay in the elevation of the insulin level. Insulin levels also tend to remain elevated for a longer period. It may be useful to measure proinsulin as well.

Other Tests. Leucine or glucagon can be administered intravenously to stimulate plasma insulin release. At times, these tests are valuable in distinguishing normal and abnormal insulin-release states. Subcutaneous epinephrine may be used for the same purpose.

A useful protocol for the *laboratory investigation* of hypoglycemia is summarized in Table 16–13.

INBORN ERRORS OF CARBOHYDRATE METABOLISM

Inborn errors of metabolism are the result of alterations in gene expression, leading to inadequate or distorted protein synthesis. Often it is not clear whether the primary injury rests with damage to DNA, instability of messenger RNA, or faulty translation. The metabolic pathways of carbohydrate metabolism that involve glycolysis and the Krebs cycle are so essential to life that any inborn deficiencies in the controlling enzyme systems are necessarily fatal in early embryonic life. Therefore, the known disorders affect glycogen breakdown or synthesis, sugar interconversion, or the hexose monophosphate shunt.

Glycogen Storage Diseases

The glycogen reserve in the liver of a healthy, well-nourished adult amounts to less than 100 g. If this store of glycogen were not readily available, severe hypoglycemia would develop in the intervals between food intake, with serious impairment of central nervous system function.

TABLE 16–13. Laboratory Investigation of Hypoglycemia

Hospital Day	Procedure
1, 2, 3	High-carbohydrate diet (300 g)
	Daily fasting blood sugar
	Clinitest four times a day (before meals and at bedtime)
	1500 h. (3 PM) blood sugar
4	5-h glucose tolerance test
5	Tolbutamide test
6, 7	48-h fast (only unsweetened tea and coffee)
	Daily blood sugar and immunoreactive insulin determinations

Glycogen storage diseases are characterized by an abnormally increased storage of glycogen, caused by deficiencies in normal glycogenolysis due to defective enzyme systems. Also, the chemical properties of the stored glycogen may vary and may be abnormal (as in Forbes' disease). The accumulation of glycogen may take place in a variety of tissues.

Several metabolic blocks are known to affect glycogen breakdown or synthesis; the enzymes responsible for most distinct types of storage disease are also known. The *classification* is presented in Table 16–14.

All types of glycogen storage disease are very rare, the relatively most common being von Gierke's. Clinical *symptoms* vary: Afflicted infants have an enlarged liver (and kidneys), fail to grow, and suffer from convulsions (probably related to hypoglycemic episodes). Pompe's disease is characterized by the predominant accumulation of glycogen in the heart and neuromuscular systems.

Diagnostic procedures involve the histochemical analysis of liver biopsy specimens; it is difficult to obtain data on the frequency of these disorders because of incomplete diagnostic results.

Galactosemia

Galactosemia, accompanied by galactosuria, results from an autosomal recessive inheritance of an inability to metabolize galactose. The defect occurs in the pathway by which galactose is converted to glucose, and may involve the enzyme *galactokinase,* but more commonly the enzyme *galactose-1-phosphate uridyl transferase.* There is less toxicity with the kinase defect, but cataracts are a problem. The heterozygous trait of transferase deficiency has a prevalence of about 1% and in homozygotes can be a serious disorder in infants. Toxicity results from the accumulation of galactose-1-PO_4 in the blood and tissues, and is possibly due to a metabolite, galactitol, which in the lens of the eye will cause cataracts.

The *untreated* patient, clinically normal at birth, starts to vomit and loses weight when given human or cows' milk. A test of the urine for galactose reveals galactosuria. If fed on a galactose-rich diet, the child will develop hepatomegaly, polyuria, aminoaciduria, and acetonuria. If the child survives, galactosuria and galactosemia persist; the victim is retarded both physically and mentally, and eventually develops cirrhosis of the liver and cataracts. Complete regression of symptoms occurs if the child is given a lactose-free diet early enough in the course of the disease. After 5 or 6 years many of these children are able to tolerate modest amounts of lactose in the diet, presumably because alternate pathways of galactose metabolism have developed. Women carrying the trait should avoid lactose during pregnancy.

Various theories try to explain the cause of *hypoglycemia* in galactosemic patients:

1. An insulinlike effect of galactose accumulated in the blood
2. A block in glucose-6-phosphatase activity
3. A block in phosphoglucomutase activity leading to a reduction in glucose-1,6-diphosphate (Fig. 16–3)

An inhibition of *phosphoglucomutase* could explain many of the signs and symptoms observed in galactosemia, including jaundice (from unconjugated bilirubinemia). Uridine diphosphoglucose

TABLE 16–14. Classification of Glycogen Storage Diseases

Type	Affected Enzyme	Principal Site of Glycogen Storage
I (von Gierke's)	Glucose-6-phosphatase	Liver, kidney, intestine
II (Pompe's)	Acid α1,4-glucosidase	Generalized disease
III (Forbes'; Cori's) (debrancher deficiency)	Amylo-1,6-glucosidase	Liver, muscle, leukocytes, kidney
IV (Andersen's)	Amylo-(1,4 $\rightarrow$ 1,6) transglycosylase	Liver
V (McArdle's)	Muscle phosphorylase	Muscle
VI (Hers')	Liver phosphorylase	Liver
VII (Tarui's)	Phosphofructokinase	Muscle (in Japanese)
VIII —	Phosphorylase-b kinase	Liver, leukocytes

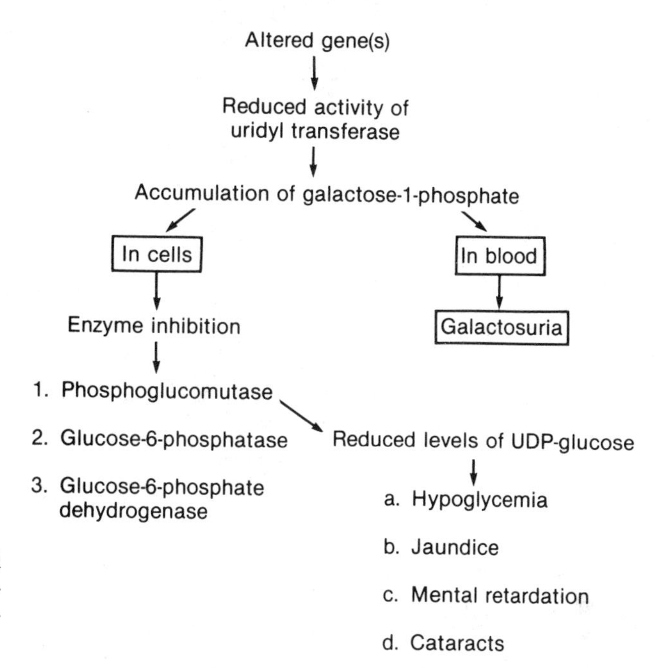

FIGURE 16–3. Pathogenesis of galactosemia and galactosuria and postulated mechanism of hypoglycemia in the galactosemic patient. UDP = uridine diphosphate.

(UDPG) is necessary for the conjugation of bilirubin to form bilirubin glucuronide. In the absence of phosphoglucomutase, the glucose-6-phosphate produced in the hexokinase reaction at the expense of glucose is no longer converted to glucose-1-phosphate and the amount of UDPG is therefore reduced. UDPG is an important precursor or factor in the biosynthesis of cerebrosides from UDP-galactose and in the biosynthesis of lens proteins. Cerebrosides contain galactose and constitute 4% to 6% of the total brain and 6% of the white matter. In the absence of UDPG, precursors of the cerebrosides are not formed, so that the mental retardation observed in galactosemia could result from an inability to synthesize basic brain components.

In addition to classic galactosemia, there is a form of transferase deficiency that seems to result from an altered regulation of the enzyme's *biosynthesis.* In this (Duarte) form of galactosemia, homozygotes have 50% and heterozygotes 75% of the normal levels of transferase activity.

Erythrocytes of galactosemic patients contain little or no galactose-1-phosphate uridyl transferase. In the past a UDPG consumption test has been used to *detect* this abnormality. At present, the formation of UDP-^{14}C-galactose from ^{14}C-galactose-1-phosphate in whole blood hemolysates is measured. Normal adults and children have enzyme activities of 1.4 to 2.8 mmol/L/hour; in heterozygotes for galactosemia the mean value is approximately half.

Lactose and Fructose Intolerance

Intolerance to dietary saccharides is caused by genetically determined enzyme deficiencies and is characterized by the excretion of the particular sugar in the urine.

Lactase is normally present in infants; its activity diminishes with age, and many adults are intolerant of milk because they have a lactase deficiency. Adult intestinal lactase levels show remarkable racial variations. In Asians, South and North American Indians, Eskimos, and Greek Cypriots, low or absent lactase activity is very common and probably normal. The prevalence of alactasia, and consequent *lactose intolerance* in adult Caucasians is approximately 6%.

A rare mendelian recessive trait is *intolerance for fructose.* Ingestion of this monosaccharide results in its accumulation in the blood; as soon as the renal threshold is reached, fructose is spilled in the urine. If the offending sugar is not removed from the diet, failure to thrive, vomiting, progressive liver damage with hepatosplenomegaly, and early death result. Profound hypoglycemia is responsible for convulsions and vomiting in infants.

Fructose appears in the urine in association with three rare conditions: essential fructosuria, hereditary fructose intolerance, and combined, familial fructose and galactose intolerance. Essential fructosuria is benign and asymptomatic; the metabolic defect is a deficiency in hepatic *fructokinase*. When fructose is ingested, it is metabolized incompletely, a significant amount remains in the circulation and is excreted in the urine. In hereditary fructose intolerance, an inborn error of metabolism, the primary defect is deficiency of fructose-1-phosphate *aldolase*. This leads to accumulation of fructose-1-phosphate, which is thought to inhibit hepatic glucose production and release and thus results in hypoglycemia.

A reduction in plasma inorganic phosphate (P_i) and hypoglycemia are both found with clinically evident fructosuria. The drop in P_i has been explained by postulating that fructose is readily converted to fructose-1-phosphate. The inhibition of phosphoglucomutase by fructose-1-phosphate is a possible mechanism for the development of hypoglycemia.

An explanation for the accumulation of fructose-1-phosphate in fructosuria is that the fructoaldolase in these patients is different from that found in normal individuals. In the *fetal* liver, fructoaldolase has an affinity for fructose-1,6-diphosphate that is two to five times greater than in the adult. It is, thus, possible that in fructosuria the fetal enzyme persists, leading to a preferential splitting of fructose-1,6-diphosphate and accumulation of fructose-1-phosphate.

There are several colorimetric methods for the *determination* of fructose in plasma and urine; all of them lack specificity. The anthrone method appears to be both simple and quick; glucose (at levels up to 11 mmol/L) scarcely influences the determinations. In alkaline urine, a positive finding of fructose must be evaluated with caution; unless the sample is fresh, chemical inversion may cause fructose formation from glucose.

Pentosuria

Pentosuria is an innocuous mendelian recessive trait, encountered more often than fructosuria but still relatively rare. The deficient enzyme is *L-xylulose reductase* (xylitol:NADP oxidoreductase). The pentose found in the urine is not identical with any known pentose contained in foods; it is L-xylulose (L-xyloketose). The condition is clinically symptomless and the urinary pentose does not react in dipstick or fermentation tests. It is determined by Bial's reaction based on conversion of the sugar to furfurol, which reacts with orcinol to form green compounds.

Physiologically, traces of pentoses may be present in the urine after ingestion of large amounts of certain fruits. Adults normally excrete 2 to 5 mg/kg/day of pentoses on a fruit-free diet. A variety of stresses (fever; drugs such as morphine, cortisone, and thyroid hormone) increase the excretion of arabinose and xylose. Pentosurics excrete 2 g to 4 g of L-xylulose/day.

SUGGESTED READING

CRAIG O: Childhood Diabetes and its Management, 2nd ed. London, Butterworths, 1981

ESCHWEGE E (ed): Advances in Diabetes Epidemiology. Amsterdam, Elsevier Biomedical Press, 1982

GRAVE GD (ed): Early Detection of Potential Diabetics. New York, Raven Press, 1979

MARBLE A, KRALL LP, BRADLEY RF, CHRISTLIEB AR, SOELDNER JS: Joslin's Diabetes Mellitus, 12th ed. Philadelphia, Lea & Febiger, 1985

MARTIN JM, EHRLICH RM, HOLLAND FJ: Etiology and Pathogenesis of Insulin-Dependent Diabetes Mellitus. New York, Raven Press, 1981

MELCHIONDA N, HORWITZ DL, SCHADE DS: Recent Advances in Obesity and Diabetes Research. Serono Symposia Publications, Vol 8. New York, Raven Press, 1984

WATTS NB, KEFFER JH: Practical Endocrine Diagnosis. Philadelphia, Lea & Febiger, 1982

17

Disorders of Calcium, Magnesium, and Bone Metabolism

Alan Pollard / Kenneth P. H. Pritzker / Marc D. Grynpas

Calcium, magnesium, and *phosphorus* are elements with important structural and functional roles. Structurally they contribute to hydroxyapatite, the crystalline mineral of bones and teeth, and functionally their ions play many roles both inside and outside the cell. Functions of Ca^{2+}, Mg^{2+}, and inorganic P (Pi, $H_2PO_4^-$, HPO_4^{2-}) include: cofactors for enzymes, including some clotting factors; control of neural transmission and muscular contraction; cell adhesiveness; hormone action (2nd messenger) and secretion. Because so many vital functions depend on these elements, it is not surprising that their concentrations are tightly controlled by complex mechanisms and that disturbances of this control lead to disease manifestations in many systems.

Ca^{2+} and Mg^{2+} in serum consist of three components: ionized, complexed, and protein bound. The complexed and protein-bound components act as a reservoir to maintain the equilibrium of the ionized fraction. The bound components may also have important roles in the configuration and function of the associated proteins. The mineral phase not only imparts rigidity to bones and teeth, but also, in bone, is a reservoir for the extracellular ions. The serum concentration, body stores, and distribution of Ca, Mg, and Pi are shown in Figures 17–1 and 17–2.

Calcium Metabolism

The Ca^{2+} (ionized Ca) concentration in extracellular fluid (ECF) is tightly regulated. The principal players in the control system are three organs: *bone, gut,* and *kidney* and three hormones: *parathyroid hormone* (PTH), *vitamin D* (1,25-dihydroxycalciferol, $1,25[OH]_2D$), and *calcitonin.* Other factors include exchange with the complexed and protein bound fractions and the influence of pH, thyroxin, cortisol, and growth hormone.

The *intestinal absorption* of Ca (from the diet and gastrointestinal secretions) is influenced by the other intestinal contents (Ca bound by phosphate, phytic acid, free fatty acids, etc.) and by transport mechanisms in the mucosa, which are stimulated by $1,25(OH)_2D$ and inhibited by cortisol. The average net absorption on a 25-mmol (1000 mg) intake is about 5 mmol. Many diets are, relatively or absolutely, deficient in Ca.

Renal excretion normally balances intestinal absorption, less minor losses through the skin. Small changes in renal handling have major effects, because there are very high fluxes across the neph-

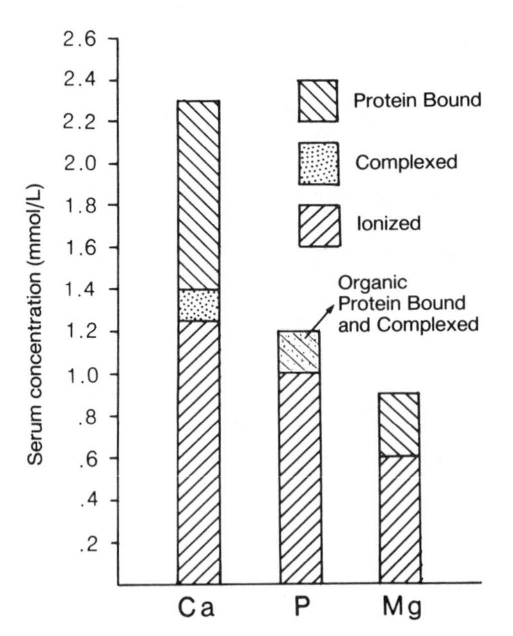

FIGURE 17–1. Normal mean concentrations and states of serum calcium (Ca), phosphate (P = Pi), and magnesium (Mg).

ron (500 mmol filtered, 495 mmol reabsorbed). *Tubular reabsorption* of Ca is complex; passive proximal reabsorption parallels sodium (Na); PTH stimulates active reabsorption down the tubule; loop diuretics (e.g., furosemide) inhibit reabsorption, and thiazide diuretics stimulate it.

Ca exchange with bone consists of accretion and resorption, two processes that are usually tightly coupled in the *bone-modeling unit.* During growth there is a net gain of bone mineral. In adults, accretion and resorption are roughly in balance, with Ca exchange very active during pregnancy and lactation. As aging ensues, the balance is tipped towards loss of bone mineral, increasing dramatically in women after the menopause.

Intracellular Ca^{2+} is involved in the control of many functions including excitation (nerve conduction, muscle contraction), hormone response, secretion, and adhesiveness. Specific *Ca-binding proteins* are involved (calmodulin, calsequestrene). Ca is transported across the cell membrane by specific transport systems that are inhibited by a class of drugs called *calcium channel blockers.* These channel blockers have a growing importance in the therapy of cardiac arrhythmias, angina pectoris, hypertension, and Raynaud's phenomenon.

Magnesium Metabolism

Unlike calcium, Mg^{2+} is predominantly an intracellular ion and is present in adequate amounts in most diets. It is absorbed with minimal regulation and stored within bone mineral. The serum concentration is regulated mainly by renal excretion. Tubular reabsorption is enhanced by low serum Mg concentrations but inhibited by diuretics, aldosterone, and hypercalcemia. The kidney can normally accommodate to an increased Mg load, but *hypermagnesemia* can ensue with renal failure. Because there is a greater fractional excretion of Mg than Ca, less Mg than Ca is stored in bone mineral, and because serum Mg is much less closely regulated than Ca, *depletion* of extracellular Mg leading to *hypomagnesemia* can occur readily.

Phosphorus (P) and Phosphate (Pi)

Phosphorus (P) is an element essential to energy storage and transfer and a constituent of substances such as nucleic acids, ATP, phospholipids, cell membranes, and skeletal mineral. Present in adequate amounts in normal diets, about 80% of P is absorbed, stimulated by $1{,}25(OH)_2D$. *Intraluminal binding* of phosphate, for example, by aluminum hydroxide, can reduce absorption substantially. Major regulation of serum Pi is renal. PTH decreases Pi reabsorption, whereas growth hormone, thyroxin, or a low serum Pi promote it. There is a maximum rate of reabsorption (threshold) above which increases in filtered load of Pi are excreted. *Pi retention* occurs when glomerular filtration rate falls, as in *renal failure.* An important aspect of Pi metabolism is its ability to shift rapidly from extracellular to intracellular compartments, stimulated particularly by glucose uptake, hyperalimentation, or alkalosis.

Parathyroid Hormone (Parathormone; PTH)

PTH, secreted by the four parathyroid glands, is an 84 amino acid single-chain polypeptide. Its main sites of action are the renal tubule and the bone-modeling unit. PTH receptors invoke *cAMP* as a second messenger. Specific PTH receptors are also found on many other cells, but their significance is not yet clear. PTH is cleaved rapidly in the circulation into an active 34 amino acid N-

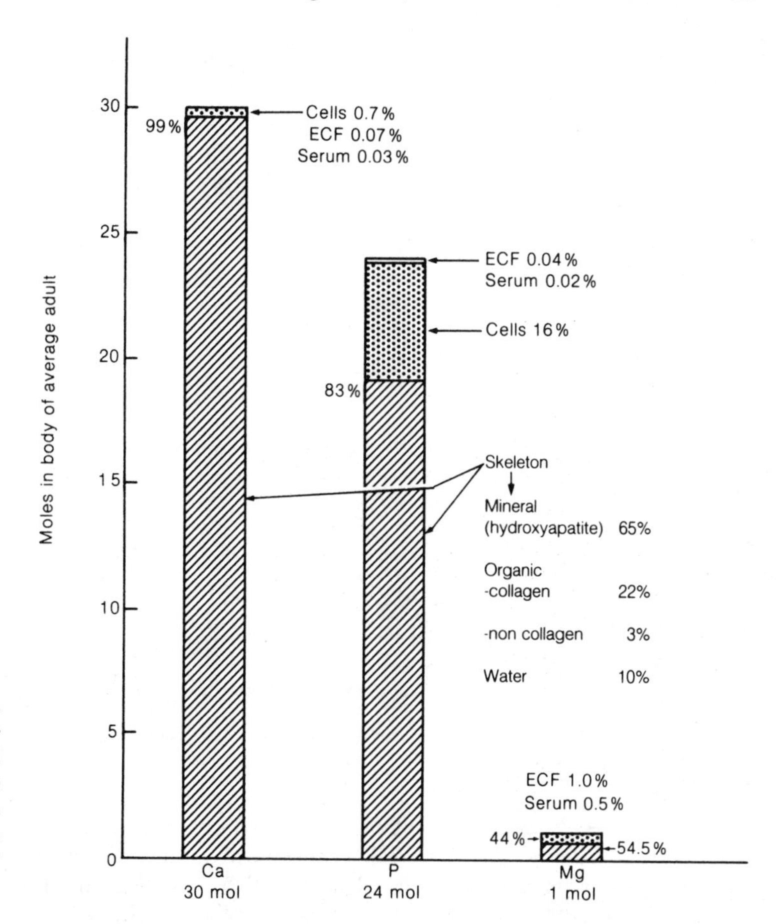

FIGURE 17-2. Distribution of Ca, P, and Mg in the different body compartments of an average adult. Total Ca = 30 moles (1200 g); Mg = 1 mole (24 g); P = 24 moles (700 g). Also shown—the mineral, organic, and water content (as %) of skeletal bone. ECF = extracellular fluid.

terminal fragment and other inactive fragments. It is not clear whether the whole molecule, the N-terminal fragment, or both are responsible for its biologic activity. The main stimulus to PTH secretion is a low serum Ca^{2+}. Parathyroid cells possess surface receptors for several different hormones which either stimulate (β-adrenergic, dopamine, histamine, secretin, and prostaglandin E1) or inhibit (α-adrenergic, prostaglandin F2, and Ca^{2+}) PTH secretion. All appear to act by modulating cAMP levels in the parathyroid cell.

Vitamin D

There are two main sources of vitamin D, the naturally occurring form from animal sources (D3) and a semisynthetic form (D2) used as a food additive. D1, the first calciferol isolated, proved to be inactive. Vitamin D2 (ergocalciferol) is produced by ultraviolet irradiation of ergosterol from certain plants. Natural D3 (cholecalciferol) occurs in fish oils, eggs, liver, and butter. D2 and D3 differ only in the side chain and are probably biologically interchangeable. Both are absorbed from the diet, and D3 is also produced in the skin from UV irradiation of 7-dehydrocholesterol. Calciferols are stored in fatty tissues and are biologically inactive until *hydroxylated* (Fig. 17–3). Both forms undergo 25-hydroxylation in the liver and the resulting *25(OH)D* is normally the main circulating form of vitamin D. It is bound to a specific carrier protein, has a low level of biologic (antirachitic) activity, and a half-life of days to weeks. Subsequent 1-hydroxylation of 25(OH)D by the kidney produces *$1,25(OH)_2D$* which is very potent biologically and has a half-life of minutes to hours. The 1-hydroxylase is stimulated by PTH and by low levels

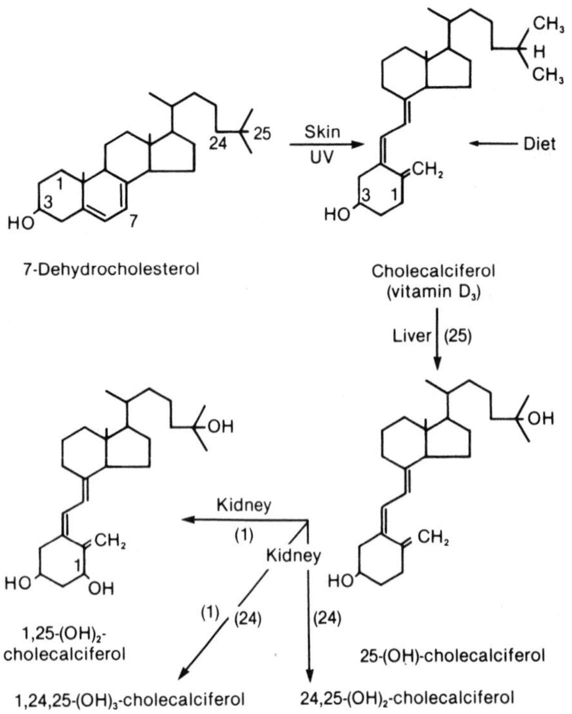

FIGURE 17–3. Sources of vitamin D_3 and metabolism of the cholecalciferols. UV = ultraviolet light. Numbers in parentheses indicate sites of hydroxylation in the liver and the kidney.

of Pi and depressed by high levels of Pi. When 1-hydroxylation is slow, *24,25(OH)$_2$D* is produced instead; other hydroxylated metabolites are also produced in small amounts. The significance of these metabolites is unclear but there is evidence suggesting that *24,25(OH)$_2$D* may be involved in bone mineralization.

The mechanism of 1,25(OH)$_2$D action is similar to that of steroid hormones. It combines with specific intracellular receptors in target tissues, is transported to the nucleus, and there regulates transcription leading to protein synthesis. Its main target tissues are the small intestine and bone, but there may also be important actions on the parathyroid cell, turning off PTH secretion, and on the kidney tubule, reducing Ca clearance. 1,25(OH)$_2$D has been shown to act on certain lymphoid cells.

Serum Calcium Homeostasis

A tendency of the Ca level to *fall* stimulates homeostatic responses by PTH and vitamin D to restore the Ca level to normal. PTH acts by:

a. Increasing renal tubular Ca reabsorption
b. Decreasing renal tubular Pi reabsorption
c. Stimulating the 1-hydroxylation of 25(OH)D
d. Stimulating osteoclastic bone resorption (with 1,25[OH]$_2$D)

The 1,25(OH)$_2$D produced acts on bone to increase resorption and on the gut to increase absorption of Ca and Pi. The net effect is to add Ca and Pi to the extracellular fluid from both gut and bone, and to excrete the resulting excess Pi via the kidney. A *high* serum Ca will inhibit PTH secretion.

Calcitonin

Calcitonin (produced by thyroid C cells) pharmacologically lowers serum Ca by inhibiting release of Ca and Pi from bone. Calcitonin, however, appears to play little role in normal or abnormal serum Ca homeostasis because hypercalcemia is not seen when calcitonin is absent (e.g., after thyroidectomy), nor is hypocalcemia found in patients with calcitonin secreting tumors. Calcitonin's main role is probably to inhibit resorption of the bone mineral when PTH and vitamin D levels are high.

ANALYTICAL ASPECTS

Calcium

Total Serum Calcium. (Reference range: 2.20 to 2.60 mmol/L; 8.8 to 10.4 mg/dL). Much effort has gone into measuring total Ca precisely and accurately, because even slight abnormalities of serum calcium may signify disease. A *definitive* method, isotope dilution/mass spectrometry, is available in selected centers to calibrate reference samples. A standardized *reference* method involves atomic absorption spectroscopy. *Routine* laboratory methods, using atomic absorption, colorimetry, or complexometric titration, can achieve a precision (1 SD) of 0.02 mmol/L, and very good accuracy. To realize this accuracy, it is important to minimize preanalytical sources of variation by taking samples with the patient fasting, in a standardized posture and with minimal forearm exercise and venous stasis.

'Corrected' Total Serum Calcium. Because roughly half the total serum Ca is reversibly bound to proteins, changes in the amount of protein and in the extent of the binding will influence both the total Ca and proportion that is free. Variations in protein concentration caused by posture, venous stasis, hemodilution, hemoconcentration, and nutritional deficiency can be corrected for (approximately) by using one of many published formulae. These formulae use either the measured total protein or the measured serum albumin to calculate the correction factor. Examples are:

1. Using *total protein* (g/L, (Rapoport and Husdan's formula)

$$\text{Corrected Ca} = \text{Total Ca}/(0.6 + \text{Total Protein}/194)$$

This assumes a mean normal total protein of 77 g/L as measured by refractometer.

2. Using *albumin* (g/L)

$$\text{Corrected Ca} = \text{Total Ca} + 0.1\ (40\text{-Albumin})/6$$

These correction methods are not perfect and will be less accurate, and even misleading, if the proteins are qualitatively abnormal, as in paraproteinemias or liver disease. Nevertheless, in routine practice, it is better to employ the *corrected* value for most patients.

Ionized (Ca_i), Complexed, and Ultrafilterable Calcium. Serum Ca not bound to protein may be measured after dialysis or ultrafiltration and includes both Ca_i and complexed Ca (bound mainly to phosphate, citrate, and bicarbonate). Ionized Ca can be measured directly, using ion-specific electrodes. The 'complexed' fraction, which is the difference between the ultrafilterable and the ionized, is much less accessible and is not measured routinely.

Ca_i-electrode systems are becoming widely available and simple to operate. They employ three main approaches:

1. Anerobic measurement in blood, serum, or plasma
2. As 1. but with simultaneous pH measurement and optional correction to pH 7.4
3. Aerobic measurement in serum after equilibration to a pCO_2 of 40 mm Hg.

Each technique and electrode type gives slightly different results, thus reference values have to be established locally. A typical reference range would be 1.15 to 1.30 mmol/L (4.6 to 5.2 mg/dL). The indications for *ionized Ca* measurement include:

a. When the corrected total Ca is borderline
b. When the proteins are qualitatively abnormal
c. When abnormal amounts of complexing anions may be present

Urine Calcium (Reference ranges on normal diet):

male: <7.5 mmol/d; <300 mg/d
female: <6.2 mmol/d; <250 mg/d
on a low Ca diet: <5 mmol/d; <200 mg/d

Urine Ca is measured to detect *hypercalciuria* which is often present in patients with *renal calculi,* without other disturbances of Ca metabolism. It is useful also in monitoring patients on vitamin D therapy. Urine for Ca measurement must be acidified and mixed thoroughly before aliquoting, because many Ca salts are insoluble at alkaline pH. Most serum Ca methods are not directly applicable to urine.

Magnesium

The serum reference range is 0.70 to 1.06 mmol/L (1.7 to 2.6 mg/dL). Serum and urine Mg can be measured by flame atomic absorption or colorimetrically with dye reagents. Both methods are capable of good accuracy and precision if performed carefully. Protein binding is less extensive than with calcium. *Hemolysis* will cause a false increase in serum concentration because Mg is mainly intracellular. Ionized Mg electrodes are not yet available.

Phosphate

The serum reference range is 0.85 to 1.45 mmol/L (2.6 to 4.5 mg/dL)—higher in children. Serum and urine Pi are measured currently by a variety of colorimetric methods without a notably high standard of performance. Hemolysis raises the value artefactually. Serum Pi falls after meals as it accompanies glucose into cells.

Newer research techniques, including plasma emission spectroscopy, mass spectrometry, and ion chromatography may in the future offer improvement in the standards of measurement of multiple elements and ions.

Alkaline Phosphatase (ALP) Isoenzymes

Differentiating the bone and liver isoenzymes is not simple because there is an overlap in their electrophoretic and kinetic properties (see also Chapters 3 and 11). It is useful to first check the serum concentration of another liver enzyme such as GGT or 5NT. A combination of heat stability and electrophoresis is often used but the sensitivity is such that, without a definite rise in the total enzyme activity, clear results are seldom obtained. Isoelectric focusing is a promising new technique. Increased bone ALP isoenzyme is a marker for *osteoblastic* activity.

Parathyroid Hormone Assays

PTH assays are usually carried out by RIA; the very sensitive cytochemical bioassays are purely research procedures. RIAs for PTH have been less than satisfactory, mainly because of the heterogeneity of PTH and its breakdown products in the plasma and the poorly defined specificity of many of the antisera used. Recently, using human PTH and its fragments prepared by gene cloning, more specific assays for intact, mid-region, and N-terminal fragments have been developed and are more promising.

PTH is secreted as intact hormone (1–84 amino acids) and possibly also as a *mid-region* fragment. In the circulation, intact PTH is cleaved rapidly into an N-terminal fragment (1–34) and one or more fragments including mid-(45–64) and C-terminal (35–84). The intact molecule and N-terminal fragments both have biologic activity and very short half lives. The mid- and C-terminal fragments are not biologically active and have much longer half lives. It follows that most of the PTH immunoactivity, determined with either mid- or C-terminal specific assays, measures mainly inactive breakdown products of PTH. If renal function is impaired the situation is worse, because these fragments are normally excreted by the kidney.

The most useful assay should be N-terminal PTH, but the assay is difficult and is working at the limits of sensitivity. Currently, the best compromise between practicality and diagnostic value seems to be the use of a mid-region specific assay. Most assays are insufficiently sensitive to distinguish clearly between 'low normal' and 'low' concentrations.

PTH assays should always be interpreted in relation to the concentration of serum Ca (preferably ionized Ca) measured on the same sample. PTH values that would be normal for a normocalcemic patient may be significantly high if serum Ca is elevated. Similarly, when the Ca is low, an elevated PTH is normally expected. An 'idealized' representation is shown in Figure 17–4. The patient's renal function must be known for accurate interpretation.

Urinary Cyclic AMP

The reference range is 340 to 570 nmol/mmol creatinine (3 to 5 μmol/g creatinine), 20 to 47 nmol/L of glomerular filtrate. One of the target organs for PTH action is the kidney tubule and because this action causes cAMP to be released into the urine, urinary cAMP output is an indirect measure of PTH action. The assay reflects not only cAMP released by PTH activity but also that from

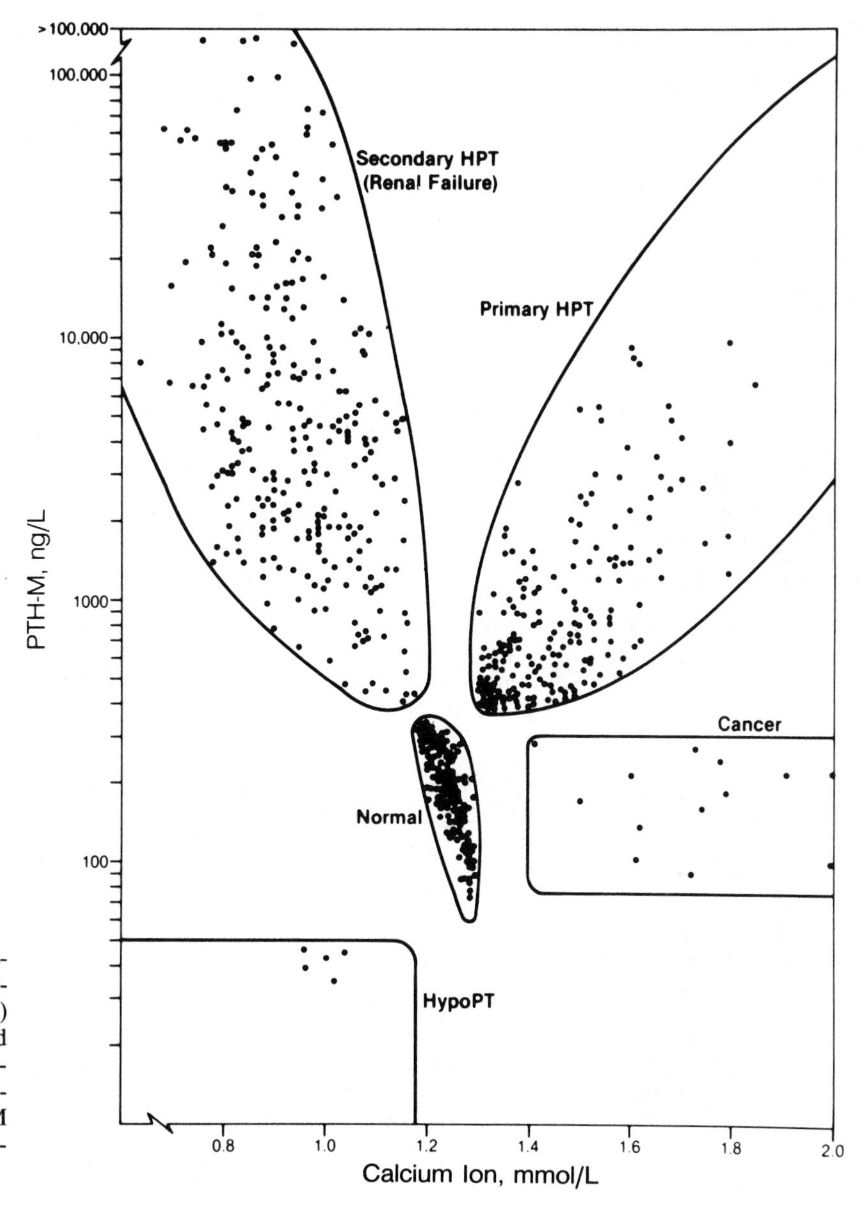

FIGURE 17–4. Idealized clustering in the relationship between parathormone (PTH) and serum Ca in health and disease. HPT = hyperparathyroidism; Hypo PT = hypoparathyroidism; PTH-M = midfragment assay of parathormone.

any other hormone (such as ADH or tumor related peptides) which acts on the kidney in a similar way. It also includes plasma cAMP excreted via glomerular filtration. The *plasma* contribution to urine cAMP can be eliminated if serum and urine cAMP and creatinine are measured simultaneously. This *nephrogenous cAMP* (NcAMP: Reference range <30 nmol/L glomerular filtrate) correlates better with PTH levels than the total urinary cAMP. It has the disadvantage that more assays are required and that serum cAMP is harder to measure than urine cAMP.

The main advantage of cAMP measurement over PTH assay is its technical simplicity and the fact that biologic rather than immunologic activity is being measured. It is used as an indicator of response to exogenous PTH in the diagnosis of *hypoparathyroidism.* The rapid response of urinary cAMP to Ca loading has been used as a test of normal parathyroid feedback control. It can also be used intraoperatively as a check on parathyroid tumor resection if a rapid assay is available.

Vitamin D

25(OH)D (Reference range in winter: 20 to 60 nmol/L) is measured by competitive protein binding. Serum concentrations correlate with dietary intake and are 30% higher in summer months because of the sunshine. $1,25(OH)_2D$ circulates in much lower concentrations (Reference range 30 to 150 pmol/L) and its assay is much more difficult. 25(OH)D measurements are used mainly to assess nutrition and malabsorption; $1,25(OH)_2D$ is mainly a research procedure.

Hydroxyproline (OH-Pr)

The reference range is 115 to 350 mmol/d. The urine output of OH-Pr, both free and peptide bound, is often measured as a marker of collagen turnover, most of which takes place as part of bone remodeling. A diet low in collagen and gelatin is taken for 2 days before the collection, because foods containing collagen will increase the urinary OH-Pr. Urinary hydroxylysine has been promoted as a more specific measurement than OH-Pr.

DISORDERS OF CALCIUM HOMEOSTASIS

HYPERCALCEMIA

Mild hypercalcemia is *common,* may persist for years, and causes few symptoms or signs. The effects, which include fatigue, malaise, nocturia, and hypertension, are subtle and nonspecific and many patients admit to no complaints. The resulting *hypercalciuria* may lead to loss of urinary concentrating power and acidification, nephrocalcinosis, and renal stones. *Severe* hypercalcemia is *rare,* often progresses, and has serious effects. It may cause anorexia and vomiting, muscular weakness, intellectual impairment, nephrogenic diabetes insipidus with resulting severe dehydration, and eventually coma and death.

Hypercalcemia is often detected accidentally by biochemical screening (1:1000 of patients screened) and should be *sought* in patients presenting with renal stone disease, bone disorders, any of the aforementioned nonspecific features, and in any patient with malignant disease. The commoner causes of hypercalcemia are listed in Table 17–1.

Hyperparathyroidism and Hypercalcemia

In *primary hyperparathyroidism* PTH is secreted in excess of the requirement to maintain calcium homeostasis. The onset is usually insidious. Patients may present with a variety of complaints that can be related to a high calcium level, including renal stones, peptic ulcer, hypertension, muscular weakness, and alteration in mood or behavior. Alternatively, patients can present with bone involvement either as a general *osteopenia* or a localized resorption presenting radiologically as a *bone cyst* (brown tumor). Most important, patients with hyperparathyroidism may be asymptomatic; recognition following detection of an elevated serum calcium during a routine laboratory assessment. *High Ca, low Pi,* and *high ALP* are typical markers for hyperparathyroidism, but are not always present. Ideally, the *laboratory* diagnosis should be made by observing permanent, or intermittent, hypercalcemia with an inappropriately

TABLE 17–1. Some Causes of Hypercalcemia

Hyperparathyroidism (primary or tertiary)
Malignant disease:
- bony secondary deposits from breast or lung tumors
- via humoral mechanisms without secondary deposits (lung and kidney tumors)
- primary tumors of bone (rare)
- myeloma and other hematologic malignancies

Vitamin D excess:
- self-administered or therapeutic
- hypersensitivity to vitamin D
  - hypercalcemia of infancy
  - granulomatous disease, e.g., Sarcoid

Hyperthyroidism
Addison's disease
Familial hypocalciuric hypercalcemia
Thiazides
Immobilization
Milk-alkali syndrome
Post renal transplant

high PTH level. The *clinical* diagnosis is made mainly by exclusion of other causes of hypercalcemia, because current PTH assays do not always come up to this standard.

Overt radiologic bone changes occur in less than 25% of patients, but *bone biopsy* is a very sensitive technique to detect the osteoid excess and the more specific active osteoclast resorption typical of the effects of PTH. In most cases, hyperparathyroidism is caused by *adenomatous hyperplasia* or *adenoma* of one gland, but sometimes hyperplasia of multiple glands is found. Hyperplasia may be familial and may be associated with adenomas in other endocrine organs, particularly the pituitary, pancreas, and adrenal medulla (MEA syndromes). The treatment is surgical. Recurrent hyperparathyroidism presents a very difficult surgical situation and neck-vein cannulation with PTH assay may be a useful preoperative guide to localization of the offending gland. Primary hyperparathyroidism must be distinguished carefully from *familial hypocalciuric hypercalcemia* (FHH) by the urinary calcium.

In chronic renal failure the parathyroids hypertrophy in response to the chronic hypocalcemia, a condition called *secondary hyperparathyroidism.* Following renal transplantation, the hypertrophied parathyroids acting on a suddenly responsive kidney frequently give rise to hypercalcemia. Prolonged secondary hyperparathyroidism can result occasionally in *tertiary hyperparathyroidism,* the chronic hypertrophy becoming autonomous or developing an adenoma.

Hypercalcemia and Malignancy

Hypercalcemia is a common feature in the course of malignant disease, occurring in over 20% of patients. Occasionally it is the presenting feature of a curable tumor but more often it arises when metastatic disease is present. Because hypercalcemia causes considerable morbidity, it is rewarding to treat even in such cases. Malignant disease can cause hypercalcemia by various mechanisms. These mechanisms include direct destruction of bone by tumor, or indirectly by a variety of humoral substances including prostaglandin E_2, osteoclast activating factor (OAF, a polypeptide secreted by some leukocytes), peptides with PTH-like activity, growth factors, or other molecules. PTH levels are generally suppressed but nephrogenous cAMP is high in some cases, responding presumably to hypercalcemic factors that simulate PTH action.

Hypercalcemia and Vitamin D Excess

Excess vitamin D activity can be caused by excessive medication, prescribed or self-administered, or by loss of regulation of the hydroxylation steps, resulting in hypersensitivity to normal amounts of vitamin D. Therapeutic use of 1-hydroxylated forms minimizes the toxicity because of their shorter half-life. Hypersensitivity is seen in some forms of *hypercalcemia of infancy,* in which the excess is seen in the 25(OH)D form, and in *sarcoidosis* in which it is seen as $1,25(OH)_2D$. In sarcoidosis, 20% of patients develop hypercalcemia. Vitamin D assays are useful here, as is therapy with cortisol, which generally reverses the effect.

Other Causes of Hypercalcemia

Hyperthyroidism sometimes causes hypercalcemia, which is related to the direct action of excess thyroid hormones which can cause bone demineralization.

Addison's disease (adrenocortical deficiency) can cause hypercalcemia although the mechanism is unclear. Lack of glucocorticoids, whose actions antagonize that of vitamin D, and dehydration, leading to retention of calcium, may be important factors.

Familial hypocalciuric hypercalcemia (FHH). This rare autosomal dominant renal tubular defect results in diminished clearance of both Ca and Mg and consequently higher serum concentrations. The disorder may present either in *infancy* (when it may be *serious*) or later in life, when it is usually mild, nonprogressive, and does not require treatment. Its main importance is that it is frequently mistaken for primary hyperparathyroidism leading to needless surgery. A very low Ca/creatinine clearance ratio distinguishes this condition from other causes of hypercalcemia.

Probably the commonest cause of mild symptomless hypercalcemia is *thiazide* therapy. The hypercalcemia is usually transient, accompanied by elevation of glucose and urate, and by depression of potassium. Thiazide therapy may occasionally unmask primary hyperparathyroidism.

Milk-alkali syndrome was associated with the treatment of peptic ulcer with large quantities of milk and absorbable alkali. The syndrome was caused by a calcium overload coupled with renal damage from chronic alkalosis. This problem should no longer be encountered, because ulcers are now treated with histamine blockers and non-absorbable alkalis.

PTH levels should be normal or suppressed in all the aforementioned causes of hypercalcemia.

Investigation of Hypercalcemia

If the hypercalcemia is only slight (<2.8 mmol/L; <11.2 mg/dL) it is first necessary to exclude artefacts (e.g., venous stasis, postprandial hypercalcemia) by repeating the measurement with the patient fasting and with minimal tourniquet. A careful history should be taken, eliciting drug or vitamin therapy and any symptoms pointing towards malignant disease, sarcoidosis, peptic ulcer, renal stones, or pancreatitis. The family history may also be very important. Physical examination and other investigations should address the same possibilities.

Many *biochemical investigations* designed specifically to rule hyperparathyroidism in or out have been devised. These investigations include measuring the renal tubular reabsorption of phosphate (reduced), the serum chloride to phosphate ratio (high), the renal phosphate response to Ca loading (diminished), the lack of response of the hypercalcemia to administered cortisone, and many others. All are now of historic interest only because they tend to be positive when the diagnosis is obvious and to let you down when most needed. Cyclic AMP measurements are not always reliable and lack specificity in distinguishing hyperparathyroidism from malignant disease. PTH assays are helpful provided that the performance of the particular assay is well established by extensive clinical correlations and that the results are always interpreted with the *serum Ca* and *renal function* in mind.

Urine Ca/creatinine clearance ratios are able to distinguish hyperparathyroidism from familial hypocalciuric hypercalcemia. As a last resort, the diagnosis of primary hyperparathyroidism is often made by exclusion of other causes, by bone biopsy, and by surgical exploration. Radiologic and bone scan signs of hyperparathyroidism are helpful but are not sensitive enough in early cases.

The diagnosis of malignant disease causing hypercalcemia is not biochemical. PTH levels should be suppressed but are often normal. Older PTH assays gave false high values in this condition because of their poor specificity.

Treatment of Hypercalcemia

Mild stable hypercalcemia requires no treatment other than the identification of its cause and its correction if possible. Severe progressive hypercalcemia requires immediate action, often before an etiologic diagnosis can be made. *Saline* infusions, to restore circulating volume and thus normalize renal sodium and calcium handling, together with *furosemide* to block Ca reabsorption, are the mainstay of treatment. *Calcitonin* is a safe and effective, if temporary, way to reduce bone resorption, and *corticosteroids* may be used to counteract excess vitamin D. The antitumor agent *mithramycin* may be used for the same purpose in the hypercalcemia of malignant disease. Oral *phosphate* supplements reduce intestinal Ca absorption and, in desperate cases, intravenous phosphate may be used. When treating severe hypercalcemia the serum levels of Mg and K often fall dangerously low and should be monitored.

HYPOCALCEMIA

Hypocalcemia (for causes, see Table 17-2) is often symptomless or may present with nonspecific symptoms of fatigue, irritability, muscular twitching, and paresthesia. The last two symptoms are caused by increased neuromuscular excitability

TABLE 17–2. Some Causes of Hypocalcemia

Hypoparathyroidism (true or pseudo)
Bone hunger (post-parathyroid surgery)
Vitamin D deficiency
Chronic renal failure
Neonatal hypocalcemia
Mg depletion
Acute pancreatitis
Abnormal Ca binding by anions

Note: Pseudopseudohypoparathyroidism and calcitonin excess do not cause hypocalcemia.

which, when more severe, presents as *tetany* or even as *seizures*. Some chronic hypocalcemic states may present with metabolic bone disease. Hypocalcemia should be *sought* in patients presenting with seizures, undiagnosed CNS disease or mental disorders, chronic renal failure, presenile cataracts, tetany, atrophy of nails, hair, or skin and after thyroid or parathyroid surgery.

Apparent hypocalcemia, caused by *low serum proteins,* is common in sick hospitalized patients and is unimportant.

Hypoparathyroidism and Hypocalcemia

Hypoparathyroidism may be acute or chronic. *Acute* forms arise from surgical or other trauma to the parathyroid glands. They often last only a few days and are treated with Ca supplements. *Chronic* hypoparathyroidism may arise from trauma or autoimmune disease, from idiopathic atrophy, or from congenital absence. Biochemically the *serum* and *urine Ca* are *low,* the *serum Pi* is *high* and the *ALP* and *bone density* are *normal.* The consequences include tetany, premature cataracts, CNS involvement (including mental disorders, seizures, and calcification of the basal ganglia), and integumentary effects including dysplasia of the nails, hair, and skin. The onset is usually insidious and the diagnosis may be missed for years.

Pseudohypoparathyroidism is a genetic disorder caused by a *receptor* defect in the *target organs* of PTH action. It causes the biochemical phenotype of hypoparathyroidism but the glands and PTH secretion are normal or even overactive. Physical stigmata include short stature, developmental delay, round facies, and short metacarpals (especially the 4th). Inheritance is X-linked dominant. The conditions can be distinguished biochemically by infusing PTH. In *true* hypoparathyroidism there is a prompt increase in urinary output of Pi and of cAMP. This response is lacking in *pseudo*hypoparathyroidism. A disorder called *pseudopseudohypoparathyroidism* exists where the physical stigmata occur without any hormonal or biochemical abnormality.

Hypocalcemia from Lack of Vitamin D

Vitamin D deficiency can arise from several causes (Table 17–3) and results in *low* serum levels of *Ca* and *Pi* and usually an *increased ALP* and *osteomalacia.*

TABLE 17–3. Some Causes of Vitamin D Deficiency

Dietary deficiency
Lack of exposure to sunlight
Malabsorption syndromes:
hepatobiliary
intestinal
pancreatic
Drug induction of hepatic microsomal enzymes (e.g., dilantin)
Lack of serum transport proteins
Lack of 1-hydroxylation:
congenital lack of enzyme (vitamin-D-dependent rickets)
hyperphosphatemia
hypoparathyroidism
renal damage
Target organ unresponsiveness (receptor defect)

Hypocalcemia from Chronic Renal Failure

Several mechanisms account for the hypocalcemia in this condition. Retention of Pi leads to reciprocal lowering of Ca because of the low solubility of calcium phosphate and because of inhibition of vitamin D 1-hydroxylase. This effect is counteracted, at first completely and later incompletely, by *secondary hyperparathyroidism*—a normal homeostatic response to the low Ca. Low $1,25(OH)_2D$ levels result also from loss of renal tissue. The lack of vitamin D causes poor absorption of Ca and this is compounded by anorexia. Metabolic acidosis increases the ionized fraction so that protein-bound Ca also falls. The complexed fraction is normal or increased. *Bone disease* in chronic renal failure is also complex, being produced by secondary hyperparathyroidism, vitamin D deficiency, and acidosis.

Neonatal Hypocalcemia

The fetus is in strong positive Ca balance *in utero,* all the Ca arriving through the placenta. This supply is suddenly cut off at birth and is not replaced until milk feeding is established. It is normal for serum Ca levels to fall 20% to 30% in the first few days of life. In some infants, especially if premature or when feeding is delayed, the fall is larger and can cause irritability or convulsions. Older infants fed cow's milk formula can become hypocalcemic because of the high ratio of Pi to Ca in cow's milk compared to human milk.

Mg Depletion and Hypocalcemia

Some hypocalcemic patients are also suffering from Mg depletion and will not respond to Ca supplements until the Mg is restored. Mg is required for the release of PTH from the parathyroids, for transmembrane Ca transport, and for the activity of several enzymes including ALP.

Acute Pancreatitis

Hypocalcemia is common from the second to the fifth day of attacks of acute pancreatitis. Its extent is roughly proportional to the severity of the attack and may be profound. Ca is bound by fatty acids released by the action of lipase on mesenteric fat and, sometime later, may become visible on abdominal x-ray. A lack of PTH response has also been suggested.

Bone Hunger

After resection of parathyroid tumors it is very common for hypocalcemia to occur for several days because the demineralization of bone, produced by the excess PTH, is suddenly reversed. Mg and Pi levels also fall. Similar 'bone hunger' for Ca may occur after treatment of hyperthyroidism and of metastatic bone cancer. Therapy for malignant disease can also cause hypocalcemia by acutely raising the serum Pi level.

Hypocalcemia due to Abnormal Calcium Binding

Alkalosis decreases the ratio of ionized to protein-bound calcium. In *chronic* alkalosis the ionized fraction is restored by endocrine homeostasis and the total Ca tends to rise. In *acute* alkalosis the ionized fraction can be quite low, even leading to tetany, with a normal total Ca. This occurs most commonly in *hyperventilation* syndromes in which many of the symptoms are caused by hypocalcemia. It can be reversed by rebreathing expired air.

Increased complexed Ca may result in a low ionized Ca, especially following multiple transfusions with blood anticoagulated with citrate. Cardioplegic solutions, which bind calcium, are infused during open heart surgery.

Treatment of Hypocalcemia

Acute hypocalcemia is treated with intravenous Ca gluconate and/or with oral Ca supplements. *Chronic* causes are treated by correcting the causative lesion when possible. Oral PTH is ineffective in hypoparathyroidism, because it is digested in the gut; parenteral PTH is antigenic, thus vitamin D is used as a substitute. Serum and urine Ca should be monitored regularly to *prevent overdosage* in all patients being treated with vitamin D. In chronic renal failure treatment is aimed at reducing elevated serum Pi and replacing missing $1,25(OH)_2D$.

DISORDERS OF MAGNESIUM

Hypomagnesemia develops quite frequently as a complication of other illnesses; it is rarely seen as an isolated problem. *Reduced intake* can result from poor diet or from malabsorption. *Increased output* occurs from abnormal gastrointestinal losses, or via the kidney as a result of diuretic excess, hyperaldosteronism, renal tubular defects, and gentamicin nephropathy. *Shifts* into *bone* and into cells occur during the treatment of hypercalcemia and of ketoacidosis. Patients most at risk for hypomagnesemia will be those with a combination of the aforementioned situations, such as alcoholic cirrhosis, post-GI surgery with complications, and intravenous feeding. Hypomagnesemia may be symptomless but may cause the same symptoms as hypocalcemia. Prolonged hypomagnesemia may lead to hypocalcemia, osteopenia, and ectopic calcification.

Hypermagnesemia causes sedation and, in high concentrations, neuromuscular depression. Mg is used therapeutically especially in pre-eclampsia and eclampsia. Therapeutic levels are 2 to 4 mmol/L; toxic >6 mmol/L. Renal failure is the commonest cause of hypermagnesemia and the moderate levels that usually result do not require treatment. Dehydration and ketoacidosis are other causes.

DISORDERS OF PHOSPHATE (Pi)

Acute hypophosphatemia is common in sick patients and is often overlooked. When moderate and not too prolonged, acute hypophosphatemia appears harmless, but when severe (<0.5 mmol/L) and prolonged (>24 h) it can have serious consequences. These consequences relate mainly to impaired ATP production. The effects are seen in the neuromuscular system as weakness, tremor, anorexia, and eventually seizures and coma, and in the blood as impaired function of red cells, white cells, and platelets. Chronic hypophosphatemia is usually caused by a renal tubular lesion and can present as osteomalacia.

Causes of acute hypophosphatemia fall into the same categories as those of hypomagnesemia. *Decreased intake* results from starvation and malabsorption, especially that caused by therapeutic use of aluminum hydroxide. *Increased losses* are usually renal, stimulated by acidosis, hyperparathyroidism, renal dialysis, and congenital or acquired renal tubular leaks. *Intracellular shifts* occur during hyperglycemia (especially in hyperalimentation), during correction of dehydration, acidosis, hyperparathyroidism, or in septic shock. An early sign of progressive Pi depletion (not due to renal causes) is a very low urinary Pi.

Vitamin D resistant rickets (or familial X-linked hypophosphatemia) is an inherited disorder of tubular phosphate transport. The serum Pi may be corrected and the bone lesions will heal if large and frequent oral phosphate supplements are given. Other proximal tubular defects, known collectively as the *Fanconi syndrome,* can have similar consequences.

Hyperphosphatemia is seen in renal failure and in hypoparathyroidism. During childhood normal levels of serum Pi are higher than in adults. The high Pi in chronic renal failure causes a low Ca and suppresses vitamin D 1-hydroxylase. It is treated by dietary restriction, by aluminum hydroxide administration, and by dialysis.

BONE

CHEMISTRY

Bone is a complex mineralizing system composed of three phases: inorganic (mineral), organic (matrix), and water (Fig. 17–2). The ***mineral phase*** is a poorly crystalline hydroxyapatite ($Ca_{10}[PO_4]_6[OH]_2$). Several other ions may replace some of the Ca, Pi, and OH in the hydroxyapatite crystals. Carbonate is the main impurity, replacing phosphate. Citrate and pyrophosphate are inhibitors of crystal growth and dissolution and probably sit on the surface of the crystals. Na and Mg can replace Ca in hydroxyapatite and this may explain the low Ca to P ratio in young bone. Fluoride can replace the OH ions, making the lattice more stable and less soluble. Bone mineral stores other physiological ions like Zn, Sr, and Fe, but also toxic metals like ^{90}Sr, Pb, Al, and Cd.

The ***organic matrix*** consists 88% of collagen Type I and 12% of various noncollagenous proteins, proteoglycans, and lipids (Fig. 17–2). The amino acid composition of *collagen* is unique in that it has a high content of hydroxyproline (10%) and contains hydroxylysine. Bone collagen fibrils are hierarchical structures 30 nm to 80 nm in diameter and several micrometers in length. They are made up of crosslinked collagen molecules, staggered at one-quarter length intervals. This arrangement gives rise to the typical banding pattern seen by electron microscopy. The individual collagen molecules are formed by three left-handed polypeptide helices coiled together in a right-handed coiled coil. This molecular structure provides specific sites and charged regions for the mineral phase to be located. The noncollagenous proteins of bone include *osteocalcin* which contains three γ-carboxy glutamic acid (Gla) residues, hence the name bone-Gla protein. This protein is probably involved in the control of mineralization. Serum Gla protein and urine Gla are used as markers of bone turnover.

BIOLOGY

Bone is laid down on cartilagenous (endochondral) and fibrous (membranous) templates during growth. It is remodeled subsequently to meet structural and mechanical stresses. The cells involved in the processes of bone formation, maintenance, and resorption are osteoblasts, osteocytes, and osteoclasts. ***Osteoblasts*** produce *osteoid,* the organic matrix of bone, and facilitate and control its subsequent mineralization. They act by synthesizing promoters of mineralization, removing inhibitors, and by raising the Ca $\times$ Pi product. One factor produced is *alkaline phosphatase,* which increases the Pi and reduces the py-

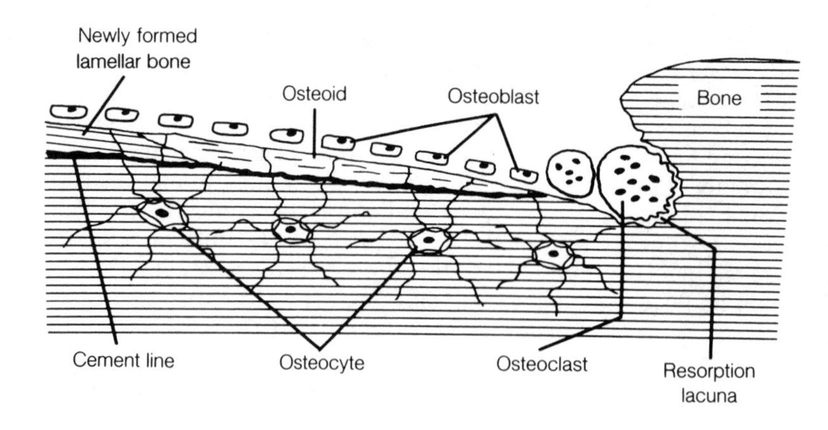

FIGURE 17–5. Cells and matrix of remodelling trabecular bone. Following resorption by osteoclasts, osteoblasts produce new matrix (osteoid) which becomes mineralized, a cycle taking 4 to 6 months. Osteocytes regulate Ca and Pi exchange with the extracellular fluid.

rophosphate in the local extracellular fluid. Osteoblasts become incorporated into the mineralized bone and continue to function as osteocytes whose cell processes extend through the bone canaliculi and remain in contact with the extracellular fluid. ***Osteocytes*** regulate the flux of calcium and phosphate between the bone mineral and the extracellular fluid. Both of these cells are of fibroblastic origin. Resorption of bone is a function of ***osteoclasts,*** multinucleated cells derived from bloodborne macrophages (Fig. 17–5, Table 17–4).

Bone remodeling adapts skeletal structure to growth and stress and repairs injury. It involves units of cellular activity, called bone modeling (or multicellular) units (***BMU***), each with an individual life cycle. The remodeling activity behaves differently in different parts of the skeleton, for example at the periosteum it results typically in a net gain of bone so that the external diameter of the long bones continually enlarges even though the bone may grow thinner. Endosteal, Haversian, and trabecular remodeling all behave differently. The degree of bone remodeling activity occurring in a given area of bone will be determined by the life span of the individual BMU as well as the activation frequency of new BMUs. The controlling factors in this will be a combination of mechanical, bioelectric, biochemical, and hormonal influences (Table 17–4).

A ***BMU cycle*** consists of the following stages: quiescence, resorption, reversal, formation, then quiescence again. The cycle begins with osteoclast activation causing resorption of bone along a cutting cone, resulting eventually in the formation of a new Haversian canal, and then osteoblastic bone formation following much more slowly behind. Each BMU turns over approximately 0.1 to 0.5 mm^3 of bone during its active life of 4 to 6 months. Typically, 7% to 10% of the skeleton is remodeled each year. PTH is the most potent stimulus to remodeling, but growth hormone, thyroxin, and physical stress are also stimuli. Calcitonin, estrogen, and corticosteroids inhibit (Table 17–4). Many osteoblasts are required to reform the quantity of bone removed by one osteoclast. The degree of new bone formation is the apposition rate, which

TABLE 17–4. Naturally Occurring Factors Affecting Bone Deposition and Resorption

Calcium regulatory hormones	Parathyroid hormone
	$1,25(OH)_2D$
	Calcitonin
Systemic hormones and growth factors	Glucocorticoids
	Insulin
	Thyroxin
	Estrogen/androgens
	Growth hormone
	Somatomedin
	Epidermal growth factor
	Fibroblast growth factor
	Platelet derived factor
Local factors	Prostaglandins
	Growth factors
	Noncollagenous proteins
	Osteoclast activating factor
	Mechanical stress
	Bioelectric potentials
Ions	Calcium
	Magnesium
	Phosphate
	Pyrophosphate

can be measured histologically by observing tetracycline uptake in bone biopsies, the distance between successive mineralizing fronts corresponding to times of tetracycline dosage. Generalized derangements of bone remodeling result in metabolic bone diseases.

BONE DISORDERS

Generalized disorders of bone are classified by their radiologic appearances, *osteosclerosis* referring to *increased* bone density and *osteopenia* to *decreased* bone density. Osteosclerosis may involve different pathologic processes (e.g., osteopetrosis, metastatic carcinoma, or myelofibrosis). Osteopenia also results from several pathologic processes. In the adult, the most common causes are osteoporosis, osteomalacia, osteitis fibrosa, and metastatic malignancy. In children the commonest cause is rickets.

Bone density is assessed radiologically at selected sites, such as vertebrae or hands. Conventional radiography is difficult to standardize and is an insensitive measure of osteopenia (at least 30% bone loss required for detection). Conventional radiography is being superseded by CAT scanning and by dual-photon absorptiometry. Neutron activation is a research technique for assessing body calcium.

TABLE 17–5. Causes of Osteoporosis

Localized
Denervation
Disuse or immobilization
Fracture
Inflammation

Generalized
Primary
(a) Postmenopausal; (b) Aging
  Probably multifactorial—possible factors include:
    Estrogen lack
    Ca deficiency
    Vitamin D deficiency
    Excessive protein intake, leading to loss of bone mineral to buffer acid production
(c) Juvenile (idiopathic)
(d) Weightlessness
Secondary
(a) Heritable bone disease, e.g., osteogenesis imperfecta
(b) Nonosseous disorders
    Nutrition—malabsorption
    Endocrine, e.g., ↑ glucocorticoids
                 ↓ estrogens
    Pulmonary, e.g., emphysema
    Renal, e.g., renal failure
    Neoplasms, e.g., myeloma

Osteoporosis

Osteoporosis is defined as decreased bone mass per unit volume, with *normally mineralized* bone. Many processes may cause this end result, as illustrated by the occurrence of high and low bone turnover osteoporosis and of local and generalized disease (Table 17–5). The *etiology* of the most common forms, postmenopausal and senescent osteoporosis, is unknown. Usually developing slowly over many years, there is no diagnostic serum or urinary biochemical marker for this disease unless, as in juvenile osteoporosis, bone resorption is very rapid. *Consequences* include bone pain, wedge fractures of the vertebral bodies leading to kyphosis, loss of height, and wrist and hip fractures.

Treatment. No well established treatments are available for osteoporosis. *Estrogen* delays the postmenopausal loss of bone but does not repair bone already lost. Therapeutic trials with *fluoride* and with *1,25(OH)$_2$D* are showing promise, as are attempts to modulate the BMU using 'coherence therapy' consisting of sequential phosphate, diphosphonate and calcium administration. Prophylaxis, ensuring well mineralized bones before aging by means of adequate Ca intake, good nutrition, and exercise, is important.

Osteomalacia

Osteomalacia is osteopenia related to deficient mineralization of bone matrix resulting in wide osteoid seams. This may be caused by a deficiency of available Ca or Pi or by an inability of the matrix to nucleate and accrete mineral (Table 17–6). Osteomalacia in *children* (rickets) is manifested by the inability to mineralize cartilage and remodel new bone at the epiphyseal growth plate. The bones do not grow in length, resulting in dwarfism, and the cartilage of long bones grows irregularly, forming lumps that are particularly noticeable on the ribs. The bone formed is soft and bends under the patient's weight. Osteomalacia in *adults* may present as bone pain, deformity, or spontaneous fracture.

TABLE 17–6. Causes of Osteomalacia

Vitamin D deficiency (Table 17–3)
Availability of phosphate
Decreased absorption
Increased loss
Renal tubular disorders (vitamin-D-resistant rickets)
Hyperparathyroidism
Availability of calcium (Table 17–2)
Abnormal losses
Dietary
Malabsorption
Inhibition of mineralization
Aluminum
Diphosphonates
Fluoride
Hypophosphatasia

Note: Hypoparathyroidism does not cause osteopenia.

Typically, serum Ca and/or Pi are low and ALP is high, but they are sometimes normal. Assessment of vitamin D and its metabolites can serve to define specific mechanisms in patients with multisystem disease or with genetic disorders. Definitive diagnosis depends on the histologic demonstration of excess osteoid together with a decreased mineralization rate. *Nutritional rickets* is rare in western countries where rickets is most often related to developmental disorders, such as vitamin D resistance, or phosphaturia. *Hypophosphatasia* is a genetic disorder with a mineralization defect similar to rickets. Alkaline phosphatase is decreased in serum and extracellular fluid and increased amounts of pyrophosphate and phosphoethanolamine are found in the urine.

Osteomalacia is a frequent but underrecognized complication of *chronic malabsorption* as from gastrectomy, inflammatory bowel disease, or chronic liver disease, especially biliary cirrhosis. Chronic *drug* ingestion can induce osteomalacia, for example, dilantin, which interferes with the 25-hydroxylation of vitamin D. Aluminum toxicity, often seen in chronic renal failure as a result of absorption of aluminum hydroxide gel administered to reduce hyperphosphatemia, causes osteomalacia possibly by inhibiting osteoid mineralization.

A specific cause can usually be discovered in a case of osteomalacia. If remediable, complete resolution can be expected although there may be residual deformity.

Osteitis Fibrosa

Osteitis fibrosa (cystica) is the accelerated removal of bone by increased *osteoclast* activity, usually stimulated by excess PTH. This process can lead to generalised osteopenia or to local bone cyst formation. It is seen in primary hyperparathyroidism and in some malignant tumors that stimulate bone resorption through PTH-like substances. *Secondary hyperparathyroidism,* a response to chronic hypocalcemia, stimulates bone resorption. On biopsy, such patients frequently show a mixed picture of osteitis fibrosa and osteomalacia.

Renal Osteodystrophy

This term is used for bone disease associated with chronic renal failure and its treatment by *dialysis.* On biopsy, osteomalacia, osteitis fibrosa, osteosclerosis, or a mixture of these processes is seen. Biochemical *assessment* is directed towards detection of osteomalacia, detection of secondary hyperparathyroidism, and evidence of excess aluminum, which may interfere with bone formation. *Treatment* endeavours to reduce serum Pi, to replace missing $1,25(OH)_2D$ in order to minimize hypocalcemia and the resulting secondary hyperparathyroidism, and to remove excess aluminum if present. Parathyroidectomy is helpful occasionally in reducing bone disease.

Paget's Disease

Paget's disease, usually occurring after age 50, is characterized by a mixed picture of bone resorption, increased vascularity, high bone turnover, and irregular bone formation. One or more bones may be affected, but even monostatic disease may cause an increased *alkaline phosphatase,* often found on a routine laboratory screen. *Clinical* features may be minimal but can include bone pain, deformities, nerve and blood vessel compression, increased cardiac output, spontaneous fractures, and osteogenic sarcoma. As with other high turnover states of bone, immobilization of patients with Paget's disease may lead to hypercalcemia, hypercalciuria, and renal calculus formation. Current *therapies* to reduce bone resorption include calcitonin, diphosphonate and, in extreme cases, mithramycin.

Fluoride

Fluoride is an ion that has complex *dose-related* effects on both bone mineral and osteoblasts. Excess fluoride ingestion results in *fluorosis,* a condition characterized by osteosclerosis and increased brittleness of bones. *Fluoride therapy* is being evaluated in the treatment of osteoporosis, because of its effect of inhibiting bone resorption and stimulating bone formation.

SUGGESTED READING

FRAME B, POTTS JT (eds): Clinical Disorders of Bone and Mineral Metabolism. New York, Excerpta Medica, 1983

RUBENSTEIN E, FEDERMAN DD (eds): Scientific American Medicine Sections 3 and 15. New York, Scientific American Inc, 1983

WILLIAMS RH (ed): Textbook of Endocrinology, Chap 19, 6th ed. Philadelphia, WB Saunders, 1981

18

Disorders of Amino Acid and Protein Metabolism

James A. Heininger / Zulfikarali H. Verjee / Michael D. D. McNeely

DISORDERS OF AMINO ACID METABOLISM

Protein is a fundamental constituent of living organisms, and the basic unit of all proteins is the ***amino acid.*** Of the 20 common amino acids, man is able to synthesize half, while the others (termed *essential*) must be derived from the diet. Elaborate synthetic pathways exist for the formation, interconversion, and metabolism of the amino acids. Derangements of these pathways account for a variety of rare, though fascinating, disorders, which are covered in the first half of this chapter. Although proteins are vitally important to the entire body, the second part of this chapter is concerned with the clinical importance of the proteins that are present in serum.

The amino acids can be linked together in chains of widely varying length known as *peptides.* Once the polypeptide chain has reached a relative mass of about 6000, it may be called a protein. The sequence of amino acids in each protein is termed its *primary* structure. Usually, the chain is arranged in a *secondary* structure of links or helices. In addition the protein can be 'formed' by bending its arrangement upon itself into a *tertiary* structure. A *quaternary* structure can be produced by assembling more than one chain or *monomer* in a particular configuration.

Disorders of amino acid metabolism (***aminoacidopathies***) are rare, genetically determined, and usually inherited in an autosomal recessive fashion. Table 18–1 lists the relatively more common aminoacidopathies and their apparent frequency. Although over 50 specific defects have been described, total-population *screening* of newborns for this wide range of amino acid abnormalities is not practical. This is due not only to a poor cost-benefit ratio but also to the inherent inefficiency of testing for rare diseases. Consider hyperprolinemia as an *example,* with an incidence of 1:50,000, and defining a positive result as one that is more than twice the standard deviation (>2 SD) from the mean. For every diseased child that is detected, over 1100 healthy children will have had a positive result, even with a test specificity of 98%.

Screening *efficiency* can be improved in two ways: the *first* is to increase the reference range considered to represent a negative result (i.e., from >2 SD to >3 SD), and the *second* is to screen selectively only populations in which the prevalence is higher. Apart from situations like tyrosinemia in the Chicoutimi region of Quebec (Table 18–1), the most effective means of selection is by clinical

TABLE 18–1. Approximate Prevalence of Amino Acid Disorders

Disease	Prevalence*
Abnormal phenylalanine levels	1:9,000
Phenylketonuria	1:12,000
Hyperphenylalaninemia	1:33,000
Cystinuria	1:7,000
Iminoglycinuria	1:20,000
Hartnup disease	1:25,000
Histidinemia	1:25,000
Hyperprolinemia	1:50,000
Argininosuccinic aciduria	1:250,000
Fanconi syndrome	1:250,000
Hyperlysinemia	1:250,000
Hypermethioninemia	1:250,000
Branched-chain ketoaciduria	1:350,000
Tyrosinemia	
General population	1:1,000,000
Chicoutimi area of Quebec	1:700

* The ease of clinically or chemically detecting a disease may have a significant effect on the apparent prevalence.

criteria. The following clinical signs and symptoms are associated with metabolic disease in the *newborn period:*

Failure to thrive, poor feeding
Seizure disorders
Hypertonicity, jitteriness, irritability
Hypotonicity, lethargy
Vomiting, diarrhea, hepatomegaly, dehydration
Metabolic acidosis, an unexplained increase in the anion gap
Hypoglycemia, hyperglycinemia, hyperammonemia
Neutropenia, thrombocytopenia

Symptoms and signs in *older children* include:

Mental retardation
Seizure disorders
Developmental delay or deterioration (mental, motor, or physical)
Hematologic abnormalities, anemias
Behavior problems, central nervous system disorders, speech defects
Hepatomegaly, splenomegaly
Renal defects, renal calculi
Blindness, cataracts, dislocated lenses, optic atrophy
Recurrent infections, fever of unknown origin

Any one or any combination of these should produce a high index of suspicion of a metabolic disease and the possibility of an aminoacidopathy should then be explored by the application of appropriate biochemical tests.

In the following section the laboratory tests commonly used for the diagnosis of amino acid disorders are reviewed briefly. Following that, some of the more common amino acid disorders are described, grouped according to metabolic pathways. For each condition, the appropriate laboratory test and the preferred biologic fluid will be indicated. Wherever possible, additional metabolic defects that are not described in the text will be indicated in the figures. References at the end of this chapter provide more comprehensive discussions of the subject.

LABORATORY METHODS FOR THE DIAGNOSIS OF AMINO ACID DISORDERS

Amino Acid Determinations

Microbiologic Assays. Several assays have been developed that use a mutant bacterial strain in which the growth rate is altered specifically by an abnormal quantity of the amino acid or other substance to be measured. Frequently referred to as the *Guthrie test,* this type of assay is used widely for total-population screening for the hyperphenylalaninemias, but it has limited value owing to the fact that only one amino acid can be measured in each assay system.

Qualitative Methods. Several screening techniques have been developed for separation of the amino acids found in body fluids. These include one- or two-dimensional paper and thin-layer partition chromatography, one-dimensional electrophoresis, and two-dimensional separations involving electrophoresis and partition chromatography. For each technique a variety of solvent systems and 'run' conditions are employed and several stains are available for visualization of the amino acids. The most widely used stains contain ninhydrin as the major constituent, because it reacts with almost all the amino acids and is very sensitive with the majority. These methods are rapid, simple, inexpensive, and permit a large number of samples to be run in a single batch. The disadvantages are that at best they are semiquantitative and that identification is only tentative.

Because amino acid levels in *plasma* are controlled within reasonably narrow limits, and few interfering substances are encountered, a one-dimensional system is all that is required. *Urine,* on the other hand, is both qualitatively and quantitatively much more complex, containing a variety of amino acids, peptides, metabolites, and drugs, with wide ranges in concentrations that are superimposed on varying concentrations of interfering salts. To correct for the wide differences in concentration, as determined by refractive index or by creatinine measurement, a varying amount of urine may be chromatographed. To assist with the resolution of the number of compounds encountered, a two-dimensional separation, or differential staining of one-dimensional runs to effect the required separation chemically, is generally recommended. All urine samples should also be screened for sulfhydryl compounds by the cyanide-nitroprusside test, and for carbonyl compounds (such as keto acids) by the dinitrophenylhydrazine test.

Quantitative Methods. The most widely employed means of quantitative amino acid analysis involves separation on a cation exchange column followed by ninhydrin color development and photometric measurement, at 570 nm for most amino acids (primary amines) and at 440 nm for proline and hydroxyproline (secondary amines).

Different analytic methods employ either continuous or discontinuous buffer elution systems, single or dual columns, rapid elution systems using three buffers but with limited resolving power, or longer elution systems requiring five or more buffers. A number of systems are commercially available that provide automatic sample injection, buffer changes, column regeneration, and peak identification and quantification. The advantages of these systems include a lack of extensive sample preparation, excellent resolution, and good precison. Disadvantages include long run times (4 to 6 hours per sample for good separation in a physiologic fluid) and expensive, complex equipment.

Any *suitably staffed hospital* with a newborn or children's ward can perform the chemical tests and amino acid chromatography in its clinical chemistry department. Interpretation of these chromatograms can be difficult, and for this reason it is best to use larger centers that frequently perform and interpret the tests. In addition, any screening or semiquantitative system, such as amino acid chromatograms, must be backed up by quantitative analyses and these are confined to larger centers or research laboratories.

Organic Acid Determinations

Organic acids are low relative mass water-soluble carboxylic acids, which are metabolic intermediates of amino acids, carbohydrates, or fats. They contain no primary amino group and hence are ninhydrin negative. Disorders that result in excessive accumulation of these acids in body fluids are generally called *organic acidemias.* Organic acids frequently have low renal thresholds, and urinary excretion can be the most sensitive reflection of their presence.

Gas–Liquid Chromatographic (GC) Screening. In this most widely used screening method, the urine is acidified, saturated with salt, and extracted with organic solvents. Water is removed, the organic solvent is blown off, and the residue is derivatized, usually by methylation or trimethylsilylation. This provides a relatively nonpolar, volatile mixture of compounds that is suitable for GC analysis. Frequently used columns include OV 17, OV 22, SE 30, and OV 101; these are relatively nonpolar packings that are stable at temperatures up to 350°C. Temperature-programming is required to elute the wide range of compounds encountered. The identity of an unknown in the organic acid profile is suggested by determining its retention time relative to an internal standard, or relative to 'methylene units' (an analogous series of straight-chain saturated hydrocarbons with increasing numbers of carbon atoms and increasing retention times).

Gas Chromatography–Mass Spectrometry (GC–MS). It has been estimated that over 300 organic acids are normally excreted in the urine. The pattern of excretion is quite variable, depending on the diet, intestinal bacteria, and metabolic state of the individual. Thus, although GC will indicate excess excretion of a compound, retention time alone cannot provide its identity. Samples that may contain abnormal peaks must be referred to the next level of analysis, GC–MS. Mass spectrometry acts as a selective detector, providing a spectrum of mass fragments that is unique to the compound being eluted. The spectrum from the unknown is compared to a library of spectra (obtained from analysis of standards) and an identification is made. Or, conversely, if no library spectrum matches the unknown, certain deductions about the molecular structure of the unknown can be made from its spectrum. GC–MS requires expensive and complex equipment and to date has been largely confined to special centers.

GC screening for organic acidurias should *exclude* the disease in 80% to 90% of cases. Large peaks found in the remainder are usually produced by drugs, food additives, or normal variation. These will require identification by GC–MS at a referral center with sufficient experience in the interpretation of mass spectra of organic acids.

Indications for Screening. Accepted indications to screen for organic acidemias include:

a. A peculiar odor (sweaty feet, tom-cat's urine, and mice)
b. Unexplained metabolic acidosis, whether persistent or intermittent, especially if associated with an elevated anion gap
c. Intractable or recurrent vomiting, especially when associated with metabolic acidosis instead of the usual alkalosis
d. Acute disease in infancy, especially when associated with metabolic acidosis and/or hyperammonemia
e. Unexplained hypoglycemia and severe metabolic acidosis with muscular hypotonia, poor feeding, lethargy, and coma in newborns
f. Progressive extrapyramidal cerebral palsy in childhood
g. Any inherited disease of obscure cause
h. Reye's syndrome, occurring in infancy.

Examples of organic acidemias that are a result of defects in amino acid metabolism are discussed in later sections.

DISORDERS OF AROMATIC AMINO ACID METABOLISM

Phenylketonuria

Phenylketonuria (PKU) is the best known of the amino acid disorders. It is an autosomal recessive deficiency of the liver enzyme *phenylalanine hydroxylase,* which normally converts phenylalanine into tyrosine (Fig. 18–1, site 1). This results in raised levels of phenylalanine in the plasma and urine. Other ketoacids that are metabolites of phenylalanine will also appear. These include phenylpyruvate, phenylacetate, phenyllactate, and *o*-hydroxyphenylacetate. The *prevalence* of the disorder averages 1:12,000, with a range from 1:4,500 (Ireland) to 1:1,000,000 (Finland). Infants appear normal; there are no consistent neonatal abnormalities but by the first month the developmental 'milestones' are noted to be delayed. The complete *syndrome* consists of mental retardation, skin lesions (eczema beginning in infancy), decreased pigmentation (blond hair and blue eyes even in Mediterranean individuals), epilepsy, agitated behavior, small head, hyperactive reflexes, tremors, and seizures. The unusual levels of phenylalanine may be toxic to the central nervous system of the growing child or may act as a competitive inhibitor of amino acid transport into cells. A third theory is that phenylalanine may alter the metabolism of other pharmacologic amines. Phenylalanine competitively inhibits the enzyme tyrosinase and decreases the development of melanin pigment.

Early diagnosis of this condition in the first few weeks of life is essential. The administration of a low-phenylalanine diet (Lofenalec) will prevent any neurologic damage or mental retardation from occurring, whereas nondiagnosis will result in irreversible damage. Phenylalanine must *not* be withheld completely from the diet. It is an essential amino acid and sufficient phenylalanine must be ingested to permit normal growth and development. In midchildhood the brain is probably less susceptible to damage from phenylalanine and the diet may be tapered off, although considerable controversy still exists on this point. Current studies indicate that although full-scale IQ is not significantly different, some aspects of school performance do decline in children taken off the diet.

An emerging problem is that of *maternal* PKU. It is now known that the fetus of the untreated PKU mother, in 97% of cases, will be severely affected. Nearly all fetuses have microcephaly and moderate-to-severe mental retardation. Many have serious congenital cardiac anomalies, and there is a high incidence of spontaneous abortion and still

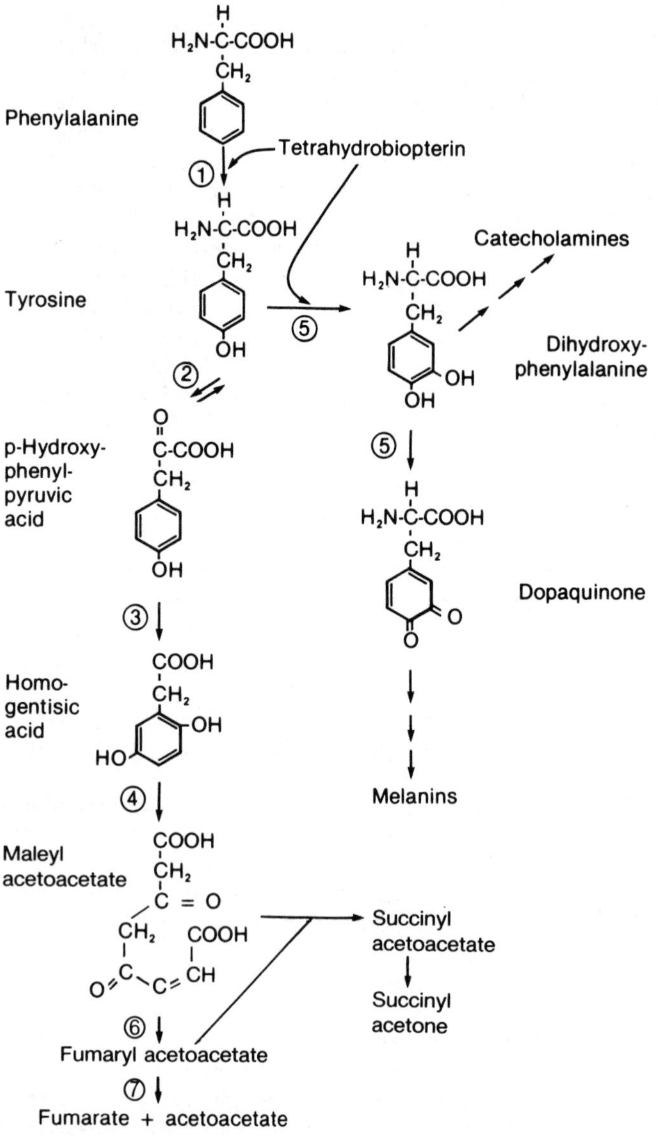

FIGURE 18–1. Some metabolic pathways of phenylalanine: (1) phenylalanine hydroxylase, defective in PKU; (2) tyrosine aminotransferase, the defect (in liver) in tyrosinemia type II; (3) *p*-hydroxyphenylpyruvic acid oxidase, a (putative primary) defect in tyrosinemia type I; (4) homogentisic acid oxidase, the defect in alcaptonuria; (5) tyrosinase, the defect (in melanocytes) in albinism; (6) maleylacetoacetate cis, trans isomerase and (7) fumaryl acetoacetate hydrolase (putative primary) defect in tyrosinemia type I.

birth. Low phenylalanine dietary therapy of these mothers during pregnancy has not been completely successful. It is essential that an appropriate diet be started before conception, a point in favor of continuing with the low phenylalanine diet in PKU patients beyond childhood.

The detection of PKU is achieved by ***screening*** all newborn infants for increased blood phenylalanine, usually by the microbiologic inhibition *Guthrie test.* For all infants found to have greater than 250 μmol/L (4 mg/dL) phenylalanine in their blood this test should be followed up by more specific quantitative assays. If the original blood test is done late in the first week of life and is normal, there is no need for a second follow-up at a later date. Under one month of age the abnormality must be detected by measurement of phenylalanine concentrations, because other metabolites may not be increased. For this reason, the *ferric chloride test* on urine, which detects elevated levels of phenylpyruvate, is not recommended. In addition, this test will give a *false negative* result if the urine is dilute, if a low-protein diet is being administered, or if the urine is old. *False positives* are produced by acetoacetate, salicylates, chlorpromazine, bile, and the byproducts of histidinemia.

Other causes of elevated phenylalanine in the newborn period are prematurity, tyrosinemia, and atypical forms of PKU, of which there are at least four. Differentiation between PKU and other types of hyperphenylalaninemia is accomplished by using quantitative phenylalanine determinations and phenylalanine loading tests at various intervals during the first few years of life. Generally speaking, blood levels greater than 1500 μmol/L (25 mg/dL) on initial testing indicate classic PKU, and levels in the region of 900 μmol/L (15 mg/dL) suggest a variant form of the disease.

An estimated 1% to 3% of infants that are diagnosed as having classic PKU deteriorate subsequently despite adequate control of their blood phenylalanine levels. They have been called *lethal PKU* or *malignant hyperphenylalaninemia,* because of the catastrophic course of the disease. The defect in these patients is not located in the protein apoenzyme but in the synthesis of *tetrahydrobiopterin,* an essential cofactor for phenylalanine hydroxylase, tyrosine hydroxylase, and tryptophan hydroxylase. The neurologic deterioration is caused by an impaired biosynthesis of the monoamine neurotransmitters dopamine and norepinephrine from tyrosine, and serotonin from tryptophan. Treatment, therefore, includes replacement of the cofactor or the administration of L-dopa, 5-hydroxytryptophan, and carbidopa. Their efficacy is not yet clearly established and may depend on early diagnosis.

Tyrosinemia

The primary enzymatic defect in hereditary tyrosinemia is not yet established but may be *fumarylacetoacetate hydrolase.* Several sites on the tyrosine pathway are affected, as well as methionine metabolism and porphobilinogen synthesis. Serum tyrosine and methionine levels are increased and there is a massive urinary excretion of tyrosine, p-hydroxyphenylpyruvate, and p-hydroxyphenyllactate. Patients with hereditary tyrosinemia have been shown to excrete large amounts of succinylacetoacetate and succinylacetone in the urine, supporting the hypothesis of a deficiency of fumarylacetoacetate hydrolase in this disease. This enzyme has also been shown to be deficient in liver tissue in patients with hereditary tyrosinemia. When the fetus is affected, abnormal amounts of succinylacetone can be detected in the amniotic fluid and thus help in the prenatal diagnosis of this disease at 15 or 16 weeks' gestation.

Clinically, the patients exhibit failure to thrive, vomiting, diarrhea, abdominal enlargement, hepatosplenomegaly, and slight or moderate mental retardation. Some cases are acute, with symptoms in the first 6 months of life; others are more chronic, with a later onset. In addition these children develop nodular cirrhosis and kidney damage that results in the Fanconi syndrome, discussed later. In tyrosinemia the diagnosis is indicated by blood findings or by the urinary tyrosine metabolites already mentioned.

Alcaptonuria

Alcaptonuria is an autosomal recessive disorder in which homogentisic acid cannot be converted to maleylacetoacetic acid because of a deficiency of the enzyme *homogentisic acid oxidase* (Fig. 18–1, site 4). Homogentisic acid builds up in the blood and is excreted in the urine. In excess this acid forms an unusual polymer which deposits within cartilage throughout the body. The slow accumulation of this material leads gradually to a darkening of the sclera, ears, and other cartilaginous areas as well as tendons and ligaments (*ochronosis*). Deposition within the cartilage of joints may cause arthritis. Alcaptonuria is a relatively benign disorder, the diagnosis of which can be made by detecting homogentisic acid in the urine. This acid is a reducing substance that will react with Benedict's sugar reagent. When homogentisic acid is present in excess amounts and the urine is alkaline, it will oxidize gradually and turn dark. These two screening tests may not always be positive and definitive diagnosis must be made by specific chemical tests for the compound.

Albinism

Albinism includes a group of inherited disorders of melanin production characterized by diminished amounts of pigment in the skin, hair, and eyes, with susceptibility to sunlight and decreased visual acuity. *Ocular* albinism is an X-linked recessive disorder of unknown biochemical nature which occurs once in every 50,000 births; pigment is normal in the hair and skin but deficient in the eyes.

Oculocutaneous albinism is an autosomal recessive disorder characterized by deficient melanin in the hair, skin, and eyes. It appears in two forms which can be differentiated by assessing the presence or absence of the enzyme *tyrosinase* in hair roots. The tyrosinase-*positive* form is thought to be due to a limitation in melanin synthesis or transport other than that involving the tyrosine-converting enzyme. Its frequency is 1:14,000 Negro births and 1:60,000 Caucasian births. The tyrosinase-*negative* variety has a frequency of 1:35,000 in all races and is probably due to a deficiency of the enzyme tyrosinase. These persons are strikingly light-skinned. There are no characteristic changes in body fluid composition in any form of albinism.

DISORDERS IN METABOLISM OF SULFUR-CONTAINING AMINO ACIDS

Cystinuria

Cystinuria is a defect of amino acid *transport* in which cystine, lysine, arginine, and ornithine are handled inappropriately. The prevalence varies in different populations but on the average is about 1:7000. There are *three subtypes* of cystinuria, all inherited with an autosomal recessive pattern. These subtypes differ in the predominant amino acid that is mishandled and in whether the defect is primarily in urinary tubular transport or in intestinal transport. The most important clinical aspect of this condition, due to a failure to reabsorb the amino acids in the proximal tubule, is the elevated urinary cystine concentration, which may form *calculi* in acid media. These stones are radiopaque owing to the presence of sulfur atoms.

As a transport defect, the condition can be detected only in urine and is screened for by using the *cyanide–nitroprusside test.* This test detects sulfur- or disulfide-containing compounds and is also positive in homocystinuria, in β-mercaptolactate-cystine disulfiduria, in a rare condition called familial pancreatitis, and with some drugs. *Diagnosis* is made by quantitative analysis of a timed (24-hour) urine collection. The four dibasic amino acids already mentioned share a common transport site and their increased excretion is pathognomonic of cystinuria. In addition, the diagnosis may be suggested from microscopic urinalysis. Cystine crystals can be observed in a concentrated, acidified urine specimen when the cystine concentration is greater than 800 μmol/L (20 mg/dL). These crystals appear as flat hexagonal rings. Any renal calculi that are presented for analysis should be examined specifically for the presence of cystine.

Cystinosis

The mechanism of this disorder is not understood but it is thought to be caused by an inability to maintain intracellular cysteine in its reduced (monomeric) form. The result is the development of abnormally high *intracellular* levels of free cystine. Cystinosis is a relatively uncommon autosomal recessive disorder and the incidence has yet to be established. Cystine crystals are deposited in numerous tissues, including the conjunctiva, bone marrow, lymph nodes, leukocytes, and various internal organs. Deposition of the crystals in the renal tubules is one of the more common causes of the Fanconi syndrome during the first year of life, with eventual progression to renal failure. *Diagnosis* can be made (after noting the characteristic chemical features of the Fanconi syndrome; see later) by discovering cystine crystals in the eye (with a slit lamp), in the bone marrow, or in circulating leukocytes. Although the names are similar, the clinical and chemical features of cystinuria and cystinosis are markedly different and no problem should exist regarding their differential diagnosis.

Homocystinuria

This syndrome is caused by at least four distinct inherited defects that affect the metabolism of homocystine. Clinically, it can produce a condition much like Marfan's syndrome. Common to both are *ectopia lentis* and other ocular disorders, skeletal abnormalities, and peculiar arterial and venous thromboses. Mental retardation is an additional frequent finding in homocystinuria. The usual cause of death is thromboembolic pneumonia. There are two basic presentations: The *mild* form may be responsive to pyridoxine (vitamin B_6); the *more severe* variety does not respond to vitamin B_6, but has been shown to respond to therapy with betaine, which promotes the synthesis of methionine from homocysteine. The skeletal abnormalities may be due to the disarray of critical cross-linking of disulfide bonds in structural molecules. The characteristic *biochemical* features include increased urinary homocystine with elevated blood methionine (Fig. 18–2, site 1) or reduced blood methionine (Fig. 18–2, site 2). Homocysteine has a low renal threshold and thus urinary *screening* for homocystine, using the cyanide-nitroprusside test, is a sensitive method of detection. A negative test excludes the condition, but a positive test must be followed up by a quantitative determination of urinary and plasma amino acids to differentiate between this and other disorders that result in increased excretion of sulfhydryl- and disulfide-containing compounds.

Cystathioninuria

Cystathioninuria is the result of a reduced activity of the enzyme *cystathionase* (Fig. 18–2, site 3), caused in some cases by a reduced affinity for its coenzyme pyridoxal phosphate (B_6). Elevated levels of blood, urine, and tissue cystathionine are found. Although the enzyme defect has been noted in association with a wide variety of clinical features, this autosomal recessive disorder may be benign.

Sulfituria

A rare disorder of *sulfite oxidase* deficiency is characterized by the increased urinary excretion of sulfite, thiosulfate, and S-sulfo-L-cysteine and by reduced urinary sulfate (Fig. 18–2, site 5). Clinically the patients have dislocating lenses, show severe neurologic abnormalities, and suffer rapid deterioration in early life. The urine nitroprusside test is negative. The condition is diagnosed by the detection of increased excretion of the urinary metabolites mentioned.

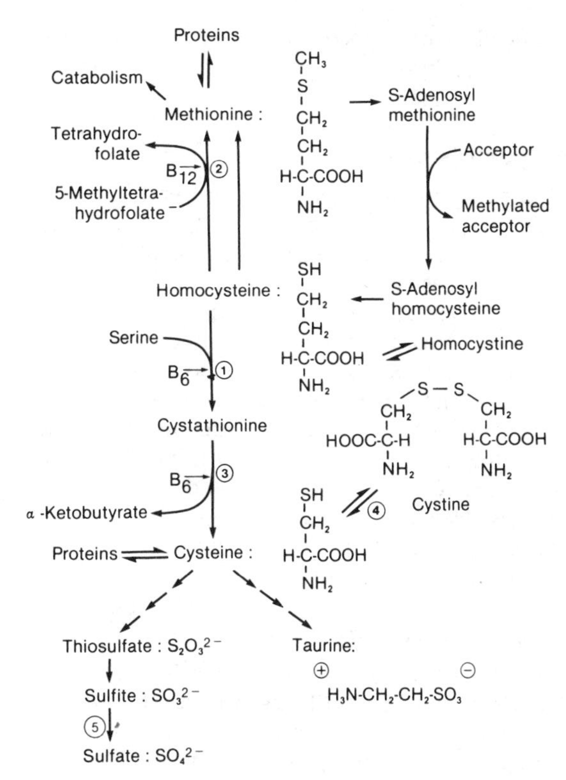

FIGURE 18–2. Some metabolic pathways of the sulfur-containing amino acids: (1) cystathionine β-synthase, the site of the defect that causes homocystinuria and hypermethioninemia; (2) homocysteine-5-methyltetrahydrofolate methyltransferase, the site of the defect that causes homocystinuria and hypomethioninemia; (3) cystathionase, the site of the defect causing cystathioninuria; (4) deficiency of intracellular cystine reduction system, postulated to explain cystinosis; (5) sulfite oxidase, the defect in sulfite oxidase deficiency.

DISORDERS OF AMINO ACID TRANSPORT

Transport defects are detectable only in the urine, and the quantities of amino acids excreted are easily recognized by using paper or thin-layer chromatographic screening techniques. Diagnosis is confirmed by quantitative analysis on timed (24-hour) urine collections. The conditions are present at birth and throughout life and should be detectable at any time. ***Cystinuria,*** the most common amino acid transport defect, has already been described.

Familial Iminoglycinuria

Familial iminoglycinuria is an inborn error of membrane transport within the renal tubule, in which there is a deletion or inactivation of a membrane transport protein that selectively binds proline, hydroxyproline, and glycine. It is autosomal recessive in nature and there is no consistent associated illness. However, it can be confused biochemically with hyperprolinemia, hydroxyprolinemia, and hyperglycinuria. The defect can be detected in urine.

Hartnup Disease

Hartnup disease is a genetic disorder in the handling of monoamino-monocarboxylic acids, which share a transport mechanism common to the renal tubule and gastrointestinal mucosa. It is inherited in an autosomal recessive fashion and is characterized chemically by massive aminoaciduria involving alanine, serine, threonine, asparagine, glutamine, valine, leucine, isoleucine, phenylalanine, tyrosine, tryptophan, histidine, and citrulline. The syndrome is characterized *clinically* by a pellagra-like rash, cerebellar ataxia, and psychologic changes. Hartnup disease is *diagnosed* by demonstrating increases in the urinary excretion of the specific amino acids listed.

Fanconi Syndrome

The Fanconi syndrome is a generalized disorder that affects amino acid transport, resulting in abnormal excretion of all amino acids, as well as affecting the transport of glucose, phosphate, and protein. It is not a disease in itself but reflects an underlying pathologic process which has affected the reabsorptive function of the proximal renal tubule. It may be acquired, as in heavy metal poisoning, vitamin deficiency, or acute tubular necrosis; or it may be secondary to an inherited disease, as seen in galactosemia, Wilson's disease, or cystinosis. These are but a few examples, by no means intended as a comprehensive list of the causes of the Fanconi syndrome. In any case, every attempt should be made to identify and treat the underlying primary disease.

DISORDERS OF BRANCHED-CHAIN AMINO ACID METABOLISM

As Fig. 18–3 indicates, several metabolic defects have been described in the branched-chain amino acid metabolic pathways. Disorders other than branched-chain ketoaciduria (BCKA), hypervalinemia, and hyperleucinemia–isoleucinemia (Fig. 18–3, sites 1, 2, and 3) are generally called *organic acidemias*. They are usually suspected clinically from an unexplained acidosis with an increased *anion gap* and are detected by gas chromatography of a derivatized urine extract. Positive identification of the accumulating metabolite is accomplished with GC–MS and definitive diagnosis is made by estimating the activity of the nonfunctioning enzyme. With the exception of BCKA, the prevalence of the conditions to be described is not known, because systematic screening for these defects is technically unrealistic. They are presumed to be quite rare and are probably all inherited in an autosomal recessive manner.

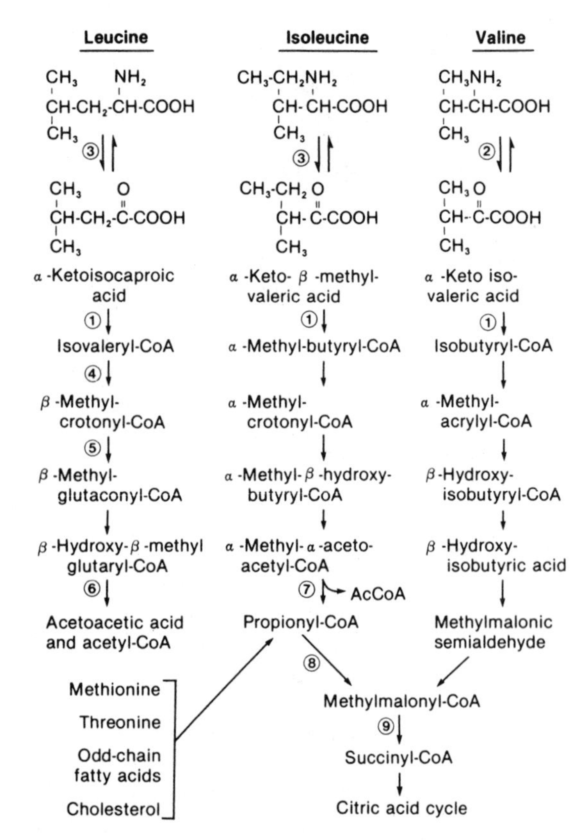

FIGURE 18–3. Some metabolic pathways of the branched-chain amino acids. The following defects (in the metabolic steps indicated) have been described: (1) branched-chain ketoaciduria; (2) hypervalinemia; (3) hyperleucinemia, isoleucinemia; (4) isovaleric acidemia; (5) β-methylcrotonic acidemia; (6) β-hydroxy-β-methylglutaric acidemia; (7) α-methylacetoacetic acidemia; (8) propionic acidemia; and (9) methylmalonic acidemia.

Branched-Chain Ketoaciduria

Branched-chain ketoaciduria is a rare autosomal recessive disorder, with a prevalence of about 1:250,000, except in the Mennonite population where it is as high as 1:200. There is a defect in the group of enzymes responsible for the oxidative *decarboxylation* of the α-keto analog of leucine, isoleucine, and valine (Fig. 18–3, sites 1, 2, and 3). Shortly after birth these branched-chain amino acids and keto acids build up to very high levels in the plasma and urine. Their presence may act to poison neurologic tissue or may possibly inhibit the transfer of essential amino acids into the brain. Whatever the cause, in the *classic case,* infants are normal at birth but begin to feed poorly by the end of the first week and may develop vomiting, become lethargic, and exhibit muscular hypertonicity and convulsions. Death can occur as early as the first week and is usually the result of an intercurrent infection. Mild and intermittent variants have also been described, however.

The most characteristic *clinical* feature is the unusual smell of maple syrup in the urine, thus the common name of *maple syrup urine disease.* This condition, like phenylketonuria, responds to dietary restriction of the offending amino acids but is more commonly fatal and harder to treat than PKU. Its presence may be suggested by a positive urinary *ferric chloride* or *dinitrophenylhydrazine test,* or by increased branched-chain amino acids on urine or plasma chromatographic screen. These amino acids are reabsorbed efficiently from the glomerular filtrate, so that the most sensitive medium for their detection is *plasma.* The condition must be differentiated from the effects of diet (since branched-chain amino acids are modestly elevated in starvation) and from two exceedingly rare conditions that affect the transaminase step of amino acid metabolism (Fig. 18–3, sites 2 and 3).

Isovaleric Acidemia

Isovaleric acidemia is a disorder of amino acid metabolism in which the enzyme *isovaleric-CoA dehydrogenase* is deficient (Fig. 18–3, site 4). The conversion of isovaleric acid into β-methylcrotonic acid is hindered. The most characteristic *clinical* feature is the unusual cheesy or 'sweaty feet' smell of the patient and the breath, which is due to the unusual metabolic products. Isovalerate accumulates in the serum and, along with its glycine conjugate (formed in the liver) and another metabolite, β-hydroxyisovalerate, is excreted in the urine.

Propionic Acidemia

Until recently this condition was called *ketotic hyperglycinemia.* The *clinical* presentation (metabolic acidosis leading to dehydration, lethargy, and coma) ranges from a very severe to a mild intermittent type. Signs, including neutropenia, are episodic, exacerbated by protein intake and catabolic states (infection, etc.). They can be moderated with dietary restriction. *Diagnosis* is made by finding increased urinary propionic acid, its glycine conjugate, and a byproduct, methylcitrate. Long-chain ketones also accumulate (butanone, pentanone, and hexanone), the result of ketone-body formation that has utilized propionate and its precursor α-methylacetoacetate. Increased blood propionate can also be demonstrated. The mechanism for the hyperglycinemia is unknown. The defective enzyme is *propionyl-CoA carboxylase* (Fig. 18–3, site 8), which requires biotin as a cofactor. Two variants have so far been described, one of which responds to pharmacologic doses of the cofactor.

Methylmalonic Aciduria

This includes at least five distinct inherited disorders that result in an inability to convert methylmalonyl-CoA to succinyl-CoA (Fig. 18–3, site 9). There are two enzymic conversions involved and defects in both enzyme proteins have been demonstrated. In addition, the second apoenzyme requires a vitamin B_{12} coenzyme, adenosylcobalamin. Three defects in the transport or synthetic pathway of the coenzyme also result in methylmalonic aciduria. Four of the five disorders present with severe ketoacidosis at birth or within the first year of life. Developmental retardation and hyperglycinemia with or without hyperglycinuria are also consistent findings, with recurrent infections and osteoporosis occurring in about 50% of cases.

Diagnosis is made by demonstrating large amounts of methylmalonate in the urine. Increased plasma levels are confirmatory, so that plasma and urine organic acids should be measured quantitatively. One of the variants involves a defect in cellular or subcellular uptake or early metabolism of cobalamin and tetrahydrofolate methyltransferase (Fig. 18–2, site 2). This results in increased excretion of homocystine and cystathionine and in decreased plasma methionine, as well as the characteristic methylmalonic aciduria. A specific diagnosis is accomplished by ruling out B_{12} deficiency and demonstrating the decreased enzymatic activity in liver biopsy tissue. Defects in the cofactor pathway may respond to pharmacologic doses of vitamin B_{12}. In addition, those with the B_{12}-responsive trait run a milder clinical course prior to initiation of therapy.

DISORDERS OF UREA CYCLE METABOLISM

Figure 18–4 depicts the metabolic pathways of the urea cycle. Biochemical defects have been described in all steps. Their prevalence is unknown but presumed to be very low. With the exception of ornithine transcarbamylase deficiency, which is inherited in an X-linked dominant fashion, all others seem to be inherited in an autosomal recessive fashion. The hallmark of the urea cycle disorders is an elevation in blood *ammonia* levels. Hyperammonemia also occurs as a transient neonatal disorder, in severe liver disease (chronic or acute, including Reye's syndrome), periodic hyperlysinemia, methylmalonic acidemia, and propionic acidemia.

FIGURE 18–4. The metabolic pathway of the urea cycle. Defects have been described for each enzymatic step: (1) carbamyl phosphate synthase deficiency; (2) ornithine transcarbamylase deficiency; (3) citrullinemia; (4) argininosuccinic aciduria; and (5) argininemia.

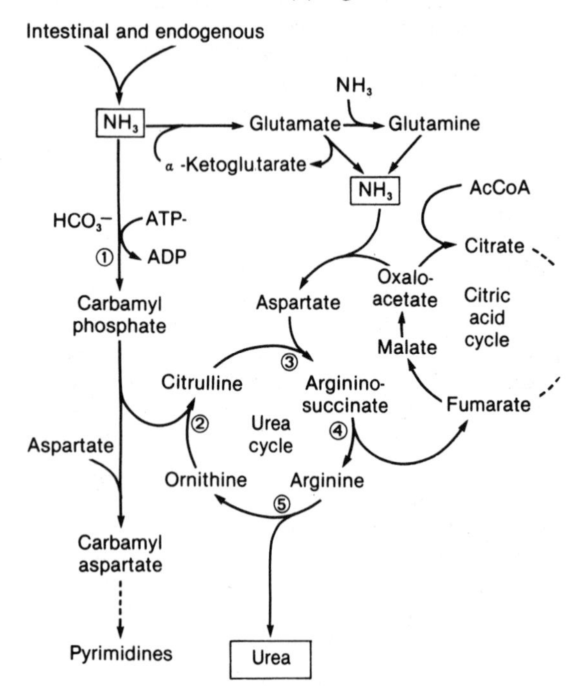

Carbamyl phosphate synthase deficiency (site 1, Fig. 18–4) and *ornithine transcarbamylase* deficiency (site 2) do not exhibit urine or plasma amino acid patterns that are helpful in the diagnosis. The latter does present with increased excretion of orotic acid, a pyrimidine precursor that is synthesized from carbamyl aspartate, but these two conditions must be diagnosed primarily by a process of elimination of other causes of hyperammonemia, and finally by a demonstration of the deficient enzyme activity in liver biopsy tissue.

Plasma and urinary amino acid determinations suggest the diagnosis in the other urea cycle disorders. *Citrullinemia* is diagnosed by finding markedly elevated plasma and urine citrulline. Argininosuccinic acid is poorly reabsorbed from the glomerular filtrate and thus *argininosuccinic aciduria* is detected most readily by urine chromatography, because it and its two cyclic anhydrides are excreted in thousand-fold excess. *Argininemia* is detected by finding increased plasma arginine. In each of these conditions the enzyme deficiency has been characterized.

A condition presenting with hyperornithinemia, homocitrullinemia, and hyperammonemia, and which is responsive to limitation of protein intake, has also been described. Patients with this syndrome exhibit mental retardation, seizures, and episodic attacks of ataxia and lethargy. The basic defect is in the transport of ornithine by hepatic mitochondria, across the inner mitochondrial membrane.

Another condition presenting with *hyperornithinemia* has also been described. Several patients with gyrate atrophy of the choroid and retina have been identified with hyperornithinemia, caused by a deficiency of *ornithine aminotransferase,* an enzyme of the mitochondrial matrix.

DISORDERS OF METABOLISM OF OTHER NEUTRAL AMINO ACIDS

Histidinemia

Histidinemia is an autosomal recessive disorder in which a lack of the enzyme *histidinase* results in elevated levels of histidine in the blood and urine. Mass screening surveys for the presence of histidinemia have shown that the prevalence of the disease may approach that of phenylketonuria (Table 18–1). The initial conclusion was that histidinemia caused mental retardation and/or speech defects. Cases detected by mass screening have not demonstrated any consistent impairment, so the condition may be benign. An alternative view is that impairment may depend on the severity and timing of the histidine elevations.

The condition is *detected* by measuring increased blood and urine histidine. Blood levels are more *consistently* elevated than urine levels but the elevation is not dramatic and may not be recognized with chromatographic screening techniques. Urine levels are often more *dramatically* elevated, but are normal with a low protein intake. In addition, the variance in the excretion of histidine is the greatest of all the amino acids in children and is second only to the variance of glycine excretion in infants. Imidazolepyruvic acid is also excreted in the urine and may be detected with the ferric chloride test; it is reported to be negative to the dinitrophenylhydrazine test.

Increased excretion of histidine occurs also during pregnancy and in a rare condition known as *imidazole aminoaciduria,* which is associated with cerebral macular degeneration. Care must be taken not to confuse histidine with its methyl derivatives, the excretion of which is entirely related to the rate of endogenous protein catabolism and dietary intake.

The Imino Acids

Two types of ***hyperprolinemia*** have been described. Type I is due to a deficiency of proline oxidase, the first catabolic step. Type II is the result of a deficiency of the second enzymatic step, which is the conversion of Δ′-pyrroline-5-carboxylic acid to glutamate. Both conditions are autosomal recessive and are characterized by hyperprolinemia and iminoglycinuria; that is, by increased urinary excretion of proline, hydroxyproline, and glycine. These three compounds share a renal transport mechanism which is saturated by an increased proline concentration. Type II is also characterized by increased excretion of Δ′-pyrroline-5-carboxylic acid.

Clinically, type I was once thought to be related to renal disease, but a careful study of the involved families revealed that the congenital nephropathy was inherited in a dominant fashion, whereas the prolinemia was inherited in a recessive fashion. Both type I and type II may be associated with neurologic symptoms. Seizures or mental retardation are not consistently found, however, and it is therefore difficult to ascribe these to the hyperprolinemia.

Hydroxyproline is an imino acid found primarily in collagen and elastin (connective tissue proteins) and is synthesized posttranslationally by the hydroxylation of certain peptide-bound proline residues. ***Hydroxyprolinemia*** is a rare disorder caused by a deficiency in the first enzyme in its catabolism, *hydroxyproline oxidase.* It is inherited as an autosomal recessive trait and about half of the cases described have had mental retardation. As with hyperprolinemia, a cause-and-effect relationship is uncertain. The two conditions may be unrelated, or may be the result of phenotypic heterogeneity. *Hydroxyprolinuria* is also seen, but without prolinuria and glycinuria. Apparently the excretion of hydroxyproline is not great enough to saturate the common transport mechanism. Detection is possible by urine or plasma screening. *Diagnosis* of hydroxyprolinemia is made by quantification of plasma and urine amino acids.

The three conditions just described, if detected by urine screening, must be differentiated from the iminoglycinuria that occurs in the newborn period (up to 9 to 12 months of age) and is a result of an immature renal tubular resorptive mechanism. Plasma amino acids are normal in this situation, however, and provide an easy means of differentiation.

Hyperlysinemia

Several clinical and biochemical presentations in this rare group of disorders are the result of blocks at different sites in the catabolic pathways of lysine. *Periodic hyperlysinemia* presents in the neonatal period with spells of vomiting, seizures, and coma, which are episodic and related to a hyperammonemia induced by protein feeding. Plasma lysine and arginine are elevated and urinary lysine excretion is increased. The urea cycle enzymes are normal, but the fact that arginase (Fig. 18–4 site 5) is strongly inhibited by elevated lysine levels has been offered as an explanation for the accumulation of arginine. Against this reasoning is the description of several patients with high plasma lysine levels but no increase in plasma ammonia. This condition should be considered along with the urea cycle disorders in the differential diagnosis of hyperammonemia.

Persistent hyperlysinemia is detected by demonstrating elevated plasma and urine lysine levels in the absence of elevated ammonia. Ornithine, ethanolamine, and α-aminobutyric acid are also excreted in excess. About half the patients described have been mentally retarded, but this may be only a coincidental finding. A defect in the enzyme that catalyzes the condensation of lysine with α-ketoglutarate has been found in one pedigree, but it is entirely possible that other metabolic defects are the cause of this disease in other patients.

Another patient has been described who excreted large quantities of lysine and citrulline. Saccharopine, an intermediate of lysine catabolism, was also elevated in plasma and urine. The metabolic defect, although not characterized, would appear to be further along the pathway than in the hyperlysinemias already described, in which elevations of saccharopine were not found.

Glycine Disorders

Nonketotic ***hyperglycinemia*** is a rare disorder that is caused by a deficiency of the glycine cleavage system (*glycine decarboxylase*) which converts glycine to CO_2, ammonia, and a single-carbon tetrahydrofolate derivative. The enzyme system is composed of four protein components: P-protein (a pyridoxal phosphate enzyme), H-protein (a lipoic acid-containing enzyme), T-protein (a tetrahydrofolate-requiring enzyme), and L-protein (a lipoamide dehydrogenase). Absence of P-protein or T-protein have been shown to be associated with this disease.

Glycine utilization is greatly impaired and glycine levels rise in the blood, urine, and cerebrospinal fluid. All patients are severely retarded from the time of birth and suffer from seizure disorders. The disease is detected by either plasma or urine amino acid screening techniques. Diagnosis is made by the quantitative measurement of CSF, plasma and urinary glycine. Secondary hyperglycinemia, resulting from various disorders of branched-chain amino acid catabolism, must be ruled out by organic acid determinations and the absence of ketoacidosis. Neonatal iminoglycinuria is easily distinguishable from this condition by plasma analysis.

Hypersarcosinemia is a rare autosomal recessive disorder of *sarcosine dehydrogenase,* the enzyme responsible for converting sarcosine (*N*-methylglycine) into glycine. Initially it was thought to be associated with mental retardation, but a patient's sibling with normal intelligence and sarcosinemia, and subsequent cases with normal intelligence, suggest that the condition is benign. The defect is characterized by detecting large elevations in blood and urine sarcosine levels.

Two distinct conditions with a grossly increased excretion of ***oxalate,*** a metabolic end-product of glycine, are known. The patients characteristically develop renal disease early in life, caused by intracellular and extracellular deposits of oxalate. Because calcium oxalate stones are regularly found in patients without metabolic disease, the diagnosis must be made by demonstrating excessive excretion of oxalate.

INTERPRETATION OF TEST RESULTS

There are other extremely rare disorders of amino acid and organic acid metabolism described in the literature which have not been included in this chapter. In addition, there are undoubtedly others as yet undescribed. With any unusual result, however, the laboratory scientist or clinician can classify the condition and order the subsequent investigations by posing three sets of alternatives:

1. Is it a metabolic defect or a transport defect?
2. Is it a primary (inherited) or a secondary (acquired) event?
3. Is it a benign or a pathologic condition?

The appropriate sample(s) can be selected, the appropriate test(s) chosen, and a diagnosis sought. Finally, treatment or genetic counseling, if necessary, can be instituted.

From a retrospective, historic look at the aminoacidopathies, the danger of assuming a cause-and-effect relationship is evident. After the discovery that PKU resulted in mental retardation and its treatment could prevent this sequel, large populations of mentally retarded patients were screened for amino acid disorders. The seemingly logical conclusion that mental retardation was the result of a metabolic aberration found in the patient has subsequently been proven incorrect in many instances.

DISORDERS OF PROTEIN METABOLISM

THE SERUM PROTEINS

More than 100 different proteins can be found in the circulation. Together, they amount to between 60 and 80 g/liter of serum and are therefore its most abundant constituent. The serum proteins act to transport metals, hormones, lipids, vitamins, and various noxious substances; they are enzymes, enzyme inhibitors, antibodies, hormones, nutritive materials, buffers, and clotting factors. Some proteins have critical life-supporting functions; others are merely 'in transit,' having leaked out of body tissues; while the purpose of many is not understood. Almost all the proteins of significance are synthesized in the liver, the main exception being the immunoglobulins, which are manufactured throughout the body in plasma cells of the immune system.

The ***laboratory evaluation*** of serum proteins can be carried out in two basic ways:

1. Individual components can be quantified specifically in order to derive information about some distinct disorder.
2. Alternatively, the serum can be separated on a buffer-impregnated inert medium (paper, agar, cellulose acetate) through which an electric current is passed. The individual serum components migrate at different rates from a common starting point. The proteins cease to migrate when the electric current is stopped. Following a staining process a number of bands (usually five) will appear (Fig. 18–5). This analytic process is known as *electrophoresis* and is useful as a starting point or as a screening procedure that may shed light on a variety of conditions.

It is helpful to consider the more important components of serum proteins on a band-by-band basis. They are summarized in Table 18–2 and will be described in more detail.

Prealbumin (M_r 54,000; 0.3 g/L, 30 mg/dL)

Prealbumin is a stable tetramer, the monomer having a M_r of 14,000. Its exact function is not known, although it is recognized to bind thyroxin and to form a complex with retinol transport protein. Prealbumin is depressed in hepatocellular damage, tissue necrosis, and low-protein diets, and it is raised in the nephrotic syndrome. Its measurement has been advocated as a means of assessing short-term protein nutritional status.

Albumin (M_r 69,000; 35–50 g/L, 3.5–5 g/dL)

Albumin is the largest single fraction of the serum proteins. It is a completely defined single

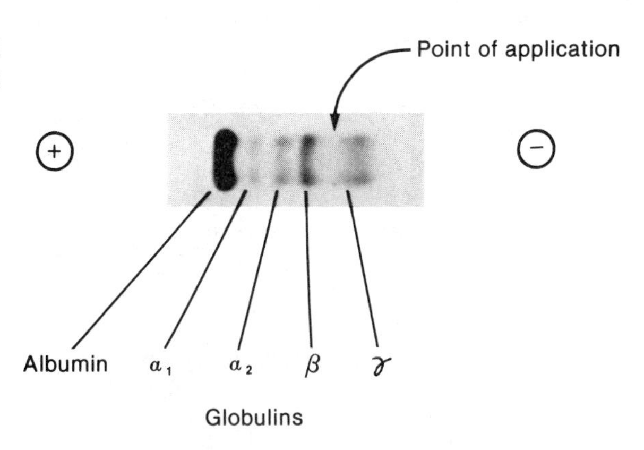

FIGURE 18–5. Separation of the proteins in normal serum by cellulose acetate electrophoresis. ⊕ = Anode; ⊖ = cathode.

TABLE 18–2. Main Components of the Plasma Proteins

Fraction	Reference Range	
	g/L	*g/dL*
Prealbumin	Normally not detected by electrophoresis	
Albumin*	35–50	3.5–5
α_1-Globulin	3–8	0.3–0.8
α_1-antitrypsin,* α_1-lipoprotein (HDL),* α_1-acid glycoprotein, thyroxin-binding globulin,* transcortin*		
α_2-Globulin	4–8.5	0.4–0.85
α_2-macroglobulin, haptoglobin,* α_2-lipoproteins,* ceruloplasmin*		
β-Globulin	5.5–10.5	0.55–1.05
β-lipoprotein (LDL),* transferrin,* hemopexin, complement,* C1 esterase inhibitor,* fibrinogen,* some immunoglobulins		
γ-Globulin	7–17	0.7–1.7
IgG,* IgA,* IgM,* IgE,* IgD*		

HDL = High-density lipoprotein; LDL = low-density lipoprotein.
* Specific protein assay is useful clinically.

polypeptide chain of 580 amino acid residues, arranged in three domains, each with three similar but not identical loops. This structure accounts for its stability, solubility, and binding affinity. Albumin plays several important physiologic roles, including binding and transportation of a variety of molecules, regulation of intravascular osmotic pressure, supplying amino acids for nutritive purposes, providing buffer capacity in the serum, and protecting thrombin from inactivation.

Binding and Transport. Albumin is well designed for its role as a binding agent. It has a net negative charge at physiologic pH, but can interact with both positive and negative ions, organic compounds, and drugs. The attraction between albumin and the substances that it binds can be purely electrostatic; however, the interactions may be enhanced by hydrogen bonding, hydrophobic bonding, van der Waals forces, and occasionally by covalent bond formation. When binding is strong enough, a *molecular complex* may be formed which moves at a different rate from normal albumin and may appear as a separate fraction on electrophoresis. This phenomenon is recognized to occur with penicillin and ampicillin when given in large doses. Binding to albumin has a number of clinical advantages. Harmful materials such as drugs, ions such as mercury, and natural waste products such as bilirubin can be transported in an inert state. Albumin is an excellent transport device for water-insoluble molecules such as fatty acids and bilirubin. Moreover, albumin provides a low-affinity, high-capacity transport for a number of hormones, which can thus be held in quantity within the vascular space.

Colloid Osmotic Pressure. Another major function of albumin is the maintenance of colloid osmotic pressure. The movement of fluid across capillary walls is determined by a delicate balance between vascular hydrostatic pressure (which tends to force fluid out of the vessels into the tissue), tissue pressure, and osmotic forces. The last-named are produced by molecules within the plasma that do not pass the semipermeable capillary wall. Albumin provides about 80% of the intravascular colloid oncotic pressure; it is therefore the major contributor to this effect and is mainly responsible for holding fluid within the vascular compartment. If albumin concentrations are low, interstitial fluid is not reabsorbed effectively and edema results.

Clinical Abnormalities. Clinical abnormalities of albumin concentration are common. A *primary* *increase* in albumin level is never found. It can be raised only through dehydration, and because this phenomenon is rarely greater than 15% of the vascular volume the albumin increase is barely perceptible. It can, however, make a somewhat low serum albumin appear normal. *Decreases* are seen in a variety of conditions. Albumin is normally low at birth but rises to adult concentrations by age one year. A decrease occurs normally after age 70 and during pregnancy. Another normal cause of lowered serum albumin is the supine position. Almost invariably, when a patient is admitted to hospital and confined to bed, orthostatic pressure will be altered in such a way that the serum albumin will show a drop of 2 to 5 g/L (0.2 to 0.5 g/dL).

In addition to hereditary analbuminemia (discussed later), a number of *acquired* disorders produce a reduced serum albumin. Such pathologic causes are usually of great concern. Normally, between 150 mg and 250 mg of albumin per kg of body weight are produced each day. In starvation, the serum albumin concentration will fall when the protein intake is reduced to less than 5 g/day, or after prolonged deprivation. Alcoholics often suffer albumin depletion owing to a combination of malnutrition and poisoning of the hepatic synthetic process. A similar mechanism may be responsible for the albumin depletion observed in chronic infections. Decreases seen in association with acute stress are caused by a catabolic effect on protein metabolism.

Liver disease is a common cause of hypoalbuminemia. It should be recognized that the reserve capacity of hepatic synthesis is great. A considerable reduction of liver function is compatible with normal serum albumin levels. Thus, when the albumin concentration falls as the result of liver disease we can be sure the patient is suffering from a severe and widespread disorder. Albumin levels may fall as a result of *direct* loss (hemorrhage with saline replacement, burns, and exudation), *renal* loss (nephrosis), or *gastrointestinal* loss (notably from Crohn's disease or villous adenoma, which cause weeping of serum into the intestinal lumen). Again, because of the liver's capacity for increased synthesis, such losses must be large for serum depletion to occur. For example, 5 g of albumin per day may be lost into the urine of a nephrotic patient without the serum level being affected, at least for a considerable time.

Effects of Albumin Deficiency. There are several effects of decreased albumin. First, when the albumin becomes low enough (less than 20 g/L [2 g/dL]) the colloid osmotic pressure can no longer be maintained and tissue *edema* will form for this reason alone. Second, the binding capacity will be reduced. This will have a number of notable effects. Metabolic *toxins,* which would otherwise be transported efficiently, may attain higher free concentrations. Also, *kernicterus* in the neonate may occur at lower levels of total bilirubin if albumin is deficient or its binding capacity is impaired.

A decreased level of serum albumin may cause confusion in *interpreting* some laboratory tests. For example, because half of the serum calcium is bound to albumin, a deficiency of this protein will cause the serum *calcium* to appear abnormally low. Formulas have been derived empirically to correct the serum calcium for the amount of albumin (see Chap. 17). When measuring therapeutic levels of *drugs* that are strongly bound to protein (such as phenytoin and digitoxin), one must appreciate that when the albumin concentration decreases, the concentration of the drug will decrease by an equivalent amount without the therapeutic effect of the medication being altered.

Hereditary Disorders of Albumin Synthesis. There are now many reports of two distinct albumin species in family clusters. The trait is rare in European populations (estimates range from 1:1000 to 1:10,000), but polymorphic variants have been reported in as many as 25% of certain North American Indian tribal groups. In these populations homozygotes for a single abnormal albumin are occasionally found. It is because of the homozygote that the term bisalbuminemia has been abandoned for the more proper designation *alloalbuminemia.* Individuals with alloalbuminemia have approximately half of their albumin in the unusual variety and half in the normal form, though this is not a rigid distribution.

Eighteen patients with congenital *absence* of albumin (<0.24 g/L, 0 to 24 mg/dL), inherited as an autosomal recessive trait, have been reported. Curiously, patients with hereditary analbuminemia do not develop severe oncotic edema. Part of the explanation for this may be the compensatory increase in other protein fractions (ceruloplasmin, transferrin, and fibrinogen). Kernicterus is a potential, though not yet recognized feature of this disease. Serum calcium concentrations are low but tetany does not occur because the free calcium is maintained at normal levels. Hypercholesterolemia is common, as are a variety of miscellaneous symptoms.

Acquired bisalbuminemia can be caused by penicillin (potentially by any substance normally bound to albumin) and has been reported in association with pancreatic pseudocysts draining into a body cavity.

α_1 Region Proteins

The α_1 region is the smallest of the five major serum protein fractions. It may be increased in a variety of disorders, including myocardial infarction, acute and chronic infections, collagen diseases, and hepatic or renal malfunction. Decreases are seen in hepatic necrosis, nephrosis, and α_1-antitrypsin deficiency (see later).

α_1-Lipoprotein (M_r approximately 200,000; 1.5 to 3.5 g/L, 150–350 mg/dL). α_1-Proteins consist mainly of high-density lipoproteins (HDL), which are almost 50% protein. HDL can transport lipids, hormones, and fat-soluble vitamins, but their major function is apparently to mobilize tissue cholesterol for transport to the liver. It can be assessed by lipoprotein electrophoresis, where it will be found elevated in association with abetalipoproteinemia. It is absent or very low in Tangier disease (see the Hypolipoproteinemias, Chap. 20).

α_1-Acid Glycoprotein (M_r 44,000; <1 g/L, 30 to 100 mg/dL). α_1-Acid glycoprotein is also known as *orosomucoid* or *acid seromucoid.* A number of hereditary forms of this protein have been recognized, but its function in serum is not known. It has been observed to increase in stress, in the acute-phase reaction (see α_2-Region Proteins), and in chronic infections. It is depressed by hepatocellular injury, the nephrotic syndrome, and malnutrition.

Thyroxin-binding Globulin (M_r 54,000; <20 mg/L, 1 to 2 mg/dL). Thyroxin-binding globulin (TBG) is a single-chain glycoprotein (23% carbohydrate) which serves to transport thyroxin (T_4) and to a lesser extent triiodothyronine (T_3) throughout the body. Hormone-binding proteins such as this are important because they allow a relatively large amount of the potentially active molecule to be transported in an inactive state. Furthermore, being bound to protein, the hormones are lost much less rapidly from the circulation. Genetic absence of TBG is compatible with normal health because free T_4 remains normal.

Corticosteroid-binding Globulin (Transcortin, CBG). CBG serves to bind cortisol in a fashion similar to thyroxin binding by TBG.

α_1-Antitrypsin (M_r 45,000; 1.0 to 2.0 g/L, 100 to 200 mg/dL). α_1-Antitrypsin is found in tears, milk, duodenal fluid, bile, cervical mucus, semen, amniotic fluid, megakaryocytes, and platelets. The name *antitrypsin* is not entirely accurate because the protein is known to have an inhibitory action not only on trypsin but also on elastase, skin collagenase, urokinase, chymotrypsin, plasmin, thrombin, renin, and leukocyte proteases. Thus, the name *α_1-protease inhibitor* is more appropriate. The biologic function of α_1-antitrypsin is to inhibit the action of leukocyte proteases, thereby protecting the lung from the damage which would occur as a result of the multiple inflammatory processes that are constantly taking place in the respiratory tissue. It appears that without α_1-antitrypsin the supporting connective tissue of the lung degenerates, leading to the destruction of alveolar walls and the ultimate development of emphysema.

α_1-Antitrypsin is present as the result of *autosomal codominant inheritance.* Everyone inherits a gene for the protein from each parent, and in turn everyone passes on one of his or her two genes to each offspring. Twenty-six allelic forms of the gene for α_1-antitrypsin are known, and therefore over 200 phenotypes of this particular protein are possible. To date, fewer than 50 have been identified.

A ***genetic classification*** known as the proteinase inhibitor (Pi) system has been formulated. The molecular variants have been designated by various letters: M, F, P, S, W, Z. These combine in groups of two, and the normal form has been designated PiMM. *Reference ranges* vary with the method used, so the result is often expressed as a percent of a normal pool, which has been determined to contain 1.3 g/L (130 mg/dL). The range is 80% to 120% of this value. The *homozygous* PiZZ defect has a prevalence of 1:7000 persons in North America and α_1-antitrypsin values are always <50% (<65 mg/dL). *Heterozygous* PiMZ values tend to be low (60 to 80 mg/dL) but are hard to establish in a patient who is ill. Low values can occur from serum protein loss, in some forms of renal disease, and occasionally in normal individuals. Several other genetic combinations can result in low levels of serum α_1-antitrypsin. These include PiSS and PiMP variants which exhibit about 60% of normal activity, PiNULL, with no detectable activity, and PiMW with near normal activity. Most of these variants are rare and all are found with different frequencies in the world populations that have been studied.

Alpha-1 antitrypsin deficiency is known to cause pulmonary disease, hepatic disease, and several associated illnesses. Of people with *emphysema,* 2% to 5% have been found to be homozygous PiZZ, and as many as 80% of homozygous PiZZ individuals will ultimately develop emphysema of the panacinar type, especially in their lower lobes. Usually the emphysema has a very early onset, often becoming debilitating before the age of 30. Cigarettes appear to be particularly devastating to individuals with this condition. Preliminary reports offer hope of treatment with α_1-antitrypsin injections.

α_1-Antitrypsin deficiency is also associated with *cirrhosis* of the liver. The hepatocytes of these patients contain intracytoplasmic inclusions consisting of a protein which is immunologically the same as α_1-antitrypsin and has the same relative mass. It has been suggested that hepatocellular cancer occurs with increased frequency in these individuals.

There have been a number of *other diseases* associated with α_1-antitrypsin deficiency, including malabsorption, thyroiditis, and membranoproliferative glomerulonephritis. These disorders may be due to the same kind of tissue destruction by white blood cell proteases as is seen in the lung.

The plasma protein α_1 region may be *increased* in acute and chronic infections, in pregnancy, following estrogen or corticosteroid therapy, during malignancy, postoperatively, and following typhoid vaccinations. These factors are important, for they may cause a false-normal level of α_1-antitrypsin to appear in a person who suffers from α_1-antitrypsin deficiency. Although the deficiency can be detected by routine serum electrophoresis, it is best to identify the disease with one of several specific methods. Abnormal findings must be investigated further by conducting Pi typing on the entire family.

α-Fetoprotein (AFP) (M_r 65,000). This protein is a major glycoprotein constituent of *fetal* plasma, reaching levels of about 3 g/L at midpregnancy. It is found in amniotic fluid and maternal serum, but has no established physiologic function, and normal *adult* serum contains only a few micrograms per liter. AFP is *increased* in a variety of hepatic disorders and in some congenital diseases, but in most disorders levels above 500 μg/L are unusual. Values exceeding 500 μg/L are regularly found with teratocarcinoma of the testes, yolk sac tumors of the ovary and some other malignancies; very high values are associated with primary hepatomas, but the value of AFP as a specific marker is limited when levels are <500 μg/L. Its value in screening for open neural tube defects is discussed in Chapter 23.

α_2 Region Proteins

Elevations of the α_2 region proteins are seen in conjunction with a depression of albumin in a number of acute situations, including infections, trauma, and tissue necrosis. This is known as the *acute-phase reaction* (see also Protein Pattern Studies).

Haptoglobin (M_r 85,000 to 160,000; 0.3 to 1.7 g/L, 30 to 170 mg/dL). Haptoglobins are proteins composed of two α chains and one β chain. There are three possible α chains and one form of β chain, which can be combined together in six different ways. Six subtypes of haptoglobin can be identified, hence the wide range in size. The *function* of haptoglobin is to bind to the α chain of hemoglobin and form a hemoglobin–haptoglobin complex which is taken up by reticuloendothelial cells and hepatocytes and removed from the circulation. Some unusual hemoglobins, such as hemoglobin H or Barts, lack an α chain and cannot be bound. Measurements of haptoglobin have been advocated for detecting chronic hemolysis, because the serum haptoglobin level falls as hemoglobin is liberated. Haptoglobin, however, may also be lowered by genetic variation and liver disease and may be increased as much as 10-fold by infections, inflammation, neoplasia, and steroid administration. Thus, there is no good correlation between haptoglobin concentration and the severity of hemolysis. Haptoglobin measurements can help to distinguish myoglobinuria from hemoglobinuria, because it will be raised in the former and lowered in the latter.

Ceruloplasmin (M_r 151,000, 0.2 to 0.4 g/L, 20 to 40 mg/dL). Ceruloplasmin is a blue, copper-containing protein, which acts as a ferrooxidase, enabling iron transport across the intestinal wall, ensuring release of iron from the liver, and facilitating copper's transit through the hepatocyte. It is not a copper-transport protein, because this metal is an intrinsic part of the molecule. Ceruloplasmin is low in liver disease, Wilson's disease, and Menke's kinky-hair syndrome (see Chap. 22). It is raised in the acute-phase reaction and by estrogen therapy. Extreme elevations of this protein can impart a blue-green color to the serum.

α_2-Macroglobulin (M_r 820,000; 1.5 to 3 g/L, 150 to 300 mg/dL). A plasmin and trypsin inhibitor that also has insulin-binding properties, α_2-macroglobulin is a somewhat heterogeneous group of molecules whose exact function is not known. It is elevated in hypoalbuminemia (especially in the nephrotic syndrome), in ataxia telangiectasia, pregnancy, estrogen therapy, and diabetes. It is decreased in some cases of multiple myeloma, peptic ulcer disease, intravascular coagulation, and preeclampsia without proteinuria. No special significance has been attached to its measurement.

α_2-Lipoproteins (M_r 5–10 $\times$ 10^6). These probably include mainly the very-low-density lipoproteins (VLDL) which are found in the *prebeta* region in lipoprotein electrophoresis. Plasma levels are usually <1 g/L (30 to 100 mg/dL), and they contain less than 10% protein. They are a major vehicle for triglyceride transport from the liver to peripheral tissues, and are increased in type 4, type 2B, and type 5 hyperlipoproteinemias.

Other α_2-Globulins. These include cholinesterase, α_2-HS glycoprotein, zinc α_2-glycoprotein, α_2-neuraminoglycoprotein, and erythropoietin.

β-Globulin Region

An elevation of proteins in this region occurs in many situations, including pregnancy in the third trimester, liver disease, and a variety of generalized disorders such as carcinoma and autoimmunity. Dramatic elevations, in association with an abnormal serum cholesterol, are highly suggestive of primary biliary cirrhosis. Compensatory increases are seen in severe albumin loss (nephrosis). Fibrinogen (if plasma is studied) and hemolysis during specimen collection (free hemoglobin) will also produce increases. Depressions are rarely observed, but may occur when there is marked activation of complement (removal of C3).

β-Lipoproteins (M_r 2–2.4 $\times$ 10^6). These include the low-density lipoproteins (LDL), which contain almost 50% total cholesterol. Although they contain only 20% to 25% protein, the high proportion of apoprotein B gives them *beta* mobility. Estimated plasma levels are 1.5 to 4 g/L (150 to 400 mg/dL); they are increased in type 2 hyperlipoproteinemias.

Transferrin (M_r 76,000; 2 to 4 g/L, 200 to 400 mg/dL). The role of transferrin is to bind and transport iron (see Chap. 22), each transferrin molecule having the capacity to carry two molecules of iron. Its concentration is increased in the third trimester of pregnancy, in acute hepatitis, and in iron-deficiency anemia. It is depressed as a result of production failure in various malignancies, liver disease, chronic infections, genetic atransferrinemia, and not infrequently in the elderly. Being about the size of albumin it is lost from the circulation in burns, effusions, nephrosis, and protein-losing enteropathy. A decrease in transferrin may be reflected in a low β-globulin.

Hemopexin (M_r 80,000; 0.7 to 1.3 g/L, 70 to 130 mg/dL). Hemopexin (Hx) binds heme but will not bind hemoglobin. It serves as a scavenger for heme following intravascular red cell breakdown. Hemopexin levels fall in liver disease, burns, and other protein–losing situations and, of course, in hemolytic conditions. Curiously, it is increased in newborn children of diabetic mothers, and in patients with multiple myeloma, Hodgkin's disease, or acute myelomonocytic leukemia. The measurement of hemopexin adds little to the diagnosis of hemolytic disease, but when haptoglobin is low and Hx is normal, hemolysis is unlikely to be serious.

Other β-Globulins. These include fibrinogen, plasminogen, β_2-glycoprotein I, and β_2-glycoprotein II. The complement components and C1 esterase inhibitor are also found in this region (see Chap. 4, Diagnostic Immunology).

γ-Globulin Region

The γ-globulin region is composed of the antibodies, immunoglobulin classes IgA, IgM, IgG, and, to a minor degree, IgD and IgE (see Chap. 4). There are some immunoglobulins that migrate in the α_2 and β regions.

Generalized Hypergammaglobulinemia. Because the γ-globulin region is composed of antibodies it is not surprising that an elevated γ-globulin level will be found in a wide variety of inflammatory conditions. Because inflammatory responses are generalized in nature, many different clones of B lymphocytes will be stimulated. This results in an increased protein content throughout the γ-globulin region. This generalized increase is observed as the result of tissue necrosis and in acute and chronic infections, collagen disorders, hepatic disease, and sarcoidosis. In cirrhosis of the liver the elevation is often so extensive that the distinction between the β and γ regions is obscured. This is known as β–γ bridging (Fig. 18–6, band 4).

Monoclonal Increases. A monoclonal increase in an immunoglobulin of the γ-globulin region is frequently observed and is believed to be produced by a single clone of abnormal cells which synthesize an unusual amount of a specific, homogeneous (i.e., monoclonal) immunoglobulin (also known as an *M component* or *paraprotein*). Because all the abnormally produced protein molecules are identical, they migrate to exactly the same position on the electrophoretic strip and a

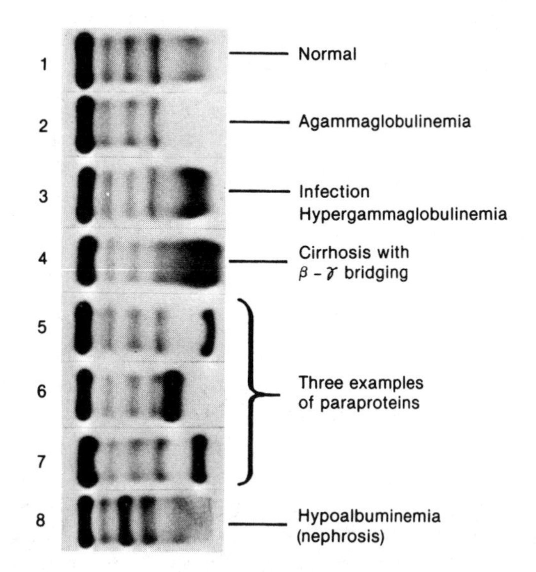

FIGURE 18–6. Representative electrophoretic patterns of normal serum and serum from patients with the disorders shown. The paraproteins represent monoclonal abnormalities referred to in the text.

narrow, discrete band is observed (Fig. 18–6; see also Fig. 4–6A).

To prove the monoclonal nature of a discrete protein abnormality, the serum must be subjected to *immunoelectrophoresis* or *immunofixation.* In these techniques, the serum is first separated by electrophoresis. Next, antisera to the immunoglobulin chains are allowed to react with the separated proteins. A monoclonal abnormality is established when the increased protein is identified as having a single immunoglobulin heavy chain (α, γ, μ, δ, or ε), a single light chain (λ or κ), or a heavy chain or light chain alone.

Heavy-chain diseases in which only heavy chains are produced, are rare. Monoclonal increases in the γ region (and also in the α_2 and β regions) are caused by multiple myeloma, macroglobulinemia, lymphoma, leukemia, other types of cancer, connective tissue disease, and primary generalized amyloidosis (Fig. 18–6 and Table 18–3). *Waldenström's macroglobulinemia* is a generalized malignancy characterized by IgM paraprotein. A number of monoclonal increases are designated 'benign' since they are found in otherwise normal individuals. The term 'benign monoclonal

TABLE 18–3. Prevalence of Paraproteins in Various Disease States

Clinical Disorder	% of Cases Showing Serum Paraprotein
Multiple myeloma	50.9
Nonmyelomatous malignancy (macroglobulinemia, lymphoma, cancer)	18.1
Connective tissue diseases	4.3
Primary generalized amyloidosis	2.5
Benign conditions	24.2

(Data from Ameis A, Ko HS, Pruzanski W: M. components—a review of 1242 cases. Can Med Assoc J 114:889, 1976)

gammopathy' is being replaced by 'monoclonal gammopathy of unknown significance' (MGUS), because many of these otherwise benign conditions undergo conversion to a malignant status at some later time.

The discovery of a monoclonal increase, or *gammopathy,* in a patient's serum is a significant event that warrants further investigation. The first dilemma is whether the monoclonal gammopathy is caused by a specific disorder or whether it is merely a benign, incidental finding. It should be recognized that benign monoclonal gammopathy rises in prevalence with age, being found in 1% of the normal population at age 60 and in 5% or more at age 75. Criteria for distinguishing *benign* from *malignant* paraproteins are shown in Table 18–4.

TABLE 18–4. Criteria for Distinguishing Benign from Malignant Disorders

Criterion	Frequency in Benign Conditions (100 Cases) %	Frequency in Malignant Conditions (500 Cases) %
Bence–Jones (light-chain) protein in the urine	<1	100
Reduction in normal immunoglobulins	10	98
Paraprotein >10 g/L (1 g/dL)	17	92
Increase of paraprotein with time	1	99

The age dependent aspect of the monoclonal gammopathies is not shown. (Hobbs JR: Bence–Jones proteins. Essays Med Biochem 1:105, 1975)

The discovery of *Bence Jones protein* (free immunoglobulin light chains) in the urine means a high probability of a malignant condition. It is detected by electrophoresis of urine that has been concentrated 200-fold. The traditional test for Bence Jones protein involved heating the urine to the boiling point and then allowing it to cool, but this test is not sufficiently sensitive and should not be used. The 'dipstick' test for urine protein will not detect Bence Jones protein.

In *light-chain disease,* either λ or κ light chains are produced. These pass directly into the urine. The serum usually appears to have a deficiency of γ-globulin but the urine contains light chains.

The diagnosis of ***multiple myeloma*** may be considered for clinical reasons or because a monoclonal gammopathy has been discovered. A clear diagnosis requires the presence of a monoclonal gammopathy, anemia, characteristic bony changes on skeletal survey, plasmacytosis in the bone marrow, and Bence Jones protein. Some (or, in fact, all) of these characteristic features may be absent, rendering the diagnosis extremely difficult. Approximately 70% of multiple myeloma patients have IgG paraproteins and 15% have IgA. IgD and IgE myelomas are very rare. More than one paraprotein may occasionally be encountered.

The discovery of a monoclonal gammopathy should initiate analyses of serum *calcium* and serum *viscosity* to give warning of two life-threatening complications of multiple myeloma—high values of either can lead to brain damage, and renal failure. With a viscosity of >4 (caused by the M-protein) plasmapheresis becomes mandatory. Because some benign monoclonal gammopathies are *premalignant,* patients discovered to have such an abnormality should be reexamined every 4 months for one year and then annually. An increase in the paraprotein level suggests malignant change.

Reduced γ-Globulins. A number of syndromes associated with reduced γ-globulins are recognized. Routine electrophoresis is not sufficiently sensitive for the diagnosis of these conditions, and quantification of the individual immunoglobulins must be performed. These tests must be interpreted with caution because the reference ranges vary significantly with age. Low immunoglobulin levels are found in transient hypogammaglobulinemia of infancy, a relatively benign condition.

A selective ***IgA deficiency*** is present in 1:700 persons, some of whom appear normal, while others are susceptible to upper and lower respiratory tract infections and to intestinal malabsorption. If serum IgA is normal, but urinary IgA is low, it indicates a problem in coupling the secretory endpiece. Total absence of IgA results in antibody formation on exposure to non-self IgA, and a subsequent transfusion can result in a severe reaction.

Isolated ***deficiencies of IgG, IgM, IgE, or IgD*** are rare and may be associated with susceptibility to infections. A combined deficiency of IgA and IgG, with normal or increased IgM, produces a syndrome of infections, thrombocytopenia, neutropenia, renal lesions, and aplastic or hemolytic anemia.

X-linked agammaglobulinemia (*Bruton's disease*) is the result of a complete absence of plasma cells. The infant is well for the first 9 months because it possesses maternal antibodies, but after this time susceptibility to pyogenic organisms becomes a serious problem.

Severe combined immunodeficiency results in early death; it is due to an absence of both T and B immunologic cell lines. Globulins and cellular immunity are both absent. In the Wiskott–Aldrich syndrome, over half the patients have low IgM and increased IgE levels. IgA and/or IgE deficiencies are found in ataxia telangiectasia.

CLINICAL VALUE OF SERUM PROTEIN MEASUREMENTS

Specific Protein Determination

Total serum protein quantification is a straightforward assay which can be used to detect (or monitor) gross abnormalities in protein concentration. A decrease in albumin, however, can be countered by an increase in γ-globulins, resulting in no apparent change in total protein.

Albumin measurements are readily available and very useful. When albumin is measured in conjunction with total protein, an estimate of total globulins can be derived. In the past, the albumin and globulin values have been expressed as an A/G ratio, which is much less useful clinically than considering the components separately. When abnormalities are discovered, a protein electrophoresis should be undertaken to define the nature of the abnormality.

Other individual protein determinations can be conducted, usually without difficulty. These tests, and their clinical utility, include the following:

1. Lipoprotein electrophoresis, for the definition of lipid abnormalities
2. TBG, an indirect estimation, for the interpretation of T_4 levels
3. α_1-Antitrypsin, for its deficiency (and genetic typing)
4. Haptoglobin, for detecting hemolysis
5. Ceruloplasmin, to assist in the diagnosis of Wilson's disease
6. Transferrin, for elucidating anemias
7. Clotting factors, α-fetoprotein, immunoglobulins, complement (particularly components C3 and C4), and a variety of enzymes

Protein Pattern Studies

When serum electrophoresis became available, it was soon apparent that certain diseases would create distinct protein patterns (Fig. 18–6). The combination of protein alterations—some rising and others falling—was at first thought to be disease-specific. Early enthusiasm was soon followed by the realization that very few disorders could be diagnosed by this approach. Protein electrophoretic studies should be used to follow-up the discovery of a gross abnormality of albumin or total protein concentration and to initiate the investigation of an immunoglobulin disorder. Electrophoresis is not a disease specific indicator and has no justification as a screening procedure. Interpretation of electrophoretic scans without accompanying clinical information is attempted by some laboratories. Such effort is unwarranted because it rarely contributes to diagnosis or treatment. Electrophoretic separations are best evaluated visually by an experienced worker. The densitometric scan contributes very little to the interpretation and, by itself, is not sufficiently sensitive to allow detection of minor abnormalities. However, a number of fundamental protein patterns have been recognized and classified by Kawai as follows:

Hypoproteinemia Pattern. All protein fractions display depletion in malnutrition, exudative skin losses, exudative pulmonary disease, or essential hypoproteinemia. Often the α_1 and α_2 globulin fractions are slightly increased.

Nephrotic Pattern. The increased permeability of the nephrotic glomerulus allows proteins of increasing relative mass to escape into the urine. Serum albumin is most dramatically affected, and the α_2 fraction (because of the macroglobulin) is commonly raised. Transferrin is a low-relative-mass protein and may be depleted. As the disease progresses, molecules of increasing size will be lost and the α-region proteins may decline. This pattern is seen not only in nephrosis but also in *protein-losing enteropathy.*

Diffuse Acute Hepatodegenerative Pattern. The total serum protein level is usually normal. Albumin is only slightly low, and the α- and β-globulins show variable increases. In severe cases the α_2 fraction will fall. The γ region usually shows a generalized broad increase which commonly merges with the β zone (β–γ bridging).

Hepatic (Alcoholic) Cirrhosis Pattern. A low albumin level and a broad, generalized increase in the γ region with β–γ bridging are the key features of this pattern.

Acute-phase Reaction. In response to acute inflammation or to stress, certain specific proteins are increased. These include α_1-glycoprotein, α_1-antitrypsin, ceruloplasmin, α_2-glycoprotein, α_2-macroglobulin, haptoglobulin, complement, and fibrinogen. Known as the *acute-phase reactants,* these proteins cause an elevation of the α_1 and α_2 regions and, in conjunction with a slight decrease in albumin, yield this pattern.

Chronic Inflammatory Pattern. A decrease in albumin, with increased α_1 and α_2 and a broad, generalized rise in γ-globulins, is seen. This pattern is quite similar to the diffuse, hepatodegenerative pattern. It is observed in chronic infections, allergies, and malignancies.

Broad Hypergammaglobulinemia. The entire γ region is increased as a result of a general stimulus to the production of antibodies. This can occur in a wide spectrum of infections, autoimmune disorders, carcinomas, and degenerative disorders. Indeed, almost any physical disorder can produce such a pattern.

Monoclonal Gammopathies. These have been discussed earlier.

Hyperlipidemia Pattern. Low-density lipoproteins, if dramatically elevated, may produce a monoclonal-appearing band between the α_2 and β zones.

Pregnancy Pattern. In pregnancy the pattern may be normal or may be associated with a decrease in albumin (particularly in the first two trimesters). The α_2 and β regions rise because of a hormonal induction of ceruloplasmin, β-lipoprotein, and transferrin. The γ region is often mildly elevated.

Defect Dysproteinemias. The absence of a major protein type may be detected on routine electrophoretic separation. These conditions have been mentioned earlier; they include analbuminemia, α_1-antitrypsin deficiency, atransferrinemia, lipoprotein deficiencies (not often detected with standard protein electrophoresis), afibrinogenemia (plasma must be studied), and various hypogammaglobulinemias.

In spite of the fact that a variety of patterns have been defined, the reader must appreciate that in a particular patient the pattern may be very difficult to interpret.

Batteries of specific protein analyses, sometimes employing automated analyzers and computers, have been used to develop much more refined diagnostic 'fingerprints.' This is a promising approach which, one hopes, will evolve into a diagnostic technique of considerable utility. Another method for future clinical application is two-dimensional, high-voltage electrophoresis. This technique produces extremely complex protein patterns and may open a whole new concept of protein diagnosis.

SUGGESTED READING

AMINO ACID DISORDERS

DURAN M, WADMAN SK: Organic acidurias. In Alberti KAMM, Price CP (eds): Recent Advances in Clinical Biochemistry, p 103. London, Churchill Livingstone, 1981

GOODMAN SJ, MARKEY SP: Diagnosis of organic acidemias by gas chromatography-mass spectrometry. In

Laboratory and Research Methods in Biology and Medicine, Vol 6. New York, Alan R Liss, 1981

ROSENBERG LE, SCRIVER CR: Disorders of amino acid metabolism. In Bondy PK, Rosenberg, LE (eds): Metabolic Control and Disease. Philadelphia, WB Saunders, 1980

STANBURY JB, WYNGAARDEN JB, FREDRICKSON DS, GOLDSTEIN JL, BROWN MS (eds): The Metabolic Basis of Inherited Disease, 5th ed. New York, McGraw-Hill, 1983

PROTEIN DISORDERS

ALPER CA: Plasma protein measurements as a diagnostic aid. N Engl J Med 291:287–290, 1974

BUEHLER BA: Hereditary disorders of albumin synthesis. Ann Clin Lab Sci 8:283–286, 1978

DANKS DM: Hereditary disorders of copper metabolism in Wilson's disease and Menkes' disease. In Stanbury JB, Wyngaarden JB, Fredrickson DS (eds): The Metabolic Basis of Inherited Disease, 5th ed., pp 1251–1268. New York, McGraw-Hill, 1983

DENNIS PM: The electrophoresis and assay of serum proteins—a changing scene. Pathol 13:651–654, 1981

MORSE JO: Alpha 1-antitrypsin deficiency. N Engl J Med 299:1045–1048, 1099–1105, 1978

RENNERT OM: The hypogammaglobulinemias. Ann Clin Lab Sci 8:276–282, 1978

SOLOMON A: Bence-Jones proteins: malignant or benign? N Engl J Med 306:605–607, 1982

TARNOKY AL: Genetic and drug-induced variation in serum albumin. Adv Clin Chem 21:101–146, 1980

19

Disorders of Purine and Pyrimidine Metabolism

Alan Pollard

Purines and pyrimidines, the nitrogenous components of the nucleic acids DNA and RNA, are also constituents of many other compounds important in intermediary metabolism, including adenosine and guanosine triphosphate (ATP, GTP), cyclic adenosine 3′,5′-monophosphate (cyclic AMP), nicotinamide adenine dinucleotide and NAD phosphate (NAD, NADP), and flavin adenine dinucleotide (FAD). Disorders of cellular control and differentiation such as cancer, may perhaps become explicable in terms of disordered nucleic acid structure or function (Xeroderma pigmentosum is one example). Inborn errors of metabolism and other genetically determined diseases may be viewed as disorders of nucleotide sequencing. At present, however, the most important conditions to understand, both because of their prevalence and because of the light they throw on purine metabolism, are *hyperuricemia* and *gout.* Hypouricemia and orotic aciduria will also be discussed.

PURINES

Excessive quantities of *urate* or *uric acid* in the body can produce one or more of the following manifestations:

1. Hyperuricemia
2. Acute arthritis
3. Chronic arthritis
4. Tophi (tissue deposits of sodium urate)
5. Nephropathy
6. Urinary uric acid calculi

Collectively, conditions 2 through 5 are known as *gout.* Gout without hyperuricemia is very rare, but hyperuricemia without other manifestations is very common. Some gouty patients have uric acid calculi but this condition too can occur in the absence of gout.

Man and the higher apes are extremely hyperuricemic compared to other mammals, with levels more than ten times higher. This is because of a loss of the enzyme uricase and the development of efficient renal tubular reabsorption of urate during the course of evolution. This high level of urate may have important antioxidant and anticarcinogenic functions and may partially replace the high level of plasma ascorbate which was lost at about the same evolutionary stage.

Hyperuricemia should be defined empirically as that level of serum urate at which clinical problems due to insolubility may arise. Instead of establishing reference values by population statistics (e.g., percentile limits, or ±2 SD) it would be preferable to use a physicochemical definition, based on the solubility limit of urate in serum ($\approx$420 μmol/L). Serum urate concentrations in adults are unimodally distributed and skewed to the right. In childhood, mean values are low and show no sex dif-

TABLE 19–1. Prevalence of Hyperuricemia and Gout

	Males*		Females*	
Study	*Hyperuricemia†*	*Gout*	*Hyperuricemia‡*	*Gout*
Ambulant Persons				
Framingham (one occasion)	4.8	0.4		<0.1
Framingham (over 14 years)	9.3	2.3		0.4
Tecumseh	6.0		6.0	
Wensleydale§	2.3		2.3	
3 English towns‖		5–12		
Finland	5.2	0.13	1.2	
France	17			
Maori§	40	10		1.8
Hospitalized patients				
Veteran's hospital	13 (>420 μmol/L; >7 mg/dL) 1.1 (>595 μmol/L; >10 mg/dL)			

* Percent of individuals tested.
† >420 μmol/L; >7 mg/dL.
‡ >360 μmol/L; >6 mg/dL.
§ Male and female.
‖ Males aged 54–74, lifetime prevalence by postal survey.

ference. After puberty they rise to adult levels and then stay relatively constant in men. Values in women are about 60 μmol/L (1 mg/dL) lower than in men until after the menopause, when they rise to approximate the male levels.

Table 19–1 shows that the reported prevalence of ***hyperuricemia*** varies widely in different populations. The determinants of hyperuricemia are multifactorial. They include both genetic and environmental factors (especially diet, alcohol, and drugs), and there is a strong correlation with obesity. Many diseases (especially renal disease and myeloproliferative disorders) cause secondary hyperuricemia. The epidemiology of *gout* is the same as that of hyperuricemia. The incidence of both gout and nephrolithiasis is closely correlated with the extent and the duration of the hyperuricemia, although additional factors are also important.

Chemistry and Biochemistry of Purines

Structure. The structures of uric acid and the urate ion are shown in Figure 19–1, together with the numbering of the purine nucleus. The substituent groups of related compounds are indicated in the legend. The weakly acidic nature of uric acid is due to the ionization of hydrogen atoms at positions 9 (pKa = 5.75) and 3 (pKa = 10.3).

Solubility of Urate and Uric Acid. At the pH and sodium concentration of serum the predom-

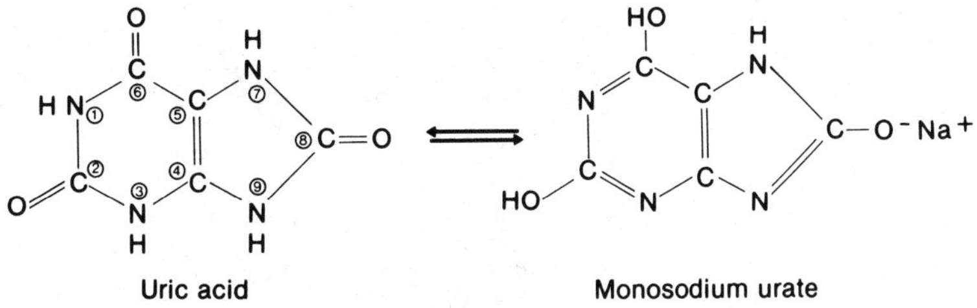

Uric acid ⇌ Monosodium urate

FIGURE 19–1. Structure and ionization of uric acid, 2,6,8-trioxy purine. Related compounds include: xanthine, 2,6-dioxy purine; hypoxanthine, 6-oxy purine; adenine, 6-amino purine; guanine, 2-amino,6-oxy purine.

inant form is urate(⁻). Tissue deposits consist of monosodium urate. A saturating concentration of urate in *serum* at 37°C is about 420 μmol/L (7 mg/dL) but supersaturated solutions can exist, probably aided by protein binding and the presence of other molecules. Solubility is markedly reduced as the temperature falls. In tissues, urate binds to and is solubilized by proteoglycans. In *urine,* uric acid and urate are both present. The proportion that is in the uric acid form varies with the pH (50% at pH 5.7, 90% at pH 4.7), and uric acid is only 1/20 as soluble as sodium urate. This has important implications for uric acid stone formation.

Purine Biosynthesis. The first, and rate-limiting, step in the biosynthesis of purines is the amination of an energy-rich derivative of ribose-5-phosphate, 5-phosphoribosyl-1-pyrophosphate (PRPP), by glutamine, catalyzed by an amidotransferase (Fig. 19–2). This amino group becomes the 9-*N* atom of the purine ring, which is then built up step by step on the phosphoribosylamine foundation by successive reactions with glycine, tetrahydrofolic acid, glutamine, bicarbonate, and aspartate. At least five molecules of ATP are consumed and Mg^{2+} is a cofactor for several steps. The first purine formed is inosinic acid (hypoxanthine-ribose-phosphate).

Purine Interconversion and Salvage Pathways. Inosinic acid is the central compound of purine interrelationships, since from it are made adenylic and guanylic acids (Fig. 19–3). Adenylic acid is broken down to hypoxanthine which, under the influence of hypoxanthine-guanine phosphoribosyl transferase (HGPRT) is largely reconverted to inosinic acid (the 'salvage' pathway), although a small proportion is oxidized further to xanthine and uric acid. Guanylic acid is broken down to guanine, which is also largely salvaged by HGPRT.

Purine Degradation and Excretion. Hypoxanthine and guanine that escape reutilization by the salvage pathway are each degraded to xanthine and thence to uric acid. *Xanthine oxidase* is the enzyme involved for converting both hypoxanthine → xanthine and xanthine → urate. This enzyme is found mainly in the liver and in the small

FIGURE 19–2. Biosynthesis of purine. Ⓟ = phosphate.

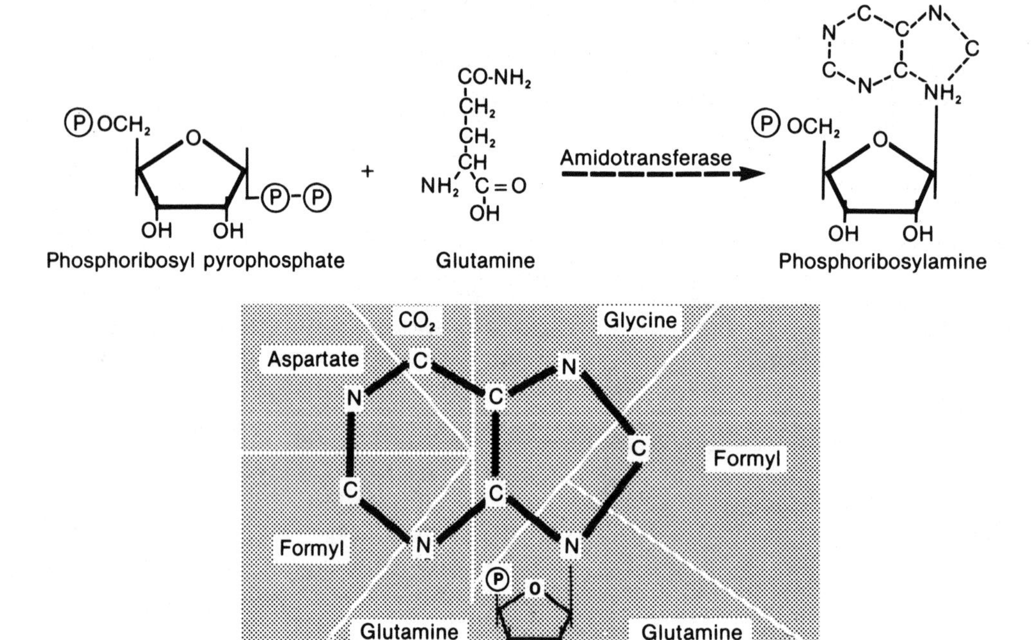

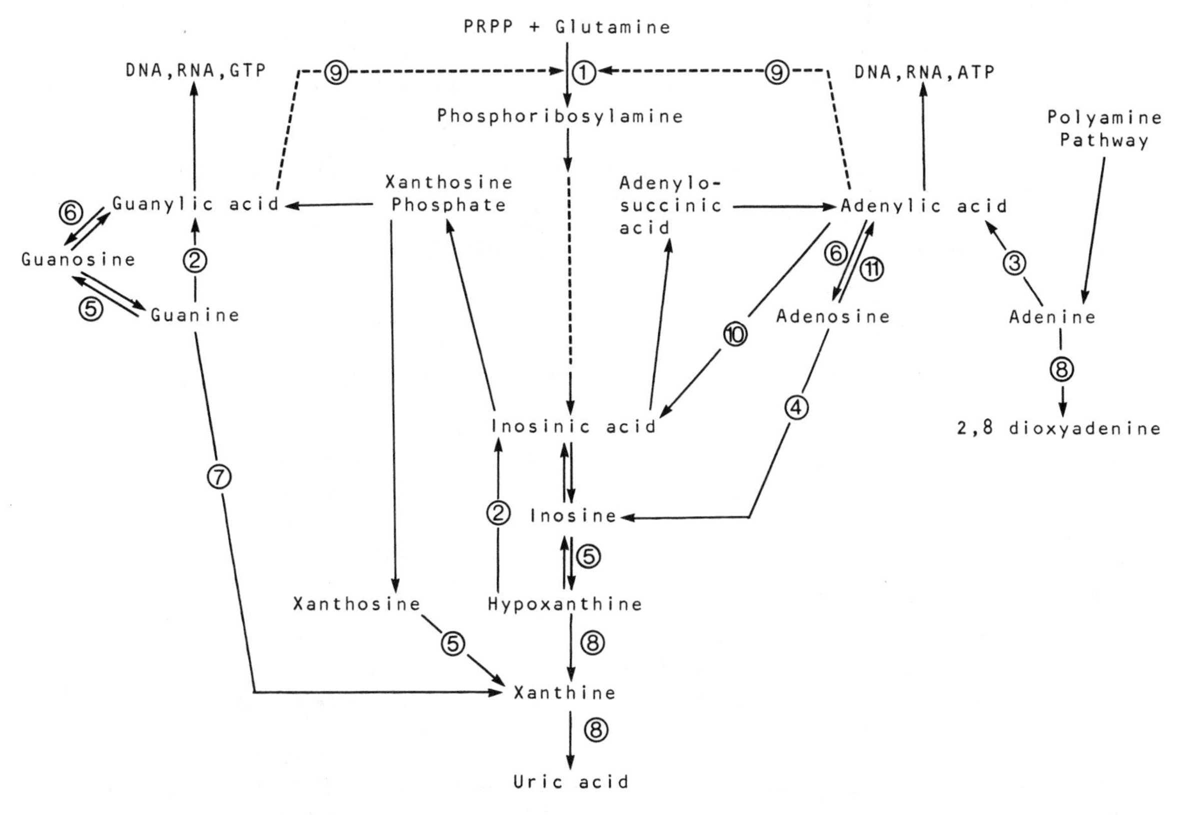

FIGURE 19–3. Purine biosynthesis and degradation—interconversion and salvage pathways. PRPP = 5-phospho-α-D-ribosyl-1-pyrophosphate. The enzymes involved at the numbered sites are: (1) amidophosphoribosyl transferase; (2) hypoxanthine guanine phosphoribosyl transferase (HGPRT); (3) adenine phosphoribosyl transferase (APRT); (4) adenosine deaminase; (5) nucleoside phosphorylase; (6) 5′-nucleotidase; (7) guanase; (8) xanthine oxidase; (9) feedback inhibition of 1; (10) AMP deaminase; (11) adenosine kinase.

intestinal mucosa. Both hypoxanthine and xanthine are more soluble than urate.

Renal Handling of Urate. Earlier it was thought that urate was completely filtered at the glomerulus and then almost fully reabsorbed in the first part of the proximal tubule, with urate appearing in the urine largely as a result of distal secretion. Today it is believed that secretion occurs higher up the nephron and reabsorption lower down, causing a considerable flux of urate as it undergoes first tubular secretion and then reabsorption. This has implications for an understanding of the pharmacology of drugs affecting the renal handling of urate. The renal clearance of urate in man is roughly 10% of the creatinine clearance; thus, the predominant effect is reabsorption (Fig. 19–4). The importance of urate binding by plasma proteins in the solubilization of serum urate and its excretion by the kidney is controversial, but is unlikely to be very great.

A variety of physiologic, pharmacologic, and pathologic effects influence renal urate handling. A water or osmotic diuresis will increase urate clearance. Salt and water depletion, with its resulting decreased tubular flow, allows a greater reabsorption of urate. Many organic anions, typically lactate, β-hydroxybutyrate, and many drugs inhibit urate secretion and so lead to urate retention. In chronic renal failure urate retention occurs,

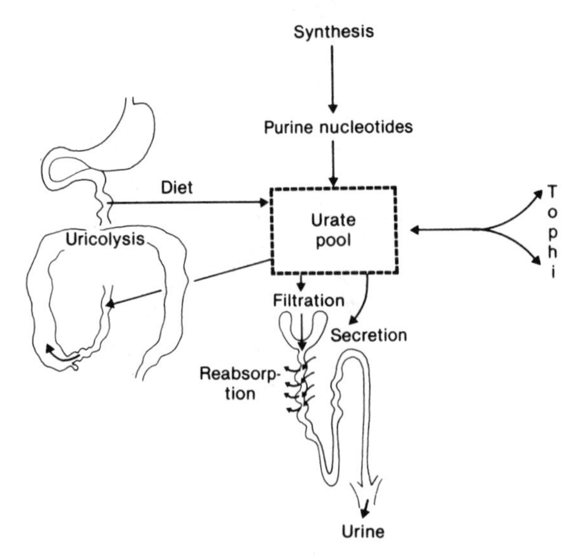

FIGURE 19–4. Urate metabolism and renal handling.

but to a much smaller extent than the retention of urea or creatinine. This is due to an increased urate clearance, perhaps because of the osmotic diuresis, and also increased intestinal uricolysis.

Studies of renal clearance of urate have shown that most gouty patients have diminished clearances and are therefore '*underexcretors*' of urate. This applies also to some of those patients who excrete supranormal quantities of urate (the '*overproducers*' of urate). The reasons for this renal dysfunction are not clear but, in at least some patients, it is not the cause of the hyperuricemia, because it is reversible if the hyperuricemia is abolished.

Uricolysis. Although man lacks uricase, studies of the excretion of labeled urate have shown that 25% to 30% of the urate turned over must be broken down. It can be shown, by sterilization of the gut with antibiotics, that most of this urate is eliminated by intestinal secretion and is broken down by microorganisms in the gut. A small proportion of uric acid turnover occurs by tissue uricolysis, in which urate may be acting to 'mop up' nonspecific oxidative systems. Alterations in uricolysis are not significant causes of hyperuricemia.

Dietary Purines. Removing purines from the diet causes a fall in both serum urate and urinary urate excretion, demonstrating that dietary purines contribute to uric acid production. On switching from an average diet to a purine-free diet the fall in serum urate averages about 60 μmol/L (1 mg/dL) and urine output decreases by about 1.2 mmol/day (200 mg/day). The dietary contribution is almost eliminated by oral allopurinol, a xanthine oxidase inhibitor, suggesting that intestinal xanthine oxidase plays a role in the absorption of purines.

Urate Pools. Serum urate is part of the rapidly miscible pool of urate. Input to the pool is by endogenous synthesis and from dietary purines. The output is via urinary excretion and intestinal uricolysis (Fig. 19–4).

The pool size varies from 860 to 1600 mg (5.1 to 9.5 mmol) in *normal* men and from 550 to 700 mg (3.3 to 4.2 mmol) in normal women (11 to 20 mg/kg body weight). Turnover rates range from 45% to 88% per day (500 to 1100 mg/day). Because such a large fraction of the pool is turned over each day, changes in input to, or output from, the pool can result in rapid and major changes in serum urate concentration. In gouty subjects without tophi, pool sizes range from 1600 to 4000 mg. In chronic gout, tissue deposits of less rapidly miscible uric acid form a second compartment communicating with the miscible pool. This second compartment, the *tophi,* may contain tens to hundreds of times as much urate as the miscible pool. An increase of the urate pool, resulting in hyperuricemia and eventually in gout, is generally the result of an increase in endogenous synthesis and/or a decrease in renal clearance, each from a variety of causes. An increased dietary intake is rarely the cause of gout, but will aggravate it in susceptible subjects. A decrease in uricolysis is not an important contributing factor.

Control of *de novo* Synthesis. The rate of reaction 1 in Figure 19–3 is controlled both by the intracellular concentration of PRPP, which is well below its Michaelis constant (Km) value and hence rate-limiting, and by the concentration of purine nucleotides, which exert feedback inhibition. They appear to do this by promoting the dimerization of the amidotransferase enzyme to an inactive form. PRPP, on the other hand, promotes the splitting of the dimer and activation of the enzyme. The glutamine concentration may also be rate-limiting in some circumstances, but there is less evidence for this.

An increased availability of PRPP, with a resulting increase of purine synthesis, could possibly be due to any of a number of metabolic abnormalities in the pentose phosphate or purine interconversion pathways. Three of these are well documented in man and are described later. In the vast majority of cases of purine *overproduction,* however, no specific metabolic lesion has been found.

Serum and Urine Urate

Measurement. The most specific method uses uricase and ultraviolet spectrophotometry. A variety of colorimetric adaptations of the uricase method are available and correlate well, but some are not suitable for urine. Nonenzymatic AutoAnalyzer methods using phosphotungstate reduction are relatively specific if combined with a preincubation step with carbonate to destroy interfering substances. Other methods, notably that for the SMA 12/60, give values that are 30 to 60 μmol/L (0.5 to 1 mg/dL) higher than the uricase methods.

Table 19–2 gives *reference ranges* for serum urate, of which many studies have been published. One of the most extensive, the Tecumseh study, which used the uricase method, included values for 6000 people; the distributions were clearly nongaussian and were age- and sex-related. Therefore the central 95th percentiles, rather than 2 standard deviation limits, are given in Table 19–2 (see comments on p. 444). An individual's serum urate values are not under tight homeostatic control, being affected by diet, by many drugs, and by season, being higher in summer. Certain ethnic groups (e.g., Maoris) have higher than average uric acid levels, but studies of transplanted ethnic groups (e.g., Philippinos who have moved to the United States) suggest that these differences may be environmental as well as genetic.

Interpretation of Serum Urate Levels. Hyperuricemia is not synonymous with gout and, unless extreme, does not demand treatment. The *prognosis* of hyperuricemia, in terms of the likelihood that gout will develop, depends on the extent and duration of the elevation. In the Framingham study only 15% of men with an initial level between 420 and 470 μmol/L (7.0 to 7.9 mg/dL) developed acute gout over the next 14 years, and thus most people with mild to moderate hyperuricemia never develop gout. Five out of six men with levels over 535 μmol/L (9 mg/dL) developed gout within 14 years. In the presence of acute arthritis, and the absence of drug therapy, normal levels of urate would not absolutely rule out the diagnosis of gout but would make it very unlikely. Most patients with acute gout have only mild to moderate elevations of their serum urate, such levels being much commoner than really high ones.

Interpretation of Urine Urate. With an average diet the upper limit of normal urate excretion is about 4.2 mmol (700 mg)/day. About 20% of gouty patients, the urate 'overproducers,' consistently excrete more than this. It is among this subgroup that the few cases of specific enzyme disorders have been identified. The urate output usually falls by 100 to 300 mg/day with a purine-free diet, which should be tried before the patient is classified as an overproducer. The incidence of uric acid *stones* correlates with the 24-hour urine excretion of uric acid, but also depends on the extremes of acidity of the urine and on the maximum concentration reached at any time in the 24-hour period. This

TABLE 19–2. Serum Urate Reference Values

	Male		Female	
	μmol/L	*mg/dL*	*μmol/L*	*mg/dL*
Tecumseh study (central 95th percentile)	150–440	2.5–7.4	130–405	2.1–6.8
Other studies*	410–460	6.9–7.7	340–395	5.7–6.6
Children (mean value)	220	3.6	220	3.6

* Range of values for upper reference interval (mean ± 2 SD).

correlation, therefore, is far from exact and many patients with uric acid stones have neither gout nor hyperuricosuria.

Urinary urate outputs cannot be interpreted in the presence of drugs that affect uric acid metabolism or excretion, and the analytic method used should be specific.

Pathology of Gout

Pathogenesis of Acute Crystal Arthritis. The essential diagnostic feature of acute gouty arthritis is the presence, in the *synovial fluid,* of sodium urate crystals, often engulfed by leukocytes. This may be demonstrated by examining the fluid microscopically under polarized light, using a first-order red compensator. Needlelike crystals are characteristic of urate. They appear yellow when their long axis is parallel to that of the compensator and blue when at right angles to it. This distinguishes gout from a similar acute arthritis (pseudogout) caused by intraarticular crystals of other substances, most commonly calcium pyrophosphate dihydrate (see Chap 13, Arthritic and Rheumatic Disorders) which show the opposite color characteristics.

Typical acute arthritis can be produced *experimentally* by the injection of sodium urate crystals of the right size into a normal joint. The crystals cause a leukocyte response; the leukocytes ingest the crystals and release chemotactic factors, kinins, and lysosomal enzymes. These enzymes have been shown to attack a variety of cells, to break down cartilage, and to increase capillary permeability. The resulting inflammatory response constitutes the attack of acute gouty arthritis. In *spontaneous* acute gouty arthritis it is not clear why the initial precipitation of urate occurs, but several clinical triggering events have been recognized. These include trauma or intercurrent illness, alcohol excess, and any sudden change in the serum urate level (even a reduction). It is likely that deposits of urate crystals (microtophi) exist silently in the synovial tissues of affected joints and the rupture of a microtophus into the joint space may be the precipitating event. Extracellular urate crystals can also be found in the nonaffected first metatarsophalangeal joints of many patients with gout and in a minority of patients with hyperuricemia but without clinical gout.

Pathogenesis of Chronic Gouty Arthritis. Pathogenesis of chronic gouty arthritis appears to be a separate process from the acute attack and is related to the deposition of tophi in the joint and periarticular tissues. Experimentally, cultured synovial fibroblasts will produce large quantities of collagenase (which attacks cartilage) and of prostaglandin E2 (which decalcifies bone) when stimulated with sodium urate crystals. Tophi occur in rare instances without a history of acute arthritis.

Connective Tissue Metabolism and Urate Deposition. The characteristic distribution of tophi in chronic gout shows their marked predilection for cartilage and other connective tissues, particularly tissues that are relatively avascular. (An exception to this rule is the kidney, which has very high urate fluxes). *Proteoglycans,* which make up a major part of connective tissue, have a strong solubilizing effect, producing supersaturated solutions of uric acid. Local damage to such tissues could cause a sudden withdrawal of this effect, with a resulting precipitation of urate. Urate solubility is also strongly temperature-dependent, so that local cooling of an avascular area could be a factor in the classic location of acute gout in the big toe starting in the middle of the night.

Urate and the Kidney. In chronic gout the deposition of microcrystals of sodium urate in the renal medulla (especially near the loops of Henle) is an early change. Later, microscopic tophi appear, with a surrounding giant cell reaction and fibrosis. The microscopic changes elsewhere in the kidney parenchyma resemble those of pyelonephritis but without the signs of infection, and vascular changes are also accelerated. In renal failure from any other cause, secondary hyperuricemia develops in the advanced stages and urate deposits may occur in the kidney. There is controversy as to whether prolonged hyperuricemia, in the absence of gout, renal calculi, or hypertension, causes nephropathy. Recent evidence suggests that deterioration of renal function is measurable but very slight.

A rare disease called *familial gout with renal failure* has been described in which renal deterioration occurs early in the disease and may be very severe. The condition also affects girls, even in childhood, and the hyperuricemia may be underestimated unless reference values appropriate for age and sex are used. Strict control of hyperuricemia and of hypertension must be implemented if a disastrous outcome is to be averted.

Pharmacology of Gout and Uric Acid

Drugs Used in the Treatment of Acute Gouty Arthritis. ***Colchicine*** has been used for centuries in the treatment of gout, yet its mode of action is still not clearly understood. It is rapidly taken up from the plasma by cells, especially leukocytes, and it is known to have many different actions, ranging from inhibition of motility, through diminished release of chemotactic factor, to prevention of mitosis. It is not clear which of its many effects is responsible for the specific action in suppressing acute gouty arthritis, which it does without affecting serum urate levels or urate pool size, but the prevention of chemotactic factor release is a likely candidate. Large doses are usually required in acute attacks and then gastrointestinal side effects become a problem. In small doses it is a useful prophylactic against acute attacks during the first few months of therapy with uricosuric agents or allopurinol.

Indomethacin is an example of the many nonsteroidal antiinflammatory drugs which are available to treat acute attacks. They vary in their side effects and some are uricosuric, so should not be used before the diagnosis is established. *Colchicine* and *phenylbutazone,* being toxic, are no longer advised for treating acute attacks. All these drugs are most effective when given early in the attack and may be ineffective if delayed, in which case intraarticular steroid may be needed.

Drugs Used in the Reduction of Hyperuricemia. The principal uricosuric drugs are ***probenecid*** (Benemid), ***sulfinpyrazone*** (Anturan), and ***benzbromarone.*** They act by inhibiting renal tubular reabsorption of urate and thus reduce the urate pool size and plasma urate. They have no role to play in the treatment of acute gouty arthritis—in fact, they may precipitate acute attacks during the early months of treatment, and this should be prevented by prophylactic doses of colchicine. Probenecid affects the metabolism of many other drugs, including penicillin, sulfobromophthalein (BSP), and heparin, and should be used with care in combination therapy. Sulfinpyrazone also has the effect of reducing platelet adhesiveness, independent of its uricosuric action, and is used for this purpose in a variety of vascular disorders. Uricosuric drugs are indicated only when renal function is good and when urate excretion is not already high. There is a danger of urate deposition in the tubules early in treatment, which should therefore begin with small doses and be accompanied by a high fluid intake.

Xanthine Oxidase Inhibitors. ***Allopurinol*** is an isomer (7-C, 8-N) of hypoxanthine and is a potent inhibitor of xanthine oxidase. It is also a substrate of xanthine oxidase and the product, *oxypurinol* (an isomer of xanthine) is another inhibitor of the enzyme. The half-life of oxypurinol is much longer than that of allopurinol, and so it is quantitatively the more important inhibitor *in vivo.*

With xanthine oxidase inhibition, hypoxanthine and xanthine accumulate and, being more soluble than uric acid, are readily excreted in the urine. The increase in hypoxanthine and xanthine is substantially less than the corresponding decrease in uric acid production, indicating a decrease in *de novo* synthesis. HGPRT is required for this effect (Fig. 19–3, site 2), which results from the recycling of the excess hypoxanthine to inosinic acid with the utilization of PRPP, which could otherwise be used for *de novo* synthesis. (PRPP may also be reduced by combination with allopurinol to form its nucleotide). Allopurinol may also reduce the absorption of dietary purines.

Allopurinol also potentiates the action of other drugs by its inhibition both of xanthine oxidase (which destroys azathioprine and 6-mercaptopurine) and of hepatic microsomal enzymes, which are involved in the metabolism of such drugs as the coumarins and probenecid. Allopurinol and oxypurinol are excreted by the kidney, so the dose should be reduced in renal impairment.

Other Drugs with Effects on Urate Handling

Salicylates in low dosage (<2 g/day) raise the serum urate level by reducing tubular secretion, whereas higher doses are uricosuric, reducing reabsorption. In any dose, salicylates interfere with the uricosuric effects of probenecid and sulfinpyrazone, and the nature of this interaction is not well understood. *Gentisic acid,* a metabolite of salicylate, may interfere with colorimetric urate measurements, giving falsely high values (as will *homogentisic acid,* which is excreted in alkaptonuria).

Thiazides and most other ***diuretics*** commonly cause hyperuricemia, although promising new diuretic compounds which are actively uricosuric, are being evaluated. The mechanism is probably complex and is partly a nonspecific consequence of sodium and water depletion. Uricosuric drugs are active against thiazide-induced urate retention. Diuretics cause as many as 50% of new cases of hyperuricemia in some studies, particularly among hospitalized patients.

The antituberculous drug ***pyrazinamide*** has a potent antiuricosuric effect, mainly because of its metabolite pyrazinoic acid, which blocks tubular secretion of urate and has been widely used experimentally for this property. The uricosuric drugs are blocked by pyrazinamide, but this effect in turn is reduced by salicylates (including *p*-aminosalicylic acid, another antituberculous drug). It is intriguing that pyrazinoic acid is also a substrate for xanthine oxidase.

Nicotinic acid, used in large doses in the treatment of hypercholesterolemia, raises the serum uric acid level by a complex mechanism, including a decreased urate clearance. The effect is not blocked by salicylate nor overcome by uricosuric agents.

There is an association between ***ethanol*** ingestion in gouty subjects and the incidence of acute attacks. Experimentally, administration of ethanol causes an increased serum urate, even in normal subjects, and 25% of alcoholics are hyperuricemic (although fewer than 1% have gout). The mechanisms are complex and may include hyperlactatemia, increased *de novo* synthesis of urate, fluid and electrolyte depletion, and the high purine content of some alcoholic beverages. Serum urate levels may be useful in monitoring alcohol ingestion in individual patients.

Cytotoxic drugs can cause extremely high levels of urate (up to 3000 μmol/L (50 mg/dL) when given in the treatment of malignancy. The rapid breakdown of malignant cells gives rise to a vastly increased load of purine nucleotides to be degraded to uric acid. In anticipation of this effect, which is likely to cause acute renal failure, such patients are pretreated with allopurinol, fluid loading, diuretics, and alkalinization of the urine.

Other drugs which are known to cause hyperuricemia by a variety of mechanisms include ethambutol (antituberculous agent), levodopa (antiparkinsonian agent), α-methyldopa (antihypertensive), methoxyflurane (anesthetic), ethyl aminothiadiazole (antitumor agent), and intravenous fructose.

Enzyme Disorders and Purine Metabolism

Hypoxanthine-Guanine Phosphoribosyltransferase (HGPRT) Deficiency (site 2 Fig. 19–3). Virtually complete absence of HGPRT, or of its activity, occurs as a rare X-linked genetic disorder known as the ***Lesch–Nyhan syndrome.*** This is a disease of male children in which overproduction of uric acid leads to hyperuricemia, hyperuricosuria, and often uric acid lithiasis. A striking retardation of growth and mental development occurs, and also a bizarre neurologic syndrome of choreoathetosis, spasticity, and compulsive self-mutilation by biting and scratching. Gouty arthritis is rare.

The overproduction of uric acid is related to an increase of PRPP as a result of the HGPRT deficiency and may also be caused by reduced feedback inhibition by purine nucleotides. In this syndrome allopurinol causes a marked increase in hypoxanthine and xanthine production because its usual effect of reducing *de novo* synthesis is lacking. Allopurinol is still very beneficial in reducing the uric acid load and preventing arthritis, and uric acid stones, although xanthine stones may occur. However, it has no beneficial effects on the neurologic disorder.

Because there is no effective treatment for this devastating condition genetic counseling is most important. Female relatives can be typed as possible heterozygous carriers with the use of cultured skin fibroblasts, and cultured fetal cells from amniocentesis can be typed for sex and for enzyme activity.

A partial deficiency of HGPRT has been demonstrated in a few kindred with gout and marked overproduction of uric acid. Most of these kindred have been neurologically normal but a few have had mild neurologic syndromes.

Phosphoribosylpyrophosphate (PRPP) Synthetase Overactivity. This enzyme produces PRPP from ribose-1-phosphate and ATP. PRPP is both a substrate for the rate-limiting step of purine biosynthesis and an activator of its catalyzing enzyme, PRPP amidotransferase. Among the many gouty patients with urate overproduction who have been screened, a very few families have shown an increased activity of PRPP synthetase. It is extraordinary that among these no fewer than three different biochemical mechanisms have been defined for the overactivity, each running true in family members. The gene for PRPP synthetase is also on the X chromosome.

Glucose-6-Phosphatase Deficiency. This causes type I ***glycogen storage disease*** (von Gierke's disease; see Table 16–14), resulting in hypoglycemia and increased intracellular glucose-6-phosphate (G6P) concentrations. Hypoglycemia leads to chronic increases in levels of free fatty acids, lactate, pyruvate, and ketone bodies, which in turn reduce renal clearance of urate. It also leads to glycogenolysis with depletion of intracellular ATP leading to increased purine synthesis. Continuous infusions of glucose over several days will normalize the serum urate. Increased G6P concentrations might be expected to lead to overproduction of PRPP, via the pentose phosphate shunt pathway, and hence to increased *de novo* synthesis of urate, but this remains to be proved.

Hyperuricemia and gout are frequent accompaniments of glycogen storage disease type I and usually warrant allopurinol treatment. Uricosuric drugs tend to be ineffective in the presence of the high levels of lactate and other organic acids.

Adenine Phosphoribosyl Transferase (APRT) Deficiency. Heterozygotes for this mutation (Fig. 19–3, enzyme 3) are common but show no clinical disorder. Homozygotes develop renal calculi consisting of 2,8-dioxyadenine which may be mistaken for uric acid on chemical testing. They can be distinguished readily by infrared or x-ray crystallographic analysis.

Other Enzymes. From current knowledge of purine metabolic pathways, several other enzyme abnormalities have been proposed as likely causes of purine overproduction in gout, but to date have not received experimental confirmation in man. These include deficiencies of glutaminase and glutamate dehydrogenase and an increase in glutathione reductase. It has been suggested that a defect in the regulatory properties of AMP deaminase (Fig. 19–3, enzyme 10) could be the prime cause of hereditary gout, and this hypothesis is now being tested. In most patients with urate overproduction, however, the cause is unexplained.

Adenosine Deaminase, Nucleoside Phosphorylase, and Immune Deficiency Disease. Adenosine deaminase (site 4 in Fig. 19–3) converts adenosine to inosine, and nucleoside phosphorylase (site 5) converts inosine and guanosine to hypoxanthine and guanine, respectively. In the rare congenital deficiencies of these enzymes uric acid production is reduced, but more importantly, severe defects in T- and B-cell function occur. Intracellular accumulation of deoxyATP, deoxyGTP, and of S-adenosylhomocysteine, which are toxic to DNA synthesis and to methylation reactions respectively, are probably responsible. Experimentally, provision of the missing enzyme in the case of adenosine deaminase has caused a temporary remission of the immune deficiency. Bone marrow transplantation may provide a cure. 5′Nucleotidase deficiency has also been implicated in some patients with immune deficiency. Patients presenting with immune deficiency syndromes should now be checked for these autosomal recessive enzyme disorders, which have both etiologic and therapeutic implications.

Muscle Adenylate Deaminase Deficiency. This autosomal recessive disorder results in a myopathy with easy fatigability following exercise. The enzyme defect is confined to muscle and the clinical features are relatively benign.

Xeroderma Pigmentosum. Xeroderma pigmentosum is an inherited disorder of DNA repair mechanisms. Several different specific enzyme defects have been described. Affected patients develop skin cancers on exposure to sunlight and/or a variety of neurologic disorders.

Differential Diagnosis of Gout by Laboratory Tests

Acute Arthritis. The most important differentiation is from acute *septic arthritis,* because the treatments for the two conditions are mutually harmful. The distinction is best made by aspirating joint fluid and examining it bacteriologically and by polarized light microscopy. This will also separate gout from pseudogout (in which the crystals have an opposite birefringence from urate), although gout and pseudogout may coexist.

Other conditions in which acute arthritis can be mistaken for gout must be distinguished on clinical grounds and by synovial fluid examination. These disorders include psoriatic and sarcoid arthropathy (in both of which the serum uric acid is often raised), Reiter's syndrome, and rheumatic fever. Many patients with joint pains take salicylates, which may result in hyperuricemia, so that the coincidence of arthritis and hyperuricemia is not sufficient to make a diagnosis of gout. Conversely, therapy with phenylbutazone or high-dose salicylate may mask hyperuricemia in cases of gout.

Chronic Gouty Arthritis. On superficial examination this may be confused with chronic rheumatoid arthritis or osteoarthritis. Biopsy of a periarticular infiltrate will reveal sodium urate crystals.

Tophi. These urate deposits may discharge spontaneously through the skin. The chalky material should be examined chemically and by microscopy. Failing a discharge, a tophus can be biopsied with a hypodermic syringe and needle.

Renal Calculi. Uric acid calculi must be distinguished from stones containing Ca and Mg, and from cystine stones. Radiolucency is one distinguishing feature. Spectroscopic examination of the calculus, if it is passed, will clinch the matter. In the presence of hyperuricosuria the incidence of calcium-containing stones is also increased, and uric acid crystals may form the nidus on which such stones develop. Whether or not this is the case, treatment with allopurinol may be indicated whenever stones containing uric acid occur, or when calcium-containing stones occur in a patient with hyperuricemia or hyperuricosuria. There is also an increased incidence of uric acid stones in cystinuria.

Chronic Renal Failure. If a patient presents with chronic renal failure and hyperuricemia it may be difficult to decide which came first. It is important to know whether hyperuricemia is the cause because allopurinol treatment might alleviate the condition. Measuring the uric acid/creatinine ratio in the urine may help. A ratio >0.7 (measured in mmol/L) or >1.0 (measured in mg/L) would indicate a primary urate nephropathy, while a ratio <0.35 (measured in mmol/L) or <0.5 (measured in mg/L) would indicate primary renal disease. Intermediate values are less helpful.

Summary of Clinical Conditions Associated with Hyperuricemia

Primary gout
- No predisposing causes
- Positive family history
- Male or postmenopausal female
- Urate overproducers (20%; a very small percentage of these will have an identifiable enzyme defect)
- Familial gout with renal failure

Increased nucleic acid turnover
- Myeloproliferative disorders and neoplasia
- Paraproteinemias
- Psoriasis
- Sarcoidosis
- Sprue

Drugs and toxins
- Beryllium poisoning
- Cytotoxic drugs
- Diuretics
- Ethanol
- Intravenous fructose
- Lead poisoning (especially plus alcohol)
- Pyrazinamide
- Salicylates

Renal causes
- Renal failure from any cause
- Preeclampsia (serum urate is useful prognostically and in following progress)
- Chronic hemodialysis (arthritis is usually pseudogout)

Endocrine, metabolic, and miscellaneous causes
- Hypo- and hyperparathyroidism
- Hypothyroidism

Obesity
Ketoacidosis (including diabetes, exercise, and starvation)
Alcoholism
Hyperlipoproteinemia (type IV)
Down's syndrome
Bartter's syndrome
Paget's disease of bone
Primary oxaluria
Cystinuria
Hereditary fructose intolerance

Hypouricemia

Hypouricemia, which may be defined arbitrarily as a serum urate level of less than 120 μmol/L (<2 mg/dL), occurs in about 1% of the population. There are many possible causes, a few of which will be described briefly. In many cases of hypouricemia a cause cannot be found, and some cases are transient.

Congenital Xanthine Oxidase Deficiency. In this very rare genetic disorder the enzyme is missing or inactive. Serum urate levels are usually less than 60 μmol/L (<1 mg/dL) and urinary urate is very low. Urine hypoxanthine and xanthine are very high and xanthine stones may occur. Xanthine deposits may also occur in muscle. The lack of serious consequences in this disorder gave early encouragement to the use of allopurinol in gout to blockade xanthine oxidase. The lifelong use of such an enzyme inhibitor might otherwise have been anticipated to cause serious side effects.

Isolated Renal Tubular Defect in Urate Reabsorption. In this rare genetic disorder the serum urate is low but the urine urate is high, clearances from 30% to 50% of the creatinine clearance being observed.

Other Causes. These include:

1. The Fanconi syndrome, with increased urate clearance
2. Advanced liver disease, which causes an increased renal clearance of urate and may also decrease production
3. Some neoplastic diseases

Drug Ingestion. The commonest cause of hypouricemia is the ingestion of drugs, including salicylates, allopurinol, uricosurics, and guaifenesin, and from the use of x-ray contrast media, which are strongly uricosuric.

PYRIMIDINES

The pyrimidine bases, thymidine, uracil, and cytidine, complement the purines in DNA and RNA. Biosynthesis of pyrimidines is from glutamine, aspartic acid, CO_2, and NH_3 (Fig. 19–5). Control of the pathway is through feedback inhibition by uridylic acid on the first, rate-limiting step (site A in Fig. 19–5). The requirement for PRPP in uridylic acid synthesis interlinks the control of purine and pyrimidine metabolism.

Hereditary Orotic Aciduria

This very rare genetic disease is associated with enzyme deficiencies at steps B and C in Fig. 19–5. This results in a deficiency of pyrimidine compounds and an excess of orotic acid from the uninhibited pathway. The pyrimidine deficiency leads to megaloblastic anemia and physical and mental retardation. The orotic acid excess causes crystal-

FIGURE 19–5. Biosynthesis of pyrimidines, PRPP = 5-phospho-α-D-ribosyl-1-pyrophosphate. The circled letters A, B, and C are referred to in the text. Ⓟ = phosphate.

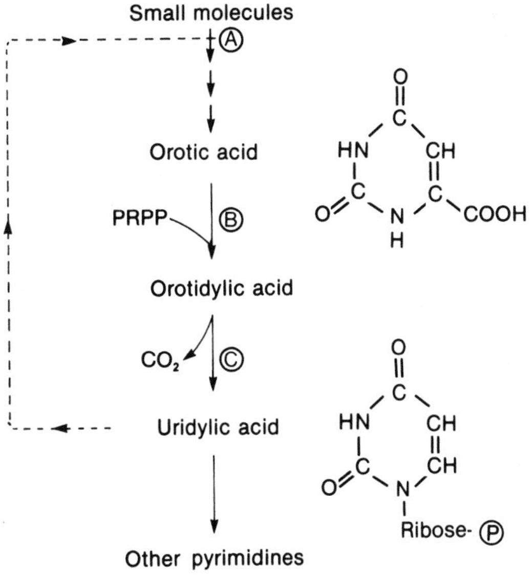

Small molecules
Ⓐ
Orotic acid
PRPP Ⓑ
Orotidylic acid
CO_2 Ⓒ
Uridylic acid
Other pyrimidines
Ribose-Ⓟ

luria which may lead to urinary obstruction. Uridine therapy results in improvement in all features of the disease. Orotic acid crystals may be *identified* by their spectral properties after recrystallization, and the orotic acid excretion may be *quantified* by chromatographic or enzymatic methods. The specific enzyme deficiencies should be identified in hemolysates. Increases in urinary orotic acid are also seen in pregnancy, in allopurinol therapy, after certain antitumor drugs, and in other enzyme defects of the pyrimidine pathway.

β-Aminoisobutyric Aciduria

β-Aminoisobutyric aciduria is a common and clinically benign inherited disorder of thymine catabolism.

Pyrimidine 5′-nucleotidase Deficiency

Pyrimidine 5′-nucleotidase deficiency is a rare cause of inherited hemolytic anemia. Basophilic stippling is caused by deposits of undegraded pyrimidine nucleotides. In *lead poisoning* this enzyme (among others) is inhibited and hemolytic anemia with basophilic stippling is also seen.

SUGGESTED READING

AMES BN, CATHCART R, SCHRIVERS E, HOCHSTEIN P: Uric acid provides an antioxidant defense in humans against oxidant- and radical-caused aging and cancer: A hypothesis. Proc Natl Acad Sci USA 78:6858–6862, 1981

BONDY PK, ROSENBERG LE (eds): Metabolic Control and Disease, 8th ed (formerly Duncan's Diseases of Metabolism). Philadelphia, WB Saunders, 1980

BOSS GR, SEEGMILLER JE: Hyperuricemia and gout. N Engl J Med 300:1459, 1979

HERS H-G, VAN DEN BERGH G: Enzyme defect in primary gout. Lancet I:585–586, 1979

STANBURY JB, WYNGAARDEN JB, FREDRICKSON DS, GOLDSTEIN JL, BROWN MS (eds): The Metabolic Basis of Inherited Disease, 5th ed. New York, McGraw-Hill, 1983

WYNGAARDEN JB, KELLEY WN: Gout and Hyperuricemia. New York, Grune & Stratton, 1976

20

Disorders of Lipid Metabolism

W. Carl Breckenridge / Robert L. Patten

Disorders of lipid metabolism include the hyperlipoproteinemias, hypolipoproteinemias and lipid storage diseases. The ***major lipids*** of the body are triglycerides (triacylglycerols), cholesterol and its esters, phospholipids, and glycolipids. The triglyceride in adipose tissue serves as the major energy reservoir for the body. The other lipids are important structural components of membranes, and cholesterol also serves as the precursor for steroid hormones and bile acids. In the blood stream, lipids are transported in the form of complex particles called *lipoproteins.* There are several lipoproteins, of differing composition, and an abnormal elevation of one or more of these characterizes the hyperlipoproteinemias. *Free fatty acids* are also transported in the blood stream, but they are carried by albumin rather than by lipoproteins.

BIOCHEMISTRY OF LIPOPROTEINS

Present physicochemical data on normal plasma lipoproteins indicate that they consist of a neutral lipid core of triglyceride and cholesteryl ester that is surrounded and stabilized by free cholesterol, amphipathic phospholipids, and apolipoproteins which are intercalated between the surface lipids. The relative proportions of nonpolar lipid, protein, and polar lipid determine the physicochemical properties of density, size, and electrostatic charge of the macromolecules. These physicochemical differences have aided in the resolution and naming of various lipoprotein families.

Apolipoproteins are proteins which are found in association with the plasma lipoproteins. Those which have been characterized range in relative mass (M_r) from 8000 to about 35,000, and all have an avidity for phospholipid. Although there are no hydrophobic regions in the primary structure of the apolipoprotein, such regions occur as amphipathic α-helices which are formed as a result of the specific arrangement of charged and hydrophobic amino acids in the primary sequence of the protein. Certain apolipoproteins are important in the synthesis and release of lipoproteins, while others are activators of plasma enzyme systems and provide recognition sites for cell receptors which are important in lipoprotein catabolism.

The convention for identifying apolipoproteins is evolving. *Letter* identification (e.g., apoC), which was originally used to designate major apolipoprotein components of specific lipoprotein groups, has been retained in this review. *Roman numerals* are added (e.g., apoC-I) to identify apolipoproteins that come under the same letter category but possess distinct immunologic determinants. *Arabic numerals* are used (e.g., apoE1) for apolipoprotein isomorphs which have immunologic cross-reactivity but different chemical characteristics that allow them to be resolved by electrophoretic or chromatographic techniques.

PHYSICOCHEMICAL CHARACTERISTICS OF LIPOPROTEINS

Many of the characteristic differences between plasma lipoproteins are determined by the relative proportions of lipid and apolipoproteins, which have very different hydrated densities. Lipoproteins containing triglyceride as the major component have a much lower density and larger particle size than those containing primarily phospholipid and protein (Table 20–1). The electrostatic charge of the particles is determined primarily by the protein constituents. Plasma lipoproteins of normal subjects have been classified, according to this property, into four *major families:*

1. Chylomicrons, with zero mobility
2. Very-low-density lipoproteins (VLDL), preβ-lipoproteins with the electrophoretic mobility of α_2-globulins
3. Low-density lipoproteins (LDL), β-lipoproteins with the mobility of β-globulins
4. High-density lipoproteins (HDL), α-lipoproteins with the mobility of α_1-globulins.

Several variants of these families, as well as abnormal lipoproteins, occur in disease states.

Chylomicrons. When viewed in the electron microscope, chylomicrons are seen as large spheres with diameters ranging from about 75 to >200 nm and M_r estimated at $1–10 \times 10^9$. Because of their large size, they impart a milky appearance to plasma. Since they contain large amounts of triglycerides and only small amounts of phospholipid, cholesterol, cholesteryl ester, and protein, they have a density < 0.95 g/mL (lower than saline), and float to the surface of plasma on standing. On paper, agarose, and acrylamide gel electrophoresis, they remain at the origin. Chylomicrons contain apolipoproteins A, B, C, and E, but lose apoA and apoC during passage through the capillary bed. Chylomicron remnants contain apoB and apoE, the latter being recognized by specific receptors in the liver.

Very-Low-Density Lipoproteins (VLDL). This family of macromolecules has a density range of 0.95 to 1.006 g/mL. They are spherical, with diameters of 28 to 75 nm and M_r of $5–10 \times 10^6$. While they impart opalescence or turbidity to plasma, they do not float on standing. Although they overlap in size and composition with chylomicrons and intermediate-density lipoprotein, the average particle has a composition of triglyceride (50% to 60%), phospholipid (15% to 20%), cholesteryl ester (15%), free cholesterol, and apolipoprotein (10%).

The apolipoproteins of VLDL are composed of at least five immunologically distinct proteins. ApoB accounts for 35% of the protein mass, while a second group (apoC) is composed of three species: C-I (5%), C-II (10%), and C-III (35%). ApoC-I and C-II have important functions as activators of the enzymes lecithin:cholesterol acyl transferase (LCAT) and lipoprotein lipase (LPL) respectively, while C-III can inhibit LPL, at least *in vitro.* Another peptide, termed apoE (15%), functions in the interaction of lipoproteins with cell surface receptors. This combination of apolipoproteins results in a lipoprotein with a mobility ahead of β-lipoproteins (termed preβ-lipoprotein).

Intermediate-Density Lipoproteins (IDL). While very low levels of intermediate-density lipoproteins are found in *normal* plasma, the concentration may be elevated in certain disease states such as type 3 hyperlipoproteinemia. The density of IDL ranges from approximately 1.006 g/mL to 1.019 g/mL, while the electrophoretic mobility is similar to that of β-lipoprotein. The particle size and relative mass of IDL are intermediate between VLDL and LDL, and IDL appear to be remnants of VLDL near the final stage of catabolism. Triglycerides (35%) and cholesteryl esters (25%) are the major components, with smaller amounts of phospholipid (17%), protein (16%), and free cholesterol (7%). All protein components of VLDL are present in IDL; however, apoB (60%) and apoE (30%) comprise a much larger percentage of the protein mass than in VLDL. The apoC content is approximately 10%.

Floating-β-Lipoprotein. In the rare condition of type 3 hyperlipoproteinemia, an abnormal VLDL with β mobility is present in large quantities. This abnormal lipoprotein overlaps with preβ- and β-lipoproteins in the electrophoretic pattern to produce the *broad-β* lipoprotein pattern. When the lipoproteins are resolved from normal preβ-lipoprotein by preparative electrophoresis or gel filtration chromatography, they are found to have

TABLE 20–1. Physicochemical Characteristics of Human Plasma Lipoproteins

Lipoprotein Classes	Density for Isolation	Particle Size (nm)	Electro-phoretic Mobility*	Composition; % by weight								Apolipoproteins	
							Phospholipid						
				Cholesterol	*Cholesteryl Ester*	*Triglyceride (Triacyl-glycerol)*	*PC†*		*Sph‡*	*Protein*	*CHO§*	*Major*	*Minor*
Chylomicrons	<1.006	75–>200	Origin	2	6	83		7		2		B, C, E	A
Very low density	<1.006	28–75	Pre-β	6	12	56	16		2	9		B, C, E	
VLDL$_1$		54		5	7	65		17		5		B, C, E	
VLDL$_2$		43		6	10	57		19		7		B, C, E	
VLDL$_3$		37		7	18	42		22		10		B, C, E	
Low density	1.006–1.063	19–26	β	10	34	18		23		14		B, E	C
LDL$_1$	1.006–1.019	22–26	β	10	34	18		23		14		B, E	C
LDL$_2$	1.019–1.063	19–22	β	9	40	6	19		6	20	1	B	E, C
Lipoprotein(a)	1.050–1.090	17–28	Pre-β	9	34	2	13		6	27	9	B, (a)	
High density	1.063–1.21		α	5	18	2	22		3	53		A_I, A_{II}	C, E
HDL$_1$	1.075–1.095	13–19	α	16	22	8		14		40		A_I, $A_{II}E$	C
HDL$_2$	1.063–1.125	9–11	α	4	19	3		29		46		A_I, A_{II}	C, E
HDL$_3$	1.125–1.21	7–9	α	2	15	2		22		61		A_I, A_{II}	C, E

* Mobility in agarose gel.

† PC = phosphatidyl choline.

‡ Sph = sphingomyelin.

§ CHO = carbohydrate.

a composition of cholesteryl ester (26%), cholesterol (7%), phospholipid (19%), triglyceride (38%), and protein (10%), which is very similar to that of IDL. In fact the abnormal lipoproteins appear to be due to a marked accumulation of IDL and chylomicron remnants. The apolipoproteins consist of apoB (60%), apoE (25%), and apoC (15%). In normal individuals apoE can be resolved into several isomorphs by isoelectric focusing, as will be described. Although the amount of apoE is increased in persons with type 3 hyperlipoproteinemia, two of the isoforms of apoE are absent.

Low-Density Lipoproteins (LDL). More than 75% of the *cholesterol* in normal plasma is transported by this lipoprotein class. LDLs have a fairly narrow density range (1.019 to 1.063 g/mL) and an M_r of $2.1–2.2 \times 10^6$. They are also visualized as spheres with a diameter of 19 nm to 25 nm. Cholesterol (10%) and its ester (38%) account for a major part of the lipoprotein mass. Similar amounts of phospholipid (20%) and protein (22%) are present, while triglyceride is a minor component (10%). The protein component is mostly apoB (96%), with only trace amounts of apoC (2%) and apoA-I (1%). Thus, the characteristic β mobility of LDL is due to apolipoprotein B.

Lipoprotein(a). Antigenic determinants that are found in some preparations of human LDL are not detected in other preparations. A variant of LDL called lipoprotein(a), which is easily identified in about 35% of the population, is due to an additional apolipoprotein known as the Lp(a) determinant. Lp(a) is identical to sinking preβ-lipoprotein which has a density of 1.050 to 1.090 g/mL and lipid chemical characteristics of LDL, but an electrophoretic mobility of preβ-lipoprotein. The rapid migration of these lipoproteins is caused by the Lp(a) apolipoprotein which has a large negative charge owing to the presence of sialic acid.

High-Density Lipoproteins (HDL). These small lipoproteins are produced by the liver and intestine and are isolated from plasma between densities of 1.063 and 1.21 g/mL. They possess a mobility equivalent to the α_1-globulins. The M_r is $0.25–0.4 \times 10^6$, while the range of diameters is 7 to 13 nm. The major components are protein (45% to 50%) and phospholipid (25% to 30%), with smaller amounts of cholesteryl ester (15% to 20%), cholesterol (5%), and triglyceride (3%). The protein is composed largely of apoA-I (70%) and apoA-II (20%), which impart an α mobility to HDL. Small amounts of apoC (10%) are present.

Lipoprotein X. This lipoprotein is a complex of phospholipid (66%), unesterified cholesterol (25%), protein (6%), and small amounts of triglyceride (3%). The protein consists of albumin (40%), apoC (50%), and small amounts of apoE and apoA-I (10%). Because cholesterol and phospholipids do not readily take up fat-soluble dyes the lipoproteins are not easily detected in routine electrophoresis. Lipoprotein X may be found in the plasma of patients with cholestasis (obstructive jaundice) and of individuals possessing the rare deficiency of LCAT (lecithin:cholesterol acyltransferase) activity, or of patients receiving infusions of lipid emulsions.

HDL_T or α_T-Lipoproteins. In the rare *Tangier disease* (T) HDL levels are extremely low and the HDL is abnormal. Only small amounts of apoA-I are immunologically detectable. ApoA-II is the major component of HDL_T but the absolute amount of A-II is only 10% of the concentrations in normal plasma. It has been suggested that HDL_T may be taken up rapidly by reticuloendothelial cells, resulting in the accumulation of cholesteryl esters in these tissues.

METABOLIC PATHWAYS OF LIPIDS AND LIPOPROTEINS

A major function of the plasma lipoprotein spectrum is the transport of triglyceride from sites of synthesis in the intestine or liver to sites of storage (adipose tissue), energy use (muscle), or metabolism (liver). Because triglyceride is hydrophobic it must be packaged in phospholipid and combined with protein for transport.

Chylomicrons. The chylomicrons are produced in the intestinal mucosa following the ingestion of dietary triglycerides containing long-chain fatty acids (see Fig. 12–5). In the intestinal lumen triglyceride is broken down by pancreatic lipase to 2-monoglycerides and free fatty acids, which are absorbed into the intestinal epithelial cell from bile acid micelles. During the process of reacylation of the 2-monoglycerides with fatty acids, particles that are similar to the chylomicrons in size appear and move toward the base of the cell. The process of assembly and release of chylomicrons requires the small amount of apoB protein associated with the lipoprotein. In the rare familial disorder, *abetalipoproteinemia,* in which apoB protein is lacking, the fat is absorbed into the intestinal cell until the cell appears filled with lipid, but no chylomicrons are released from the cell. Immunohistochemical studies and intestinal perfusion systems have revealed that apoA-I and apoA-II are also synthesized in the intestinal mucosa, while very little apoC and apoE are made in this tissue.

The chylomicrons move from the intestine to the thoracic lymph duct and the blood stream. Apolipoproteins A-I and A-II are rapidly lost, while apoC and apoE are acquired from HDL. Chylomicrons are removed from plasma with a half-life of 5 to 15 min. Most of the triglycerides are hydrolyzed by lipoprotein lipase in the capillary bed of adipose tissue and muscle. ApoC-II, an activator of lipoprotein lipase, is required for normal catabolism of the chylomicrons. After removal of most of the triglyceride, *chylomicron remnants* are rapidly cleared by the liver through a chylomicron remnant receptor specific for apoE. The defect in *type 3 hyperlipoproteinemia* has been attributed to mutant changes in apoE which lower its affinity for the apoE receptor. The disorder involves an accumulation of these remnant particles, along with IDL derived from VLDL, to give rise to the abnormal floating-β-lipoprotein.

Very-Low-Density Lipoproteins (VLDL). Very-low-density lipoproteins serve to transport endogenously synthesized triglyceride from the liver and intestine. Lipid particles that are similar morphologically and biochemically to VLDL have been isolated from the Golgi apparatus of the rat liver. Persons with abetalipoproteinemia also fail to synthesize VLDL, and inhibitors of protein synthesis block the synthesis of VLDL. Thus, as with chylomicrons, apoB is essential for VLDL synthesis. In addition to apoB the liver also synthesizes apoC and apoE. The major precursors for triglyceride synthesis are glucose and plasma free fatty acids. The synthesis of VLDL can be stimulated by carbohydrate feeding, and by a variety of metabolic stimuli that cause an increase in plasma free fatty acids.

VLDL are catabolized fairly rapidly with a half-life of 6 to 12 hours. The process appears to involve mechanisms similar to chylomicron catabolism. Studies with protein-labeled VLDL have shown that radioactive apoB of VLDL first appears in the IDL and later in LDL, indicating that LDL, a major lipoprotein group in the plasma, are remnants of VLDL catabolism. In contrast to apoB, the apoC proteins associated with VLDL readily exchange with the same components in HDL.

The catabolism of VLDL can be envisaged as a cascade mechanism. As triglycerides are removed, apoC, free cholesterol, and phospholipid leave the particle, which becomes smaller and richer in cholesteryl ester, apoB, and apoE to give the IDL. The cholesterol and phospholipids that leave the particles appear to be incorporated into HDL with the aid of transfer proteins in the plasma. They are potential substrates for LCAT, which transfers the fatty acid at position 2 of phosphatidylcholine to the hydroxyl group of cholesterol. ApoA-I and apoC-I are activators of LCAT, and HDL is the preferred substrate. The products of the enzyme activity are *cholesteryl ester,* which can be stored in the lipid core of lipoproteins, and *lysolecithin,* which is bound to albumin. In humans, cholesteryl esters are redistributed from HDL to other lipoproteins by an exchange protein.

Although the exact mechanism of catabolism of IDL to LDL is unknown, much has been learned. IDL contains apolipoprotein E and apoB-100. About half the IDL formed is taken up in 2 to 6 hours by the liver, because apoE has a higher affinity for hepatic LDL receptors than apoB-100. The remaining IDL loses its apoE as it converts to LDL, a process which appears to involve a lipase on the sinusoidal endothelium of the liver (hepatic lipase). The resulting LDL is taken up by LDL receptors throughout the body over the next 2 to 3 days, much of it by the liver, which has a high concentration of receptors and uses the cholesterol to form bile acids. On the basis of the rate of triglyceride synthesis (VLDL production) and LDL protein catabolism, it can be calculated that all LDL may be derived from VLDL through IDL in normal persons. Figure 20–1 summarizes the main features of these metabolic steps of lipoprotein catabolism.

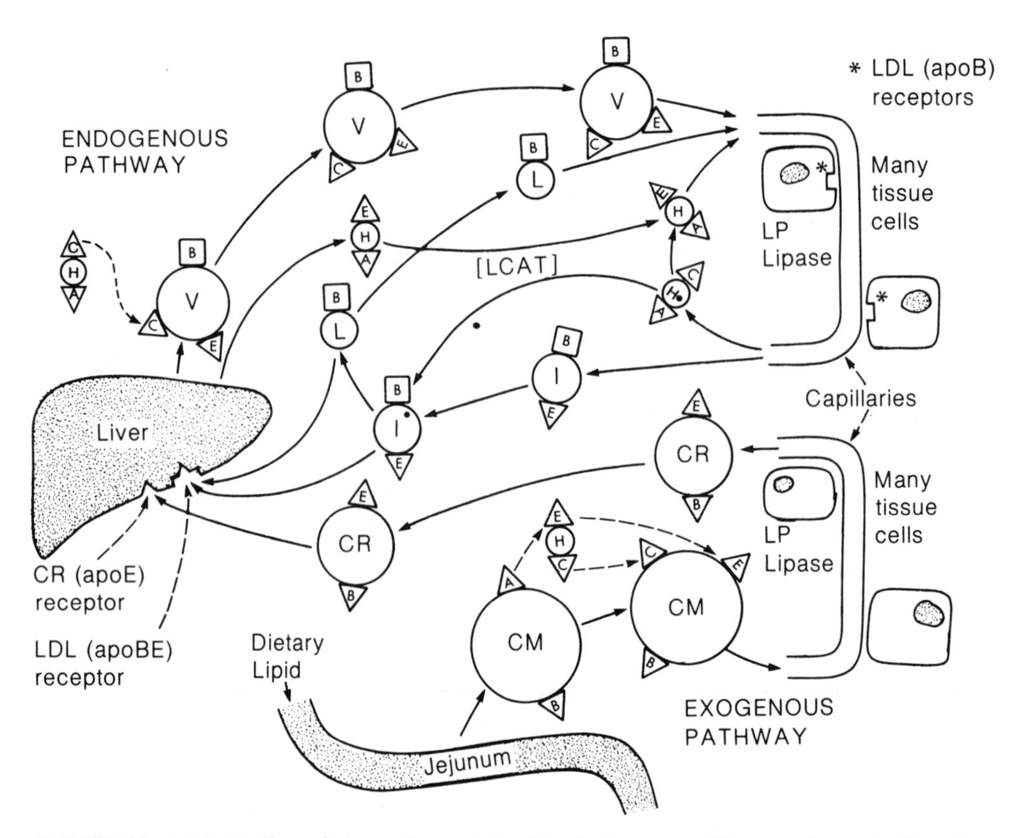

FIGURE 20–1. Metabolism of plasma lipoproteins. Metabolic routes of lipoproteins are indicated by **solid-line arrows;** specific interlipoprotein movement of apoproteins is shown by **dashed arrows.** The major function of plasma lipoproteins involves the transport of energy in the form of triglyceride from sites of synthesis (intestine, liver) to sites of storage or energy use. Chylomicrons (**CM**) are produced in the intestine, while very-low-density lipoprotein (VLDL; **V** in the figure) is synthesized primarily by the liver and to a small extent by the intestine. Apolipoprotein B is required for the transport and release of these lipoproteins. Although both the intestine and liver appear capable of synthesizing several of the apolipoproteins, the intestine is a major source of apoA-I (**A**), while the apoC (**C**) and apoE (**E**) proteins are synthesized largely by the liver. While lymph chylomicrons are rich in apoB-48 (**B** in triangles) and apoA-I, the latter apoprotein is lost rapidly to high-density lipoprotein (HDL; **H** in the figure). Nascent VLDL from the liver is rich in apoB-100 (**B** in squares) and apoE. In the plasma both VLDL and chylomicrons acquire apoC peptides from HDL. These lipoproteins are substrates for lipoprotein lipase (**LPL**) in the capillaries of the periphery. As triglycerides are broken down to free fatty acids, which can enter cells for storage or energy use, the chylomicrons or VLDL become smaller. ApoC peptides and some free cholesterol and phospholipid are lost to HDL. Chylomicrons give rise to chylomicron remnants (**CR**), which are removed by receptors that are specific to the liver. VLDL gives rise to intermediate-density lipoprotein (IDL; **I** in the figure) and subsequently to low-density lipoprotein (LDL; **L**). Some of the IDL is cleared rapidly from the plasma because of a high affinity of apoE receptors in the liver. The remainder loses triglyceride and apoE, becomes LDL, and is cleared more slowly because of a lower affinity of apoB-100 receptors in the tissues and liver. ApoB has a unidirectional flow from triglyceride-rich particles to LDL whereas apoC peptides move between the triglyceride-rich particles and HDL. Some exchange of apoE probably occurs as well. HDL is a substrate for lecithin cholesterol acyl transferase (**LCAT**), resulting in the formation of cholesteryl ester. The black dot in HDL represents cholesterol on its way from the tissues to IDL and the liver.

High-Density Lipoproteins (HDL). Liver and intestine synthesize nascent HDL which are composed of phospholipid, cholesterol, some apoA-I and A-II and major amounts of apoE. The particles have a flat lamellar appearance under electron microscopy. In plasma, LCAT converts the nascent HDL into spherical HDL, with a normal complement of cholesteryl ester and apoA-I and A-II. Nascent lymph chylomicrons contain apoA-I and A-II, which transfer rapidly to HDL on entry of the chylomicron into plasma. Finally, the smaller molecules of the HDL spectrum can readily incorporate cholesterol and phospholipid that are released from the surface of triglyceride-rich lipoproteins during their catabolism. The process, which is facilitated by phospholipid-transfer factors, causes an increase in the size of HDL particles and provides substrates for LCAT. HDL may also promote the removal of cholesterol from peripheral tissues by a similar mechanism. ApoC and apoE also appear to cycle between HDL and triglyceride-rich lipoproteins. The half-life of HDL apolipoproteins is 3 to 4 days. Although HDL can donate cholesterol to the liver it is not clear whether these lipoproteins are catabolized as a unit by the liver, or whether cholesteryl ester may be transferred to other lipoproteins for removal.

Low-Density Lipoproteins and Cholesterol Metabolism. As described previously LDLs are remnants of VLDL catabolism. There is very little evidence to suggest that LDLs are synthesized directly by the liver. Recent studies with cell cultures indicate that LDL catabolism may be linked with cholesterol catabolism.

Cholesterol is obtained from the diet or synthesized in the liver or the intestinal mucosa for release in association with lipoproteins. In the adult most other cells, with the exception of certain endocrine glands, do not have active cholesterol synthesis. In the synthesis of cholesterol (see Fig. 11–3) two molecules of acetyl CoA condense to give acetoacetyl CoA. A further residue of acetyl CoA is added to give β-hydroxy-β-methylglutaryl · CoA (HMG · CoA). This compound is reduced to mevalonate, which is converted to isopentenyl pyrophosphate. A number of isoprenoid intermediates are formed and cyclized, eventually to yield cholesterol. The enzyme responsible for reducing HMG · CoA (HMG · CoA reductase) is under feedback control by cholesterol.

Recent studies suggest that *LDL* may be the biologic vehicle to provide cholesterol to the cell for the *feedback inhibition.* Working with skin fibroblasts, investigators have discovered that the activity of HMG · CoA reductase is suppressed in the cultured cell by serum. When serum is removed the activity increases several fold. The major components responsible for suppressing the activity are apoB- and apoE-containing lipoproteins, which are bound by a receptor for these apolipoproteins on the cell surface. The lipoproteins are taken up into the lysosomes by endocytosis. ApoB is degraded and the cholesteryl esters are hydrolyzed to free cholesterol, which inhibits HMG · CoA reductase activity, and the synthesis of receptors.

In the liver the cholesterol is converted to *bile acids.* These bile acids are transported to the gallbladder for release into the small intestine, where they aid in fat absorption. They are reabsorbed mainly in the lower ileum and transported via the portal vein to the liver, where they exert a feedback inhibition on the further metabolism of cholesterol. Any interruption of bile acid reabsorption leads to increased catabolism of cholesterol via the bile acids (see Fig. 11–3).

STRUCTURE AND FUNCTION OF NORMAL OR MOLECULAR VARIANTS OF APOLIPOPROTEINS

Although the consumption of dietary fats influences the concentration of plasma lipoproteins, possibly by saturation of metabolic pathways, it is now apparent that apolipoproteins are important metabolic 'programmers' of lipoprotein metabolism. Analyses of the amino acid sequence of apolipoproteins have established that they possess regions of amphipathic helices that are responsible for binding of the proteins to phospholipids to allow for their interaction with lipoproteins. Specific regions of apolipoproteins are responsible for activation of enzymes or for recognition of lipoproteins by cell surface *receptors.* Furthermore, it has been established that mutations in specific regions of the apolipoproteins result in profound changes in functional capacity of some of the apolipoproteins. The characteristics and functions of the major apolipoproteins are given in Table 20–2. Some of the more significant functions of apolipoproteins are now described along with important genetic variants which influence their function.

TABLE 20–2. Characteristics of Human Apolipoproteins and Variants

Apolipoprotein	Relative Mass (M_r)	Isoelectric* Point	Biologic Functions
AI†	27,000	5.3–5.4	Activates LCAT
AII	17,400 (dimer)	5.0	Undefined
B-100	5.4×10^5	Undefined	Lipid transport from liver; receptor recognition site
B-48	2.6×10^5	Undefined	Lipid transport from intestine
CI	6630	6.5	Activates LCAT
CII†	8835	4.8	Activates lipoprotein lipase
$CIII_{0,1,2}$	9960	4.5–4.9	Inhibits lipoprotein lipase; inhibits receptor recognition of apoE
E†	34,145	5.4–6.1	Lipid transport Receptor recognition

* Estimated in urea (7–8 M) acrylamide gels. Range of values given for major isomorphs.

† These apolipoproteins have been shown to have variants (usually single amino acid substitutions) which influence the normal function of the apolipoprotein.

Apolipoprotein A-I. Apolipoprotein A-I, which accounts for approximately 20% to 30% of the mass of HDL, is present in the largest quantities of all apolipoproteins. It is an important activator of LCAT. Furthermore it appears to promote efflux of cholesterol from peripheral tissues. Epidemiologic studies indicate that individuals with low HDL or low apoA-I have a much greater risk of coronary heart disease than those with high concentrations of these constituents. Subjects who lack apoA-I, because of a change in the structure of the apoA-I gene, have very low HDL and severe coronary heart disease. Patients with Tangier disease clear apoA-I and HDL rapidly into reticuloendothelial tissues. HDL levels are low but they do not suffer from premature coronary heart disease. A number of genetic mutations in apoA-I have been described. However these subjects, who are heterozygotes and thus possess some normal HDL, have no severe complications.

Apolipoprotein B. The exact molecular structure of apoB remains undefined because of its extreme insolubility following removal of the lipids of the lipoprotein. The M_r appears to be in excess of 250,000. Two forms of apoB, called apoB-48 and apoB-100, have been reported. In humans apoB-48 is synthesized by the intestine whereas apoB-100 is synthesized in the liver. Partial or total apoB deficiency may occur. In total apoB deficiency essentially no lipid is transported from the intestine and the liver produces no VLDL, thus no LDL is present. In normotriglyceridemic abetalipoproteinemia apoB-100 is absent. Lipid absorption is normal and apoB-48 is detectable in chylomicrons. No VLDL and LDL are present in the plasma.

Although apoB is important in the transport of triglyceride from cells it is also important in the delivery of cholesteryl ester-rich lipoproteins to cells via cell surface receptors. Specific regions of apoB containing basic amino acids interact with specific receptors on cells. This results in endocytosis and catabolism of the lipoprotein as described in the previous section.

Apolipoprotein E. ApoE is a protein containing 299 amino acids. The apolipoprotein exists in three major forms (designated apoE2, E3, E4) with several minor forms which are produced by the addition of sialic acid to the major isoforms. An analysis of the amino acid sequence has shown that the three variants are a result of specific amino acid substitutions. Isoform E2 contains 2 cysteine residues at position 112 and 158 of the polypeptide chain. Isoform E3 has a cysteine residue at 112 but an arginine residue at position 158. Both cysteine residues are replaced with arginine in isoform E4. These changes result in single charge shifts in the apoE4, E3, and E2, as observed by isoelectric focusing in polyacrylamide gels. The substitutions have a dramatic effect on the function of apoE, which also possesses the ability to bind to the same cell surface receptor as apoB. The arginine residues at positions 112 and 158 are important in the binding process. Thus apoE2 binds very poorly, whereas apoE3 and E4 have much higher affinities for the receptor. The mode of inheritance of apoE involves three alleles ($\epsilon4$, $\epsilon3$, and $\epsilon2$) at a single genetic locus which provide for six phenotypes; E4/4, E3/3, E2/2, E4/3, E3/2, and E4/2. Subjects with E2/2 show accumulation of IDL because of poor removal of this apoE-rich lipoprotein by receptor dependent processes. These subjects, however, may or may not be hyperlipidemic. *Type 3 hyperlipoproteinemic* subjects have the E2/2 phenotype as well as some other underlying defect in lipid or lipoprotein metabolism.

Apolipoprotein C-II. ApoC-II is a small apolipoprotein with 78 amino acids. The amino terminal portion of the protein possesses the amphipathic helices while the carboxyl terminal portion is responsible for the marked stimulation of the activity of *lipoprotein lipase,* which is responsible for hydrolysis of lipoprotein triglyceride. Subjects with a deficiency of apoC-II have severe elevations of triglyceride-rich lipoproteins because of inactive lipoprotein lipase. ApoC-II deficiency has been shown to result from either a lack of apoC-II or a defective apoC-II. There is considerable potential for genetic variants in this apolipoprotein which could influence the ability of the protein to activate lipoprotein lipase and thereby influence the metabolism of triglyceride rich lipoproteins.

It is apparent that considerable genetic variation exists for apolipoprotein amino acid sequences. These variations can influence the function of the apolipoprotein. It is probable that many more variants of the apolipoproteins will be found which may provide a better appreciation of polygenic mechanisms influencing lipoprotein concentrations in the population.

PRINCIPLES OF LIPID AND LIPOPROTEIN DETERMINATIONS

Most cases of hyperlipoproteinemia are discovered as the result of a routine measurement of plasma cholesterol and triglyceride and a qualitative or quantitative estimate of lipoprotein levels.

Cholesterol

Approximately 30% of the total plasma cholesterol is present as the free alcohol, while the remainder is esterified. Only total cholesterol is measured routinely, because the ratio of *free* cholesterol to *total* cholesterol does not vary extensively, except in cases of severe liver disease, bile secretory failure, and LCAT deficiency in which the plasma levels of LCAT activity are low or absent.

Until recently, most methods for measuring ***cholesterol*** have relied on the polyunsaturated chromogen produced from the sterol nucleus when sulfuric acid is added to dilute solutions of cholesterol in organic solvent (*Liebermann–Burchardt reaction*). The most reliable methods involve a preliminary hydrolysis of the esters to free cholesterol and an extraction of sterol into an organic solvent prior to the color reaction. In methods in which the reagents are added directly to serum, the results are usually higher than those obtained with the hydrolytic procedure because of differences in the extinction coefficients for cholesterol and cholesteryl ester. Bilirubin interference is also significant in direct procedures.

An *enzymatic* method for determining cholesterol uses cholesteryl esterase to hydrolyze the esters. Cholesterol oxidase in the presence of oxygen converts cholesterol to cholestenone with the stoichiometric production of hydrogen peroxide. The hydrogen peroxide is measured colorimetrically in the presence of a peroxidase and a suitable dye.

Triglycerides

Plasma neutral glycerides are composed almost entirely (97%) of triglycerides, with minor amounts (1% to 3%) of mono- and diglycerides. A *reliable* estimate of normal circulating levels of plasma triglycerides can be obtained only after a 12- to 14-hour fast.

All routine methods for the determination of triglycerides estimate the amount of *glycerol* after chemical saponification with alkali, or after enzymatic hydrolysis with lipases. In *chemical* assays the glycerol is oxidized to formaldehyde which is combined with ammonium acetate and acetyl acetone to yield a fluorescent chromogen (3,5-diacetyl-1,4-dihydrolutidine). Triglycerides are extracted from lipoproteins with isopropanol, and interfering phospholipids, free glycerol, and glucose are removed by an adsorbent (zeolite mixture). In *enzymatic* procedures the glycerol is converted to glycerophosphate in the presence of glycerol kinase and adenosine triphosphate (ATP). The reaction product, adenosine diphosphate (ADP), reacts with phosphoenolpyruvate in the presence of pyruvate kinase to yield ATP and pyruvic acid. The pyruvic acid is converted to lactic acid in the presence of lactate dehydrogenase with reduced nicotinamide adenine dinucleotide (NADH). The amount of NADH converted to NAD, which is stoichiometric for the amount of glycerol present, is measured by the decrease in optical density at 340 nm.

Lipoproteins

Samples for lipoprotein analysis should be maintained at 4°C and analyzed within 5 days of sampling. Plasma containing ethylene diamine tetraacetic acid (EDTA) is preferred to serum, since EDTA prevents the entrapment of chylomicrons in the clot and the peroxidation of lipoproteins by heavy metal ions. Most methods for the separation and quantification of lipoproteins depend on differences in the surface charge, density, and size of the macromolecular complexes resulting from the relative proportions of lipid and apolipoproteins. A *simple assessment* of lipoproteins can be made by observing the plasma after allowing it to stand at 4° for 18 hours. Chylomicrons give a milky appearance to plasma and will float to the surface, forming a distinct chylomicron layer. Increased levels of VLDL give an opalescence or turbidity to plasma but these lipoproteins do not readily float to the surface of plasma.

Electrophoresis

Zone Electrophoresis. Paper, cellulose, and agarose gel are suitable media for resolving the lipoprotein families into α (HDL), preβ (VLDL), and β (LDL) lipoproteins, with chylomicrons remaining at the origin (Fig. 20–2). In all procedures the lipids of the lipoproteins are stained with an aqueous emulsion of a fat-soluble dye. The resolution of β- and preβ-lipoproteins on paper is enhanced by the addition of albumin to the buffers. Agarose gives the best resolution of preβ- and β-lipoproteins and the highest level of sensitivity. The presence of floating-β-lipoprotein should be considered when a broad-β pattern is observed with no resolution between the β- and preβ-lipoproteins. The presence of floating-β should be confirmed by ultracentrifugation of plasma.

Polyacrylamide Gel Electrophoresis. In this medium lipoproteins are resolved primarily on the basis of molecular size. Three different gels (sample, stacking, and running gels) are prepared in a glass tube. The chylomicrons remain at the origin while VLDL enter the stacking but not the running gel. Both HDL and LDL enter the running gel and migrate in a fashion similar to α and β mobility on paper. By virtue of their abnormal characteristics, all VLDL and LDL particles in samples containing floating-β-lipoprotein remain at the interface of the stacking and running gels. Thus, a diagnosis of floating-β-lipoprotein can be made from the presence of a β band on agarose or paper, but the absence of an LDL band in polyacrylamide gel electrophoresis.

Urea-polyacrylamide gels containing ampholytes are particularly useful for the resolution of delipidated apolipoproteins. The apolipoproteins are resolved on the basis of distinct isoelectric points (Table 20–2). The method has been used extensively to detect deficiencies of apoA-I and C-II and to characterize the isoforms of apoE. It has also been useful in characterizing homozygous E2/2 in type 3 hyperlipoproteinemia.

Ultracentrifugation

When plasma (density 1.006 g/mL) is subjected to high centrifugal forces (100,000 × *g* for 16 hours) VLDL will float to the top of the centrifuge tube while LDL and HDL will sediment. Chylomicrons can be isolated at lower centrifugal forces (30,000 × *g* for 30 min). The density of the medium can be altered by the addition of soluble salts such as sodium bromide. At a density of 1.063 g/mL, LDL floats after centrifugation at 100,000 × *g* for 20 hours, while HDL are isolated at density 1.21 g/mL at 100,000 × *g* for 30 to 40 hours. The quantities of lipoprotein in each fraction are expressed on the basis of cholesterol content. Ultracentrifugation provides the most reliable technique for the demonstration of floating-β-lipoprotein, which floats with VLDL but migrates in zone electrophoresis with β mobility.

Polyanion Precipitation

Because clinical laboratories may not have an ultracentrifuge, other methods have been devel-

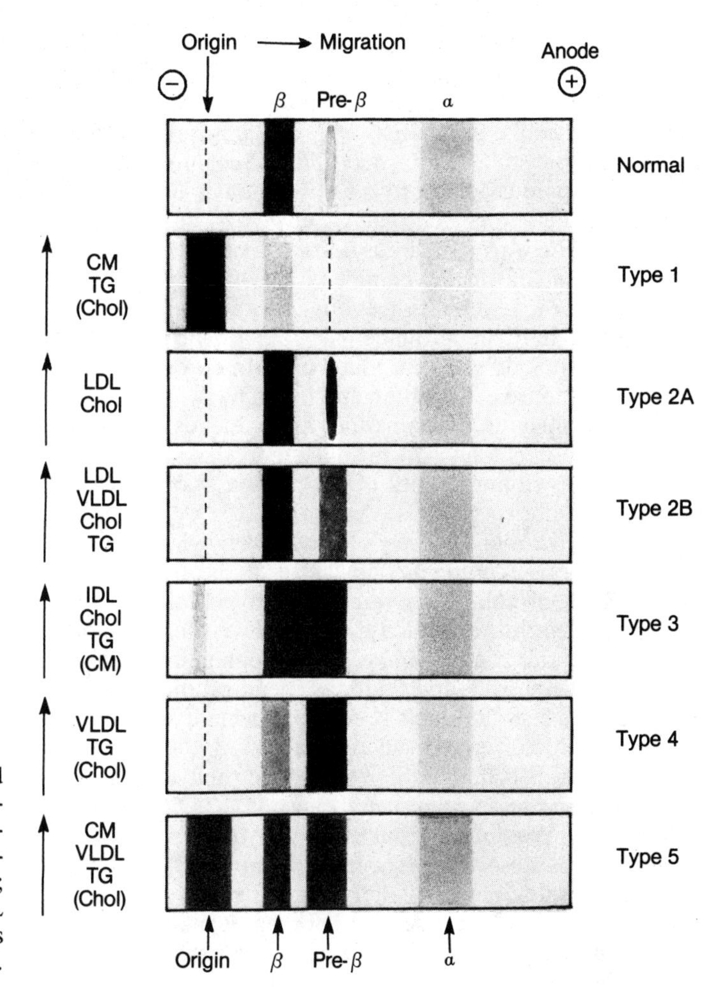

FIGURE 20–2. Electrophoretic patterns and typical biochemical abnormalities in the lipoproteinemias. ↑ = Increased; CM = chylomicrons; TG = triglyceride; Chol = cholesterol; LDL = low-density lipoprotein; VLDL = very-low-density lipoprotein; IDL = intermediate-density lipoprotein. Symbols in parentheses may or may not be increased.

oped for quantifying lipoproteins. Both VLDL and LDL, as a result of their apoB content, readily associate with heparin and divalent cations such as Mg^{2+} or Mn^{2+}. The lipoprotein-heparin complex precipitates, leaving only HDL in solution. The amount of *HDL cholesterol* is determined quantitatively, while *LDL cholesterol* is determined by the formula:

$$\text{LDL cholesterol} = \text{plasma cholesterol} - (\text{plasma triglyceride}/2.2 + \text{HDL cholesterol})$$

(all values in mmol/L; for mg/dL values, use TG/5)

Thus an estimate of the lipoprotein concentrations is obtained by analyzing the total plasma cholesterol and triglyceride, and the HDL cholesterol. It has been found that the amount of cholesterol in VLDL is equal to approximately $^{5}/_{11}$ the total plasma triglyceride, as long as there are no chylomicrons or floating-β-lipoproteins in the sample and the total triglyceride levels do not exceed 4.5 mmol/L (or <400 mg/dL; here the fraction is $^{1}/_{5}$).

LIPID DISORDERS

THE HYPERLIPOPROTEINEMIAS

The hyperlipoproteinemias are disorders of metabolism in which one or more of the plasma lipoproteins are increased. Each can be inherited (*primary hyperlipoproteinemias*) and more than one heritable disorder may cause a given lipoprotein abnormality. These abnormalities may be caused by other diseases as well, or by the influence of other factors such as improper diet in patients who have a predisposition to the development of hyperlipoproteinemia (*secondary hyperlipoproteinemias*). There are several classifications of these disorders, none of which is completely satisfactory at this time. Genetic classifications are ideal for the primary hyperlipoproteinemias, but do not adequately include the secondary hyperlipoproteinemias or those of uncertain etiology. The phenotypic classification remains the most practical for routine use as long as it is understood that it describes only abnormalities of the plasma lipoproteins and not specific diseases (Table 20–3).

Although the identification of a *type* of hyperlipoproteinemia requires a knowledge of the lipoprotein abnormality, the *presence* of a hyperlipidemia is determined mainly by the levels of plasma ***cholesterol*** and ***triglyceride.*** If either of these lipids is elevated, then further investigation may be necessary to elucidate the abnormality. There is a problem, however, in establishing the reference range of cholesterol or triglyceride above which a value may be considered elevated.

It must be remembered that the concentrations of serum cholesterol and triglyceride form a continuous spectrum in the population, and that there is no clearly defined cutoff point between normal and abnormal lipid levels. Various *upper reference values* that have been chosen, whether they are 95% limits in defined populations or not, must be recognized as being arbitrary. Furthermore, the major interest in plasma lipids is due to the association that exists between elevated lipids and the development of premature atherosclerosis. There is clearly an increased risk of developing coronary heart disease with increasing cholesterol levels and the risk increases almost as the *cube* of the cholesterol concentration, even within a plasma cholesterol range of 5.2 to 7.8 mmol/L (200 to 300 mg/dL). It is not clear whether there is an increased risk of atherosclerosis due to an elevated triglyceride level *per se.*

TABLE 20–3. Types of Hyperlipoproteinemias

Type	Increased Lipoprotein
1	Chylomicrons
2a	LDL
2b	LDL and VLDL
3	IDL
4	VLDL
5	VLDL and chylomicrons

IDL = Intermediate-density lipoprotein; LDL = low-density lipoprotein; VLDL = very-low-density lipoprotein.

Type I Hyperlipoproteinemia

Primary, Type 1 hyperlipoproteinemia (exogenous hyperlipemia) is a rare disorder which is caused by either an inherited deficiency of *lipoprotein lipase* (LPL), the enzyme necessary for the clearance of chylomicrons from the blood stream, or by the inherited absence of *normal apoC-II,* an apoprotein necessary for LPL activity. Both disorders are autosomal recessive; parents of patients with LPL deficiency usually have no lipid abnormality, whereas parents of patients with apoC-II deficiency often have mild hypertriglyceridemia. Type 1 hyperlipoproteinemia has been reported to develop *secondary* to uncontrolled diabetes, probably because insulin deficiency inhibits the synthesis of LPL. Type 1 has also appeared in the course of diseases with abnormal plasma globulins. In some of these it appears that the abnormal proteins have the ability to bind heparin, a mucopolysaccharide necessary for LPL activity.

Clinical Features. LPL deficiency usually manifests itself in early childhood. Attacks of *abdominal pain* are a frequent complication and can vary from mild to very severe. The pathogenesis of milder episodes is not clear, but the more severe episodes are usually caused by attacks of acute pancreatitis. The occurrence of abdominal pain shows a rough correlation with the level of triglycerides.

Enlargement of the liver and spleen is common in patients with grossly elevated triglyceride values, owing to uptake of chylomicrons by the reticuloendothelial system. The liver and spleen are often tender, and this may be responsible for some episodes of mild abdominal discomfort. Patients may develop *eruptive xanthomas.* The lesions that appear in the skin, usually on the extensor surfaces, consist of small round creamy lipid deposits surrounded by reddish areas of inflammation. Examination of the optic fundi may reveal a pale pink appearance of the retinal blood vessels, *lipemia retinalis.*

The less common disorder, apoC-II deficiency, is more often diagnosed in adolescence or adulthood. Pancreatitis is frequently seen but hepatomegaly and eruptive xanthomas do not appear to be features of this condition. Secondary Type 1 hyperlipemia may appear at any age.

Biochemical Features. Because of the large number of chylomicrons present, the blood has the appearance of cream of tomato soup. The plasma is grossly lipemic and on standing overnight at 4°C develops a layer of fat at the top, while the infranatant becomes clear. The plasma cholesterol may be normal but is usually elevated, owing to the cholesterol contained in the chylomicrons. The plasma triglyceride is greatly elevated, sometimes in the order of 20 to 40 mmol/L (1.8 to 3.6 g/dL). On electrophoresis a chylomicron band is apparent at the origin and there is a decrease in the amount of LDL and HDL. Postheparin triglyceride lipase activity (plasma LPL after the administration of heparin) is low in LPL deficiency, whereas apoC-II can be shown to be absent in apoC-II deficiency. In either LPL or apoC-II deficiency VLDL may increase as patients become older, or because of alcohol or estrogen use, changing their apparent phenotype to Type 5.

Course, Prognosis, and Treatment. In spite of very high levels of triglyceride, patients with type 1 hyperlipoproteinemia do not have an increased risk of atherosclerosis. Recurrent episodes of pancreatitis are common, however, and constitute the main threat to survival.

The principle of treatment is to reduce the levels of circulating chylomicrons by reducing dietary fat. Medium-chain triglycerides may be substituted for ordinary fats, as these are not incorporated into chylomicrons but are absorbed via the portal venous system.

Type 2 Hyperlipoproteinemia

Type 2a hyperlipoproteinemia (hypercholesterolemia) is characterized by an increase in LDL, and ***type 2b*** by an increase in LDL and VLDL (combined hyperlipidemia). Both types 2a and 2b can be inherited. The classic inherited disorder is *familial hypercholesterolemia,* which occurs in heterozygous and homozygous forms. The usual lipid pattern in this disease is type 2a, but type 2b may appear as a phenotypic variant in involved families. A disorder called *familial combined hyperlipoproteinemia* (familial multiple lipoprotein-type hyperlipidemia) has been described in which types 2a, 2b, 4, and 5 may occur in a single family. *Secondary* causes of type 2 (2a and 2b) hyperlipoproteinemia include hypothyroidism, the nephrotic syndrome, obstructive disease of the biliary tree, and acute intermittent porphyria. In addition to the above conditions, many individuals have moderately elevated plasma cholesterol and LDL without a significant family history of hypercholesterolemia. This may be the result of polygenic or sporadic disorders of cholesterol metabolism, or it may reflect an individual response to a diet high in cholesterol and saturated fats.

Clinical Features. Clinical findings in familial hypercholesterolemia include *xanthelasma* (deposits of cholesterol around the eyes) and *xanthomas* (deposits of cholesterol in the skin and tendons). Cutaneous xanthomas may be rounded (tuberous) or flat (planar) and usually appear on the elbows, knees, and hands. Tendon xanthomas appear most frequently in the Achilles tendon and in the extensor tendons of the hands. Patients often have a premature *arcus senilis,* a greyish ring around the outside of the cornea of the eye. If hypercholesterolemia is severe, aortic stenosis may develop.

The clinical features of type 2 hyperlipoproteinemia differ with the cause of the disorder. Patients who are *homozygous* for familial hypercholesterolemia may have xanthomas at birth or develop them in early childhood. These patients develop severe atherosclerosis in childhood and usually die from coronary artery disease or aortic stenosis before the age of 30. Patients with the *heterozygous* form begin to develop the clinical features of the disorder about the age of 20, and usually have evidence of coronary artery disease in their 40s. Patients with milder forms of type 2 hyperlipoproteinemia are less prone to the development of xanthomas, but they have a higher than normal risk of developing atherosclerosis. Patients with secondary forms of type 2 hyperlipoproteinemia may develop xanthomas, but this is unusual.

Biochemical Features. In type 2a hyperlipoproteinemia the plasma is clear; in type 2b the plasma may be turbid owing to increased amounts of VLDL. The plasma remains turbid on standing overnight, and does not develop a chylomicron layer. The plasma cholesterol is elevated in type 2a, and both plasma cholesterol and plasma triglyceride are elevated in type 2b. Electrophoresis shows increased β-lipoproteins in type 2a and increased β-lipoproteins and preβ-lipoproteins in type 2b. LDL cholesterol values are elevated in both 2a and 2b.

The highest plasma ***cholesterol*** values are found in patients with *homozygous* essential familial hypercholesterolemia. In these patients the total plasma cholesterol is usually between 15.6 and 26.0 mmol/L (600 to 1000 mg/dL). LDL cholesterol usually exceeds 13.0 mmol/L (500 mg/dL). In patients with the *heterozygous* form, the total plasma cholesterol is usually in the range of 7.8 to 13.0 mmol/L (300 to 500 mg/dL) and the LDL cholesterol is usually between 4.4 and 10.4 mmol/L (170 to 400 mg/dL).

At the present time there is no clear demarcation between patients with 'normal' plasma cholesterol and those with elevated plasma cholesterol levels. *Reference ranges* based on 95% limits, whether or not they are subdivided by age and sex groupings, define upper reference limits which are much higher than is considered healthy. It has been suggested that anyone with a plasma *total cholesterol* greater than 6.2 mmol/L (240 mg/dL) should be considered hypercholesterolemic. It has also been argued that the upper limit should be 5.2 mmol/L (200 mg/dL), based on the observations that populations with a low incidence of coronary atherosclerosis have a mean cholesterol of approximately 4.7 mmol/L (180 mg/dL), and that the incidence of coronary artery disease increases dramatically as the cholesterol level increases above 5.2 mmol/L (200 mg/dL). At present a level of *LDL cholesterol* greater than 4.3 mmol/L (165 mg/dL) is used to define type 2 hyperlipoproteinemia.

Pathogenesis of Familial Hypercholesterolemia. Human fibroblasts maintained in cell culture ordinarily have receptor sites that bind LDL, permitting the internalization and degradation of the lipoprotein by the cell. In familial hypercholesterolemia *three* different *receptor defects* have been identified, presumably specified by three different alleles at the same locus. One allele specifies a receptor unable to bind LDL, another a receptor with greatly reduced binding ability, and the third a receptor which binds LDL normally but cannot internalize the lipoprotein. Patients who are heterozygotes for familial hypercholesterolemia have *one* of these abnormal alleles, whereas patients who are homozygous have *any* combination of *two* abnormal alleles.

Treatment. The primary treatment of patients with type 2 hyperlipoproteinemia is *dietary*. Diets containing large amounts of cholesterol and saturated fat elevate the plasma cholesterol; hence the diet for these patients should be low in total fat and cholesterol, with some of the saturated fats replaced by polyunsaturated fats.

If dietary therapy alone is not sufficient, then *drug therapy* may be added. The drug of choice at the present time is a resin (e.g., cholestyramine) which has the ability to bind bile acids in the intestine. The binding of bile acids prevents their reabsorption, and thereby increases the amount of cholesterol required for bile acid synthesis. Other medications such as neomycin, nicotinic acid, clofibrate, or gemfibrozil may also be used.

Dietary therapy of milder forms of hypercholesterolemia is often successful, even without very severe dietary modifications. In many patients with *heterozygous* essential familial hypercholesterolemia, significant lowering of plasma cholesterol levels can be achieved with strict diets and the use of medication. In *homozygotes,* although some lowering of the plasma cholesterol can be achieved, the resulting levels remain very elevated. In some of these patients, an experimental form of treatment—the surgical creation of a portacaval shunt—has resulted in dramatic improvement. It is not known why this therapy is effective. Plasmaphoresis will also lower cholesterol concentrations in homozygotes, but this procedure must be repeated at frequent intervals.

Type 3 Hyperlipoproteinemia

Type 3 hyperlipoproteinemia (*dysbetalipoproteinemia*) is a rare disorder characterized by the presence in the plasma of a lipoprotein having β mobility on electrophoresis but the density of VLDL. This lipoprotein is an intermediate in the formation of LDL from VLDL and accumulates in this disorder because of a *defect in IDL catabolism.* The disorder appears in individuals who are homozygous for an isoform of apolipoprotein E (E2/E2) which binds poorly to hepatic receptors. This apoprotein disorder has a prevalence of about 1%, but most individuals who have the defect do not have hyperlipidemia. The development of type 3 hyperlipidemia appears to require the presence, in addition, of a gene for another form of hyperlipidemia such as familial multiple lipoprotein-type hyperlipidemia or familial hypertriglyceridemia. Type 3 hyperlipoproteinemia also occurs because of hepatic triglyceride lipase deficiency, and secondary to hypothyroidism.

Clinical Features. Manifestations do not usually appear until age 30 in men and age 50 in women. Tuberous and tuberoeruptive *xanthomas* may occur on the knees, elbows, and buttocks. Palmar xanthomas, if present on the palmar surfaces of the hands and fingers, are almost diagnostic of this condition. Premature *arcus senilis* is fairly common, but xanthelasma is rare. There is a high incidence of peripheral vascular disease in these patients, and ischemic heart disease is common.

Biochemical Features. The plasma is usually turbid, and on standing overnight at 4° it shows a thin layer of chylomicrons over a turbid infranatant. The plasma cholesterol and triglyceride are elevated, usually in a ratio approximating 1:1. On electrophoresis a broad-β lipoprotein band may be observed. Unfortunately the absence of a broad-β band does not rule out type 3 hyperlipoproteinemia, nor does its apparent presence establish the diagnosis. Definitive *diagnosis* of type 3 hyperlipoproteinemia requires ultracentrifugation and the subsequent demonstration of lipoproteins in the VLDL fraction having β mobility on electrophoresis. Diagnosis of familial dysbetalipoproteinemia requires the analysis of apoE isoforms by isoelectric focusing.

Many patients will have glucose intolerance, and many will have an elevated serum urate.

Treatment. The initial treatment of patients with type 3 hyperlipoproteinemia is *dietary.* Many of these patients are overweight, and it is known that obesity worsens the disorder. It is therefore necessary to restrict total calories to ensure that the patient maintains a slow steady weight loss. Cholesterol intake should be reduced, and polyunsaturated fats substituted for saturated fats as much as possible. Simple carbohydrates and alcohol should be restricted.

In almost all patients with type 3 hyperlipoproteinemia this regimen will lower the plasma cholesterol and triglyceride, and in some the plasma lipids will approach acceptable levels. The addition of *clofibrate* or *gemifibrozil* to the dietary regimen will often return the lipids to normal. Regression of xanthomas usually occurs with effective treatment.

Type 4 Hyperlipoproteinemia

Type 4 hyperlipoproteinemia (*endogenous hyperlipemia*) is characterized by an increase in VLDL without other lipoprotein abnormalities. It is the most frequent of the hyperlipoproteinemias.

Pathogenesis. This lipoprotein disorder also may be primary, or secondary to other disorders. The *primary* form appears to be inherited, at least in some families, as an autosomal dominant trait (familial hypertriglyceridemia) with other family members exhibiting the type 4 pattern. It has been reported in familial combined hyperlipoproteinemia, in which other family members have the type 4, type 2a, type 2b, or type 5 patterns. It may appear as a sporadic disorder. *Secondary* type 4 hyperlipoproteinemia occurs fairly frequently; the commoner causes of this condition are:

Obesity
Alcohol
Diabetes mellitus
Oral contraceptives
Steroid therapy
Chronic renal failure
Hypothyroidism
Nephrotic syndrome

Some patients who have the type 4 pattern secondary to other conditions may have a genetic predisposition to develop this pattern. It can be very difficult in such patients to decide whether the disorder is inherited, is induced by indiscrete dietary habits, or is caused by a combination of both factors.

The increase in VLDL levels can be the result of an increase in the rate of hepatic VLDL triglyceride synthesis or of a decrease in the rate of VLDL catabolism. It has been suggested that the primary defect in familial combined hyperlipidemia is an increase in hepatic apoB synthesis.

Clinical Features. Most patients with type 4 hyperlipoproteinemia have no symptoms and no physical signs. Many will be overweight. Several studies have shown an increased incidence of coronary artery and peripheral vascular disease, but this relationship is not considered proven. Nor is it clear whether there is an increased incidence of cerebral vascular disorders. Some patients with type 4 hyperlipoproteinemia may develop very high triglyceride levels with hyperchylomicronemia (type 5). Under these conditions, eruptive xanthomas, lipemia retinalis, or pancreatitis may occur.

Biochemical Features. The plasma will be turbid if the *triglyceride* level is over approximately 2.8 mmol/L (250 mg/dL). On standing overnight at 4° the plasma remains turbid and does not develop a layer of chylomicrons. As with cholesterol, the level of plasma triglycerides that is considered abnormal is to a certain extent arbitrary. Some use age- and sex-adjusted 95th percentiles to define *reference limits;* others state that any adult with a plasma triglyceride level greater than 2.3 mmol/L (200 mg/dL) should be considered to have hypertriglyceridemia. The plasma total cholesterol may be increased, and this is more common with higher levels of triglyceride because of the cholesterol carried in the VLDL. The LDL cholesterol should be normal; if LDL cholesterol is increased the disorder is considered type 2b.

Lipoprotein electrophoresis shows an increase in preβ-lipoproteins with no abnormality of other lipoproteins. Patients, however, who have severe type 4 may develop a type 5 pattern with the presence of chylomicrons. In addition to the lipid abnormalities, hyperuricemia and mild to severe glucose intolerance may be present.

Treatment. The most important aspect of treatment is modification of the *diet.* Obesity worsens type 4, so that overweight patients should have their total caloric intake restricted to ensure a slow but steady weight loss to ideal body weight. Alcohol also worsens type 4, and intake of alcoholic beverages should be restricted or eliminated. The condition is thought by some to be worsened by a high carbohydrate intake, possibly more by simple sugars than by complex carbohydrates. Because dietary saturated fats elevate triglycerides, these should be reduced, and replaced with polyunsaturated fats.

Dietary modification alone may be adequate to lower plasma triglycerides sufficiently in many patients. In some, however, even faithful adherence to a strict diet does not achieve normal triglyceride levels and the addition of *drugs* such as clofibrate, gemfibrozil, or nicotinic acid is necessary.

Type 5 Hyperlipoproteinemia

Type 5 hyperlipoproteinemia (*mixed hyperlipoproteinemia,* hyperchylomicronemia with hyperprebetalipoproteinemia) is characterized by the presence of chylomicrons in the fasting state and an increase in VLDL.

This disorder may be inherited (*primary*) but the genetics are not clearly understood. It has been reported in familial combined hyperlipoproteinemia, and it also occurs sporadically. It appears as a complication of severe type 4 hyperlipoproteinemia and may represent a combination of increased hepatic VLDL triglyceride synthesis and decreased triglyceride-rich lipoprotein removal. As a group, patients tend to have low but not absent postheparin lipase activity, but this is not diagnostic. It is suspected that those patients with severe type 4 hyperlipoproteinemia, who then develop the type 5 pattern, have saturated their clearing mechanisms with VLDL and that chylomicrons accumulate because of this.

Clinical Features. The clinical manifestations of type 5 hyperlipoproteinemia are the same as in type 1 hyperlipoproteinemia except that the disease presents in *adulthood,* with abdominal pain being the usual first symptom.

Biochemical Features. When large numbers of chylomicrons are present, the blood will have the same appearance as in type 1 hyperlipoproteinemia. Plasma kept overnight at 4° will develop a chylomicron cake at the top of the tube, but remains turbid owing to the presence of VLDL. The plasma triglycerides may be very high, and cholesterol is often increased owing to the cholesterol contained in VLDL and chylomicrons. Electrophoresis reveals the presence of both chylomicrons and increased VLDL.

As in type 4 hyperlipoproteinemia there is an increased incidence of hyperuricemia and glucose intolerance. Occasionally, the level of triglycerides becomes high enough to cause an *apparent* hyponatremia. In this situation the sodium concentration in plasma water is in fact normal, but as whole plasma or serum is used for electrolyte determinations in most analyzers, the large quantity of chylomicrons reduces the observed sodium concentration (approximately 3 mmol/L for each 10 mmol/L of triglyceride).

Treatment. The management of type 5 hyperlipoproteinemia is mainly *dietary*. An important step is the attainment of ideal body weight, and this by itself may cause a drastic reduction in the level of triglycerides. Many of the patients, particularly those who actually have type 4 and who go on to develop type 5, are very sensitive to the triglyceride-elevating effects of alcohol, and alcohol ingestion should be restricted or, preferably, prohibited. In patients with primary type 5 lipoproteinemia, dietary fat of all types may have to be restricted in order to reduce the level of chylomicrons.

Drug therapy of this form of hyperlipoproteinemia is not encouraging. Nicotinic acid may have some effect. Clofibrate or gemfibrozil can be tried but neither is very effective in most patients.

Investigation and Management of the Hyperlipoproteinemias

The prevalence of hyperlipoproteinemia in the North American population depends on the reference values chosen to define hypercholesterolemia and hypertriglyceridemia. By most criteria used now, a significant percentage of the population will have mild hyperlipoproteinemia; severe hyperlipoproteinemia is less frequent.

Patients with severe hyperlipidemia may present with symptoms, but many patients with hyperlipoproteinemia are identified only by determining *plasma cholesterol and triglyceride* concentrations. An important matter that has not been resolved is the question of which asymptomatic individuals should be screened for hyperlipidemia by plasma cholesterol and triglyceride determinations. Suggested guidelines are as follows:

Clinical manifestation of hyperlipidemia
Atherosclerosis (except in older age groups)
Hypertension
Diabetes mellitus
Hyperuricemia or gout
Obesity
Diets with excessive cholesterol or saturated fats
High alcohol intake
Other diseases known to cause hyperlipoproteinemia
Family history of atherosclerosis
Family history of hyperlipidemia

Plasma lipids should be *measured* following an overnight fast and patients should have been following their usual diet without weight loss for approximately 2 weeks prior to the test. Plasma lipid determinations obtained during a period of acute illness may not reflect the patient's usual lipid levels. Initially, plasma cholesterol and triglyceride are determined; if the levels are borderline or elevated, the determinations should be repeated on one or preferably two occasions. It is prudent to determine HDL-cholesterol concentrations as well, because mild elevations of total plasma cholesterol may be caused by increased HDL or by increased LDL. The levels of LDL cholesterol may then be calculated from the formula on page 467. Elevations of HDL-cholesterol require no treatment.

Once the presence of hyperlipidemia has been established, the definitive *classification* of the disorder into a type of hyperlipoproteinemia may require further investigation, including the electrophoretic and ultracentrifugal steps described previously. However, this is impractical and would be too costly to apply to the majority of patients with *mild* hypertriglyceridemia or hypercholesterolemia. For most patients, identifying the type of hyperlipoproteinemia is not necessary for successful management. If the patient does not respond to appropriate treatment, or if the disorder appears unusual or complicated, then the more detailed investigation is warranted. In centers where the problem of hyperlipoproteinemia is receiving special study, it may be appropriate to undertake a detailed investigation of most patients referred. In addition, if secondary hyperlipoproteinemia is suspected on clinical grounds, then investigations appropriate to the diagnosis of the underlying disease must be undertaken.

Recommended Approach. In patients with hypercholesterolemia it is important to establish whether the hypercholesterolemia is a result of increased LDL, because it is now clearly established that lowering even moderate elevations of plasma cholesterol in these patients can decrease their risk of coronary artery disease. It has not been proved whether lowering elevated plasma triglycerides can improve the risk from coronary heart disease, but, because mild to moderate hypertriglyceridemia often responds to a change to proper nutritional habits, it is reasonable to attempt dietary therapy in these patients.

Should phenotyping of patients with hyperlipidemia be desired, the following guidelines may be followed. Patients with *hypercholesterolemia* caused by increased LDL, *without hypertriglyceridemia,* have type 2a hyperlipoproteinemia. If *hypertriglyceridemia without hypercholesterolemia* is present and no chylomicrons are observed, then the disorder can be assumed to be type 4 hyperlipoproteinemia. If there is *hypercholesterolemia and hypertriglyceridemia without chylomicrons,* the disorder may be type 2b, type 3, or type 4. Estimation of LDL concentrations as already described will distinguish type 2b from type 4, but identification of the rare type 3 disorder requires further investigation.

The *presence of chylomicrons* can readily be detected by allowing plasma to stand at 4°C. If they are present the disorder is type 1 or type 5, although a small chylomicron layer may be observed in type 3. A clear infranatant suggests type 1, and a turbid infranatant type 5. *Lipoprotein electrophoresis* may establish the disorder as type 5 by showing the presence of increased preβ-lipoproteins in addition to chylomicrons, but if increased preβ-lipoproteins are not demonstrable, the condition cannot be assumed to be type 1. A *lipoprotein lipase* assay is then necessary to distinguish the two conditions.

In the definitive diagnosis of type 3 hyperlipoproteinemia, *ultracentrifugation* at a density of 1.006 is necessary, followed by *electrophoresis* of the supernatant and infranatant fractions. In type 3 there will be a lipoprotein having β mobility in the supernatant while types 2b and 4 will show lipoproteins having only preβ mobility in the supernatant.

HIGH-DENSITY LIPOPROTEINS

There is considerable evidence that high-density lipoproteins, measured in clinical practice as ***HDL cholesterol,*** have a protective effect against the development of *coronary artery disease.* Families with familial hyper*alpha*lipoproteinemia have a very low incidence of coronary heart disease, and population studies indicate that *low* HDL cholesterol carries an independent increase in the risk of coronary heart disease. The factors that regulate HDL cholesterol are not well known, but heredity may have a role in determining its levels. HDL cholesterol is *higher* in women than in men, and it has been reported to be increased in marathon runners. Specific diseases that *lower* HDL cholesterol are:

Obstructive and hepatocellular liver disease
The nephrotic syndrome
Chronic renal failure
Familial lecithin:cholesterol acyltransferase deficiency

HDL cholesterol is reported to be decreased in untreated diabetes and in most hyperlipoproteinemias.

THE HYPOLIPOPROTEINEMIAS

These disorders may be secondary to other diseases, or they may be inherited.

The ***primary, inherited hypolipoproteinemias*** are very rare and include the following:

Abetalipoproteinemia (Bassen-Kornzweig syndrome, acanthocytosis). This disorder is inherited as an autosomal recessive trait. It appears to be caused by a lack of ability to synthesize *apoB.* Chylomicrons, LDL, and VLDL are absent and HDL is decreased. Plasma cholesterol, triglycerides, and phospholipids are low. Patients have acanthocytosis of the red blood cells, and they develop retinitis pigmentosa and ataxia. The cells of the intestinal mucosa become loaded with dietary fat that cannot be formed into chylomicrons, and a malabsorption syndrome develops.

Hypobetalipoproteinemia. This may be inherited but is genetically unrelated to abetalipoproteinemia. It may also occur *secondary* to malabsorption, hyperthyroidism, or terminal hepatic failure. Plasma LDL and cholesterol are decreased but not absent and the other lipoproteins are normal. Mild clinical findings similar to those observed in abetalipoproteinemia may be present.

Analphalipoproteinemia (Tangier disease). This disorder appears to be due to synthesis of an abnormal apoA-I, which is rapidly destroyed by reticuloendothelial tissues. α-Lipoproteins cannot be detected on electrophoresis, but small amounts of an abnormal HDL (HDL_T) are present. LDL and total cholesterol are reduced but VLDL and chylomicrons are increased. The clinical manifestations are mainly due to the accumulation of cholesteryl ester in the reticuloendothelial system and consist of hypertrophied, yellow-orange tonsils and enlargement of the liver and spleen and sometimes of the lymph nodes. In addition, a peripheral neuropathy often develops.

LIPID STORAGE DISEASES

Lipid storage diseases (Table 20–4) are uncommon inherited disorders in which lipids accumulate in various tissues. The different types are caused by a deficiency of different enzymes necessary for the catabolism of certain lipids, chiefly the sphingolipids. The major symptoms are usually from involvement of the nervous system. They vary in their severity from relatively benign, to fatal in infancy. The diagnosis of many of these disorders can be established by finding the specific enzyme deficiency in plasma, leukocytes, or cultured fibroblasts. Some can be identified in the prenatal period by study of the amniotic fluid. (See also Chap. 15, Neurologic and Psychiatric Disorders, and Chap. 23, Prenatal Diagnosis and Biochemical Assessment of High-Risk Pregnancy.)

TABLE 20–4. Lipid Storage Diseases

Clinical Disorder	Lipid Abnormality
Tay–Sachs disease	GM_2 ganglioside
Gaucher's disease	Glucosylceramide
Fabry's disease	Ceramide trihexoside
Niemann–Pick disease	Sphingomyelin
Metachromatic leukodystrophy	Cerebroside sulfate
Krabbe's disease	Galactocerebroside
Generalized gangliosidosis	GM_1 ganglioside
Refsum's disease	Phytanic acid
Wolman's disease	Cholesteryl ester and triglyceride
Cerebrotendinous xanthomatosis	Cholestanol

SUGGESTED READING

BROWN MS, GOLDSTEIN JL: The hyperlipoproteinemias and other disorders of lipid metabolism. In Petersdorf RG, Adams RD, Braunwald E et al (eds): Harrison's Principles of Internal Medicine, 10th ed, pp 547–549. New York, McGraw-Hill, 1983

BROWN MS, GOLDSTEIN JL: How LDL receptors influence cholesterol and atherosclerosis. Sci Am 251:58, 1984

HAVEL RJ (ed): The Medical Clinics of North America, Vol 66, No. 2, Symposium on lipid disorders. Philadelphia, WB Saunders, 1982

MILLER NE: Plasma lipoproteins, lipid transport, and atherosclerosis: Recent developments. J Clin Pathol 32: 639, 1979

RIFKIND BM, SEGAL P: Lipid research clinics program reference values for hyperlipidemia and hypolipidemia. JAMA 250:1869–1872, 1983

SCHAEFER EJ: Clinical, biochemical, and genetic features in familial disorders of high density lipoprotein deficiency. Atherosclerosis 4:303, 1984

Part Four

Special Topics

21

Pediatric Clinical Biochemistry

J. Gilbert Hill

Of the various subspecialty areas, pediatric clinical biochemistry is one of the broadest and most rewarding. Many of the biochemical abnormalities found in adults are obviously also found in children, but to these must be added a whole host of clinical conditions associated primarily with the pediatric age group. It is not possible to provide a comprehensive review in a single chapter of this nature, and thus a selection of topics has been made. The reader is referred to chapter 18 in particular for discussions of inborn errors of metabolism.

REFERENCE (NORMAL) VALUES

It is important to recognize that the 'normal' values for a variety of common components of body fluids are dependent on the age and sex of the individual, as well as on the method of analysis. Tables 21–1 to 21–3 provide some examples, but the reader is cautioned that these data are offered mainly to indicate the complexity of the problem and should be used only with adequate knowledge of the manner in which they were obtained.

PEDIATRIC BLOOD SPECIMENS

To those interested in the welfare of tiny infants, the volume of blood collected for laboratory testing and the method of collection are very important. It is clear that procedures suitable for use in a 80-kg adult male, with a plasma volume of 4000 mL, are not appropriate for use in a 800-g (0.8-kg) premature infant, with a total plasma volume of less than 40 mL. To emphasize this point: it is not uncommon to find that in the course of a week in an intensive care nursery an infant may have given up an amount of blood equivalent to half its original blood volume. The importance of discretion in requesting tests, and of using micro testing procedures, is self-evident.

In older children, *venipuncture* is a rapid convenient way to collect blood for all tests except pO_2. For interpretable pO_2 results, arterial or arterialized capillary specimens must be obtained.

In infants and young children, ***capillary specimens*** are desirable and, when carefully collected, are quite suitable for all tests except pO_2 (which, again, requires an arterial specimen). However, in interpreting the results of tests performed on capillary blood, it is important to recognize that potassium, total protein, and calcium values are marginally higher than in venous blood. In the fasting patient, glucose values are almost identical in venous and capillary specimens, but differences as great as 4 mmol/L (70 mg/dL) may be observed in the nonfasting patient. In addition, capillary specimens are particularly prone to hemolysis, with the result that the actual amounts of potassium, aspartate aminotransferase (AST), lactate dehydrogenase (LDH), and aldolase measured in the serum may be higher than the 'true' value. Finally, some of the erythrocyte components released during hemolysis (e.g., hemoglobin) may interfere with certain test procedures (e.g., the measurement of bilirubin), resulting in false values.

TABLE 21–1. Reference Values for Serum Constituents That Do *Not* Show Sex Differences

Substance	Age (Years)	Percentile 5th	Percentile 95th	SI Units	Percentile 5th	Percentile 95th	Conventional Units
Sodium	4–20	136	142	mmol/L	136	142	mEq/L
Potassium	4–20	3.8	5.0	mmol/L	3.8	5.0	mEq/L
Chloride	4–20	100	109	mmol/L	100	109	mEq/L
Urea nitrogen	4–20	3.2	6.4	mmol/L	9	18	mg/dL
Calcium	4–20	2.40	2.72	mmol/L	9.6	10.9	mg/dL
Magnesium	4–20	0.77	0.93	mmol/L	1.54	1.86	mEq/L
Total protein	4–11	66	79	g/L	6.6	7.9	g/dL
	12–20	68	84	g/L	6.8	8.4	g/dL
Albumin	4–20	38	50	g/L	3.8	5.0	g/dL
α_1-Globulin	4–20	1	3	g/L	0.1	0.3	g/dL
α_2-Globulin	4–20	6	11	g/L	0.6	1.1	g/dL
β-Globulin	4–20	8	12	g/L	0.8	1.2	g/dL
γ-Globulin	4–20	6	14	g/L	0.6	1.4	g/dL

(Cherian AG, Hill JG: Percentile estimate of reference values for 14 chemical constituents in sera of children and adolescents. Am J Clin Pathol 69:24, 1978)

TABLE 21–2. Reference Values for Two Serum Constituents That Vary Significantly with Age and Sex

	Age (Years)	Percentile 5th	Percentile 95th	SI Units	Percentile 5th	Percentile 95th	Conventional Units
Inorganic phosphorus							
Females	4–8	1.28	1.74	mmol/L	3.9	5.4	mg/dL
	9–13	1.13	1.64		3.5	5.1	
	14–20	0.90	1.39		2.8	4.3	
Males	4–9	1.23	1.75		3.8	5.4	
	10–15	1.10	1.71		3.4	5.3	
	16–20	0.81	1.43		2.5	4.4	
Creatinine							
Females	4–5	35	62	μmol/L	0.4	0.7	mg/dL
	6–9	44	71		0.5	0.8	
	10–14	53	80		0.6	0.9	
	15–18	62	88		0.7	1.0	
	19–20	71	97		0.8	1.1	
Males	4–6	35	62		0.4	0.7	
	7–9	44	71		0.5	0.8	
	10–11	53	80		0.6	0.9	
	12–14	62	88		0.7	1.0	
	15–17	71	97		0.8	1.1	
	18–20	80	106		0.9	1.2	

(Cherian AG, Hill JG: Percentile estimate of reference values for 14 chemical constituents in sera of children and adolescents. Am J Clin Pathol 69:24, 1978)

TABLE 21-3. Reference Values for Three Serum Enzymes That Vary with Age and Sex

	Alkaline Phosphatase (ALP; IU/L)		Aspartate Aminotransferase (AST; IU/L)		Creatine Kinase (CK; IU/L)	
	Percentiles		*Percentiles*		*Percentiles*	
Sex and Age	*5th*	*95th*	*5th*	*95th*	*5th*	*95th*
Males						
5–8	146	291	16	27	32	108
9–10	145	389	16	26	33	109
11–12	159	355	15	23	33	108
13–14	162	497	14	22	30	130
15–16	92	365	13	23	45	247
17–20	70	174	12	24	30	180
Female						
5–8	155	358	16	28	29	101
9–10	165	355	15	23	32	88
11–12	121	372	13	22	28	87
13	80	334	11	20	26	85
14	69	203				
15–16	57	140	11	19	21	74
17–20	40	97	10	23	18	66

(Cherian AG, Hill JG: Age dependence of serum enzymatic activities [alkaline phosphatase, aspartate aminotransferase, and creatine kinase] in healthy children and adolescents. Am J Clin Pathol 70: 783, 1978)

SPECIAL PROBLEMS IN PEDIATRICS

RESPIRATORY DISTRESS SYNDROME

The respiratory distress syndrome (sometimes called *hyaline membrane disease*) is the most common cause of severe acute respiratory distress in the first few days of life. It is said to occur in approximately 1% of neonates and to be the most common cause of death in the neonatal period. Fortunately, the development of neonatal intensive care units and the ready availability of biochemical monitoring procedures have had a striking effect in reducing the mortality rate in these infants. Continued improvements are anticipated as knowledge of the pathogenesis and treatment of the syndrome increases.

The syndrome occurs primarily in infants delivered prematurely and appears to be the direct consequence of an insufficient supply of the ***pulmonary surfactant*** which normally facilitates gas exchange by reducing surface tension in the alveoli. Two surfactants, both lecithins, have been described. The first is α-palmitoyl-β-myristoyl phosphatidyl choline and is the product of a biosynthetic process that begins in the fetal lung at 22 to 24 weeks' gestation. This pathway is very sensitive to acidosis, hypothermia, hypoxia, and hypercapnia and so may be 'turned off' during labor and delivery. An alternate pathway, which begins to function at about the 35th week of pregnancy, produces the second surfactant, dipalmitoyl phosphatidyl choline. Should pregnancy be interrupted before term, continuing activity of the first pathway becomes extremely important, and any inhibition sets the stage for development of the syndrome.

In a high-risk pregnancy, measurement of the *lecithin/sphingomyelin ratio* in the amniotic fluid is a useful device to predict the likelihood that a fetus will develop respiratory distress syndrome (see also Chap. 23) and permits the development of a plan to try to prevent this occurrence. Prevention may include strategies to delay delivery, thus permitting lung maturation, or attempts to stimulate surfactant synthesis through the use of glucocorticoid therapy.

In neonates with respiratory distress syndrome the most characteristic *clinical features* are rapid, labored breathing associated with expiratory grunting and increasing cyanosis. The infant is often relatively unresponsive to changes in the amount of inspired oxygen. As the lungs become more difficult to ventilate, arterial pO_2 falls, pCO_2 rises, and severe acidosis develops. *Treatment* is directed at improving oxygenation and managing the secondary hydrogen ion and electrolyte disturbances. Correction of the oxygen deficit may or may not require the use of a respirator, but in any case the objective is to achieve an arterial pO_2 value in the range of 50 to 70 mm Hg. Excessive oxygen carries with it the risk of retrolental fibroplasia or bronchopulmonary dysplasia, so that oxygen therapy must be closely monitored by frequent measurements of the arterial or capillary oxygen tension.

NEONATAL HYPERBILIRUBINEMIA

Under normal circumstances a rise in the concentration of unconjugated bilirubin in the blood leads to jaundice in the first week of life in approximately 60% of full-term infants and 80% of preterm infants. In *full-term* infants, this so-called ***physiologic jaundice*** becomes visible on the second or third day; the serum bilirubin peaks between the second and fourth day at 70 to 100 μmol/L (4 to 6 mg/dL), but may reach 200 μmol/L (12 mg/dL), and declines to less than 35 μmol/L (2 mg/dL) between the fifth and seventh days. Among *prematures* the rise in serum bilirubin tends to appear a little later (between 3 and 4 days) and has a higher peak (up to 260 μmol/L; 15 mg/dL), and the jaundice may last longer (up to 9 days; see Fig. 10–9).

Factors contributing to physiologic jaundice in the newborn include the following:

1. The rate of *production* of bilirubin in the neonate averages 7 mg/kg/day, in contrast to 2.5 mg/kg/day in adults, owing to the shorter life span of fetal erythrocytes.
2. The *transfer* of bilirubin from extracellular fluid into the hepatocyte depends on the availability of Y (and Z) receptor-carrier proteins, and the amount of Y is low in the neonate.
3. The *conversion* of bilirubin to bilirubin glucuronide depends on the availability of glucuronic acid and the activity of uridine diphosphate glucuronyl transferase (UDPGT), and both are relatively deficient in the newborn.
4. The amount of bilirubin *reabsorbed* from the gut is greater because the newborn lacks the bacterial enzymes which normally reduce bilirubin to urobilinogen.

Although the hyperbilirubinemia found in most neonates is of the physiologic type, many forms of ***pathologic jaundice*** have been described. Among them, the most important are related to these conditions:

1. Bilirubin *overproduction* secondary to hemolysis from
   a. Rh or ABO incompatibility
   b. Viral or bacterial infections
2. Impaired *transport* of bilirubin due to abnormalities in albumin binding caused by:
   a. Hypoxia
   b. Acidosis
   c. Drugs (e.g., sulfonamides)
3. Impaired *conjugation* of bilirubin in the hepatocyte associated with:
   a. The presence of pregnane-$3\alpha,20\beta$-diol in maternal milk (breast milk jaundice)
   b. Drugs (e.g., chloramphenicol)
4. Impaired *excretion* of conjugated bilirubin due to biliary atresia

In addition, neonatal hyperbilirubinemia is seen in association with a wide variety of *inherited* metabolic diseases, such as hypothyroidism, galactosemia, cystic fibrosis, and α_1-antitrypsin deficiency.

CYSTIC FIBROSIS

Cystic fibrosis is a relatively common inherited disease in which a defect in exocrine gland function leads to chronic lung disease, general malnutrition secondary to malabsorption, and failure to thrive. The disease may present in the neonatal period as intestinal obstruction, but the diagnosis may be delayed several years until severe lung disease, gross steatorrhea, or growth failure becomes apparent. A decade ago 25% of patients with cystic fibrosis died in the first year of life, and 60% between the ages of 1 and 9 years. Now, however, through the use of more vigorous and effective treatment with antibiotics and pancreatic enzymes, the life span of patients with cystic fibrosis has significantly lengthened and an increasing number are reaching adulthood.

The ***diagnosis*** of cystic fibrosis requires that at least two of the following four criteria are met:

a. The presence of the characteristic lung disease
b. The presence of pancreatic insufficiency
c. The presence of an increased level of sweat electrolyte
d. A positive family history

The most useful test to establish pancreatic insufficiency is the quantitative measurement of the *total fat* content of a 5-day *fecal* collection. The same specimen may be used to demonstrate a deficiency of trypsin and chymotrypsin as well, but this information is of relatively little additional value in the presence of proven steatorrhea.

The demonstration of an increased *electrolyte* content of *sweat* is probably the most important element in the diagnosis of cystic fibrosis. Confirmed values for sweat chloride exceeding 60 mmol/L or sweat sodium exceeding 70 mmol/L are generally considered diagnostic. It is recognized, however, that normal values vary with age, and that abnormally high values may occasionally be found in conditions other than cystic fibrosis (e.g., glucose-6-phosphatase deficiency, glycogen storage disease, hypothyroidism, untreated renal insufficiency, and malnutrition).

On the *first* day of life up to 8% of infants will show abnormally elevated levels of sweat electrolytes, but most will subsequently be found to be normal. On the other hand, abnormally high sweat electrolyte values on the *third* day of life are unlikely in the absence of cystic fibrosis. Finally, a very small number of patients with cystic fibrosis have been shown to have intermittently normal sweat electrolytes during the course of their illness.

In the differential diagnosis of cystic fibrosis, the *Schwachman syndrome* must be considered. This syndrome, which is characterized by pancreatic insufficiency, neutropenia, bone abnormalities, and failure to thrive, is second only to cystic fibrosis as a pancreatic cause of malabsorption in infancy. The clinical presentation of the two conditions may be very similar, but the *normal sweat chloride* of the Schwachman syndrome is a reliable distinguishing feature.

Technical problems are frequent in sweat analyses, and most attempts to simplify the test have led to an increase in incorrect results. As a consequence, it is essential to use a method which includes pilocarpine iontophoresis to stimulate sweating, followed by gravimetric quantification of the amount of sweat produced and the coulometric amperometric measurement of chloride.

The merits of neonatal screening for cystic fibrosis are still being debated. Proponents argue that early detection and treatment can significantly prolong life, whereas others argue that this has not been proved. Until recently, the arguments have been somewhat academic, because reliable screening procedures were not available. Now, however, evidence is accumulating that the measurement of circulating *immunoreactive trypsin* may identify infants who will eventually develop cystic fibrosis. These infants appear to begin life with an abnormally *high* concentration of immunoreactive trypsin, which gradually falls to abnormally *low* levels over a period of weeks, months, or years. The test is attractive because it can be performed on blood spots obtained for phenylketonuria screening, but the specificity and sensitivity have still to be established.

NEUROBLASTOMA

Neuroblastoma is one of the most common malignant tumors in infancy and early childhood, with approximately 25% of cases presenting in the first year of life and 75% by the fifth year. The usual presenting sign is an abdominal mass, although in many patients the first complaint arises as a consequence of tumor metastases to the liver or the cervical or axillary lymph nodes. Less frequently, systemic manifestations of a catecholamine-secreting tumor may be present—hypertension, tachycardia, excessive sweating, headache. Rarely, persistent diarrhea is the sole complaint. Treatment with surgery and/or radiation now offers these patients significant hope for survival, so that a high premium is placed on accurate early diagnosis.

Diagnostic procedures include radiologic techniques to demonstrate a soft tissue mass displacing the kidney downward and laterally, and/or the presence of foci of calcification in the kidney. Histologic examination of the bone marrow may reveal the characteristic cells of metastatic tumor.

Biochemical examinations may be extremely important in both the diagnosis and the monitoring of treatment in neuroblastoma, and the metabolic pathways for the catecholamines (Figs 14–9 and 14–10) provide the rationale for test selection. Unfortunately, the urinary excretion patterns of the catecholamines, their precursors, and their metabolites are extremely variable in patients with neuroblastoma, so that no single measurement can be relied on to rule the diagnosis in or out. The interpretation of test results is further complicated by the inadequacy of reference values, the difficulty of collecting timed urine specimens from infants, and the lack of precision and accuracy of some of the test procedures. Nevertheless, most studies indicate that approximately 80% of patients with neuroblastoma excrete two to ten times the expected amount of *vanillylmandelic acid* (VMA), and the determination of the 24-hour excretion of this compound is probably the most useful test in the diagnosis of neuroblastoma. A significant but smaller percentage of patients will be found to have an excessive excretion of *homovanillic acid* (HVA), total *catecholamines,* total *metanephrines,* and *dopamine.* It is important to note, however, that almost all possible permutations and combinations of normal and abnormal results have been reported for these five tests.

Plasma dopamine β-hydroxylase and urine cystathionine have also been suggested as markers for neuroblastoma but in neither case has sufficient specificity or sensitivity been demonstrated to justify their routine use. Finally, mention should be made of various 'spot tests' designed to screen for neuroblastoma through the demonstration of abnormal amounts of catecholamine metabolites in the urine. None of these tests has proved to be a reliable alternative to a careful quantitative assay for VMA or related compounds, and these spot tests should be abandoned.

GASTROINTESTINAL DISEASE

The evaluation of chronic gastrointestinal disease in children is a challenging problem and frequently requires comprehensive testing of some or all of the metabolic processes involved in digestion, absorption, and excretion of carbohydrate, fat, and protein. Approaches to this problem have been considered in detail previously (Chap. 12), but three topics are of special interest in pediatrics and merit comment here.

Fat Absorption

Fat absorption may be assessed directly or indirectly. In the *direct* approach a diet of known fat content is provided for 3 to 5 days; feces are collected for the same time period, and the fecal fat content is measured by the method of Van de Kamer (or equivalent). Fat excretion may be as high as 15% of intake (20% in prematures) for patients up to 3 months of age. Thereafter, the upper limit of normal is generally considered to be 10%, gradually falling to 5% in adults.

Various *indirect* approaches to the assessment of fat absorption have been used, such as the measurement of carotene in fasting serum, but none has proved to be as reliable as the direct measurement of fecal fat.

Disaccharide Intolerance

Milk and other dairy products are important components of the diet of many children and are the principal source of the disaccharide *lactose.* In normal circumstances, the nonabsorbable dietary lactose is hydrolyzed by the small-intestinal enzyme lactase, to yield the absorbable monosaccharides glucose and galactose. In some circumstances, there may be an absolute or relative absence of the enzyme, so that lactose passes unchanged to the large intestine. Under the influence of bacteria normally resident in the colon, the lactose is then fermented, leading to symptoms which may include abdominal pain, distention, and diarrhea.

Lactase activity in the intestinal brush border is at its highest level in the newborn period, declining gradually after 3 years. In some children the natural decrease in the capacity to digest lactose is rapid enough to result in symptoms as early as 4 or 5 years of age. This so-called *primary* lactose intolerance is particularly prevalent in Asian and black children. *Secondary* lactose intolerance is more common and may be seen following acute gastroenteritis or in chronic gastrointestinal conditions such as celiac disease.

The *diagnosis* of lactase deficiency has been based historically on an oral lactose tolerance test, demonstration of excess reducing substance in the stool, or the direct assay of enzymes in a small bowel mucosal biopsy. Currently, the procedure of choice is the breath hydrogen test, which is noninvasive and is a sensitive discriminator between normal and abnormal.

In the ***breath hydrogen test,*** the patient is given an oral load of lactose, and the content of hydrogen in the expired air is measured by gas-liquid chromatography. In lactase deficiency there is a demonstrable increase in breath hydrogen between the baseline sample and samples collected at intervals up to 120 minutes thereafter. The test depends on the ability of colonic bacteria to metabolize lactose, with the production of hydrogen, followed by the passage of hydrogen to the blood stream, and eventually to the expired air.

Interpretation of the test requires knowledge of several important factors that may influence results. It has been estimated that there may be up to 2% *false negatives,* caused by the absence of hydrogen producing bacteria in the colon. This possibility can be ruled out by showing an increase in breath hydrogen excretion following ingestion of a substrate known not to be absorbed, such as lactulose. *False-positive* tests are rare, but can occur with bacterial colonization of the upper small bowel. Variations in breath hydrogen may be related to other factors, such as delayed gastric emptying or short intestinal transit time, but these factors do not reduce the ability of the test to discriminate between lactase sufficient and lactase deficient individuals.

It should also be noted that measurement of breath hydrogen following ingestion of *other sugars* permits the identification of patients with other forms of sugar intolerance, such as isolated sucrase-isomaltase deficiency.

Protein-Losing Enteropathy

Excessive loss of plasma protein across the intestinal mucosa is described as protein-losing enteropathy and is a frequent complication of conditions such as celiac disease, regional ileitis, allergic gastroenteropathy, colitis, and lymphangiectasia.

The *diagnosis* has traditionally been difficult and for many years depended on the use of a radioactively labeled substrate such as albumin-^{51}Cr. Recently, however, the diagnosis has been greatly simplified by development of the concept of *protein clearance* by the intestine, analogous to creatinine clearance by the kidney. The plasma glycoprotein *α_1-antitrypsin* has proved to be a very useful endogenous marker for this purpose, in that it is representative of the other circulating plasma proteins and is resistant to degradation in the gut.

The clearance of α_1-antitrypsin can be calculated by measuring its concentration in plasma and feces, and then applying a standard clearance formula. Reference values will depend on the actual analytical technique used, but a clear distinction can usually be drawn between normal and abnormal.

HEAVY METAL POISONING

A brief account of the most common examples of heavy metal poisoning in children is given here (see also Chap. 22, Nutrition, Vitamins, and Trace Elements).

Iron

Exposure to excess iron is almost always the result of a child's ingestion of a parent's medication. The minimum lethal dose has not been clearly established, but children have died after the ingestion of as little as 1000 mg of ferrous sulfate (200 mg of elemental iron). Severe systemic toxicity can occur as soon as 30 min after ingestion, and may manifest as vomiting, diarrhea, severe acidosis, convulsions, coma, and circulatory collapse. Although the mechanisms leading to these signs and symptoms are not clearly understood, they are related to the presence in the circulatory system of excess *unbound* iron, and do not occur unless the serum iron concentration exceeds the serum iron-binding capacity.

Along with conventional supportive measures, specific treatment with the chelating agent *deferoxamine* has been found to be very effective in managing patients with iron intoxication. The drug itself, however, is potentially toxic and for this reason should be administered only when serum iron levels exceed the serum iron-binding capacity. Traditionally, the measurement of iron and iron-binding capacity are tedious, time-consuming procedures, and are not readily available on an emergency basis. A *semiquantitative test* is available, however, to determine whether the serum iron concentration exceeds the iron-binding capacity, and this has proved to be an important adjunct in patient care.

Lead

Although public health measures designed to reduce exposure to lead are almost universally practiced, acute and chronic lead poisoning are still major problems in many parts of the world. Because there are no completely specific signs and symptoms of this disorder, the possibility of lead intoxication must be considered in any child presenting with unexplained anorexia, pallor, apathy, hyperirritability, vomiting, regression in development, or behavior disorder.

There is no known biologic requirement for lead, and all the known effects are undesirable. These effects are first noted in the erythroid cells of the bone marrow, but in time direct injury to the nervous system and the kidneys will occur.

Lead appears to exert its *toxic effects* by interfering with the biosynthesis of ***heme,*** and it is likely that this is a consequence of the blocking of thiol groups in a number of the enzymes in this complex pathway (see Fig. 10–1). In particular, heme synthetase, which catalyzes the insertion of ferrous iron into protoporphyrin IX to form heme, and porphobilinogen (PBG) synthetase, which facilitates the conversion of aminolevulinic acid (ALA) to PBG, are very sensitive to inhibition by lead.

The definitive ***diagnosis*** of lead poisoning requires laboratory assistance, and the available tests fall into two main categories: those measuring lead itself, and those measuring the metabolic consequences of excess lead exposure.

The direct measurement of *whole blood lead* is the most widely used test to assess lead exposure, and in children values above 1.5 μmol/L (30 μg/dL) should be viewed with concern. Because of problems in obtaining accurately timed collections, the measurement of lead excreted in the urine is less useful. The *metabolic effects* of toxic amounts of lead may be assessed by demonstrating (a) in erythrocytes, excess protoporphyrin or decreased PBG synthetase, or (b) in urine, increased excretion of coproporphyrin or ALA.

SUGGESTED READING

CLAYTON BE, JENKINS P, ROUND JM: Paediatric Chemical Pathology: Clinical Tests and Reference Ranges. Oxford, Blackwell Scientific Publications, 1980

HICKS JM, BOECKX R (eds): Pediatric Clinical Chemistry. Philadelphia, WB Saunders, 1983

JOHNSON TR, MOORE WM, JEFFRIES JE (eds): Children Are Different, 2nd ed. Columbus, Ross Laboratories, 1978

MEITES S (ed): Pediatric Clinical Chemistry. A Survey of Normals, Methods, and Instrumentation with Commentary, 2nd ed. Washington, American Association for Clinical Chemistry, 1981

22

Nutrition, Vitamins, and Trace Elements

Michael D. D. McNeely

NUTRITIONAL PROBLEMS

Nutrition is the process by which food is assimilated and converted into the tissues of living organisms. Inappropriate nutrition is a major cause of illness in the world today. While the inhabitants of underdeveloped nations suffer from a lack of food, many citizens of economically developed countries eat improperly or in excess.

Malnutrition

Underdeveloped Nations. In stark contrast to the affluence of Europe and North America is the suffering of millions who are deprived of normal nutrition. Specific deficiency syndromes such as rickets, beriberi, scurvy, and vitamin A blindness remain endemic. Protein-calorie malnutrition of some degree affects up to two thirds of preschool-aged children in 'third world' countries and, every year, thousands of people die from starvation. Several identifiable syndromes of protein-calorie malnutrition are recognized.

Kwashiorkor is caused by a diet that is low in protein but contains some carbohydrate. Intercurrent infections and socioeconomic problems complicate the picture. It develops usually at the time of weaning but may occur in older children. Clinically, the main features are edema, misery, growth retardation, wasted muscles, some subcutaneous fat, an enlarged fatty liver, light-colored dyspigmented skin and hair, anemia, infections, and cardiac failure. Biochemical changes are widespread but nonspecific.

Marasmus is the condition produced by a diet which is grossly inadequate in both protein and carbohydrate. There is serious wasting of muscle and fat. In contrast to kwashiorkor, there is no edema, the liver is normal in size, the skin color does not change, and no subcutaneous fat is present. For every case of diagnosable marasmus there are many milder cases of starvation.

Developed Nations. Overt malnutrition still occurs in economically developed nations. The extremely poor, the old and feeble, and the ill are often prone to starvation and vitamin deficiency.

In considering the apparently 'normal' population several observations can be made. Calcium intake is frequently inadequate and this is probably part of the reason for the high incidence of osteoporosis in older women. In some groups vitamin C intake may be inadequate. Certainly, the North American diet is low in fiber content. Though not absorbed, fiber is needed to facilitate the normal function of the bowel. Low-fiber diets have been implicated in the genesis of carcinoma of the bowel, derangement of the enterohepatic circulation of bile (leading to increased cholesterol levels and gallstones), and impaired glucose tolerance.

Clearly, the major nutritional problems in North America are those of *excess.* Too much alcohol, in addition to its obvious involvement in social and economic problems, is a metabolic poison and a source of 'empty' (nonnutritive) calories. An overabundance of meat and dairy products, high in saturated fatty acids, has characterized our diet in the past to the detriment of our lipid metabolism, but this trend is reversing as the public becomes more aware of nutrition and as the economy changes. Better dental health and possibly better carbohydrate metabolism would result from a reduction of dietary sugar. The most disturbing nutritional problem in North America, however, is obesity.

Obesity

Obesity is a disease affecting 15% of the population, characterized by too much adipose tissue, which results from a continued positive caloric balance. It is recognized that psychosocial and behavioral factors play a prime role in the ***development*** of this condition. In a few cases, obese individuals are found to have excessive numbers of fat cells. Obese persons move less than nonobese persons. An ongoing search for *biochemical* factors, however, has generated a number of observations. Triglyceride formation is increased in the fat cells of the obese. Hypertriglyceridemia is commonly seen and high-density-lipoprotein (HDL) cholesterol is lower. Hyperinsulinemia is often found and leads to decreased numbers of insulin receptors and thus to reduced glucose tolerance. This may also explain the lack of a feeling of satiety. There is generally no problem in *diagnosing* obesity, although occasionally Cushing's syndrome, myxedema, or hypothalamic disease must be ruled out.

Treatment of obesity requires a reduction of calories. The loss of fat depends on the negative caloric balance achieved by the diet. A diuresis will occur early in caloric restriction, possibly caused by urea reduction and an alteration of renal tubular concentrating power. This diuresis and weight loss may give false hope to the dieter. Ketosis will occur if the deficiency of carbohydrate in the reducing diet is extreme. There are several successful, reputable self-help and group-help approaches to weight loss, which are based on selecting a diet that produces a negative caloric balance while providing all the essential nutritive requisites. These plans usually provide nutritional education and social assistance in maintaining a new eating pattern. At the same time, fad diets, pills, and unusual foods enjoy immense popularity; they are to be viewed with scientific scepticism.

A surgical treatment for morbid obesity has been the *intestinal bypass.* Though successful in preventing calorie absorption, the derangement of intestinal physiology, in conjunction with the additional operative risk associated with obesity, has created an extensive list of *complications.* These include a 4% operative mortality rate, pulmonary emboli, wound infections, gastrointestinal hemorrhage, and renal failure. Later problems include hyperoxaluria and renal calculi caused by increased intestinal absorption of oxalate precursors, a dilated colon from bacterial overgrowth, anemia, acute cholecystitis resulting from deranged enterohepatic bile circulation, diarrhea, electrolyte abnormalities, hypoproteinemia, and fatty infiltration of the liver. Adequate hepatic function must be established prior to this procedure and frequent postoperative monitoring with liver, oxalate, electrolyte and hematologic tests must be carried out.

Currently, an operation that reduces the size of the stomach by *surgical stapling* has been introduced to avoid these functional derangements. This procedure still allows noncompliant patients to overeat, by continual eating or by ingesting high caloric liquid foods (alcohol, ice cream). Vomiting, abdominal pain, carbohydrate absorption derangements, and even perforation are the risks of this approach.

Dental 'splinting' has been used as a temporary aid. Total starvation regimes have also been used. Patients undergoing this radical therapy must be observed constantly for specific nutritional disorders, and a multivitamin supplement is usually needed. Cosmetic fat removal is a popular approach in some centers but does nothing to correct the primary problem.

Surgical treatment of obesity must be reserved for those whose obesity is intractable and represents a seriously debilitating or life-treatening problem.

Parenteral Nutrition

Total parenteral nutrition (TPN) is now an accepted form of therapy when food cannot be taken by mouth. Details of this treatment are beyond the scope of this book but several comments are in order.

The *complications* of TPN are caused by management problems (intravenous sites, infections) and biochemical or nutritional abnormalities (dehydration, overhydration, hyperosmolality, hypoglycemia, allergic reactions, specific nutritional deficiencies and excesses, metabolic acidosis, and hepatic failure). The development of commercially available micronutrient and vitamin mixtures should reduce the incidence of the bizarre deficiencies observed in the past.

Patients are monitored by clinical assessment, including weighing and the measurement of urinary output. *Biochemical monitoring* of the urine is most useful and may include measurements of: (1) urinary osmolality, to assess hydration; (2) sodium and potassium, to indicate electrolyte balance; (3) urea concentration, as a rough guide to overall nitrogen balance; and (4) ketones and glucose to indicate poor carbohydrate control and caloric loss. Blood analyses can be useful but should be interpreted with caution. The intravenous solution being administered at the time of blood collection may influence the results. For example, concentrated glucose solutions may give a false picture of carbohydrate metabolism, and lipid emulsions generally interfere with spectrophotometric tests and may produce a false lowering of electrolyte values.

Nutritional Assessment

The clinical assessment of nutritional state is important for epidemiologic studies, clinical evaluation (particularly presurgically) and during refeeding or TPN programs. Unfortunately, no single definitive laboratory test for nutritional status exists. Clinical evaluation should consist of a detailed nutritional history by a trained dietitian, and a physical examination which evaluates general status and identifies specific nutrient deficiencies. It should include measurements of percent ideal body weight, recent weight change, triceps skin-fold thickness, and arm circumference. Evaluation is hampered by wide confidence limits for the measurable parameters and because concomitant disease confounds the picture. Serum albumin concentrations are low in serious, long-standing nutritional deficiency but this test lacks sensitivity. Prealbumin may become a useful measurement of recent low protein intake. Ketonuria is a sign of negative caloric balance and a low serum and urine urea and uric acid signal protein deficiency.

VITAMINS

The term *vitamin* comes from the words 'vital amine' and refers to polyatomic trace substances, vital to normal growth and development, which the body cannot synthesize and must obtain exogenously. Although the essential amino acids conform to this definition, they are not considered in this category.

There is no other area of medicine which is less understood, engenders more debate, or encourages more unverified claims in the popular press than vitamin therapy. The administration of vitamin supplements as a health aid to normal persons ingesting a balanced diet has no secure scientific support. Megavitamin therapy, as a panacea for a variety of illnesses, has not been generally recognized and indeed has been the cause of toxicity to vitamins A, C, and D.

In the light of such widespread controversy, the material presented in this chapter is conservative and only those facts which have been demonstrated by generally accepted scientific techniques are presented. To condone unproved vitamin therapy on grounds that 'It can't hurt and it might do some good' represents intellectual dishonesty that can only impair rational medical therapy.

Vitamin A

Vitamin A is a generic term referring to a group of compounds that have the biologic activity of retinol and are essential for vision, reproduction, growth, epithelial differentiation, and mucus secretion. The major sources of vitamin A are plant carotenoid pigments (mainly β-carotene) and retinyl esters from animals. These compounds are converted into retinol and esterified with long-chain fatty acids. The resulting retinyl esters are absorbed in close association with fats and transported to the liver for storage.

Depending on tissue demands, retinyl esters are reconverted to retinol and transported by retinol-binding protein (RBP: M_r 20,000), which forms a complex with prealbumin. Serum concentrations of RBP vary directly with the availability of vitamin A; they decrease in protein deficiency and liver disease and increase with kidney disease and estrogen administration.

Vitamin A maintains the integrity of epithelial tissue by its influence on cell division, RNA synthesis, protein glycosylation, lysosomal membrane stability, and prostaglandin biosynthesis. Through these mechanisms, vitamin A determines the keratinization and differentiation of the epithelial layers. It plays a vital role in *vision.* Retinal is the prosthetic group of the photosensitive pigment in the eye which enables incident light to be converted into nerve impulses.

Vitamin A *deficiency* is endemic in several third-world countries. Incredibly, it remains the principal cause of blindness. Deficiency is also seen in fat malabsorption syndromes. Clinically, typical ocular signs, beginning with night blindness and terminating with total blindness, are seen. In addition, widespread epithelial changes develop. General malnutrition often accompanies the syndrome. The diagnosis is made clinically.

Serum vitamin A measurements have limited diagnostic use because the concentration is maintained at a fairly constant level until liver reserves are exhausted. Nevertheless, a fasting value below 20 μg/dL suggests vitamin A deficiency and less than 10 μg/dL is almost diagnostic. Hepatic tissue vitamin A measurement is the most accurate (though least practical) test. Liver disease is associated with low levels.

Vitamin A *toxicity* results in a cerebral syndrome with epithelial, liver, and bone involvement and raised serum vitamin A concentrations. It was once seen only in infants but in recent years has become more common because of overenthusiastic self-medication. Another syndrome, *carotenemia,* is caused by excessive ingestion of carotenoids (in lettuce and carrots), which discolor the serum and skin (notably the palms of the hands). Serum carotene levels become raised while serum bilirubin remains normal.

Vitamin A has gained attention as a potential anticancer agent. The theory is that retinoids may reverse the loss of cellular differentiation which is associated with cancer. Other unsubstantiated uses of vitamin A are the treatment of schizophrenia, learning disabilities, and autism.

Zinc metalloenzyme (retinol dehydrogenase) is responsible for the conversion of retinol to retinaldehyde (the photochemically active form of vitamin A). Thus, a form of vitamin A deficiency may be caused by zinc deficiency.

Vitamin B_6

Vitamin B_6 is the term used to refer to pyridoxine, pyridoxal, and pyridoxamine. Pyridoxal phosphate serves as a cofactor in a number of enzyme reactions, particularly those involving the synthesis and catabolism of amino acids; it is vital for protein synthesis.

Vitamin B_6 *deficiency* has been implicated in a variety of disorders, including skin changes, anemia, convulsions, and neurologic degeneration. Isoniazid, hydralazine, penicillamine, and cycloserine all interfere with B_6 and without vitamin replacement their use may lead to polyneuritis. There are also a number of pyridoxine-responsive microcytic anemias.

A deficiency of vitamin B_6 can be *evaluated* biochemically by the tryptophan loading test (2 g per os); normal persons will excrete more than 50 μmol of xanthurenic acid in the urine. The test is not generally available and has limited use.

Vitamin B_{12} and Folate

Vitamin B_{12} (cobalamin) is a tetrapyrrole porphyrin-like structure surrounding a cobalt atom. Following its ingestion, vitamin B_{12} is bound by a glycoprotein known as intrinsic factor (IF) which is secreted by the stomach. The vitamin B_{12}–IF complex becomes attached to specific receptors in the terminal ileum. Here, the vitamin B_{12} is absorbed through the intestinal epithelium and transported in the bloodstream by the protein transcobalamin II. Although it is distributed throughout the body, most of the vitamin is stored in the liver. Under usual conditions the body contains several years' supply of vitamin B_{12}.

Vitamin B_{12} *deficiency* may be produced by:

Total gastrectomy (loss of IF)
Pernicious anemia (essential IF deficiency)
Surgical removal or inflammatory disease of the terminal ileum
Infestation by the fish tapeworm (which consumes vitamin B_{12})
Dietary deficiency
Malabsorption syndromes (bacterial overgrowth)
Pregnancy
Selective vitamin B_{12} malabsorption

Folates are found in many vegetables as well as in animal liver. This natural folate is conjugated and prior to its absorption must be freed of all but a single glutamate residue by intestinal deconjugases. The body is able to store several months' supply of folic acid (pteroyl glutamic acid). The active form of folic acid is tetrahydrofolic acid, formed by dihydrofolic acid reductase. The analogs aminopterin and amethopterin inhibit this enzyme and have been used to halt the growth of cancer cells, particularly in leukemias.

Folate *deficiency* may result from:

Inadequate intake
Generalized malabsorption
Increased utilization (in malignancy and hyperthyroidism)
Congenital folate malabsorption
Abetalipoproteinemia

Both vitamin B_{12} and folate act in DNA synthesis. The active forms of vitamin B_{12} are deoxyadenosylcobalamin and methyl-B_{12}. The former is the cofactor essential for the N^5-methyltetrahydrofolate–homocysteine methyltransferase reaction that produces methionine. The active forms of vitamin B_{12} and folic acid are involved in purine and pyrimidine synthesis, and hence are necessary for normal DNA synthesis during cell replication.

Deficiency of either vitamin B_{12} or folate leads to the development of *megaloblastic anemia.* Vitamin B_{12} deficiency (*pernicious anemia*) may also result in a severe degenerative neurologic syndrome of the lateral and dorsal columns of the spinal cord. Folate deficiency does not produce this latter disorder. When a case of megaloblastic anemia is encountered, it is important to determine whether it is caused by deficiency of folate or vitamin B_{12} or both, because the treatment of pernicious anemia with folate rather than B_{12} will correct the hematologic picture but allow the neurologic degeneration to proceed.

The prevalence of pernicious anemia increases after the age of 40. The pathophysiology is obscure. Ninety percent have antibodies to parietal cells, 50% have antibodies to intrinsic factor, thyroglobulin, and microsomes. In pernicious anemia, the Schilling's test is usually diagnostic, the gastric juice is alkaline and the serum gastrin and LDH are dramatically elevated.

Laboratory measurements of *serum folate and vitamin B_{12}* are readily available. The type of assay used to determine the vitamin B_{12} level is important. Some methods measure non-active cobalamin compounds and therefore overestimate the vitamin B_{12} concentration. Administration of vitamin B_{12} and folate is common and results in very high values. Patients who have received folate may have normal serum folate levels. Such patients are more accurately assessed with *red cell* folate determinations. The concentration of *methylmalonic acid* in the urine is a sensitive indicator of vitamin B_{12} deficiency. The test is not routinely available and is rarely required.

If vitamin B_{12} deficiency is diagnosed, an effort must be made to determine its cause. This is done primarily from the clinical history but can, and should, be confirmed in the laboratory. The ***Schilling test*** is very useful for this purpose. In this procedure, the patient is given an intramuscular (IM) injection of vitamin B_{12} (flushing dose) to saturate the body's B_{12} binding sites. This step prevents subsequent sequestration of orally-administered vitamin B_{12}. Oral radioactive vitamin B_{12} is given in conjunction with the IM dose. Urine is collected for 24 hours and its radioactivity is measured. Normal individuals will absorb and then excrete at least 10% of the ingested dose. The patient is judged to have vitamin B_{12} malabsorption if less than 10% is excreted. The test may be repeated using an oral dose of radioactive vitamin B_{12}-intrinsic factor complex. If absorption is *normal* with the complex then intrinsic factor deficiency is the probable cause. If absorption remains *abnormal* then a lesion of the terminal ileum is implied.

Vitamin C

Vitamin C (ascorbic acid) is necessary for hydroxylation of proline and lysine during collagen synthesis. It has an undefined role in the maintenance of blood vessel integrity, which may be related to collagen production.

Deficiency of vitamin C leads to *scurvy.* The deficiency is usually caused by a diet devoid of citrus fruits, leafy vegetables, and other plant foods. It is manifested by swollen, bleeding gums, arthralgias, fatigue (in adults), and tender, swollen legs and subcutaneous hemorrhages (in children).

The diagnosis is usually clinically clear. Serum ascorbic acid levels less than 11 μmol/L (0.2 mg/dL) are suggestive of vitamin C deficiency. Such a finding should be followed by 24-hour urinary ascorbic acid measurements, carried out before and during the second day following 2 days of oral administration of 200 mg ascorbic acid. Vitamin C deficiency is associated with the excretion of less than 50 mg per day under these conditions.

A concept widely popularized is that large daily doses of vitamin C will prevent the common cold, or reduce its severity. There is no consistent evidence in favor of such megadoses. Only about 500 mg/day can be absorbed from the intestine and even such amounts have been suggested as detrimental to health. A large amount of vitamin C in the urine has been associated with an increased incidence of uric acid and cystine stones. It will also interfere with tests for glucose and oxalate in the urine.

Vitamin D

The physiology and pathology of vitamin D (1,25-dihydroxycholecalciferol) have been covered in Chapter 17. Plasma assays of vitamin D are now available. Deficiency (rickets, osteomalacia) is usually associated with a low serum calcium and inorganic phosphorus, and a raised alkaline phosphatase.

Vitamin E

Vitamin E is the name given to a group of molecules known as *tocopherols;* it acts as an antioxidant, particularly in adipose tissue and it protects membrane lipids. Vitamin E may have an effect in reducing oxidative damage to lungs, red cells, and neurons. It is required for reproduction in animals (never demonstrated in humans) and this attribute has given it a reputation as a fertility agent.

Premature *infants* may develop vitamin E *deficiency,* a syndrome which consists of edema, anemia, thrombocytosis, and an erythematous, papular, erupting skin lesion. A balance of iron, vitamin E, and polyunsaturated fatty acids is necessary to protect infants from hemolytic anemia. *Adult* vitamin E deficiency is rare, despite the popular attribution of this etiology to a spectrum of ills. Plasma tocopherol levels (generally an unobtainable assay) are low in vitamin E deficiency.

Vitamin K

Dietary vitamin K is absorbed in conjunction with fat and acts within the liver in the production of prothrombin and clotting factors V, VII, IX, and X. Additional vitamin K is synthesized by intestinal bacteria.

Newborn infants and persons with fat malabsorption are often deficient in vitamin K; a lack of sufficient bile acids is the most frequent cause. *Deficiency* of this vitamin may be gauged by measuring the *prothrombin time.* If a prothrombin time abnormality is exclusively caused by vitamin K deficiency then it should be corrected within hours by 1 mg of parenteral vitamin K.

Niacin (Nicotinic Acid)

Niacin is derived from the essential amino acid tryptophan. *Deficiency* of niacin will lead to *pellagra,* a disorder which begins insidiously but develops ultimately into the full-blown syndrome of 'four Ds': diarrhea, dermatitis, dementia, and death. Several forms of niacin deficiency are recognized.

Populations that depend on *corn* as the main protein staple are susceptible to pellagra. The tryptophan content of corn is low and its niacin is in a combined form that is not readily available. Diagnosis is primarily clinical, although the urinary measurement of *N^1-methylnicotinamide* and its *pyridone* (the principal metabolites of niacin) is often helpful. Normally more than 12 mg of these products will be excreted. In pellagra, less than 2 mg is common.

A pellagralike disorder is found in *Hartnup disease* (see Chap. 18), in which massive urinary loss of monoamino-monocarboxylic acids leads to tryptophan deficiency. It can be diagnosed by its characteristic urinary amino acid excretion.

Isoniazid is a pyridoxine antagonist. Pyridoxine is necessary in the conversion of tryptophan to niacin and therefore isoniazid administration may lead to niacin deficiency.

Malignant *carcinoid tumors* occasionally produce such vast amounts of 5-hydroxytryptamine (serotonin) that the entire body reserve of tryptophan is consumed, resulting in a pellagralike syndrome. This disorder has a unique clinical presentation (see Chap. 12, Gastrointestinal and Pancreatic Disorders) and is characterized by the excretion of unusually large amounts of the serotonin metabolite, *5-hydroxyindoleacetic acid,* in the urine.

Riboflavin

Riboflavin (vitamin B_2) is the precursor of flavin mononucleotide (FMN) and flavin adenine dinucleotide (FAD) and is involved in numerous biochemical reactions throughout the body, including electron transport, liver enzymes and corticosteroid production. A *deficiency* of riboflavin results in oropharyngeal changes, dermatitis, neuropathy, and anemia. Deficiency is usually found in association with low-B-vitamin syndromes. Blood and urine levels of riboflavin are of questionable diagnostic value. The measurement of red cell *glutathione reductase* activity before and after the addition of FAD to the reaction mixture, is probably the best laboratory indicator of riboflavin deficiency.

Thiamine

Thiamine (vitamin B_1) is necessary for the integrity of the central nervous system. It may be essential for hexose monophosphate shunt activity in myelin lipid maintenance. It has a major, undefined role in the myocardium.

Thiamine *deficiency* occurs in populations that depend on rice as their staple: If the rice is milled, the outer portion of the grain is lost, along with its vitamins. The condition is seen in children in underdeveloped countries and in alcoholics in developed nations. The latter has led to advocacy of thiamine-fortified beer. Combined vitamin B_1,B_6 deficiency has a detrimental effect on the transport of calcium, cadmium, and zinc across the intestinal wall.

Thiamine deficiency has several clinical presentations. *Dry beriberi,* or chronic thiamine deficiency, is characterized by neurologic involvement. *Wet* beriberi (a form with severe edema) and an *acute* form with heart failure are also seen. Thiamine antagonists are present in coffee, tea, betel, and raw fish, and are a complicating factor in the cause of beriberi (particularly in Thailand). Diagnosis is largely clinical. Urinary thiamine levels, whole blood thiamine, and tests of lactate and pyruvate production are of little value. Measurement of *erythrocyte transketolase* activity (a thiamine-dependent enzyme of the hexose monophosphate shunt), made before and after the addition of thiamine pyrophosphate to the reaction mixture, is probably the most suitable laboratory indicator of thiamine deficiency but cannot be depended upon completely.

Biotin

Biotin is a coenzyme of CO_2 fixation, the synthesis and oxidation of fatty acids, and the deamination of some amino acids. It is also related to cholesterol metabolism and storage. Biotin-synthesizing bacteria have been found in normal gastrointestinal tracts, but their role is not known.

Biotin deficiency can cause *alopecia totalis* and a fine scaly dermatitis. A disease of generalized carboxylase deficiency which can be corrected by biotin administration has been described. This condition is characterized by vomiting and ketoacidosis at birth and acrodermatitis at 3 to 6 months.

Choline

Although the body can synthesize choline from methionine and serine, the amount is not enough to sustain normal function. Choline acts as a source of labile methyl groups in several biologic reactions and is involved in the detoxification of toxins, synthesis of phospholipids, lecithins, sphingomyelins, and acetylcholine and in the metabolism and transportation of fat from the liver.

Deficiency is associated with fatty infiltration of the liver (chronic alcoholics, Kwashiorkor).

TRACE ELEMENTS

A *trace element* is one which makes up less than 1% of the organism's mass. It is an evolutionary puzzle why man and other higher animals should have a dependence on a series of elements which, although ubiquitous, are found in very low concentrations in the environment. Recent developments in analytic techniques, including atomic absorption spectrometry and anodic stripping polarography, have enabled clinical laboratories to assess metal levels reliably in a variety of biologic fluids.

Unfortunately, trace element measurements in both clinical and research settings have been plagued by two *problems.* First, collection techniques are a common source of *contamination.* Needles, syringes, storage containers, and rubber stoppers all may alter a specimen so much that the results obtained on it are worthless. Second, even when the measurement is carried out perfectly it may have no *clinical relevance.* Trace elements are distributed in a very unequal fashion throughout the body, so that knowledge of the concentration of an element in one body space (e.g., serum) may or may not reflect the concentration in another (e.g., brain), or in the body in general.

There are two basic disorders associated with trace element metabolism—*excess* and *deficiency.* Our knowledge of these conditions is still very incomplete. Dramatic claims concerning the relationship between trace elements and cancer, neurologic disease, behavioral disorders, and so forth must be viewed with scientific scepticism until clearly proven.

ESSENTIAL TRACE ELEMENTS

Iron

Iron is the metal required for the formation of hemoglobin, myoglobin, and the cytochromes. Even although body iron is recycled in an efficient manner, humans maintain a precarious iron balance and can lapse quite easily into iron deficiency, which leads to *anemia.* The normal requirement in men is approximately 1 mg iron/day. Menstruating women need almost twice this amount, and pregnant women require four times as much. Normally, iron is lost from the body through urinary excretion (0.1 mg/day), skin losses (desquamation, sweat; 0.2 mg/day), and gastrointestinal losses (0.7 mg/day).

Iron is *absorbed* from the intestine in the ferrous state. In the acid environment of the stomach iron interacts with a number of ligands (fructose, ascorbic acid, citric acid), as well as with gastroferrin from the stomach mucosa, to form complexes which maintain the solubility of iron in the alkaline fluid of the duodenum. These interactions are important, and dietary phytates and phosphates may actually impair iron absorption.

The intestinal mucosal cells absorb ferrous iron, which combines with a protein known as apoferritin (M_r 450,000) to form *ferritin.* If iron is needed by the body it is easily transferred to the serum from the ferritin in the mucosal cells. If not required, the ferritin remains within the mucosal cells. Eventually these cells are lost by desquamation and the body is protected from an oversupply of iron.

Iron within the circulation is in *equilibrium* with the iron in all body tissues. Although it is but a fraction of the total body store, a depression or elevation of the serum iron is roughly correlated with body iron balance.

Within the circulation, iron is *transported* in association with a specific iron-binding protein known as *transferrin* (siderophilin). This protein is a β-globulin (M_r 76,000). Normally only 30% to 40% of the potential binding sites on transferrin are occupied by iron. Eventually, iron is delivered to the bone marrow and other tissues to serve in the production of hemoglobin, myoglobin, and cytochromes.

Serum iron measurements are conveniently performed by manual or automated spectrophotometric techniques, or by atomic absorption spectrometry. *Transferrin* levels may be measured directly by immunologic quantification or, more often, indirectly, as the *iron-binding capacity* (IBC). In this latter approach the patient's serum is mixed with an excess of iron. Unbound iron is removed and the iron taken up by transferrin in the patient's serum is measured. The result is expressed as the total amount of iron that the serum is capable of holding. Serum iron and IBC measurements should be considered together (Table 22–1).

Serum ferritin assays are now commonly available and in some centers have replaced the measurement of serum iron and IBC. Radioimmunoassay (RIA) methods are used. *Reference values* for men average 120 ± 70 μg/dL, and for women 60 ± 30 μg/dL. As mentioned earlier, ferritin is the primary available storage form of iron. Plasma levels reflect the destruction of erythrocytes in the reticuloendothelial cells and correlate with iron stores. Serum ferritin levels increase in hepatocellular necrosis and hemolytic disorders. Ferritin levels below 10 μg/dL are almost certain evidence of iron deficiency and those between 10 and 20 μg/dL are highly suspicious. In rheumatoid arthritis, or infection, serum ferritin values may be 50 to 100 μg/dL, even though there is no stainable bone marrow iron. The significance of ferritin levels in iron-overload is a matter of considerable debate, but they are useful in monitoring the effects of phlebotomy.

Iron *deficiency* may result from a wide variety of dietary and gastrointestinal abnormalities. Chronic blood loss (including menorrhagia) is an important cause. Chronic illness impedes iron utilization and renal disease interferes with hematopoiesis.

Excess body iron may result in *hemosiderosis* (an increase in tissue iron without alteration of tissue structure or function), or *hemochromatosis* (in which tissue is damaged). Causes include

TABLE 22–1. Serum Iron, Iron-Binding Capacity, and Ferritin in Various Disorders

Disorder	Serum Iron	Iron-Binding Capacity	Serum Ferritin
Iron deficiency	↓↓	↑	↓
Chronic disease	N to ↓	N to ↓	N to ↓
Hemolysis	↑	N	↑
Pernicious anemia	↑	N	↑
Iron overload	↑↑	N	N or ↑
Nephrotic syndrome	↓	↓	N or ↓
Liver necrosis	N to ↑	N to ↓	↑

N = Normal; ↑ = high; ↓ = low; ↑↑ = very high; ↓↓ = very low.

increased parenteral intake from repeated transfusions, alcoholic cirrhosis, oral iron therapy, Kaschin–Beck disease, and (in African Bantus) iron-cooking-pot disease. *Primary* hemochromatosis is a genetic disorder characterized by widespread accumulation of iron in the liver, pancreas, and myocardium. It leads to the classic triad of pigmentation, hepatic cirrhosis, and diabetes mellitus (bronzed diabetes) and is treated by repeated venesection and sequestering agents (desferroxamine). Iron poisoning in children is discussed in Chapter 21.

Iodine

This essential element is an intrinsic part of the thyroid hormones whose physiology has been covered elsewhere (see Chap. 14).

Iodine is absorbed from the gastrointestinal tract as inorganic iodide and distributes itself quickly throughout the extracellular fluid. The kidney and particularly the thyroid are extremely efficient scavengers of iodide. After being taken up by the thyroid, the iodine is used in the synthesis of thyroid hormones. After their release, thyroid hormones are deiodinated by peripheral tissues, and the iodine is reutilized efficiently.

Deficiency may occur as a result of inadequate intake or, in rare instances, from insufficient dehalogenase to allow reutilization. This results in thyroid hormone deficiency and thyroid enlargement (*goiter*) caused by thyrotropin (TSH) stimulation.

Cobalt

Vitamin B_{12} is the only biologic compound known to be complexed with cobalt in humans. Cobalt can substitute for zinc in certain zinc metalloenzymes; it can stimulate erythropoiesis and has been considered as a possible treatment for anemia.

Cobalt was once added to beer as a foam stabilizer but can cause polycythemia, pericardial effusion, thyroid hyperplasia, and neurologic abnormalities. The combination of cobalt with ethanol can cause a dramatic cardiomyopathy (beer drinkers' cardiomyopathy). The direct measurement of cobalt is not clinically useful.

Copper

Following intestinal absorption, copper is transported bound to albumin. Within 2 hours 66% to 95% is found in the liver. In the liver some of the copper is incorporated into the oxidase enzyme ceruloplasmin, while the excess is excreted in the bile. Copper is found in cytochrome C oxidase, lysyl oxidase (collagen, elastin), tyrosinase, superoxide dismutase (erythrocoprein), and dopamine-β-hydroxylase. *Raised* serum copper levels are seen in a spectrum of disorders, because ceruloplasmin levels are raised as part of the acute-phase stress reaction. Acute copper toxicity has occurred in hemodialysis and following the ingestion of copper salts.

Lowered serum copper levels are associated with several syndromes. The best known of these is ***Wilson's disease*** (hepatolenticular degeneration). In this condition the liver copper-transporting mechanism is impaired so that it fails to synthesize ceruloplasmin and to excrete copper in the bile. Copper builds up in the body and is deposited in the cornea, the brain, the liver, and the renal tubules. The diagnosis is suggested by low serum copper and ceruloplasmin levels and a raised urine copper, although a low ceruloplasmin is not invariable. It is confirmed by finding raised hepatic tissue copper levels in a biopsy specimen.

Menke's kinky hair syndrome (trichopoliodystrophy) is due to an inability to absorb copper, or handle it intracellularly. The deficiency results in retarded growth and development, pili torti (kinky hair), arterial tortuosity, scorbutic bone changes, and cerebral degeneration. Biochemically the serum copper, ceruloplasmin, and hair copper are low. Copper therapy does not cure the condition.

Dietary copper deficiency of obscure cause in premature infants and in parenterally nourished individuals may result in neutropenia, scorbutic bone changes, and anemia. Treatment with iron has no effect because copper is needed to incorporate iron into heme. A central nervous system disorder is characterized by spinal demyelination, low brain catecholamines, hypotonia, psychomotor retardation, and decreased visual acuity. Depigmentation is seen in these patients caused by inactive tyrosinase.

Manganese

Manganese is an essential cofactor in many biochemical processes. Manganese deficiency may cause impaired growth, skeletal abnormalities, depressed reproductive function, and ataxia in animals. Deficiency has been suspected in one man who had mild dermatitis, slight reddening of his hair, and depressed vitamin K-dependent clotting factors which did not respond to vitamin K administration. Manganese dust inhalation can produce basal ganglion impairment.

Molybdenum

Although molybdenum has a recognized essential role in almost every form of life studied thus far, its function in man is not well understood. It is known that xanthine oxidase is a molybdenum-containing metalloenzyme. Claims that it has a protective influence on dental caries are not conclusive. Combined deficiency of sulfite oxidase and xanthine dehydrogenase is the result of an absent molybdenum cofactor. This produces a syndrome of mental retardation, seizures, bilateral lens dislocation, sulfaturia, and thiosulfaturia.

Zinc

Zinc is an essential element found as part of many enzymes including carboxypeptidases A and B, carbonic anhydrase, and alkaline phosphatase. Low levels of plasma zinc are a nonspecific finding in a number of conditions including liver and lung disease, myocardial infarction, uremia, increased catabolism, malnutrition, some cancers, pernicious anemia, leukemia, pregnancy, and steroid therapy. Cadmium and mercury compete with zinc for intestinal absorption. Ingested zinc is known to protect against cadmium toxicity.

Body zinc *deficiency* is suggested by low levels of plasma zinc and may cause several bizarre disorders. *Juvenile* zinc deficiency was first reported in Egypt and Iran as the result of low dietary zinc, the chelation of zinc by dietary phytic acid, chronic fecal zinc loss because of intestinal parasitism, and the ingestion of clay which prevents zinc absorption. Boys suffer dwarfism and delayed sexual maturation which is responsive to zinc administration. Zinc deficiency may also cause hypogeusia (impaired taste). A loss of zinc commonly occurs in burns and with total parenteral nutrition or chronic penicillamine therapy. Depleted body zinc stores impair wound healing and may encourage the formation of decubitus and other chronic ischemic ulcerations. Night blindness has also been attributed to zinc deficiency.

Acrodermatitis enteropathica (Danbolt's disease) is a lethal, autosomal recessive disorder characterized by progressive bullous, pustular lesions, which may be due to inadequate body zinc. *Maternal* zinc deficiency has also been implicated in the production of congenital abnormalities.

Zinc *toxicity* has been reported in children who ingested zinc from toys, or as an acute pulmonary syndrome in adults following exposure to hot zinc vapor. Single dose poisoning may cause necrosis and atrophy of the testes. Industrial exposure to zinc salts is not known to be toxic. Raised zinc levels have been reported in hypothermia but the most common cause of raised plasma zinc is contamination of the specimen.

Chromium

Chromium plays a role in glucose and lipid metabolism and its deficiency has been suggested as a causal factor in atherosclerosis. It has been demonstrated to improve glucose tolerance, possibly by facilitating insulin release or tissue binding. However, the exact relationship is far from clear. Chromium also stimulates fatty acid and cholesterol synthesis. Deficiency of chromium results in reduced growth, impaired glucose tolerance, and the formation of aortic atheromas.

Selenium

Selenium is the least abundant and most toxic of the essential elements. Selenium *deficiency* has been seen in animals in which hepatic necrosis, retarded growth, muscular degeneration, and infertility have been reported. It has caused a congestive cardiomyopathy in children (Keshan disease) in China. Incapacitating muscle pain, reversed by selenium, was observed in a patient receiving TPN. Selenium deficiency increases platelet aggregation.

Excess selenium is teratogenic, hepatotoxic, and neurotoxic. Selenium accumulates in plants in high-selenium soils. Ingestion of such food is the cause of 'blind staggers' and 'alkali disease' in cattle. Man is most exposed to selenium in industrial settings. Selenium is carcinogenic in animals. Its oncogenic influence in man is not established.

Nickel

The role of nickel in human physiology is not known, but evidence is growing that this element is essential. It is poorly absorbed and is transported throughout the circulation loosely bound to albumin and incorporated within an α_2-macroglobulin known as *nickeloplasmin.* It is excreted by the kidney. Serum nickel (26 to 76 nmol/L; 0.15 to 0.45 μg/dL) rises following myocardial infarction, cardiovascular accidents, and burns. It is depressed in hepatic cirrhosis and chronic uremia. Serum and urine nickel levels are a reflection of environmental exposure and such measurements may be useful in monitoring industrial workers. Most prone to exposure are those working in refineries using the Mond nickel carbonyl process. Acute exposure to nickel carbonyl [$Ni(CO)_4$] can cause a potentially fatal cardiopulmonary syndrome. Chronic exposure to organic nickel compounds (oxides and sulfides) has been implicated in the genesis of neoplasia.

For *analysis* serum samples must be collected in absolutely clean containers, using nonmetallic intravenous cannulas.

Vanadium

Vanadium is probably essential. Considerable investigation is focused on the role of vanadate as an inhibitor of Na-K-ATPase and a modifier of sodium pump activity. Vanadium may play a role in cholesterol and fatty acid metabolism, calcium flux, bone development, and cardiac and renal function. Its measurement has no current clinical value.

NONESSENTIAL TRACE ELEMENTS

Aluminum

There is no demonstrated biologic role for aluminum in humans. The average oral intake is between 10 and 100 mg each day, but a very small percentage of this is actually absorbed. Toxicity from aluminum compounds is not seen in normal persons. Patients with renal failure who receive aluminum-containing antacids to reduce phosphate absorption may experience a positive phosphate balance. Other patients on TPN, or renal dialysis, have experienced *aluminum toxicity.* Dialysis dementia may present with telangiectases and progress to myoclonus, dementia, behavior changes, and dysphagia. Bone pain and osteomalacia are seen in aluminum toxicity. After aluminum deposits in bone it blocks the incorporation of calcium into osteoid, which leads to osteomalacia. The blocking also causes hypercalcemia and then low parathyroid hormone. Aluminum normally increases in cerebral tissue with age, but does not cause Alzheimer's syndrome as some people have suggested.

Arsenic

Long-term, low-dose ingestion of arsenic can produce palmar and plantar keratoses, skin cancer, and neuropathy. Measurements of urinary arsenic have been used to identify such cases. The recent ingestion of seafood can result in spuriously high levels. Hair and fingernail arsenic determinations have been used to detect past exposure. Occupational exposure (pesticides and herbicides), gold refining, nefarious deeds, contaminated water supplies, and medicine (Fowler's solution) are the prime causes of poisoning.

Cadmium

Cadmium is extremely toxic and long-term exposure has been implicated in the production of testicular tumors, renal damage, hypertension, arteriosclerosis, growth inhibition, premature aging, and cancer. Geographic cadmium levels correlate directly with cardiovascular mortality. Exposure to its fumes can cause metal 'fume fever.' Such exposure may occur during smelting as the molten cadmium is poured into ingots, but is more common in welders who cut through a metal object using a torch and encounter a cadmium alloy structure.

Acute exposure is associated with nausea, vomiting, diarrhea, and renal and hepatic failure. *Chronic* industrial exposure causes pulmonary fibrosis and renal damage. Although cadmium can be measured in the blood and urine, workers are best *monitored* by pulmonary function testing and urine protein measurements (particularly β_2-microglobulin). An environmental disease has been reported in the Jintsu River valley in Japan. Here, cadmium contamination of farms has resulted in the development of extremely brittle bones in many of the inhabitants (mainly the women). The disorder is known as "Itai-Itai" ("Ouch, Ouch"). Cadmium normally increases in the tissues with age.

Fluoride

Fluoride monitoring in drinking water is useful, because supplementation with this element reduces dental caries. Fluoride has also been advocated in the treatment of osteoporosis. In industrial workers it may be necessary to monitor fluoride. Too much fluoride will cause mottling of the teeth.

Gold

Gold, administered parenterally in the form of gold sodium thiomalate and gold sodium thioglucose, is sometimes of benefit in the treatment of rheumatoid arthritis. The mechanism is possibly an inhibitory effect on lysosomal enzymes. Patients on gold therapy should have a complete blood count (CBC), platelet count, urine protein, and serum AST at the outset. Each week thereafter a CBC with differential and platelets, and a urine protein determination, should be carried out.

Serum and urine measurements of gold have not proved worthwhile because there is an extensive overlap among patients with and those without a therapeutic response to the treatment. An ability to estimate the half-life of the administered dose may possibly be useful in designing therapy.

Lead

Lead poisoning (see also Chap. 10, Porphyrins, the Heme Proteins, Bile Pigments, and Jaundice, and Chap. 21, Pediatric Clinical Biochemistry) is currently a major public health concern. Although fatal cases are few in number, the serious and frequently widespread effects on health are well established. The long-term effect of chronic, subtoxic exposure is not known. The effect of lead on children is suspected of causing reduced intelligence, although this is not proven. Lead poisoning may produce anemia, encephalopathy, neuropathy, gastrointestinal disturbances, and porphyria.

Exposure to lead is mainly industrial, or from automobile exhaust, or from eating lead paint (pica, in children). Lead smelter workers often transport the contamination home on their clothing.

The most practical test for the *diagnosis* of lead poisoning is the level of this metal in whole blood. Values from 0.5 to 1.5 μmol/L (10 to 30 μg/dL) are found in a normal urban population. In the rural population values generally remain below 1 μmol/L (20 μg/dL). Values above 1.5 μmol/L are considered potentially harmful for pregnant women and children, and above 2 μmol/L (40 μg/dL), potentially toxic. The legal limit for industrial workers varies from 70 to 100 μg/dL, and should probably be 3 μmol/L (60 μg/dL). Overt toxicity is not usually seen in adults until levels of 5 μmol/L or more are obtained. Seizures have developed in children at slightly greater than 2.5 μmol/L.

Lead interferes with porphyrin synthesis at a number of sites. Red cell aminolevulinic acid (ALA) dehydrase, an enzyme necessary for the conversion of ALA into porphobilinogen, is inhibited. As a result the urinary ALA concentration will increase.

Lead also interferes with the incorporation of iron into the tetrapyrrole ring of hemoglobin, allowing its replacement by zinc. Zinc protoporphyrin (free erythrocyte protoporphyrin) levels rise. *Zinc protoporphyrin estimations* are easy to perform using a specially designed fluorimetric device suitable for on-site testing. This may become the preferred screening technique.

Mercury

Mercury has no known biologic function. It is important for its insidious and devastating toxicity. Elemental mercury (as a vapor) passes the alveolar membrane and can diffuse into lipid-containing tissues (the central nervous system in particular). Liquid mercury *per se* is not hazardous. However, if spilled it may easily be vaporized by heat or converted into organic mercury compounds. *Chronic exposure* to mercury causes oral cavity disorders, a fine tremor of the extremities, and psychologic disturbances. Somewhat different effects appear when mercury is covalently linked to an organic molecule. Those compounds that are metabolized in the body (phenyl and methoxyethyl mercury) yield inorganic mercury. Those that resist metabolism (alkyl mercury compounds) are much more easily absorbed and become extensively distributed throughout the body.

Dramatic episodes of grievous harm to entire populations from environmental mercury have been seen in Iraq, Guatemala, Russia, Japan (Minamata and Niigata), and Canada. Industrial and sporadic poisonings have been recorded in all countries.

Mercury *analyses* are notoriously hard to perform and almost impossible to interpret. In addition, the clinical picture is very nonspecific. The definitive diagnosis of mercury poisoning is therefore exceedingly difficult.

SUGGESTED READING

ANDERSON RA: Nutritional role of chromium. Sci Total Environ 17:13–29, 1981

BAKER JP, DETSKY AS, WESSON DE et al: Nutritional assessment; a comparison of clinical judgement and objective measurements. N Engl J Med 306:969–972, 1982

BIERI JG, CORASH L, HUBBARD VS: Medical uses of vitamin E. N Engl J Med 308:1063–1071, 1983

BLAKE DR, WATERWORTH RF, BACON PA: Assessment of iron stores in inflammation by assay of serum ferritin concentrations. Br Med J 283:1147–1148, 1981

BOYCE BF, ELDER HY, ELLIOT HL et al: Hypercalcaemic osteomalacia due to aluminum toxicity. Lancet ii:1009–1013, 1982

CANNATA JB, JUNOR BJR, BRIGGS JD et al: Effect of acute aluminum overload on calcium and parathyroid-hormone metabolism. Lancet i:501–503, 1983

CHAN WAI-YEE, RENNERT OM: Cadmium nephropathy. Ann Clin Lab Sci 11:229–238, 1981

Committee on Dietary Allowances, Food and Nutrition Board, National Research Council (ed): Recommended Dietary Allowances, 9th ed. Washington DC, National Academy of Sciences, 1979

COULEHAN JL: Ascorbic acid and the common cold; reviewing the evidence. Postgrad Med 66:153–160, 1979

ELIAS PM, WILLIAMS ML: Retinoids, cancer, and the skin. Arch Dermatol 117:160–180, 1981

FAIRBANKS VF, KLEE GG: Ferritin. Prog Clin Pathol 8: 175–203, 1981

FINCH CA, HUEBERS H: Perspectives in iron metabolism. N Engl J Med 306:1520–1528, 1982

GOLDEN MHN, GOLDEN BE: Trace elements; potential importance in human nutrition with particular reference to zinc and vanadium. Br Med Bull 37:31–36, 1981

GRANT JP, CUSTER PB, THURLOW J: Current techniques of nutrition assessment. Surg Clin North Am 61:437–463, 1981

HERBERT V: The vitamin craze. Arch Intern Med 140: 173–176, 1980

JANDHYALA BS, HOM GJ: Minireview of physiological and pharmacological properties of vanadium. Life Sci 33:1325–1340, 1983

JOHNSON JL, WAUD WR, RAJAGOPALAN KV et al: Inborn errors of molybdenum metabolism: Combined deficiencies of sulfite oxidase and xanthine dehydrogenase in a patient lacking the molybdenum cofactor. Proc Natl Acad Sci 77:3715–3719, 1980

KITAY DZ: Folic acid and reproduction. Clin Obstet Gynecol 22:809–817, 1979

KOPP SJ, GLONEK T, PERRY HM, ERLANGER M, PERRY EF: Cardiovascular actions of cadmium at environmental exposure levels. Science 217:837–839, 1982

LAU KS: Editorial: Cobalamins in man. Pathology 13: 189–195, 1981

LIU VJK, ABERNATHY RP: Chromium and insulin in young subjects with normal glucose tolerance. Am J Clin Nutr 35:661–667, 1982

MARKESBERY WR, EHMANN WD, HOSSAIN TIM, ALAUDDIN M, GOODIN DT: Instrumental neutron activation analysis of brain aluminum in Alzheimer disease and aging. Ann Neurol 10:511–516, 1981

MERANGER JC, CONACHER HBS, CUNNINGHAM HM, KREWSKI D: Levels of cadmium in human kidney cortex in Canada. Can J Public Health 72:269–272, 1981

MULLIN GT, JR: Gold for rheumatoid arthritis; current perspectives. Postgrad Med 72:205–217, 1982

Nutritional oedema, albumin and vanadate (editorial). Lancet i:646–647, 1981

ROTH KS: Biotin in clinical medicine—a review. Am J Clin Nutr 34:1967–1974, 1981

SHAW JCL: Trace elements in the fetus and young infant. I. Zinc. Am J Dis Child 133:1260–1268, 1979; II. Copper, manganese, selenium and chromium. Am J Dis Child 134:74–81, 1980

WEISBERG HF: Evaluation of nutritional status. Ann Clin Lab Sci 13:95–106, 1983

WILLIAMS DM: Copper deficiency in humans. Semin Hematol 20:118–128, 1983

YOUNG DS: Effect of vitamin C on laboratory tests. Lab Med 14:278–282, 1983

YOUNG VR: Selenium: A case for its essentiality in man. N Engl J Med 304:1228–1230, 1981

23

Prenatal Diagnosis and Biochemical Assessment of High-Risk Pregnancy

Lynn C. Allen

PRENATAL DIAGNOSIS

The objective of prenatal diagnosis is to determine if a fetus is affected with a disorder for which it may be at risk. A diagnosis is now possible for more than 50 genetic abnormalities, allowing at-risk families, unwilling to chance reproduction, the opportunity of electing to have only unaffected offspring. Indications for prenatal diagnosis are presented in Table 23–1.

Genetic counselling is a fundamental part of any prenatal diagnosis program. The couple must be informed of the risk of occurrence or recurrence of the defect, the severity of the disease, the reproductive options, the risks of the procedure for obtaining specimens (amniotic fluid, fetal blood, chorion) and the reliability of results obtained. The specific diagnosis must be confirmed in the affected family member and, in disorders with autosomal recessive inheritance, the heterozygosity of the parents must be confirmed. The specific defect must be expressed in a fetal tissue (fibroblasts, chorionic villi, plasma, red cells). Follow-up of prenatal diagnosis includes confirmation of the diagnosis, either following termination of an affected fetus or birth of an affected or normal child.

Amniotic Fluid

Amniotic fluid surrounds and protects the fetus, and its constituents change throughout pregnancy. In early pregnancy, amniotic fluid may be regarded as an ultrafiltrate of maternal serum; later contributions come from the fetal circulation, fetal urination, and secretion from the amnion and fetal respiratory tract. Amniotic fluid contains both maternal and fetal cells. By 16 weeks of gestation there are 100 to 200 mL of amniotic fluid, and 15 to 17 mL can be removed at *amniocentesis* (transabdominal aspiration of fluid from the amniotic sac) for cytogenetic and biochemical analyses. At this stage amniocentesis carries a risk about 0.5% above that of the general population of spontaneous abortion or miscarriage.

Fetal Blood

Fetal blood required to diagnose certain disorders can be obtained at about 18 weeks of gestation by fetoscopy (direct endoscopic visualization of the fetus). A fetal vessel on the placental surface is punctured and fetal blood (mixed with some amniotic fluid) is aspirated. The risk of spontaneous abortion following fetal blood sampling is 2% to 5%.

TABLE 23–1. Indications for Prenatal Diagnosis

Chromosomal
Maternal age > 35 years
Previous chromosomal anomaly, e.g., Trisomy
Parental chromosome rearrangement (translocation/inversion)
Hemoglobinopathies
Inborn error of metabolism
Previous neural tube defect, ↑ maternal serum AFP
X-linked disorders

CLINICAL DISORDERS

Chromosomal Abnormalities

The most frequent indication for amniocentesis is exclusion of a chromosomal disorder; this accounts for 80% to 90% of patients referred for prenatal diagnosis. About 95% of these patients are referred for reasons of maternal age >35, because the risk of trisomy 21 (Down's syndrome) increases from about 1:350 at age 35 to 1:25 at age 45. The other 5% of referrals are for such indications as a previous child with trisomy 21, 13, or 18, or a parental chromosome rearrangement.

Laboratory Diagnosis. *Cytogenetic analysis* of fetal cells involves growing the cells in nutritive medium for 2½ weeks, arresting growth, disrupting the nuclei, and staining, banding, and photographing the chromosomes. The karyotype, prepared by arranging the chromosomes from *one cell* by size and morphology, is examined for abnormalities. The time required for the entire process (about 3 weeks) causes anxiety, because amniocentesis is performed at 16 weeks, and many hospitals restrict therapeutic abortions to gestations under 20 weeks.

Techniques for obtaining *chorionic villi* (early placenta) at 7 to 8 weeks of gestation have now been developed. If the risk of the procedure is sufficiently low, this will allow direct analysis of chromosomal patterns and other analyses requiring fetal cells, much earlier in pregnancy.

Neural Tube Defects

α_1-Fetoprotein (*AFP*) is synthesized in the yolk sac during embryonic life and later in the parenchymal cells of the fetal liver. It is similar in structure to albumin, but is a glycoprotein with 4% carbohydrate. It appears in *fetal serum* by the sixth week, reaches a peak concentration of about 3 g/L at 14 weeks' gestation, and subsequently declines (Fig. 23–1A). AFP probably enters the amniotic fluid by fetal urination and leaves by fetal swallowing and digestion. The peak level of AFP in amniotic fluid is about 50 mg/L at 14 weeks' gestation, decreasing to around 15 mg/L by 20 weeks' gestation (Fig. 23–1B). AFP also enters the *maternal* circulation. Here, however, the level rises from about 10 μg/L at 8 weeks' gestation to about 100 μg/L at 20 weeks' gestation, and continues to increase until term (Fig. 23–1C).

Clinical Features. The common open neural tube defects are anencephaly and spina bifida. *Anencephaly* is a fatal condition in which the brain is amorphous and the vault of the skull is absent. In *spina bifida,* protrusion of the meninges and neural tissue (meningomyelocele) or of the meninges alone (meningocele) occurs. About 10% of spina bifida are closed (skin-covered) defects.

About half of the infants with spina bifida die in the first month of life. More than half of the survivors are severely handicapped, often lacking bladder and bowel control; about one fifth are mentally retarded.

The *incidence* of neural tube defects is about 2:1000 births in Canada and the United States. The recurrence risk is 2% (10× the incidence) if the mother has had a child with a neural tube defect. The highest incidence occurs in Scotland, Wales, and Ireland (about 5:1000) and the recurrence risk is 5%.

Pathophysiology. Neural tube disorders are developmental defects in which the neural groove fails to close, at the cranial level in anencephaly, and at the spinal level in spina bifida. In the presence of an *open* neural tube defect, an increased concentration of AFP is seen in the amniotic fluid, caused by transudation of fetal serum across capillaries at the site of the defect. The AFP concentration in maternal serum is also increased. Acetylcholinesterase, found in spinal fluid, enters the amniotic fluid as well. *Closed* defects generally do not cause an AFP elevation in amniotic fluid or maternal serum.

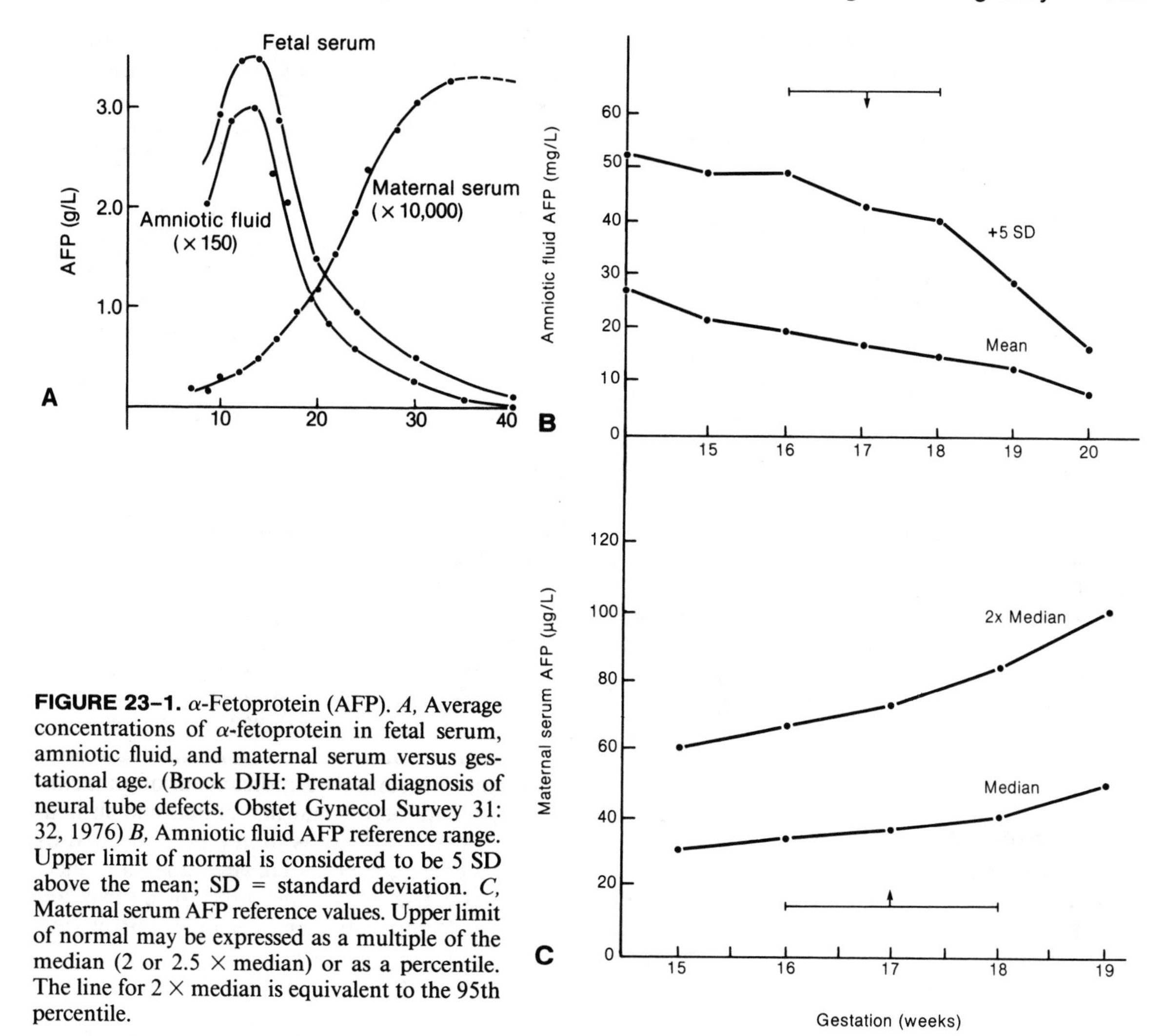

FIGURE 23–1. α-Fetoprotein (AFP). *A,* Average concentrations of α-fetoprotein in fetal serum, amniotic fluid, and maternal serum versus gestational age. (Brock DJH: Prenatal diagnosis of neural tube defects. Obstet Gynecol Survey 31: 32, 1976) *B,* Amniotic fluid AFP reference range. Upper limit of normal is considered to be 5 SD above the mean; SD = standard deviation. *C,* Maternal serum AFP reference values. Upper limit of normal may be expressed as a multiple of the median (2 or 2.5 × median) or as a percentile. The line for 2 × median is equivalent to the 95th percentile.

Laboratory Assessment. High risk patients have a family history of a neural tube defect and this indication accounts for about 7% of patients seen for prenatal diagnosis. A routine ***ultrasound*** examination is performed at 14 to 16 weeks to establish gestational age (by measurement of the fetal biparietal diameter), placental location, and presence of a normal skull outline. ***Amniocentesis*** is carried out at 16 weeks' gestation to obtain an amniotic fluid specimen for analysis of *AFP,* which is measured by electroimmunoassay (rocket immunoelectrophoresis) or by radioimmunoassay (RIA) on diluted specimens. A detailed ultrasound study of the fetal spine is made at 18 weeks. In anencephaly, the skull outline is incomplete, whereas in spina bifida, the two parallel lines observed with a normal spine flare at the site of the defect. On transverse section the normal circle of the spinal canal appears instead as an open U-shape. The gestational age must be known accurately to interpret the AFP result. A level above the reference range (Fig. 23–1B), in conjunction with an abnormal ultrasound study, confirms the diagnosis of an *open* neural tube defect.

Contamination of the amniotic fluid by fetal blood gives a *false* elevation of the amniotic fluid

AFP. The assay of *acetylcholinesterase* can be used as an alternate test, or as an adjunct when the fluid is blood-stained. With an open neural tube defect, agarose-gel electrophoresis will demonstrate the presence of an acetylcholinesterase band in addition to the cholinesterase band present normally in amniotic fluid. Other causes of false-positive results are congenital nephrosis, exomphalos (umbilical hernia), and duodenal atresia.

Screening programs were begun in the mid-1970s because of the severity of the handicap suffered by children with spina bifida. Screening of the low-risk population for open neural tube defects is based on the increase in *maternal serum AFP* (measured by RIA) in the presence of such a defect (Fig. 23–1C). A blood specimen taken at 16 to 18 weeks of gestation gives the best detection efficiency for spina bifida. After ruling out other causes of an increased result, such as wrong dates, twins, and threatened abortion, the patient is offered a thorough ultrasound examination and amniocentesis for amniotic fluid AFP. False-positive results are common. Depending on the prevalence of the disorder, a decision level 2 to $2\frac{1}{2}$ times the median is chosen so as to attain an acceptable detection efficiency, but still avoid an excessive call-back rate and undue maternal anxiety.

Inborn Errors of Metabolism

Prenatal Diagnosis. More than 50 inborn errors of metabolism have been diagnosed prenatally and diagnosis of many others awaits identification of a pregnancy at risk. Most inherited disorders are autosomal recessive, which means a 25% chance of a child being affected. An *affected family member* is generally the first indication of a genetic problem because carriers are usually clinically normal. For some diseases, such as Tay-Sachs, the carrier status of the parents can be established by screening them. Biochemical disorders that have been diagnosed prenatally are listed in Table 23–2. Some diseases (e.g., Tay-Sachs) are invariably fatal at an early age, whereas others can be treated by diet (e.g., galactosemia). Ethical considerations arise concerning how serious a disorder must be before prenatal diagnosis is justified.

The *steps in prenatal diagnosis* are as follows: the first objective is to assess the risk of occurrence of the disorder in a subsequent pregnancy, then to determine whether the disorder can be identified prenatally so the parents can be counselled accordingly. The exact nature of the defect must be confirmed in the affected family member and the level of enzyme or substrate in both parents should be measured to facilitate the interpretation of the fetal assays. The appropriate fetal sample is obtained and analyzed, after which the parents are informed and counselled. Finally, the diagnosis must be confirmed after abortion of an affected fetus, or after birth of the child. Fewer than 1% of patients undergoing prenatal diagnosis are seen because of the risk of an inborn error of metabolism.

Laboratory Diagnosis. With defects that are expressed in amniotic fluid cells, prenatal diagnosis requires a demonstration of the specific enzyme deficiency in cultured cells. In other cases, diagnosis may be made by analysis of amniotic fluid metabolites (e.g., tryosinemia), characterization of enzymes or proteins in fetal blood, linkage analysis, and ^{35}S-cystine uptake (cystinosis). Such analyses should be performed only in specialized laboratories familiar with the problems encountered. Cells in amniotic fluid are heterogeneous (fibroblasts, epithelial, and other cells) and the deficiency may not be expressed in each type. Enzyme activity depends on culture conditions, for example, contamination (blood, maternal cells, mycoplasma) and the medium. Results should be confirmed by two procedures, such as enzyme assay and electrophoresis, use of two substrates, or by two laboratories. Both positive and negative control samples must be run.

Hemophilia

Clinical Features. Hemophilia A (classic hemophilia) and B (Christmas disease) are X-linked disorders; that is, female carriers transmit the defect to half their sons and the gene to half their daughters. The prevalence of the two disorders is 1:10,000 and 1:50,000 respectively.

Recurrent painful bleeding into joints and muscles, with progressive joint deformity and crippling, dominates the clinical course. Post-traumatic and operative surgery are life-threatening. Most patients can be treated by replacement of factor VIII or IX, and treatment at home has largely reduced the need for hospital care.

Pathogenesis. *Hemophilia A* is associated with a deficiency of factor VIII coagulant activity, re-

TABLE 23–2. Prenatal Diagnosis of Metabolic Disease*

Disorder	Diagnostic Test
Lipid Disorders	
Tay-Sachs disease (GM_2-gangliosidosis)	β-Hexosaminidase A
Sandhoff's disease	β-Hexosaminidase A and B
Gaucher's disease	Acid β-glucosidase
GM_1-gangliosidosis	GM_1-ganglioside β-Galactosidase
Nieman-Pick disease	Sphingomyelinase
Fabry's disease	α-Galactosidase A
Metachromatic leukodystrophy	Cerebroside sulfate (Arylsulfatase A)
Krabbe's disease	Galactocerebroside β-galactosidase
Farber's disease	Ceramidase
Type II hypercholesterolemia	Low-density-lipoprotein receptor
Wolman's disease	Acid esterase
Adrenoleukodystrophy	Long chain fatty acid (C_{26})
Glycoprotein and Mucopolysaccharide Disorders	
Fucosidosis	α-L-Fucosidase
Mannosidosis	Acidic α-mannosidase
Sialidosis	α-N-acetylneuraminidase and β-galactosidase
Mucolipidosis II	Lysosomal hydrolases
Mucolipidosis III	Lysosomal hydrolases
MPS IH (Hurler)	α-L-iduronidase
MPS IS (Scheie)	α-L-iduronidase
MPS II (Hunter)	α-L-iduronate sulfatase
MPS IIIA (Sanfilippo A)	Heparan N-sulfatase
MPS IIIB (Sanfilippo B)	N-acetyl-α-D-glucosaminidase
MPS IV (Morquio)	Hexosamine-6-sulfatase
MPS VI (Maroteaux-Lamy)	N-acetylhexosamine-4-sulfatase (Arylsulfatase B)
Carbohydrate Disorders	
Galactosemia	Galactose-1-phosphate uridyl transferase
Glycogen storage disease type II (Pompe's)	Acid α-1,4-glucosidase
Glycogen storage disease type IV	α-1,4-glucan:β-1,4-glucan-6-glycosyl transferase
Pyruvate decarboxylase deficiency	Pyruvate decarboxylase
Urea Cycle and Amino Acid Disorders	
Ornithine carbamyl transferase deficiency	Ornithine carbamyl transferase
Argininosuccinic acidemia	Argininosuccinic acid lyase
Glutaric acidemia type I	Glutaryl-CoA dehydrogenase
Cystinosis	^{35}S-cystine uptake
Nonketotic hyperglycemia	Glycine/serine ratio
Maple syrup urine disease	Branched-chain ketoacid decarboxylase
Tyrosinemia	Succinylacetone
Isovaleric acidemia	Isovaleryl-CoA dehydrogenase
Methylmalonic acidemia	
(B_{12}-nonresponsive)	Methylmalonic acid
(B_{12}-responsive)	Adenosylcobalamin synthesis
Propionic acidemia	Propionyl-CoA carboxylase
Other Disorders	
Lesch-Nyhan disease	Hypoxanthine-guanine phosphoribosyltransferase
Congenital erythropoietic porphyria	Uroporphyrinogen III cosynthetase
Acute intermittent porphyria	Uroporphyrinogen I synthetase
Congenital adrenal hyperplasia type III (21-hydroxylase deficiency)	HLA B linkage; 17-hydroxyprogesterone
Combined immunodeficiency	Adenosine deaminase
Hypophosphatasia	Alkaline phosphatase isoenzymes
Congenital nephrosis	α-Fetoprotein

Modified from Mennuti MT in Bolognese RJ, Schwarz RH, Schneider J (eds): Perinatal Medicine. Management of the High Risk Fetus and Neonate. Baltimore, Williams and Wilkins, 1982.

* This list is expanding constantly as new diagnostic procedures become available.

sulting in defective clotting. Sensitive immunoradiometric assays show that factor VIII coagulant antigen is deficient, but that the related antigen (associated with platelet function) is normal, indicating that there is a normal production of an abnormal protein. *Hemophilia B* is associated with deficient factor IX coagulant activity and antigen. Many patients produce an abnormal protein with defective coagulant activity.

Laboratory Diagnosis. Before the prenatal diagnosis of hemophilia was possible, known carriers could have amniocentesis for fetal sex and elect to terminate the pregnancy if the sex was male. This carried a 50% risk of aborting a normal male fetus.

The development of sensitive immunoradiometric assays for factor VIII and IX antigens and the improvement in the technique of obtaining fetal blood now make prenatal diagnosis possible in the fetuses at risk. In the fetus with *hemophilia A,* factor VIII coagulant antigen is deficient while factor VIII related antigen is at normal levels. In *hemophilia B,* factor IX coagulant antigen is deficient. Antigen assays are necessary because coagulant activity assays are subject to interference by thromboplastin-like material in amniotic fluid.

Hemoglobinopathies

Clinical Features. The two main disease groups are *sickle cell* (Hb S) and related diseases (Hb C, E), resulting in production of abnormal globin chains, and the *thalassemias,* resulting in impaired production of normal globin chains. Their inheritance follows classic mendelian genetics.

About 9% of black Americans are heterozygous for Hb S, whereas the prevalence in Africa may be 20% to 40%. The gene has persisted because heterozygotes have some protection against malaria. Patients who have a sickle cell *trait* (heterozygotes) have few clinical problems, but those who have sickle cell *disease* (homozygotes) develop severe hemolytic anemia (with crises precipitated by infection, dehydration, and other stresses), impaired growth, and a shortened life-span. About 2% of black Americans are heterozygous for Hb C. Homozygous Hb C and the double heterozygous states of Hb SC and Hb S/β-thalassemia have clinical manifestations similar to sickle cell disease.

β-Thalassemia is the classic Mediterranean anemia. In Italy, Sicily, and Greece, about 10% of the population are heterozygous for β-thalassemia, and it is encountered in central Africa and southeast Asia. In northern Italy and Sardinia, the thalassemia is predominantly of the β^0 type, in which synthesis of β chain is completely suppressed. In southern Italy and Greece, β-thalassemia is commonly of the β^+ type, in which β-globin is produced in small amounts. Patients with β-thalassemia *minor* (trait) have few clinical manifestations, but those with β-thalassemia *major* (deletion of both β chains) have severe anemia and hepatosplenomegaly; iron overload from repeated blood transfusions results eventually in death from myocardial damage.

α-Thalassemia occurs primarily in southeast Asia. Because the gene for α-globin is duplicated (four α-genes are normally inherited), deletions of one to four α-genes can occur. Homozygous α-thalassemia, or α-thalassemia *major* (deletion of four α-genes), results in *hydrops fetalis* and death *in utero;* whereas moderately severe microcytic hypochromic anemia with splenomegaly occurs in hemoglobin H disease (deletion of three α-genes). The excess γ- and β-chains form the unstable tetramers γ_4 (Hb Barts) and β_4 (Hb H). α-Thalassemia *minor* (trait) (deletion of one or two genes) is not associated with anemia.

Pathogenesis. *Sickle cell* and related diseases arise from a single mutation in the DNA of the structural gene, causing the substitution of one amino acid for another. In Hb S, a single point mutation (β^6 Glu $\rightarrow$ Val) causes unoxygenated Hb S molecules to aggregate at low O_2 tensions; as a result the cells elongate to a sickle shape which can plug the capillaries. Although many single point mutations occur, only one third are associated with clinical manifestations.

β^0-Thalassemia has several causes: generation of a termination codon in the coding region, deletion of part of the β-globin structural gene, or impairment of β-globin m-RNA transcription. *β^+-Thalassemia* results from a single nucleotide change (G $\rightarrow$ A) in the first intervening sequence within the β-globin gene, leading to an unstable m-RNA. Some normal mRNA is produced in small amounts.

Almost all cases of *α-thalassemia* result from deletion of structural genes coding for α-globin.

Laboratory Diagnosis. The two approaches available are globin chain synthesis and, more recently, restriction endonuclease techniques. Diagnosis by *globin chain synthesis* has been the most frequent indication for fetal blood sampling. The fetal cells are washed, incubated with ^{3}H-leucine, and then lyzed. The newly synthesized globin is separated by chromatography into α, β, and γ chains. At 18 weeks' gestation normal fetal reticulocytes contain about 95% $\alpha_2\gamma_2$ and 5% $\alpha_2\beta_2$ hemoglobins. Comparison of the amount of β^s, β or α chain to the γ chain allows diagnosis of the respective trait or disease to be made.

Restriction endonuclease techniques use amniotic fluid cells and eliminate the need for fetal blood samples. After digestion of nuclear DNA with a specific restriction endonuclease, the DNA fragments are separated according to size by agarose gel electrophoresis and transferred to nitrocellulose filters (Southern blot technique). The distribution of specific sequences among the DNA fragments is determined by molecular hybridization with a DNA probe (^{32}P-DNA complementary to the gene sequence), and autoradiography (see also Chap. 28).

Indirect assessment uses restriction fragment length polymorphisms (normal inherited variations in DNA) linked to the abnormal gene. Absence of a restriction site at the locus of a polymorphism produces a larger DNA fragment than that produced when the restriction site is present. Indirect diagnosis of sickle cell disease depends on a polymorphism linked to the β^s gene. Although the β-gene is associated generally with a 7.6 kb (kilobase) fragment of DNA, in 50% to 70% of black Americans, it is linked to a 13.0 kb fragment. Family studies are necessary to determine whether the linkage is present and if diagnosis is possible. Similarly, diagnosis of β^0-thalassemia may be made using several polymorphic restriction endonuclease sites in the $^G\gamma$ and $^A\gamma$-genes and on the 3′-side of the β-gene.

New specific techniques, employing synthetic probes and specific restriction endonucleases, are being developed rapidly and are simplifying diagnosis greatly. In the diagnosis of sickle cell disease, the specific restriction endonuclease *Mst II* cleaves the gene sequence of the normal β-gene but does not recognize the abnormal DNA sequence of the β^s-gene. The β^A-gene gives two DNA fragments of 1.15 and 0.2 kb whereas the β^s-gene gives one 1.35 kb fragment. A more recent technique uses synthetic 19-oligonucleotide probes which specifically recognize either the β^A or β^s gene sequences.

Similarly, in the diagnosis of β^+-thalassemia and β^0-thalassemia caused by a termination codon, synthetic probes that recognize the normal and abnormal sequences of DNA have been produced. Family studies must be done to ascertain the type of thalassemia.

Homozygous α-thalassemia is diagnosed by digestion of the DNA with *Eco R1* and hybridization with an α-gene probe. Absence of the large DNA fragment which contains the α-genes indicates homozygous α-thalassemia.

New procedures are being developed and other genetic diseases will soon be diagnosed by recombinant DNA techniques.

BIOCHEMICAL ASSESSMENT OF HIGH-RISK PREGNANCY

Laboratory testing in a high-risk pregnancy is directed toward assessment of fetal risk and fetal maturity. The fetus is inaccessible for routine biochemical testing. Measurements must be made on the fluids available—maternal blood, maternal urine, and amniotic fluid. *Biochemical* analyses play a role in diagnosing ectopic pregnancy, in assessing the severity of Rh immune disease, in determining fetal lung maturity for the prevention of respiratory distress syndrome, in assessing fetal well-being in the presence of maternal diabetes, or toxemia, and in the assessment of intrauterine growth retardation.

LABORATORY ASSESSMENT OF FETAL DISTRESS AND FETAL LUNG MATURITY

Fetoplacental Synthesis of Estriol

Adequate placental and fetal function are both necessary for the synthesis of ***estriol*** (Fig. 23–2). The precursors cholesterol and pregnenolone are obtained from the mother and the placenta. The *fetal adrenals* lack 3β-hydroxysteroid dehydrogenase and cannot convert pregnenolone to progesterone. Instead, they convert pregnenolone, via 17-

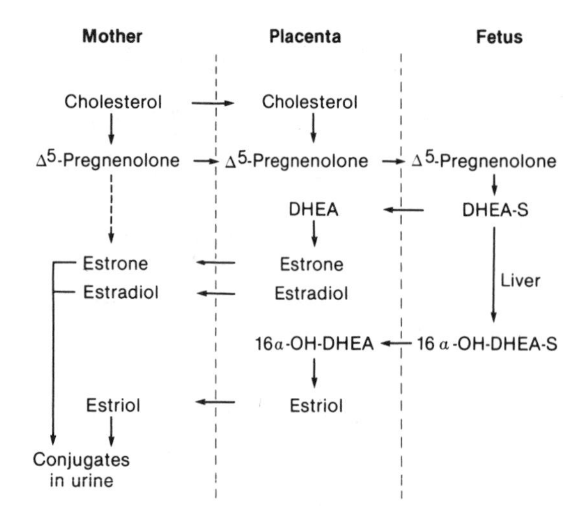

FIGURE 23–2. Fetoplacental synthesis of estriol. DHEA-S = dehydroepiandrosterone (sulfate).

hydroxypregnenolone, to dehydroepiandrosterone (DHEA), which in turn is converted to 16-OH-DHEA by the *fetal liver.* The fetus also rapidly and extensively conjugates steroids with sulfate. The sulfated derivatives of DHEA and 16-OH-DHEA pass to the placenta because the fetus lacks sulfatases and aromatases.

The *placenta* depends on the 19-carbon precursors (DHEA and 16-OH-DHEA) from the fetus for estrogen synthesis, because it lacks 17-hydroxylation and desmolase activities needed to form C-19 steroids from C-21 precursors. The placenta has sulfatase activity to cleave the sulfate conjugates and aromatase activity to convert both DHEA to estrone and estradiol, and 16-OH-DHEA to estriol. Estriol is conjugated in the maternal liver to the 3-sulfate, 16-glucuronate and 3-sulfate-16-glucuronate, in the intestinal mucosa to the 3-glucuronate, and to a minor extent in the kidney to the 16-glucuronate. The conjugates are excreted by the kidneys and the liver; the liver excretes them by way of the bile into the intestine, where they are hydrolyzed by gut bacteria and estriol is reabsorbed (enterohepatic circulation). Only about 10% of maternal blood estriol is unconjugated. Pregnancy is characterized by a great increase in placental estrogen production. Because estriol makes up 90% of estrogens in normal pregnancy, and its production depends to a large extent on the fetal precursor, 16-OH-DHEA sulfate, either total urinary estrogen or serum estriol level has been considered a reflection of fetal well-being and placental function.

Assessment of Fetal Distress

Serial measurement of estriol concentrations is still used occasionally in the third trimester of pregnancy to assess fetal growth and well being. Fetal heart-rate monitoring, however, is steadily replacing estriol determinations in the assessment of fetal distress. Estriol determinations must be assessed serially, because the reference range is wide. Day to day variations may exceed 30%.

In most centers the assay of ***serum unconjugated estriol*** (about 10% of the plasma total estriol level) has replaced that of 24-h total urinary estrogens. Unconjugated estriol is derived almost exclusively from the fetal adrenals and liver and is cleared by the maternal liver, rarely affected in high-risk pregnancies. Because its half-life is about 20 minutes, unconjugated estriol rapidly reflects changes in estriol production. The assay procedure (radioimmunoassay) is not subject to drug interferences.

Rising concentrations of serum unconjugated estriol and levels within the reference range (Fig. 23–3) reflect a *healthy* fetus, whereas low or falling levels indicate fetal *jeopardy* or fetal death. A decrease in estriol of 40% from the average of the previous three results is associated with severe risk to the fetus. Very low estriol levels occur with an anencephalic fetus, and in a clinically normal pregnancy in which placental sulfatase deficiency is present.

In many centers ***non-stress testing*** (fetal heart-rate monitoring) has replaced estriol measurements in the assessment of fetal distress. The fetal heart rate is measured over a period of 20 minutes. The basis of the test is the physiologic response of the fetus to a decreased fetal pO_2, which occurs when the fetus undergoes stress (fetal movement, small contractions prior to labor). With a *normal* result (appropriate acceleration of the fetal heart rate) the fetus is considered to be capable of coping with a greater stress in the future. With an *abnormal* result (lack of expected accelerations or continuous decelerations) the fetus is judged unable to cope with even small stresses and immediate delivery must be considered. Non-stress tests are performed generally once a week on high-risk patients, except

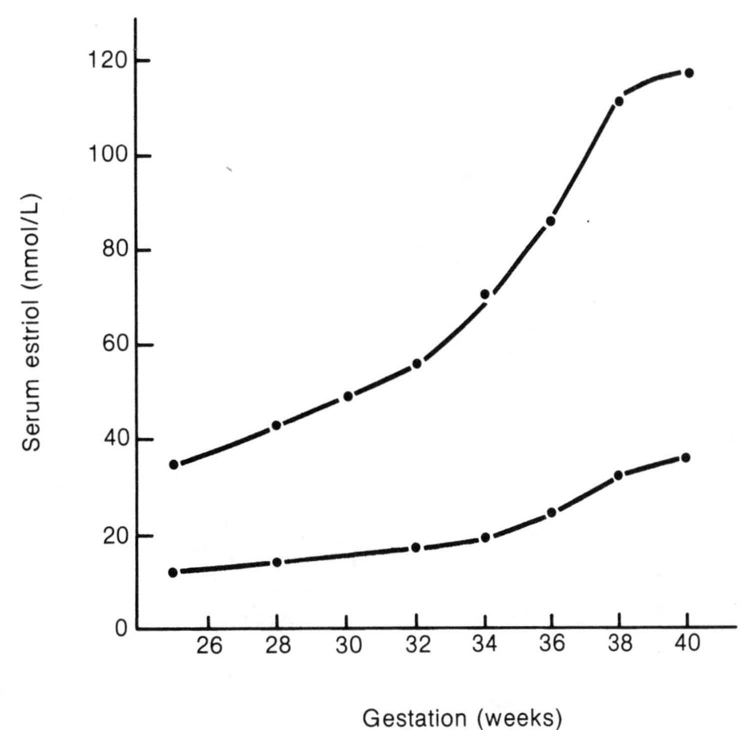

FIGURE 23–3. Reference range for maternal serum estriol.

in diabetes, intrauterine growth retardation, and postmaturity (gestation >42 weeks), when the tests may be done every 1 to 3 days.

Lung Maturation

A surface-tension-lowering agent is necessary to sustain normal respiration after delivery. *Pulmonary surfactant* acts in the normal lung: (1) to reduce surface tension as the alveolar radius decreases, thus preventing collapse of the alveoli with expiration; and (2) to maintain alveolar stability by varying surface tension with alveolar size.

The surfactant material is a phospholipid-protein complex produced in the type-II epithelial cells of the lung. Surfactant is approximately 90% lipid and 10% protein. About 90% of the lipids are phospholipids, with lecithin being the major component, and phosphatidyl glycerol a very active minor component. *Dipalmitoyl lecithin* is more effective as a surfactant than lecithin containing unsaturated fatty acids.

Most *de novo* ***lecithin synthesis*** appears to occur by the choline incorporation pathway (Fig. 23–4), which forms lecithin containing predominantly a

FIGURE 23–4. Synthesis of lecithin. The choline incorporation pathway, showing the synthesis of 1,2-diacyl lecithin, its conversion to 2-lysolecithin, and the synthesis of 1,2-dipalmitoyl lecithin by transacylation or reacylation. CDP-choline = cytidine diphosphocholine.

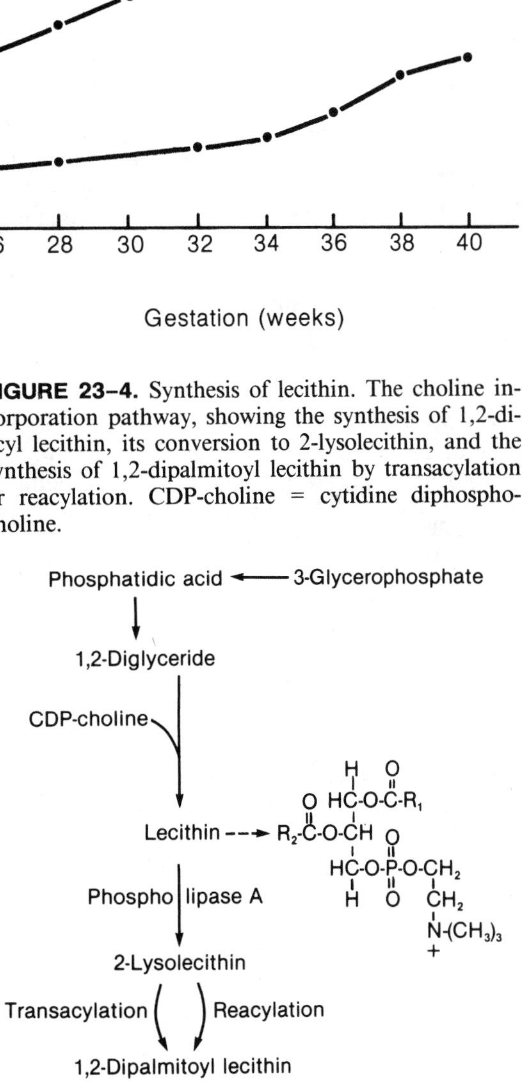

Phosphatidic acid ← 3-Glycerophosphate
↓
1,2-Diglyceride
↓ CDP-choline
Lecithin --→ R_2-C(=O)-O-CH with HC-O-C(=O)-R_1 and HC-O-P(=O)(O)-O-CH_2-CH_2-$N^{+}(CH_3)_3$
↓ Phospholipase A
2-Lysolecithin
Transacylation ↓ Reacylation
1,2-Dipalmitoyl lecithin

saturated fatty acid, palmitic acid, at the 1 position and an unsaturated fatty acid at the 2 position. Synthesis of dipalmitoyl lecithin appears to occur by the hydrolysis of this lecithin to 2-lysolecithin, followed by transacylation or reacylation. In transacylation, two molecules of 2-lysolecithin, each with a saturated fatty acid at the 1 position, form one molecule of disaturated lecithin. Reacylation of lysolecithin may be less important.

The activity of the choline incorporation pathway increases rapidly at about 35 weeks of gestation, causing a rise in the concentration of dipalmitoyl lecithin in the lung and in the amniotic fluid and resulting in sufficient amounts of surfactant to permit normal breathing after delivery.

Assessment of Fetal Lung Maturity

Clinically, the most serious consequence of fetal immaturity is the *respiratory distress syndrome* (RDS), which remains a major cause of death in the neonatal age group. The infant developing RDS is almost invariably premature and at birth or shortly thereafter begins to show signs of tachypnea, expiratory grunting, chest wall retraction, see-saw respiration, and cyanosis. An x-ray of the lungs shows a ground-glass appearance.

Pathogenesis. The lack of *surfactant* in the lung of the premature infant initiates a sequence of atelectasis, pulmonary hypoperfusion, right-to-left shunting, shock, and hyaline membrane formation. Acidosis and hypoxia in the newborn appear to inhibit the synthesis of surfactant, and such factors as hypothyroidism and maternal diabetes retard its production. By contrast, factors such as prolonged rupture of the membranes, chronic fetal distress, and maternal glucocorticoid promote its synthesis.

Laboratory Assessment. The fetal lung secretes fluid which mixes with amniotic fluid in the nasopharynx, and the surfactant lecithin of the pulmonary secretions thus enters the amniotic fluid. The amount of lecithin in amniotic fluid has been shown to reflect the degree of maturity of the fetal lung. It rises sharply at about 35 weeks' gestation at the same time as a slight fall in the sphingomyelin concentration occurs.

The amount of lecithin in amniotic fluid is generally determined by measuring the ***lecithin/sphingomyelin ratio*** using thin-layer chromatography. Values above a critical reference level (which depends on the method used) are predictive of fetal pulmonary maturity (Fig. 23–5). The predictive value of a *negative* result (i.e., L/S ratio above the reference value, indicating maturity) is very high. The predictive value of a *positive* result (a low ratio, indicating immaturity) is only moderate; one can not have a high degree of confidence that the fetal lung is immature. This problem of false-positive results exists presently for all tests of fetal lung maturity.

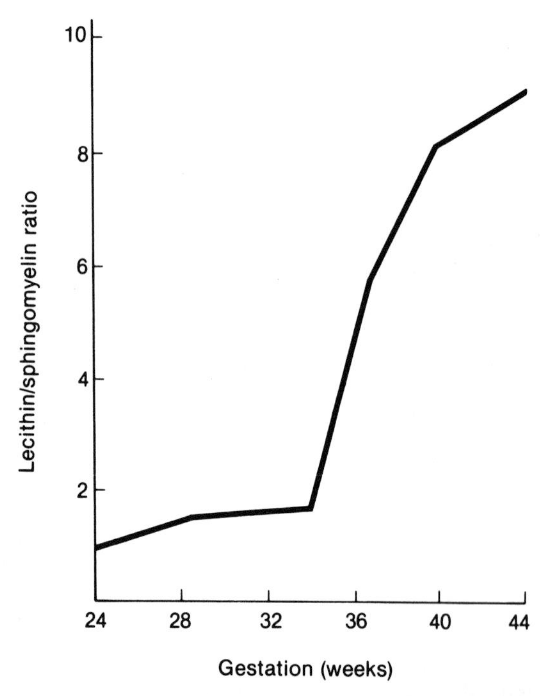

FIGURE 23–5. Mean amniotic fluid lecithin/sphingomyelin (L/S) ratio versus gestational age. A ratio of >3 indicates a low risk of respiratory distress syndrome. Our procedure omits the acetone precipitation step of Gluck, whose critical level is a ratio of 2.

The presence of *phosphatidyl glycerol,* measured by thin-layer chromatography, is a good predictor of fetal lung maturity. Alternatively, a simple test of fetal lung maturity is the *bubble stability* (or shake) *test,* in which amniotic fluid is mixed with an equal volume of 95% ethanol and shaken (final ethanol concentration 47% to 48% by volume). A complete ring of bubbles will persist at the meniscus if sufficient surfactant is present to prevent RDS.

Ectopic Pregnancy

Human chorionic gonadotropin (hCG, M_r 36,000), a glycoprotein secreted by trophoblastic tissue, is comprised of an α-chain, also found in TSH, LH, and FSH, and a β-chain which confers specificity. It can be detected in the maternal circulation about 10 days after ovulation if fertilization has occurred, by using a sensitive immunoassay procedure specific for the β-chain. It continues to rise to a broad peak from 6 to 12 weeks of pregnancy and then falls to about 1% of the peak concentration for the remainder of gestation. Its detection in urine, about a week following the last expected menstrual period, forms the basis of the urine pregnancy test.

Clinical Features. About one in every 90 pregnancies is *ectopic* (extrauterine), and of these pregnancies, 95% occur in the fallopian tube. Ruptured ectopic pregnancies with intra-abdominal hemorrhage cause 7% to 10% of maternal deaths. Early diagnosis and adequate treatment minimizes tissue damage and increases the likelihood of future normal pregnancy. Predisposing factors are pelvic inflammatory disease, previous abdominal or tubal surgery, and previous ectopic pregnancy.

The *classic triad* of pain, bleeding, and adnexal mass (mass by the ovary) occurs in only 30% to 40% of ectopic pregnancies. The presenting symptoms are often nondiagnostic, and the differential diagnosis includes rupture of an ovarian cyst, pelvic inflammatory disease, intrauterine pregnancy, early spontaneous abortion, and dysfunctional uterine bleeding.

Laboratory Assessment. A patient suspected of having an ectopic pregnancy first has a qualitative *serum β-hCG screen* performed. The screening test utilizes a rapid immunoassay (e.g., RIA, or sensitive enzyme immunoassay using monoclonal antibodies), with results being reported as negative, inconclusive, or positive. An 'inconclusive' test should be repeated in a few days. The patient is investigated for other causes of her symptoms if the screen is negative. If the screen is *positive,* the patient is pregnant and an ultrasound examination is done to determine whether the pregnancy is intrauterine (gestational sac seen) or ectopic (no gestational sac). Ultrasound does not always demonstrate an adnexal mass (by the ovary) in women with ectopic pregnancies. In some centers quantitative hCG assays are being performed 2 days apart to assess the hCG doubling time, which is normally about 2 days. It is generally shorter if twins are present, and longer if the pregnancy is ectopic.

Management. A patient with a ruptured ectopic pregnancy and in shock requires immediate surgery, whereas a patient who is hemodynamically stable and who wishes to remain fertile may have surgery with conservation of the tubes. Nonetheless, a significant infertility risk remains.

Rh Isoimmunization

Clinically, *erythroblastosis fetalis* is an immune hemolytic disorder of the fetus and newborn, characterized by hydrops fetalis (fetal edema), severe jaundice, and anemia. The incidence is about 1:150 live births.

Mildly affected infants are not anemic at birth but may develop severe hyperbilirubinemia with concentrations of unconjugated bilirubin up to 270 to 340 μmol/L (16 to 20 mg/dL). Moderately affected infants have anemia at birth, and as the placenta is no longer available to remove bilirubin, the unconjugated bilirubin concentration may rise to 340 μmol/L or more. This concentration will exceed the bilirubin-binding capacity of the infants' albumin and is then associated with a high risk of kernicterus (deposition of bilirubin in the brain, and associated degenerative lesions). The mortality rate of severely affected infants with fetal hydrops (hypoalbuminemia, edema, and ascites), severe anemia, and heart failure, is high. Above bilirubin concentrations of 200 μmol/L and up to about 300 μmol/L, infants are treated with phototherapy at 440 nm to 470 nm (see Chap. 10 and Chap. 21); above 300 μmol/L (if not sooner) the baby should receive an exchange transfusion.

Pathophysiology. Erythroblastosis fetalis results from *sensitization* of an Rh-negative mother to Rh-positive fetal red cells. The resulting maternal antibodies cross the placenta and cause the hemolysis of fetal red cells. Sensitization is usually due to Rh_0(D) incompatibility, but can also occur (in about 2% of cases) with other Rh and minor blood group antigens such as E, c, Kell, Kidd, and Duffy. Rh sensitization occurs by transplacental hemorrhage during delivery in 1% to 2% of first pregnancies, and 10% of subsequent pregnancies. It can occur also during a spontaneous or induced abortion, following an ectopic pregnancy, or at amniocentesis. Occasionally it occurs during pregnancy before delivery without any known cause.

About 8 to 9 weeks after sensitization, a weak antibody of the IgM type is produced. Then follows production of IgG, which crosses the placenta. On reexposure of the Rh-negative mother to Rh-positive fetal cells, anti-D IgG levels rise rapidly. The antibody crosses the placenta and initiates hemolysis of the fetal Rh-positive cells. Fetal anemia, which stimulates the production of erythropoietin, is countered to a variable extent by bone marrow hyperplasia and extramedullary hematopoiesis (causing hepatosplenomegaly). Normoblasts and erythroblasts (red cell precursors) are released into the fetal circulation, hence the name of the disease. Released hemoglobin is metabolized to *bilirubin,* some of which enters the amniotic fluid, apparently via lung secretions.

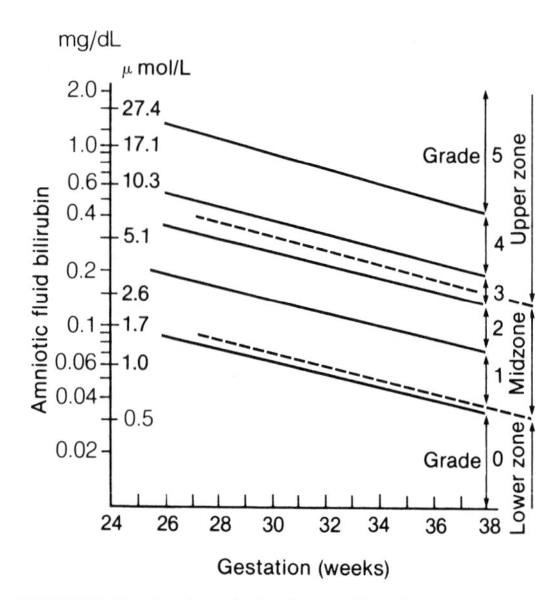

FIGURE 23–6. Amniotic fluid bilirubin versus gestational age, showing grades 0–5 (see text). Superimposed are Liley's lower, mid, and upper zones. (Bjerre S et al: Amniotic fluid spectrophotometry, urinary estrogen estimations, and intrauterine transfusion and severe Rh isoimmunization. Am J Obstet Gynecol 102:275, 1968)

Laboratory Assessment. An Rh-negative patient should be screened for anti-D ***antibodies*** at fairly frequent intervals during pregnancy if the father is Rh-positive. If antibody titers rise above a critical titer level (determined by the laboratory) amniocentesis is indicated. If there has been a previous affected pregnancy or if the patient was sensitized by blood transfusion, amniocentesis is necessary to assess the severity of the fetal disorder.

The assessment of fetal status is provided by the measurement of ***bilirubin*** in the amniotic fluid. Normal amniotic fluid contains a small amount of bilirubin, maximum at about 24 weeks and falling until term. Increased levels are seen in erythroblastosis and are a reflection of the degree of hemolysis. A sustained elevation of bilirubin indicates a greater degree of disease as gestation advances. The graph of bilirubin versus gestational age (Fig. 23–6) is divided into several zones. Grades 0 and 1 indicate an unaffected fetus and a mildly affected fetus, respectively. Grades 2 and 3 indicate moderate severity, and grades 4 and 5 indicate a severely affected fetus, usually with impending fetal death. Similarly Liley's lower, mid, and upper zones correspond to a mildly, moderately, and severely affected fetus.

Monitoring Rh Sensitization. Serial amniocenteses are performed to follow the changes in *bilirubin* concentration. In severe disease, between 24 and 34 weeks of gestation, an intrauterine fetal transfusion is indicated, while after 34 weeks the fetus is delivered and given exchange transfusions. Measurement of maternal *estriol* is of limited value in assessing erythroblastosis, nor is it used any longer in assessing fetal status after an intrauterine transfusion.

Prevention. Since about 1967, Rh sensitization in unsensitized women has been prevented by the administration of $Rh_0(D)$ immune globulin within 72 hours of delivery and after abortion, amniocentesis, or ectopic pregnancy. This has reduced the incidence of sensitization from about 15% of Rh-negative patients to less than 2%. If unsensitized women are given $Rh_0(D)$ immune globulin at 28 weeks of gestation, the incidence of Rh sensitization will decrease even further.

Diabetes

Clinically, the complications in pregnant women with diabetes mellitus include the risk of sudden intrauterine death and of neonatal death from prematurity, trauma, or congenital malformations. Perinatal mortality has been reduced to under 4% by careful control of maternal diabetes, monitoring with frequent estriol determinations, fetal heart-rate monitoring, and amniocentesis for assessment of fetal lung maturity before preterm delivery, and by improved postnatal care.

The *prevalence* of overt diabetes in pregnancy is about 0.5%. Complications in the mother include ketoacidosis, toxemia, and polyhydramnios, whereas in the neonate the prevalence of respiratory distress syndrome is about five times normal.

Pathology. Placental weight varies with fetal growth. In the recent diabetic without vascular complications, the placenta is hyperplastic and the fetus is often large, whereas in the diabetic with vascular complications the placenta is small and fibrous and fetal growth is retarded. No specific anatomic lesion in the placenta has been observed.

Laboratory Assessment. The risks of sudden intrauterine death and of respiratory distress syndrome caused by prematurity must be balanced in deciding when to deliver the diabetic patient. Non-stress testing is frequently performed on a daily basis. Serial estriol measurements are still considered to be useful here, when sudden fetal death may occur despite a normal non-stress test. A fall in estriol concentration of more than 40% is a predictor of fetal morbidity and death.

Fetal lung maturity is assessed by measurement of the L/S ratio and phosphatidyl glycerol concentration in amniotic fluid. In diabetic pregnancies false-negative L/S ratio results do occur; thus, the presence of *phosphatidyl glycerol* in the amniotic fluid may be a better test of fetal lung maturity.

Hypertensive States of Pregnancy

Clinically, the hypertensive states of pregnancy include preeclampsia/eclampsia (toxemia) and essential hypertension. *Preeclampsia* is characterized by the appearance of hypertension, albuminuria, and edema, whereas *eclampsia* is the occurrence of convulsions and coma in a preeclamptic patient. Patients who have longstanding hypertension may be misdiagnosed as preeclamptic.

The prevalence of preeclampsia in the third trimester is estimated to be 5% to 7%. Risk factors include primigravidity (first pregnancy), twins, poor nutrition, diabetes, hydramnios, renal disease, and hypertension. The maternal mortality rate is under 10%, whereas the perinatal mortality rate is somewhat higher (10% to 25%).

Pathology. Preeclampsia affects the kidneys and vascular system, but the etiology is unknown. Uteroplacental blood flow is compromised and may be associated with severe fetal growth retardation. Intravascular fibrin deposition may contribute to the development of glomeruloendotheliosis (reversible swelling of the endothelial cytoplasm), which may result in increased proteinuria.

Laboratory Assessment. Management of the pregnancy includes 24-h urinary protein measurements, assessment of fetal well-being by non-stress testing and serial estriol measurements, and assessment of fetal lung maturity by measurement of the L/S ratio. Infants born to hypertensive mothers have accelerated pulmonary maturation, so that pregnancies with severe preeclampsia are generally delivered early.

Intrauterine Growth Retardation

Clinically, growth retarded or 'small for gestational age' infants are those whose weights at delivery are below the tenth percentile for gestational age and must be distinguished from premature infants whose weights are appropriate for gestational age. Growth retarded infants appear wasted, lack subcutaneous fat, and have dry skin which appears to hang. Perinatal problems include asphyxia, fasting hypoglycemia, polycythemia, and temperature instability.

Growth retardation is seen by ultrasound to follow two patterns: slowing of growth in the *second* trimester related to an intrinsic fetal problem (e.g., infection, chromosomal disorder), and growth arrest in the *third* trimester caused by impaired uteroplacental function or nutritional deficiency (e.g., hypertension, toxemia).

Pathogenesis. Growth retardation results in decreased numbers of cells in the newborn and failure to deposit depot fuels (glycogen, fat) required postnatally. The placenta reflects abnormalities of uteroplacental blood flow and maternal nutritional state, because placental blood volume and trophoblastic mass are decreased in 'small for gestational age' and poorly nourished infants.

Genetic and environmental factors affecting fetal growth include low maternal weight and low weight gain, multiple pregnancy, fetal sex, chromosomal defects, nutrition, use of drugs (cigarettes, alcohol, and heroin), infection, placental insufficiency, and maternal disease (e.g., hypertension, diabetes).

Laboratory Assessment. Low *estriol* concentrations are associated with pregnancies complicated by growth retardation. Low but consistently rising concentrations are reassuring, whereas declining levels indicate fetal jeopardy. Non-stress testing is being used more and more in place of serial estriol assays. Elevated concentrations of maternal serum *AFP* at 16 weeks of gestation, when not indicative of another disorder, are frequently associated with low birth-weight infants.

SUGGESTED READING

BOLOGNESE RJ, SCHWARZ RH, SCHNEIDER J (eds): Perinatal Medicine. Management of the High Risk Fetus and Neonate. Baltimore, Williams and Wilkins, 1982

BROCK DJH: Early Diagnosis of Fetal Defects. Edinburgh, Churchill Livingstone, 1982

GABROWSKI GA, DESNICK RJ: Prenatal diagnosis of inherited metabolic diseases. In Lott SA, Darlington GJ (eds): Methods in Cell Biology 26, pp 95–179. New York, Academic Press, 1982

GERBIE AB (ed): Clinics in Obstetrics and Gynecology: Antenatal Diagnosis of Genetic Defects: Vol 7(1). London, WB Saunders, 1980

MAKOWSKI EJ (ed): High risk obstetrics. Clin Obstet Gynecol 21(2), 1978

24

Cancer-Associated Biochemical Abnormalities

Kenning M. Anderson / Philip D. Bonomi

Cancers may present clinically as an apparently localized disease; or metastases from an unsuspected primary cancer may provide the first evidence of its presence. Most often, the physician is confronted by a primary cancer complicated by metastases, some of which (micrometastastases) are clinically occult.

Cancers are associated with such varied clinical and biochemical manifestations that no single biochemical test exists which can alert the clinician to the presence of the many different malignancies. Although probably of diverse etiology, cancers exhibit a major common feature, namely, abnormal control of growth and replication. Increasing knowledge of the mechanisms responsible for cellular differentiation may yet lead to the discovery of one or more basic biochemical abnormalities associated with cancer cells which could be used to signal their presence.

CANCERS AS EXAMPLES OF DEFECTIVE CELLULAR DIFFERENTIATION

During normal cellular differentiation, integrated sets of genes are believed to be turned 'on' or 'off,' in a sequence-dependent pattern characteristic of the particular tissue. In cancer cells, *control* of these events is abnormal, resulting in the proliferation of cells beyond any physiologic requirement of the host.

Many cancers synthesize proteins that are *inappropriate* for the final differentiated form of the cell line from which they arose. These include changes in enzymes or isoenzymes, the ectopic synthesis of hormones, the appearance of oncofetal proteins and antigens, and the expression of proteins similar to those of oncogenic retroviruses. The release of such materials into the circulation or their evocation of a biologic response by the host could provide biochemical *markers* for the detection and monitoring of the inciting cancer. However, a number of these proteins are synthesized normally during fetal development or at times of cell division. Whether cancers contain tumor-specific proteins that are truly unique to their cell of origin is unknown. It is reasonably certain that many of the biochemical abnormalities to be discussed represent late sequelae or epiphenomena of the more fundamental defects in gene regulation that are responsible for the aberrant differentiation of cancer cells.

STRATEGY OF CANCER DETECTION BY BIOCHEMICAL MEANS

Unless a clinically occult cancer synthesizes a biologically active product such as a hormone, an enzyme, or a macromolecule that can be detected immunologically, or impairs the formation of such materials leading to a deficiency syndrome, or evokes a biochemical response by the host, its detection will have to be accomplished by nonbiochemical studies. Once local extension or distant metastases occur, and the function of a vital organ is compromised, biochemical studies can help to define the extent of spread and of remaining function.

A definitive *biochemical diagnosis* of cancer can rarely be made. Some exceptions include: (1) the detection of sufficiently increased amounts of 5-hydroxyindoleacetic acid (5HIAA) secreted from carcinoid tumors, (2) human chorionic gonadotropin (hCG) secreted by choriocarcinoma, (3) M protein in multiple myeloma, and (4) melanuria in malignant melanoma. Prostatic acid phosphatase is one of the more cancer-specific diagnostic enzymes in common use, but the presence of metastatic prostatic cancer must still be confirmed by biopsy. In addition to the secretion (or loss due to cellular necrosis) of hormonally active peptides or cellular enzymes, the detection of immunogenic proteins synthesized by the tumor, or humoral or cell-mediated responses to them by the host, represent promising approaches to the early detection and diagnosis of cancer. Given the present state of the art, clinicians and biochemists concerned with the diagnosis of cancer are obliged to function within these constraints.

MONOCLONAL ANTIBODIES, RECOMBINANT DNA, AND 'ONCOGENES'

There have been three developments which in time should profoundly affect the clinical and laboratory aspects of oncology.

An ability to generate large amounts of homogeneous antibody against a single epitope, or antigenic determinant, has many applications in oncology. Antigen-sensitized lymphocytes are generally fused with a hypoxanthine-guanine-phosphoribosyl transferase (HGPRT), or thymidine kinase (TK)-deficient cell line in selective 'HAT' medium, and a proportion of 'immortalized' cells, secreting a single type of antibody is cloned. Usually sensitized human or mouse lymphocytes are fused with mouse or human myeloma cells, with B-lymphocyte cell lines, or clones obtained by transformation of sensitized human lymphocytes with Epstein-Barr virus. Monoclonal antibodies to carcinoembryonic antigen (CEA), α-fetoprotein (AFP), and human tumor-specific or tumor-associated colorectal, melanoma, breast and lung cancer, and glioma antigens have been developed. ***Monoclonal antibodies*** have been used to assay the serum concentration and determine the tissue localization of cancer cell products (e.g., CEA); to localize metastases by radioimmunoscintography with anti-CEA and antiprostate acid phosphatase antibodies; for targeting of chemotherapeutic or cytotoxic agents against cancer cells; in immunotherapy, therapeutic drug monitoring, identification of new surface antigens on tumor cells, measurement of steroid hormone receptors, identification of B and T cells and their subsets, and removal of T cells involved in the graft-versus-host response prior to transplantation of bone marrow.

Detection of monoclonal antibodies is usually by radioimmunoassay (RIA) or various forms of enzyme-linked immunosorption assay (ELISA). To detect an antigen of interest by ELISA a tissue section is incubated with the monoclonal antibody, exposed to an anti-IgG antibody to which peroxidase is linked, or to a second enzyme-linked anti-IgG antibody; the enzyme substrate is then added and incubated. Deposition of an insoluble precipitate denotes the location of the antigen of interest.

Although many of the cancer-associated antigens against which monoclonal antibodies have been raised are probably 'differentiation' antigens, present in embryonic and some adult cells, such antibodies have proved especially useful in defining antigens present in adult lymphocytes. The monoclonal antibodies designated as OKT_4 and OKT_8, for example, identify helper and suppressor T-cell populations, respectively. CALLA (common ALL) antigen is present in 75% to 80% of non-T, non-B (or 'null') cell acute lymphoblastic leukemia (ALL) in children (20% are T-cell and about 2% B-cell derived). Prognosis is more favorable when CALLA antigen is detected in conjunction with the enzyme *terminal deoxynucleotidyl transferase* (TDT). Lymphomas that are difficult to identify histologically can often be identified by their expression of one or more leukocyte-common antigens. Other monoclonal antibodies that are cytotoxic against certain human cancer cells *in vitro* have been described. Identification of gene-protein products from recombinant DNA studies is another powerful use of monoclonal antibodies.

The development of ***techniques for cloning DNA*** has been even more remarkable. A complementary DNA (cDNA)-synthesizing clone is represented by a bacterial (or less often eukaryotic) cell transformed by a plasmid (in the case of bacteria) containing a DNA copy of the gene being sought. One sequence for developing a recombinant DNA plasmid requires: (1) the synthesis from a messenger RNA of interest (e.g., interferon mRNA) of a double stranded DNA copy, using reverse transcriptase and *E. coli* DNA polymerase, (2) insertion into the plasmid at a restriction enzyme cleavage site (as by dCTP-tailed double-stranded DNA to a dGTP-tailed pBR322 vector), and (3) transformation of a bacterial host by the plasmid (see also Chap. 28).

Restriction endonucleases are able to 'recognize' and cleave specific base sequences within double-stranded DNA. Identification of plasmid or bacteriophage carrying eukaryotic DNA can be accomplished by *in situ hybridization* of colonies with ^{32}P-labeled DNA or RNA. Alternatively, messenger RNA may be used to hybridize with complementary DNA immobilized on a nitrocellulose filter. The DNA:RNA duplexes are heated to release the mRNA and its presence is determined by translation in a cell-free rabbit reticulocyte or wheat germ system, followed by immunoprecipitation if antibody to the product is available, and/or by SDS gel electrophoresis. Selective 'blotting procedures' have been used to detect restriction-nuclease-digested DNA fragments, separated electrophoretically by size, then transferred to nitrocellulose filters for ease in identification with known complementary ^{32}P-DNA or RNA probes. Such procedures, and the ability to sequence cloned DNA, have been prerequisites for the discovery and characterization of 'oncogenes.'

Some 18 oncogenes (or better, *proto-oncogenes*) have been described in a variety of eukaryotic cells. The presence of 'transforming' genes in oncogenic C-type RNA (retro) viruses, the association of Epstein-Barr virus (a DNA virus) with Burkitt's lymphoma and nasopharyngeal carcinoma, of hepatitis B virus with subsequent development in a few chronically infected patients of primary hepatoma, and the retrovirus associated with a particular T-cell lymphoma, all bespeak a role for virus infection in at least some human cancers. Most interestingly, DNA that is structurally related (complementary) to the RNA of many different retroviruses is widely dispersed in mammalian cells. These homologies are demonstrated by hybridization of radioactive viral 'probes' to eukaryotic DNA using 'blotting' procedures.

These *proto-oncogenes* represent *normal* vertebrate cellular genes. Human cells, for example, contain DNA related to avian myeloblastosis virus on chromosome 6, avian myelocytomatosis, and sarcoma virus on chromosome 8. Viral homologues of these genes possibly represent eukaryotic DNA 'captured' by viruses; indeed viruses can be regarded as distantly related to the eukaryotic 'mobile' genetic elements of moderately repetitious DNA. Additional proto-oncogenes may eventually be described, perhaps different ones related to various pathways of cellular commitment and differentiation. Transformation *in vitro,* by the animal DNA virus SV-40, is associated with synthesis of a viral gene product that functions as a *protein kinase,* phosphorylating tyrosine rather than the usual serine or threonine. Homology between the amino acid sequence of *platelet-derived growth factor* and that of a simian *sarcoma oncogene protein product,* based on its DNA sequence, has been found. In some Burkitt's lymphoma cell lines, *translocation* of the terminal portion of the long arm of chromosome 8, which contains an avian myelocytomatosis virus-related 'myc' gene, to the long arm of chromosome 14, near the gene that regulates *Ig class switching,* suggests that juxtaposition of the switch region and the 'myc' gene on chromosome 14 somehow 'activates' the oncogene. The gene product 'myc' is a 47-K protein; in some mouse myelomas the concentration of its messenger RNA is increased 20- to 40-fold. Translocation of the 'myc' gene to chromosomes 2 or 22, the loci of genes for immunoglobulin light chains, has been found in other Burkitt's lymphomas that secrete light chains. Although the amount of specific oncogene product is not markedly increased in the particular bladder cancer cell line T 24, a single change of 1 DNA base pair (dG to dT), reflected in a single altered amino acid (GGC to GTC, glycine to valine) in its final 21 K 'ras' gene product, characterizes that cancer. This constitutes a 'unique' gene product associated with these bladder cancer cells.

It has been possible to transform reproducibly an unusually susceptible 3T3 fibroblast line with DNA extracted from human cancers, a process termed *'transfection.'* This finding suggests that provision of 'transformed' DNA (and its presumed activated oncogene) is sufficient to convert this permanent nonmalignant cell line to a 'malignant' state. The DNA of different types of cancer contains different activated oncogenes, although the same type(s) of oncogenes are believed to be functioning in the same type of malignancy. Different proto-oncogenes can be activated in cancers from the same line of cell, depending on their stage of differentiation. Finally, although cellular transformation is considered to be a multi-step process, it seems likely that qualitative or quantitative changes in the function of endogenous proto-oncogenes will represent a *central biochemical feature* of malignant change.

The karyotypic changes associated with some cancers, viewed in this context, provide a rational framework for understanding them. If the product of an activated oncogene is a form of cellular growth factor required during embryonic development, and which normally is repressed during subsequent maturation, attempts to diagnose cancer can be focused on its presence, and to treat cancer will require abrogating its effects. This may be another way of saying that most cancers represent a form of 'blocked' differentiation, and rational therapy would seek to correct or to circumvent this presumptive defect(s). As a corollary to this view, a variety of physical or chemical carcinogens may have as their final cellular 'target' the activation of normally repressed cellular proto-oncogenes. We seem to contain within ourselves the 'seeds' of our own destruction. Application of this new knowledge, which depended on the use of powerful new methods for its discovery, must turn now to the diagnosis and treatment of cancer.

USE OF CLINICAL BIOCHEMISTRY BY THE ONCOLOGIST

The clinical oncologist employs biochemical tests for any of at least five reasons:

a. To detect its presence
b. In a few instances to determine its biochemical nature
c. To assess its extent (or stage)
d. To estimate the prognosis
e. To monitor response to therapy (including recurrence)

USE OF TUMOR PRODUCTS AND ENZYMES IN THE DIAGNOSIS AND MONITORING OF CANCER

Oncofetal Gene Products

One of the characteristics of tumor tissue is a reappearance of embryonic metabolic activity which represents a reactivation of genes that had been repressed in adult tissues. Different cancers may synthesize particular *oncofetal* gene products, such as embryonic proteins or isoenzymes, which can be useful in diagnosis and in monitoring the total body burden of cancer. Monoclonal antibodies for some of these antigens (CEA and AFP) are available commercially as assay 'kits.'

Carcinoembryonic Antigen (CEA). This is a protein–polysaccharide β-globulin (M_r 1–2 $\times 10^5$) with a 7S–8S sedimentation coefficient. The protein can be extracted from intestine with 0.6 M perchloric acid, purified by chromatography, and used to elicit antibodies in rabbits. Radioimmunoassay (RIA) of CEA is available, and small quantities can be found in the sera of apparently healthy individuals, in pregnant women, in the elderly, in patients who smoke, and in those with uremia and nonmalignant pulmonary and hepatic disease. The upper *reference limit* is 2.5 μg/L. With the use of a more sensitive RIA, increased levels of serum CEA are found in 60% to 80% of patients with carcinoma of the gastrointestinal tract (including the pancreas), in a lesser number of patients with breast, lung, and various other cancers, and in some individuals with nonmalignant disorders.

Small tumors rarely increase the CEA level; extensive local lesions or metastatic tumor are more often positive. Determination of the CEA concentration does not constitute a general screening test for gastrointestinal or other cancers, and a value within reference limits (negative) cannot exclude serious cancer. However, elevated (positive) values, especially if persistently present without evident cause (such as regeneration of the liver after acute hepatitis), require a thorough evaluation of the patient for primary or metastatic cancer. CEA is used primarily to *monitor recurrence* after excision of a CEA-positive cancer; increased CEA may precede clinically evident disease by weeks or months. In general, the CEA level reflects the total body burden of CEA-producing tumor.

α-Fetoprotein (AFP). This is a glycoprotein (M_r $5–7 \times 10^4$) containing 4.3% carbohydrate, of which about 20% is sialic acid. A sensitive RIA is available; serum values range from 2 to 16 μg/L in normal adults and up to 550 μg/L during the third trimester of pregnancy (see Fig. 23–1). Initially the test was used as an aid in the diagnosis of hepatomas and, with less sensitive immunodiffusion techniques, 50% to 70% positive results were obtained. With RIA 90% of patients with hepatomas will have an elevated AFP, which can reach thousands of micrograms per liter. About 60% of adults with testicular teratocarcinoma have an elevated AFP. A few other common cancers, usually with hepatic metastases, increase the AFP level; these include carcinoma of the stomach, gallbladder, prostate, and bronchi; nonmalignant conditions associated with regeneration of the liver will also elevate the AFP concentration.

Human Chorionic Gonadotropin (hCG). This is a sialic acid–containing glycoprotein of which the β subunit can now be measured in serum and urine by a highly sensitive RIA. Except during pregnancy, hCG in body fluids indicates the presence of a tumor such as a choriocarcinoma, hydatiform mole, or malignant testicular or extragonadal teratoma. In choriocarcinoma, synthesis of 10^6 IU or more daily has been reported, and with successful treatment of the tumor, resulting in cure, hCG is no longer detected by RIA. Recurrence of the tumor or response to chemotherapy can be *monitored* by serum or urine levels of hCG. A low or normal AFP in the presence of an elevated hCG helps to distinguish hydatiform mole or choriocarcinoma from late pregnancy, in which both proteins are increased.

Ferritin and Casein. ***Ferritin*** is increased in the serum of some patients with various malignancies, in a minority of patients with cirrhosis, and occasionally in apparently normal subjects. The ferritin apoprotein originates in the liver and not the tumor. ***Casein*** can be detected in the serum of lactating women and has been reported in 72% of women with breast cancer, but is found occasionally in control patients with benign disease of the breast.

Test Batteries. In an effort to detect human cancer serologically, RIA for CEA, AFP, hCG β subunit, and κ-casein have been used to examine patients with and without cancer. Among 194 patients with early symptoms of cancer, 72% had one or more abnormal antigen levels, but only 40% were identified when CEA alone was used. Patients with metastases had more abnormal antigens in larger amounts, although 15% false positives occurred. Antigen levels declined or disappeared in 50% of patients treated surgically but normal levels were achieved less often in patients treated with chemotherapy or irradiation. The persistence of abnormal antigen concentrations is taken to indicate incomplete removal of the cancer, or the existence of unsuspected micrometastatic disease.

Other Antigens. Many other antigens are being evaluated by immunologic assay. These include several fetal and placental proteins or enzymes, leukemia- and Hodgkin's-associated antigens, pancreatic carcinoma antigens, and T-antigens. Monoclonal antibodies to most common human cancers have been described and their eventual role in diagnosis and treatment is under intense study. These tests represent an effort to detect different cancers with a level of sensitivity and specificity that should approach and may surpass the assay of tumor products such as hormones. For purposes of detection and diagnosis, class-specific tumor antigens which do not simply reflect the total body burden of tumor or the sequelae of metastases, would be most helpful. Until such antigens are discovered, and their clinical utility assessed, assays for oncofetal gene products that are normally absent, or that occur in very low concentrations in dividing cells, represent an important effort to diagnose cancer biochemically, before the onset of clinical symptoms.

Oncologic Enzymology

Most enzymes are intracellular in location and unless histochemical procedures are resorted to, must be released from cells in order to be diagnostically useful. Few enzymes are specific for a particular organ, but some discrimination can be achieved by considering their concentration and electrophoretic properties (see also Chap. 3).

Acid Phosphatase (ACP). Prostatic ACP is used clinically as a measure of metastases from carcinoma of the prostate. Bone, liver, spleen, and kidney contain about 0.1% as much ACP as the prostate, but sufficient enzyme is present in erythrocytes to require the determination of tartrate-labile and formaldehyde-stable ACP activity (relatively specific for the prostate), especially in the presence of hemolysis.

Transient elevations of ACP were thought to occur after rectal examination of a normal prostate, but this is now disputed. If cancer is contained within the capsule of the gland, serum ACP should not be consistently elevated. About 75% of patients with *metastatic* cancer of the prostate have increased serum ACP. The principal uses of the assay are to supplement the diagnosis and to follow the course of this disease. Increased serum ACP may also be seen in carcinomas of the breast, myeloproliferative disease, lymphoblastic leukemias, Gaucher's disease, primary hyperparathyroidism, and osteoporosis.

In a recent study, polyclonal antisera against a purified *prostatic antigen* (PA) and a *prostate-specific acid phosphatase* (PSACP) were developed. PSACP was assayed chemically by counterimmunoelectrophoresis (CIEP), solid-phase fluorescent immunoassay (SPIF), solid-phase immunoadsorbent assay (SPIA), and an enzyme-linked immunosorbent assay (ELISA). An ELISA was used to detect serum PA. Sera from patients with localized (stage A) to widely metastatic (stage D) disease were examined. PA was detected singly in 25% (stage A) to 64% (stage D) of patients, and PSACP in 33% to 73%. When combined, 58% to 91% (stage D) of the sera contained both antigens, and 80% of all patient's sera contained at least one of the antigens sought. Use of both assays in early stage disease doubled the number of abnormal sera detected, which was a significant improvement in the biochemical diagnosis of this cancer.

Alkaline Phosphatase (ALP). Determination of ALP is an aid in evaluating metastatic bone disease and intra- and extrahepatic obstruction of bile flow; the differentiation between an elevated ALP caused by bone or liver disease is a common clinical problem. A concomitantly elevated ACP suggests carcinoma of the prostate, which is a frequent primary source of osteoblastic metastases.

Determination of other serum enzymes can help to distinguish the source of an elevated ALP. In metastatic carcinoma of the liver, 80% of nonjaundiced patients have an increased ALP; changes in the aminotransferases reflect the extent of liver necrosis. An increased ALP in the presence of metastatic deposits may be caused by increased enzyme synthesis in hepatic tissue adjacent to cancer cells, or to obstruction of bile flow from infiltration and pressure by cancer cells. Some patients with extensive hepatic metastatic disease, however, have normal liver function studies.

Extremely high values of ALP are seen in Paget's disease of bone (osteitis deformans), which some consider a low-grade malignancy of the osteoclasts. Osteitis fibrosa cystica caused by primary hyperparathyroidism, osteoblastic osteogenic sarcoma, and metastatic invasion of bone with stimulation of the osteoblasts are associated with an elevated ALP, whereas osteolytic lesions, Ewing's sarcoma, chondrosarcoma, giant cell tumors, and benign bone tumors are not. Following successful hormone therapy of breast cancer, lytic lesions recalcify and ALP values fall.

Regan isoenzyme (*placentallike ALP*) is synthesized by a few tumors, especially those of the colon, breast, and female genital tract. In one study, one in seven cancer patients, and one in three cancer patients with elevated total ALP activity, were found to have this isoenzyme, although an overall frequency of less than 5% has been observed by others. Increased levels of Regan isoenzyme can occur in clinically normal controls and in patients with non-neoplastic liver disease. Regan isoenzyme in patients with cancer is thought to represent ectopic enzyme production by the tumor.

Lactate Dehydrogenase. Serum ***lactate dehydrogenase*** (LDH) is often elevated in cancer patients, reflecting the body burden of tumor, and declines with regression of the underlying cancer. In addition to fluctuations in total serum levels, selective increases in the LDH isoenzymes LDH_3 and LDH_5 (liver enzyme) may occur. An isomorphic increase in all five LDH isoenzymes can be seen in leukemias and lymphomas, and in hemolysis and pernicious anemia. Cancer-specific changes in the distribution of LDH isoenzymes have been reported but not universally confirmed.

Peritoneal, pleural, or pericardial ***effusions*** due to metastases can develop in patients with cancer. An *exudate* of fluid collecting in a body cavity, when due directly to cancer, is characterized by:

a. A fluid protein/serum protein ratio >0.5
b. A fluid LDH level greater than 200 IU
c. A fluid LDH/serum LDH ratio in excess of 0.6

The proportion of LDH_5 in malignant effusions may be increased, relative to the level in the serum. Exudates can be caused by either inflammatory or malignant disease. Consequently, such samples should also be cultured for bacteria, including tubercle bacilli, and subjected to stained smear and other cytologic examinations for bacteria and cancer cells. In some instances, malignant effusion with the properties of a *transudate* may collect, owing to obstruction of the venous or lymphatic system. A frankly bloody or xanthochromic effusion is more likely to be a result of cancer. Electrophoresis of the proteins and determination of CEA, AFP, or hCG may help to achieve a diagnosis.

Other Enzymes. The glycolytic enzymes ***phosphohexose isomerase*** and ***aldolase*** have been used to assess the response of various cancers to treatment, and in those tumors that synthesize either enzyme the serum levels paralleled the activity of the disease. Compared with serum aminotransferases, serum ***isocitrate dehydrogenase*** may be increased to a greater extent in metastatic carcinoma of the liver.

Serum ***ribonuclease*** is increased in several different cancers, including prostatic carcinoma. When it was measured along with ACP, a detection rate of 85% was obtained, compared to 71% and 78% for either enzyme alone. ***Histaminase*** has been reported to be associated with medullary carcinoma of the thyroid and the MEA II syndrome. Serum ***lysozyme*** probably originates in part from the lysis of leukocytes; increased concentrations are found in myelocytic and monocytic but not lymphocytic leukemia. Tuberculosis, sarcoidosis, and extensive burns can also increase serum lysozyme. Serum levels of ***α_1-antitrypsin*** are increased in pregnancy, myocardial infarction, inflammatory diseases, and malignancy. In the absence of the first three conditions increased values could reflect the presence of cancer. Serum ***amylase*** is increased in about 25% of patients with pancreatic carcinomas, and isoenzymes can be used to distinguish salivary from pancreatic enzymes; serum ***trypsin*** and ***lipase*** are less constantly increased. ***Leucine aminopeptidase,*** once thought to reflect pancreatic carcinoma, originates chiefly in the biliary tree, and is correlated more with biliary obstruction. ***Terminal deoxynucleotidyl transferase (TDT)*** is an enzyme marker for lymphoid cells; its presence often correlates with more differentiated cells, response to steroids, and a better prognosis.

OTHER BODY FLUID CONSTITUENTS OF USE IN ONCOLOGY

A variety of biochemical abnormalities occur in diverse cancers. None of them is of established clinical value in cancer diagnosis, but they form part of the total picture and are dealt with briefly.

Decreased serum *albumin* can be an early finding in many solid tumors, owing to (1) decreased protein intake, (2) the hypermetabolic state of cancer and reduced hepatic synthesis, (3) competition for nutrients by the tumor, and (4) increased external or internal losses such as ascitic fluid. In solid tumors α_1- and *α_2-globulins* and *glycoproteins* may be increased, while in hematologic malignancies α_1-, α_2-, and *γ-globulins* often are reduced and other nonspecific changes in γ-globulin occur. *Cryoglobulins,* including both IgG and IgM, are seen most often in cancers associated with increased immunoglobulins. Elevated IgA has been found in some patients with breast cancer.

Plasma *histamine,* correlating with the basophil count, is elevated in acute and chronic myelogenous leukemia, and in systemic mastocytosis with urticaria, pigmentation, and duodenal ulceration. Serum and urinary *hydroxyproline* are increased in metastatic disease of bone, and in any condition with sufficient destruction of collagen. Serum *copper* is elevated in a number of acute and chronic disorders. In Hodgkin's disease and the leukemias increased serum copper is associated with active disease, while remission is accompanied by a fall in concentration. Provided other causes of these changes could be excluded, serum copper has been used to help assess the activity of these diseases.

Polyamines, detected with an amino acid analyzer, are sometimes increased in the urine of patients with various cancers, or other causes of rapid tissue proliferation. Increased amounts of *transfer*

RNA nucleosides, detected by gas phase chromatography of urine from patients on low-purine diets, and including pseudouridine, N^2,N^3-dimethylguanosine, and 1-methyl inosine, have been found in patients with leukemias and other cancers. Such substances may reflect the presence and extent of various cancers, but this remains to be established.

Patients with increased parathormone (PTH) secretion excrete greater amounts of urinary *cyclic AMP,* while individuals with hypercalcemia that is not caused by increased PTH excrete normal, increased, or decreased amounts. Thus, a low level of urinary cyclic AMP in the presence of hypercalcemia favors a non-PTH, cancer-related cause. The measurement of plasma and urinary cyclic AMP and the renal glomerular filtration rate (GFR) allows one to calculate the cyclic AMP derived from glomerular filtration. The nephrogenic contribution to the total urinary cyclic AMP has been used to distinguish hypercalcemia due to parathyroid overactivity, or to ectopic formation of PTH, from hypercalcemia due to other causes such as direct destruction of bone.

STEROID HORMONE RECEPTORS AND RESPONSES TO THERAPY

Classic steroid hormone target cells contain steroid hormone receptors. For example, cytosolic *estradiol-receptor* (E · R) can be detected in the uterus and mammary glands, both organs that require estradiol for normal growth and development. Some cancers occurring in steroid-dependent target cells retain, at least for a time, an obligatory requirement for steroid hormone in order to achieve maximal rates of growth, and will regress after the removal of the appropriate steroid-producing endocrine gland, or other endocrine manipulation.

Receptor-steroid *interactions* involve specific high-affinity ($K_D = 10^{-9}$ M) binding of the steroid to a low-capacity cytosol protein. The receptor-steroid complex can undergo transformation to an 'activated' form which binds to specific chromatin acceptor sites within the nucleus. This results in the initiation of RNA synthesis at a number of these sites.

To *measure* cytosolic steroid-receptor, the tissue of interest is homogenized, particulate matter is removed by centrifugation, the supernatant is incubated with a radioactive steroid hormone (with or without a competing steroid), and receptor-bound radioactivity is analyzed, generally by sucrose gradient centrifugation or charcoal assay. The use of indirect, or eventually direct, RIA may supplant these methods for routine clinical work, because a battery of steroid hormone receptors could be analyzed conveniently. However, detection of immunoreactive substances may or may not correlate closely with results of steroid-binding assays; neither assay assesses whether the receptor being detected is biologically active.

The presence or absence of estradiol receptors is used to ***predict the response*** of breast cancer to hormone manipulations. If all premenopausal patients with breast cancer are castrated, only 30% undergo a clinical remission. When castration is limited to patients with E · R (+) tumors, 60% of them respond. In cancers that are also positive for *progesterone receptors* (P · R), the rate of response is about 80%. Generally, the greater the absolute values for E · R and P · R the more predictable the response. About 30% of patients with E · R (+) and P · R (−) cancers respond, and possibly 40% to 50% with P · R (+) and E · R (−) cancers. A response rate of 5% or less can be expected in receptor-negative patients. The minimum concentration of cytosol receptor likely to be associated with a clinical remission is 5 to 10 fmol (10^{-15} M)/mg of cytosol protein.

Thirty percent of *receptor-positive* mammary cancer patients fail to respond to hormonal manipulations. Possible reasons include the presence of some cancer cells that are receptor negative, receptors that are continuously activated independent of any steroid hormone, or defects in the interaction of receptor with nuclear binding sites. Mammary cancers may also contain receptors for dexamethasone, dihydrotestosterone, aldosterone, and vitamin A, but their significance is less well established.

Other receptors currently being studied include androgen, estrogen, and progesterone receptors in carcinoma of the *prostate,* estrogen and progesterone receptors in *endometrial* cancer, possible estrogen receptors in *malignant melanoma* (which may be an artifact), progesterone receptors in *renal* cancer, and the ability of *acute lymphocytic leukemia* cells to bind dexamethasone. In all these diseases it is hoped that the presence of steroid hormone receptors reflects a more differentiated cancer, and will correlate with clinical responses to hormonal manipulation. Evidence indicates that in general this is true for a number of these cancers.

Other attempts to predict the probable response of mammary cancer to endocrine manipulations or chemotherapy include:

1. Calculations based on the ratio of etiocholanolone to 17-hydroxycorticosteroids in the urine
2. The ability of mammary cancers to sulfate steroids
3. The presence of an active hexose monophosphate shunt in mammary cancer

None of these is used clinically.

Receptor-positive, steroid-hormone–responsive cancer represents one of the few human tumors whose growth can be reduced by manipulating the plasma concentration of physiologic substances. Converting a hormone-independent to a hormone-dependent cancer represents a rational goal, and a thorough understanding of hormone receptors and especially the newer knowledge concerning the control of gene function, may eventually help to achieve it. For example, in a patient with β^+ thalassemia, and one with sickle cell anemia, the agent azathioprine, which interferes with methylation of DNA, has been used to increase the synthesis of normally regressed fetal hemoglobin (HbF), presumably by reducing the extent of methylation of its gene, resulting in some transient alleviation of the anemia. DNA that is extensively methylated is rendered unsuitable for transcription by RNA polymerases.

PRIMARY NEOPLASTIC ENDOCRINOPATHIES

Clinical syndromes associated with primary cancers of the endocrine glands are caused by the synthesis and release of hormones appropriate for the cell of origin (e.g., excessive secretion of insulin by pancreatic islet β-cell adenomas). These syndromes can usually be diagnosed by using available assays of hypothalamic, pituitary, or target gland hormones, or by measuring their metabolic effects, and by the use of various suppression and stimulation tests to assess the autonomy of feedback mechanisms and end-organ responses. Because these tests are discussed in Chapters 14, 16, and 17, we emphasize here the *para*neoplastic syndromes, some of which closely mimic primary neoplastic endocrinopathies.

PARANEOPLASTIC SYNDROMES

Classification, Prevalence, and Pathophysiology

Paraneoplastic syndromes include a diverse group of systemic effects that occur distant from the primary cancer or its metastases. Ectopic hormone synthesis by nonendocrine cells, or the synthesis of a hormone not normally associated with particular endocrine cells, represent *paraendocrine* disorders distinct from primary cancers of the endocrine glands. The distinction between these categories is not always precise, as exemplified by the multiple endocrine adenopathy syndromes, MEA I and II (see the section on APUD Tumors). Symptoms due to a paraneoplastic syndrome may provide the patient's only presenting complaint, and can be life-threatening.

Paraneoplastic syndromes are recognized in 20% of cancer patients, but their actual prevalence may be as high as 50%. Several ***mechanisms*** can be responsible for their expression. *Derepression* of the cell genome is one, with inappropriate synthesis of normally repressed oncofetal gene products, including the aberrant synthesis of hormones and immunologically active proteins. Cancer cells may induce *cell-mediated or humoral antibody immune responses,* leading to autoimmune phenomena. Because of the disorganized vasculature and the sinusoidal spaces within many tumors, normally 'forbidden' cell-to-cell contacts may occur, leading to cell- or antibody-mediated cytotoxicity or other manifestations of autoimmune disease. The phagocytosis of dead cancer cells may be impaired, resulting in prolonged antigenic stimulation by tumor-associated antigens. *Neurovascular reflex mechanisms* are probably responsible for part of the *hypertrophic pulmonary osteoarthropathy syndrome* (HPO syndrome).

One third of patients with paraneoplastic syndromes exhibit effects of ectopic hormone production, another *third,* symptoms of connective tissue and dermatologic disease; *one sixth* express neurologic and psychiatric syndromes, and the *remainder* show hematologic, vascular, gastrointestinal, or direct immunologic complications. Recognition of these syndromes is important not only in order to treat them properly, but also in favorable instances to provide a diagnosis, at times early in the clinical course of the underlying cancer.

To ***diagnose*** a paraneoplastic endocrine syndrome requires, ideally:

1. An appropriate endocrine syndrome associated with an increased concentration of the hormone in both the tumor and the venous blood draining from it
2. Demonstration of *in vitro* labeling (synthesis) of the hormone by the tumor
3. Remission of the clinical syndrome following effective therapy, and its return in the event the cancer recurs

This biochemical equivalent to Koch's postulates (used to prove the etiology of an infectious disease) is rarely achieved in clinical practice.

APUD Tumors and the Multiple Endocrine Adenopathy Syndromes

The high ***a***mine content, amine ***p***recursor ***u***ptake, amino acid ***d***ecarboxylation (APUD) concept has been developed as an aid in classifying the cellular origin of paraneoplastic and primary endocrine products observed in cancer. During formation of the embryonic neural tube, clusters of neuroectodermal cells migrate from the neural crest to the primitive ectoderm and endoderm, where they differentiate into sensory and sympathetic neuroganglia, melanocytes, Schwann cells, and APUD cells. The last provide ectodermal endocrine components of the parathyroid, thyroid ultimobranchial cells (C cells), and hormone-producing cells of the pancreatic islets (α and β cells), stomach (G cells), duodenum and intestine (enterochromaffin cells), and the lung.

APUD cells are characterized by their high content of catecholamines or 5-hydroxytryptamine, amine precursors such as levodopa or 5-hydroxytryptophan, amino acid decarboxylases, and neurosecretory granules (seen by electron microscopy), and by their synthesis of polypeptide or amine hormones. These hormones can include corticotropin (ACTH), calcitonin, gastrin, insulin, glucagon, melanocyte-stimulating hormone (MSH), and vasoactive peptides. 'Apudomas' are also seen in the *familial paraendocrine syndrome* ***MEA type I,*** characterized by multiple endocrine adenomas (anterior pituitary, parathyroid, and islet cell), and frequently associated with intractable peptic ulceration (the *Zollinger-Ellison syndrome*). ***MEA type II,*** also a hereditary disorder, can include medullary (C cell) carcinoma of the thyroid, pheochromocytoma, and parathyroid adenomas or hyperplasia. A variant is associated with a marfanoid habitus. Some of the hormones secreted by isolated apudomas and in MEA are of APUD cell origin (α and β islet cells) but others are not (pituitary somatotropes). Whether a hormonal syndrome associated with a particular APUD tumor should be considered paraneoplastic depends on whether elaboration of the hormone is *inappropriate* for its cell of origin.

General Systemic Effects of Paraneoplastic Syndromes

Fever, anorexia, asthenia, loss of body fat and protein, increased basal metabolic rate, and increased energy expenditure in the face of reduced dietary intake are general ***metabolic*** consequences of advanced cancer. This differs from starvation, with its *reduced* metabolic rate and energy expenditure. Although the pathogenesis of cancer cachexia is not established, the inability to reduce caloric expenditure, excessive gluconeogenesis with degradation of host proteins, and the release of lactic acid from the tumor via the Cori cycle, all contribute to a hypermetabolic state. Patients with cancer often exhibit diabeticlike glucose tolerance curves. These metabolic events are associated with changes in key regulatory glycolytic enzymes toward fetal or tumor-related (isoenzyme) forms which are not subject to normal allosteric regulation.

Fever in cancer patients may be secondary to infection or directly caused by the malignancy. Hypernephroma, metastatic liver tumors, and lymphomas are frequent causes of fever not associated with infection. Various cells, including polymorphonuclear leukocytes, monocytes, macrophages, and reticuloendothelial cells, contain endogenous pyrogens (e.g., interleukin I) which may have a role in fever that is directly or indirectly related to cancer.

Some General Metabolic Syndromes

Lactic Acidosis. This disorder, reflecting the shift toward intracellular anaerobic glycolysis, can complicate the clinical course of cancer patients. Elevations of lactate to 4 to 8 mmol/L (36 to 72 mg/dL) may occur in severe anemia, hypoxemia, cardiovascular shock, and septic shock. In cancer patients, hyperlacticacidemia with hyperventilation can occur in the absence of these conditions. With small increases in lactate, serum pyruvate is not markedly elevated, and the normal lactate:pyruvate ratio of 8:1 to 15:1 is maintained. In some cancer patients, with no acute precipitating cause, marked lactic acidosis without increased pyruvate will be found. Many of these patients have a leukemia or lymphoma in relapse, at times with infiltration of the liver by cancer cells. The neoplastic tissue releases amounts of lactate which exceed the capacity of the liver and renal cortex to metabolize it.

Hypoglycemia. This has been observed in patients with retroperitoneal sarcomas, mesotheliomas, and other nonpancreatic tumors. Mechanisms invoked to explain the hypoglycemia include excess utilization of glucose by the tumor, ectopic secretion of insulin or an insulinlike substance, and increased release of immunoreactive insulin from the pancreas or its reduced degradation. Sensitivity to tolbutamide but not to leucine may be present. Extensive metastases to the liver or destruction of both adrenal cortices can also cause profound hypoglycemia.

Hyperuricemia. This may occur in malignant disease, especially in lymphomas and with myeloproliferative disorders such as leukemias. It is caused by an acute or subacute overproduction of uric acid and is especially common in response to the cytolytic effects of chemotherapy. Secondary ***gout*** similar to the acute attacks seen in patients with primary hyperuricemia may result. Use of a xanthine oxidase inhibitor, hydration, and alkalinization of the urine are essential measures for *prophylaxis,* especially before cytolytic therapy, to prevent uric acid nephropathy. Serum levels of urate greater than the upper reference limit of 420 μmol/L (7 mg/dL), or 500 μmol/L (8.5 mg/dL) when determined on most AutoAnalyzers, may be present without symptoms.

Hypocalcemia. This can be seen in patients with extremely active osteoblastic lesions, in occasional carcinomas of the lung or breast, and in leukemia. It may cause muscle weakness and twitching, a positive Chvostek's sign, mental confusion, and myocardial irritability, leading to coma and death. Mesenchymal tumors associated with vitamin D-resistant osteomalacia, bone tenderness, pathologic fractures, hypophosphatemia, reduced tubular reabsorption of phosphate and normocalcemia occur. Medullary carcinoma of the thyroid (whether a component of MEA syndromes or not), producing increased serum calcitonin, can be associated with hypocalcemia and hypophosphatemia because it inhibits resorption of bone mineral.

Fluid, Electrolyte, and Hydrogen Ion Derangements. During the course of many cancers, various derangements of *fluid balance* occur, at times precipitated by a complication such as hemorrhage, infection, obstruction, or actual destruction of organs responsible for homeostasis. *Hypernatremia* may occur in cancer patients. Frequent causes include dehydration, vomiting, diuretic therapy, and adrenocortical insufficiency, all of which result in reduced intravascular volume, a decreased glomerular filtration rate (GFR), and increased blood urea nitrogen (BUN). In order to conserve free water in the renal collecting duct system, increased secretion of antidiuretic hormone (ADH) occurs, and if any sodium deficiency is not corrected, this can result in *hyponatremia.* Massive destruction of cancer cells by effective therapy may cause the release of potassium to the point of *hyperkalemia,* resulting in cardiotoxicity.

Secondary Amyloidosis. This can complicate multiple myeloma, Waldenström's macroglobulinemia, and, less often, lymphomas. Variable fragments of IgG light chains, which are insoluble in physiologic fluids, are deposited in the musculoskeletal system, kidney, liver, and heart. This distribution is typical of primary amyloidosis, although the disorder is secondary to the underlying cancer. *Diagnosis* requires biopsy of an affected region, staining with Congo red dye, and microscopic examination, preferably by polarized light, to detect amyloid deposits. Microscopic examination of the bone marrow, electrophoresis and immunoelectrophoresis of the serum for myeloma M protein and of the urine for Bence Jones protein, may uncover an underlying malignancy in patients with secondary amyloidosis.

Syndromes Due to Ectopic Hormone Synthesis

Because some of these syndromes are described elsewhere, only a brief synopsis is given here.

Hypercalcemia. This is the most common biochemical manifestation of ectopic neoplastic endocrinopathy, and malignant disease is the most common cause of hypercalcemia. Tumors, particularly of the breast or lung, and multiple myeloma, cause resorption of bone, probably by one or more of several mechanisms. Direct activation of the patient's osteoclasts by osteoclast-activating factors occurs in multiple myeloma. Calcium-mobilizing sterols distinct from vitamin D have been reported in human breast cancer. Prostaglandins secreted by some tumors that have metastasized to bone may activate the osteoclasts. Other tumors also may cause hypercalcemia without evidence of osseous metastases. Some cancers synthesize a parathyroid-stimulating or parathormonelike peptide. The hypercalcemia, hypophosphatemia, hypercalciuria, and increases in alkaline phosphatase seen in these patients regress with effective treatment of the primary tumors, which include carcinomas of almost every variety, sarcomas, and lymphomas. In some patients, coexisting primary hyperparathyroidism has also been diagnosed.

Ectopic ACTH Syndrome. This is the second most frequent paraneoplastic endocrine disorder. Extraadrenal sites of the tumor include the lung (especially oat-cell or small-cell carcinoma), thymus, and pancreatic islet cell and less often the thyroid, ovary, testes, adrenal medulla, kidney, or salivary gland. Classic physical signs of Cushing's syndrome may not be striking, but hyperpigmentation caused by the concurrent secretion of MSH may be present. Hypokalemic alkalosis, muscle weakness, and glucose intolerance are often more marked than in other forms of Cushing's syndrome, and both plasma cortisol and urinary 17-hydroxycorticoids and 17-ketosteroids are increased (see Chap. 14).

Inappropriate ADH Secretion. This can result from excessive secretion from an apparently normal pituitary or can occur in some cancer patients, especially those with oat-cell carcinoma of the lung (*Schwartz–Bartter syndrome*). Patients present with hyponatremia, a normal BUN, decreased serum osmolality, and an increased urinary sodium concentration. The urine is hyperosmolar in relation to the serum, despite the absence of dehydration, oliguria, edema, or hypertension. If other causes of *hyponatremia,* including cardiac, renal, hepatic, pituitary, adrenal, and thyroid disease, are excluded, ectopic formation of ADH can be diagnosed (see Fig. 6–3). ADH can be identified in tumor tissue, plasma, or urine by radioimmunoassay. Characteristically, plasma renin is suppressed and aldosterone levels are normal.

Ectopic Gonadotropins. During normal pregnancy the placenta secretes large amounts of human chorionic gonadotropin (hCG) and human chorionic somatomammotropin (hCS; human placental lactogen, hPL). Both have actions similar to those of pituitary luteinizing hormone (LH), and hCS in addition is lactogenic. Excess gonadotropins are often secreted by trophoblastic cancer in women (causing menstrual irregularities), teratomas and lung cancer in men (causing gynecomastia), and hepatoblastoma in young boys (causing precocious puberty). Testicular tumors are also a common source of gonadotropin. The concentration of hCG in the urine may be insufficient to cause a positive pregnancy test, so that radioimmunoassay of the β subunit of hCG in body fluids is to be preferred.

Somatotropin (STH); **the HPO Syndrome.** Patients with *adenocarcinoma of the lung* can synthesize somatotropin (growth hormone, STH), and there may be an association between the hypertrophic pulmonary osteoarthropathy (HPO) syndrome and ectopic formation of hCG. The HPO syndrome includes clubbing of the distal phalanges, pulmonary osteoarthropathy, gynecomastia, and pachydermoperiostosis, which need not coexist. Abnormal levels of gonadotropins, estrogen, or

STH may be present, but a neuroendocrine mechanism may be responsible for the clubbing.

Thyrotropin (TSH). Increased thyroid function may be related to the ectopic synthesis of TSH in some patients with *trophoblastic tumors,* including choriocarcinoma, hydatiform mole, and embryonal carcinoma of the testes. Symptoms of hyperthyroidism are minimal; tachycardia is present but eye signs, thyroid enlargement, and tremor are often absent. Thyroxin (T_4) levels and radioiodine uptake are increased. Plasma or tumor 'TSH' is elevated when measured by bioassay, but normal by radioimmunoassay, suggesting that this thyrotropic hormone differs from the chorionic thyrotropin produced normally by the placenta.

Syndromes Associated with Pancreatic Islet Cells. Indigenous islet-cell hormones include insulin (beta, B cells), glucagon (alpha, A cells) and somatostatin (delta, D cells). Ectopic vasoactive intestinal peptide (VIP) and gastrin come from D cells, and ectopic serotonin, ACTH, and MSH from non-beta cells. The ***Zollinger–Ellison syndrome*** is associated with a non-β-cell pancreatic islet tumor, which secretes gastrin. Diagnosis is confirmed if plasma gastrin levels are above 150 ng/L. Increased levels of gastrin can be found in pernicious anemia, but not in patients with peptic ulceration unassociated with an ulcerogenic tumor. Up to one third of patients with the Zollinger–Ellison syndrome may also have MEA I involving the parathyroid, adrenal, or pituitary glands, and additional studies are warranted. Whether this syndrome is paraneoplastic will be decided once the cell of origin and its biosynthetic properties are settled.

Whether ***hyperglycemia syndrome*** should be considered paraneoplastic is debatable, because α cells of the pancreas are considered the physiologic source of glucagon. In any event, α-cell tumors of the pancreatic islets secrete excess glucagon, which is associated with diabetes, skin rashes, hemolytic anemia, and edema. Hyperinsulinism, hyperglucagonemia, and hyperglycemia with an abnormal glucose tolerance support the diagnosis.

Non-β-cell islet tumors of the pancreas can also give rise to the ***WDHA syndrome*** (watery diarrhea, hypokalemia, and hypo- or achlorhydria), associated with the synthesis and release of vasoactive intestinal peptide (VIP). In some patients with reduced gastric secretion, a gastric inhibitory polypeptide (GIP) has also been identified (see Chap. 12). The biochemical work-up requires measurement of gastric secretion and of maximum acid output after stimulation with pentagastrin, because achlorhydria or hypochlorhydria would be compatible with the WDHA syndrome. In some centers, levels of VIP and GIP can be measured.

Carcinoid Syndrome. Carcinoid syndrome results from tumors that originate in the enterochromaffin cells, present throughout of the alimentary tract, including the liver, gallbladder, and pancreas. Carcinoid tumors in the small intestine tend to secrete large quantities of serotonin and kallikrein (see Chap. 12). If these are metabolized by a normal liver, little reaches the general circulation. However, carcinoids draining directly into the general circulation (e.g., bronchial or ovarian carcinoids) may release sufficient vasoactive amines to cause severe symptoms. The typical carcinoid syndrome includes flushing, diarrhea, bronchial constriction, and telangiectasia (usually attributed to serotonin). Excretion of 5-hydroxyindoleacetic acid (5HIAA) in a 24-hour specimen of urine exceeds 100 μmol (19 mg) and may reach hundreds of micromoles daily (reference range: 10 to 75 μmol/d).

Ectopic Calcitonin. Medullary carcinomas of the thyroid, arising in C cells, usually secrete large amounts of calcitonin without associated hypocalcemia, probably because of a compensatory secretion of parathormone (parathyroid hormone, PTH). About 10% of the patients exhibit other signs of MEA II syndrome. Paraendocrine secretion of calcitonin has been observed in tumors of neural crest origin, in oat-cell carcinomas of the lung, in intestinal and bronchial carcinoids, and in pheochromocytomas. Calcitonin may prove to be the most common hormone elaborated by APUD cells.

Erythrocytosis and Erythropoietin. Paraneoplastic *erythrocytosis* is distinguished from polycythemia vera by the absence of pancytosis and splenomegaly, and from secondary erythrocytosis by a normal arterial oxygen saturation. Half of the malignant tumors responsible for this condition originate in the kidney, while a mixed group of other tumors are responsible for the rest. Benign renal disease, including cysts, hydronephrosis, and adenoma, can also produce this syndrome. An increased serum or urine concentration of erythropoietin is usually found. The erythrocytosis of benign renal disease may be secondary to renal anoxia with increased synthesis of erythropoietic factor by normal renal tissue.

Cancer Cell Growth Factors

In addition to classic endocrine secretions, cellular growth factors can be extracted from normal cells and from the culture medium of established nontransformed cell lines. Factors derived from normal cells include epidermal, fibroblast, muscle, bone, platelet, red and white blood cell precursor growth factors, and a panoply of T and B immune-cell growth factors (e.g., interleukin 1 and 2). Growth factors have also been detected in serum-free culture medium of transformed cell lines. Cancer-derived growth factors are of great interest because they may stimulate division of the parent cell (*autocrine secretion*) or modify the growth and differentiation of contiguous cells (*paracrine secretions,* e.g., an induced synthesis of angiogenesis factor that promotes the local growth of the vascular system into the cancer). Such factors include a polypeptide similar to platelet-derived growth factor obtained from cultured human osteosarcoma cells, an epidermal growth-factor-like substance from murine sarcoma virus-transformed rat kidney, multiplication stimulating activity by a human fibrosarcoma line, and a nerve growth-factor-like polypeptide from a human melanoma line. Other proteins (TGF, *transforming growth factors* alpha and beta, and other TGFs from murine sarcomas) confer a 'transformed' phenotype on cells cultured in their presence. Tumor growth factors might provide a point of attack by immuno or other therapy, while assays for them might be used to assess the 'benign' or 'malignant' character of a cancer.

EFFECTS ON OTHER ORGAN SYSTEMS

Gastrointestinal Abnormalities

Malabsorption syndromes resembling nontropical sprue can be a direct or indirect consequence of gastrointestinal or other malignancies. Steatorrhea is the most obvious sign of malabsorption, which can also be associated with diarrhea, weight loss, anemia, and evidence of multiple vitamin and mineral deficiencies.

Obstruction of intestinal lymphatics and portal vessels presents a barrier to normal absorption, whereas submucosal infiltration of the bowel can lead to malabsorption, fistula formation, and intestinal obstruction, with hemorrhage, perforation, and generalized peritonitis. Exudative enteropathy with excessive *loss of protein* (notably albumin) can occur after ulceration of the esophagus, stomach, or colon because of carcinoma, lymphoma, or lymphatic obstruction, and may cause hypoproteinemia and edema. Hypertrophic gastritis and villous adenoma or carcinoma of the rectum are additional causes of *protein-losing enteropathy.*

Renal Disorders

Renal disorders associated with various cancers have included:

1. The nephrotic syndrome, in some patients with various carcinomas and Hodgkin's disease, without amyloidosis or renal vein thrombosis
2. Renal tubular acidosis and the Fanconi syndrome, in hypergammaglobulinemic states
3. Hypokalemic, hypercalcemic, or hyperuricemic nephropathies, in cancers responsible for these metabolic disorders
4. Renal vein thrombosis and cortical necrosis, secondary to intravascular coagulopathies, radiation nephritis, and postrenal obstruction. Secondary amyloidosis can also involve the kidney.

Hematologic Disorders With Distinctive Biochemical Characteristics

Leukemias. Many primary hematologic and nonhematologic cancers have profound effects on the composition and function of formed elements and soluble components of the blood. Results of a blood count, differential smear, and examination of the bone marrow permit the diagnosis of many leukemias. Leukocyte *alkaline phosphatase* (ALP), determined histochemically and distinct from the other forms of serum ALP, is reduced in chronic myelogenous leukemia, and less often in acute myelogenous leukemia.

Enzymatically determined *terminal deoxynucleotidyl transferase* (TDT) is a unique form of DNA polymerase found only in the thymus and bone marrow of vertebrates. Some 95% of patients with *acute lymphocytic* leukemia have had a large fraction of TDT-positive cells, while only 5% of patients with *acute myelogenous* leukemia are TDT positive. *Chronic myelogenous* leukemia, during blast crisis, may exhibit the enzyme, suggesting that the blast cells share features of acute lymphocytic leukemia. This information is important, because therapy differs depending on the cellular origin of the tumor. The blood level of *uridyl transferase* is reduced in hemolyzed erythrocytes from patients with chronic lymphocytic leukemia, and this has been proposed as a screening test.

Anemias. Red cell and hemoglobin deficiencies are common in cancer patients for several reasons, including: (1) blood loss, (2) 'toxic' depression of the bone marrow similar to that seen in many illnesses, (3) nutritional inadequacies, (4) hemolysis, and (5) results of chemotherapy. The ***anemia of chronic disease*** (ACD) is usually normocytic and normochromic, with a reduced red cell mass; increased plasma volume, normal absolute reticulocyte, WBC, and platelet counts; reduced serum iron; reduced saturation of transferrin; evidence of moderate hemolysis; and a normocellular marrow with an essentially normal ratio of erythroid to myeloid elements and reduced sideroblasts. Decreased plasma iron in the presence of *increased* iron stores in the bone marrow distinguishes this entity from iron-deficiency anemia, which is a microcytic, hypochromic anemia with *reduced* body iron stores and depressed serum ferritin levels. Shortened red cell survival, impaired utilization of iron bound in the reticuloendothelial system, and a failure of compensatory erythrocytosis account for much of the ACD syndrome.

Megaloblastic hyperchromic anemia can occur in carcinoma of the stomach with reduced vitamin B_{12} uptake, or folic acid deficiencies secondary to lymphoma invading the small bowel, and can complicate other cancers as well. *Leukemoid* reactions, with a WBC count in excess of 100,000/mm^3 (100×10^9/L) and circulating myeloblasts and promyelocytes, occur in patients with breast or lung cancer, often in the presence of hepatic or bone marrow metastases. *Eosinophilia,* at times associated with Hodgkin's disease and monocytosis, and a lymphocytic type of leukemoid disease may be seen with various cancers. *Myelophthisic anemia* following invasion of bone marrow by cancer cells, frequently with supervening fibrosis, also occurs, along with pancytopenia. *Erythrocyte agenesis* (pure red cell aplasia or erythroblastic anemia) has been seen in 2% of patients presenting with thymomas. Here an absence of reticulocytes in the peripheral blood and of erythroblasts in the bone marrow is associated with normal WBC and (usually) platelet counts. Acquired agammaglobulinemia and positive Coombs and LE (lupus erythematosus) tests occur in some patients with this syndrome, which suggests that it represents an autoimmune disease with an IgG antibody to erythrocyte precursors.

Disseminated Intravascular Coagulation (DIC). Coagulation and defibrination syndromes with abnormal fibrinolysis are serious hazards of a number of cancers, particularly those of the lung, pancreas, and liver. DIC syndromes are caused by the release of thromboplastic substances from such tumors, with activation of the coagulation pathway and continued generation of thrombin, resulting in deficiencies of coagulation factors (especially V, VIII, and fibrinogen) and of platelets. To the extent that fibrinolysis occurs, fibrin split products are generated, and accelerated fibrinogen turnover, decreased platelet and plasminogen survival, and reduced plasminogen levels are observed. Hemorrhage due to *thrombocytopenia* can result from chemotherapy, hypersplenism, destruction of platelets by autoimmune mechanisms, DIC, and reduced formation of megakaryocytes by the bone marrow. Nonbacterial *endocarditis,* leading to arterial embolization and infarction, may be a related phenomenon, associated most often with mucinous adenocarcinomas of the stomach, lung, or pancreas.

DIC can present acutely or be present chronically. The one-stage prothrombin time and partial thromboplastin time are prolonged owing to consumption of coagulation factors; the thrombin clotting time is increased because of decreased serum fibrinogen and inhibition by fibrinogen/fibrin degradation products, and the platelet count is low. Fibrinogen is diminished, fibrinopeptide A is increased, and fibrinogen degradation products may be present. In *chronic* DIC, these tests can give close to normal results.

Paraproteinemias. These include multiple (plasma cell) myeloma and Waldenström's macroglobulinemia, both *monoclonal gammopathies,* each derived from a single clone of cells.

Multiple myeloma is characterized by plasma cell infiltration of bone and other tissues. Biochemically, M protein, a monoclonal γ-globulin, is demonstrated in the serum of a majority of patients, often at concentrations in excess of 20 g/L; it is IgG in 70%, IgA in 18%, and IgD or IgE in the remaining cases. About 20% of all myeloma patients excrete only Bence Jones protein (either κ or λ chains) in the urine, as do 75% of patients with M protein in the serum. Serum albumin, IgA, and IgM are often reduced, the latter two rendering the patients more susceptible to bacterial infection. Hypercalcemia, hyperuricemia, and anemia are common, and secondary amyloidosis (with a primary pattern of distribution) occurs in 15% of patients during some stage of the disease.

Waldenström's macroglobulinemia is a less common malignant lymphoreticular disorder associated with the monoclonal synthesis of a high-relative mass IgM with a Svedberg constant of 19S. Increased rouleaux formation and an increased sedimentation rate are also characteristic. *Heavy-chain disease,* with formation of incomplete γ, α, or μ chains, has been described. The diagnosis is made by immunoelectrophoretic analysis of serum and immunologic analysis of the purified protein.

Serum hyperviscosity, which most commonly results from elevated serum M protein in multiple myeloma and Waldenström's macroglobulinemia, can be measured in an Ostwald viscometer. A viscosity greater than 4 compared with water is *diagnostic* and plasmapheresis is then imperative. This contrasts with polycythemia vera in which hyperviscosity due to increased numbers of red blood cells is treated by phlebotomy.

Muscle and Connective Tissue Disorders (see also Chap. 13)

There are several paraneoplastic syndromes in which the role of clinical biochemistry is distinctly limited. Autoimmune responses triggered by tumor-specific or tumor-associated factors, or the development of autoantibodies to various tissues, may be responsible for many of the disorders that develop secondary to a cancer, but can occasionally precede detection of the underlying malignancy by months or years. Thus, 20% of patients with *polymyositislike myopathy* have a malignancy, and symptoms of muscle disease can be the first sign of occult cancer. Depending on the extent of muscle necrosis, the serum enzymes lactate dehydrogenase (LDH), aspartate aminotransferase (AST), creatine kinase (CK), and aldolase, and the sedimentation rate, may be elevated. *Dermatomyositis* is a rare disorder that can develop concurrently with or subsequent to carcinoma of the lung or gastrointestinal tract, again suggesting that the tumor initiates the secondary disease. *Systemic lupus erythematosus* is also associated with some cancers.

Neurologic Syndromes (see also Chap. 15)

Neurologic syndromes, some of which may represent an autoimmune response by the patient against components of his own nervous tissue, can develop in cancer patients. Dementia, cerebellar degeneration, myelopathies, posterior root degeneration, and peripheral neuropathy have all been described. The cerebrospinal fluid (CSF) is often abnormal in a nonspecific way, with pleocytosis, increased protein, an elevated γ-globulin, and a 'paretic' colloidal gold curve. *Progressive multifocal leukoencephalopathy,* a form of demyelinating disease of the central white matter usually seen in immunosuppressed patients, is probably due to subacute infection with a polyomalike virus of the papova group. *Carcinomatous myasthenia* (Eaton–Lambert syndrome) mimics myasthenia gravis in several respects. However, serum 'muscle' enzymes are normal, and electromyography reveals a block in neuromuscular transmission at rest and an increase in response with continued electrical stimulation; in true myasthenia gravis the response decreases.

Nonmetastatic neurologic syndromes may occur as a consequence of hypoxia, hypoglycemia, hyperlacticacidemia, hepatic coma, hypercalcemia, hyponatremia, the hyperviscosity syndrome and uremia, in addition to cerebral vascular accidents due to thrombosis or hemorrhage, uncommon central nervous system infections in immunosuppressed patients, and damage from radiation or chemotherapy. The clinical chemist may be called on to provide supporting evidence for one or more of these complications.

The concentration of *protein in the CSF* of patients with brain tumors is usually increased, exceeding 1 g/L in about one third of the patients. In the presence of diffuse neoplastic involvement of the meninges, *CSF glucose* may be less than 2.2 mmol/L (40 mg/dL).

Since acute or chronic bacterial or fungal *infections* of the central nervous system (CNS) also increase CSF protein and reduce glucose, the fluid should be subjected to microscopic examination and cultured for bacterial, fungal, or other pathogens. In immunosuppressed patients, infections due to *Toxoplasma, Cryptococcus neoformans, Candida,* or *Aspergillus,* or disseminated viral diseases, are more common. CSF interferon could be determined in suspected virus disease; in demyelinating illness, serum cholinesterase can increase and so may the enzyme in the CSF. In some instances, it will be diagnostically useful to obtain CSF for enzyme or immunologic studies such as LDH, hCG, or carcinoembryonic antigen (CEA) assays, which may correlate with malignancy within the cranial cavity.

Dermatologic Syndromes

Certain dermatoses can alert the clinician to the presence of an underlying malignancy. Malignant *acanthosis nigricans* is characterized by darkly pigmented hyperkeratotic acneform lesions, especially involving the body folds, and is often associated with cancer of the gastrointestinal tract or lung. Secretion of an MSH-like hormone probably contributes to its development. A sudden increase in facial or body *hair* ('malignant down'), or the appearance of *pachydermoperiostosis* and osteoarthropathy, suggest malignant production of hormonally active materials. Episodes of cutaneous *flushing,* associated with carcinoid tumors, and of migratory *thrombophlebitis* in pancreatic or other neoplasias may precede diagnosis of these cancers.

A variety of dermatoses including ichthyosis, exfoliative dermatitis, pemphigoid and psoriatic lesions, toxic erythemas, vesicular and bullous eruptions, and Raynaud's phenomenon occur with different malignancies, and while their pathogenesis usually is not understood, they may include immunologic and neuroendocrine mechanisms and responses to vasoactive peptides. Cryoglobulinemia and macroglobulinemia can result in cold urticaria, cyanosis, and gangrene caused by vascular occlusion.

Porphyria cutanea tarda, with small blisters on areas of the skin exposed to the sun, has been reported in patients with hepatoma, and excretion of uro- and coproporphyrins may be increased. *Pruritus* occurs in lymphoma and other malignancies but the pathogenesis, including local infiltration of tumor cells, is different. Increased tissue bile acids in obstructive jaundice due to cancer may also cause pruritus.

CANCER ASSOCIATED WITH FAMILIAL AND CONGENITAL DISORDERS

Cancer occurs in several familial and congenital disorders more often than in the general population. *Examples* include:

1. Chromosomal disorders such as Down's syndrome (C21 trisomy), with a 20-fold greater risk of leukemia
2. Mendelian recessive disorders such as Bruton type agammaglobulinemia, with increased leukemia and lymphoreticular cancer
3. Familial tumor syndromes, mendelian dominant in some pedigrees, such as MEA I, MEA II, and familial polyposis of the colon, with benign and malignant colon tumors

The *diagnosis* of some of these familial disorders is not difficult once the cancer is far advanced, or if the family history is known. Unless the latter is obvious, many are not detected until the former occurs. No means of preventing them are known, other than removing an organ at risk, when that is appropriate. Patients with certain hereditary forms of intestinal multiple polyposis, who have a high probability of malignant change, are being treated with retinoids. The hope is that this differentiation-promoting factor will reduce the incidence of polyps and hence cancer—a form of *biologic response modifier.* The existence of so many neoplastic disorders to which clinical biochemistry makes only a limited contribution at present, should provide continuing incentive for the future development of the discipline.

Histologically Indeterminate Cancers and the Unknown Primary

The solid cancers of from 3% to 5% of all patients and from 5% to 10% of medical oncology patients seen at a typical referral center are *histologically undifferentiated;* about half of them are never identified. A metastatic deposit that cannot be diagnosed is frequently the first evidence of the occult primary cancer. If the ontogeny of cancer cells significantly recapitulates phylogeny, it can be expected that many proteins expressed by histologically ambiguous cancer cells would exhibit homology with the protein profiles of identified cancer cells, or even of normal cells originating from *cognate 'stem' cells.*

Two-dimensional protein electrophoresis, in which proteins are first separated according to their isoelectric points, and subsequently in a second dimension by SDS polyacrylamide gel electrophoresis according to their relative mass, permits the analysis of a complex mixture of proteins. For example, when the distribution of extracted proteins from human liver, colon, and prostate cancer are compared, the patterns are so dissimilar that they can easily be distinguished. Most of these proteins are present in the normal tissue of origin. Although only 5% to 10% of the total complement of proteins is believed to be detected by staining with Coomassie Blue, their distribution among different organs is distinctive. *Digitization* of such patterns and their *computer-assisted analysis* may lead to the development of 'libraries' of representative cancer cell patterns, with which histologically ambiguous cancer patterns can be compared. A similarity of protein patterns should denote a comparable stem cell origin.

Use of monoclonal antibodies against a variety of *intermediate filaments* may provide an alternative means of identifying histologically undifferentiated cancers. Fluorescent ligand-labeled antibodies to vimetin (mesenchymal cells), desmin (muscle cells), cytokeratins (epithelial cells), and neurofilaments (neural cells) are available for use in immunofluorescent identification of these cells. One caveat is evident; for this procedure to be useful, the synthesis of these or other 'marker' proteins must remain constitutive in all poorly differentiated histologically ambiguous cancers.

BIOCHEMICAL TEST SELECTION, MONITORING OF CANCER THERAPY, AND DIAGNOSIS OF COMPLICATIONS

Biochemical markers such as ACP, hCG or other hormonally active agents, CEA, AFP, myeloma proteins, and melanin provide a means of assessing the extent and activity of those cancers synthesizing them. But clones of tumor cells lacking the particular synthetic ability may evolve spontaneously, or at times be promoted in response to therapy. The measurement of urinary cyclic AMP, polyamines, and nucleosides may be combined with immunodiagnostic procedures to provide diagnostic *profiles* that can be expected to assume increasing importance as a means of determining the presence and estimating the extent of a cancer. Certainly more sensitive, selective, and less cumbersome approaches to earlier detection and identification of cancer will be needed.

The development of reliable *assays* (in the past involving the labeling of recently synthesized DNA with ^{3}H-thymidine) for estimating the response of a patient's tumor to different combinations and scheduling of ***chemotherapeutic agents*** may help to individualize therapy for each patient. *Tumor-cell cloning assays,* somewhat similar to bacterial antibiotic sensitivity testing, have been used to assess the sensitivity of a patient's cancer to various drugs. Cancer cells are dispersed into solid agar (which limits the growth of nonmalignant cells) in the temporary presence of various chemotherapeutic agents, and the number of resulting colonies compared. These studies have been fraught with difficulties, such as inadequate growth of cancer cells, and seem useful chiefly in identifying agents to which the cells are insensitive. However, discovery of hitherto unknown 'growth factors' or other modifications of culture conditions might change this assessment. An expanding role can be envisioned for biochemical procedures in the more rational selection of therapy and prediction of response.

Clinical chemists will be called on to monitor the serum levels of certain chemotherapeutic agents. During high-dose *methotrexate* treatment, when a patient must be 'rescued' from the potentially lethal effects of the drug by its 'antidote,' citrovorum factor, it is essential to follow the concentration of methotrexate in the patient's serum by RIA or chromatography. Serum concentrations of the cardiotoxic chemotherapeutic drug *adriamycin* can also be measured by RIA.

The efficacy of chemotherapeutic drugs and of radiation depends on the differential responses of cancer cells and normal cells to their cytotoxic effects (selective toxicity). Tissues with a high rate of replication, including bone marrow and lymphoid, gastrointestinal, reproductive, and embryonic tissues, are susceptible to damage and this limits the use of either form of therapy. Less constant *acute* effects occur in other organ systems, as for example:

1. Vincristine-induced peripheral neuropathy or inappropriate ADH syndrome
2. Cytoxan-induced hemorrhagic cystitis
3. Acute hypercalcemia, hyperuricemia, and hyperkalemia in treated patients with breast cancer, leukemias, and lymphomas, respectively

Chronic complications, some of which may become evident long after therapeutic intervention, include:

1. The cardiotoxicity of adriamycin
2. Pulmonary fibrosis associated with the use of bleomycin or methotrexate
3. Temporary or permanent sterility
4. Teratogenic effects
5. The occurrence of second malignancies such as
   a. blood dyscrasias in patients treated with various alkylating agents
   b. bladder cancer in patients receiving cytoxan for protracted periods of time

SUMMARY AND FUTURE TRENDS IN ONCOLOGIC CLINICAL BIOCHEMISTRY

There is evidence that from 10^3 to 10^5 cancer cells can, under favorable circumstances, be suppressed by host defenses. An ideal diagnostic marker would be detectable at concentrations approaching the amount synthesized by this number of cells, and released into the circulation in amounts proportional to the number of cancer cells present in the patient. Currently, very few tests approach this ideal. The quantification of multiple myeloma M protein in the serum has been used to estimate the total body burden of myeloma cells. Development of an RIA for the β subunit of hCG, to which a highly selective antibody can be induced, achieves a sensitivity capable of detecting nanogram to picogram quantities. Release, from a cancer, of a substance (e.g., a hormone) with measurable effects on the patient may amplify the ability to detect it and may alert the clinician to its presence.

Despite the sensitivity of enzymatic reactions, the detection of oncofetal or other ***enzymes*** by classic methods has not led to definitive diagnostic tests for cancer that are capable of identifying clinically occult disease. The release of prostate-specific ACP or prostatic antigen, by metastatic carcinoma of the prostate, approaches the ideal of a tissue-specific marker which, if measured by RIA or ELISA, increases the ability to diagnose clinically occult prostatic carcinoma. Immunologically active, enzymatically inactive forms of ACP are elaborated by some prostatic cancers which, if detectable, might also serve as tumor markers. Indeed, it seems likely that cancers synthesize a variety of abnormal proteins, including nonfunctional *precursors* of hormones and enzymes, often without producing a characteristic clinical syndrome, although many of them are immunologically active.

Cancer-associated and a lesser number of putative cancer-specific ***antigens*** have been detected in virtually every common type of human cancer. The extent to which synthesis of a particular cancer antigen (against which a monoclonal antibody has been developed) is constitutive, no matter how undifferentiated the cancer, may depend on the antigen studied. Since there are major differences in the structure of cancer cell plasma membranes, tumor-associated oncofetal membrane components, whether protein, glycoprotein, carbohydrate, or lipoprotein, may provide a source of material released into the circulation, and may be capable of being detected by a monoclonal antibody at an early stage of the disease, in amounts proportional to the body burden of cancer cells. The identification of such materials in the serum, possibly in some instances by measurement of the patient's response to them, is an important area for the future development of oncologic biochemistry. Putting emphasis on the surface properties of cancer cells may circumvent difficulties that may or may not be inherent in using intracellular macromolecules as tumor markers. Two-dimensional electrophoresis of proteins in the serum (including acute phase reactants, such as C-reactive protein) or in extracts from tumor tissue, capable of resolving several hundred individual proteins, could provide cancer-specific patterns that will facilitate the detection and diagnosis of various malignancies. If the degree of cellular differentiation can be assessed from the cancer protein profile, inferences concerning prognosis should follow.

The *measurement* of urinary polyamines, transfer RNA nucleosides, serum β-chain hCG and CEA, have been used to follow the course of human mammary cancer. Urinary putrescine, spermidine, spermine, and cadaverine were measured with an amino acid analyzer; pseudouridine N^2,N^2-dimethylguanosine (M_2^2G), and methylinosine, by gas chromatography. All but 1 of 69 patients with metastatic breast cancer had increased values of one or more of the different substances. CEA, βhCG, and M_2^2G, either singly or in combination, were increased in 94% of the patients. In 16%, M_2^2G was the only elevated marker. CEA, βhCG and M_2^2G, with or without determination of polyamines, were also used to compare pre- and postoperative patients and those with metastases. Greater discrimination occurred with all four tests and abnormal results were obtained in 69%, 58%, and 99%, respectively, of these patients. CEA and hCG tended to parallel the tumor burden and could provide premonitory information about clinical recurrences. However, use of such a disparate group of tests involving different methodologies is cumbersome and expensive.

High-resolution gas chromatographic-mass spectroscopic analysis of body fluids for virtually any soluble small-relative-mass component, including amino acids, carbohydrates, lipids, nucleic acid, and trace elements, may generate disease-associated *diagnostic profiles* reflecting metabolites characteristic of underlying cancer or other disorders. Hundreds of constituents have been detected and partially quantified in urine from normal individuals. The determination of complex patterns in cancer patients and the application of computers for their analysis holds promise of being useful. The development of panels of reagents for organ-specific detection of multiple oncofetal proteins is another possibility. A role for nuclear magnetic resonance spectroscopy, capable of measuring fluxes in tissue water and phosphorus content, may prove to be useful in locating the primary or metastatic cancer, or measuring the response of a solid cancer to therapy. In time, it should become clear whether any approach alone, or some combination of them, will facilitate the subclinical diagnosis and subsequent monitoring of human cancer.

Particles with features of RNA tumor ***viruses*** containing reverse transcriptase activity have been reported in human T-cell lymphomas. The exploitation of antigenic relationships between human cancer-associated proteins, proto-oncogene products and oncogenic viruses, may facilitate the production of antisera with diagnostic or therapeutic roles against the former. The malfunction of cellular genes that exhibit homology with genes of certain oncogenic viruses may represent a *sine que non* of malignant transformation; if so, the implications for clinical oncology should be profound.

It is our belief that the really significant advances in oncologic biochemistry are still to be made, and that they will depend on the application of fundamental discoveries in the basic biologic and physical sciences in ways that cannot yet be foreseen. It could be anticipated that much of currently accepted clinical oncologic biochemistry will eventually be rendered obsolete by the application of newer knowledge concerning monoclonal antibodies, recombinant DNA techniques, the evident ubiquity of proto-oncogenes in all cells, and their apparent malexpression in specific human cancers.

SUGGESTED READING

BECKER FF (ed): Cancer, a Comprehensive Treatise 2nd ed. New York, Plenum Press, Vol I, 1982

CARTER SK, GLATSTEIN E, LIVINGSTON RB: Principles of Cancer Treatment. New York, McGraw-Hill, 1982

DEVITA V, HELLMAN S, ROSENBERG SA: Cancer Principles and Practice of Oncology. Philadelphia, JB Lippincott, 1982

HOLLAND J, FREI E (eds): Cancer Medicine, 2nd ed. Philadelphia, Lea & Febiger, 1982

MIHICH E: Immunologic Approaches to Cancer Therapeutics. New York, Wiley, 1982

MITCHELL MS, OTTIGEN HF: Hybridomas in Cancer Diagnosis and Treatment, Vol 21 of Progress in Cancer Research Therapy. New York, Raven Press, 1981

O'CONNOR TE, RANSHER FJ: Oncogenes and Retroviruses: Evaluation of Basic Findings and Clinical Potential. New York, AR Liss, 1983

ROBBERSON DL, SAUNDERS GF: Perspectives on Genes and the Molecular Biology of Cancer. New York, Raven Press, 1982

SELL S: Cancer Markers. Diagnostic and Developmental Significance. Clifton, NY, Humana Press, 1980

SPORN MB, ROBERTS AB: Autocrine growth factors and cancer. Nature 313:745–747, 1985

SYMINGTON T, CARTER RL (eds): Scientific Foundations of Oncology. Chicago, Year Book Medical, 1976

WATERS H: Handbook of Cancer Immunology, Vol 1–9. New York, Garland, STM Press, 1978–1980

Yearly monographs, such as Advances in Clinical Chemistry and Advances in Cancer Research, are valuable sources of information on a variety of specific topics.

25

Therapeutic Drug Monitoring

Steven J. Soldin

Physicians have a responsibility to be adequately informed about the risks, limitations, and use of the drugs they prescribe and to acquaint their patients with possible adverse effects. Some doctors may never use a particular drug, or prescribe only small doses, for fear of inducing disease. Others prescribe drugs with insufficient attention to possible consequences. Improvement in the quality of medical care is dependent on physician involvement in drug monitoring in order to establish appropriate prescribing practices.

The risk of using a drug is not always clearly established. When the incidence of serious *drug-induced* disease is only one in every 20,000 or more patients, as for example, the devastating aplastic anemia caused by chloramphenicol, the recognition of a causal relationship is sometimes difficult and delayed. Methods for the detection and investigation of adverse drug reactions are required, as are studies of the pathogenesis of drug-induced diseases.

THEORETICAL CONSIDERATIONS FOR A THERAPEUTIC DRUG MONITORING SERVICE

Therapeutic drug monitoring (TDM) is based on the assumption that there is a better correlation between the plasma drug *concentration* and the pharmacologic effect than between the drug *dosage* and the pharmacologic effect. A typical therapeutic decision process is the following:

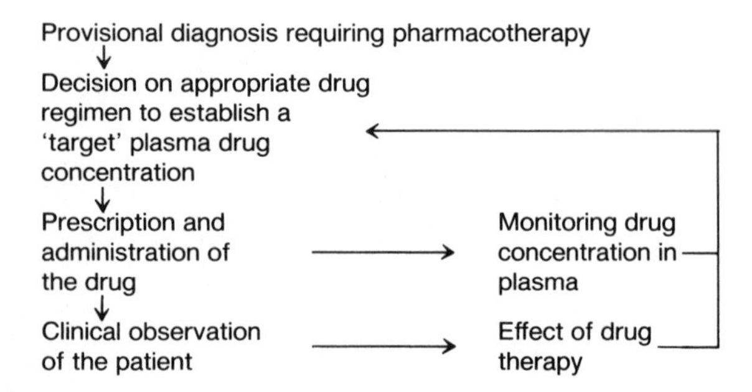

Thus, the plasma drug concentration *and* clinical observation are used together to optimize therapy. But there can be many pitfalls such as:

a. Physician error in diagnosis, drug choice, and dosage
b. Errors in drug dispensing and in compliance by the staff or the patient
c. Rate and extent of absorption, distribution, biotransformation, and excretion of the drug
d. Presence of liver or kidney disease
e. Effect of plasma proteins on binding and tissue uptake of the drug (and its metabolites)
f. The 'free' drug concentration and the number of 'receptors' in target tissues
g. Possibility of interaction when more than one drug is given

Adults, Infants, and Children

Although the aforementioned problems pertain to both adult and pediatric populations, important differences exist between the two groups:

1. Absorption is altered for many drugs in the neonatal period because of differences in gastric pH and emptying time.
2. There are differences in the apparent volume of distribution because of differences in body composition (the neonate has proportionately less body fat and more water).
3. The *clearance* of drugs is often low in neonates and premature infants (due to immature hepatic and renal function), and the biotransformation of many drugs is slow (due to immaturity of the hepatic microsomal enzyme system), however
4. The activity of this latter system is often greater in *children* than in adults, requiring a higher mg/kg dosage to achieve a comparable steady-state concentration.

Rationale for Therapeutic Drug Monitoring

The pharmacologic response to a drug is a result of the interaction of the drug with tissue receptors. The intensity and duration of the response for most drugs is proportional to the drug concentration at the receptor site. This in turn depends on several factors, including drug dose and the pharmacokinetic properties of the drug administered. Factors affecting the pharmacokinetics of drugs include genetic differences in drug metabolism, disease, age, drug interactions, diet, and life style (alcohol, nicotine, and caffeine). For many drugs it is therefore *impossible to predict* a serum concentration for any given mg/kg dose.

For most drugs there is a plasma concentration below which the clinical response is unsatisfactory (subtherapeutic). The drug elicits a therapeutic effect at higher concentrations, while at still higher concentrations unwanted toxic side-effects occur. The aim of drug dosage design is to maintain the plasma concentration in the *therapeutic range.* For the drug *theophylline* this is between 55 and 110 μmol/L (10 and 20 mg/L); <55 μmol/L is probably subtherapeutic, >110 μmol/L is potentially toxic.

For therapeutic monitoring to be effective in the clinical management of patients, it is necessary that the drugs of interest fulfil most of the following:

1. A reliable and rapid method for drug analysis must be available.
2. The relationship between serum concentration and pharmacologic effect must be close. This usually implies a strong correlation between serum concentration and concentration in the target tissue (Fig. 25–1).
3. The range between serum concentrations that provide therapeutic effects and those that cause toxic effects should be relatively narrow (e.g., digoxin, theophylline, aminoglycoside, and antibiotics).
4. The correlation between serum concentration and drug dosage, caused by interindividual differences in drug absorption, metabolism, and excretion should be poor (as shown for clomipramine in Fig. 25–2).
5. Pharmacologic effects of the drug are not readily measurable, for example, anticonvulsants, in which the suppression of seizure activity is difficult to monitor.

Factors Affecting the Serum Concentration of Drugs

Drug Formulation. The bioavailability of drugs can vary markedly with drug formulation. Any change in the brand of a product should therefore be approached cautiously. The clinical status of the patient should be followed closely at such times, and drug concentrations should be measured before and after any adjustments in drug regimen.

FIGURE 25–1. Relationship between serum digoxin concentrations and postmortem (6) or surgical (2) myocardium values in infants and young children.

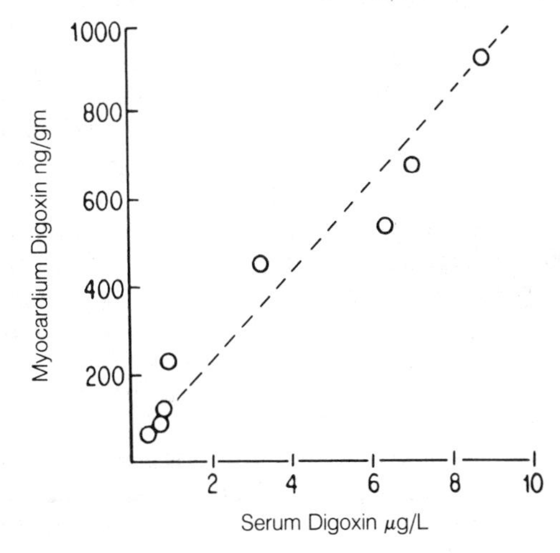

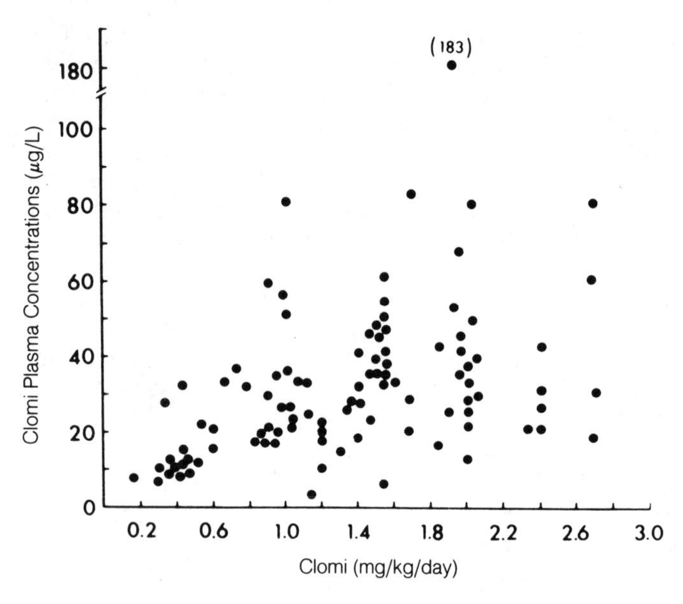

FIGURE 25–2. 'Relationship' between oral daily doses of clomipramine (Clomi) and plasma drug concentrations *12 h* after taking the drug.

Time of Sampling. The interpretation of laboratory data depends not only on drug dosage, but also on the time of the last dose relative to the time of blood sampling. For a drug administered orally at intervals equal to its half-life (e.g., 4 hours), it takes approximately 4 to 5 times the half-life to achieve steady-state plateau concentrations (Fig. 25–3).

Samples for analysis should not be drawn until sufficient time has elapsed to enable a steady-state concentration to be achieved (unless, of course, toxicity is suspected at an earlier stage). For drugs with a long half-life (such as phenobarbital) administered at intervals shorter than the half-life, there is little difference between the steady stage 'peak' and 'trough' drug concentrations. However, for drugs with a short half-life (such as the aminoglycosides, theophylline, and primidone), differences between peak and trough concentrations may be considerable, and it is often advisable to measure both. As a general rule, however, the ideal sample is that which would provide the *steady-state trough* serum concentration. This is the sample drawn just prior to the next dose.

Steady-state drug concentrations can be achieved rapidly by administering oral, intramuscular, or intravenous *loading* doses of a drug. Loading doses circumvent the necessity of waiting five half-lives to achieve a maximum therapeutic effect. For a drug administered intravenously (e.g., digoxin), it is necessary to wait a fixed time interval to allow for drug distribution to occur. In the special case of digoxin this is approximately 6 hours (Fig. 25–4).

In general, *intravenous* medications should be sampled 30 to 60 min after administration. For example, a sample representing the 'peak' serum concentration for gentamicin is most appropriately drawn after distribution is complete (30 minutes after the infusion of the drug ends).

Absorption. Drug absorption is affected by numerous factors including route of administration, drug formulation, age of recipient, and concomitant administration of other drugs or food. Administration by the intravenous route is rapid and 'absorption' is complete. In contrast, absorption (e.g., of phenytoin or diazepam) after intramuscular administration is slower, less complete, and therefore to be avoided when possible. Most drugs are administered orally, and factors that influence the amount of drug absorbed from the gastrointestinal tract (the bioavailability of a drug) include drug formulation, drug solubility and pK, concomitant administration of other drugs, and simultaneous ingestion of food.

Most drugs are absorbed from the gastrointestinal tract by a process of passive diffusion, pri-

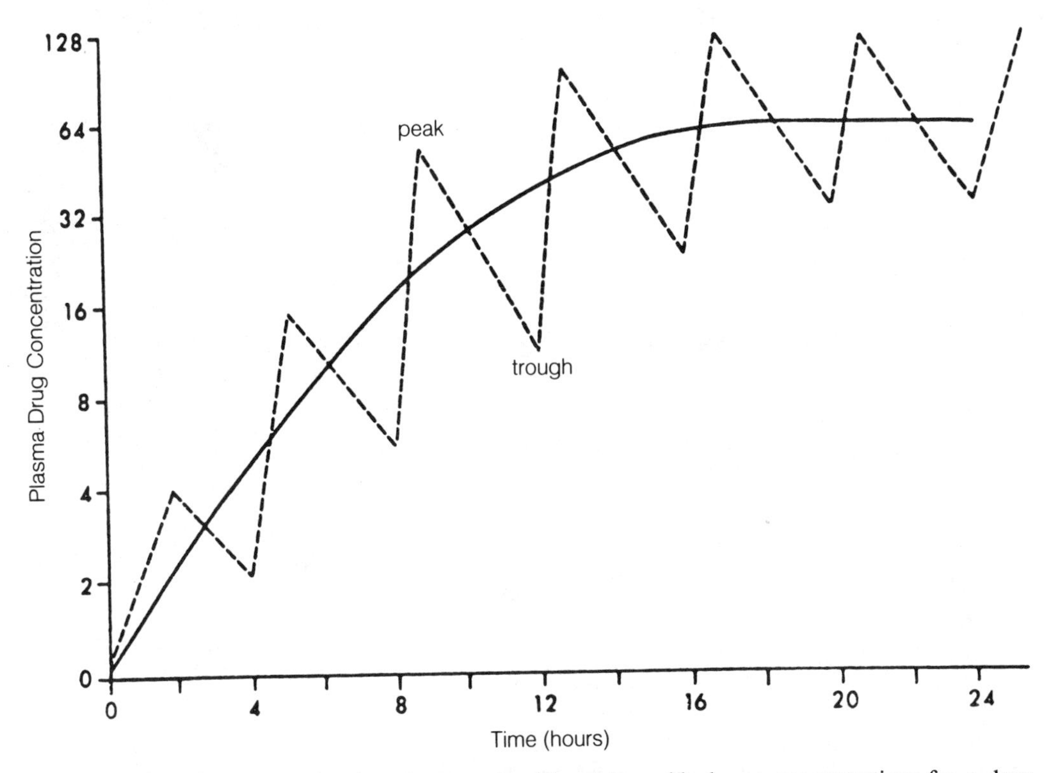

FIGURE 25–3. Fluctuations between 'peak' and 'trough' plasma concentrations for a drug administered orally at intervals equal to the drug half-life (in this case 4 h).

marily in the small intestine. Variables include gastric emptying time, which is considerably prolonged in the neonate and approaches adult values only after 6 months of age. Food or any other factor that delays gastric emptying will delay drug absorption. Gastric pH affects the state of ionization of some drugs and hence their absorption across lipid membranes. Erythromycin and ampicillin are acid-labile and delayed retention in the stomach results in less being available for absorption. Gastric pH is close to neutral at birth and remains neutral for 1 to 2 weeks. Adult values are reached only after 2 years of age.

Drug Metabolism. Drug metabolism is influenced by genetic and dietary factors and also by age and the activity of drug metabolizing enzymes. In addition, hepatic, renal, and cardiac function can markedly affect drug metabolism and lead to serious problems if dosage regimens are not tailored accordingly. Genetic factors exert a significant control over rates of drug metabolism; for example, identical twins have very similar plasma half-lives for antipyrine.

Isoniazid and procainamide are metabolized by *acetylation* in the liver. The capacity for rapid acetylation occurs in families as a Mendelian dominant. Individuals lacking the dominant gene for this acetylase are 'slow acetylators' (a trait found in approximately half the population of North America). Plasma concentrations of these drugs will be higher because of this reduced capacity and will remain elevated longer than in the fast-acetylator group.

Humans are exposed to many chemicals, some of which alter the rates of drug biotransformation by enhancing the activity of the hepatic *microsomal enzyme system.* The half-life of theophylline in smokers, for example, is considerably shorter than in non-smokers. The effect of feeding a charcoal-broiled beef diet to patients on long-term theophylline therapy was to decrease theophylline

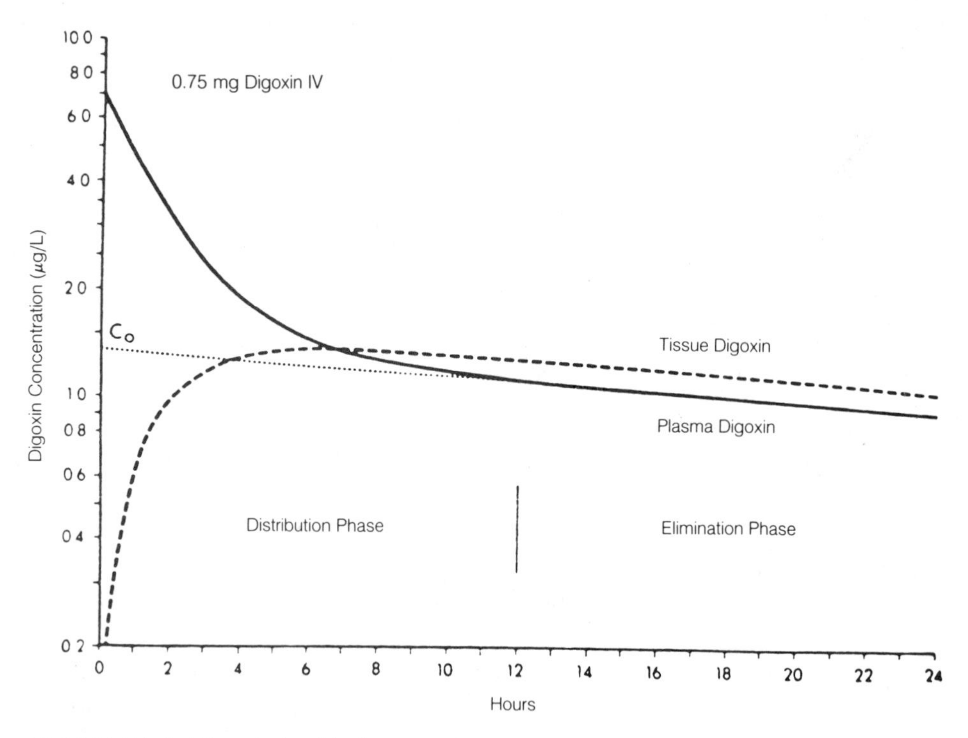

FIGURE 25–4. The relationship between tissue and plasma digoxin concentrations after administration of an IV dose of the drug. C_0 = drug concentration at zero time.

half-life. The ratio of protein to carbohydrate in the diet can also affect the rate of drug metabolism. The half-life of both antipyrine and theophylline, for example, is markedly reduced when the diet is changed from a low protein/high carbohydrate to a high protein/low carbohydrate content.

The hepatic microsomal enzymes are active at birth although their titers are considerably reduced in comparison to adult values. Activity increases with advancing postnatal age, for example, the theophylline half-life in premature infants has been quoted as 14.4 to 57.7 hours, in children 1 to 4 years of age as 1.9 to 5.5 hours and in adults as 3.0 to 9.5 hours.

Clearly, the effects of *disease* on drug elimination are also important. Clearance (C) can be regarded as a composite of renal clearance (C·R) and hepatic clearance (C·H) so that C = C·R + C·H. In patients with impaired renal function C·R is decreased, whereas in patients with advanced liver disease, C·H may be reduced. Reduction of either C·R or C·H will increase the elimination half-life according to the equation:

$$t_{1/2} = \frac{0.69\ \text{Vd}}{\text{C}\cdot\text{R} + \text{C}\cdot\text{H}}$$

where Vd = apparent volume of distribution. Renal and hepatic blood flow is a major determinant of drug clearances and significantly impaired cardiac output will, therefore, affect drug elimination.

The extent of *protein binding* can also affect drug elimination. It is the 'free' drug that is pharmacologically active. In disease states characterized by hypoalbuminemia (e.g., hepatic or renal failure, nephrotic syndrome, and protein-losing enteropathy), the concentration of the free, active drug will be higher at any given total drug concentration. This can give rise to toxicity in patients who have a total serum concentration within the therapeutic range. Enhanced protein binding slows the elimination of drugs that are removed from the serum by glomerular filtration or uptake by the liver, thus increasing the duration of action of such drugs.

In addition to albumin, other blood constituents such as red blood cells and α_1-acid glycoprotein (α_1AGP, orosomucoid) are capable of binding drugs. The concentration of α_1AGP in plasma, a protein that binds many basic drugs, increases with infectious, inflammatory, and malignant diseases and after surgery. Clearly the binding of drugs such as propranolol and chlorpromazine to α_1AGP is dependent on the concentration of this protein in serum. The binding of lidocaine and disopyramide and the concentration of α_1AGP, increase significantly in patients with myocardial infarction. Serum α_1AGP is low in the neonate and the binding of lidocaine and propranolol is reduced in cord serum as compared to adult controls.

The routine measurement of 'free' drug concentrations would be ideal, but is still impractical. Equilibrium dialysis is time-consuming and the various membranes available on the market (e.g., the Millipore ultra-free membrane system), that allow the free drug concentration to be measured in a protein-free ultrafiltrate, are costly. Both procedures require large sample volumes, which is always a problem in a pediatric center. Nevertheless, measurement of the 'free' drug should allow a more meaningful evaluation of dosage requirements and will eventually probably replace the present practice of correlating the 'total' serum drug concentration with the clinical effect.

Knowledge of the 'free' drug concentration is important especially in those instances in which the drug has a high association constant (Ka) for plasma proteins. For many drugs with a high Ka value (phenytoin, primidone, ethosuximide, and carbamazepine), the concentration in *saliva* is equal to the 'free' serum drug concentration. This finding has led to the suggestion that in many instances saliva should be substituted for the plasma sample. A further advantage is that saliva can be collected by noninvasive techniques.

Patient Compliance. Noncompliance is failure to follow a prescribed medication regimen and can occur in different ways and to varying degrees. When noncompliance occurs the effectiveness of therapy may be misjudged. Unsuspected noncompliance can lead to unnecessary tests, additional medications, and even to hospital admissions. Examples as high as 82% noncompliance have been reported with short-term penicillin therapy. In pediatric populations noncompliance is often the result of parental neglect, or unwillingness to follow the prescribed drug regimen.

PRACTICAL CONSIDERATIONS FOR A TDM SERVICE

Prelaboratory Considerations

Optimal *timing* of the sample is imperative. A service is therefore recommended that includes several nurses whose sole task is to ensure that blood samples are drawn at appropriate time intervals relative to the drug dose. A special TDM *requisition* is also necessary, listing the patient's age, sex, weight, height, dose, time of last dose, time of sampling, clinical status (especially with regard to renal, hepatic, and cardiac function) and a list of other medications received by the patient. Fig. 25–5 and Table 25–1 show the front and the back of the requisition employed currently at the Hospital for Sick Children. A means to convey the sample rapidly to the laboratory for analysis is also required.

Laboratory Considerations

The analytical procedure chosen should be accurate and precise, rapid, easy to perform, readily automated and inexpensive (see also Chap. 5). No single technique meets all these requirements all of the time. For example, although HPLC and GLC would provide for probably the most accurate and precise analysis of many drugs, they require specialized equipment and trained personnel and cannot even today be regarded as 'easy to perform.' With a high laboratory workload the reagent cost per analysis with these techniques is lower by an order of magnitude than with the alternative methodologies (enzyme immunoassay; fluorescence immunoassay; fluorescence polarization immunoassay; radioimmunoassay) available. Furthermore, GLC and HPLC techniques often per-

THE HOSPITAL FOR SICK CHILDREN
Department of Clinical Biochemistry

THERAPEUTIC DRUG MONITORING REQUISITION

STAT ☐ INITIAL REQUEST ☐ REPEAT ☐ (within 7 days)

IF OUTPATIENT SEND RESULT TO

ROOM................ LOCAL.................

DATE

PATIENT NAME

BIRTH DATE DAY MO YR SEX HISTORY NO

PARENT NAME

ADDRESS

IMPRINT OR ENTER ABOVE DETAILS BY HAND

FOR LABORATORY USE ONLY

VEN ☐ CAP ☐ LAB NO

THE ANALYSIS WILL NOT BE PERFORMED UNLESS THIS INFORMATION IS PROVIDED.

TYPE OF SPECIMEN.................................. (if other than blood) TIME OF COLLECTION hrs DATE OF COLLECTION Yr Mo DAY

PATIENT'S WEIGHT kg

	DRUG A	DRUG B
DRUG ANALYSIS REQUESTED	....	....
FOR THEOPHYLLINE, SPECIFY DRUG PREPARATION USED	....	....
TIME OF LAST DOSE (Prior to collection)	hrs	hrs
LAST DOSE	mg μg	mg μg
THERAPY WITH ABOVE DRUG COMMENCED ON	....	....
DOSING SCHEDULE	q4h ☐ q6h ☐ q8h ☐ q12h ☐ continuous infusion ☐ other	q4h ☐ q6h ☐ q8h ☐ q12h ☐ continuous infusion ☐ other ....
ROUTE OF ADMINISTRATION	p.o. ☐ s.c. ☐ i.m. ☐ i.v. ☐ START hrs FINISH hrs	p.o. ☐ s.c. ☐ i.m. ☐ i.v. ☐ START hrs FINISH hrs

TO BE COMPLETED BY MEDICAL STAFF/RESIDENTS

REASON FOR REQUEST QUERY TOXICITY ☐ QUERY SUBTHERAPEUTIC ☐
QUERY COMPLIANCE ☐ CHECK DOSING INTERVAL ☐

CONCURRENT MEDICATIONS RECEIVED

Cimetidine ☐	Quinidine ☐	Erythromycin ☐	Ticarcillin ☐	Chloramphenicol ☐
Phenobarbital ☐	Phenytoin ☐	Carbamazepine ☐	Valproic Acid ☐	Corticosteroids ☐

RESEARCH FUND NUMBER (if applicable) NAME OF PHYSICIAN .... (Please print)

IN EMERGENCY CONTACT CLINICAL CHEMIST ON CALL

13842 REV 02 83

FIGURE 25–5. Requisition currently employed in the therapeutic drug monitoring program at the Hospital for Sick Children, Toronto.

mit the simultaneous analysis of several drugs or drug metabolites in a single sample—an advantage not offered by any of the immunoassay techniques. In contrast, analyses performed with the enzyme-multiplied-immunoassay system (EMIT, Syva) or a fluorescence polarization immunoassay procedure (TDX, Abbott) are easier, and provide adequate specificity, sensitivity, accuracy, and precision. For *smaller* laboratories, therefore, the increased reagent cost per analysis is offset by the ease of drug quantification and the ability to provide a fairly comprehensive drug analytic service with minimal technical expertise and equipment cost.

TABLE 25-1. General Guidelines for Therapeutic Drug Monitoring

Drug	Optimal Sample Time	Therapeutic (Effective) Range	Interacting Drugs*
Theophylline	Trough prior to next dose Peak 1–2 hours post dose (4 hrs post dose if sustained release product is used)	55–110 μmol/L	Erythromycin Cimetidine
Phenobarbital	Trough prior to next dose	65–130 μmol/L	Other anticonvulsants Chloramphenicol Corticosteroids
Phenytoin	Trough prior to next dose	40–80 μmol/L	Other anticonvulsants Chloramphenicol Corticosteroids
Carbamazepine	Trough prior to next dose	17–50 μmol/L	Other anticonvulsants Chloramphenicol Corticosteroids
Ethosuximide	Trough prior to next dose	280–710 μmol/L	Other anticonvulsants Chloramphenicol Corticosteroids
Valproic acid	Trough prior to next dose	350–700 μmol/L	Other anticonvulsants Chloramphenicol Corticosteroids
Primidone	Trough prior to next dose	23–55 μmol/L	Other anticonvulsants Chloramphenicol Corticosteroids
Gentamicin	Trough (<2 mg/L) prior to next dose Peak (<10 mg/L) 30 minutes after completion of IV	5–10 mg/L	Ticarcillin
Tobramycin	Trough (<2 mg/L) prior to next dose Peak (<10 mg/L) 30 minutes after completion of IV	5–10 mg/L	Ticarcillin
Amikacin	Trough (<10 mg/L) prior to next dose Peak (<35 mg/L) 30 minutes after completion of IV	15–25 mg/L	Ticarcillin
Chloramphenicol	Trough (<15 mg/L) prior to next dose Peak (<30 mg/L) 30 minutes after completion of IV	10–25 mg/L	
Digoxin	Trough prior to next dose	1.0–2.5 nmol/L	Quinidine
Methotrexate (High dose)	24 hr post dose 48 hr post dose	<10 μmol/L <0.9 μmol/L	

* Altering disposition of drug listed.

Note: Some patients may require a more individualized approach.

Postlaboratory Considerations

Once the analysis has been performed it is imperative that the results be conveyed rapidly to the physician who requested the test. Ideally, an interpretative arm of the TDM service should interrelate between the laboratory and the wards to ensure that appropriate adjustments in drug regimen have been made as a result of the analytic service provided. Such a function can probably be carried out in the most cost-effective manner by clinical pharmacists. The official report form should include the drug concentration found, the desired therapeutic concentration range, and recommendations as to how the latter may be achieved. In special instances it may be desirable to carry out pharmacokinetic studies in order to permit dosage adjustment for optimal patient management.

SUGGESTED READING

AVERY GS (ed): Drug Treatment, Principles and Practice of Clinical Pharmacology and Therapeutics, 2nd ed. Sydney and New York, Adis Press, 1980

GOODMAN LS, GILMAN A (eds): The Pharmacological Basis of Therapeutics, 7th ed. New York, MacMillan, 1985

KALANT H, ROSCHLAU WHE, SELLERS EM: Principles of Medical Pharmacology, 4th ed. Toronto, Department of Pharmacology, 1985

MOYER TP, BOECKX RL (eds): Applied Therapeutic Drug Monitoring, Vol I: Fundamentals. Washington, DC, American Association for Clinical Chemistry, 1982, Vol II, 1984

MUNGALL DR (ed): Applied Clinical Pharmacokinetics. New York, Raven Press, 1983

26

Receptors and Disease

Allan G. Gornall

Virtually all neurotransmitters, hormones, and drugs exert their biologic effects by binding to specific cellular recognition sites called 'receptors.' The characteristic response is an alteration of intracellular metabolic events, often involving enzyme activation and/or ion fluxes. An ability to study and assay receptors has led to their assuming a growing importance in our understanding and possible management of disease. The homology that exists between growth factor receptors and certain oncogenic products offers a new insight into the problems of cancer.

Receptors may occur in a cell plasma membrane, or intracellular membrane, in the cytosol, or in the nuclear chromatin. They are a characteristic of the molecular structure of specific proteins, glyco-, lipo-, or glycolipoproteins, that may range in size from M_r 50 to 500 K (kilodaltons). The fact that a protein (even in a cell membrane) binds a particular substance (ligand) does not make it a receptor. Receptors have two characteristic functions in a cell: (a) to recognize a specific external signal, and (b) to relay a message that will elicit an appropriate response by the cell. Thus,

$$\text{Ligand (L)} + \text{Receptor (R)} \rightleftharpoons \text{LR} \dashrightarrow \text{RESPONSE}$$

Criteria for a *true receptor* require the demonstration of: (a) saturability; (b) reversibility; (c) ligand and tissue specificity, and (d) a correlation between binding affinity and physiologic potency. Definitive proof of a receptor requires solubilization, purification, chemical characterization, reconstitution of a functional receptor, and elucidation of the mechanism leading to the biologic response.

Receptor recognition sites exist in large numbers on and in all cells, enabling them to recognize among the myriad of substances in the plasma, specific signals to which they must respond. In the hypothalamus there are receptors not only for messages from higher neural centers but also for feedback control by circulating hormones. The anterior pituitary cells have receptors for hypothalamic peptides and amines, as well as for target gland hormones. Various organs and tissues have receptors that are specific for anterior and posterior pituitary hormones. Appropriate tissues possess specific receptors for the target gland hormones. Chronic degenerative disease in any tissue or organ will result in a depletion of its receptor population. The degree of autonomy of endocrine adenomas and carcinomas may be related to the loss of receptors and therefore of feedback restraint.

An *understanding* of *receptor binding* requires some simple derivations based on the law of mass action.

$$\text{Free ligand [L]} + \text{free receptor [R]} \underset{k_d}{\overset{k_a}{\rightleftharpoons}} \text{ligand occupied receptor [LR]}$$

The rate of association is k_a [L][R]; the rate of dissociation is k_d [LR]. At *equilibrium* the two rates are the same, $k_a[L][R] = k_d[LR]$ and the equilibrium dissociation constant K_D can be expressed as:

$$\frac{k_d}{k_a} = K_D = \frac{[L][R]}{[LR]}$$

When [R] is equal to [LR], that is, at 50% occupancy, $K_D = [L]$; which means that the concentration of free ligand [L] necessary to half-saturate its receptors is equal to K_D. This is a useful working definition of *affinity,* the lower this concentration the higher the affinity.

Typically receptors exhibit high affinity and low capacity for their specific ligand, which means easy saturability, rapid kinetics, and a degree of reversibility. These properties are investigated using ligands tagged with a radionuclide (L*). Binding studies are made with fixed quantities of the receptor preparation, brought to equilibrium with increasing concentrations (at least 0.1 to 10 times the K_D) of ligand. The free L* is then separated from the bound L*R.

The *total number* of receptor sites (R_T) cannot be measured directly, but can be expressed as

L^*R_{max}. Unoccupied sites, $R = L^*R_{max} - L^*R$.

Now if $K_D = \frac{[L^*][R]}{[L^*R]}$ then the concentration of bound/free ligand $\frac{[L^*R]}{[L^*]} = \frac{[R]}{K_D}$, and substituting for R, $\frac{[L^*R]}{[L^*]} = \frac{[L^*R]_{max} - [L^*R]}{K_D}$

$$= -\frac{[L^*R]}{K_D} + \frac{[L^*R]_{max}}{K_D}$$

Because K_D and $[L^*R]_{max}$ are constants, this assumes the form of the algebraic expression $y = mx + b$. There is a straight line relationship between bound/free and bound ligand. Slope m is $-1/K_D$ and $K_D = -1/m$. This is the basis of the *Scatchard plot* (Fig. 26–1*A*). The ratio of bound to free ligand $[L^*R]/[L^*]$ is plotted on the *ordinate* against bound ligand [L*R] on the *abscissa.* The intercept where the extrapolated line cuts the abscissa gives the receptor concentration; numbers may range from <1500 to >50,000 per cell. When the plot is curvilinear with an upward concavity (dashed line), it means at least two different slopes and usually indicates heterogeneity of receptor sites.

The proportion of receptors occupied by a specific ligand is called *Fractional occupancy* (FO) and is measured as LR/LR_{max}.

From the equilibrium equation:

$$K_D = \frac{[L]([LR]_{max} - [LR])}{[LR]}$$ it follows that

$$\frac{LR_{max}}{LR} = \frac{K_D}{L} + 1$$ and therefore

$$\frac{LR}{LR_{max}} = \frac{1}{1 + K_D/L} = \text{F.O.}$$

Nonspecific binding is a problem that can be circumvented by varying the ratio of inert ligand (L) to tagged ligand (L*). When [L*] is high and the receptors are fully occupied, K_D/L^* approaches zero and F.O. $\simeq$ 1.0. When [L] is high relative to [L*], K_D/L^* will be much greater than 1.0 and F.O. $\simeq$ 0. This enables specific binding to be measured in the presence of nonspecific binding (Fig. 26–1*B*).

In competition-binding studies the use of synthetic analogues of ligands as agonists or antagonists has been a powerful tool in defining the properties of receptors.

MEMBRANE RECEPTOR MODELS

Many models of receptor structure have been proposed; four of the best characterized are shown in Figure 26–2. Some models are simple transmembrane proteins and function mainly to effect the internalization of the ligand by endocytosis. In many cases there is a *system complex* consisting of a ligand recognition, or *binding* site, on the outer surface of the cell membrane, an intramembrane *transducer* (a GTP-dependent protein), and an *effector* moiety on the inner surface of the membrane. Best known is the receptor, G-protein, adenylate cyclase system that generates cyclic AMP as a second messenger. The so-called 'Ca^{2+}-mobilizing' receptors probably act through an analogous GTP-dependent protein to activate a phos-

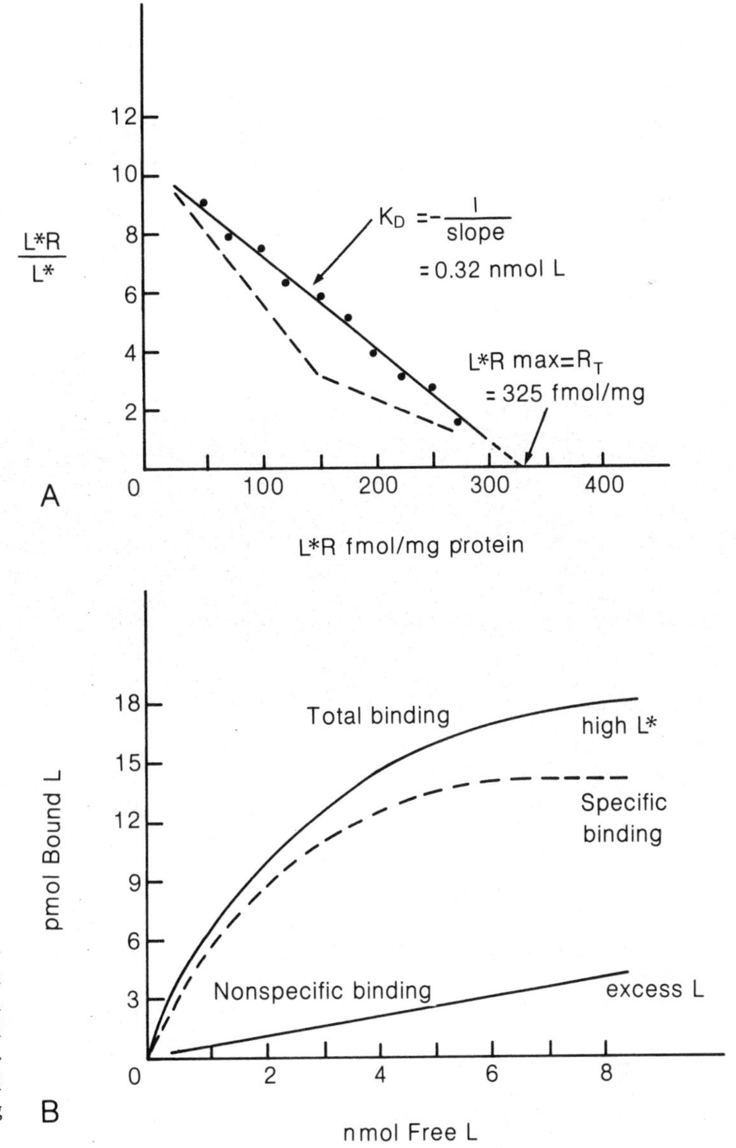

FIGURE 26–1. *A,* A Scatchard plot analysis of receptor affinity (K_D), from which the concentration of receptors (R_T) can be determined. L* = radionuclide labeled ligand. R = receptor. *B,* A graph to illustrate the use of labeled (L*) and unlabeled (L) ligand in measuring specific binding to a receptor.

phodiesterase that hydrolyzes phosphatidylinositol 4,5-bisphosphate (PIP_2) to yield two second messengers, diacylglycerol (DG) and inositol triphosphate (IP_3). DG activates a membrane Ca^{2+}-dependent protein kinase; IP_3 mobilizes intracellular Ca^{2+}. Insulin receptors phosphorylate their own β-subunits and probably other cell proteins, but details are still obscure. The second messenger may be a small peptide.

RECEPTOR METABOLISM AND REGULATION

Receptor protein synthesis is determined by a portion of the nuclear genome and occurs on ribosomes of the endoplasmic reticulum. Plasma *membrane* receptors must be translocated in small vesicles and inserted into the cell membrane. These receptors have a *half-life* measured usually in hours

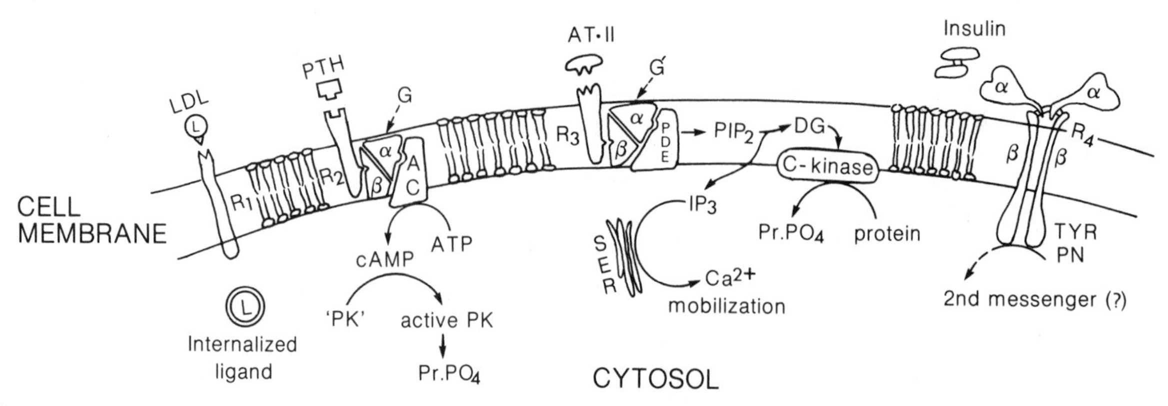

FIGURE 26–2. Diagrammatic illustrations of four known types of receptor. R_1 = receptor concerned primarily with internalization of its ligand. R_2 = receptor complex in which the recognition moiety binds the β-component of the G-protein complex (G), allowing the α-component to activate adenylate cyclase (AC). R_3 = receptor complex in which a similar, or different, G-protein complex (G′) is involved in the activation of phosphodiesterase (PDE). R_4 = present understanding of the insulin receptor complex. LDL = low-density lipoprotein; PTH = parathormone; AT-II = angiotensin II; PIP_2 = phosphatidylinositol diphosphate; DG = diacylglycerol; IP_3 = inositol triphosphate; Pr = protein; PK = protein kinase; C = calcium dependent; PN = phosphorylation; SER = smooth endoplasmic reticulum.

(10 to 20 h) but *affinity* for their ligand can be altered in minutes by a variety of factors. Some receptors exist in reversible states of high and low affinity that depend on phosphorylation and dephosphorylation of the receptor protein. When ligands bind to their specific *cytosolic* receptor there is a conformational change that allows the complex to translocate and bind to another receptor in the nuclear chromatin.

Receptors that are internalized (usually in vesicles) fuse with lysosomes. The receptor complex is dissociated and its components may be degraded further, or in some cases the receptor may be recycled to the membrane. A reduction in receptor numbers or affinity has the effect of desensitizing the cell to further ligand stimulation. Little is known about the factors that control transcription and synthesis of new receptors.

Receptor affinity (and hence cell sensitivity) for a specific ligand can be affected by several factors, both endocrine and metabolic, including existing states of energy, electrolyte and H^+ ion balance. Commonly, a hormone tends to *downregulate* its receptor population, acting thus to modulate the effect of continuous presence of the hormone. This is why pulsatile release of tropic hormones is important for a sustained effect. A few hormones appear to *upregulate* (increase) the concentration of their receptors. Hormone antagonists may increase the receptor population resulting in a 'hypersensitivity withdrawal syndrome.' Some hormones influence the concentration of receptors for other hormones.

RECEPTORS AND DISEASE

Tissue cell receptors are, with a few exceptions, not accessible to direct assay. They can be evaluated, however, where a biopsy is justified (e.g., a breast or prostate tumor). Congenital receptor defects may be expressed in circulating blood cells as well as in the primary tissue. Cultured skin fibroblasts also lend themselves to study.

A defect or dysfunction of some component of the receptor-transducer-effector complex has now been identified for a significant number of diseases. The better known examples will be described briefly, based on the nature of the underlying defect.

ABNORMAL NUMBERS OR AFFINITY OF MEMBRANE RECEPTORS

Insulin Receptors

The insulin receptor is coded by a single gene and its 1370 amino acid sequence is now established. It is an integral plasma membrane glycoprotein having two exterior α-subunits (each M_r 125K) linked by disulfide bonds to two transmembrane β-subunits (each M_r 90K). Insulin binds to the glycosylated α-subunits and stimulates autophosphorylation of tyrosine residues in the β-subunits. Other specific cell proteins are also phosphorylated and perhaps thereby activated to express the action of insulin. Tyrosine kinase activity is found also in at least two growth factor receptors and in certain oncogenic protein products.

The insulin-receptor complex is modulated by β-adrenergic stimulation, ketoacidosis, and other factors. There is a steady turnover of membrane receptors by internalization and degradation, and by synthesis and insertion of new membrane proteins. Chronic excess of insulin decreases and a deficiency increases the receptor population. *Growth hormone* (GH; somatotropin, STH) tends to decrease the concentration of insulin receptors. *Glucocorticoids* (cortisol) appears to decrease the affinity of receptors for insulin, but may act mainly on postreceptor events.

In normal individuals, 5 hours after ingesting 100 g glucose, there is a modest increase in receptor affinity and insulin binding. Neonates of normal mothers have somewhat greater numbers of receptors with increased affinity, but neonates of diabetic mothers have greatly increased numbers of insulin receptors, as well as increased affinity and binding.

Receptor defects may involve a mix of decreased numbers, affinity and/or activity of the insulin receptor complex. Although present in the membrane of almost all cells it is most accessible to study in monocytes or erythrocytes.

Obesity is the most common state of insulin resistance and probably has the following *pathogenesis:* Overeating produces a state of relative hyperinsulinism that leads to a decreased concentration of insulin receptors and results in insulin resistance. A paucity of insulin receptors in the hypothalamus may account for the lack of satiety in most obese people. Fasting has the effect of increasing receptor binding. Non-insulin-dependent (type 2) diabetics, among whom 75% are obese, also have a subnormal receptor population with normal affinity.

Insulin-dependent (type 1) diabetics, who lack insulin, actually have increased numbers of receptors with normal affinity. When they develop severe *ketoacidosis,* both affinity and numbers of receptors are depressed and they become refractory to insulin therapy.

Defects of the insulin receptor/effector complex have been implicated in the following disorders:

Leprechaunism is a congenital form of extreme insulin resistance characterized by an increased affinity of the receptors for insulin but failure to internalize the message. A structural abnormality of the receptor is suspected. These patients die in infancy.

Acanthosis nigricans (type ***A***) is a rare disorder in young women who exhibit signs of excess androgens. Insulin binding may decrease markedly due to a receptor defect or deficiency; whatever the cause the cell fails to respond.

Lipoatropic diabetes is characterized by almost complete absence of adipose tissue and an insulin-resistant diabetes mellitus. Whether the disorder is caused by receptor or postreceptor defects is not yet clear.

Insulinomas, as might be expected, are associated with decreased numbers of insulin receptors.

Whether receptors are involved in the relative insulin resistance of late *pregnancy,* or the impaired glucose uptake found in the *elderly,* remains to be clarified.

Growth Factor Receptors

Rare cases of growth failure may be caused by a somatoliberin receptor defect in the somatotropes, but this has not been established. Some cases of impaired growth are associated with low concentrations of somatomedin-C (IGF-I) in spite of apparently adequate amounts of growth hormone (STH) in the plasma. This has been attributed to absent or defective STH receptors in the liver.

Of special interest are epidermal growth factor (EGF), platelet derived growth factor (PDGF), and the fact that viral oncogenes may code for a protein that shows considerable homology with known growth factors or GF receptors. Tyrosine phosphorylation appears to be involved in all these growth-promoting stimuli. They all stimulate phosphodiesterase hydrolysis of polyphosphoinositides to release diacylglycerol, which activates a C-kinase phosphorylation pathway.

The EGF receptor is a glycoprotein (M_r 170-180 K) containing 1186 amino acids, with an external N-terminal region, a hydrophobic transmembrane domain, and a cytoplasmic C-terminal region. The cytoplasmic C-terminal region has tyrosine kinase activity and a striking sequence homology with the v-*erb*-β gene product. The full significance of this link with pathogenic mechanisms in cancer remains to be determined.

Low-Density Lipoprotein (LDL) Receptors

It is now established that an absence or deficiency of LDL receptors is the primary cause of *familial hypercholesterolemia* and a factor in the development of *atherosclerosis.* There are at least two types of lipoprotein receptors, identified by their affinity for particular ligand apolipoproteins (see also Chap. 20). ApoE-specific receptors are found only in the liver, show little variation in numbers or activity, and function mainly in the removal of chylomicron remnants from the circulation. ApoB,E-specific receptors are found mainly in organs with a high requirement for cholesterol (liver, adrenals, and gonads) but also in smooth muscle, adipose tissue, lymphocytes, macrophages, and fibroblasts. These receptors are involved in cholesterol homeostasis.

The liver synthesizes very low density lipoprotein (VLDL) rich in triglyceride and carrying apoB-100, apoE, and apoC. Passage through the capillaries results in a loss of triglyceride and of apoC, yielding intermediate density lipoprotein (IDL). Almost half the IDL is taken up by apoB,E receptors in the liver, the rest is processed further to LDL, which has only apoB-100 and a lower affinity for receptors (half-life 2 to 3 days). Tissue apoB,E receptors normally take up LDL particles by binding, clustering, and endocytosis. Cholesterol is removed and the remainder is discharged into the circulation, possibly as high-density lipoprotein (HDL). Intracellular cholesterol acts to repress HMG·CoA reductase (cholesterol synthesis), to stimulate acylation (cholesterol storage), and to inhibit transcription of mRNA for synthesis of LDL receptors.

Normal tissue LDL receptors have an M_r around 160,000. The apoE moiety of LDL has a central cluster of lysyl and arginyl residues. A mutation involving the substitution of *one* of these amino acids is sufficient to reduce affinity for the receptor and may result in type 3 *hyperlipoproteinemia.* Hepatic apoB,E receptors are downregulated by excess cholesterol and bile acids and upregulated by estrogen and cholestyramine therapy.

Familial hypercholesterolemia (FHC) is an autosomal dominant deficiency of the gene coding for the LDL receptor. Three different receptor defects have been identified (see Chap. 20). Heterozygotes that have *one* mutant gene are common, 1/500 people in most populations. Their plasma LDL concentrations are about twice normal and they tend to suffer coronary heart disease in the third or fourth decade. If two heterozygotes marry, each child has a 25% chance of inheriting the mutant gene from *both* parents (prevalence 1/1,000,000). Such children have plasma LDL concentrations about six times normal and experience heart attacks before adulthood. Homozygotes cannot synthesize any LDL receptors; heterozygotes synthesize half the normal number. The deficiency of such receptors in the liver prolongs the half-life of IDL and increases LDL production. In the peripheral tissues there will be a decreased rate of removal.

Treatment of the disorder in heterozygotes is aimed at increasing the synthesis of LDL receptors. This can be achieved by lowering cholesterol intake (diet), increasing bile acid excretion (cholestyramine), and inhibiting HMG·CoA reductase (drugs).

Acquired deficiency of LDL receptors can result from excessive intake of foods that increase tissue cell cholesterol and inhibit the synthesis of LDL receptors. The remedy for this situation is better dietary habits.

Dopamine Receptors

Several types of receptors for dopamine (DA) have been identified in the brain and other organs and the effector in most cases is adenylate cyclase.

DA receptors are found in blood vessels, especially of the kidney, mesenteric, cerebral, and coronary vascular beds. DA_2 receptors are found on postganglionic sympathetic nerves and act to inhibit norepinephrine release. At low-dose infusion rates, DA lowers heart rate and blood pressure and increases renal blood flow. At somewhat higher doses DA activates β-adrenergic receptors, then at still higher doses both α_1- and α_2-adrenergic receptors resulting in vasoconstriction. Judicious use of dopamine can be valuable in the treatment of shock. Drugs that serve as DA agonists or antagonists have been effective in the selective management of cardiac, hypertensive, neurologic, and renal disorders. DA receptors are implicated in the pathogenesis of several diseases, for example *idiopathic hyperprolactinemia* may result from a deficiency of DA receptors in the lactotropes.

Parkinson's Disease. Dopamine receptors are found in several areas of the brain, including the median eminence, as well as in the anterior pituitary. Some of these receptors exhibit states of high and low affinity as judged from binding studies. In Parkinson's disease, the caudate and putamen nuclei of the brain contain respectively only about 15% and 5% of the normal content of DA. Concentrations of the metabolite homovanillic acid (HVA) are found to be low in the cerebrospinal fluid, but this occurs in other brain disorders as well.

From the perspective of striatal DA receptor sites, Parkinson's disease is viewed as a disorder of presynaptic dopaminergic mechanisms, with decreased activity of DA receptors. The resulting low DA input is believed to explain the signs and symptoms of the disease. Treatment, either indirectly by providing large amounts of DA precursor (L-DOPA), or directly by DA agonists (apomorphine, bromocriptine) is presumed to involve stimulation of DA receptors.

Tardive Dyskinesia. Tardive dyskinesia is marked by orolingual facial movements which include chewing, tongue protrusions, and facial grimacing. Tardive dyskinesia can occur in 3% to 6% of psychiatric patients and is common in elderly institutionalized patients, especially after stopping antipsychotic drug medication. The disorder is attributed to the activity of dopamine at striatal DA receptors. A postsynaptic hypersensitivity to DA probably results from neuroleptic blockade of DA receptors.

Two types of DA receptors have been described in brain, excitation-mediating (DA_e) and inhibition-mediating (DA_i). An imbalance in the activity of these receptors is believed to result in tardive dyskinesia. Because a relative DA excess is responsible for the dyskinesia, drugs that deplete brain dopamine (e.g., tetrabenazine) or block DA receptors (e.g., primozide) or a combination of the two have been effective in treatment of the disorder.

Schizophrenia. Of the many theories that have surrounded this baffling and relatively common disorder the possibility of increased numbers of postsynaptic DA receptors, with an underactive antagonist system, has to be considered. Much of the evidence comes from the favorable response to antischizophrenic drugs that are known to block the effect of DA on postsynaptic adenylate cyclase. More research is necessary to establish excess DA as a major factor in schizophrenia.

Muscarinic, Cholinergic, and GABA Receptors

Muscarinic, cholinergic and GABA receptors, as well as DA receptors, have been implicated in *Huntington's disease,* which is an autosomal dominant disorder characterized by premature tissue degeneration in the corpus striatum of the brain. Its prevalence is about 5/100,000 in Caucasians, occurring usually between ages 30 to 60. The patient survives about 15 years. The disease produces increasing involuntary and eventually incapacitating movement of skeletal muscles, along with personality changes, depression, and finally dementia. The defect has been localized on the short arm of chromosome 4. Almost all neurotransmitter functions involving the striatum are impaired, correlating with decreased concentrations of receptors. These are probably all *epiphenomena* resulting from the death of neurons. Recombinant DNA techniques with suitable probes should produce a presymptomatic diagnostic test within a few years.

The neurologic signs of acute *porphyrias* may result from excess aminolevulinic acid acting on GABA receptors.

Adrenergic Receptors

There are at least four different receptors for norepinephrine and epinephrine, which explains the diversity of responses to these catecholamines.

α_1-Adrenergic receptors affect smooth muscle contraction in the blood vessels and urinary tract and stimulate glycogenolysis in the liver. They activate the Ca^{2+}-mobilizing phosphodiesterase system. Other adrenergic receptors act through cAMP.

α_2-Receptors effect smooth muscle relaxation in the gastrointestinal tract and smooth muscle contraction in the vascular bed. They suppress norepinephrine release in the SNS and renin release by renal juxtaglomerular cells. They inhibit adenylate cyclase, depress lipolysis and insulin release.

β_1-Receptors increase the rate and force of cardiac contractions. They stimulate adenylate cyclase and increase lipolysis. β_2-Receptors increase norepinephrine and promote smooth muscle relaxation in the bronchi and blood vessels. They stimulate adenylate cyclase, increase insulin, glucagon, glycogenolysis, and gluconeogenesis. They also increase renin release.

Adrenergic receptors exhibit states of high affinity and low affinity; GTP converts the former to the latter by phosphorylation. They are also downregulated by internalization. Upregulation in response to treatment with antagonists offers an explanation for the hypersensitivity of the *propranolol-withdrawal syndrome.*

Thyroid hormones (T_3, T_4) increase the concentration of both α- and β-adrenergic receptors, probably explaining the symptoms of increased SNS activity in hyperthyroidism and the opposite in hypothyroidism.

Adrenergic receptors are presumed to be involved in cardiovascular disorders, the ventricular arrhythmias of myocardial ischemia, and the loss of SNS support in congestive heart failure. They are believed to play a role, still undefined, in hypertension.

Bronchial asthma correlates in severity with a reduction in β-adrenergic receptors and a shift from β- to α-receptors in lungs and lymphocytes. Appropriate treatment includes β-agonists (epinephrine) and glucocorticoids, which reverse the shift. *Autoantibodies* to β_2-adrenergic receptors have been found in sera from patients with allergic rhinitis and bronchial asthma.

Disorders of micturition, such as stress incontinence, bladder neck dysfunctions, and benign prostatic obstruction have been attributed to α_1-adrenergic receptor defects in the smooth muscle of the lower urinary tract.

Histamine Receptors

Histamine receptors are widespread in the body and appear to be involved in anterior pituitary regulation of sex related hormones. H_1 receptors are probable mediators of allergic manifestations in the bronchioles, which respond to specific antagonists (e.g., chlorpheniramine). H_2 receptors are involved in gastric acid secretion and respond to receptor blocking agents (e.g., cimetidine).

Angiotensin II Receptors

Angiotensin (AT) II receptors are found in the adrenals, kidney, liver, brain, platelets, and in smooth muscle membranes of arteries, bladder, and uterus. Receptor numbers (and perhaps affinity) are modulated by K^+, Na^+, ACTH (adrenals), and estradiol (lactotropes and adrenals). Receptor activation is linked to the Ca^{2+}-mobilizing phosphodiesterase system. A defect or deficiency of AT II receptors has been postulated in disorders of blood pressure, aldosterone production, and glomerular filtration.

Vasopressin (Antidiuretic Hormone, ADH) Receptors

The rare disorder of *nephrogenic diabetes insipidus* has been ascribed to a deficiency of receptors for ADH in cells lining the collecting ducts of the kidney. Nephrogenic diabetes insipidus can be a congenital or an acquired defect. A blunting of hypothalamic thirst perception (receptors detecting increased osmolality) is the probable defect that occurs in the elderly, and in chronic asymptomatic hypernatremia syndrome.

Gonadoliberin (GnRH, LHRH) Receptors

GnRH appears to regulate its own receptor numbers in pituitary gonadotropes, although gonadectomy increases receptor concentration and appropriate gonadal steroids prevent the increase. GnRH receptors act through a Ca^{2+}-calmodulin mechanism. Endogenous opioids appear to modulate GnRH receptors and gonadotrope function.

Prolactin (PRL) and Luteinizing Hormone (LH) Receptors

Leydig cells possess PRL and LH receptors. Excess prolactin (caused by a prolactinoma) can cause infertility, possibly by downregulation of LH receptors of the Leydig cell membrane in males, and of luteal tissue LH receptors in females.

Opiate Receptors

Although the effects of exogenous opiates have been known for many years, physiologic receptors for endogenous opiates have only recently been identified in the brain. The effect of *enkephalins* on pain perception and emotional control is now well established. The role of pituitary *endorphins* is less well understood.

ANTIBODIES TO RECEPTORS

Disorders of immune mechanisms have been implicated in several endocrine and neuroendocrine diseases. Antibodies may develop to hormones themselves (e.g., insulin) or to endocrine tissues (thyroid, adrenal). The target of antibody production appears to be cell membrane *receptors* in at least *four* diseases, which are described briefly.

Antibodies to Insulin Receptors

Acanthosis nigricans, type ***B,*** is a rare syndrome of profound insulin resistance resulting from anti-insulin-receptor antibodies. There is a chronic inflammatory thickening and darkening of the skin with severe glucose intolerance and often ketosis. Insulin antibody titers are high and therapy, even with huge doses of insulin, is rarely successful. Receptor concentrations (in monocytes) are usually normal but receptor affinity is very low, probably as a generalized defect. Remissions are rare, the clinical course may stabilize but usually leads to renal complications. Immunosuppressive therapy, corticosteroids, and plasma exchange have had variable effects.

Ataxia-Telangiectasia is a multisystem disease with progressive ataxia, abnormalities of the immune system, insulin-resistant diabetes mellitus, and a tendency to neoplasia. Ataxia-telangiectasia is an autosomal recessive trait carried by about 1% of the population in North America; prevalence is 1/40,000. The presenting symptom is unsteadiness as the child learns to walk; the child is usually confined to a wheelchair by the age of 12. Dilatation of groups of capillaries (telangiectasia) appear at age 3 to 6, followed by sinopulmonary infections and a relentless downhill course. They develop an insulin-resistant diabetes with an abnormal function of the liver and thymus.

Insulin receptors appear to be structurally normal but show a striking decrease in affinity. The sera contain antibodies of an IgM type that cause the binding defect and the insulin resistance. It is not clear what renders the insulin receptor antigenic and this may be only part of a more generalized abnormality in antibody production.

Graves' Disease

The commonest cause of hyperthyroidism is Graves' disease, a genetically determined autoimmune disorder characterized by the production of thyroid-stimulating IgG antibodies (TSIg) directed at thyroidal cell-membrane *TSH-receptors.* These immune globulins can be demonstrated in the serum. They bind to (or close by) the receptor and exert a TSH-like effect on the cell. Adenylate cyclase is activated and the increase in cAMP leads to the production and release of thyroxin. This proceeds in excess of requirements because of the failure of normal feedback suppression of pituitary TSH to prevent continued stimulation. Receptor numbers are actually increased and the TSIg · R complex is not rapidly internalized but continues to activate the cell.

The TSH receptor is a large glycoprotein (M_r 280K) made up of several subunits and contains 30% carbohydrate and 10% neuraminic acid, which is required for binding. Gangliosides provide recognition specificity. Although not yet commonly available, TSIg can be measured in serum by a radioassay using a thyroid membrane receptor. Results correlate with natural remissions and responses to drug therapy.

Myasthenia Gravis

Myasthenia gravis is a disease of severe muscle weakness and fatigue caused by antibodies (IgG) directed against the *acetylcholine receptors* (AcCh · R) of the neuromuscular junction. The signal for muscle contraction involves the release of AcCh, its diffusion across the synapse, and a change in permeability of the postsynaptic membrane. The depolarization and end-plate potentials are abnormally small in amplitude in myasthenia gravis. This is caused by reduced numbers of AcCh · R (10% to 30% of normal) resulting from accelerated internalization and degradation initiated by cross-linking and clustering of receptors by the divalent antibody. The rate of new receptor synthesis is unable to maintain a normal population in the plasma membrane.

Antireceptor antibodies can be demonstrated in >90% of patients with clinical myasthenia gravis. The concentrations vary widely and do not always correlate with the severity of symptoms, but are clearly of value in diagnosis and in monitoring treatment.

Other Antireceptor Antibody Diseases

It is likely that the development of antireceptor antibodies is more frequent than presently recognized. These disorders are more common in women than in men and may result from a variety of factors. Genetic predisposition to Graves' disease and myasthenia gravis is higher in patients with HLA types B8 and DRw3. Antibodies to ACTH receptors may explain the development of adrenal insufficiency in Addison's disease. Antibodies to gastrin receptors may be an etiologic factor in pernicious anemia.

Treatment of receptor *antibody* diseases is usually indirect, although selective control of antibody production is a hope for the future. Graves' disease responds usually to antithyroid drugs and myasthenia gravis to anticholinesterase therapy. Management of severe cases has included plasmapheresis, glucocorticoid and immunosuppressive drugs.

TRANSDUCER DEFECTS

In several membrane receptor complexes a guanine-nucleotide-dependent G-protein functions as a 'transducer,' converting the recognition-site signal into a stimulus of the effector unit. The *G-protein,* when complexed with GTP (Mg^{2+}), dissociates into α-(M_r 45K) and β-(M_r 35K) subunits. Depending on the signal ligand the active GTP-bound α-subunit stimulates either adenylate cyclase to generate cAMP, or phosphodiesterase to generate DG and IP_3. There are two recognized clinical disorders in which the 'transducer' is involved.

Pseudohypoparathyroidism

Parathormone (PTH) receptors are found in kidney and bone cells and act through the adenylate cyclase system. Pseudohypoparathyroidism is characterized by a reduction in PTH-receptor activity caused by a defect in the G-protein, which appears to be locked in the *inactive* G/GDP form. As a result there is decreased production of cAMP. This leads to decreased activation of 1α-hydroxylase, which is necessary to convert 25(OH)D to $1,25(OH)_2D$. The defect can be demonstrated in erythrocytes and in cultured skin fibroblasts.

There is defective bone resorption and reduced entry of Ca^{2+} into extracellular fluid and bone because of the reduced effectiveness of parathormone. Intestinal Ca^{2+} absorption is impaired. Renal loss of Ca^{2+} occurs along with PO_4^{3-} retention.

Diagnosis of pseudohypoparathyroidism should be considered when hypocalcemia and hyperphosphatemia occur in a patient with normal renal function and a *normal* or *elevated* serum PTH. If serum PTH is low the patient probably has primary hypoparathyroidism. Definitive diagnosis is based on the lack of a urinary cAMP response to infused PTH. Treatment with $1,25(OH)_2D$ bypasses the deficiency created by the receptor defect.

Watery Diarrhea Due to Cholera

When, due to ingestion of contaminated water or food, the cholera *vibrios* binds to specific sites on the membrane of intestinal epithelial cells, a sequence of events leads to ADP-ribosylation of the G-protein which becomes stuck in the *active* G/GTP form. As a consequence, adenylate cyclase is continually activated to produce excess cAMP, which stimulates water and electrolyte secretion producing diarrhea. The condition persists until the infected cell is shed into the intestinal lumen.

DEFECTS INVOLVING THE EFFECTOR, 2ND MESSENGER, AND ION MOBILIZATION

No disorders in these aspects of the receptor-response mechanism have yet been defined. Although highly speculative, it is tempting to suggest that certain uncontrolled hypermetabolic states could result from massive activation of the polyphosphoinositol receptor mechanism.

Malignant hyperthermia* is an autosomal dominant pharmacogenetic disorder characterized by an acute, potentially fatal, hypermetabolic reaction in skeletal, heart, and possibly smooth muscle. The incidence may be as high as 1 in 2000 surgical cases. It correlates with muscle mass and severe stress, and is most often induced by certain drugs and anesthetics. Mild reactions may occur postoperatively and consist of tachycardia, moderate fever, muscle pains, and malaise. Acute reactions developing soon after the induction of anesthesia exhibit muscle rigidity, tachycardia, and hyperventilation to excrete massive amounts of CO_2 being produced by the muscles. The pathogenesis is not yet clear, but evidence points to a flooding of sarcoplasm by Ca^{2+}, which probably induces the hypermetabolic state.

Quick diagnosis and prompt treatment are imperative. Anesthetic agents are terminated; the patient is ventilated with oxygen, given dantrolene sodium to lower myoplasmic calcium, sodium bicarbonate to counter lactic acidosis, and sometimes insulin to lower serum K^+. Awareness, early diagnosis, and proper treatment result in 100% survival. The mortality rate has fallen in the last 20 years from 60% to about 7%.

An *in vitro* bioassay involving *platelet* muscle filaments remains to be validated. At present the only definitive way of demonstrating the hyperthermia trait requires a skeletal *muscle biopsy* and a caffeine-halothane contracture test.

DEFECTS IN CYTOSOL RECEPTORS

Unoccupied receptors for steroids are found in the cell cytoplasm. Steroid hormones are in an equilibrium state of easy transfer across cell membranes; those inside the cell bind to specific receptors. The steroid-receptor complex undergoes transformation or 'activation' and then translocates to the nucleus, where it binds to a specific site on the chromatin. This leads to transcription and synthesis of a protein that expresses the action of the hormone on that cell.

* Condensed from a personal communication by Dr. Beverley A. Britt FRCP(C), staff anesthetist, Toronto General Hospital, Toronto, Ontario.

Androgen Receptors

Receptors that bind testosterone have been identified in kidney, brain, and other tissues and exist in both sexes. The more significant androgen receptor is a protein found only in male reproductive tissues. The molybdate-stabilized cytosolic receptor sediments at about 8–9S and has a high affinity for dihydrotestosterone (DHT). The steroid-receptor complex that binds to the nuclear chromatin contains a 3.5S protein. Receptor studies are generally performed on prostate biopsy tissue or cultured genital skin fibroblasts.

Two types of *inherited resistance* to *androgens* are found in 46XY males with testes. The *first* type is an autosomal recessive defect in the intracellular 5 α-reductase that converts testosterone to DHT. The *second* type is an X-linked defect that is seen in: (a) phenotypic 'women' with or without some virilization, and (b) phenotypic men with or without ambiguous external genitalia. In some cases there may be decreased numbers or lower affinity of androgen receptors, but in most of these patients there appear to be qualitatively abnormal receptors that will not bind DHT.

In the complete ***testicular feminization syndrome*** there are no DHT-binding receptors and the individual develops both pre- and postnatally under the influence of estrogen. The patient appears as a normal woman, but has internal testes and only a small vaginal pouch. Gonadectomy is mandatory and the patient is maintained as a woman. In the *incomplete* form of this disorder there is some, but very low DHT binding. Female characteristics may be less well developed as a result and there are varying degrees of virilization.

Reifenstein's syndrome is a type of male pseudohermaphrodism associated with hypospadias, small testes and sterility, absence of beard, short stature, and often gynecomastia. Present evidence implicates deficient numbers or function of androgen receptors as the commonest cause.

The ***infertile male syndrome*** has been found, in about 40% of cases, to be associated with decreased numbers or impaired function of androgen receptors.

Hyperplasia of the ***prostate*** is known to be associated with an increase in cytoplasmic DHT. This finding is explained by an increase in 5α-reductase activity in the prostate and a decrease in DHT-degrading enzymes. Present evidence does not support an hypothesis that increased numbers of androgen receptors could be the cause of benign prostatic hypertrophy. In ***cancer*** of the prostate assays of biopsy specimens has provided some evidence that the best responses to anti-androgen therapy occur when receptor assays are high; results are poor when assays are low. More studies of both cytoplasmic and nuclear receptors are needed.

Estrogen and Progesterone Receptors

Estrogen receptors are present in a number of tissues but the major sites are the endometrium, vagina, mammary glands, and the liver. Feedback receptors occur in the hypothalamus and pituitary. Several forms of estrogen receptors have been described, but basically it is a 4S protein that is active only in its phosphorylated form. When the estrogen receptor binds estradiol it is no longer accessible to phosphatase inactivation and moves quickly to the nucleus and binds to a specific receptor on the chromatin. Transcription and protein synthesis then follow. Little is known about the fate of the nuclear receptor complex, but some receptor is probably recycled to the cytoplasm. In addition to its well-known sex-tissue effects, estrogen has been found to increase the synthesis not only of its own receptors but also of oxytocin receptors in the uterus, progesterone receptors in the ovary, endometrium and breast, and gonadotropin receptors in granulosa cells.

The *progesterone* receptor is known to be a dimer composed of subunits A (M_r 79K) and B (M_r 108K). Its major sites are the endometrium, ovary, vagina, and liver, but receptors occur also in the brain and kidney.

A deficiency of estrogen and progesterone receptors may be a factor in some forms of infertility and virilization, but more studies in these areas are needed.

Receptors in Mammary Cancer. Reliable assay procedures are now available for estrogen and progesterone receptors. There is abundant evidence that breast cancer patients whose tumors contain these receptors are much more likely to respond favorably to hormone manipulation than those who lack receptors for these hormones. A summary of nine reports gave the following results:

Receptors:	ER(−) PgR(−)	ER(−) PgR(+)	ER(+) PgR(−)	ER(+) PgR(+)
Survivors:	9/89	4/11	30/105	102/132
	10%	30%	29%	78%

Treatment may be ablative (ovariectomy, hypophysectomy, or adrenalectomy) or additive (androgen, glucocorticoid, estrogen, and progesterone).

The alternative use of *cytotoxic drugs* correlates poorly with receptor data, much better with the rate of *thymidine incorporation.* Clearly, the less well differentiated and more rapidly growing the tumor the better the response to such drugs.

Glucocorticoid Receptors

Glucocorticoid (GC) receptors occur in most tissue cells; two classes have been reported. Type I has a high affinity for corticosterone, low for dexamethasone; type II has a high affinity for dexamethasone, moderately high for cortisol. Type I in the kidney favors occupancy by aldosterone, in the brain by cortisol. Type II exhibits two modes: (a) a high affinity receptor that mediates GC effects on mood, appetite, and sleep/wake patterns, and (b) a low affinity form occupied by cortisol only when plasma levels are high in a stress situation.

A component (M_r 90K) of the GC receptor can be autophosphorylated in the presence of Ca^{2+} and then exhibits protein kinase activity. A form sedimenting as a 9S protein is converted to an active 5S form by dephosphorylation, probably a reversible process.

A syndrome of *primary cortisol insensitivity* (end-organ resistance) occurs in humans as a rare autosomal defect of glucocorticoid receptors. It is caused by a decrease in or instability of the 'activated' form of the receptor. The patients have high circulating concentrations of free cortisol and ACTH. They show no signs of Cushing's syndrome, but in severe cases may develop hypertension and hypokalemia caused by excess deoxycorticosterone.

Aldosterone Receptors

Two aldosterone (AO) receptors have been described. Type I (M_r 85K) has a high affinity

($Ka = 5 \times 10^{10}$ M) for aldosterone; type II ($Ka = 3 \times 10^{8}$ M) will bind glucocorticoids as well. AO receptors are found in the kidney, liver, and the smooth muscle cells of the smaller arteries and the heart.

Clinical disorders of AO receptors are not yet clearly defined, but are suspected in some forms of hyponatremia and hypotension. *Pseudohypoaldosteronism* is a rare congenital insensitivity of the renal collecting ducts to AO, possibly caused by a receptor defect.

Vitamin D (1,25[OH]$_2$D) Receptors

A rare syndrome of hereditary resistance to vitamin D, associated with rickets and baldness, is attributed to a deficiency of cytosolic receptors for 1,25(OH)$_2$D. The defect is noted in childhood and results in hypocalcemia with high circulating concentrations of 1,25(OH)$_2$D. A definitive diagnosis may require receptor studies on cells cultured from bone biopsy or skin fibroblasts.

NUCLEAR RECEPTORS

Two types of disorders involving nuclear receptors can be envisaged: (a) receptors for hormone-receptor complexes formed in the cytosol and translocated to the nucleus (knowledge in this area is at present very sparse) and (b) primary target tissue receptors.

Thyroid Hormone (T_3) Receptors

Thyroid hormones enter the cell freely and the active hormone (T_3) diffuses directly to a receptor located on the nuclear chromatin.

A rare disorder of *thyroid-hormone resistance* has been attributed to a defect or *deficiency* of T_3 receptors. *Familial thyrotoxicity* has been attributed to an *increase* in nuclear T_3 receptors. The patients show symptoms of thyrotoxicity in spite of normal thyroid function.

SUGGESTED READING

BAXTER JD, FUNDER JW: Hormone receptors. N Engl J Med 301:1149–1161, 1979

BERRIDGE MJ: Inositol triphosphate and diacylglycerol as second messengers. Biochem J 220:345–360, 1984

BLECHER M, BAR RS: Receptors and Human Disease. Baltimore, Williams and Wilkins, 1981

BROWN MS, GOLDSTEIN JL: How LDL-receptors influence cholesterol and atherosclerosis. Sci Am 251:58–67, 1984

CLARK GM, OSBORNE CK, MCGUIRE WL: Correlations between estrogen receptor, progesterone receptor, and patient characteristics in human breast cancer. J Clin Oncol 2:1102–1109, 1984

DRACHMAN DB, ADAMS RN, JOSIFEK LF, SELF SG: Functional activities of autoantibodies to acetylcholine receptors and the clinical severity of myasthenia gravis. N Engl J Med 307:769–775, 1982

LEFKOWITZ RJ, CARON MG, STILES GL: Mechanisms of membrane-receptor regulation. Biochemical, physiological and clinical insights derived from studies of the adrenergic receptors. N Engl J Med 310:1570–1579, 1984

STOBO JD: Autoimmune antireceptor diseases, Chap. 25 in Dixon FJ, Fisher DW: Biology of Immunologic Diseases. Sunderland, Sinauer Assoc Inc, 1983

TITELER M: Multiple Dopamine Receptors. New York, Marcel Dekker, 1983

ULLRICH A, BELL JR, CHEN EY et al: Human insulin receptor and its relationship to the tyrosine kinase family of oncogenes. Nature 313:756–761, 1985

ULLRICH A, COUSSENS L, HAYFLICK JS et al: Human epidermal growth factor receptor cDNA sequence and aberrant expression of the amplified gene in A431 epidermoid carcinoma cells. Nature 309:418–425, 1984

27

Biochemical Aspects of Aging

Choong-Chin Liew

Aging has been defined as a progressive unfavorable loss of adaptation leading to increased vulnerability, decreased viability, and decreased life expectancy. *Gerontology* is the study of this deterioration process in the human body. *Geriatrics* is a branch of general medicine that deals with remediable and preventable clinical problems and with the social consequences of illness in elderly patients. In general, patients over 65 years of age with chronic illnesses requiring medical treatment can be considered as geriatric problems. As shown in Fig. 27–1, the human life span has been influenced greatly by advances in health care and improvements in the environment. For example, antibiotics and vaccination for infectious diseases, and good sanitation, have significantly extended the mean and median life span in recent decades. The *maximum* life span, however, has not been prolonged.

Little is known about the origins of senescence, and the ***mechanism of aging*** remains an enigma. During fetal development all cells divide and differentiate. Cells of the central nervous system and muscle cells have ceased to replicate by late childhood. By middle age some cells show a declining capacity to regenerate, for example, fibroblasts and hepatocytes. Cells of the gastrointestinal mucosa continue to replicate throughout life. The processes of *development* and *aging* may be distinguished on the basis that during aging there are no proteins, matrices, or organs with essentially new structures or functions.

There are three clinical syndromes that can be considered to represent an early onset of aging:

1. ***Werner's syndrome*** is characterized by shortness of stature; early graying and loss of hair; juvenile cataracts; a tendency to develop diabetes, atherosclerosis, calcification of blood vessels, and osteoporosis; and a high prevalence of cancer. Such patients carry an autosomal recessive gene linked to the loss of HLA-A2 (see HLA Antigens, Chap. 4) and their fibroblasts exhibit an impaired capacity to grow.
2. ***Progeria*** occurs at an earlier age than Werner's syndrome. It is a term that describes several rare disorders in which the cells are defective in repairing breaks in DNA induced by ultraviolet (UV) or cosmic radiation. An example is *Cockayne's syndrome,* in which children stop growing after a few years, then age rapidly and die of 'old age' before their midteens.
3. ***Alzheimer's disease*** and senile dementia exhibit similar microscopic pathologic abnormalities. Cortical atrophy is seen, with a high density of neurofibrillary plaques and tangles and neurons that are clogged with insoluble protein. Alzheimer's disease occurs in 2% to 5% of the elderly population. Clinically, this disease presents with loss of memory (*amnesia,* a common complaint) but progresses to loss of speech or verbal comprehension (*aphasia*), loss of motor coordination

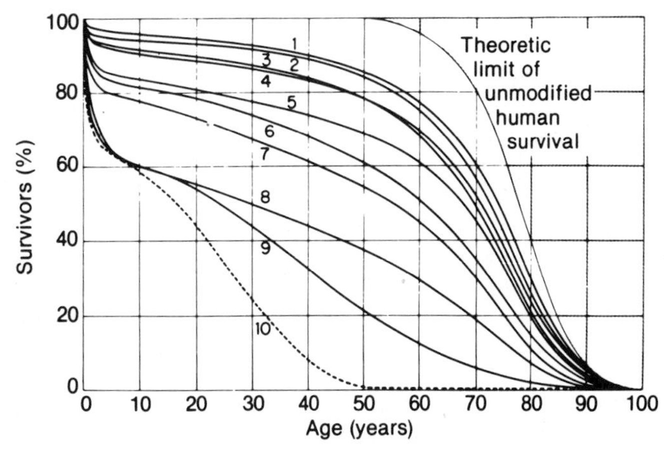

1. New Zealand, 1934-1938
2. US (whites), 1939-1941
3. US (whites), 1929-1931
4. England and Wales, 1930-1932
5. Italy, 1930-1932
6. US (whites), 1900-1902
7. Japan, 1926-1930
8. Mexico, 1930
9. British India, 1921-1930
10. Stone age man

FIGURE 27-1. Historical changes in the human survival curve. (Comfort A [ed.]: The Biology of Senescence. New York, Elsevier, 1956).

(*apraxia*), and loss of comprehension of sounds, sight, or feeling (*agnosia*). Onset can be from age 40 onwards, but more commonly after age 65; the course may be 5 to 15 years with periods of apparent remission. Diagnosis is based on tests of mental state with deficits in two or more areas of cognition. Definitive diagnosis at present requires a biopsy for confirmation.

The etiology of Alzheimer's disease is still an enigma. Up to 50% are usually found to have a familial aspect. Although increased concentrations of aluminum have been found in brain chromatin, this is probably not the cause of Alzheimer's disease. The possibility of a 'slow' virus has been suggested. A deficiency of choline acetyltransferase accounts for a marked reduction in acetylcholine, particularly in the region of the hippocampus. Cerebrospinal fluid acetylcholinesterase tends to be low in Alzheimer's disease, but the test lacks both sensitivity and specificity. Whether the metabolic deficiency of this disease is associated primarily with the degeneration and death of certain neurons, or *vice versa,* remains to be established.

THEORIES OF AGING

Several theories have been proposed to explain the natural mechanism or causes of aging. In general, according to current concepts, the biochemistry of aging can be summarized as an interference with the flow of cellular information somewhere in the sequence from DNA to RNA to protein synthesis and degradation. The loss of cellular information could occur either (1) through processes of *random deterioration* of cell metabolism related to errors or damage, or (2) via a *genetic program* related to the continuity of cell differentiation and development. These two aspects could also be interdependent.

Random Deterioration

The theories that are related to the processes of random deterioration can be outlined briefly as follows:

Free Radical Involvement. It is known that free radicals (e.g., $O_2^{(-)\cdot}$, $^{\cdot}OH$) react easily and could damage both DNA and other cell structures (e.g., cell membranes). The radicals may act directly or indirectly by generating strong oxidizing agents within the cells as a result of metabolic processes or local environmental factors. Antioxidant substances (e.g., vitamin E) have been shown to interfere with reactions mediated by free radicals. It has also been shown that cells cultured in the presence of vitamin E show delayed senescence and increased doubling times.

Somatic Mutation. It has been known for years that an increase in chromosomal aberrations, including aneuploidy, occurs with advanced age. Biochemically, these changes are caused mainly by the imbalanced reactions that occur between breakage and repair systems of DNA, which cause an eventual accumulation of defective DNA. Normal DNA is essential for normal regulatory functions. As a result, genes of somatic cells that are inactivated by continuous subtle breaks in DNA become potentially lethal to those cells. This situation is particularly serious for cells that do not divide after they have become highly differentiated to their mature forms. For example, when brain and muscle cells function poorly or die they are irreplaceable. Other cell types like hepatocytes or epithelial cells of the gastrointestinal tract, which have the capacity to divide, may be less seriously affected.

A relationship may exist between the *stability of DNA* and the *longevity* pattern of a given species. The average life span of man, mouse, and rat are, roughly, inversely proportional to their estimated rates of spontaneous somatic mutation. If somatic mutation is involved in senescence, the efficiency of DNA repair mechanisms may be the determining factor, because the rate of DNA repair is proportional to longevity. It has been reported that fibroblasts derived from human, elephant, and bovine tissues are five times more active in repairing DNA than those of the rat, mouse, and shrew. It has also been postulated that cosmic and UV radiation may accelerate the processes of aging. No clear evidence exists as yet to substantiate this.

Error Catastrophe of Protein Synthesis. This hypothesis is based mainly on the proposition that a random accumulation of errors (i.e., abnormal protein) occurs within the cell. An accumulation of changes in the amino acid sequences of one or more proteins, in particular, errors that affect the specificity of enzymes required for protein synthesis, would lead to cell deterioration and death. Such errors could result either during RNA transcription, which specifies the amino acid sequence of a protein, or during the translational processes by a miscoding of amino acid residues. For example, when amino acid analogs such as *p*-fluorophenylalanine and ethionine are fed to larvae of the fruit fly, they are incorporated into proteins in place of phenylalanine and methionine, respectively. As a result, the activities of glucose-6-phosphate dehydrogenase and 6-glucophosphogluconatase are affected, and these enzymes then become sensitive to heat.

Autoimmunity. Many of the original concepts regarding the development of autoimmunity with advancing age were attributed to somatic mutations, leading to incorrectly specified cells. Burnet has suggested that mutated lymphocyte clones may lose their property of inhibiting the production of antibodies for correctly specified body tissues. An important feature of this theory is the support it receives from the *autoimmune diseases* that are commonly associated with senescence, such as various forms of arthritis, myasthenia gravis, probably diabetes mellitus, and possibly even cancer. The cardinal features of these diseases include:

1. Infiltration of a homogeneous substance known as *amyloid* (chondroitin sulfuric acid–protein complex)
2. Accumulation of lymphocytes and plasma cells
3. Death of parenchymal cells
4. Perivascular inflammation and changes in blood vessels, with thickening and hyalinization of their walls

Genetic Programming

An alternative theory that the life span of each species is programmed genetically was derived from the following observations:

Characteristics of Cell Types. All tissues and organs gradually become differentiated in the course of embryogenesis and by adolescence all cell types are completely differentiated. Based on their mitotic characteristics, cells can be classified into three different types:

1. Those with a continuous mitotic capability that allows cell division to continue throughout the life span of the organism (e.g., gastrointestinal and hematopoietic cells)
2. Cells with an intermittent mitotic capability that are active in turnover at times of stimulation (e.g., partial hepatectomy), but otherwise have a very low turnover rate; hepatic parenchymal cells, renal tubular cells, and bone cells are examples
3. Nonmitotic cells that have lost their mitotic capability by early adulthood (e.g., muscle and nerve cells)

Tissue studies *in vitro* have shown that the growth rates of the liver, kidney, and spleen are the most active, whereas those of the brain and heart are the least active. This provided strong evidence that the phenotypic expression of each cell type is genetically programmed.

Diploid Cell Cultures. All cells cultured *in vitro* exhibit three characteristic phases. Following the primary culture and the rapid growth phases, the cells gradually lose their mitotic activity and eventually become senescent and die, regardless of the technique employed. It has been postulated that the life span of diploid cells cultured *in vitro* is genetically programmed. Diploid cells derived from human embryo and adult lung have been shown to possess a finite life span. The average number of cell doublings is 48 for embryonic cells and 20 for adult cells. These cells retain a normal diploid character until the later stages in their life span, when nuclear abnormalities and chromosome aberrations appear. Recent work on explanted tissues from patients with diabetes mellitus or Werner's syndrome showed a more restricted life span for these than for normal tissues.

Turnover Rate of Protein. It has been shown biochemically that the turnover rate of plasma protein is closely correlated with the life span of mammals. As shown in Table 27–1, the highest turnover rate of proteins such as albumin and γ-globulin is in the mouse, whereas the lowest rate is in humans. The biologic half-life of these proteins is roughly proportional to the maximum life span of the species.

TABLE 27–1. Turnover of Plasma Proteins and Homologous Proteins of Various Species

Species	Maximum Life Span (years)	Half-life (days)	
		Albumin	*γ-globulin*
Mouse	3		1.9
Rat	5	3–4.4	
Dog	18	5–6.6	8.0
Man	110–120	7–10	13.1

(Tarver H: In Neurath H, Bailey K (eds): The Proteins, p 1199–1296. New York, Academic Press, 1954)

Chromatin Proteins. The biochemistry of aging within the nucleus is relatively unexplored. If senescence is the result of continuous error accumulation at the transcriptional level, and has a genetic basis, then the components most susceptible to changes in the genes would be the chromatin proteins. The functional aspects of chromatin proteins have been studied extensively in recent years. In general, chromatin is composed of DNA, proteins, and trace amounts of RNA. The proteins are separated into two major fractions: The basic proteins are known as *histones,* and the acidic proteins as *nonhistone chromatin proteins.* The histones have a mass equal to that of DNA and are regarded mainly as structural proteins. On the other hand, it has been suggested that the nonhistone chromatin proteins act both as structural proteins and as potential regulators (see Fig. 14–2).

The most discernible differences between ***histone*** and ***nonhistone chromatin protein*** are as follows:

1. The structure of histone is conserved in evolution and has five major fractions in most species and in most eukaryotes; the chromatin structure is constituted as repeating core particles containing two molecules of each of the four histones (H_{2a}, H_{2b}, H_3, H_4) and 140 base pairs of DNA. The nonhistones are highly heterogeneous and exhibit limited tissue and species specificity.
2. Histone inhibits RNA synthesis *in vitro,* but nonhistone chromatin protein does not.

3. Histone binds nonspecifically to DNA fragments, whereas nonhistone chromatin protein has some specific binding capacity.
4. The synthesis of histone occurs mostly at the time of DNA synthesis (S phase of the cell cycle) and its turnover rate is very low; the synthesis of nonhistone chromatin protein occurs in all phases and it has a high turnover rate.

Both histone and nonhistone chromatin protein can be modified covalently by acetylation, ADP-ribosylation, phosphorylation, and methylation. There is evidence that increased *covalent modification* of these proteins is associated with the processes of aging, possibly reflecting a deterioration in the efficiency of homeostatic mechanisms.

BIOCHEMICAL OBSERVATIONS IN GERIATRIC PATIENTS

Aging is associated with a decline in the function of most organs and an accelerated incidence of many diseases. Some of the better known aspects of these problems are considered briefly.

Organ System Functions that Decline with Age

The maintenance of homeostasis is less efficient with advancing age, and there is a decrease in cell water, a reduction in muscle mass, and a gradual decline in:

1. Respiratory function (vital capacity; maximum breathing capacity)
2. Cardiovascular function (cardiac index; blood pressure regulation)
3. Kidney function (creatinine clearance; glomerular filtration rate; renal plasma flow); renal concentrating capacity (functioning nephrons; response to antidiuretic hormone)
4. Liver function (sulfobromophthalein clearance)
5. Immune system function (T-cell function; immunologic surveillance)
6. Neurologic function (coordination; thirst mechanism)
7. Endocrine system function (see Endocrinologic Aspects)

Endocrinologic Aspects of Aging. The capacity of the hypothalamus–anterior pituitary system seems to be well maintained in elderly people, but the efficiency of homeostatic regulation may decline. There is no clear evidence of any significant age-related posterior pituitary dysfunction. Low urine osmolality may occur, but is more likely to be a renal problem. Elderly patients, however, can have quite marked increases in plasma osmolality without experiencing thirst.

The ***thyroid*** can present serious problems in older people. *Hypothyroidism* is often missed because the changes mimic normal aging. There is little evidence of thyroid failure as part of the normal aging process. There is no increase in thyrotropin (TSH) and the response to TSH administration is essentially normal. Plasma thyroxin (T_4) is usually normal; triiodothyronine (T_3) levels decline above age 55 and more T_4 is converted to reverse T_3 (rT_3), but the significance of this observation is not clear. Elderly patients with signs of hypothyroidism should be screened with T_4, T_3, rT_3 and TSH assays. A significant number of elderly patients who have hypothyroidism can be 'rejuvenated' with treatment. About 2% of senile dementia cases are in fact hypothyroidism. Hyperthyroidism occurs in the elderly, but the signs relate primarily to cardiovascular disturbances.

Adrenal disorders are not more frequent in old age. Glucocorticoid production is well maintained, aldosterone levels may decline, and androgen production definitely falls. Age has a selective effect on androgen synthetic pathways in the adrenal cortex and the balance may shift to catabolic dominance. Urinary 17-ketosteroids decline to about half by age 75, caused mainly by a fall in dehydroepiandrosterone production. Cortisol production is fairly well maintained. There may be some blunting of the response of the adrenal medulla, for example, to hypoglycemia.

The prevalence of *hypertension* increases progressively with age; about 30% of men and 40% of women become definitely hypertensive between ages 65 and 80. Renal factors are most likely involved. Adrenal causes are rare, although increased norepinephrine levels and hyperaldosteronism do occur in elderly patients.

Gonadal failure is the cause of significant problems in females beyond the age of 50. In addition to the menopausal symptoms of vasomotor and psychic instability, the loss of the anabolic effects of estrogen leads to bone and tissue wasting and atrophy of the genital tract. The male climacteric occurs usually between ages 60 to 80 and the effects are less notable. At this time the ratio of testosterone to estradiol will have fallen from a normal 12:1 to about 2:1.

Metabolic Functions that Decline with Age

In establishing criteria for identifying deviations from 'normal' it is quite apparent that reference ranges must be adjusted for patients above the age of 60. Furthermore, the coefficient of variation is greater and reference limits are harder to define. A significant number of samples must therefore be analyzed prior to drawing any conclusions.

There is experimental evidence that restriction of food intake, so long as good nutrition is maintained, acts by some metabolic mechanism to slow the aging process.

Carbohydrate Metabolism. Glucose tolerance declines with age, the 2-hour value increasing approximately 0.55 mmol/L (10 mg/dL) per decade from age 40. ***Diabetes mellitus*** is a common problem in the aged, with a prevalence of about 17% (1 in 6) at age 65 and 26% (1 in 4) by age 85. A clinical picture seen frequently is a mild to moderate hyperglycemia in a somewhat obese patient with no tendency to ketosis. The response to insulin appears blunted, possibly because of a decreased number of insulin receptors. There may be difficulty in converting proinsulin to insulin. A serious problem is the insidious development of *hyperosmolar nonketotic coma,* which may be the first indication of the disease. Blood glucose may reach levels of 55 mmol/L (1000 mg/dL) and serum osmolality may exceed 350 mosm/kg. The severe dehydration can result in shock and death unless treated effectively. Insulin levels may be 15 to 30 mU/L, enough to prevent ketosis but not hyperglycemia.

Protein Metabolism. The protein wasting and decline in muscle mass and bone protein that occur in elderly people is attributed usually to the reduction in anabolic hormones. In addition to a fall in anabolic steroids, the anabolic effectiveness of insulin may decline. In general, protein metabolism is the sum of protein synthesis and degradation. This measurement is based usually on the nitrogen turnover in the entire body. In a study in which ^{15}N-glycine was administered orally to normal individuals, it was demonstrated that protein synthesis in the elderly (69 to 91 years) was about 60% of that in the young adult (20 to 30 years). However the basal energy metabolism, which is closely associated with protein metabolism, is within the normal range in the elderly.

Lipid Metabolism. The ratio of fat to lean mass is increased with aging. Serum cholesterol increases slowly in both sexes from age 30, and more rapidly in women after age 50. Triglyceride levels rise more rapidly in males aged 30 to 50, but in parallel with females after age 50. The relationship of lipids to lipoprotein and apolipoprotein metabolism is still being worked out.

Atherosclerosis is a complex disease that affects 1 in 8 males at age 40 and 1 in 4 at age 60. Patients with lipoproteinemia type 2 (familial hypercholesterolemia) are prone to cardiovascular disease. Homozygotes die before the age of 30; heterozygotes develop coronary artery disease in their forties. The most significant index is the low-density lipoprotein (LDL) cholesterol (see Chap. 20).

High-density lipoprotein (HDL) levels are inversely proportional to plasma triglyceride. The main functions of HDL are to facilitate the transport of cholesterol from the cells and to provide substrate for the formation of cholesteryl ester. HDLs also inhibit the binding of LDL by cell surface receptors for subsequent uptake and degradation, which increases with age.

Calcium Metabolism. The total mineral mass of the body declines from about age 45. ***Osteoporosis*** is a major problem in aging women, associated with a sharp increase in the incidence of bone fractures after age 50. The administration of cortisone accentuates the bone loss. Estradiol and calcitonin inhibit the action of parathyroid hormone and the postmenopausal deficiency of estrogen may be the etiologic factor. Bone loss can be delayed by estrogen administration but a decision on replacement therapy becomes a problem of balancing the morbidity of bone fractures against the risk of cancer.

Blood Chemistry in the Elderly

Automation of clinical chemistry services has provided a convenient examination of at least 18 blood constituents. Reports on the effects of both sex and age in healthy blood donors have been published. In general, the concentrations of plasma or serum creatinine, urea, cholesterol, globulin, potassium, alkaline phosphatase, and glucose show a tendency to *increase* with age. Total serum proteins, albumin, calcium, inorganic phosphate, and iron levels tend to *decline* with age. The most notable changes relating to sex are those that occur in women following the menopause.

OTHER METHODS OF EVALUATION

Many approaches have been taken in an attempt to evaluate the natural causes of aging, but sound methods dealing with geriatrics on a biochemical basis remain to be established. A few non-biochemical approaches that have met with some success will be described briefly.

Tests of Mental Status. Three functional tests of organic brain syndromes in elderly patients are as follows:

1. The face–hand test (FHT) evaluates the organicity of mental dysfunction by recording problems in identifying two simultaneous points at times of stimulation, for example, the face and the ipsilateral or contralateral hand.
2. Memory for design (MFD) examines the ability to recall geometric designs.
3. The Short Portable Mental Status Questionnaire (SPMSQ) and the Mental Status Questionnaire (MSQ) deal with the degree of orientation as measured by the recognition of time, place, person, and other generalized data. The SPMSQ is a concise form of the MSQ.

It has been demonstrated that the SPMSQ yields the best correlation with a clinical diagnosis of organic brain disorder. The FHT helps to confirm the identification of mental status, thus serving as a useful second test.

Hematologic Examination. Blood cell counts and hemoglobin concentration may bear some relationship to the course of natural aging. A significant gradual decrease in the erythrocyte count and hemoglobin content has been observed during the life span of mice, and probably occurs in humans.

Statistical Analysis. The evaluation of a gerontologic survey should be based on the following criteria:

1. ***Population sampling.*** The planning of the population sampling should preclude biased conclusions. For example, in a city or town with a large population, varied characteristics of different regions can usually be identified. Any sample must be highly heterogeneous to be unbiased.
2. ***Sample size.*** The number of persons studied should be in the range of 1% to 10% of the population. It is also necessary to correlate the sample size to the population sampling.
3. ***Longitudinal and cross-sectional studies.*** Conclusions made from sampling performed only once may be quite erroneous. In gerontology, factors that contributed in the past and those relevant in the present may be different (e.g., the height of the individual). Therefore, studies should consider the patient's history as well as the present conditions.

Finally, any study must be subjected to a proper biometric analysis. A well-recognized problem in gerontology is the *survivor effect* on biochemical statistics. Persons who deviate most from normal tend to be at greater risk and die earlier, whereas those remaining in the study will have mean values closer to the reference range. This places greater significance on moderate deviations, but makes it more difficult to separate those who are 'normal for age' from 'diseased' populations.

SUGGESTED READING

ADELMAN RC, ROTH GS (eds): CRC Series in Aging, p 82. Boca Raton, CRC Press, 1981

BROCKLEHURST JC, HANLEY T: Geriatric Medicine for Students, 2nd ed. Edinburgh, Churchill-Livingstone, 1981

BURNET M: Somatic mutation and chronic disease. Br Med J 1:338–342, 1965

CARTWRIGHT IL, ABMAYR SM, FLEISCHMANN G et al: In CRC Critical Reviews in Biochemistry, p 1. Boca Raton, CRC Press, 1982

GOLDSTEIN S: The biology of aging. N Engl J Med 285:1120–1128, 1971

HAGLUNG MJR, SCHUCKIT MA: A clinical comparison of tests of organicity in elderly patients. J Gerontol 31: 654–659, 1976

HAYFLICK L: The cell biology of human aging. Sci Am 242:58–65, 1980

HAYFLICK L: Theories of biological aging. Exp Gerontol 20:145, 1985

KORENMAN SG: Endocrine Aspects of Aging. New York, Elsevier, 1982

MASORO EJ: Nutrition as a modulator of the aging process. The Physiologist 27:98, 1984

ROTHSTEIN M: Biochemical Approaches to Aging. New York, Academic Press, 1983

ROWE JT, BESCHIN RW: Health and Disease in Old Age. Boston, Little Brown, 1982

WURTMAN RJ: Alzheimer's disease. Sci Am 252:62, 1985

28

Molecular Diagnosis of Genetic Defects

Choong-Chin Liew

The number of diseases known to be related to heritable factors has increased since 1960 from about 500 to more than 3000. Thus, genetics plays a major role in determining the likelihood of many people developing illnesses. Genetic disorders can be identified successfully using conventional methods, such as clinical signs and symptoms, serum enzyme levels, and so forth, but in most cases diagnosis occurs only after the onset of the disease. The defect can rarely be established early or carriers of the trait identified.

The development of recombinant DNA technology, beginning in the early 1970s, has had a major impact on our understanding of the molecular basis and biochemistry of human diseases, particularly in the following areas:

a. The molecular basis of heritable disease
b. Analysis of human gene linkage
c. Mapping of the human genome
d. Regulation and expression of specific genes
e. Manipulation of gene transfer

The application of recombinant DNA technology to the study of gene structure and function has opened a new era of clinical chemistry diagnostic services, namely the molecular diagnosis of genetic defects.

The principle of molecular diagnosis is relatively simple, because the human genome contains many cleavage sites for restriction endonucleases. Restriction endonucleases are microbial enzymes that cleave double stranded DNA at specific and known sequences. The type, number, and location of these restriction enzyme sites can vary in a population. If two different samples of DNA are cut with a particular restriction endonuclease and those DNA samples have a different number or placement of restriction sites for that endonuclease, then two different sets of DNA fragments will be generated from the two samples. The differences in the fragments can be detected by methods that will be described later. These differences are called *restriction fragment length* (RFL) *polymorphisms,* or DNA markers. If specific markers can be associated with particular genetic disorders at a high level of probability, then an individual's chances of having a gene defect can be calculated and hence his chances of developing a genetic disease.

The advantage of this method is that, potentially, it can provide a large number of new genetic markers (DNA as opposed to protein markers) that can be detected before the disease produces either clinical or biochemical (i.e., protein polymorphism) effects. These DNA markers can be detected whether or not they lie within a coding region of the genome. This technique also makes it possible to map the location of the defective gene(s) in the genome. One can then clone, isolate, and characterize the defective gene(s). The sequence change in the DNA that causes the genetic disorder can eventually be determined.

BASIC PRINCIPLES OF MOLECULAR DIAGNOSIS

Three fundamental requirements must be met in order to establish a molecular diagnosis:

1. ***Human DNA.*** An intact and highly purified DNA sample must be prepared from the patient and from a normal control. In practice, DNA is isolated from a tissue or from nucleated leukocytes fractionated from the cells in 20 to 40 mL of blood. DNA is obtained by amniocentesis or from a biopsy of chorionic villi for prenatal diagnosis. DNA from the cells in 10 to 20 mL of amniotic fluid, which can be obtained during the 15th to 17th week of pregnancy, is sufficient for diagnosis. Diagnosis using a chorionic villi DNA sample can be achieved as early as the 6th week of gestation. The isolation of the DNA requires several stages of purification, including digestion with proteinase K, which removes proteins associated with nucleic acid, and digestion with ribonuclease A which removes contaminating RNA. The DNA must be very pure and in the form of a continuous double strand of very high relative mass, so that the restriction enzymes will cut in a predictable manner. Sheared or broken DNA would lead to a misinterpretation of the results.
2. ***Restriction Enzymes.*** These endonucleases are highly specific. Each recognizes a particular double-stranded sequence of four or more nucleotides. Although more than 150 restriction enzymes are known, only about 20 to 30 are commonly used for recombinant DNA work. Nine enzymes used regularly for RFL polymorphism studies are listed in Table 28–1. After restriction endonuclease digestion, the DNA of the patient generates a set of fragments whose size (relative mass) distribution differs from that of a normal person's DNA cut with the same restriction enzyme(s). Thus, the patient's DNA must have some polymorphic (compared with the normal) restriction endonuclease sites so that it will exhibit characteristic RFL polymorphisms. In addition, these DNA markers must be associated with the disease gene locus (or loci) so that they can be used for predictive purposes. The marker site must therefore occur in the *disease gene locus itself,* or be linked with the locus in such a way that it appears only in a high proportion of diseased or 'carrier' individuals.
3. ***Molecular Probe.*** In order to identify the aforementioned RFL polymorphisms, it is necessary to have a molecular probe. Human DNA cut with a restriction enzyme, separated by electrophoresis and stained with ethidium bromide, will appear as a smear because there are so many fragments and so much DNA.

   A molecular probe is a piece of DNA or RNA between 15 and several thousand nucleotides in length, labeled with a radionuclide, that can bind to or 'hybridize' with complementary sequences of the DNA from both the patient and the control. The conditions of hybridization are made sufficiently stringent that

TABLE 28–1. Restriction Enzymes Commonly Used for RFL Polymorphism Studies

Restriction Enzyme	Recognition Sequence
Bam HI	5′G/GATC C3′ 3′C CTAG/G5′
Bgl I	GCCN NNN/NGGC CGGN/NNN NCCG
Eco RI	G/AATT C C TTAA/G
Hae III	GG/CC CC/GG
Hind III	A/AGCT T T TCGA/A
Msp I	C/CG G G GC/C
Pst I	C TGCA/G G/ACGT C
Taq I	T/CG A A GC/T
Xba I	T/CTAG A A GATC/T

A = adenine; C = cytosine; G = guanine; T = thymine. N can be any nucleotide base. The slash marks indicate the sites of cleavage.

the probe will bind only with sequences that are highly homologous to it.

If an RFL polymorphism exists between that patient and the normal, the probe will bind to fragments of different relative masses. This can be detected by a technique called Southern blot transfer (described later) and autoradiography.

Molecular probes can be obtained by chemical synthesis or from a DNA library. A DNA library may be of complementary DNA or genomic DNA. A *complementary* DNA (cDNA) library is made from messenger RNA from a specific tissue. An enzyme called reverse transcriptase is used to synthesize a single stranded piece of DNA using the mRNA as a template. This single stranded DNA is then made double stranded using DNA polymerase I (Klenow fragment). The double stranded DNA is then inserted into a vector (e.g., plasmid pBR322). The resulting recombinant DNA is used to transform *E. coli* cells so that many copies of the cDNA can be made.

To construct a *genomic* library, DNA is isolated from a specific tissue and cut into fragments of about 15 to 20 kilobases (kb) using a restriction endonuclease such as Mbo I. These fragments are then inserted into lambda phage (very large pieces of DNA are not as easily incorporated into bacteria as plasmids), which is transfected into a host (*E. coli*). DNA fragments up to 40kb can be cloned using 'cosmid' cloning. This technique takes advantage of the fact that, in order to be packaged into lambda phage, a piece of DNA need only have the lambda 'cos' sites (complementary stretches of single stranded DNA). These *cos* sites have been cloned into the tetracycline resistance gene of pBR322 (the ampicillin resistance gene is left intact). The plasmid is cut at a restriction enzyme site adjacent to the *cos* site so that the 40kb fragments can be ligated to the *cos* site; concatamers (long stretches of DNA linked end to end) are formed so that the DNA fragments are flanked by *cos* sites. Lambda packaging enzymes and proteins are then added; these will cut out the 40kb fragments flanked by the *cos* sites and package them into lambda phage. When this DNA is injected into *E. coli* it will behave like a plasmid and will confer ampicillin resistance on its host, thus enabling it to be detected on the basis of conferred antibiotic resistance.

With these three elements in hand (human DNA, restriction enzymes and molecular probes) one has the basic tools with which to *establish* the *presence* of a ***genetic defect.*** The purified DNA of the control and the patient is cut with the restriction enzyme that produces the RFL polymorphism for the disease in question. The cut DNA is then separated by electrophoresis on an agarose gel. The DNA is transferred from the gel to a sheet of nitrocellulose by a process called Southern blot transfer. This transfer to nitrocellulose (or a material with similar properties, e.g., positively charged nylon), which has a very high affinity for DNA and binds it tightly, is necessary because the molecular probe cannot bind to DNA lodged in a gel matrix.

The labeled probe is then hybridized to the DNA on the nitrocellulose. The nitrocellulose is exposed to x-ray film in order to observe the banding pattern produced by the probe.

If the probe has hybridized to different fragments (having different relative masses) in the DNA of the patient as compared with the control, the banding patterns for the two samples will be different, that is, the bands will appear at different distances from the origin of migration or different numbers of bands will appear.

This would indicate that the patient's DNA exhibits RFL polymorphism and the patient's chances of having the disease can then be calculated. Figure 28–1 illustrates the procedure. Table 28–2 is a list of some diseases currently under investigation, the restriction enzymes being used to generate RFL polymorphisms, and the probes used to detect the markers.

APPROACHES TO MOLECULAR DIAGNOSIS

Simple Mendelian Inheritance

Diseases in Which the Biochemical Defect is Known. In general, RFL polymorphisms are detected by random testing of patient and family DNA with different restriction enzymes or, if one knows the sequence abnormality that

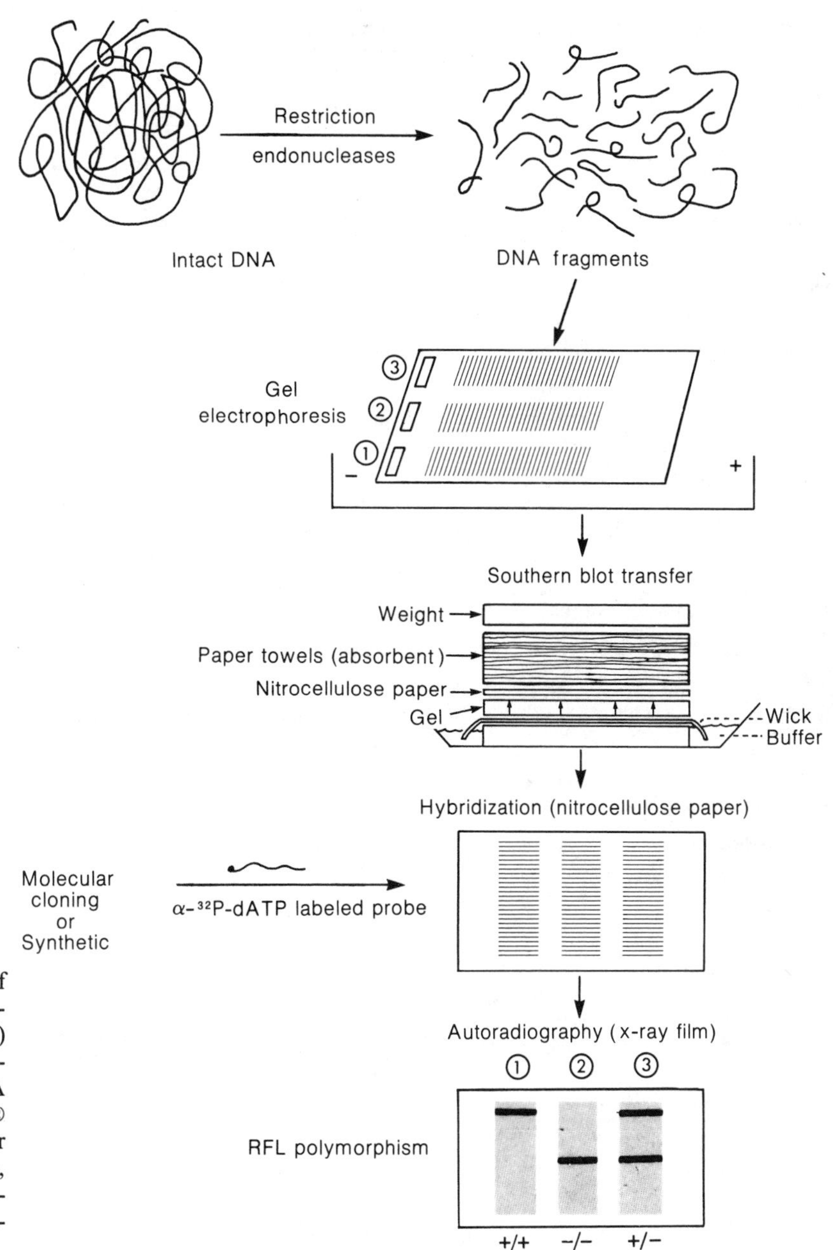

FIGURE 28–1. Technique of molecular diagnosis by restriction fragment length (RFL) polymorphism. Restriction endonucleases fragment the DNA from ① a normal individual, ② a person carrying two alleles for the genetic defect (homozygous), and ③ a person carrying one allele for the genetic defect (heterozygous).

leads to the disorder, it may also be possible to predict the polymorphic enzyme sites.

In disorders in which there is a direct correlation between the defective enzyme (or protein structure) and physiologic dysfunction in the patient, a molecular probe may be obtained by synthesis, or from a cDNA library. A short (15 to 20 nucleotides) probe can be synthesized using the amino acid sequence of the defective protein. Alternatively, a cDNA clone of the gene for the protein

TABLE 28–2. Genetic Disorders Under Active Investigation

Disease	Gene Defect	Restriction Enzyme	Probe
Coronary artery disease (e.g., hypercholesterolemia	Apo A-1	Pst I	Apo A-1
Cystic fibrosis	Paraoxonase(?)	Msp I Taq I	pJ3·11 (chromosome 7)
Diabetes:			
Type I	Insulin(?)	Bgl II	pHLA-DR α-chain
Type 2	Unknown	Bgl I Sac I	Insulin gene
Duchenne type muscular dystrophy	Unknown	BstXI	Xp·21 (X-chromosome)
Dwarfism:			
Achondroplastic	Collagen Type II(?)	Bam HI Eco RI	pgHCol (II) (genomic clone)
Thanatophoric hereditary	Growth hormone	n/d	GH gene
Hemophilia A	Factor VIII deficiency	Taq I	St14
Hemophilia B (Christmas disease)	Factor IX deficiency	Eco RI	Factor IX
Huntington's chorea	Unknown	Hind III	G8 fragment of chromosome 4
Lesch–Nyhan syndrome	Hypoxanthine-guanine phosphoribosyl transferase (HGPRT)	Bam HI	HGPRT cDNA clone
Osteogenesis imperfecta	Procollagen (C-peptides)	n/d	Pro 1(I)
Phenylketonuria	Phenylalanine hydroxylase	Hind III	pPH gene
Pulmonary emphysema or infantile liver cirrhosis	α_1-antitrypsin	Hind III Xba I Bam HI	α_1-antitrypsin
Sickle cell anemia	β-globin	Mst II	β-globin gene
Thalassemias:			
alpha	α-globin	Eco RI	α-globin gene
beta	β-globin	Hinc II	β-globin gene
Thromboembolism	Antithrombin deficiency	Pst I	Antithrombin III

n/d = not defined.

may be selected from a library using the synthesized probe.

Diseases in Which the Biochemical Defect is Unknown. The strategy used to approach this category of diseases depends fundamentally on *linkage analysis.* First, one must determine from family pedigrees, whether the disorder is sex-linked or autosomal. One can then select probes, from a chromosome specific library, that have no repetitive sequences in them. This is a genomic DNA library in which one first sorts the chromosomes into discreet size classes (using fluorescence-activated cell sorting). The DNA is isolated from each size class and then cloned into a lambda phage or cosmid vector, which is then used to transform *E. coli.*

RFL polymorphisms are found by random testing of family and patient DNA with restriction enzymes and probes. Duchenne muscular dystrophy, myotonic dystrophy, cystic fibrosis, and Huntington's chorea are examples of this class of disorder that can be diagnosed.

Multifactorial Inheritance

This group of genetic defects is the most difficult to detect by molecular diagnosis. One must proceed with three lines of investigation:

1. Obtain evidence that will at least suggest what set of genes are involved.
2. Trace inheritance in large families in which the disease is expressed.
3. Improve methods of clinical diagnosis to enhance pedigree studies.

MOLECULAR DIAGNOSIS IN PRACTICE

Hemoglobinopathies

Hemoglobinopathies are disorders of globin synthesis and include the thalassemias and sickle-cell anemia.

Thalassemia is a condition in which there is an insufficient production of one of the polypeptides of globin. This condition is caused by point mutations and/or short deletions which result in translational mutations (nonsense and frameshift), transcriptional mutations (faulty initiation), and mRNA-processing mutations (faulty splicing).

Fortunately, the globin genes have been cloned and large regions have been sequenced, thus making it possible to synthesize short probes for specific segments of the globin genes. In addition, single base changes, approximately every 200 base pairs in the globin genes, provide a number of polymorphic sites that can generate RFL polymorphisms.

The application of molecular diagnosis in Sardinia, where there is a high prevalence of thalassemia, has already reduced the incidence in newborns.

Sickle-cell anemia is a disease in which the β-globin gene is defective. Hemoglobin in sickle-cell patients tends to polymerize, causing the red blood cells to deform. The defect in the β-globin gene is the result of a single point mutation (A to T); this causes the triplet at position 6 to code for a valine residue instead of glutamic acid.

This single base change removes a restriction enzyme site (Dde I), which can be detected using a cDNA clone of the β-globin gene. Several other DNA markers have also been discovered. The RFL polymorphism used for routine testing is generated by the enzyme Mst II. It is now possible to diagnose at least 80% of families at risk for this type of anemia.

Diabetes

Diabetes is a complex disorder characterized usually by hyperglycemia and, in some cases, by an abnormal production of insulin. There are, however, several forms of the disease, each differing in its symptoms and severity. The standard classification is type 1 insulin-dependent diabetes (IDD) and type 2 noninsulin dependent diabetes (NIDD).

The precise biochemical defect that causes diabetes has yet to be elucidated. Extensive studies that have been carried out so far have not produced any evidence to indicate that the insulin gene locus is involved directly in the etiology of diabetes.

For example, a highly polymorphic area found in the 5′ flanking region of the insulin gene (i.e., a region contiguous with a structural gene at the 5′ end but not transcribed) provides markers for linkage analysis. But in a study of a large family with noninsulin dependent diabetes (type 2) manifesting at a young age, linkage analysis indicated, with a high probability, that there was no linkage between the insulin locus and the onset of type 2 diabetes. Thus, the defect does not lie in the insulin gene but must involve another gene.

This finding illustrates *two* important points. First, that the association between a particular gene and a disease does not necessarily mean that the gene is defective. Second, polymorphism may not result in genetic abnormality, or even be related to it. It has been estimated that random base changes occur in many noncoding sequences, for example, within introns, changes occur once in every 200 base pairs along a chromosome when the same intron sequences are compared between individuals in a population. Thus, while polymorphisms are necessary to molecular diagnosis, one must find polymorphic sites that are usefully linked to the defective gene(s). This process can be very laborious.

Although molecular diagnosis is a promising new technique, there are many obstacles, not least of which is the sheer size of the human genome (3×10^9 base pairs). The research required to identify suitable RFL polymorphisms and to select the right probes is tedious and difficult. Ultimately the defective gene(s) themselves will have to be mapped, isolated, and characterized. Only such a fundamental approach will be successful in delineating genetic defects and providing a definitive diagnosis of the disease at the molecular level. The road will be long, but the work will be exciting and the results rewarding.

SUGGESTED READING

BOEHM CD, ANTONARAKIS SE, PHILLIPS JA, STETTEN G, KAZAZIAN HH JR: Prenatal diagnosis using DNA polymorphisms: Report on 95 pregnancies at risk for sickle-cell disease or β-thalassemia. N Engl J Med 308:1054–1058, 1983

BOTSEIN D, WHITE RL, SKOLNICK M, DAVIS RW: Construction of a genetic linkage map in man using restriction fragment link polymorphisms. Am J Hum Genet 32:314–331, 1980

GUSELLA JF, WEXLER NS, CONNEALLY PM: A polymorphic DNA marker genetically linked to Huntington's disease. Nature 306:234–238, 1983

KAN YW, DOZY AM: Polymorphism of DNA sequence adjacent to human β-globin structural gene: Relation to sickle mutation. Proc Natl Acad Sci USA 75:5631–5635, 1978

LAWN RM, VEHAR GA: The molecular genetics of hemophilia. Sci Am 254:48–54, 1986

LEWIN B: Genes. New York, John Wiley and Sons, 1983

MANIATIS T, FRITSCH EF, SAMBROOK J: Molecular Cloning, A Laboratory Manual. Cold Spring Harbor, Cold Spring Harbor Laboratory, 1982

ORDOVAS JM, SCHAEFER EJ et al: Apolipoprotein A-I gene polymorphism associated with premature coronary artery disease and familial hypoalphalipoproteinemia. N Engl J Med 314:671–677, 1986

WOO SLC, LIDSKY AS, GÜTTLER F, CHANDRA T, ROBSON KJH: Cloned human phenylalanine hydroxylase gene allows prenatal diagnosis and carrier detection of classical phenylketonuria. Nature 306:151–155, 1983

Index

The letter *f* following a page number indicates a figure; the letter *t* following a page number indicates a table.